T0180254

# Graduate Texts in Physics

## Graduate Texts in Physics

Graduate Texts in Physics publishes core learning/teaching material for graduate- and advanced-level undergraduate courses on topics of current and emerging fields within physics, both pure and applied. These textbooks serve students at the MS- or PhD-level and their instructors as comprehensive sources of principles, definitions, derivations, experiments and applications (as relevant) for their mastery and teaching, respectively. International in scope and relevance, the textbooks correspond to course syllabi sufficiently to serve as required reading. Their didactic style, comprehensiveness and coverage of fundamental material also make them suitable as introductions or references for scientists entering, or requiring timely knowledge of, a research field.

More information about this series at http://www.springer.com/series/8431

Massimo V. Fischetti • William G. Vandenberghe

# Advanced Physics of Electron Transport in Semiconductors and Nanostructures

Massimo V. Fischetti
Department of Materials Science and Engineering
University of Texas at Dallas
Richardson, TX, USA

William G. Vandenberghe
Department of Materials Science and Engineering
University of Texas at Dallas
Richardson, TX, USA

ISSN 1868-4513 ISSN 1868-4521 (electronic)
Graduate Texts in Physics
ISBN 978-3-319-79126-5 ISBN 978-3-319-01101-1 (eBook)
DOI 10.1007/978-3-319-01101-1



Printed on acid-free paper

This Springer imprint is published by Springer Nature
The registered company is Springer International Publishing AG Switzerland

*To the memory of my father Nico, who initiated my lifelong love affair with physics, and to Alessandra, Sebastian, and Alice.*

MF

*To my family*

WGV

# Preface

This book is aimed at second-year graduate students in physics, electrical engineering (EE), or materials science (MS). Its main goal is to present a rigorous introduction to electronic transport in solids, especially at the nanometer scale, within a self-contained text (and course) and with a not-so-hidden emphasis on computational aspects.

Understanding electronic transport in solids requires some basic knowledge of Hamiltonian classical mechanics, quantum mechanics, condensed matter theory, and statistical mechanics. Thus, usually, four "heavy" courses are required to acquire this background knowledge. Unfortunately, in practice, EE and MS students may only attend a quick course on quantum mechanics at best; they are unlikely to have ever seen the Hamiltonian formulation of classical mechanics or to have been exposed to statistical mechanics or condensed matter beyond some quick preview, especially as far as electronic (band) structure is concerned. MS students may be required to enroll in a class on the electrical and optical properties of solids but this is often only a superficial introduction. Finally, not all physics students will have seen these subjects at the depth needed. Our aim is to provide a deep discussion of those specific subtopics of these four disciplines which are required to deal with electronic transport, so that a single, self-contained class may suffice. This will be useful for students who intend to work in academia or the nano-/microelectronics industry.

We assume some basic knowledge of classical mechanics (of course) and of quantum mechanics. Therefore, only the Lagrangian and Hamiltonian formulations of classical mechanics are reviewed. Similarly, the principles of quantum mechanics are revisited only formally, with emphasis on canonical quantization, since this leads to its generalization to systems with infinitely many degrees of freedom (fields) and so to second quantization, elementary excitations in solids, and scattering processes. A bird's-eye view of the structure of atoms, bonds, and molecules serves the purpose of introducing some useful mathematical tools and concepts used in the text.

Topics covered in this book are: the theory of energy bands in crystals, second quantization and elementary excitations in solids, dielectric properties of semiconductors, with an emphasis on dielectric screening, electron scattering with phonons, plasmons, electrons, and photons, the derivation of transport equations in semiconductors and semiconductor nanostructures, both at the quantum and semiclassical level. The text presents examples relevant to current research, not only about Si, but also III–V compound semiconductors, nanowires, graphene, and graphene nanoribbons. In particular, the text gives major emphasis to plane-wave methods regarding the electronic structure of solids, both Density Functional Theory (DFT) and empirical pseudopotentials, always paying attention to their effect on (and numerical implementation in a description of) electron transport. The core of the text deals with electronic transport, as we said above, with ample discussions of the transport equations derived both in the quantum picture (the Liouville–von Neumann equation) and semiclassically (the Boltzmann transport equation, BTE). Several methods for solving the semiclassical BTE are also reviewed, including the method of moments, expansions into orthogonal functions, iterative methods, Monte Carlo, cellular automata, and direct matrix inversion.

Four appendices conclude the text. The first one presents time-independent and time-dependent perturbation theory, as well as the Born approximation. The second appendix provides full information, and references, about the empirical pseudopotentials that are commonly used for fcc semiconductors. The third appendix, a quick-and-dirty introduction to the principles of special relativity, is required to understand the "minimal" electromagnetic coupling between electrons and photons. It also introduces the relativistic wave equation for spin-1/2 particles which, in its massless version, describes approximately the electron dispersion in graphene. The fourth appendix pays tribute to our emphasis on computational aspects and lists the source code of a simple computer program to compute the band structure of Si using empirical pseudopotentials. We hope that students will enjoy running and modifying this program.

The breath and length of the subjects presented in this text makes it impossible to cover the totality of material presented here in a single one-semester course (typically 28 lectures of 75 min each) and even less so in a one-quarter course. In our experience, only the basic subjects can be covered in such a class. Subjects that may be covered "optionally," if time allows or in a sequel of this course, have been highlighted by an asterisk. Some are simple historical remarks or curiosities, but those that deal with advanced topics are intentionally treated at a slightly deeper, more research-oriented level. In these chapters and/or sections, more frequent reference is made to journal papers than to textbooks. In the second year of a graduate program, we find it appropriate to start exposing graduate students to the more terse and concise style used in research papers, while the students are preparing to perform research on their own.

Richardson, TX, USA
July 6, 2015

Massimo V. Fischetti
William G. Vandenberghe

# Acknowledgments

For one of the authors (MVF), this book is the result of more than 30 years spent playing with the physics of electron transport in solids, after a precious apprenticeship as a Ph.D. student of Professor James Hartle. The "golden years" spent at the IBM Thomas J. Watson Research Center in Yorktown Heights, New York, have contributed to my enjoyment of physics in a way that cannot be overstated. While there are too many colleagues to thank individually, particular gratitude and appreciation are extended to: Drs. Edward Cartier, Dan DiMaria, Alan Fowler, John Kirtley, Paul Solomon, Frank Stern, Thomas Theis, Sandip Tiwari, and the late Drs. Markus Büttiker and Rolf Landauer. A very special acknowledgement goes to Dr. Steve Laux, who contributed some of the figures reproduced here, as well as many of the concepts presented in the last chapter. Of the many other mentors and colleagues that have managed to teach (and stimulate interest in) many of the fundamental ideas of condensed matter physics and electronic transport, Professors Roberto Car, David Ferry, Chihiro Hamaguchi, Karl Hess, Carlo Jacoboni, Gerhard Klimeck, Mark Lundstrom, Wolfgang Porod, Lino Reggiani, Massimo Rudan, Nobuyuki Sano, and the late Marcello Anile deserve special thanks and gratitude. In particular, results of calculations performed by professor Sano are included in this text (Fig. 15.6). Faculty and colleagues at the University of Massachusetts, Amherst, and at the University of Texas at Dallas have made it possible to teach the graduate class on whose lecture notes this book is based. Special thanks are due to Dr. Shela Aboud for her many tutorials on density functional theory, for providing Figs. 7.29, 7.32, and 7.33, and for her strong encouragement to convert those lecture notes into a book, a task whose difficulty was grossly underestimated. Finally, many students and postdoctoral fellows have contributed not only to a better understanding of many issues with their, occasionally embarrassing questions, but also to the technical content of this text, often contributing ideas and results of their own. So, many thanks are due to Drs. Jingtian Fang, Bo Fu, Seonghoon Jin, Jiseok Kim, Alexander Kirk, Sudarshan Narayanan (who performed the Monte Carlo calculations shown in Figs. 13.5, 13.4, 13.10, and 13.11), Zhun-Yong Ong, Terrance O'Regan, and Yan Zhang. Ms. Catherine Sachs calculated the phonon spectrum of graphene shown in Fig. 13.6 and, most important, has been patient enough to spot mistakes and typographical errors in the lecture notes on which this book is based. We have done our best to introduce new ones.

The second author (WGV) is first and foremost indebted to the first author (MVF) for presenting him the opportunity to work as a research scientist at the University of Texas at Dallas and freely roam the frontier of nanoelectronics research topics. WGV graduated from KU Leuven, Belgium, and thanks his promoters Guido Groesenken (KU Leuven) and Wim Magnus (UAntwerpen, Belgium) for supporting him during that time. He performed his research at the nanoelectronics research center imec, located in Leuven, Belgium, where he was exposed to studying electronic transport in state-of-the-art electronic devices. From the many great colleagues at imec, WVG extends special thanks to Professors Marc Heyns and Bart Sorée as well as his master thesis advisor Dr. Anne Verhulst for crystallizing many of his ideas into publications. Further thanks go to fellow Ph.D. students

Kuo-Hsing Kao and Geert Hellings as well as his own master thesis students Maarten Van de Put and Sergej Fischer and Dr. Verhulst's master thesis student Devin Verreck. At the University of Texas, the second author also had the pleasure of having interesting discussions with many of the same people as the first author, and wishes to recognize the help of Terrance O'Regan, Bo Fu, Jiseok Kim, Zhun-Yong Ong, and Jingtian Fang, as well as Professors Robert Wallace and Christopher Hinkle.

# Contents

# Acronyms

| | |
|---|---|
| 1DEG | One-dimensional electron gas |
| 2DEG | Two-dimensional electron gas |
| AGNR | Armchair-edge graphene nanoribbon |
| aSiNR | Armchair-edge silicane nanoribbon |
| APW | Augmented plane waves |
| BE | Bose–Einstein |
| BH | Brooks–Herring |
| BTE | Boltzmann transport equation |
| BZ | Brillouin zone |
| CA | Cellular automata |
| CAR | Canonical anti-commutation rules |
| CCR | Canonical commutation rules |
| CME | Constant matrix element approximation |
| CMOS | Complementary metal-oxide-semiconductor |
| CNT | Carbon nanotube |
| CW | Conwell–Weisskopf |
| DFPT | Density functional perturbation theory |
| DFT | Density functional theory |
| DH | Debye–Hückel |
| DOS | Density of states |
| FD | Fermi–Dirac |
| FET | Field-effect transistor |
| FFT | Fast Fourier transform |
| GGA | Generalized gradient approximation |
| GNR | Graphene nanoribbon |
| HEMT | High electron-mobility transistor |
| HOMO | Highest occupied atomic orbital |
| IQHE | Integer quantum hall effect |
| KS | Kohn–Sham |
| LA | Longitudinal acoustic |
| LCAO | Linear combination of atomic orbitals |
| LCBB | Linear combination of bulk bands |
| LDA | Local density approximation |
| LER | Line-edge roughness |
| LO | Longitudinal optical |
| LUMO | Lowest unoccupied atomic orbital |

| | |
|---|---|
| MBE | Molecular beam epitaxy |
| MC | Monte Carlo |
| MESFET | Metal-semiconductor field-effect transistor |
| MOS | Metal-oxide-semiconductor |
| MOSFET | Metal-oxide-semiconductor field-effect transistor |
| NEGF | Non-equilibrium Green's functions |
| NW | Nanowire |
| OPW | Orthogonalized plane waves |
| PBE | Perdew-Burke-Ernzerhof |
| PDE | Partial differential equation(s) |
| QFT | Quantum field theory |
| QM | Quantum mechanics |
| OTBM | Quantum transmitting boundary method |
| QW | Quantum well |
| RPA | Random phase approximation |
| SBE | Semiconductor Bloch equations |
| SCF | Self-consistent field |
| SHE | Spherical harmonics expansion |
| SMA | Scattering matrix approach |
| TA | Transverse acoustic |
| TMD | Transition metal dichalcogenides |
| TO | Transverse optical |
| VCA | Virtual crystal approximation |
| VLSI | Very large-scale integration |
| WS | Wigner–Seitz |
| ZGNR | Zigzag-edge graphene nanoribbon |
| zSiNR | Zigzag-edge silicane nanoribbon |

# About the Notation

The use of a consistent and correct notation is always a major issue, since compromises must be found between correctness and clarity. Here we have adopted the notation that is "conventional" in the physics community. However, having to deal with subjects spanning a variety of subfields, we have encountered some difficulties due to the use of the same symbols to denote different quantities in many of these subfields. Listing some examples may also help the readers:

- The symbol $q$ is used as electric charge in electromagnetism, as canonical position coordinate in Lagrangian and Hamiltonian classical mechanics, and as phonon wavevector—or, more generally, wavevector transfer—in condensed matter physics. We have avoided its use to denote charge, but we have decided to use it both as canonical coordinate and wavevector. We felt that referring to a position coordinate with a symbol we have not used, such as $\mathcal{M}$, would have created even more confusion.
- The symbol $V$ usually denotes potential energy, electrostatic potential, bias applied to contacts, and volume. We have opted for the use of $\Omega$ for volume and $\varphi$ for the electrostatic potential, reserving $V$ to potential energy and voltages, and liberally used subscripts and superscripts to clarify its meaning. Having done that, in one of the appendices, having used $\Omega$ for volume, we had to choose a symbol to denote solid angles, almost universally denoted also by $\Omega$. We have used $\omega$ for the solid angle, since no radian frequencies are used in that appendix. But the fear of having created confusion remains.
- The letter $H$ denotes Hamiltonians. So, we have used $\mathscr{H}$ to denote Hamiltonian density. We were forced to look for a different-looking font, $\mathcal{H}$, when a symbol was needed to denote Hilbert spaces.
- The electric field is usually denoted by $\mathbf{E}$. However, we have denoted it by $\mathbf{F}$ in some subfields, in order to avoid confusion with energy, denoted by $E$. We have followed our usual idea of using the symbol that is most commonly used in the literature dealing with the subject we discuss. For example, dealing with the quantization of the electromagnetic field (Chap. 9), we have used the symbol $\mathbf{E}$. However, in chapters that treat electron transport and make frequent use of the energy $E_i$ or $E(\mathbf{k})$ of electronic states, such as Chaps. 18 and 19, we have followed the usual convention and employed the symbol $\mathbf{F}$.

In summary, we have tried to adopt the notation and conventions that are most commonly used in the literature specific to a particular subfield, even though this has forced us to use a slightly different notation in different chapters of the book. We think that students, who are encouraged to read the scientific literature, should become familiar with these conventions.

Some additional comments are necessary:

1. Scalar quantities are indicated by italic symbols, vectors by bold symbols. So, for example, the vector $\mathbf{v}$ has magnitude $v$. The vectors for the position and wavevector are denoted by lower-case symbols ($\mathbf{r}$ and $\mathbf{k}$) in three dimensions and by upper-case symbols ($\mathbf{R}$ and $\mathbf{K}$) in two dimensions.
2. Hermitian operators acting on Hilbert spaces are denoted by a "hat," $\widehat{A}$.
3. In the Dirac notation we use, $|a\rangle$ is a general element of a Hilbert space. Two commonly used "bases" are the plane-wave basis $|\mathbf{k}\rangle$ of eigenstates of the momentum operator and the eigenstates $|\mathbf{r}\rangle$ of the position operator. Obviously and ironically, in standard courses of quantum mechanics, having defined Hilbert spaces and such, these basis vectors are commonly used, although neither basis belongs to a Hilbert space. Indeed, the representation of these basis states in terms of "functions" is $\langle\mathbf{r}'|\mathbf{r}\rangle = \delta(\mathbf{r}-\mathbf{r}')$, which is not even a "function," and $\langle\mathbf{r}'|\mathbf{k}\rangle = e^{i\mathbf{k}\cdot\mathbf{r}'}$, which is not square integrable and so does not belong to the usual single-particle Hilbert space $L^2(\mathbb{R}^3)$. The problem is either ignored, or bypassed by defining eigen-packets, but the best answer, the definition of "rigged" or "equipped Hilbert spaces," is seldom discussed. We follow this "easy way out" as well. Readers who would like to understand the mathematically correct solution could consult, for example, I.M. Gelfand and N.J. Vilenkin, *Generalized Functions, vol. 4: Some Applications of Harmonic Analysis. Rigged Hilbert Spaces* (Academic Press, New York, 1964).
4. The notation $\sum_{\mathbf{q}}$ represents the sum over the "quantum numbers" $\mathbf{q}$. Depending on whether $\mathbf{q}$ represents a quantum number in the discrete or continuum spectrum, a fact that often depends on the normalization chosen, it should be interpreted as follows: For normalization over a finite volume $\Omega$,

$$\sum_{\mathbf{q}} f(\mathbf{q})\ \delta(\mathbf{q}-\mathbf{q}_0) \rightarrow \frac{1}{\Omega} \sum_{\mathbf{q}} f_{\mathbf{q}}\ \delta_{\mathbf{q},\mathbf{q}_0} \ .$$

So, we tacitly imply a division by the normalization volume, and the "Dirac delta-function" should be interpreted as a Kronecker delta. For normalization over an infinite volume in $n$ dimensions, instead,

$$\sum_{\mathbf{q}} f(\mathbf{q})\ \delta(\mathbf{q}-\mathbf{q}_0) \rightarrow \frac{1}{(2\pi)^n} \int d\mathbf{q}\, f(\mathbf{q})\ \delta(\mathbf{q}-\mathbf{q}_0) \ .$$

So, we imply a division by the phase-space factor $(2\pi)^n$ and the Dirac delta-function should be interpreted as a genuine tempered functional over a Hilbert space (to be pedantic). Therefore, we shall be "sloppy" and consider expressions like $f(\mathbf{q})$ and $f_{\mathbf{q}}$ as equivalent, much to a mathematician's horror, but we have already horrified mathematicians by writing $\int f(\mathbf{q})\ d\mathbf{q}$ as $\int d\mathbf{q}\, f(\mathbf{q})$, as is common in the physics community. Note, however, that occasionally it will be necessary to show explicitly the normalization volume, for example, when decomposing an integral over the entire normalization volume into identical integrals over unit cells of the crystal.

# List of Symbols

Special symbols are defined in the text when first introduced. We list here only symbols, not defined in the text, that have an almost universal meaning, including physical constants:

$\epsilon_0$ Permittivity of vacuum, $8.854187817 \times 10^{-12}$ F/m
$\mu_0$ Permeability of vacuum, $4\pi \times 10^{-7}$ N/A$^2$
$\mu_B$ Bohr magneton, $e\hbar/(2m_{el}) = 9.27400968(20) \times 10^{-24}$ J/T
$c$ Speed of light in vacuum, $2.99792458 \times 10^8$ m/s
$e$ Magnitude of the electron charge, $1.602176565(35) \times 10^{-19}$ C
$h$ Planck constant, $6.62606957(29) \times 10^{-34}$ J·s
$\hbar$ Reduced Planck constant, $h/(2\pi) = 1.054571726(47) \times 10^{-34}$ J·s
$k_B$ Boltzmann constant, $1.3806488(13) \times 10^{-23}$ J/K
$m_{el}$ Electron mass, $9.109\,38291(40) \times 10^{-31}$ kg
Ry Rydberg unit of energy, $[e^2/(4\pi\epsilon_0)]^2 m_{el}/(2\hbar^2) = 13.60569253(30)$ eV

# Part I
# A Brief Review of Classical and Quantum Mechanics

# Chapter 1
# Canonical Quantization of Physical Systems

## 1.1 Overview

The electronic properties of crystals, of their dielectric behavior, of the dynamics of electrons in solids, and of the elementary excitations with which electrons (or photons) interact, are the topics treated in the main chapters of this text. Treating these topics requires a formal use of Quantum Mechanics that goes beyond what is usually presented in introductory courses. In this chapter we set the foundations of the formalism used in following chapters: We first discuss why we need a formulation of Quantum Mechanics that goes beyond the simple single-particle Schrödinger equation. We then review the Lagrangian and Hamiltonian formulation of Classical Mechanics, since this leads to the formal Canonical Quantization; that is, the formalism that allows us to "quantize" any physical system starting from its classical description. Whereas in this chapter this formalism is applied only to systems with a finite number of degrees of freedom (so, excluding fields), in later chapters this limitation will be lifted, following a natural extension of what is discussed here. Having quickly reviewed the main mathematical structure on which Quantum Mechanics rests, Hilbert spaces and bound Hermitian operators on Hilbert spaces, we finally briefly touch upon the problem of interpreting the mathematical formalism in physical terms and present formally the interpretation commonly known as the "Copenhagen interpretation." We shall proceed very formally, assuming previous knowledge of Quantum Mechanics in its simple formulation. As a result, in contrast to the other chapters of this book, here we will use a more terse, concise, less discursive style, since we intend to set the foundations in terms of concept and terminology used later.

As a matter of notation, operators will be denoted by symbols with "hats," the usual mathematical notation. In some mathematical texts they are denoted by bold symbols. However, we shall follow the conventional notation adopted by the physics community of reserving bold characters for vectors.

## 1.2 Formalization of Quantum Mechanics: Why?

Quantum Mechanics, as a formal "theory" of the microscopic world as we know it today, is the fruit of the efforts made in the 1930s to formalize the patchwork picture that had emerged in the few decades before. As one can read from any textbook on Quantum Mechanics, the end of the nineteenth century broke the attitude of certainty that had inspired the confidence of physicists up to that point: On the one hand, the fact that Maxwell's equations are not invariant under Galilean relativity, as Newton's equations are, together with the Michelson and Morley experiment, related to this Galilean-relativity



M. Fischetti, W.G. Vandenberghe, *Advanced Physics of Electron Transport in Semiconductors and Nanostructures*, Graduate Texts in Physics, DOI 10.1007/978-3-319-01101-1_1

issue, pointed at the need to revisit the concept of ether. This eventually led to Einstein's theory of special relativity [1]. On the other hand, several puzzles posed by experiments dealing with light and matter at the microscopic scale questioned the validity of the theory of electromagnetism and classical Statistical Mechanics to explain the spectrum of a black body, the photoelectric effect, Compton scattering between light and electrons, and, of course, the structure and stability of the hydrogen atom, as postulated by Bohr's theory [2]. Planck's work on the black-body spectrum (see, for example, [3]) [4], Einstein's explanation of the photoelectric effect (and, implicitly, of Compton scattering) [5], and Heisenberg's [6] and Schrödinger's [7] work published in 1925 and 1926, respectively, gave each explanations of the phenomena. But these explanations were somewhat ah hoc and partial. For example, Schrödinger's equation could be applied successfully to explain the spectrum of the hydrogen atom. But how to apply the same type formalism to a vibrating or rotating molecule? Einstein's work explained the wavelength dependence observed in the photoelectric effect. But how to treat the electromagnetic field on the same "quantum" footing used to deal with the electrons?

In order to build a formal more general theory applicable, in particular, to crystals, we must revisit the Hamiltonian formulation of Classical Mechanics. This will lead to a complete and consistent procedure to "quantize" any physical system. In this chapter this will be limited to systems with a finite number of degrees of freedom (so, system constituted by a finite number of "discrete" particles, each of them, in turn, possibly with a finite number of discrete internal degrees of freedom, such as spin up or down). More general systems with an infinite (actually, a continuum) number of degrees of freedom, or "fields," will be treated in later chapters.

## 1.3 Lagrangian and Hamiltonian Formulation of Classical Mechanics

### *1.3.1 Hamilton's Principle*

Consider a system with $N$ degrees of freedom. $N$ may be thought of as equal to $3n$, where $n$ is the number of classical particles in three dimensions. Let $q_i$ be their "generalized" coordinates, with $i = 1, N$. By "generalized" we mean that these coordinates are related to the position of particle $\alpha$ via relations of the form $\mathbf{r}_\alpha(q_1, q_2, \ldots, q_N)$. Thus, the coordinates $q_i$ may be the particles positions in another coordinate system (polar, cylindrical, etc.), or may be the coordinates on the subspace on which the system is forced to be by constraints (such as particles on a surface of a sphere), so that, in this case $N < 3n$.

Now assume that forces induced by interactions among the particles, or due to an external source (such as gravity or electromagnetic fields, for example), can be described by a potential energy $V(\mathbf{q})$ [where the $\mathbf{q}$ labels the $N$-dimensional vector $(q_1, q_2, \ldots, q_N)$] and let's define the kinetic energy $T$ of the system:

$$T(\dot{\mathbf{q}}) = \frac{1}{2}\sum_\alpha m_\alpha v_\alpha^2 = \frac{1}{2}\sum_\alpha m_\alpha \left( \sum_i \frac{\partial \mathbf{r}_\alpha}{\partial q_i} \dot{q}_i + \frac{\partial \mathbf{r}_\alpha}{\partial t} \right)^2 . \tag{1.1}$$

Defining the *Lagrangian* function:

$$L(\mathbf{q}, \dot{\mathbf{q}}) = T(\dot{\mathbf{q}}) - V(\mathbf{q}) , \tag{1.2}$$

the time integral of the Lagrangian,

$$S[\mathbf{q}] = \int_{t_1}^{t_2} L(\mathbf{q}, \dot{\mathbf{q}}) \mathrm{d}t = 0 , \tag{1.3}$$

is called the "action" and should be considered as a "functional" of all possible trajectories $[\mathbf{q}(t), \dot{\mathbf{q}}(t)]$ of the system. Hamilton's principle states that the motion of the system, $[\mathbf{q}(t), \dot{\mathbf{q}}(t)]$, during the time interval $(t_1, t_2)$ is such that

$$\frac{\delta S[\mathbf{q}]}{\delta \mathbf{q}} = 0 . \tag{1.4}$$

This simply states that the action is an extremum when we substitute the physical trajectory $[\mathbf{q}(t), \dot{\mathbf{q}}(t)]$ in the integral above of the Lagrangian functional.

### *1.3.2 A Primer on Variational Calculus*

How do we solve the variational problem described by Eq. (1.4)? Let $f(y, \dot{y}, x)$ (with $\dot{y} = \mathrm{d}y/\mathrm{d}x$) be a functional of $y$ and $\dot{y}$ and let's look for the particular function $y(x)$ for which the integral

$$\mathscr{I} = \int_{x_1}^{x_2} f(y, \dot{y}, x)\, \mathrm{d}x \tag{1.5}$$

is an extremum, with the boundary conditions $y(x_1) = y_1$ and $y(x_2) = y_2$. Let's parametrize all possible "trajectories" $y(x)$ by a parameter $\lambda$, so that the integral $\mathscr{I}$ becomes a function of $\lambda$,

$$\mathscr{I}(\lambda) = \int_{x_1}^{x_2} f[y(x,\lambda), \dot{y}(x,\lambda), x]\, \mathrm{d}x . \tag{1.6}$$

Then, our problem consists in solving

$$\frac{\partial \mathscr{I}(\lambda)}{\partial \lambda} = 0 . \tag{1.7}$$

Now:

$$\frac{\partial \mathscr{I}(\lambda)}{\partial \lambda} = \int_{x_1}^{x_2} \left\{ \frac{\partial f}{\partial y}\frac{\partial y}{\partial \lambda} + \frac{\partial f}{\partial \dot{y}}\frac{\partial \dot{y}}{\partial \lambda} \right\} \mathrm{d}x . \tag{1.8}$$

The second integral can be integrated by parts:

$$\int_{x_1}^{x_2} \frac{\partial f}{\partial \dot{y}} \frac{\partial^2 y}{\partial x \partial \lambda}\, \mathrm{d}x = \left. \frac{\partial f}{\partial \dot{y}} \frac{\partial y}{\partial \lambda} \right|_{x_1}^{x_2} - \int_{x_1}^{x_2} \frac{\mathrm{d}}{\mathrm{d}x}\left(\frac{\partial f}{\partial \dot{y}}\right) \frac{\partial y}{\partial \lambda}\, \mathrm{d}x . \tag{1.9}$$

Our boundary conditions require all functions to take the same values $y_1$ and $y_2$ at $x_1$ and $x_2$, so the first term vanishes. Thus,

$$\frac{\partial \mathscr{I}(\lambda)}{\partial \lambda} = \int_{x_1}^{x_2} \left( \frac{\partial f}{\partial y} - \frac{\mathrm{d}}{\mathrm{d}x}\frac{\partial f}{\partial \dot{y}} \right) \frac{\partial y}{\partial \lambda}\, \mathrm{d}x . \tag{1.10}$$

Now let's consider the variation of $y$ around an arbitrary value $\lambda_0$. Then:

$$\left( \frac{\partial \mathscr{I}}{\partial \lambda} \right)_0 \mathrm{d}\lambda = \int_{x_1}^{x_2} \left( \frac{\partial f}{\partial y} - \frac{\mathrm{d}}{\mathrm{d}x}\frac{\partial f}{\partial \dot{y}} \right) \left( \frac{\partial y}{\partial \lambda} \right)_0 \mathrm{d}x\, \mathrm{d}\lambda . \tag{1.11}$$

Now set

$$\delta\mathscr{I} = \left(\frac{\partial\mathscr{I}}{\partial\lambda}\right)_0 \mathrm{d}\lambda\ , \tag{1.12}$$

$$\delta y = \left(\frac{\partial y}{\partial\lambda}\right)_0 \mathrm{d}\lambda\ , \tag{1.13}$$

and

$$\delta\dot{y} = \left(\frac{\partial\dot{y}}{\partial\lambda}\right)_0 \mathrm{d}\lambda\ . \tag{1.14}$$

Thus, Eq. (1.11) becomes

$$\delta\mathscr{I} = \int_{x_1}^{x_2} \left(\frac{\partial f}{\partial y} - \frac{\mathrm{d}}{\mathrm{d}x}\frac{\partial f}{\partial\dot{y}}\right) \delta y\, \mathrm{d}x = 0\ . \tag{1.15}$$

But now note that $\delta y$ is an arbitrary variation of the trajectory. The only way Eq. (1.15) can be satisfied is when the term within parentheses vanishes, that is

$$\frac{\partial f}{\partial y} - \frac{\mathrm{d}}{\mathrm{d}x}\frac{\partial f}{\partial\dot{y}} = 0\ . \tag{1.16}$$

This is known as the Euler equation (although there are so many "Euler equations" in so many areas of mathematics and physics!).

### 1.3.3 Euler–Lagrange Equations

Equation (1.4) can be solved in a very similar way, leading us to the *Euler–Lagrange equations*:

$$\frac{\mathrm{d}}{\mathrm{d}t}\frac{\partial L}{\partial\dot{q}_i} - \frac{\partial L}{\partial q_i} = 0\ . \tag{1.17}$$

Note immediately that this formulation of Classical Mechanics exhibits explicitly some conservation laws: If the Lagrangian $L$ does not depend explicitly on the variable $q_j$ for some $j$, then from Eq. (1.17) we see that

$$p_j = \frac{\partial L}{\partial\dot{q}_j} \tag{1.18}$$

is a constant of motion, i.e., is a conserved quantity during the time evolution of the system. In this case, $q_j$ is said to be a "cyclical coordinate" while $p_i$ is called the "canonical generalized conjugate momentum."

### 1.3.4 Hamiltonian and Hamilton's Equations

This observation, as well as the definition of the conjugate momentum, gives the idea of transforming the Lagrangian $L$ into a functional of $\mathbf{q}$ and $\mathbf{p}$, rather than of $\mathbf{q}$ and $\dot{\mathbf{q}}$. This can be done by the Legendre transform

$$L(\mathbf{q},\dot{\mathbf{q}},t) \to H(\mathbf{q},\mathbf{p},t) = \sum_i p_i \dot{q}_i - L(\mathbf{q},\dot{\mathbf{q}},t) . \tag{1.19}$$

This is the familiar "Hamiltonian." In order to derive the equations of motion, note that

$$dH = \sum_i \frac{\partial H}{\partial q_i} dq_i + \sum_i \frac{\partial H}{\partial p_i} dp_i + \frac{\partial H}{\partial t} \mathrm{d}t . \tag{1.20}$$

On the other hand, from the definition (1.19), we also have

$$dH = \sum_i \dot{q}_i dp_i + \sum_i p_i \mathrm{d}\dot{q}_i - \sum_i \frac{\partial L}{\partial \dot{q}_i} \mathrm{d}\dot{q}_i - \sum_i \frac{\partial L}{\partial q_i} dq_i - \frac{\partial L}{\partial t} \mathrm{d}t . \tag{1.21}$$

Note that the second and third terms in this equation cancel, thanks to the definition (1.18). Since $\dot{p}_i = \partial L/\partial q_i$, Eq. (1.21) becomes

$$dH = \sum_i \dot{q}_i \mathrm{d}p_i - \sum_i \dot{p}_i \, \mathrm{d}q_i - \frac{\partial L}{\partial t} \mathrm{d}t . \tag{1.22}$$

Comparing Eqs. (1.20) and (1.22) we arrive at the *canonical Hamilton's equations*:

$$\dot{q}_i = \frac{\partial H}{\partial p_i} , \tag{1.23}$$

$$\dot{p}_i = -\frac{\partial H}{\partial q_i} , \tag{1.24}$$

together with

$$\frac{\partial H}{\partial t} = -\frac{\partial L}{\partial t} . \tag{1.25}$$

### 1.3.5 Poisson Brackets

A very abstract reformulation of Hamilton's equations relies on the use of the *Poisson brackets* between any two functions $A(\mathbf{q},\mathbf{p})$ and $B(\mathbf{q},\mathbf{p})$:

$$\{A,B\} = \sum_{i=1} \left( \frac{\partial A}{\partial q_i} \frac{\partial B}{\partial p_i} - \frac{\partial A}{\partial p_i} \frac{\partial B}{\partial q_i} \right) . \tag{1.26}$$

We shall call any function of $\mathbf{q}$ and $\mathbf{p}$ a *dynamic variable*. Using this definition, it is trivial to see that Eqs. (1.23) and (1.24) can be written as

$$\frac{\mathrm{d}q_i}{\mathrm{d}t} = \{q_i, H\} , \tag{1.27}$$

$$\frac{\mathrm{d}p_i}{\mathrm{d}t} = \{p_i, H\} , \tag{1.28}$$

while for a general dynamic variable $A(\mathbf{q}, \mathbf{p})$:

$$\frac{\mathrm{d}A}{\mathrm{d}t} = \{A, H\} + \frac{\partial A}{\partial t} . \tag{1.29}$$

Note also the following relations:

$$\{q_i, p_j\} = \delta_{ij} , \quad \{q_i, q_j\} = 0 , \quad \{p_i, p_j\} = 0 . \tag{1.30}$$

This reformulation of the laws of Classical Mechanics was developed in the late seventeenth [8] and early nineteenth century [9]. At that time, it was part of the attempts to re-express Newton's laws in a way that made it possible to employ the symmetries of the system in order to identify constants of motion and, as a consequence, simplify the task of finding the motion of the system. Together with the search for a proper transformation of coordinates that rendered the symmetry transparent (see the text by Goldstein [10] for an outstanding discussion of symmetries, generating functions, and the Hamilton–Jacobi equations), this formal structure of Classical Mechanics represents one of the most elegant formulations in mathematical physics. However, here we are interested in seeing how this abstract formalization allows us to treat a general physical system from a quantum-mechanical perspective. We shall see shortly how those "dry" and abstract equations, employing the Hamiltonian and Poisson brackets, can be converted to a complete formulation of quantum dynamics with a simple change of perspective and interpretation of the mathematical symbols. This change of perspective is formally simple, but conceptually deep.

## 1.4 Formal Structure of Quantum Mechanics

### *1.4.1 Hilbert Spaces*

We now introduce the basic ideas about the mathematical structure on which Quantum Mechanics is based. Again, we shall first present in "terse" terms the mathematical concepts and we shall later discuss the physical picture. Let's start with some basic definitions of "abstract algebra."

**Some Definitions**

- A *ring* $\mathcal{R}$ is a set in which two binary operations are defined: *addition* and *multiplication*. The ring is an abelian (i.e., multiplication is commutative) group under addition and a semigroup (i.e., the inverse does not necessarily exist) under multiplication.
- A *field* (or *corpus*) $\mathcal{F}$ is a ring whose nonzero elements form a group under multiplication. Real numbers, complex numbers, and rational numbers, for example, form well-known fields.
- A *vector* (or *linear*) *space* $\mathcal{V}$ over a field $\mathcal{F}$ is a set for which a binary operation ("+", vector addition) and a multiplication by an element of the field (scalar product) are defined such that

1. $\forall \mathbf{u} \in \mathcal{V}\,; \forall \alpha \in \mathcal{F} \mid \alpha\mathbf{u} \in \mathcal{V}$ .
2. $\forall \mathbf{u}, \mathbf{v} \in \mathcal{V}\,; \forall \alpha, \beta \in \mathcal{F} \mid \alpha\mathbf{u} + \beta\mathbf{v} \in \mathcal{V}$.
3. The vector addition satisfies the conditions of *commutativity*, *associativity*, and admits an identity and an inverse element, that is

$$\begin{array}{ll} \mathbf{u}+\mathbf{v}=\mathbf{v}+\mathbf{u} & \text{(commutativity)} \\ \mathbf{u}+(\mathbf{v}+\mathbf{z})=(\mathbf{u}+\mathbf{v})+\mathbf{z} & \text{(associativity)} \\ \exists \mathbf{0} \mid \forall \mathbf{u} \mid \mathbf{0}+\mathbf{u}=\mathbf{u} & \text{(existence of identity element)} \\ \forall \mathbf{u}\ \exists(-\mathbf{u}) \mid \mathbf{u}+(-\mathbf{u})=\mathbf{0} & \text{(existence of inverse element)} \end{array}$$

4. The scalar product and the vector addition are compatible in the sense that

$$\begin{array}{ll} \alpha(\mathbf{u}+\mathbf{v})=\alpha\mathbf{u}+\alpha\mathbf{v} & \text{(distributivity I)} \\ (\alpha+\beta)\mathbf{u}=\alpha\mathbf{u}+\beta\mathbf{u} & \text{(distributivity II)} \\ (\alpha\beta)\mathbf{u}=\alpha(\beta\mathbf{u}) & \text{(compatibility)} \\ 1\mathbf{u}=\mathbf{u} & \text{(where 1 is the identity element in } \mathcal{F}\text{)} \end{array}$$

5. *Linear independence*: Two elements $\mathbf{u}$ and $\mathbf{v}$ of $\mathcal{V}$ are said to be linearly independent when $\mathbf{u}$ cannot be expressed as scalar multiple of $\mathbf{v}$ and vice versa. A set of more than two vectors are said to be linearly independent when none of them can be expressed as a linear combination of the other vectors.
6. *Dimension of* $\mathcal{V}$: A *basis* in $\mathcal{V}$ is a minimal set of $n$ linearly independent elements $\mathbf{u}_i$ $(i=1,n)$ such that every element of $\mathcal{V}$ can be expressed as a linear combination of them. The integer $n$ is the "dimension" of the vector space.

- A *normed vector space* is a vector space in which a "norm" $\|\mathbf{u}\| \in \mathbb{R}$ (or "length") of a vector $\mathbf{u}$ is defined. This defines a "metric" which allows us to define the "distance" between two vectors $\mathbf{u}$ and $\mathbf{v}$ as $\|\mathbf{u}-\mathbf{v}\|$ which can be used to define a topology.
- An *inner-product vector space* is a vector space in which a new binary operation ("inner product" or "dot product") is defined such that $\forall \mathbf{u}, \mathbf{v} \in \mathcal{V} \mid (\mathbf{u}, \mathbf{v}) \in \mathcal{F}$.
- A *normed inner-product* vector space is a vector space in which the "norm" $\|\mathbf{u}\|$ is defined by the inner product: $\|\mathbf{u}\| = \sqrt{(\mathbf{u},\mathbf{u})}$.
- A *topological vector space* is a normed, inner-product vector space in which the topology is defined by the norm, that is, the notion of "$\mathbf{u}$ and $\mathbf{v}$ being close" is determined by the norm $\|\mathbf{u}-\mathbf{v}\|$.
- A *complete topological space* is a topological space in which every Cauchy sequence has a limit. (A Cauchy sequence is a sequence $s_n$ $(n=1,\infty)$ such that $\forall \epsilon \in \mathbb{R}\ \exists N \in \mathbb{Z} \mid \|s_{n+1}-s_n\| < \epsilon\ \forall n > N$).
- A *Banach space* $\mathcal{B}$ is a complete normed topological vector space.
- A *Hilbert* space $\mathcal{H}$ is a Banach space in which the norm is given by the inner product.

In simple words, Hilbert spaces are rich mathematical entities: Their structure is well suited to Quantum Mechanics because their linear (vector) structure captures automatically the linear superposition principle, as we shall see below. Their topological nature inherited by a metric in turn defined by a norm—that is, the possibility of defining distances, "proximity," completeness, "size" of elements—leads to simplified calculus. The additional fact that the norm derives from an inner product adds to this richness, making possible to define "projections" in a natural way. An intuitive way to view Hilbert spaces is to regard them as conventional three-dimensional space $\mathscr{R}$ (i.e., the usual vector space $\mathbb{R}^3$ built on the field $\mathbb{R}$) extended to a continuum of dimensions. Regarding the vector $\psi$ as an infinite-dimensional version of a conventional vector $\mathbf{v}$, the inner product $(\psi_1, \psi_2)$ corresponds to the dot-product $\mathbf{v}_1 \cdot \mathbf{v}_2$; the norm $||\psi||$ to the "length" $|\mathbf{v}|$.

As in the space $\mathscr{R}$, one can introduce a set of "coordinates" $(x_1, x_2, x_3)$ by defining three linearly independent orthogonal vectors of unit length, $\mathbf{e}_1$, $\mathbf{e}_2$, and $\mathbf{e}_3$, and defining the $i$th component $v_i$ of the vector $\mathbf{v}$ by projecting it onto the vector $\mathbf{e}_i$, $v_i = \mathbf{e}_i \cdot \mathbf{v}$, so one can define a "complete basis" of elements of the Hilbert space as a set $\{\phi_i\}$ of orthonormal vectors (that is, such that $(\phi_j, \phi_i) = \delta_{ji}$) and such that any element of the Hilbert space can be represented by its coordinates $\psi_j = (\phi_j, \psi)$, so that

$$\psi = \sum_j \psi_j \, \phi_j = \sum_j \, (\phi_j, \psi) \, \phi_j \, . \tag{1.31}$$

In its most commonly used and simplest realization, the space of single-particle wavefunctions, the Hilbert space is the space of all square-integrable complex functions of three real variables, $L^2(\mathbb{R}^3)$ with the norm of a "vector" $\psi$ given by

$$\|\psi\| = \int_{\mathbb{R}^3} |\psi(\mathbf{r})|^2 \, d\mathbf{r} \, , \tag{1.32}$$

and inner product between two vectors $\psi_1$ and $\psi_2$

$$(\psi_1, \psi_2) = \int_{\mathbb{R}^3} \psi_1^*(\mathbf{r}) \; \psi_2(\mathbf{r}) \, d\mathbf{r} \, . \tag{1.33}$$

In the following we will use Dirac "bra-ket" notation: A vector in a Hilbert space will be denoted by $|\psi\rangle$, and the inner product between $|\psi_1\rangle$ and $|\psi_2\rangle$ by $\langle\psi_1|\psi_2\rangle$. "Kets" $|\psi\rangle$ appear on the right, "bras" $\langle\psi|$ appear on the left. For the special case of $L^2(\mathbb{R}^3)$, "bras" can be viewed as entering the integral Eq. (1.32) via the "complex conjugate" of the function $\psi$. In this notation, Eq. (1.31) becomes

$$|\psi\rangle = \sum_j \, |\phi_j\rangle\langle\phi_j|\psi\rangle \, . \tag{1.34}$$

Formally, we can identify $\sum_j \, |\phi_j\rangle\langle\phi_j|$ as the identity mapping $\widehat{I}$ (or "operator," as defined below). It is usual to write

$$\widehat{I} = \sum_j \, |\phi_j\rangle\langle\phi_j| \, . \tag{1.35}$$

The right-hand side of this equation is usually called a "completeness."

### 1.4.2 Operators on Hilbert Spaces

As we shall see shortly, classical dynamic variables—functions of the generalized coordinates and their conjugate momenta, such as linear or angular momentum—are mapped to a particular class of operators acting on Hilbert spaces—namely, bound or continuous Hermitian operators—according to Canonical Quantization. Some mathematical definitions must be given before explaining why and how these operators are associated with physical observables.

- *Linear operators*. A linear operator $\widehat{A}$ on a Hilbert space $\mathcal{H}$ is a linear map $\mathcal{H} \to \mathcal{H}$, that is

$$\forall \mathbf{u} \in \mathcal{H}, \; \widehat{A}\mathbf{u} = \mathbf{v} \in \mathcal{H}$$
$$\forall \mathbf{u}, \mathbf{v} \in \mathcal{H}, \; \forall \alpha, \beta \in \mathcal{F}, \; \widehat{A}(\alpha\mathbf{u} + \beta\mathbf{v}) = \alpha\widehat{A}\mathbf{u} + \beta\widehat{A}\mathbf{v}$$

Note that linear operators form a vector space themselves: If $\widehat{A}$ and $\widehat{B}$ are linear operators, then also $\alpha\widehat{A}+\beta\widehat{B}$ is a linear operator. Moreover, a product $\widehat{A}\widehat{B}$ may be defined as $\widehat{A}\widehat{B}\mathbf{u}=\widehat{A}(\widehat{B}\mathbf{u})$. The inverse element corresponding to $\widehat{A}$ will be denoted by $\widehat{A}^{-1}$. Combining the operator product and the inverse element with the definition of the "identity operator" $\widehat{I}$ such that $\forall\mathbf{u}\in\mathcal{H},\widehat{I}\mathbf{u}=\mathbf{u}$, this defines a group. However, note that this defines a "nonabelian" group, since in general $\widehat{A}\widehat{B}\neq\widehat{B}\widehat{A}$. The operator equal to the difference between $\widehat{A}\widehat{B}$ and $\widehat{B}\widehat{A}$ is called the "commutator," $[\widehat{A},\widehat{B}]=\widehat{A}\widehat{B}-\widehat{B}\widehat{A}$.

- A *bounded* or *continuous* linear operator is an operator $\widehat{A}$ such that

$$\forall\epsilon\in\mathbb{R}\ |\ \|\mathbf{u}-\mathbf{v}\|<\epsilon\,,\exists\delta\in\mathbb{R}\ |\ \|\widehat{A}(\mathbf{u}-\mathbf{v})\|<\delta$$

- *Hermitian adjoint* operator: The Hermitian adjoint (or *conjugate* or *transpose*) $\widehat{A}^{\dagger}$ of a (bounded) operator $\widehat{A}$ is defined as the operator such that

$$(\mathbf{v},\widehat{A}\mathbf{u})=(\widehat{A}^{\dagger}\mathbf{v},\mathbf{u})\ ,$$

$\forall\mathbf{v},\mathbf{u}\in\mathcal{H}$.

- A *Hermitian* or *self-adjoint* operator is an operator which is identical to its adjoint, so that

$$(\mathbf{v},\widehat{A}\mathbf{u})=(\widehat{A}\mathbf{v},\mathbf{u})\ ,$$

$\forall\mathbf{v},\mathbf{u}\in\mathcal{H}$.

- *Spectrum* of an operator: The spectrum of a linear operator $\widehat{A}$ is the set of all its eigenvalues $\lambda_i$, that is, those elements of $\mathcal{F}$ such that

$$\widehat{A}\ \mathbf{u}_i=\lambda_i\ \mathbf{u}_i\ ,$$

where the vectors $\mathbf{u}_i$ are the eigenvectors of $\widehat{A}$.
*A few very important theorems*: (1) The spectrum of a Hermitian operator on a Hilbert space $\mathcal{H}$ on the field $\mathbb{C}$ is real; (2) Eigenvectors corresponding to different eigenvalues are orthogonal; and (3) The set of its eigenvectors spans the entire Hilbert space, i.e., it forms a (complete) basis.
Let's prove the first two theorems.
If $\widehat{A}\mathbf{u}=\lambda\mathbf{u}$ with $\mathbf{u}\neq\mathbf{0}$, then:

$$\lambda(\mathbf{u},\mathbf{u})=(\mathbf{u},\widehat{A}\mathbf{u})=(\widehat{A}\mathbf{u},\mathbf{u})=\lambda^*(\mathbf{u},\mathbf{u})$$

which proves that $\lambda$ is real. Moreover, let's assume that $\widehat{A}\mathbf{v}=\mu\mathbf{v}$. Then:

$$\lambda(\mathbf{v},\mathbf{u})=(\mathbf{v},\widehat{A}\mathbf{u})=(\widehat{A}\mathbf{v},\mathbf{u})=\mu(\mathbf{v},\mathbf{u})\ ,$$

but if $\lambda\neq\mu$, we must have $(\mathbf{v},\mathbf{u})=0$.
*Another very important theorem*: Two commuting Hermitian operators have common eigenvectors. This can be easily proved as follows: Consider an eigenvector $\mathbf{a}$ with eigenvalue $\alpha$ of the operator $\widehat{A}$ that commutes with the operator $\widehat{B}$. Thus:

$$\widehat{A}\ \mathbf{a}=\alpha\ \mathbf{a}\ . \tag{1.36}$$

Applying the operator $\widehat{B}$ to both sides of this equation, we have

$$\widehat{B}\,\widehat{A}\,\mathbf{a} = \widehat{B}\,\alpha\,\mathbf{a} = \alpha\widehat{B}\,\mathbf{a} = \alpha\,\mathbf{b}\,, \tag{1.37}$$

having set $\mathbf{b} = \widehat{B}\,\mathbf{a}$. On the other hand, thanks to the fact that the two operators commute:

$$\widehat{B}\,\widehat{A}\,\mathbf{a} = \widehat{A}\,\widehat{B}\,\mathbf{a} = \widehat{A}\,\mathbf{b}\,. \tag{1.38}$$

Comparing the right-hand sides of Eqs. (1.37) and (1.38), we see that $\mathbf{b}$ is also an eigenvector of $\widehat{A}$ with the same eigenvalue of $\mathbf{a}$. Therefore, either $\mathbf{b}$ coincides with $\mathbf{a}$ up to a constant $\gamma$, so that $\widehat{B}\,\mathbf{a} = \mathbf{b} = \gamma\,\mathbf{a}$, which shows that $\mathbf{a}$ is an eigenvector also of $\widehat{B}$ with eigenvalue $\gamma$; or, alternatively, $\mathbf{b}$ belongs to the same subspace of degenerate eigenvectors to which $\mathbf{a}$ belongs. In either case, this shows that $\widehat{A}$ and $\widehat{B}$ have common eigenvectors.

The importance of the first theorem above cannot be overstated: In order for an operator to correspond to a physical observable and for its eigenvalues to correspond to results of an experiment aimed at measuring the value of that observable for a system in a given state, it is absolutely necessary that all eigenvalues of the operator be real numbers, obviously.

In Dirac's notation, an operator (observable) $\widehat{A}$ acting on a vector $|\psi\rangle$ will be denoted as $\widehat{A}|\psi\rangle$. The inner product between $\widehat{A}|\psi_1\rangle$ and $|\psi_2\rangle$ by $\langle\psi_2|\widehat{A}|\psi_1\rangle$. The quantity $\langle A\rangle = \langle\psi|\widehat{A}|\psi\rangle$ is called the "expectation value" of $\widehat{A}$ on $|\psi\rangle$. Note that, since the observable $\widehat{A}$ is Hermitian, there is no confusion, so one can view $\langle A\rangle$ either as the inner product of $\widehat{A}|\psi\rangle$ with $\langle\psi|$ or of $\langle\psi|\widehat{A}$ with $|\psi\rangle$. We shall call $\langle\psi_1|\widehat{A}|\psi_2\rangle$ the *matrix element* of $\widehat{A}$ between $|\psi_1\rangle$ and $|\psi_2\rangle$ and denote it by $A_{12}$. Note that $A_{21} = A_{12}^*$, that is, $\langle\psi_2|\widehat{A}|\psi_1\rangle^* = \langle\psi_1|\widehat{A}|\psi_2\rangle$.

### 1.4.3 Canonical Quantization

We are finally ready to describe in detail how any physical system described classically in Hamiltonian terms can be treated quantum mechanically. Let's start by describing this procedure, called "canonical quantization," very schematically. We shall then discuss how to interpret physically the mathematical construct.

1. Define a Hilbert space $\mathcal{H}$ associated with the system. Vectors of this space are state vectors of the system. Observables (or dynamical variables) are ("bounded" or "continuous") Hermitian operators on this space.
2. Provide a Hamiltonian classical description of the system, identifying the canonical coordinates $\mathbf{q}$ and momenta $\mathbf{p}$.
3. Promote the canonical variables to operators acting on the Hilbert space $\mathcal{H}$,

$$q_i \to \hat{q}_i\,, \quad p_i \to \hat{p}_i = -\mathrm{i}\hbar\frac{\partial}{\partial q_i}\,, \tag{1.39}$$

and convert Poisson brackets $\{\ldots,\ldots\}$ to commutators $[\ldots,\ldots]$, so that

$$[\hat{q}_i,\hat{p}_j] = \mathrm{i}\hbar\delta_{ij}\,, \quad [\hat{q}_i,\hat{q}_j] = 0\,, \quad [\hat{p}_i,\hat{p}_j] = 0\,. \tag{1.40}$$

* *The idea of associating the classical momentum with the spatial derivative originates from the theory of continuous groups called* Lie groups. *In rough terms, we can associate transformations with "generating operators" and generating functions. Translations in time are associated with energy and their infinitesimal generating operator is: "take the time derivative." Similarly,*

*translations in space are associated with momentum along the direction of the translation and the corresponding operator is the spatial derivative along that direction. Note the analogy with conservation laws and symmetries, by the way. One may speculate that Schrödinger "derived" his equation starting from $E = H$. He then interpreted the energy $E$ as the generating operator of translations in time and the momentum entering the Hamiltonian $H$ as the generating operator of translations in space. Hence, the* $\mathrm{i}\hbar \mathrm{d}/\mathrm{d}t$ *for $E$, the* $-\mathrm{i}\hbar\nabla$ *replacing* $\mathbf{p}$ *inside $H$. The imaginary units come from Lie groups, since transformations are expressed as complex exponentials of the infinitesimal generating operators. Some other constant with the units of action ($\hbar$) is demanded by dimensional considerations but, clearly, Schrödinger must have known that something related to the Planck constant had to enter the picture.*

4. In the so-called *Heisenberg representation*, state vectors do not evolve in time, but dynamical variables (operators) do and they evolve according to equations of motion that are similar to the Hamilton equations in their Poisson brackets form, Eq. (1.29):

$$\frac{\mathrm{d}\widehat{A}}{\mathrm{d}t} = \frac{\mathrm{i}}{\hbar}[\widehat{H},\widehat{A}] + \frac{\partial \widehat{A}}{\partial t} . \tag{1.41}$$

5. In the so-called *Schrödinger representation*, instead, dynamic variables do not evolve in time but state vectors do. The equation of motion for the state vectors is given by Schrödinger equation which, in its most abstract and general formulation takes the form:

$$\mathrm{i}\hbar \frac{\mathrm{d}\mathbf{u}}{\mathrm{d}t} = \widehat{H}\,\mathbf{u} . \tag{1.42}$$

6. *Equivalence of the Heisenberg and Schrödinger representations.* What we can "know" about a physical system consists of the information contained in the expectation values of the observables. Thus, the entire physical content of the theory is captured by the time evolution of the expectation value $(\mathbf{u},\widehat{A}\mathbf{u})$ of the observable associated with the operator $\widehat{A}$, knowing that the system was initially in the state $\mathbf{u}$. Therefore, consider first the Heisenberg picture in which the state vector is fixed in time and the observable evolves, Eq. (1.41):

$$\begin{aligned}
\frac{\mathrm{d}}{\mathrm{d}t}(\mathbf{u},\widehat{A}(t)\mathbf{u}) &= (\mathbf{u},\frac{\mathrm{d}}{\mathrm{d}t}\widehat{A}(t)\mathbf{u}) \\
&= \frac{\mathrm{i}}{\hbar}(\mathbf{u},[\widehat{H},\widehat{A}(t)]\mathbf{u}) \\
&= \frac{\mathrm{i}}{\hbar}\,[(\mathbf{u},\widehat{H}\widehat{A}(t)\mathbf{u}) - (\mathbf{u},\widehat{A}(t)\widehat{H}\mathbf{u})] \\
&= \frac{\mathrm{i}}{\hbar}\,[(\widehat{H}\mathbf{u},\widehat{A}(t)\mathbf{u}) - (\widehat{A}(t)\mathbf{u},\widehat{H}\mathbf{u})] .
\end{aligned} \tag{1.43}$$

Now let's stop for a moment and change perspective, freezing the observable at time $t$, but letting now the state vector $\mathbf{u}$ evolve in time. So, let's rewrite the last term of Eq. (1.43) as

$$\frac{\mathrm{i}}{\hbar}\,[(\widehat{H}\mathbf{u}(t),\widehat{A}\mathbf{u}(t)) - (\widehat{A}\mathbf{u}(t),\widehat{H}\mathbf{u}(t))] ,$$

and use now Schrödinger equation, Eq. (1.42), to express $\widehat{H}\mathbf{u}(t)$ as $\mathrm{i}\hbar\,\mathrm{d}\mathbf{u}(t)/\mathrm{d}t$. Then the expression above becomes

$$\frac{\mathrm{i}}{\hbar}\left[-\mathrm{i}\hbar\left(\frac{\mathrm{d}\mathbf{u}(t)}{\mathrm{d}t},\widehat{A}\mathbf{u}(t)\right)-\mathrm{i}\hbar\left(\widehat{A}\mathbf{u}(t),\frac{\mathbf{u}(t)}{\mathrm{d}t}\right)\right]=\frac{\mathrm{d}}{\mathrm{d}t}(\mathbf{u}(t),\widehat{A}\mathbf{u}(t))\,, \tag{1.44}$$

which is once more the time evolution of the expectation value of the observable $\widehat{A}$ when the system is on the state $\mathbf{u}$, now in the Schrödinger representation. This shows that the two representations provide identical physical information about the dynamical evolution of the system.

### *1.4.4 The Copenhagen Interpretation*

In his textbook "The Principles of Quantum Mechanics," Dirac introduced the superposition principle as the basic principle of Quantum Mechanics [11]. Gordon Baym followed the same approach [12]. According to this principle, a physical system can be in a linear superposition of two distinct states (two eigenvectors of an observable corresponding, in general, to different eigenvalues of that observable). A photon, for example, can be in a linear superposition of clockwise and counterclockwise polarization, or an electron can simultaneously pass through the top *and* the bottom slit of a double slit.

Certainly, this principle embodies the "weirdness" of the quantum world, as it emerged from those experiments (polarization, double slit) of the early days, and it has emerged even more strikingly in recent experiments, especially Alan Aspect's "proof" [13] (why this is in quote will be discussed shortly) of the Bell inequality [14] that was originally aimed at supporting the EPR paradox [15]. We shall return briefly to these issues. However, the superposition principle per se does not provide an interpretation of what a "state" is, of what result one will obtain from the measurement of an observable. At a more philosophical level, it does not provide a clear "intuitive" picture of how to define "reality" in the quantum world.

That such a satisfactory picture was in doubt emerged right from the early work by Heisenberg [6] and Schrödinger [7]. "What is the wavefunction?" was a question that triggered a period of confusion and intense discussion. It is fair to say that this discussion continues today.

This text is not the right venue to present an overview of all the possible interpretations that have been offered: From Everett's "many-worlds" interpretation [16] to Bohm's semiclassical trajectories in a quantum potential [17, 18], from Griffith's and Omnes' "consistent histories" [19, 20] to classical models resulting in quantum behavior in a stochastic environment [21], from the postulation of a new universal physical constant inducing spontaneous decoherence [22], all the way to the formulation in absence of "observers," as demanded in the cosmological context [23]. We are interested in computational aspects (as the saying goes: We belong to the "shut-up and compute" philosophical school). We adopt the so-called Copenhagen interpretation since, so far, it has proven more than adequate to tackle computational problems providing results that can be successfully compared to experimental observations.

There is no unique way to define "the Copenhagen interpretation." While Niels Bohr was its champion (hence the name, from the institute he directed in Copenhagen), it emerged from ideas originating in the late 1920s with contributions from all of the major players of the time. Its most formal crystallization is due to Jordan [24, 25] together with Born and Heisenberg, and especially to John von Neumann [26]. We list below the main "axioms" that most people would agree define the Copenhagen interpretation. Violating a correct historical perspective, we formulate these axioms in terms of the Canonical Quantization framework we have outlined above.

1. Any physical system can be in any of the states $|\psi\rangle$ belonging to the Hilbert space associated with the system. This Hilbert vector embodies the totality of information we can obtain about the system and there is no additional information, like "hidden variables," local or not. The state itself, however, is not observable.

2. Although one may define and, in principle, measure many dynamical variables associated with the system, only commuting observables can be measured simultaneously. Complementary observables can be measured simultaneously only within the limitations posed by the Heisenberg uncertainty principle: Defining the "standard deviation" $\sigma_A = [\langle \widehat{A}^2 \rangle - \langle \widehat{A} \rangle^2]^{1/2}$ for an observable $\widehat{A}$ in terms of expectation values on a state $|\psi\rangle$, if $\widehat{A}$ and $\widehat{B}$ are two observables, then

$$\sigma_A \sigma_B \geq \frac{1}{2} \left| \langle [\widehat{A}, \widehat{B}] \rangle \right| \tag{1.45}$$

the equal sign being achievable only for a suitably selected "minimum uncertainty" state. For the well-known case of position and momentum of a particle in one dimension, $[\hat{x}, \hat{p}] = \mathrm{i}\hbar$, so Eq. (1.45) takes the well-known form $\sigma_x \sigma_p \geq \hbar/2$.
3. (*Schrödinger picture*). An isolated system in the state $|\psi\rangle$ will evolve in time according to the "generalized" Schrödinger equation

$$\frac{\mathrm{d}}{\mathrm{d}t}|\psi\rangle = \widehat{H}|\psi\rangle \,, \tag{1.46}$$

where $\widehat{H}$ is the Hamiltonian operator.
(*Heisenberg picture*). Alternatively, an isolated system initially in a state $|\psi\rangle$ will remain in this state, but observables $\widehat{A}$ evolve in time according to the Hamilton equation (1.41):

$$\frac{\mathrm{d}\widehat{A}}{\mathrm{d}t} = \frac{\mathrm{i}}{\hbar}[\widehat{H}, \widehat{A}] + \frac{\partial \widehat{A}}{\partial t} \,. \tag{1.47}$$

In terms of the dynamics of the results expected from measurements, as described below, we have seen that the time evolution of the expectation value $\langle A \rangle$ is the same in both the Schrödinger and Heisenberg pictures.
4. Measurements are performed by an observing apparatus (laboratory device) that obeys the laws of classical physics.
5. Consider an observable $\widehat{A}$ with eigenvector $|a_i\rangle$ corresponding to eigenvalues $a_i$, that is

$$\widehat{A}|a_i\rangle = a_i|a_i\rangle \,. \tag{1.48}$$

Express now the state $|\psi\rangle$ (assumed normalized, $\langle \psi|\psi\rangle = 1$)) over the complete basis $\{|a_i\rangle\}$ of the eigenvectors of the observable $\widehat{A}$,

$$|\psi\rangle = \sum_j \langle a_j|\psi\rangle \, |a_j\rangle \,. \tag{1.49}$$

Then, the coefficients $\alpha_j = \langle a_j|\psi\rangle$ are called the amplitudes of the state $|\psi\rangle$ on the eigenvectors $|a_j\rangle$ and their squared magnitudes $|\alpha_j|^2$ represent the probability that a measurement of the observable $\widehat{A}$ on the system in the state $|\psi\rangle$ will yield the result $a_j$.
6. Upon a measurement of the observable $\widehat{A}$ on the system in the state $|\psi\rangle$, if the result of the measurement is $a_n$, then the system instantaneously "collapses" irreversibly into the state $|a_n\rangle$, an eigenstate of the observable $\widehat{A}$.
7. Once a measurement has collapsed the system into a state $|a_n\rangle$, successive measurements of the same observable $\widehat{A}$ will always yield the same results, $a_n$.
8. From Eqs. (1.36)–(1.38), a system can be in a state that is an eigenvector of two commuting observables, $\widehat{A}$ and $\widehat{B}$. Therefore, the two observables can be measured simultaneously and, after

the measurement, the system can be said to possess definite values for both observables. In this case, the observables are said to be "compatible." If, instead, the observables do not commute, they are called "complementary" and the system can be in an eigenstate of only one of them, not of both simultaneously. Accordingly, "pure states" (that is, excluding mixtures due to "classical ignorance") that are eigenstates of an observable can be considered as linear superpositions of other pure states that are eigenstates of compatible observables. A state cannot be thought of as a superposition of eigenstates of complementary observables. (This, of course, allows the mathematical representation of the systems, states, and observable, in terms of the Hilbert-space formalism used to perform the Canonical Quantization.)

9. If two isolated systems are brought together, the total state of the larger combined system can be thought as a "tensor product" of the states of the two systems. A fortiori, the Hilbert space associated with the larger system is the tensor product of the Hilbert spaces associated with each sub-system.
10. As the size of a system (often defined as the number of degrees of freedom) grows, the quantum-mechanical laws approach those of Classical Mechanics. This is known as the "correspondence principle" of Bohr and Heisenberg.
11. The "inner workings" of the system are hidden. Reality is exclusively what we measure and the act of measuring defines it. In the usual paradoxical terms, 'a tree falling in a forest does not make any noise if there is no one to hear it.'

As a simple corollary, note that the expectation value of an observable $\widehat{A}$ for a system in a state $|\psi\rangle$ has been defined before as $\langle A\rangle = \langle\psi|\widehat{A}|\psi\rangle$. Using the expansion (1.49), we have

$$\langle A\rangle = \langle\psi|\widehat{A}|\psi\rangle = \sum_{ij} \alpha_i^* \alpha_j \, \langle a_i|\widehat{A}|a_j\rangle \, . \tag{1.50}$$

Because of the orthonormality of the eigenvectors $\{|a_j\rangle\}$,

$$\langle A\rangle = \sum_{ij} \alpha_i^* \alpha_j \, a_j \, \langle a_i|a_j\rangle = \sum_{i} |\alpha_i|^2 \, a_j \, . \tag{1.51}$$

The coefficients $\alpha_i$, as we saw before, express the probability that a measurement of $\widehat{A}$ performed on the system in a state $|\psi\rangle$ will yield the result $a_i$. Therefore, Eq. (1.51) shows that the expectation value of the observable $\widehat{A}$ is simply the average of all possible eigenvalues, weighted by the probability of obtaining each one of them.

### *1.4.5 The "Copenhagen Controversy"*

The Copenhagen interpretation we have just outlined has emerged in the late 1920s amidst confusion and controversy. Indeed, the picture of the physical world ("reality") that emerges from this interpretation leaves many people unhappy and also presents some paradoxes (some would say: "contradictions").

At a general "intuitive" or philosophical level, "reality" ceases to exist as a concept independent of the observer. This is what is usually known as the "logic positivist" view. The first two axioms above imply that it is meaningless to assign a property $b$ (say: the position of a particle) to a system when we have measured a complementary property $a$ (say: its momentum), even only conceptually, regardless of measurements: Not only can we not assign a position to the particle; the particle cannot be thought of as being endowed with a property like "position" once we have measured its momentum.

In other words, the ultimate nature of reality is determined by the observer. Realists, like Einstein or Schrödinger himself, never accepted this view and fought back. Einstein, in particular, claimed that Quantum Mechanics is an incomplete theory [15] that, eventually, will need to be replaced by a better picture that will uncover and deal explicitly with those "hidden variables" that Copenhagen dismisses. Attempts to re-introduce realism are at the roots also of Bohm's interpretation [17, 18], among others.

At a more technical level, axioms 6–8 coupled to axiom 9 present an intrinsic problem. A measurement apparatus must obey the laws of classical physics. Yet, axiom 9 tells us that, since Quantum Mechanics is assumed to be a complete theory, we could describe quantum mechanically the larger system $\{\text{original system}\}+\{\text{measuring apparatus}\}$. This larger system, associated with the tensor-product Hilbert space of the Hilbert spaces of both systems, would then be subject to measurements performed by a third system, a new measuring apparatus. This new apparatus must obey the laws of classical physics. Clearly, this results in a recursive process, known as the "von Neumann chain." Where does the quantum world end and does the classical world begin? Particularly puzzling are the paradoxes that emerge as soon as we are willing to include the observer in the quantum description, since we are brought into the realm of defining human consciousness and how this affects reality [27].

Despite the severity of these problems, from a very pragmatic perspective, so far the Copenhagen interpretation has withstood all possible tests. John Bell, another realist, intended to show how absurd this interpretation is, by emphasizing differences between classical probabilities and quantum probabilities that include "interference" terms. His famous "Bell inequality" [14], finally tested experimentally [13], has confirmed the validity of the Copenhagen interpretation, raising even more puzzles. Skeptics remain and loopholes are constantly found in the experiments to disprove Aspect's result, which why we wrote "proof" in quotes. Yet, it is fair to say that, as new experiments are performed, a realist's viewpoint becomes ever harder to defend. Theories that introduce local hidden variables, in an attempt to restore "intuitive" reality, seem to be inconsistent with these experimental results. Theories that attempt to achieve the same goal by assuming nonlocal hidden variables, such as Bohm's nonlocal quantum potential [17, 18], appear to be still consistent (or, at least, "not inconsistent") with experiments. However, this nonlocality introduces a feeling of "weirdness" via "spooky actions at a distance" (a statement attributed to Einstein) that feels uncomfortable to the realists themselves.

Richard Feynman said that Quantum Mechanics with its Copenhagen interpretation works well "for all practical purposes." In the rest of this text we shall leave philosophy on the side: After all, the mathematical machinery of Quantum Mechanics works well indeed (that is, it provides results consistent with experimental observations available at present), the rest we shall leave to after-dinner discussions... However, we must admit that some issues that recently have become popular and of practical interest (such as quantum entanglement, quantum computing, and quantum teleportation, to mention just a few) make the problem more pressing, more practically relevant, and more severe than we would like to believe.

## Problems

**1.1. (Newton's Law)** Consider a particle of mass $m$ in three dimensions in the presence of a potential $V(\mathbf{r})$. Write the Lagrangian, the Hamiltonian, Euler–Lagrange, and Hamilton equations for this simple system. Convince yourself that you recover Newton's law $\mathbf{F} = m\mathbf{a}$.

**1.2. (Conservation Laws)** Consider the same situation of the previous problem, but now assume that $V(\mathbf{r})$ has cylindrical symmetry; that is, that it depends only on the cylindrical coordinates $r$ and $z$, but not on the angle $\phi$. Write the Lagrangian, Hamiltonian, Euler–Lagrange, and Hamilton equations in cylindrical coordinates. Finally, verify directly that the $z$-component, $L_z$, of the angular momentum of the particle is a constant of motion.

**1.3. (Tunneling Across a Square Potential Barrier)** Consider the tunneling problem with the potential barrier:

$$V(z) = \begin{cases} 0 & \text{for } z \leq 0 \\ V > 0 & \text{for } 0 < z < L \\ 0 & \text{for } z \geq L \end{cases} .$$

Write the wavefunction as

$$\psi(z) = \begin{cases} Ae^{ikz} + Be^{-ikz} & \text{for } z \leq 0 \\ Ce^{\kappa z} + De^{-\kappa z} & \text{for } 0 < z < L \\ Fe^{ikz} & \text{for } z \geq L \end{cases} ,$$

with $k = (2mE)^{1/2}/\hbar$ and $\kappa = [2m(V-E)]^{1/2}/\hbar$.

(a) Write the system of four equations expressing the continuity of the wavefunction and its derivative at $z = 0$ and at $z = L$.
(b) Find the transmission coefficient, $T = |F|^2/|A|^2$. There's no need to solve the full system. Be creative. Hint: Multiply the equation expressing continuity of $\psi$ at $z = 0$ by $ik$ and add and subtract it from the equation expressing continuity of the derivatives at $z = 0$. Do a similar thing with the other two equations (by multiplying one by $\kappa$). Now it should be relatively easy to solve for $F$ in terms of $A$ alone. This gives you $T$.

**1.4. (The WKB Approximation)** The Wentzel–Kramers–Brillouin (WKB) approximation to solve the Schrödinger equation consists in writing the solution of the time-independent problem:

$$-\frac{\hbar^2}{2m}\frac{d^2\psi(x)}{dx^2} + V(x)\psi(x) = E\psi(x) ,$$

as

$$\psi(x) \approx \frac{1}{k^{1/2}} \exp\left\{ i \int^x k(x') \, dx' \right\} ,$$

where $k(x) = \{2m[E - V(x)]\}^{1/2}/\hbar$. This is a good approximation if the potential $V(x)$ varies slowly (that is, it does not change much compared to the electron energy $E$ when $x$ varies over several de Broglie wavelengths). If $E - V(x) < 0$, the WKB wavefunction becomes

$$\psi(x) \approx \frac{1}{k^{1/2}} \exp\left\{ - \int^x \kappa(x') \, dx' \right\} ,$$

where now $\kappa(x) = \{2m[V(x) - E]\}^{1/2}/\hbar$.
Let's now ignore the factor $k^{-1/2}$ (which simply ensures continuity of probability current). Consider now the previous tunneling problem (Problem 1.2.) and identify the WKB approximation to the transmission coefficient as

$$T_{\text{WKB}} = |\psi(L)|^2 = \exp\left\{ -2 \int_0^L \kappa(x') \, dx' \right\} . \tag{1.52}$$

Compare $T_{\mathrm{WKB}}$ with the "exact" transmission coefficient $T$ of the previous tunneling problem, in the limit in which $\kappa L >> 1$.

**1.5. (Probability Density and Current Density)** From the Schrödinger equation derive the continuity equation:

$$\frac{\partial \rho}{\partial t} + \nabla \cdot \mathbf{S} = 0 \, ,$$

where $\rho = |\Psi|^2$ is the "probability density" and $\mathbf{S} = \frac{i\hbar}{2m}[\Psi \nabla \Psi^* - \Psi^* \nabla \Psi]$ is the "probability density current." It is this conservation law that makes it possible to interpret consistently the squared amplitude of the single-particle wavefunction as a probability density.

**1.6. (Some Useful "Matrix Elements")** Calculate the matrix element between two wavefunctions of the form

$$\psi(\mathbf{k},\mathbf{r}) = \frac{1}{V^{1/2}} \mathrm{e}^{\mathrm{i}\mathbf{k}\cdot\mathbf{r}} \quad \text{and} \quad \psi(\mathbf{k}',\mathbf{r}) = \frac{1}{V^{1/2}} \mathrm{e}^{\mathrm{i}\mathbf{k}'\cdot\mathbf{r}} \, ,$$

and the perturbation potentials of the form:

(a) $\widehat{H} \propto \mathrm{e}^{\mathrm{i}\mathbf{q}\cdot\mathbf{r}}$
(b) $\widehat{H} \propto \delta(\mathbf{r})$
(c) $\widehat{H} \propto |\mathbf{r}|^{-2}$
(d) $\widehat{H} \propto \mathrm{e}^{-|\mathbf{r}|/r_0}$ Polar coordinates are useful in **c** and **d**.

**1.7. (Second-Order Time-Independent Perturbation Theory)** Extend the procedure followed in Appendix A to find the perturbed eigenvalues $E_n^{(2)}$ to second order in $\alpha$ using time-independent perturbation theory.

## References


1. A. Einstein, Zur Elektrodynamik bewegter Körper (On the electrodynamics of moving bodies). Ann. Phys. **17**, 891 (1905)
2. N. Bohr, On the constitution of atoms and molecules, part I. Philos. Mag. **26**, 1 (1913)
3. T. Kuhn, *Black-Body Theory and the Quantum Discontinuity: 1894–1912* (Clarendon Press, Oxford, 1978)
4. M. Planck, Über das Gesetz der Energieverteilung im Normalspektrum (On the law of distribution of energy in the normal spectrum). Ann. Phys. **4**, 553 (1901). Translated in K. Ando, http://theochem.kuchem.kyoto-u.ac.jp/Ando/planck1901.pdf
5. A. Einstein, Über einen die Erzeugung und Verwandlung des Lichtes betreffenden heuristischen Gesichtspunk (Concerning an heuristic point of view toward the emission and transformation of light). Ann. Phys. **17**, 132 (1905)
6. W. Heinsenberg, Über quantentheoretishe Umdeutung kinematisher und mechanischer Beziehungen (Quantum-theoretical re-interpretation of kinematic and mechanical relations). Z. Phys. **33**, 879 (1925)
7. E. Schrödinger, Quantisierung als Eigenwertproblem. (Erste Mitteilung) (Quantization as an eigenvalue problem (Part I)). Ann. Phys. **70**, 261 (1926)
8. J.L. Lagrange, *Mécanique Analytique*, Paris (1788) [English translation: *Analytical Mechanics* (Kluwer, Dordrecht, 1998)]
9. S.W.R. Hamilton, On a general method of expressing the paths of light, & of the Planets, in *Coefficients of a Characteristic Function* (P.D. Hardy, Dublin, 1833)
10. H. Goldstein, *Classical Mechanics* (Person Education, Upper Saddle River, 2002)
11. P.A.M. Dirac, *The Principles of Quantum Mechanics* (Oxford University Press, Oxford, 1930)
12. G. Baym, *Lectures on Quantum Mechanics* (Westview, New York, 1990)
13. A. Aspect, P. Grangier, G. Roger, Experimental realization of Einstein-Podolsky-Rosen-Bohm Gedankenexperiment: a new violation of Bell's inequalities. Phys. Rev. Lett. **49**, 91 (1982)

14. J.S. Bell, On the Einstein–Poldolsky–Rosen paradox. Physics **1**, 195 (1964)
15. A. Einstein, B. Podolsky, N. Rosen, Can quantum-mechanical description of physical reality be considered complete? Phys. Rev. **47**, 777 (1935)
16. H. Everett, Relative state formulation of quantum mechanics. Rev. Mod. Phys. **29**, 454 (1957)
17. D. Bohm, A suggested interpretation of the quantum theory in terms of "hidden variables". I. Phys. Rev. **85**, 166 (1952)
18. D. Bohm, A suggested interpretation of the quantum theory in terms of "hidden variables". II. Phys. Rev. **85**, 180 (1952)
19. R.B. Griffith, *Consistent Quantum Theory* (Cambridge University Press, Cambridge, 2002)
20. R. Omnes, *The Interpretation of Quantum Mechanics* (Princeton University Press, Princeton, 1994)
21. E. Nelson, Derivation of the Schrödinger equation from Newtonian mechanics. Phys. Rev. **150**, 1079 (1966)
22. G.C. Ghirardi, A. Rimini, T. Weber, A model for a unified quantum description of macroscopic and microscopic systems, in *Quantum Probability and Applications*, ed. by L. Accardi et al. (Springer, Berlin, 1985)
23. M. Gell-Mannn, J.B. Hartle, Classical equations for quantum systems. Phys. Rev. D **47**, 3345 (1994)
24. M. Born, P. Jordan, Zur Quantenmechanik (Quantum mechanics). Z. Phys. **34**, 858 (1925)
25. M. Born, W. Heisenberg, P. Jordan, Zur Quantenmechanik. II (Quantum mechanics II). Z. Phys. **35**, 557 (1926)
26. J. von Neumann, *Mathematical Foundations of Quantum Mechanics* (Princeton University Press, Princeton, 1996)
27. E. Wigner, M. Henry, Remarks on the mind body question, in symmetries and reflections, scientific essays. Am. J. Phys. **35**, 1169 (1967)

# Chapter 2
# The Periodic Table, Molecules, and Bonds

## 2.1 Atoms: Building up the Periodic Table

In this chapter we show how the Schrödinger equation—together with a few additional concepts—can allow us to build (conceptually) the periodic table of the elements. In turn, this will give us some ideas about the nature of atomic orbitals, concept that constitutes the basic building block underlying the formation of atomic bonds in molecules and crystals.

First, we need to introduce the concept of *quantum number*. The example of a particle in a three-dimensional (3D) box will illustrate this. For simplicity, we treat the simpler case of a free particle. Any text on Quantum Mechanics could—and should—be consulted to have a more complete picture and set of examples [1–5].

### 2.1.1 *Free Particles in Three Dimensions*

The Schrödinger equation describing a particle of mass $m$ in absence of any potential is

$$-\frac{\hbar^2\nabla^2}{2m}\,\Psi(\mathbf{r},t) = \mathrm{i}\,\hbar\frac{\partial}{\partial t}\,\Psi(\mathbf{r},t). \tag{2.1}$$

In order to solve this partial differential equation we use a technique called "separation of variables." Let us look for a solution of the form

$$\Psi(\mathbf{r},t) = \psi(\mathbf{r})\;T(t), \tag{2.2}$$

that is, we look for a solution expressed as the product of a function $\psi$ that depends only on the spatial coordinates $\mathbf{r}$, and a function $T$ that depends only on time. Thanks to the linearity of the Schrödinger equation, the most general solution satisfying the required boundary and initial conditions can always be written as a linear combination of solutions of this type. Therefore, the particular choice given by Eq. (2.2) does not restrict the generality of the solution we can obtain.

Inserting Eq. (2.2) into Eq. (2.1) and dividing by $\Psi$, we obtain

$$-\frac{1}{\psi(\mathbf{r})}\,\frac{\hbar^2\nabla^2}{2m}\,\psi(\mathbf{r}) = \mathrm{i}\,\hbar\frac{1}{T(t)}\,\frac{\mathrm{d}T(t)}{\mathrm{d}t}. \tag{2.3}$$



M. Fischetti, W.G. Vandenberghe, *Advanced Physics of Electron Transport in Semiconductors and Nanostructures*, Graduate Texts in Physics, DOI 10.1007/978-3-319-01101-1_2

Note now that the left-hand side (lhs) of this equation is a function of the spatial coordinates **r** only, whereas the right-hand side (rhs) depends only on $t$. Since **r** and $t$ are independent variables, we are free to change them as we please. Therefore, the equality expressed by Eq. (2.3) can hold only if both sides of the equation are equal to a constant. Therefore, calling this constant $E$, must we have

$$\frac{dT(t)}{dt} + \mathrm{i}\frac{E}{\hbar}\,T(t) = 0, \tag{2.4}$$

and

$$-\frac{\hbar^2\nabla^2}{2m}\,\psi(\mathbf{r}) = E\,\psi(\mathbf{r}). \tag{2.5}$$

Equation (2.4) has a general solution $T(t) = A\mathrm{e}^{-\mathrm{i}\,\omega t}$, where $A$ is a constant to be determined by the initial conditions, and $\hbar\omega = E$. Note that, since the lhs of Eq. (2.5) is just the Hamiltonian of the system, this equation is exactly the eigenvalue equation for the operator $\widehat{\mathbf{H}}$ of the simple system. Therefore, we can identify the search for the solution $\psi$ as the search for the eigenvectors of the Hamiltonian, and $E$ as their eigenvalues. Therefore, $E$ is just the energy of the particle.

We conclude that our solution has the form $\Psi(\mathbf{r},t) = A\mathrm{e}^{-\mathrm{i}\,\omega t}\psi(\mathbf{r})$. This represents an eigenvector of the Hamiltonian, $|\psi\rangle$, in the Hilbert space $L^2(\mathbb{R}^3)$, that rotates in time with radian frequency $\omega$. The wavefunction $\psi(\mathbf{r}) = \langle\mathbf{r}|\psi\rangle$ satisfies the "time-independent" Schrödinger equation, Eq. (2.5).

Assuming that the potential energy of the particle is just a constant, $V_0$, we can determine $\psi(\mathbf{r})$ by following the same procedure of "separation of variables." We express the Laplacian and the spatial coordinate **r** in terms of their components:

$$-\frac{\hbar^2}{2m}\left[\frac{\partial^2}{\partial x^2}\psi(x,y,z) + \frac{\partial^2}{\partial y^2}\psi(x,y,z) + \frac{\partial^2}{\partial z^2}\psi(x,y,z)\right] + V_0\,\psi(x,y,z) = E\,\psi(x,y,z). \tag{2.6}$$

Now, let us look for a solution of the form:

$$\psi(x,y,z) = \psi_x(x)\,\psi_y(y)\,\psi_z(z), \tag{2.7}$$

where $\psi_x$, $\psi_y$, and $\psi_z$ are three different functions in one variable only. Inserting this into Eq. (2.6), we obtain

$$-\frac{\hbar^2}{2m}\left[\psi_y(y)\,\psi_z(z)\,\frac{\mathrm{d}^2\psi_x(x)}{\mathrm{d}x^2} + \psi_x(x)\,\psi_z(z)\,\frac{\mathrm{d}^2\psi_y(y)}{\mathrm{d}y^2} + \psi_x(x)\,\psi_y(y)\,\frac{\mathrm{d}^2\psi_z(z)}{\mathrm{d}z^2}\right] = E\,\psi_x(x)\psi_y(y)\psi_z(z). \tag{2.8}$$

Dividing both sides by $\psi_x(x)\psi_y(y)\psi_z(z)$:

$$-\frac{\hbar^2}{2m}\left[\frac{1}{\psi_x(x)}\frac{\mathrm{d}^2\psi_x(x)}{\mathrm{d}x^2} + \frac{1}{\psi_y(y)}\frac{\mathrm{d}^2\psi_y(y)}{\mathrm{d}y^2} + \frac{1}{\psi_z(z)}\frac{\mathrm{d}^2\psi_z(z)}{\mathrm{d}z^2}\right] = E. \tag{2.9}$$

This equation tells us that the sum of three functions of different variables ($x$, $y$, and $z$) must always be equal to a constant ($E$) for all possible values of $x$, $y$, and $z$. This is possible only if each function is equal to a constant, that is

$$-\frac{\hbar^2}{2m}\frac{1}{\psi_x(x)}\frac{\mathrm{d}^2\psi_x(x)}{\mathrm{d}x^2} = E_x, \tag{2.10}$$

$$-\frac{\hbar^2}{2m}\frac{1}{\psi_y(y)}\frac{\mathrm{d}^2\psi_y(y)}{\mathrm{d}y^2} = E_y, \tag{2.11}$$

$$-\frac{\hbar^2}{2m}\frac{1}{\psi_z(z)}\frac{\mathrm{d}^2\psi_z(z)}{\mathrm{d}z^2} = E_z. \tag{2.12}$$

We have indicated with $E_x$, $E_y$, and $E_z$ three arbitrary constants subject to the condition $E_x + E_y + E_z = E$. Rewriting Eq. (2.10) as

$$-\frac{\hbar^2}{2m}\frac{\mathrm{d}^2\psi_x(x)}{\mathrm{d}x^2} = E_x\psi_x(x), \tag{2.13}$$

its general solution has the form:

$$\psi_x(x) = A_x\,\mathrm{e}^{\mathrm{i}k_x x}, \tag{2.14}$$

where $A_x$ is an arbitrary multiplicative constant that will be determined below by "normalizing" the wavefunction. In this equation

$$k_x = \frac{1}{\hbar}\sqrt{2mE_x}. \tag{2.15}$$

is the wavenumber along the $x$-axis. We can proceed in an analogous way for the $y$- and $z$-components, so that the full wavefunction has the form:

$$\psi(\mathbf{r}) = A_xA_yA_z\,\mathrm{e}^{\mathrm{i}(k_x x + k_y y + k_z z)} = A\,\mathrm{e}^{\mathrm{i}\mathbf{k}\cdot\mathbf{r}}, \tag{2.16}$$

having defined the *wavevector* **k** as the vector with components $(k_x, k_y, k_z)$, and A is an arbitrary constant.

Recalling that $\Psi(\mathbf{r},t) = A\mathrm{e}^{-\mathrm{i}\,\omega t}\psi(\mathbf{r})$, Eq. (2.16) describes the particle as a "plane-wave" propagating in the direction of the wavevector **k** with wavenumber (number of wavelengths per unit length along the direction of propagation) given by the magnitude $k$ of **k** and with phase velocity $\omega/k$. Note that its momentum **p** will be $\hbar\mathbf{k}$, and the energy of the particle is $E = \hbar^2k^2/(2m)$.

### 2.1.2 Particles in a Three-Dimensional Box

Consider again Eq. (2.5) or (2.6), but assume now that the particle is confined in a cubic box with side length $L$; i.e., the potential vanishes for $-L/2 < x < L/2$, $-L/2 < y < L/2$, and $-L/2 < z < L/2$, but it is infinite otherwise. Therefore, we must look for solutions of a form similar to what we have just found for a free particle, but subject to the condition that the wavefunction vanishes at the boundary, and outside, of the box. Therefore, we must look for solutions $\psi_x(x)$, $\psi_y(y)$, and $\psi_z(z)$ of the form:

$$\psi_x(x) = \left(\frac{2}{L}\right)^{1/2} \sin\left(\frac{n_x\pi x}{L}\right)$$

where $n_x$ is an integer, with similar expressions being valid for $y$ and $z$. The full wavefunction will now be

$$\psi(\mathbf{r}) = \left(\frac{8}{\Omega}\right)^{1/2} \sin\left(\frac{n_x \pi x}{L}\right) \sin\left(\frac{n_y \pi y}{L}\right) \sin\left(\frac{n_z \pi z}{L}\right), \tag{2.17}$$

where $\Omega = L^3$ is the volume of the box and the energy of the particle will be

$$E_{n_x,n_y,n_z} = \frac{(n_x^2 + n_y^2 + n_z^2)\hbar^2\pi^2}{2mL^2}.$$

Note that the normalization constant $A = (8/\Omega)^{1/2}$ in this case has been fixed by requiring that there is unit probability of finding the electron inside the box:

$$\int_{\Omega} d\mathbf{r}\, |\psi(\mathbf{r})|^2 = 1.$$

If we were considering a particle in one dimension, the particle would have only one degree of freedom and we would need only a single "quantum number," say, $n_x$, to label the energy levels. In 3D, instead, the particle has 3 degrees of freedom and we now need three quantum numbers, $n_x$, $n_y$, and $n_z$ to label the possible states (that is, the possible "types" of wavefunction) of the particle.

### 2.1.3 The Hydrogen Atom

The case of the H atom is conceptually similar to the case of the particle in a cubic box. Unfortunately, it is significantly more complicated from a mathematical perspective, because the potential confining the particle (the Coulomb potential energy is $-e^2/(4\pi\epsilon_0 r)$) is not "flat," but depends on the distance $r$ from the nucleus. Indeed, the time-independent Schrödinger equation we must solve is

$$\left[-\frac{\hbar^2\nabla^2}{2m_e} + V(r)\right] \psi(\mathbf{r}) = E\psi(\mathbf{r}), \tag{2.18}$$

where $V(r) = -e^2/(4\pi\epsilon_0 r)$. We have used the notation $m_e$ for the mass of the particle, since we are specifically interested in an electron around a proton. Also, we have attempted to minimize the confusion caused by the fact that the symbol $m$ will be used below to label a quantum number. It is true that the potential is not constant, so separating variables in Cartesian coordinates would not work: We would not be able to re-express this equation as the sum of terms that depends on a single variable, as we found in Eq. (2.9). However, the potential energy $V$ depends only on the radial coordinate $r$, not the polar and azimuthal angles $\theta$ and $\phi$. This spherical symmetry can be exploited to separate variables if we employ spherical coordinates. Thus, we look for solutions of the form:

$$\psi(\mathbf{r}) = \psi(r, \theta, \phi) = R(r)\, \Theta(\theta)\, \Phi(\phi). \tag{2.19}$$

The "radial function" $R(r)$ describes the way the wavefunction spreads away from the nucleus. The "angular functions" $\Theta(\theta)$ and $\Phi(\phi)$ describe how the wavefunction is distributed as a function of the polar and azimuthal angles.

Rewriting Eq. (2.18) in spherical coordinates,

$$\frac{1}{r^2\sin\theta}\left[\sin\theta\frac{\partial}{\partial r}\left(r^2\frac{\partial\psi}{\partial r}\right)+\frac{\partial}{\partial\theta}\left(\sin\theta\frac{\partial\psi}{\partial\theta}\right)+\frac{1}{\sin\theta}\frac{\partial^2\psi}{\partial\phi^2}\right]-\frac{2m_eV(r)}{\hbar^2}\psi$$
$$=-\frac{2m_eE}{\hbar^2}\psi. \tag{2.20}$$

and inserting Eq. (2.19), we have

$$\frac{1}{r^2}\Theta\Phi\frac{\mathrm{d}}{\mathrm{d}r}\left(r^2\frac{\mathrm{d}R}{\mathrm{d}r}\right)+\frac{R\Phi}{r^2\sin\theta}\frac{\mathrm{d}}{\mathrm{d}\theta}\left(\sin\theta\frac{\mathrm{d}\Theta}{\mathrm{d}\theta}\right)+\frac{R\Theta}{r^2\sin^2\theta}\frac{\mathrm{d}^2\Phi}{\mathrm{d}\phi^2}-\frac{2m_eV(r)}{\hbar^2}R\Theta\Phi$$
$$=-\frac{2m_eE}{\hbar^2}R\Theta\Phi. \tag{2.21}$$

Now, following the same procedure we have followed before in Cartesian coordinates, let's divide by $R\Theta\Phi$ the expression above:

$$\frac{1}{Rr^2}\frac{\mathrm{d}}{\mathrm{d}r}\left(r^2\frac{dR}{\mathrm{d}r}\right)+\frac{1}{r^2\sin\theta}\frac{1}{\Theta}\frac{\mathrm{d}}{\mathrm{d}\theta}\left(\sin\theta\frac{\mathrm{d}\Theta}{\mathrm{d}\theta}\right)+\frac{1}{r^2\sin^2\theta}\frac{1}{\Phi}\frac{\mathrm{d}^2\Phi}{\mathrm{d}\phi^2}-\frac{2m_eV(r)}{\hbar^2}=-\frac{2m_eE}{\hbar^2}. \tag{2.22}$$

Multiplying by $r^2\sin^2\theta$ and rearranging the result, we obtain

$$-\frac{1}{\Phi}\frac{\mathrm{d}^2\Phi}{\mathrm{d}\phi^2}=r^2\sin^2\theta\left\{\frac{1}{Rr^2}\frac{\mathrm{d}}{\mathrm{d}r}\left(r^2\frac{dR}{\mathrm{d}r}\right)+\frac{1}{r^2\sin\theta}\frac{1}{\Theta}\frac{\mathrm{d}}{\mathrm{d}\theta}\left(\sin\theta\frac{\mathrm{d}\Theta}{\mathrm{d}\theta}\right)-\frac{2m_e[V(r)-E]}{\hbar^2}\right\}. \tag{2.23}$$

Now we follow the "usual" reasoning: The lhs is a function of $\phi$ alone, while the rhs is a function of $r$ and $\theta$. Since we can vary $r$, $\theta$, and $\phi$ independently, the only way Eq. (2.23) can hold is for both sides to be equal to a constant. Let's call this constant $m^2$ (where $m$ is to be viewed, for now, as a general complex number). Thus,

$$-\frac{1}{\Phi}\frac{\mathrm{d}^2\Phi}{\mathrm{d}\phi^2}=m^2, \tag{2.24}$$

so that the "azimuthal" factor $\Phi$ must be of the form:

$$\Phi(\phi)\sim \mathrm{e}^{\pm\mathrm{i}m\phi}, \tag{2.25}$$

up to a normalization constant. Since we require periodic (and so, single-valued in real space) solutions, we see that $m$ must be an integer. This condition should remind us of the De Broglie and Bohr's arguments to explain the discrete nature of the atomic spectra by requiring that an integer number of wavelengths must "fit" into an orbit. Note also how the eigenvalue problem given by Eq. (2.24) [and, more generally, Eq. (2.20)] is defined not only by the differential equation, but also by the boundary conditions, and so by the space of functions we intend to confine our attention to; namely, differentiable, integrable over the entire space and single-valued/periodic, in our case. This is another way of expressing physically sensible boundary conditions. Returning to Eq. (2.23) and using Eq. (2.24), we see that

$$\frac{1}{Rr^2}\frac{\mathrm{d}}{\mathrm{d}r}\left(r^2\frac{dR}{\mathrm{d}r}\right)+\frac{1}{r^2\sin\theta}\frac{1}{\Theta}\frac{\mathrm{d}}{\mathrm{d}\theta}\left(\sin\theta\frac{\mathrm{d}\Theta}{\mathrm{d}\theta}\right)$$
$$+\frac{1}{r^2\sin\theta}\frac{1}{\Phi}\frac{\mathrm{d}^2\Phi}{\mathrm{d}\phi^2}-\frac{m^2}{r^2\sin^2\theta}-\frac{2m_e[V(r)-E]}{\hbar^2}=0. \tag{2.26}$$

Multiplying this equation by $r^2$, we can rewrite it as

$$\frac{1}{R}\frac{\mathrm{d}}{\mathrm{d}r}\left(r^2\frac{dR}{dr}\right)-\frac{2m[V(r)-E]}{\hbar^2}r^2=-\frac{1}{\sin\theta}\frac{1}{\Theta}\frac{\mathrm{d}}{\mathrm{d}\theta}\left(\sin\theta\frac{\mathrm{d}\Theta}{\mathrm{d}\theta}\right)+\frac{m^2}{\sin^2\theta}. \tag{2.27}$$

Once more, we see that the lhs is a function of $r$ alone, while the rhs depends only on $\theta$. Thus, both sides must be equal to a constant, say $K$. Therefore, we reach the two equations:

$$\frac{\mathrm{d}}{\mathrm{d}r}\left(r^2\frac{dR}{\mathrm{d}r}\right)-\frac{2m_\mathrm{e}[V(r)-E]}{\hbar^2}r^2R=KR, \tag{2.28}$$

and

$$\frac{1}{\sin\theta}\frac{\mathrm{d}}{\mathrm{d}\theta}\left(\sin\theta\frac{\mathrm{d}\Theta}{\mathrm{d}\theta}\right)+\frac{m^2\Theta}{\sin^2\theta}=-K\Theta. \tag{2.29}$$

We can now proceed in two ways, the first restricted to the case at hand, the second path being more general and being related with the *angular momentum operators*. Let's discuss them both. In order to proceed from Eq. (2.29), we'll skip some details here and simply state without proof that requiring $\Theta$ to be single-valued implies that $K$ must be of the form $l(l+1)$, where $l$ is a positive integer, much like requiring $\Phi$ to be single-valued resulted in an integer $m$ before. Moreover, the integer $m$ must take values of magnitude smaller than $l$ (i.e., $m=0,\pm1,\pm2,\ldots,\pm l$).

* *We should stop here for a brief comment. The statement above is the essence of what is called the theory of "orthogonal polynomials." The text by Arfken [6] gives an exceptionally clear overview of the topic. When trying to solve many types of ordinary differential equations of the form* $Q(x)y(x)''+P(x)y(x)'+\lambda y(x)=0$, *it is convenient to expand the unknown solution,* $y(x)$, *in a power series of* $x$; *that is,* $y(x)=\sum_n a_n x^n$. *Inserting this series into the original differential equation, one can equate terms of the same power of* $x$ *and obtain a set of recursive relations that the coefficients* $a_n$ *must satisfy in order for* $y(x)$ *to be a solution. One finds that when the parameter* $\lambda$ *takes discrete values,* $\lambda_N$, *then one can terminate the series at the term* $x^N$, *all coefficients* $a_n$ *vanishing for* $n\geq N$, *for finite* $N$. *This results in the generation of polynomials* $y_N(x)$, *that constitute a basis set on the functional space considered; for example, in the Hilbert space* $L^2(\mathbb{R})$ *of all square-integrable complex functions of a real variable. These can be orthogonalized or orthonormalized using the Gram–Schmidt procedure. Hermite, Legendre, Jacobi, Laguerre, and Chebyshev polynomials belong to this class. In the context of Quantum Mechanics, usually the parameter* $\lambda$ *is related to the eigenvalue of the Hamiltonian and the resulting discrete set of values* $\lambda_N$ *yields the discrete spectrum, the polynomials* $y_N$ *being related to the eigenvectors. We will encounter shortly two examples of such functions, the Legendre and Laguerre polynomials.*

Returning to Eq. (2.29), with these considerations we see that it must take the form:

$$\frac{1}{\sin\theta}\frac{\mathrm{d}}{\mathrm{d}\theta}\left(\sin\theta\frac{\mathrm{d}\Theta}{\mathrm{d}\theta}\right)+\frac{m^2\Theta}{\sin^2\theta}+l(l+1)\Theta=0. \tag{2.30}$$

Setting $x=\cos\theta$ transforms this equation into what is known as the "associated Legendre" equation whose solutions are the "associated Legendre polynomials"

$$P_l^m(x)=(1-x^2)^{m/2}\frac{\mathrm{d}^m}{\mathrm{d}x^m}P_l(x), \tag{2.31}$$

where the functions $P_l(x)$ are the Legendre polynomials. Thus, we have found that the full angular dependence of our solution has the form:

$$Y_l^m(\theta,\phi) = P_l^m(\cos\theta)\mathrm{e}^{\mathrm{i}m\phi}. \tag{2.32}$$

Properly normalized, so that

$$\int_0^{\pi} \mathrm{d}\theta \int_0^{2\pi} \mathrm{d}\phi |Y_l^m(\theta,\phi)|^2 = 1, \tag{2.33}$$

these functions are called "spherical harmonics."

The second way to proceed from Eq. (2.29) is to consider the angular momentum operators. In Classical Mechanics the orbital angular momentum of a particle is defined as

$$\mathbf{L} = \mathbf{r} \times \mathbf{p}, \tag{2.34}$$

so that the $z$-component, for example, will be

$$L_z = xp_y - yp_x. \tag{2.35}$$

The quantum-mechanical equivalent will be the operator

$$\widehat{L}_z = -\mathrm{i}\,\hbar\left(x\frac{\partial}{\partial y} - y\frac{\partial}{\partial x}\right). \tag{2.36}$$

Note that from the basic commutation rules between the position and the momentum operators,

$$[\hat{x}_i,\hat{p}_j] = \mathrm{i}\,\hbar\delta_{ij}, \qquad [\hat{x}_i,\hat{x}_j] = [\hat{p}_i,\hat{p}_j] = 0, \tag{2.37}$$

(where we have used the notation $x_1 = x$, $x_2 = y$, and $x_3 = z$ and similarly for $p_1$, etc.) we have

$$[\widehat{L}_z,\hat{x}] = \mathrm{i}\,\hbar\,\hat{y}, \qquad [\widehat{L}_z,\hat{p}_x] = \mathrm{i}\,\hbar\,\hat{p}_y, \tag{2.38}$$

$$[\widehat{L}_z,\hat{y}] = -\mathrm{i}\,\hbar\,\hat{x}, \qquad [\widehat{L}_z,\hat{p}_y] = -\mathrm{i}\,\hbar\,\hat{p}_x, \tag{2.39}$$

and:

$$[\widehat{L}_i,\widehat{L}_j] = \mathrm{i}\,\hbar\,\epsilon_{ijk}\,\hbar L_k, \tag{2.40}$$

where $\epsilon_{ijk} = 1$ when $(i,j,k)$ is a cyclical permutation of (1,2,3), $\epsilon_{ijk} = -1$ for an anti-cyclical permutation, $\epsilon_{ijk} = 0$ otherwise. Defining the square magnitude of the total angular momentum,

$$\widehat{L}^2 = \widehat{L}_x^2 + \widehat{L}_y^2 + \widehat{L}_z^2, \tag{2.41}$$

we have

$$[\widehat{L}^2,\widehat{L}_j] = 0. \tag{2.42}$$

So, since $\widehat{L}^2$ commutes with any $\widehat{L}_i$, we can select a basis of common eigenstates of $\widehat{L}^2$ and $\widehat{L}_z$, that is

$$\widehat{L}^2|\psi\rangle = \alpha|\psi\rangle, \tag{2.43}$$

and

$$\widehat{L}_z|\psi\rangle = \beta|\psi\rangle. \tag{2.44}$$

In the spherical-coordinates representation we have

$$\widehat{L}^2 = \hbar^2 \left[ \frac{1}{\sin\theta} \frac{\partial}{\partial\theta} \left( \sin\theta \frac{\partial}{\partial\theta} \right) + \frac{1}{\sin^2\theta} \frac{\partial^2}{\partial\phi^2} \right], \tag{2.45}$$

and

$$\widehat{L}_z = -\mathrm{i}\hbar \frac{\partial}{\partial\phi}. \tag{2.46}$$

Using these expressions, we see that Eq. (2.44) reduces to Eq. (2.24), so that $m$ must be an integer (positive or negative). Then, Eq. (2.43) becomes identical to Eq. (2.29) with $K = l(l+1)$.
* *This last conclusion can also be reached in another very elegant way that bypasses the need to consider orthogonal polynomials. We present it here, because a similar, purely algebraic, technique will be used to deal with the harmonic oscillator when we shall discuss elementary excitations and second quantization. Define the "raising/lowering" operators:*

$$\widehat{L}_\pm = \widehat{L}_x \pm \mathrm{i}\,\widehat{L}_y \tag{2.47}$$

*It is easy to see, using the commutation rules we have derived above, that*

$$[\widehat{L}^2, \widehat{L}_\pm] = 0, \tag{2.48}$$

*(where, for clarity, we should recall that $\widehat{\mathbf{L}} = (\widehat{L}_x, \widehat{L}_y, \widehat{L}_z)$, so that $\widehat{L}^2 = \widehat{L}_x^2 + \widehat{L}_y^2 + \widehat{L}_z^2$), and*

$$[\widehat{L}_z, \widehat{L}_\pm] = \pm\hbar\, \widehat{L}_\pm. \tag{2.49}$$

*Applying $\widehat{L}_\pm$ on an eigenstate $|\psi\rangle$ of $\widehat{L}_z$ with eigenvalue $\beta$, using Eq. (2.49), we have*

$$\widehat{L}_z \widehat{L}_\pm |\psi\rangle = [\widehat{L}_\pm \widehat{L}_z \pm \hbar\; |\psi\rangle = (\beta + \hbar)|\psi\rangle. \tag{2.50}$$

*This expression shows that the operators $\widehat{L}_\pm$ raise (+) or lower (−) the eigenstate $|\psi\rangle$ of $\widehat{L}_z$ to another eigenstate of $\widehat{L}_z$, now with a larger/smaller eigenvalue $(\beta \pm \hbar)$. Since the total angular momentum is finite, the eigenvalues $\beta$ of $\widehat{L}_z$ must span in discrete steps the range $(-m\hbar,\, m\hbar)$, where m is an integer, for this raising/lowering process to terminate upwards or downwards. Therefore, the states $|\psi\rangle$ can be labeled as $|\alpha, m\rangle$, where $\alpha$ is a yet-to-be-determined eigenvalue of $\widehat{L}^2$. To determine $\alpha$, a little algebra using the commutation rule, Eq. (2.48), and the definitions of $\widehat{L}^2$ and $\widehat{L}_\pm$, shows that we must have $\alpha = l(l+1)\hbar^2$, where l is a positive integer, including 0, and that m must be smaller than l. Thus, we can recast Eqs. (2.43) and (2.44) as*

$$\widehat{L}^2|l,m\rangle = l(l+1)\hbar^2|l,m\rangle, \tag{2.51}$$

*and*

$$\widehat{L}_z|l,m\rangle = m\hbar|l,m\rangle, \tag{2.52}$$

*with $l = 0,1,2,3,\ldots$ and m can take the 2l+1 values $-l$, $-l+1$, …, $-1$ 0, 1, $l-1$, $l$.*

Leaving now the abstract algebraic domain and returning to the real-space representation of the eigenstates, we see that the spherical harmonics are eigenfunctions of the angular momentum operator corresponding to a total angular momentum $\sqrt{l(l+1)}\hbar$ with projection $m\hbar$ over the $z$-axis (the polar axis). Note also that the spherical harmonics satisfy the orthogonality condition:

$$\int_0^{\pi} \mathrm{d}\theta \int_0^{2\pi} \mathrm{d}\phi \; Y_l^m(\theta,\phi)\; Y_{l'}^{*m'}(\theta,\phi) = \delta_{mm'}\delta_{ll'}. \tag{2.53}$$

Returning to the radial function $R$, this must obey Eq. (2.28) with $K = l(l+1)$:

$$\frac{\mathrm{d}}{\mathrm{d}r}\left(r^2\frac{dR}{\mathrm{d}r}\right) - \frac{2m_\mathrm{e}[V(r)-E]}{\hbar^2}r^2R - l(l+1)R = 0. \tag{2.54}$$

Now we can make use of the fact that we are dealing with a Coulomb potential and set $V(r) = -e^2/(4\pi\epsilon_0 r)$. Thus, the radial wavefunction $R(r)$ must satisfy the equation:

$$\frac{\mathrm{d}}{\mathrm{d}r}\left(r^2\frac{dR}{\mathrm{d}r}\right) + \frac{2m_\mathrm{e}E}{\hbar^2}r^2R - l(l+1)R + \frac{2m_\mathrm{e}}{\hbar^2}\frac{e^2}{4\pi\epsilon_0}rR = 0. \tag{2.55}$$

Let us now introduce the new dependent variable $u(r) = rR(r)$, which obeys the equation:

$$-\frac{\hbar^2}{2m_\mathrm{e}}\frac{\mathrm{d}^2u}{\mathrm{d}r^2} + \frac{\hbar^2 l(l+1)}{2m_\mathrm{e}r^2}u - \frac{e^2}{4\pi\epsilon_0 r}u = Eu. \tag{2.56}$$

Defining the independent variable

$$\rho = \frac{(8m_\mathrm{e}|E|)^{1/2}r}{\hbar}, \tag{2.57}$$

and a measure $\lambda$ of the binding energy of the particle,

$$\lambda = \frac{e^2}{4\pi\epsilon_0\hbar}\left(\frac{m_\mathrm{e}}{2|E|}\right)^{1/2}, \tag{2.58}$$

Eq. (2.56) becomes

$$\frac{\mathrm{d}^2u}{\mathrm{d}\rho^2} - \frac{l(l+1)}{\rho^2}u + \left(\frac{\lambda}{\rho} - \frac{1}{4}\right)u = 0. \tag{2.59}$$

Clearly, at large $\rho$ the solution behaves as $u \sim \exp(\pm\rho/2)$, the minus sign being the only physically meaningful choice. Thus, setting

$$u(\rho) = F(\rho)\exp(-\rho/2), \tag{2.60}$$

Eq. (2.59) implies

$$\frac{\mathrm{d}^2F}{\mathrm{d}\rho^2} - \frac{\mathrm{d}F}{\mathrm{d}\rho} + \left[\frac{\lambda}{\rho} - \frac{l(l+1)}{\rho^2}\right]F = 0. \tag{2.61}$$

Following the general procedure outlined when discussing orthogonal polynomials, one can show that the solutions of this equation (closely related to the so-called associated Laguerre polynomials) have an acceptable behavior in the limit $\rho \to \infty$ only for integer values of $\lambda$ (say, $\lambda = n$, where $n$ is an integer) larger than $l$ so that the "radial" component of the wavefunction will be defined by the two integers $n$ and $l < n$. The index $n$ is directly related to the energy $E$ via Eq. (2.58).

As a result of this discussion, we see a situation similar to what we saw before in the case of the 3D particle in the cubic box: We need three quantum numbers to label the eigenvectors (with eigenfunctions $\psi_{nlm}(\mathbf{r}) = R_{nl}(r)\Theta_{lm}(\theta)\Phi_m(\phi) = R_{nl}(r)Y_l^m(\theta,\phi)$), and eigenvalues (energy levels) of the Hamiltonian: The "principal" (or "radial") quantum number, $n$, can take the values:

$$n = 1,\ 2,\ 3,\ \ldots\ \infty.$$

Usually (that is, in the absence of magnetic fields and ignoring weaker perturbations), the electron energy depends only on this number via Eq. (2.58): When $\lambda = n$,

$$E_n = \frac{m_e e^4}{2(4\pi\hbar)^2 n^2}, \tag{2.62}$$

in agreement with the early result of Bohr's model. The quantum numbers $l$ and $m$ are related to the angular momentum: as we saw, $l(l+1)\hbar^2$ is its squared magnitude and $m\hbar$ its component along the polar axis. These can take the values:

$$l = 0,\ 1,\ \ldots n-1, \quad (n \text{ total number of values})$$
$$m = -l, -l+1,\ \ldots 0,\ 1,\ \ldots l-1,\ l, \quad (2l+1 \text{ total number of values}).$$

For $l = 0$, we must also have $m = 0$. In this case the wavefunction does not depend on the angles $\theta$ and $\phi$. These are fully spherically symmetric wavefunctions (for every $n$) and are called $s$-waves. For $l = 1$ we can have three different wavefunctions for $m = -1, 0, +1$. These are called $p$-waves and can be expressed as functions shaped like "lobes" pointing along the $x$-, $y$-, or $z$-axis. For $l = 2$ we can have five states ($m = -2, -1, 0, +1, +2$), called $d$-waves. For $l = 3$ we have seven $f$-waves, and so on.

### *2.1.4 Periodic Table*

We are now ready to build the periodic table of the elements. We shall do it as if we were building an onion, one "shell" at a time, from the inside out. We shall call "shell" each layer defined by a common quantum number $n$, "subshells" those with common angular quantum number $l$. The word "orbital" or "state" denotes any combination of quantum numbers.

Before building atoms, though, we need to know two additional facts: The existence of an additional electronic degree of freedom ("spin") and of "Pauli's exclusion principle." They will be discussed at length later. For now, we consider them qualitatively.

Regarding spin, in addition to the three degrees of freedom that all particles have in three dimensions (3D), most particles also possess an additional "internal" degree of freedom, $s$, which one can visualize as the particle "spinning" around a polar axis, much like a spinning top or the planets. Thus, spin is an internal angular momentum of the particle. As we just saw for the orbital angular momentum, spin can also take only discrete values in integer or semi-integer multiple of the reduced Planck constant. Photons, some nuclei, and some elementary particles have integer values of spin ($s = 0$ or $n\hbar$, where $n$ is an integer.). For reasons we shall see later, these particles are called "Bosons."

Electrons, protons, and neutrons have half-integer spin ($s = (n\hbar/2)$). They are called "Fermions." Electrons have spin $\hbar/2$ (usually said "spin one-half"). Every electron can exist in two different spin states: Spin "pointing up" ($s = \hbar/2$) or "pointing down" ($s = -\hbar/2$). Both are states with angular momentum of magnitude $\hbar/2$, but they differ in the direction of rotation around their axis.

Regarding Pauli's principle, in order to explain the periodic table and the electronic structure of the atoms, the German physicist Wolfgang Pauli had to invoke a new postulate (later demonstrated rigorously): Given an energy level, characterized by a set of quantum numbers, it can be occupied only by one Fermion (so, electrons). Therefore, a given hydrogen state characterized by the quantum numbers $(n, l, m)$, can be empty or occupied by one or two electrons (in which case an electron will have spin up, the other spin down). Attempting to add a third electron would violate Pauli's principle.

Let's start with the simplest element, H. Its nucleus has a single positively charged proton and it is surrounded by a single electron. The lowest possible orbital the electron can occupy is the "ground state" ($n = 1$, $l = 0$, and $m = 0$). We use the notation $1s^1$, meaning that in the orbital $1 (n = 1)$, which is an $s$-wave ($l = 0$), we have one electron (the superscript "1"). Note that the electron is bound rather weakly to the nucleus: the ionization energy, $E_1$ in Eq. (2.62) is about 13.6 eV. This quantity is called a "Rydberg" and is indicated by the symbol "Ry". We shall use it very frequently in later chapters. Therefore, H loses its electron quite easily; that is, it is easily ionized. This trait is common to all elements with only one electron in the outermost shell. These elements are very reactive (think of hydrogen reacting with oxygen!).

The next simplest element is He. Solving the Schrödinger equation for this atom already introduces enormous complications. The wavefunction we must find is now a function of six variables, $\psi(\mathbf{r}_1, \mathbf{r}_2)$, $\mathbf{r}_1$ and $\mathbf{r}_2$ being the coordinates of the two electrons. Unlike the case of the hydrogen atom, in which the potential energy is given, now the potential energy must include also the electrostatic potential felt by each electron due to the charge of the other electron. The potential energy due to the Coulomb interaction with the charge of the nucleus and with the "time averaged" charge distribution of the other electron is called the "Hartree" energy. However, more subtle contributions to this potential energy come from the fact that Pauli's principle can also be reformulated by requiring that the total wavefunction be antisymmetric under exchange of the two electrons, that is, $\psi(\mathbf{r}_1, \mathbf{r}_2) = -\psi(\mathbf{r}_2, \mathbf{r}_1)$. This property will also be discussed at length in later chapters, since it affects very strongly the electronic properties of crystals. For now, we simply observe that this "antisymmetrization" makes electron with spins pointing in the same direction repel each other; the opposite is true for electrons with spins pointing in the opposite direction. The Coulomb repulsion between the electrons, therefore, will be modified by this effect and the additional (positive or negative) change in energy is called the "exchange" energy. An additional correction (the "correlation" energy) will be discussed in later chapters. What matters now is that the Hartree and exchange energies depend on the spatial distribution of the electrons, $|\psi(\mathbf{r}_1, \mathbf{r}_2)|^2$. So, the potential energy entering the Schrödinger equation depends on the wavefunction itself. This renders the Schrödinger equation nonlinear and only approximate or numerical solutions can be found. The case of heavier atoms, so with a much larger number of electrons, is, of course, extremely complicated. In later chapters we will see how one can handle such a situation even in crystals, systems with an extremely large number of electrons. Here, we should just keep these facts in mind and we shall deal with the problem at a qualitative level.

Returning to the He atom, its nucleus now has two protons (so that the "atomic number" of He is $Z = 2$). Two neutrons are required to prevent the Coulomb repulsion between the protons from causing a disintegration of the nucleus itself. (These are forces of a different nature, the *strong nuclear forces* that we will ignore in this book.) Two electrons orbit the nucleus and, in the configuration of the lowest possible energy, both are in the lowest-energy "shell," provided they align their spins in opposite directions. We use the notation $1s^2$. Note that we have fully populated the first "shell." If we were to ionize negatively the atom by adding another electron, this would have to occupy the next higher-energy shell, $n = 2$, because there is no more room in the $n = 1$ shell. Thus, the third electron would have to go farther away from the nucleus. So, He does not easily acquire another electron.

On the other hand, stripping one electron away from the He atom is energetically very expensive, since both electrons sit quite close to the doubly charged nucleus. Therefore, helium does not "like" to lose or gain electrons (that is, becoming ionized). He, like all elements which have a fully occupied shell, is chemically inert.

The third element, Li, has three protons in its nucleus ($Z = 3$), so it must have three electrons orbiting it in order to be charge neutral. The third electron, as we just saw, must occupy the $n = 2$ shell, so Li will have the electronic configuration $1s^2 2s^1$. It is a very reactive metal, for the reasons we discussed dealing with H: It is very easy to strip away the lone outer electrons, leaving behind a positive Li ion.

We now keep populating the $n = 2$ shell, adding another electron in the $s$ orbital (beryllium, Be, $1s^2 2s^2$ with $Z = 4$), then 6 more into the 3 $p$ orbitals, adding electrons with spin up and spin down in sequence. As we move from the reactive metals (Li and Be), we build elements with 3 "outer" electrons (boron, B, $1s^2 2s^2 2p^1$), 4 outer electrons (carbon, C, $1s^2 2s^2 2p^2$), 5 (nitrogen, N, $1s^2 2s^2 2p^3$), 6 (oxygen, O, $1s^2 2s^2 2p^4$), and 7 outer electrons (fluorine, F, $1s^2 2s^2 2p^5$). As we do so, we move from elements which like to lose electrons, to those who like to gain them. Eventually, with neon (Ne, $Z = 10$, $1s^2 2s^2 2p^6$) we complete the $n = 2$ shell and we hit another inert gas, like He, unlikely to react chemically. The eight slots we have filled while populating the $n = 2$ shell are the eight columns of the periodic table. The chemical properties of any two elements having the same number of electrons in the outer shell are very similar. So, as we fill the $n = 3$ shell, we start from sodium (column I, Na, $Z = 11$, $1s^2 2s^2 2p^6 3s^1$) which, having only one electron in the outer shell, behaves like Li. And we end up with argon (column VIII, Ar, $Z = 18$, $1s^2 2s^2 2p^6 3s^2 3p^6$) which is another inert gas like He and Ne.

Things get a little more complicated now, because the energy of the $4s$ shell is lower than that of the $3d$ shell. This is caused by the perturbation of the Coulomb energy (Hartree, exchange, and correlation) due to the many electrons present in the system, so that the energetic "ladder" expected from an analogy with the eigenlevels of the H atom fails significantly. The dependence of the energy level on the quantum number $l$, for example, is a manifestation of this fact. So, with potassium (K) and calcium (Ca) we fill the $4s$ shell, but then we go back filling the $3d$ shell. The fact remains that—up until further complications caused by the $f$ states of transition metals—each element belongs to its own column which determines the chemical properties via the number of electrons in the outer shell.

Of interest to us are silicon (Si, $Z = 14$, $1s^2 2s^2 2p^6 3s^2 3p^2$) and germanium (Ge, $Z = 32$, $1s^2 2s^2 2p^6 3s^2 3p^6 3d^{10} 4s^2 4p^2$) which, like C, belong to the IV column. Their 4 outer electrons are $s^2 p^2$ and can easily "hybridize" into 4 $sp^3$ orbitals, forming the tetrahedral structure required to bond atoms in their cubic crystal form. Elements of column III (B, Al, Ga, and In) have outer electrons arranged as $s^2 p^1$, that is, one fewer electron than in Si and Ge. Elements of column V (N, P, As, and Sb) have outer electrons arranged as $s^2 p^3$, that is, one more electron than in Si and Ge.

Finally, electrons in fully occupied shells are called "core electrons." They do not contribute to the chemical activity of the element. Electrons in the outer shell are called "valence electrons."

## 2.2 Molecules and Bonds

The nature of chemical bonds, molecular orbitals, hybridization, and such, has been subject, and still is, of too many studies and even controversy. Being interested in solids and electronic transport in crystals, we can only give here a bird's eye view. Excellent texts such as Pauling's [7] and Harrison's [8] give a comprehensive and detailed account from the perspective of a chemist (Pauling, arguably one of the fathers of the concept of chemical bond in light of Quantum Mechanics) and of a physicist (Harrison).

If we bring several atoms together, interesting things happen, depending on the electronic populations of the outer shell.

1. *Nothing*. For example, bringing 2 Ne atoms together, both atoms retain their electronic configuration unaltered, as they are both in a configuration of a completely filled outer shell with 8 electrons (2 in *s* orbitals, 6 in *p* orbitals). This is why elements of the VIII column are called "inert" elements (or gases).
2. *Formation of an ionic bond*. Let's consider Na and Cl. Na has 1 electron in the outer *s* shell, Cl 7 valence electrons (2 in the *s* shell, 5 in the *p* shell). Energetically, it is advantageous for Na to release its electron (weakly bound, as the core electrons shield the nuclear charge), while Cl "loves" to acquire that electron to complete the filling of its outer shell. Thus, the sodium atom becomes positively ionized ($Na^+$) and the chlorine atom becomes negatively ionized ($Cl^-$). Now we have a positive ($Na^+$) and negative ($Cl^-$) particle which are attracted by the strong Coulomb force. The two ions are said to have formed an "ionic bond" and stick together in the NaCl molecule (regular kitchen salt when in crystalline form). Semiconducting crystals, like GaAs (Ga in column III, As in column V, so this is a so-called III–V compound semiconductor), are bound in large part by ionic bonds. Water ($H_2O$) is similarly formed, although now the charge transfer takes place among three atoms: The 2 H atoms give their electrons to O which has 6 valence electrons, and so it fills its outer shell. Thus, we now have two positively charged $H^+$ ions bound to a doubly ionized $O^{2-}$ ion. Ionic bonds are usually quite strong with "binding energies" of several eV, i.e., of the order of 1–10 eV.
3. *Formation of a covalent bond*. Let's consider two oxygen atoms brought together. They are in the $2s^2 2p^4$ configuration in the outer shell. They would both "love" to fill their outer shells by adding two electrons. They can do that by sticking together and sharing their two outer electrons, thus forming the $O_2$ molecule. They form a "covalent bond." The situation with Si and Ge is somewhat more complicated. They have 4 electrons in their outer shells ($s^2 p^2$). They can "hybridize" to form 4 $sp^3$ orbitals. These are linear combinations of the *s* orbital and of the 3 *p* orbitals that, when Si atoms are arranged in a crystal, form a state with lower energy than the separate *s* and *p* orbitals. These hybrid orbitals are arranged in a tetragonal fashion, so that each Si atom has four neighbor Si atoms to form covalent bonds. Therefore, the $Si_2$ molecule cannot exist, but several Si atoms can form a "network" of tetragonal bonds and form a covalent crystal. Covalent bonds are, usually, the strongest bonds.
4. *Metallic bond*. When we bring together a large number of metallic atoms (say, Li or Be), the weakly bound single electrons belonging to each one of the many atoms are "shared" among all ions. The ions form a "lattice" (as we shall see below dealing with crystals), the shared electrons keeping the lattice together. The metallic bond has a strength comparable to that of ionic bonds.
5. *Hydrogen bond*. This is still a poorly understood bond present in water-ice, and, for example, binding the bases to the chain of DNA molecules. A H atom between two molecules acts as a bridge binding (somewhat weakly) the molecules together, as its lone electrons "resonates" between them.
6. *Van der Waals force*. Although generally a weaker effect, another type of interaction affects solids and some relatively chemically inactive molecules. When brought together, rather than exchanging or sharing electrons, atoms or molecules "polarize" their electronic clouds. The resulting interaction is between electric dipoles with opposite orientation. The force decays very quickly with distance. This is an interaction typical of flat surfaces (think of two sheets of glass) and polymers.

Illustrations of the ionic, covalent and Van der Vaals bond are shown in Fig. 2.1. Note also that, although we have made a sharp distinction between ionic and covalent bonds, in many cases both types of bonds contribute: For example, in III–V compound semiconductors, a large charge transfer causes ionicity of the bonds, but the fact that most of these crystallize into the face-centered cubic

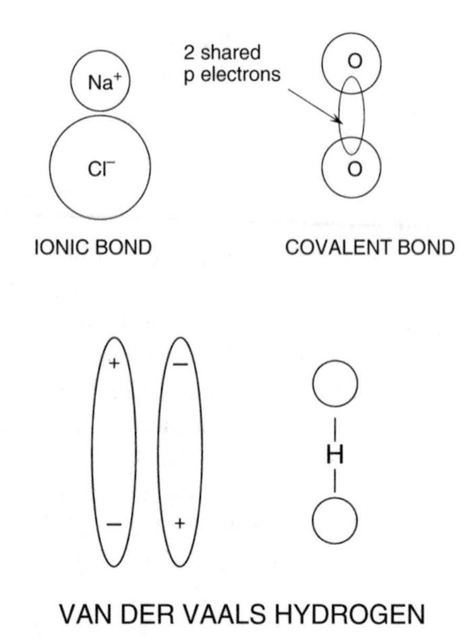


**Fig. 2.1** Cartoon representation of chemical bonds: in an ionic bond, two atoms with large electron affinities exchange charge, resulting in two ionized atoms that are attracted by the resulting Coulomb force. In a covalent bond, atoms with similar or identical electron affinities share their valence electrons in a bonding orbital

structure resulting from $sp^3$ hybridization shows that the bond is also largely covalent. Such bonds are often characterized by specifying the degree of ionicity.

We have outlined why the chemical properties of the elements are fully determined by the valence electrons. The number of electrons in the outer shell also determines the chemical valence of the atom. As a general rule, electrons with fewer than 3 electrons in the outer (incomplete) shell have a chemical valence equal to the number of electrons. For example, Li (one electron in the outer shell) has valence I, Ga (3 electrons in the outer shell) has valence III. If the outer shell contains 6 electrons or more, the valence of the element is 8 minus the number of electrons. Elements with a more complicated electronic structure of the outer shell(s) can exhibit several different chemical valence numbers.

Finally, it is important to realize that during the formation of a bond, the electronic orbitals change their structure. For example, during the formation of the $O_2$ molecule the electrons "shared" by the two oxygen atoms will "orbit" both cores (nuclei+core electrons), forming new orbitals by a "linear superposition" of the wavefunctions of the $2p$ orbitals of each atom. As shown in the figure below, out of these two "single atom" orbitals we form two molecular orbitals: A lower-energy "bonding orbital" and a higher-energy "antibonding orbital," respectively, at a slightly lower and higher energy than the original single-atom orbitals. The bonding orbital is occupied by the two shared electrons and it exhibits a larger charge density (that is, a larger $|\psi|^2$) between the atoms. In other words, there is a large probability of finding the electrons between the atoms. Thus, this orbital contributes to keeping the atoms together. The antibonding orbital is empty and it is associated with a charge density (again, $|\psi|^2$), which is larger away from the bond. A bonding and an antibonding orbital are illustrated in Fig. 2.2.

Very large molecules may consist of a very large number of atoms. For example, polymers, such as the polyparaphenylene (PPP) chain of C atom illustrated in Fig. 2.3. In such a chain, each C atom on the "center" atomic line is bound to three more C atoms in hybrid $sp^2$ orbitals (note that the notation used here differs from the notation used before: The superscript "2" now indicates that 2 $p$ orbitals make up the hybrid orbital): The outer $s$ orbital and three $p$ orbitals split into three linear combinations of the lone $s$ and two $p$ bonding orbitals, and a single $p_z$ orbital that "sticks out" of the plane of the molecule. For C atoms on the outer lines of the chain, two such $sp^2$ orbitals bond each C atom to its two neighbors; the remaining $sp^2$ orbital is "terminated" by an H atom. The out-of-plane $p_z$ orbitals hybridize into a set of bonding $\pi$ orbitals that extend throughout the chain (the use of Greek letters is briefly explained at the end of the chapter). Higher-energy orbitals (the $3s$ and $3p$)

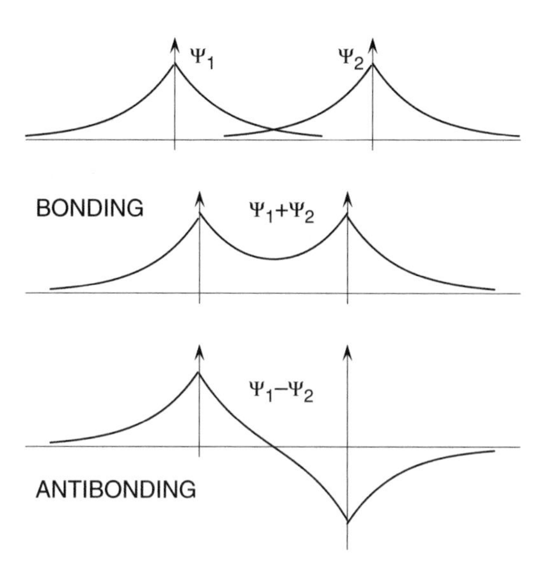


**Fig. 2.2** Schematic representation of bonding and antibonding states in a diatomic molecule: Two molecular orbitals can result from the linear superposition of the two outer-shell wavefunctions of the two atoms, a bonding orbital with a large electron charge density between the atoms, charge that contributes to the bonding and has a lower Coulomb energy; and an antibonding, higher-energy orbital. The Van der Waals interaction between two molecules stems from the mutually induced charge polarization and resulting Coulomb attraction. In a H-bond, a hydrogen atom acts as a bridge, being shared by the two atoms or molecules

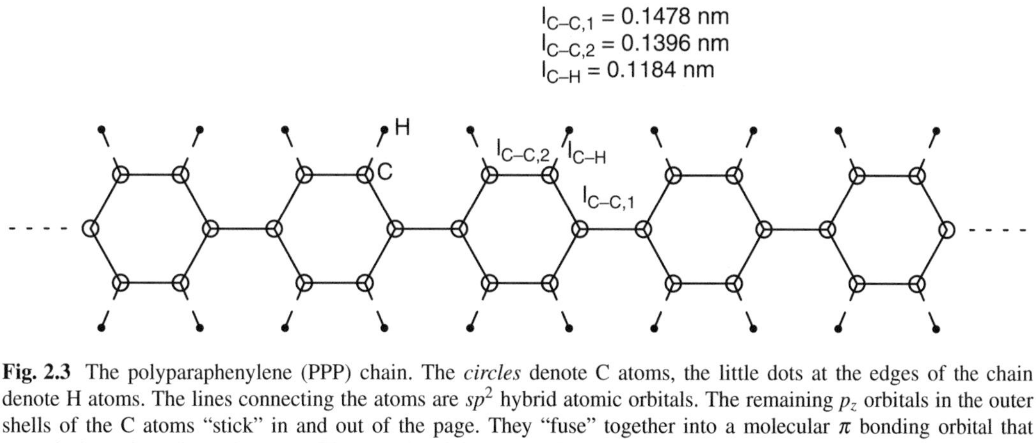

**Fig. 2.3** The polyparaphenylene (PPP) chain. The *circles* denote C atoms, the little dots at the edges of the chain denote H atoms. The lines connecting the atoms are $sp^2$ hybrid atomic orbitals. The remaining $p_z$ orbitals in the outer shells of the C atoms "stick" in and out of the page. They "fuse" together into a molecular $\pi$ bonding orbital that extends throughout the entire chain. They are the highest-occupied molecular orbitals, HOMO. The next higher-energy molecular orbitals, $\sigma^*$ and $\pi^*$, are, instead, empty antibonding orbitals. They are the lowest-unoccupied molecular orbitals, LUMO

hybridize into antibonding $\sigma$ and $\pi$ bonds. In this case, the polymers exhibit an energy difference between the highest-energy occupied (hybridized) molecular orbital (HOMO) and the lowest-energy unoccupied molecular orbital (LUMO). Such a molecule will be electrically insulating, since only electrons occupying the delocalized LUMO (that is, spread among many atoms) can move if driven by an electric field, but there are no electrons in these states. Other examples of molecules may be found in which the valence hybridized orbitals are only partially occupied, so these molecules will be good conductors, behaving as metals. These considerations will be repeated almost verbatim when dealing with insulators and conductors in the context of crystals (solids). Strands of DNA, or even larger proteins, can fall into any of these categories. Their electrical properties are now being exploited to "mark" them for medical or biological purposes, "sense" them in diagnostics, such as detecting the change of conductivity induced by a mismatched basis in a DNA sequence. Also, electronic devices based on single molecules are an extremely active area of research. A complete discussion of these topics is, clearly, outside the scope of this text, although the connections are many: understanding their physical principles, their atomic and electronic structure, and their electronic and electrical properties requires concepts, mathematical and numerical techniques that are very similar, often identical, to those discussed here. The text by Helgaker et al. [9] constitutes an excellent and relatively recent overview of this fast-developing field.

Finally, note how hybrid molecular orbitals are termed with the Greek letter(s) associated with the Latin character of the atomic orbital that enters the hybridization ($s$ atomic orbitals hybridize into $\sigma$ molecular orbitals, $p$ can enter into $\pi$-bonds as well as $\sigma$-bonds).

## Problems

**2.1. (The de Broglie Hypothesis and the Bohr Atom)** The *de Broglie's assumption* to explain atomic spectra (among other things) consists in assuming that *all* forms of matter behave as particles or waves, depending on the situation. For a particle of mass $m$, the wavelength $\lambda$ of the associated "pilot wave" will be given by

$$p = mv = \frac{h}{\lambda}$$

Consider how de Broglie's suggestion may explain some features of the hydrogen atom:

(a) Show that the de Broglie assumption, together with the "quantization condition" that the circular orbit be an integer multiple of the length of the electron wavelength (that is, $n\lambda = 2\pi r$, where $r$ is the radius of the orbit, $n$ an integer), implies that only discrete orbits are allowed.
(b) Calculate the total energy (kinetic plus potential) in each orbit characterized by an integer $n$. Show that the result is identical to what is shown in Eq. (2.62).

## References

1. D. Bohm, *Quantum Theory* (Dover, New York, 1989)
2. A. Messiah, *Quantum Mechanics* (Dover, New York, 2014)
3. L.I. Schiff, *Quantum Mechanics* (McGraw-Hill, New York, 1968)
4. R.H. Dicke, J.P. Wittke, *Introduction to Quantum Mechanics* (Addison Wesley, New York, 1960)
5. R.P. Feynman, R.B. Leighton, M. Sands, *The Feynman Lectures on Physics*, vol. 3 (Addison Wesley, New York, 1970)
6. G.B. Arfken, H.J. Weber, F.E. Harris, *Mathematical Methods for Physicists*, 7th edn. (Academic, New York, 2012)
7. L. Pauling, *The Nature of the Chemical Bond and the Structure of Molecules and Crystals: An Introduction to Modern Structural Chemistry* (Cornell University Press, Ithaca, NY, 1960)
8. W. Harrison, *Electronic Structure and the Properties of Solids: The Physics of the Chemical Bond* (Dover, New York, 1989)
9. T. Helgaker, P. Jorgensen, J. Olsen, *Molecular Electronic-Structure Theory* (Wiley, New York, 2013)

# Part II
# Crystals and Electronic Properties of Solids

# Chapter 3
# Crystals: Lattice, Reciprocal Lattice, and Symmetry

## 3.1 Overview

We are usually taught that matter can exist in three forms (or "states"): Gas, liquid, and solid. Occasionally, a fourth state is added, plasma (a gas of ionized atoms and electrons). Sir Nevil Francis Mott, however, disagreed: If we look at many of what we call "solids," a small change of the ambient temperature can induce an apparent change of their state. Butter, for example, when taken as a hard, brittle "solid" out of the freezer, becomes softer, more malleable, when put in the warmer shelves of the refrigerator. When left on the kitchen table on a warm day, it melts into a liquid. There are no sharp "phase transitions" to speak of. The microscopic structure of butter remains the same: Long organic molecules, arranged in a random fashion, are free to slide one over the other at a higher temperature; lowering the temperature simply makes it more difficult for the molecules to move and butter becomes "solid." Mott would have said that, even in the freezer, butter is a liquid, as hard as it may feel to the touch. Most people would say that butter is an "amorphous" solid, lacking an internal regular order, being made up, instead, by atoms of molecules arranged in a random network. A proper "solid," instead, exhibits a phase transition—a sudden change of its behavior—when melting. This is the property of "crystals": Regular, periodic arrangement of atoms or molecules.

Intuitively, it is convenient to think of crystals as tiles on a floor. The shape of the tile (rectangular, square, or hexagonal) is analogous to what is defined as "lattice." The design on the tile (e.g., a flower) is the analogous of what is called the "basis." In a proper solid, the lattice defines the symmetry of the crystal, each tile being called a "cell"; the arrangement of the atoms in each cell defines the basis.

Crystals are the main subject of this text. The reason why they are so important as to deserve a huge literature and extensive study is, of course, the enormous role that they play in our society and culture. Our age has been defined as the "silicon age." The cultural and financial effects that crystals (specifically: semiconducting crystals) have had to our society cannot be overstated. If crystals had entered our society only through kitchen salt (cubic NaCl crystals), interest in them would have not been so high, as much as our bodies and nervous system may need salt.

Their regular, ordered, periodic structure makes them ideal systems in which we can control the flow of charge carriers (electrons). This order also makes it possible to study them, and so to optimize their properties. In this chapter we shall define crystals and discuss their symmetries. In later chapters we shall use this "order" to study their all-important electronic properties.



M. Fischetti, W.G. Vandenberghe, *Advanced Physics of Electron Transport in Semiconductors and Nanostructures*, Graduate Texts in Physics, DOI 10.1007/978-3-319-01101-1_3

## 3.2 Crystals

*Crystals* are regular arrangements of atoms which look identical when viewed from 2 points

$$\mathbf{r} \quad \text{and} \quad \mathbf{r}' = \mathbf{r} + \mathbf{R}_l, \tag{3.1}$$

where

$$\mathbf{R}_l = l_1\, \mathbf{a}_1 + l_2\, \mathbf{a}_2 + l_3\, \mathbf{a}_3 \tag{3.2}$$

is a translation vector with $l_1, l_2, l_3 \in \mathbb{Z}$ (the group of integers, from the German *Zahlen* for "numbers") and with *fundamental translation vectors* or *primitive vectors* $\mathbf{a}_1$, $\mathbf{a}_2$, and $\mathbf{a}_3$. The vector $\mathbf{R}_l$ is called a *lattice vector*.

The *lattice* of the crystal is the set of all points $\mathbf{r}'$ generated by translation operations $T$; that is, by translating a point by a translation vector $\mathbf{R}_l$: $T : \mathbf{r} \to \mathbf{r} + \mathbf{R}_l$. The *basis* of the crystal is the set of the coordinates $\tau_\alpha$ of all atoms within a primitive cell. The position of each atom in the crystal can be uniquely written as

$$\mathbf{R}_{\text{atom}} = \mathbf{R}_l + \tau_\alpha. \tag{3.3}$$

### 3.2.1 Lattice

Lattices as defined by the set of ideal (mathematical) points with the properties given above are called "Bravais lattices," from the studies of August Bravais [1]. In two dimensions, it is easy to see that there is a small set—five in all—of possible lattices: Think of the number of ways we can cover a floor using identical tiles. These five Bravais lattices are: square, rectangular, center-rectangular, oblique, and hexagonal (or rhombic). Different lattices are considered equivalent when they can be transformed one into the other via a simple symmetry operation (isomorphism). In three dimensions, there are only 14 Bravais lattices. We shall not discuss them here: We shall be interested mainly in the face-centered cubic (fcc) lattice in three dimensions, since it is the lattice that describes most technologically significant semiconductors; and in the hexagonal lattice in two dimensions, since graphene (and other two-dimensional materials of high scientific and technological interest, at the time this is written) belong to this lattice. Exhaustive information about lattices and crystals can be found in texts on crystallography (for example, the book by Borchardt-Ott [2]) or in the introductory text on solid-state physics by Kittel [3]. This book also provides information about how crystal structures can be observed, identified, and quantified experimentally, via X-ray or neutron diffraction.

Some definitions and properties concerning the lattice are

- The parallelepiped formed by the fundamental translation vectors is a *primitive cell* and its volume is given by $\Omega_c = |\mathbf{a}_1 \times \mathbf{a}_2 \cdot \mathbf{a}_3|$. Note that repeating the unit cell, with an offset determined by all translation vectors, fills the entire space.
- Another primitive unit cell is the *Wigner–Seitz cell*. The Wigner–Seitz (WS) cell can be generated by connecting all lattice points with straight lines, bisecting these lines with planes normal to the lines, and considering the volume enclosed by these planes.
- As we have mentioned before, the symmetry properties of a lattice allow the existence of 14 types of Bravais lattices (those generated by Wigner–Seitz cells).
- An arbitrary plane will intersect the axes $\mathbf{a}_1$, $\mathbf{a}_2$, and $\mathbf{a}_3$ at distances $(A, B, C)$ (in units of the lengths $(\mathbf{a}_{1,2,3})$, the lattice constants). The *Miller indices* $(l, m, n)$ of the plane can be found by taking the reciprocal of the distances $(1/A, 1/B, 1/C)$ and finding the smallest integers $(l, m, n)$ having the same ratios as the reciprocals.

As a matter of notation, $(lmn)$ denotes the plane defined by its Miller indices; $\{lmn\}$ denotes the family of all planes that are equivalent to the plane $(lmn)$; $[lmn]$ denotes the direction of the translation vector $l\mathbf{a}_1 + m\mathbf{a}_2 + n\mathbf{a}_3$ (that, for cubic crystals, is the vector perpendicular to the planes in the family $\{lmn\}$); $< lmn >$, denotes the family of all directions equivalent to the direction of $[lmn]$. Also, usually a negative Miller index is indicated by a "bar" over the index; for example, $[1\bar{1}0]$ is the direction defined by the Miller indices $(1,-1,0)$. Unfortunately, this notation is often used loosely, crystallographers and physicists employing different conventions.

### 3.2.2 Reciprocal Lattice

Because of the translational symmetry of the crystal, any of its physical properties, described, for example, by some function $f(\mathbf{r})$, must not change under translation:

$$f(\mathbf{r}+\mathbf{R}_l) = f(\mathbf{r}), \tag{3.4}$$

for any $\mathbf{R}_l$. We can express such a periodic function $f$ as a Fourier series:

$$f(\mathbf{r}) = \sum_{\mathbf{q}} A_{\mathbf{q}}\, \mathrm{e}^{\mathrm{i}\mathbf{q}\cdot\mathbf{r}}, \tag{3.5}$$

where the Fourier coefficients $A_{\mathbf{q}}$ are given by

$$A_{\mathbf{q}} = \frac{1}{\Omega_{\mathrm{c}}} \int_{\Omega_{\mathrm{c}}} f(\mathbf{r}) \mathrm{e}^{-\mathrm{i}\mathbf{q}\cdot\mathbf{r}}\, \mathrm{d}\mathbf{r}. \tag{3.6}$$

Making use of the periodicity, Eq. (3.4), to rewrite Eq. (3.5), we have

$$\sum_{\mathbf{q}} A_{\mathbf{q}}\, \mathrm{e}^{\mathrm{i}\mathbf{q}\cdot\mathbf{r}}\, \mathrm{e}^{\mathrm{i}\mathbf{q}\cdot\mathbf{R}_l} = \sum_{\mathbf{q}} A_{\mathbf{q}}\, \mathrm{e}^{\mathrm{i}\mathbf{q}\cdot\mathbf{r}}. \tag{3.7}$$

Clearly, this equality can hold if and only if $\mathrm{e}^{\mathrm{i}\mathbf{q}\cdot\mathbf{R}_l} = 1$. This implies that the sum must extend over a discrete set of $\mathbf{q}$, that we shall denote as $\mathbf{G}_h$, such that

$$\mathbf{G}_h \cdot \mathbf{R}_l = 2\pi \text{ times an integer.} \tag{3.8}$$

To satisfy Eq. (3.8), $\mathbf{G}_h$ has the form

$$\mathbf{G}_h = h_1\mathbf{b}_1 + h_2\mathbf{b}_2 + h_3\mathbf{b}_3, \tag{3.9}$$

with

$$\mathbf{b}_1 = \frac{2\pi}{\Omega_{\mathrm{c}}}\mathbf{a}_2 \times \mathbf{a}_3, \quad \mathbf{b}_2 = \frac{2\pi}{\Omega_{\mathrm{c}}}\mathbf{a}_1 \times \mathbf{a}_3, \quad \mathbf{b}_3 = \frac{2\pi}{\Omega_{\mathrm{c}}}\mathbf{a}_1 \times \mathbf{a}_2. \tag{3.10}$$

The vectors $\mathbf{b}_1$, $\mathbf{b}_2$, and $\mathbf{b}_3$ are called *primitive reciprocal lattice vectors* and they define the *reciprocal lattice*. They are often referred to as $\mathbf{G}$-vectors.

Similar to the real lattice, a Wigner–Seitz cell of the reciprocal lattice can be constructed. This cell is called the *first Brillouin Zone (BZ)* of the lattice.

### 3.2.3 Bloch's Theorem

Bloch's theorem is an extremely important theorem that simplifies enormously the study of electronic states in crystals. We shall make use of it in the following to such an extent that it is hard to envision how one could study electrons in crystals without it. It states that a wavefunction of electrons in the (periodic) crystal potential can be labeled by a vector-index $\mathbf{k}$ and must satisfy the condition:

$$\psi_{\mathbf{k}}(\mathbf{r}+\mathbf{R}_l) = e^{i\mathbf{k}\cdot\mathbf{R}_l}\ \psi_{\mathbf{k}}(\mathbf{r}), \tag{3.11}$$

which is equivalent to the condition:

$$\psi_{\mathbf{k}}(\mathbf{r}) = e^{i\mathbf{k}\cdot\mathbf{r}}\ u_{\mathbf{k}}(\mathbf{r}), \tag{3.12}$$

where $u_{\mathbf{k}}(\mathbf{r})$ is periodic:

$$u_{\mathbf{k}}(\mathbf{r}+\mathbf{R}_l) = u_{\mathbf{k}}(\mathbf{r}). \tag{3.13}$$

Note that this result expresses the fact that electronic wavefunctions in a crystal are plane waves (the factor $e^{i\mathbf{k}\cdot\mathbf{r}}$), as for electrons of momentum $\hbar\mathbf{k}$ in free space, modulating localized wavefunctions that originate from the orbitals of the atoms that constitute the crystal (the factor $u_{\mathbf{k}}(\mathbf{r})$). As we shall see, in many cases studying the modulating "envelope" $e^{i\mathbf{k}\cdot\mathbf{r}}$ will be sufficient to derive many electronic properties that are relevant to charge transport.

To prove Bloch's theorem, we know that if $\psi(\mathbf{r})$ is a solution, then so is $\psi(\mathbf{r}+\mathbf{R}_l)$. We can choose $\psi(\mathbf{r})$ so that $\psi(\mathbf{r}+\mathbf{R}_l) = \alpha\psi(\mathbf{r})$ and requiring $|\psi(\mathbf{r})|^2 = |\psi(\mathbf{r}+\mathbf{R}_l)|^2$, we see that we must have $|\alpha| = 1$. Bloch's theorem is now straightforwardly obtained by writing $\alpha = e^{i\mathbf{k}\cdot R_1}$.

An alternative proof of Bloch's theorem uses group theory: (1) the Hamiltonian commutes with the translation operator, (2) two commutating operators can be simultaneously diagonalized, and (3) the eigenvalues of the group of translations are $e^{i\mathbf{k}\cdot\mathbf{R}_l}$.

The expression for $\psi_{\mathbf{k}}(\mathbf{r})$ can be seen as that of a plane wave (a free electron) modulating a periodic function describing the effect of the (periodic) ionic potential. However, since replacing $\mathbf{k}$ with $\mathbf{k}+\mathbf{G}$ describes the same wavefunction, $\hbar\mathbf{k}$ cannot be the true momentum of the state and $\hbar\mathbf{k}$ is referred to as *crystal momentum*, not necessarily directly related to the "real" electron momentum.

Note that, whereas $\psi_{\mathbf{k}}(\mathbf{r})$ is not periodic, the only physically meaningful quantities we can derive from it are periodic. For example, charge density, $e|\psi_{\mathbf{k}}(\mathbf{r})|^2$, is periodic:

$$|\psi_{\mathbf{k}}(\mathbf{r}+\mathbf{R}_l)|^2 = \left|e^{i\mathbf{k}\cdot(\mathbf{r}+\mathbf{R}_l)}\right|^2\ |u_{\mathbf{k}}(\mathbf{r}+\mathbf{R}_l)|^2 = |u_{\mathbf{k}}(\mathbf{r})|^2 = |\psi_{\mathbf{k}}(\mathbf{r})|^2. \tag{3.14}$$

### 3.2.4 Bragg Reflections

To better understand the meaning of the wavevector $\mathbf{k}$, consider the potential of the ions as very small, that is, as a weak perturbation of the "empty lattice." Let us consider the Fourier series of the crystal potential:

$$V^{(\mathrm{lat})}(\mathbf{r}) = \sum_{\mathbf{G}} V_{\mathbf{G}}\ e^{i\mathbf{G}\cdot\mathbf{r}}, \tag{3.15}$$

and assume that a free electron (characterized by a plane wave of wavevector $\mathbf{k}$) "enters" the crystal and feels the lattice potential as a perturbation. The effect of this perturbation on the wavefunction $|\mathbf{k}\rangle = |\psi_{\mathbf{k}}\rangle$, given by the Bloch wave defined above, will be that of "scattering" the wave to a new direction and wavelength, i.e., to a new wavevector $\mathbf{k}'$. The matrix element for this process is

$$\langle \mathbf{k}'|V^{(\mathrm{lat})}|\mathbf{k}\rangle = \sum_{\mathbf{G}} V_{\mathbf{G}}^{(\mathrm{lat})} \, \delta_{\mathbf{k}-\mathbf{k}'-\mathbf{G}}, \tag{3.16}$$

which vanishes unless $\mathbf{k}+\mathbf{G} = \mathbf{k}'$. Assuming that the collision between the electron and lattice is elastic (the recoil of the lattice, much heavier, will be ignored), we can assume $k = k'$, so that, squaring the relation $\mathbf{k}+\mathbf{G} = \mathbf{k}'$:

$$G^2 = -2\,\mathbf{k}\cdot\mathbf{G}, \tag{3.17}$$

which is the condition for Bragg reflection. It also determines the edge of the BZ. Therefore, the electron wavevector $\mathbf{k}$ can be treated as a "quasi-momentum," but it is restricted to be inside the BZ: As soon as the electron wavelength approaches the BZ boundary, Bragg reflections will "destroy" the wave.

From Bloch's theorem and the condition for Bragg reflections, we conclude that waves with crystal momentum $\hbar\mathbf{k}$ and $\hbar\mathbf{k}+\hbar\mathbf{G}$ are equivalent. This is because a propagating wave carries with it all the diffracted components $e^{i\mathbf{G}\cdot\mathbf{r}}$ as well. However, the energy of the wave will depend on which $\mathbf{G}$ we are considering, as we will see when dealing with energy bands in solids.

### 3.2.5 The Zincblende and Diamond Crystal

Most semiconductors of interest have a "tetragonal" coordination resulting from the $sp^3$ hybridization of the bonding orbitals. This yields a *face-centered cubic (fcc) lattice* with two atoms in each basis. If the basis atoms are different, as in GaAs, this is the zincblende crystal; if the atoms are the same, as in Si or C (diamond), this is the diamond lattice.

- The fcc primitive vectors are

$$\mathbf{a}_1 = a_0\left(0,\frac{1}{2},\frac{1}{2}\right) \quad , \quad \mathbf{a}_2 = a_0\left(\frac{1}{2},0,\frac{1}{2}\right) \quad , \quad \mathbf{a}_3 = a_0\left(\frac{1}{2},\frac{1}{2},0\right), \tag{3.18}$$

  where $a_0$ is the lattice constant (see Fig. 3.1 on page 44). The fcc unit cell is shown in Fig. 3.2 and the fcc Wigner–Seitz cell is illustrated in Fig. 3.3.
- fcc lattices have body-centered cubic (bcc) reciprocal lattices (and vice versa).
- There are 24 fundamental "point" symmetry transformations which map the cubic lattice onto itself (the so-called $T_d$ group): All permutations of the three coordinates (six operations) times sign-swapping two of them (four operations).
- If both atoms in the zincblende crystal are the same, the crystal gains inversion symmetry and is now the diamond crystal. Accounting for inversions, there are 48 symmetry inequivalent operations in all (the so-called $O_h$ group).

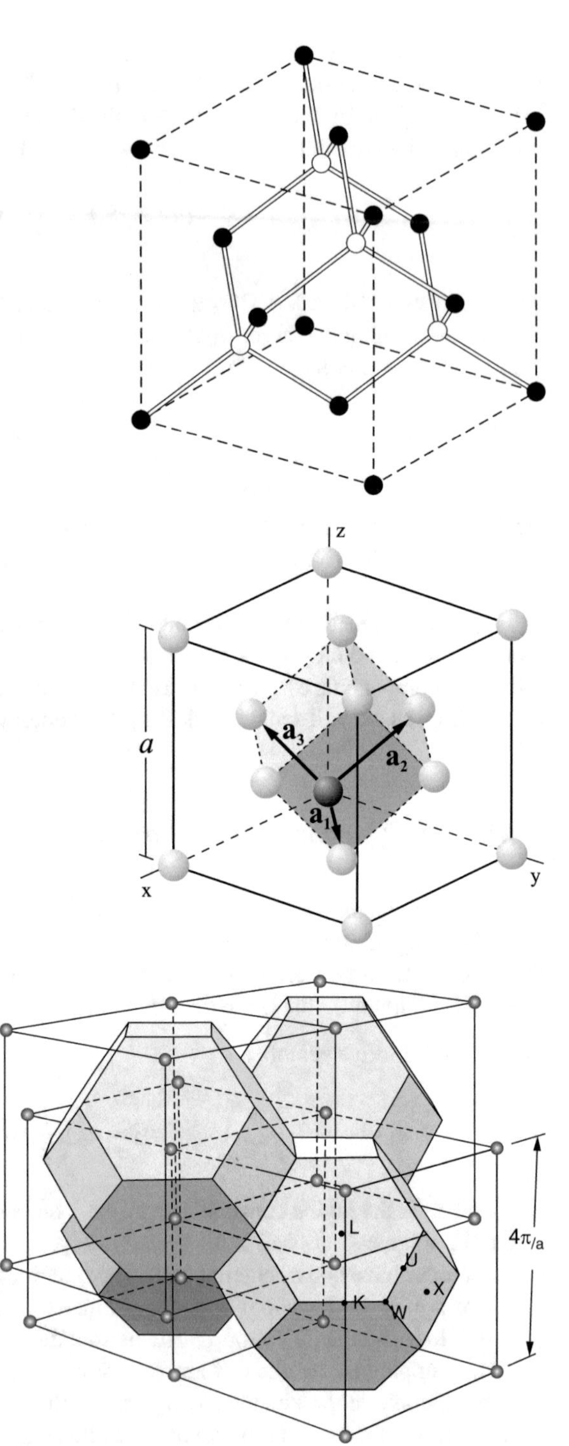


**Fig. 3.1** Illustration of the zincblende crystal in its cubic unit cell

**Fig. 3.2** Illustration of the fundamental translation vectors and the primitive unit cell

**Fig. 3.3** Four adjacent Brillouin zones of the fcc lattice

## 3.3 *Group Theory

The effects of symmetry can be formally analyzed using algebra and more specifically "group theory." Using symmetry information is often useful to prove or understand why certain degeneracies appear in the band structure or why some transitions are prohibited.

For example, that silicon is not efficient at absorbing/emitting light with an energy close to the band gap is an immediate consequence of crystal symmetry and translational symmetry in particular. Translational symmetry assigns a wavevector **k** to each state and unless translational symmetry is broken, **k** has to be conserved between the final and the initial state. Now, silicon has an indirect band gap, i.e., the valence-band minimum and conduction-band maximum have a different **k**-vector. The interaction with an electromagnetic field only breaks translational symmetry very weakly because of the long wavelength of light compared to the lattice constant. Silicon's indirect gap means that light absorption in silicon is prohibited to first order. Second-order interactions, where light absorption/emission is mediated by the electron–phonon interaction, are allowed but are much less inefficient.

However, most crystalline materials, and certainly most interesting crystalline materials, have many symmetries other than translational symmetries. These can be rotations and reflections and combinations thereof, such as inversions, screw displacements (combinations of rotations and translations), glide plane operations (combinations of reflections and translations), and improper rotations (combinations of reflections and rotations).

Group theory, as applied to crystals, is a vast subject. Here, we provide the basic elements. Texts, such as Dresselhaus' [4], should be consulted for a complete treatment in the context of solid-state physics.

### 3.3.1 *Symmetry and Groups*

When studying electronic properties of semiconductors, we are interested in those symmetry operations which preserve the energy of the system. Formally, any symmetry operation $\widehat{T}$ preserving the energy must leave the Hamiltonian unchanged

$$\widehat{T}(\widehat{H}) = \widehat{H}. \tag{3.19}$$

Since any practical Hamiltonian contains the kinetic energy ($\propto -\nabla^2$), the allowed symmetry operations are limited to Euclidean (or rigid) transformations which have a general form

$$\widehat{T} : \mathbf{r} \to M\mathbf{r} + \mathbf{a} \tag{3.20}$$

where $M$ is an orthogonal $3 \times 3$ matrix and $\mathbf{a}$ a translation vector. For pure translations, $M$ is the identity matrix, for rotations around the origin and reflections through a plane containing the origin, $\mathbf{a} = \mathbf{0}$.

In Eq. (3.19), $\widehat{T}$ is written as an operator acting on an operator, i.e., the Hamiltonian. But it is more convenient to have $\widehat{T}$ simply act on a Hilbert space, similar to other operators. If $\widehat{T}$ represents an operator on a Hilbert space, Eq. (3.19) can be rewritten as

$$\widehat{T}\widehat{H} = \widehat{H}\widehat{T} \quad \text{or} \quad [\widehat{H}, \widehat{T}] = 0. \tag{3.21}$$

Symmetry operators can be shown to have four properties:

- Closure: Taking two symmetry operators $\widehat{T}_1$ and $\widehat{T}_2$, $\widehat{T}_1\widehat{H} = \widehat{H}\widehat{T}_1$ and $\widehat{T}_2\widehat{H} = \widehat{H}\widehat{T}_2$. It immediately follows that their combination $\widehat{T}_3 = \widehat{T}_1\widehat{T}_2$ is also a symmetry of the system:

$$\widehat{T}_3\widehat{H} = \widehat{T}_1\widehat{T}_2\widehat{H} = \widehat{T}_1\widehat{H}\widehat{T}_2 = \widehat{H}\widehat{T}_1\widehat{T}_2 = \widehat{H}\widehat{T}_3. \tag{3.22}$$

- Identity: The identity symmetry operator $\widehat{E} : \mathbf{r} \to \mathbf{r}$ is always a symmetry of the Hamiltonian: $\widehat{E}\widehat{H} = \widehat{H} = \widehat{H}\widehat{E}$.
- Invertibility: If a symmetry operator $\widehat{T}_1$ is a symmetry of the system, then its inverse $\widehat{T}_1^{-1}$, satisfying $\widehat{T}_1\widehat{T}_1^{-1} = \widehat{E} = \widehat{T}_1^{-1}\widehat{T}_1$, is also a symmetry of the system. Defining an auxiliary wavefunction $\psi'$ such that $\psi = \widehat{T}_1\psi'$,

$$\widehat{T}_1^{-1}\widehat{H}\psi = \widehat{T}_1^{-1}\widehat{H}\widehat{T}_1\psi' = \widehat{T}_1^{-1}\widehat{T}_1\widehat{H}\psi' = \widehat{H}\psi' = \widehat{H}\widehat{T}_1^{-1}\widehat{T}_1\psi' = \widehat{H}\widehat{T}^{-1}\psi. \tag{3.23}$$

- Associativity: $\widehat{T}_1\widehat{T}_2\widehat{H} = \widehat{T}_1(\widehat{T}_2)\widehat{H}$ which is a general property of (Euclidean) transformations.

#### 3.3.1.1 Groups

The (abstract) algebraic structure that takes the four properties of symmetry operators (closure, associativity, identity, and invertibility) as its axioms is the mathematical *group*. In solid-state physics, we can draw on the results from group theory, but it is important to remember that group theory is a formal abstract mathematical construct. The development of group theory has yielded many interesting mathematical results and predates modern condensed matter physics and even Quantum Mechanics.

A group is defined by a set of elements $\{a, b, \dots\}$ and an operation $a \cdot b = ab$, often called the group multiplication. In group theoretical language, each symmetry operation $(\widehat{T}_1, \widehat{T}_2, \widehat{T}_3)$ of the crystal can be represented by an element $(a, b, c)$ of its space group. The result of one symmetry operation acting upon another symmetry operation $(\widehat{T}_3 = \widehat{T}_1\widehat{T}_2)$ will determine the group multiplication $(c = a \cdot b = ab)$ in the space group.

For small groups, it is convenient to characterize the group by its multiplication or Cayley table. As an example of some groups we have

- The trivial group containing only one element $\{e\}$ and its group multiplication is trivially determined as $ee = e$. Its multiplication table is

$$\begin{array}{c|c} \cdot & e \\ \hline e & e \end{array}. \tag{3.24}$$

- The group containing two elements $\{e, a\}$ with multiplication $ee = aa = 1$ and $ea = ae = a$. This is the smallest nontrivial group and is denoted as $\mathbb{Z}_2, C_2, C_i, \dots$. Its multiplication table is

$$\begin{array}{c|c|c} \cdot & e & a \\ \hline e & e & a \\ \hline a & a & e \end{array}. \tag{3.25}$$

- The cyclic group containing $n$ elements $\{e, a, aa, aaa, \dots, (a)^{n-1}\}$ is denoted as $\mathbb{Z}_n$. Each element in the cyclical group can be represented by a complex number with unit magnitude: $a = \mathrm{e}^{\mathrm{i}2\pi/n}$.

- The group of integers $\mathbb{Z}$ $\{\ldots,-2,-1,0,1,2,\ldots\}$ where the "group multiplication" operation corresponds to the conventional addition of numbers, i.e., $1 \cdot 2 = 3$ and 0 is the identity element. $\mathbb{Z}$ is an infinite group.
- The dihedral group $D_3$ with six elements $\{e,a,b,c,d,f\}$ with multiplication table

$$\begin{array}{c|c|c|c|c|c|c} \cdot & e & a & b & c & d & f \\ \hline e & e & a & b & c & d & f \\ \hline a & a & e & d & f & b & c \\ \hline b & b & f & e & d & c & a \\ \hline c & c & d & f & e & a & b \\ \hline d & d & c & a & b & f & e \\ \hline f & f & b & c & a & e & d \end{array} . \tag{3.26}$$

  Note that contrary to the above listed groups, the multiplication table is not symmetric. This means that the group multiplication of the dihedral group is not commutative for some elements, e.g., $ab = d \neq ba = f$. Any such group where group multiplication is not commutative is known as a nonabelian group.

For larger groups, the multiplication tables are not practical and alternative ways, such as character tables, are used to identify groups.

Groups can be combined into new groups by taking the direct product of the groups. The direct product of two groups $G,H$, with $n_G,n_H$ elements $\{e_G,a_G,b_G,\ldots\}$, $\{e_H,a_H,b_H,\ldots\}$ is a group $G \times H$ with $n_G n_H$ elements $\{e_G e_H, e_G a_H, e_G b_H, \ldots, a_G e_H, a_G a_H, a_G b_H, \ldots\}$. Denoting the operations in $G$ and $H$ as $\cdot_G$ and $\cdot_H$, the group operation (multiplication) in $G \times H$ is defined by $(x_G x_H) \cdot_{G\times H} (y_G y_H) = (x_G \cdot_G y_G)(x_H \cdot_H y_H)$.

### 3.3.1.2 Space Groups

All symmetry operations in a crystal form the space group of the crystal and the symmetry operations can be categorized as

- (i) Translations $T_l : \mathbf{r} \to \mathbf{r} + \mathbf{R}_l$
- (ii) Point symmetries $R$: Rotations, reflections, and combinations of rotations and reflections around a point of choice in the crystal. If we take the origin as the point of choice, then $R : \mathbf{r} \to M\mathbf{r}$. Groups composed of point symmetry operations around a point of choice are called *point groups*.
- (iii) All other operations which are combinations of translations with point symmetries including point symmetries around a point other than the point of choice.

All of these combined form the space group of the crystal $\mathcal{G}$.

The description of translation symmetry can always be done by a group $\mathcal{T}$. In three dimensions, any translation by $\mathbf{R}_l = l_1\mathbf{a}_1 + l_2\mathbf{a}_2 + l_3\mathbf{a}_3$ is a symmetry if $\mathbf{a}_{1,2,3}$ are the primitive vectors and the group of translations is isomorphic to $\mathbb{Z} \times \mathbb{Z} \times \mathbb{Z}$.

In the simplest case, any element of category (iii) can be obtained by combining a translation from category (i) with an element of the point group $\mathcal{R}$ described by category (ii). In this simple case, the space group $\mathcal{G}$ is said to be *symmorphic* and the space group is the direct product of the translation and the point group of the system $\mathcal{G} = \mathcal{T} \times \mathcal{R}$.

In the alternative case, some elements of category (iii) are a combination of a rotation or a reflection with a translation which is not an element of category (i). In this case, the space group is said to be *non-symmorphic*. The point group $\mathcal{R}$ of non-symmorphic space groups is the combination of the group formed by the elements of category (ii) together with some of the elements of category (iii).

Specifically, those elements in category (iii) that do not differ from each other by a lattice translation found in category (i). In group technical terms, the point group is the factor group of the space group and the translation group: $\mathcal{R} = \mathcal{G}\backslash\mathcal{T}$. However, the space group is not the product of the translation and the point group $\mathcal{G} \neq \mathcal{T} \times \mathcal{R}$ for non-symmorphic space groups.

### 3.3.2 Character Tables

The unique symmetry information of a crystal is found in its point group $\mathcal{R}$ and to describe these point groups, character tables are used. In this section, we will explain what a character table is and how it can be read.

A first concept when constructing a character table is the concept of a class. Intuitively, we can say that "similar" symmetry operations belong to the same class. For example, a reflection along the $x-z$ plane belongs to the same class as a reflection along the $y-z$ plane if the crystal also has a 90° rotation axis along the $z$-axis. Formally, two elements of a group $a,b$ belong to the same class if an element $c$ exists such that

$$a = cbc^{-1}. \tag{3.27}$$

The second concept when constructing a character table is the concept of a (matrix) representation. Each element in a group can be represented by a matrix $\Gamma(a)$ so that the group multiplication also holds for each matrix multiplication

$$\Gamma(a)\Gamma(b) = \Gamma(ab). \tag{3.28}$$

If each element $a \neq b$ is mapped onto a different matrix $\Gamma(a) \neq \Gamma(b)$, the representation is said to be faithful, otherwise it is an unfaithful representation. For example, for $\mathbb{Z}_2$, we can take

$$\Gamma(e) = \begin{bmatrix} 1 & 0 \\ 0 & 1 \end{bmatrix} \quad \text{and} \quad \Gamma(a) = \begin{bmatrix} 0 & 1 \\ 1 & 0 \end{bmatrix} \tag{3.29}$$

or through a similarity transformation

$$\Gamma(e) = \begin{bmatrix} 1 & 0 \\ 0 & 1 \end{bmatrix} \quad \text{and} \quad \Gamma(a) = \begin{bmatrix} -1 & 0 \\ 0 & 1 \end{bmatrix} \tag{3.30}$$

as faithful representations of dimension 2. $\Gamma(e) = 1, \Gamma(a) = -1$ as a faithful representation of dimension 1 or $\Gamma(e) = 1, \Gamma(a) = 1$ as an unfaithful representation of dimension 1.

A representation is reducible if a similarity transformation exists that brings all representation matrices in block-diagonal form at the same time

$$\Gamma(a) = \begin{pmatrix} \Gamma^{(1)}(a) & 0 \\ 0 & \Gamma^{(2)}(a) \end{pmatrix}. \tag{3.31}$$

If this is the case, $\Gamma^{(1)}$ and $\Gamma^{(2)}$ are also representations of the group and Eq. (3.31) defines the sum of two representations $\Gamma = \Gamma^{(1)} \oplus \Gamma^{(2)}$. If it is not possible to bring all matrices in block-diagonal form, or equivalently, to write the representation as the sum of two or more representations, the representation is said to be irreducible.

Equation (3.31) defines the sum of two representations. But two representations of dimension $n_1$ and $n_2$ can also be multiplied with each other, $\Gamma^{(3)} = \Gamma^{(1)} \times \Gamma^{(2)}$, yielding a representation of dimension $n_3 = n_1 n_2$. In matrix form, the elements of $\Gamma^{(3)}$ are the Kronecker product of the matrices in representations 1 and 2. The product of 2 representations may be reducible or irreducible.

The *character* $\chi^{(i)}(a)$ of an element is the trace of its matrix in representation $i$:

$$\chi^{(i)}(a) = \mathrm{Tr}(\Gamma^{(i)}(a)). \tag{3.32}$$

Some properties are noted below; required proof can be found in [5].

- The character (the trace) is invariant under similarity transformations.
- Any matrix representing two elements $a, b$ in the same class satisfies

$$\Gamma(a) = \Gamma(c)\Gamma(b)\Gamma(c)^{-1}, \tag{3.33}$$

  has the same eigenvalues, and consequently the same character $\mathrm{Tr}(\Gamma(a)) = \mathrm{Tr}(\Gamma(b))$.
- The characters of all elements in the group for different irreducible representations are orthogonal:

$$\sum_a \chi^{(i)}(a)^* \chi^{(j)}(a) = \sum_C \chi^{(i)}(C)^* \chi^{(j)}(C) N_C = N\delta_{ij}, \tag{3.34}$$

  where $C$ sums over all classes, $N_C$ is the number of elements in class $C$, and $N = \sum_C N_C$ the total number of elements.
- Characters for the same element in different representations are orthogonal:

$$\sum_i \chi^{(i)}(C)^* \chi^{(i)}(C') = \frac{N}{N_C}\delta_{CC'}, \tag{3.35}$$

  where $C, C'$ indicate the classes of the elements under consideration.
- The characters of the sum of two representations are the sum of the characters of the representations: if $\Gamma^{(3)} = \Gamma^{(1)} \oplus \Gamma^{(2)}$, then $\chi^{(3)}(C) = \chi^{(1)}(C) + \chi^{(2)}(C)$.
- The characters of the product of two representations are the product of the characters of the representations: if $\Gamma^{(3)} = \Gamma^{(1)} \times \Gamma^{(2)}$, then $\chi^{(3)}(C) = \chi^{(1)}(C)\chi^{(2)}(C)$.
- The number of (inequivalent) irreducible representations always equals the number of classes.
- Any representation can be decomposed into irreducible representations in a unique way.

A character table is a table listing all characters with its rows indicating the different irreducible representations and its columns the different classes. The first row is always the trivial representation ($\chi(C) = 1$) for all $C$. The first column is always the identity transformation $E$ whose representation matrices are always the identity matrix and its character (trace) is the dimension of the representation. Character tables for all point groups can be found in [4].

The point group of a system with only inversion symmetry is denoted as $\mathrm{C}_i$ (isomorphic to $\mathbb{Z}_2$). The character table of $\mathrm{C}_i$ is shown in Table 3.1 together with the character table of the dihedral group $D_3$. Representations of point groups denoted by a letter $A, B$ are one-dimensional representations which are symmetric and antisymmetric around their principle axis of rotation, respectively, $E$ and $T$ representations are two- and three-dimensional representations. A subindex $g, u$ indicates if the representation is even (German: *gerade*) or odd (German: *ungerade*).

**Table 3.1** Character table of the point group $C_i$ and $D_3$

| | E | i |
|---|---|---|
| $A_g$ | 1 | 1 |
| $A_u$ | 1 | −1 |

| | E | $2C_3$ | $3C_2'$ |
|---|---|---|---|
| $A_1$ | 1 | 1 | 1 |
| $A_2$ | 1 | 1 | −1 |
| E | 2 | −1 | 0 |

### 3.3.3 *Degeneracies and Selection Rules*

#### 3.3.3.1 Degeneracies

The irreducible representations of the point group of a system help to understanding degeneracies in the band structure of a crystal. If $\psi$ is an eigenstate of the Hamiltonian with an energy $E$, then so is $\widehat{T}\psi$. Taking a set of orthogonal degenerate wavefunctions $\psi_i$ with energy $E$ satisfying

$$\widehat{H}\psi_i = E\psi_i, \tag{3.36}$$

then $\widehat{T}\psi_i$ must be a linear combination of $\psi_i$:

$$\widehat{T}_a\psi_i = \sum_{i'} \Gamma_{ii'}(a)\psi_{i'}. \tag{3.37}$$

The matrices formed by the coefficients $\Gamma_{ii'}(a)$ form a representation of the space group of the crystal. If the representation is reducible, however, the degenerate states can be separated into two (or more) sets $\psi_1,\ldots,\psi_M$ and $\psi_{M+1},\ldots,\psi_N$ and the symmetry operations only carry $\psi_1,\ldots,\psi_M$ into linear combinations of $\psi_1,\ldots,\psi_M$. In this case, symmetry does not protect the degeneracy between the two sets but only within the set. For this reason, almost all degeneracies observed in theoretical or experimental band structures are governed by irreducible representations of their point group. Although accidental degeneracies are possible, most often when more degeneracy than expected is present, it is because the system has a symmetry which was not accounted for.

As an example, consider a one-dimensional system with inversion symmetry (which is the same as reflection in 1D) whose character table is given in Table 3.1. The simplest such system is a one-dimensional system without any applied external potential which has the solutions $\psi_{1,2} = \mathrm{e}^{\pm ikx}$ with an energy $\hbar^2/(2m_0)k^2$. The matrix representation showing how inversion carries $\mathrm{e}^{\pm ikx}$ into $\mathrm{e}^{\mp ikx}$ is given in Eq. (3.29). An alternative but equivalent orthogonal set of solutions is given by $\psi_{1,2} = \sin(x),\cos(x)$. The matrix representation for the inversion is now given by Eq. (3.30), which shows that the degeneracy between $\cos(x)$ and $\sin(x)$ is not protected against inversion. But $\cos(x)$ and $\sin(x)$ are degenerate in this case (with zero potential) because the system also has translational symmetry in addition to inversion symmetry. Breaking the translation symmetry by considering a particle in a box $x \in [-L/2, L/2]$, the solutions are $\cos(x(2n-1)\pi/L)$ and $\sin(x2n\pi/L)$ with $n = 1,2,\ldots$ which correspond to $A_g$ and $A_u$ irreducible representations, respectively, i.e., the "cos solutions" are even under inversion $T_i \cos(x(2n-1)\pi/L) = \cos(x(2n-1)\pi/L)$ while the "sin solutions" are odd $T_i \sin(x2n\pi/L) = -\sin(x2n\pi/L)$.

#### 3.3.3.2 Selection Rules

Selection rules are the specification of which transitions are allowed under a given symmetry and a given perturbation. The transition rates for a transition from an initial state $\psi_i$ to a final state $\psi_f$ under a perturbing Hamiltonian $\widehat{H}'$ are measured by the matrix element (see Appendix A):

$$M = \int \mathrm{d}r^3 \psi_i(\mathbf{r}) \widehat{H}' \psi_f(\mathbf{r}) = \int \mathrm{d}r^3 f(\mathbf{r}). \tag{3.38}$$

If $\psi_i(\mathbf{r})$ and $\psi_f(\mathbf{r})$ are nondegenerate and $H'$ is invariant under the crystal symmetry, $f(\mathbf{r})$ is a one-dimensional representation of the crystal symmetry and $T_a[f(\mathbf{r})] = \alpha_a f(\mathbf{r})$ for any symmetry operation $a$. In this case, it is straightforward to derive selection rules. Applying a crystal transformation $a$ leaves $M$ invariant

$$M = \int \mathrm{d}r^3 \; T_a[f(\mathbf{r})] = \alpha_a \, M, \tag{3.39}$$

since $T_a$ is a Euclidean transformation. The matrix element $M$ must vanish unless $\alpha_a = 1$ for all $a$, since $(1-\alpha_a)M = 0$. Equivalently, the representation of $f(\mathbf{r})$ must be the trivial representation, i.e., the representation with all of its characters equal to unity.

In general, by identifying how the perturbing Hamiltonian behaves under the symmetry operations of the crystal, its representation $\Gamma^{(H')}$ of dimension $n_{H'}$ can be determined. If the perturbation does not respect the symmetry of the crystal, the representation of the Hamiltonian will be a higher-dimensional representation (which may be reducible), although the specific perturbation will only be one of many equivalent perturbations. For example, evaluating if a perturbation dependent on the $x$-coordinate is allowed is equivalent to evaluating whether a perturbation dependent on $y$ is allowed if the system has a 90° rotation around the $z$-axis. If the initial or final state is degenerate because of symmetry, selection rules will be determined between any combination of the degenerate initial or final states.

Taking $\psi_i(\mathbf{r})$ and $\psi_f(\mathbf{r})$ to have $n_i$ and $n_f$-fold degeneracy and to be represented by $\Gamma^{(i)}$ and $\Gamma^{(f)}$, $n_i n_{H'} n_f$ $\psi_i^*(\mathbf{r}) H' \psi_f(\mathbf{r})$ products are possible. Applying the symmetry operations to these products, the $n_i n_{H'} n_f$-dimensional product representation $\Gamma^M = \Gamma^{(i)} \times \Gamma^{(H')} \times \Gamma^{(f)}$ is obtained. This product can subsequently be written as a sum of irreducible representations $\Gamma^M = \sum_j \Gamma^{(j)}$. For a transition to be allowed, the product of the representations $\Gamma^M$ must contain the trivial representation.

### 3.3.4 Application of Group Theory to Graphene

As a concrete example, we determine qualitative features of the band structure of graphene. Graphene has a honeycomb lattice with two primitive vectors $\mathbf{a}_{1,2}$ with an angle of 60° between them. Taking the origin at the center of the honeycomb, the point group of the graphene lattice has the following symmetries:

- $2C_6$: Clockwise and counterclockwise rotations over 60° around the $z$-axis
- $2C_3$: Clockwise and counterclockwise rotations over 120° around the $z$-axis
- $C_2$: Rotation over 180° around the $z$-axis
- $3C_2'$: 3 180° rotations around the $x$-axis and equivalent (rotated over 120° around the $z$-axis)
- $3C_2''$: 3 180° rotations around the $y$-axis and equivalent
- $i$: Inversion around the origin
- $2S_3$: Inversion and rotation over 120° around the $z$-axis
- $2S_6$: Inversion and rotation over 60° around the $z$-axis
- $\sigma_\mathrm{h}$: Mirror symmetry through the horizontal $x-y$ plane
- $\sigma_\mathrm{d}$: 3 mirror symmetries through the $x-z$ plane and equivalent
- $\sigma_\mathrm{v}$: 3 mirror symmetries through the $y-z$ plane and equivalent

The point group describing this symmetry has 24 elements and is called $D_{6h}$, its character table is shown in Table 3.2.

**Table 3.2** Character table of the graphene point group $D_{6h}$

| | E | $2C_6$ | $2C_3$ | $C_2$ | $3\,C_2'$ | $3\,C_2''$ | i | $2\,S_3$ | $2\,S_6$ | $\sigma_h$ | $3\,\sigma_d$ | $3\,\sigma_v$ |
|---|---|---|---|---|---|---|---|---|---|---|---|---|
| $A_{1g}$ | 1 | 1 | 1 | 1 | 1 | 1 | 1 | 1 | 1 | 1 | 1 | 1 |
| $A_{2g}$ | 1 | 1 | 1 | 1 | −1 | −1 | 1 | 1 | 1 | 1 | −1 | −1 |
| $B_{1g}$ | 1 | −1 | 1 | −1 | 1 | −1 | 1 | −1 | 1 | −1 | 1 | −1 |
| $B_{2g}$ | 1 | −1 | 1 | −1 | −1 | 1 | 1 | −1 | 1 | −1 | −1 | 1 |
| $E_{1g}$ | 2 | 1 | −1 | −2 | 0 | 0 | 2 | 1 | −1 | −2 | 0 | 0 |
| $E_{2g}$ | 2 | −1 | −1 | 2 | 0 | 0 | 2 | −1 | −1 | 2 | 0 | 0 |
| $A_{1u}$ | 1 | 1 | 1 | 1 | 1 | 1 | −1 | −1 | −1 | −1 | −1 | −1 |
| $A_{2u}$ | 1 | 1 | 1 | 1 | −1 | −1 | −1 | −1 | −1 | −1 | 1 | 1 |
| $B_{1u}$ | 1 | −1 | 1 | −1 | 1 | −1 | −1 | 1 | −1 | 1 | −1 | 1 |
| $B_{2u}$ | 1 | −1 | 1 | −1 | −1 | 1 | −1 | 1 | −1 | 1 | 1 | −1 |
| $E_{1u}$ | 2 | 1 | −1 | −2 | 0 | 0 | −2 | −1 | 1 | 2 | 0 | 0 |
| $E_{2u}$ | 2 | −1 | −1 | 2 | 0 | 0 | −2 | 1 | 1 | −2 | 0 | 0 |

To determine the graphene band structure, we consider the free electron in a "virtual" hexagonal lattice. We assume a confining potential in the $z$ direction so the electrons are only free in the $x$ and $y$ direction and have no $k_z$-dependence. The wavefunctions of the free electron are

$$\psi(\mathbf{r}) = \mathrm{e}^{\mathrm{i}\mathbf{K}\cdot\mathbf{R}} u_{g,u}(\mathbf{r}), \tag{3.40}$$

where $\mathbf{K}$ and $\mathbf{R}$ are two-dimensional vectors and $u(\mathbf{r})$ is a function with crystal symmetry accounting for the confinement in the $z$ direction. We will consider the case where $u(\mathbf{r})$ is even ($u_g(\mathbf{r})$) and odd ($u_u(\mathbf{r})$) under reflection through the $x-y$ plane.

The lowest-energy wavefunction is found for $\mathbf{K} = \mathbf{0}$ together with a $u_g(\mathbf{r})$. We label this representation as $[00]_g$ and its corresponding irreducible representation is $A_{1g}$. Higher-energy wavefunctions are obtained by taking $u_u(\mathbf{r})$ and the thus obtained representation $[00]_u$ corresponds to the irreducible representation $A_{2u}$.

Higher-energy wavefunctions can also be obtained by taking $u_g(\mathbf{r})$ and selecting the plane waves $\mathrm{e}^{\mathrm{i}\mathbf{G}_i\cdot\mathbf{R}}$ with

$$\mathbf{G}_1 = \frac{4\pi}{\sqrt{3}}\left(\frac{\sqrt{3}}{2}, \frac{1}{2}\right) \tag{3.41}$$

$$\mathbf{G}_2 = \frac{4\pi}{\sqrt{3}}(0, 1) \tag{3.42}$$

$$\mathbf{G}_3 = \frac{4\pi}{\sqrt{3}}\left(-\frac{\sqrt{3}}{2}, \frac{1}{2}\right) \tag{3.43}$$

$$\mathbf{G}_4 = \frac{4\pi}{\sqrt{3}}\left(-\frac{\sqrt{3}}{2}, -\frac{1}{2}\right) \tag{3.44}$$

$$\mathbf{G}_5 = \frac{4\pi}{\sqrt{3}}(0, -1) \tag{3.45}$$

$$\mathbf{G}_6 = \frac{4\pi}{\sqrt{3}}\left(\frac{\sqrt{3}}{2}, -\frac{1}{2}\right). \tag{3.46}$$

**Table 3.3** Characters for the plane waves in the hexagonal lattice

| | E | 2C$_6$ | 2C$_3$ | C$_2$ | 3 C$_2'$ | 3 C$_2''$ | i | 2 S$_3$ | 2 S$_6$ | $\sigma_h$ | 3 $\sigma_d$ | 3 $\sigma_v$ | Irreps |
|---|---|---|---|---|---|---|---|---|---|---|---|---|---|
| $[00]_g$ | 1 | 1 | 1 | 1 | 1 | 1 | 1 | 1 | 1 | 1 | 1 | 1 | A$_{1g}$ |
| $[11]_g$ | 6 | 0 | 0 | 0 | 2 | 0 | 0 | 0 | 0 | 6 | 2 | 0 | A$_{1g}\oplus$ B$_{1u}\oplus$ E$_{2g}\oplus$ E$_{1u}$ |
| $[00]_u$ | 1 | 1 | 1 | 1 | −1 | −1 | −1 | −1 | −1 | −1 | 1 | 1 | A$_{2u}$ |

Assuming $\mathbf{a}_{1,2} = (1,0),(1/2,\sqrt{3}/2)$, $\mathbf{b}_{1,2} = 4\pi/\sqrt{3}(\sqrt{3}/2,-1/2),4\pi/\sqrt{3}(0,1)$. These 6 wavefunctions will yield a 6-dimensional representation of $D_{6h}$ which we denote as $\Gamma^{[11]_g}$. Applying all symmetry operations and counting which wavefunctions remain invariant, we can determine the characters for each class of symmetry operations. $E$ leaves all wavefunctions unchanged and will give $\chi^{[11]_g}(E) = 6$, $C_6$ permutes all wavefunctions $1 \to 2, 2 \to 3, \ldots$ yielding $\chi^{[11]_g}(C_6) = 0$, $C_3$ permutes $1 \to 3, 2 \to 4, \ldots$ yielding $\chi^{[11]_g}(C_3) = 0$, a nonzero contribution comes from $C_2'$ which leaves $\mathbf{G}_{i,i+3}$ unchanged. The resulting characters for all classes is given in Table 3.3.

To have an idea of what the wavefunctions corresponding to these symmetries look like, we take the representation of $C_6$ in $\Gamma^{[11]_g}$:

$$\Gamma^{[11]}(C_6) = \begin{bmatrix} 0&0&0&0&0&1 \\ 1&0&0&0&0&0 \\ 0&1&0&0&0&0 \\ 0&0&1&0&0&0 \\ 0&0&0&1&0&0 \\ 0&0&0&0&1&0 \end{bmatrix} \tag{3.47}$$

which is the permutation matrix of dimension 6 with eigenvalues $e_i = \mathrm{e}^{\mathrm{i}j\pi/6}$ with $j = 0,\ldots,5$ and without normalization the eigenvectors are

$$V_j = [1, \mathrm{e}^{\mathrm{i}j\pi/3}, \mathrm{e}^{\mathrm{i}2j\pi/3}, \mathrm{e}^{\mathrm{i}3j\pi/3}, \mathrm{e}^{\mathrm{i}4j\pi/3}, \mathrm{e}^{\mathrm{i}5j\pi/3}].$$

The one-dimensional representations $A_{1g}$ and $B_{1u}$ have characters $\pm 1$, which must correspond to the first and the fourth eigenvalues of $\Gamma^{[11]_g}$ and have eigenvectors $V_{A_{1g}} = [1,1,1,1,1,1,1]$ and $V_{B_{1u}} = [1,-1,1,-1,1,-1,1]$. We can also immediately determine the two-dimensional representations since $\chi^{E_{2g},E_{1u}}(C_6) = \mp 1$ (from character Table 3.2) and this can only be realized by assigning the eigenvalues $\mathrm{e}^{\pm \mathrm{i}2\pi/3}$ to $E_{2g}$ and $\mathrm{e}^{\pm \mathrm{i}\pi/3}$ to $E_{1u}$.

Calculating the magnitude of the wavefunction at the position of the carbon atoms in the graphene lattice $\tau = (0,1/\sqrt{3})$, then

$$\psi(\tau) = [\mathrm{e}^{\mathrm{i}2\pi/3}, 1, \mathrm{e}^{-\mathrm{i}2\pi/3}, \mathrm{e}^{-\mathrm{i}2\pi/3}, 1, \mathrm{e}^{\mathrm{i}2\pi/3}] = \mathrm{e}^{\mathrm{i}2\pi/3k}$$

gives

$$\psi_{A_{1g}}(\tau) = \mathrm{e}^{\mathrm{i}2\pi/3} + 1 + \mathrm{e}^{-\mathrm{i}2\pi/3} + \mathrm{e}^{-\mathrm{i}2\pi/3} + 1 + \mathrm{e}^{\mathrm{i}2\pi/3} = 0,$$

and $\psi_{E_{2g}}(\tau) = 0;6$, $\psi_{B_{1u}}(\tau) = 0$ and $\psi_{E_{1u}} = 0;0$, showing that only the $E_{2g}$ has a nonzero value for its wavefunction at the carbon atoms where the potential is the lowest. The $E_{2g}$ representation will thus have the lowest energy of the four representations.

Carbon has 4 valence atoms, graphene has 2 carbon atoms per unit cell and thus 8 atoms per unit cell which means that 4 spin-degenerate bands will be occupied. We have already established

**Table 3.4** Character table of the graphene point group $D_{3h}$

| | E | $\sigma_h$ | $2C_3$ | $2S_3$ | $3\,C_2'$ | $3\,\sigma_v$ |
|---|---|---|---|---|---|---|
| $A_1'$ | 1 | 1 | 1 | 1 | 1 | 1 |
| $A_2'$ | 1 | 1 | 1 | 1 | $-1$ | $-1$ |
| $A_1''$ | 1 | $-1$ | 1 | $-1$ | 1 | $-1$ |
| $A_2''$ | 1 | $-1$ | 1 | $-1$ | $-1$ | 1 |
| $E'$ | 2 | 2 | $-1$ | $-1$ | 0 | 0 |
| $E''$ | 2 | $-2$ | $-1$ | 1 | 0 | 0 |

that the lowest-energy state has $A_{1g}$ symmetry corresponding to $[00]_g$ and the lowest-energy band corresponding to $[11]_g$ is the two-dimensional $E_{1u}$ representation, the fourth electron will occupy the $[00]_u$ state with an $A_{2u}$ representation corresponding to a $\pi$-bond between the carbon $p_z$ orbitals.

At the K-point ($\mathbf{K} = (4\pi/3,0)$), the symmetry of the lattice is reduced. For example, rotation over 60° maps $\mathbf{K}$ onto $\mathbf{K}' = 4\pi/3(1/2,\sqrt{3}/2)$ and $\mathbf{K}$ and $\mathbf{K}'$ are inequivalent since $\mathbf{K}-\mathbf{K}'$ is not a reciprocal lattice vector. Symmetries which are broken are $C_6$, $C_2$, $C_2''$, $i$, $\sigma_d$, and $S_6$. The remaining symmetry of the K-point is $D_{3h}$ whose character table is shown in Table 3.4.

Repeating our analysis of the occupied bands using the free-electron bands at $K$: we start from the 3 plane-wave functions at $\mathbf{K}$:

$$\mathbf{K}_{1,2,3} = 4\pi/3[(1,0),(-1/2,\sqrt{3}/2),(-1/2,-\sqrt{3}/2)]$$

which reduce into a two-dimensional $E'$ representation and an $A_1'$ representation. These 3 bands will be fully occupied and similar to the case at $\Gamma$, a fourth band which is odd under vertical mirror symmetry will be occupied. However, the lowest-energy band which is odd under vertical mirror symmetry corresponds to a two-dimensional $E''$ representation. So at K, the graphene band is degenerate (i.e., a two-dimensional irreducible representation) but only half-occupied and graphene is a semi-metal with a Dirac-cone at K. The semimetallic nature of graphene is protected by its symmetry.

## Problems

**3.1. (Equivalent Formulations of Bloch's Theorem)** Show in detail the equivalence between the two formulations of Bloch's theorem:

$$\psi(\mathbf{k},\mathbf{r}+\mathbf{R}_l) = \mathrm{e}^{\mathrm{i}\mathbf{k}\cdot\mathbf{R}_l}\,\psi(\mathbf{k},\mathbf{r}), \tag{3.48}$$

and:

$$\psi(\mathbf{k},\mathbf{r}) = \mathrm{e}^{\mathrm{i}\mathbf{k}\cdot\mathbf{r}}\,u_{\mathbf{k}}(\mathbf{r}), \tag{3.49}$$

where $u_{\mathbf{k}}(\mathbf{r})$ is periodic:

$$u_{\mathbf{k}}(\mathbf{r}+\mathbf{R}_l) = u_{\mathbf{k}}(\mathbf{r}).$$

**3.2. (Reciprocal Vectors of the fcc Lattice)**

(a) Find the reciprocal lattice vectors of the fcc lattice. As fundamental translation vectors use

$$\mathbf{a} = \frac{a}{2}(\hat{\mathbf{x}} + \hat{\mathbf{y}}), \quad \mathbf{b} = \frac{a}{2}(\hat{\mathbf{y}} + \hat{\mathbf{z}}), \quad \mathbf{c} = \frac{a}{2}(\hat{\mathbf{z}} + \hat{\mathbf{x}}),$$

where $\hat{x}$, $\hat{y}$, and $\hat{z}$ denote the unit vectors along the directions of the Cartesian axes.

(b) Find the volume of the primitive cell and the density of valence electrons in $\mathrm{cm}^{-3}$ for Si ($a = 0.543$ nm).

(c) Find the volume of the BZ.

(d) Using a computer program, list the first 50–100 **G**-vectors in order of increasing magnitude.

## References

1. A. Bravais, Mémoire sur les systémes formés par les points distribués réguliérement sur un plan ou dans l'espace. J. Ecole Polytech. **19**, 1 (1850) [English: Memoir 1, Crystallographic Society of America, 1949]
2. W. Borchardt-Ott, *Crystallography: An Introduction* (Springer, Berlin, 2011)
3. C. Kittel, *Introduction to Solid State Physics*, 7th edn. (Wiley, New York, 1996)
4. M.S. Dresselhaus, G. Dresselhaus, A. Jorio, *Group Theory: Application to the Physics of Condensed Matter* (Springer, Berlin, 2008)
5. M. Tinkham, *Group Theory and Quantum Mechanics* (McGraw-Hill, New York, 1964)

# Chapter 4
# The Electronic Structure of Crystals: Theoretical Framework

## 4.1 Overview

In this chapter we discuss the electronic properties of crystals from a formal theoretical perspective. These are characterized mainly by the existence of allowed "energy bands" separated by "energy gaps." Understanding energy bands quantitatively requires solving the Schrödinger equation for many electrons in the potential due to all ions in the crystal. For example, ignoring core electrons, a crystal composed of $N$ Si atoms can be viewed as a lattice of $N$ $Si^{+4}$ ions and $4N$ electrons embedded in the complicated Coulomb potential due to the presence of all of these ions and of the electrons themselves.

The problem may appear daunting. However, the periodicity of the lattice leads to major simplifications: Roughly speaking, all we have to do (so to speak...) is to study the electronic properties of a single cell subject to the periodic boundary conditions that originate from the symmetry of the lattice. Thus, we can reduce the problem to a simpler one involving "only" eight electrons and two fourfold-ionized Si ions.

General theorems are available to compute the band structure of solids: Most notably, several schemes are based on some approximation of the potential of the ions and all "other electrons," reducing them to a fixed *pseudo*potential which includes the effect of all other electrons, so that we can use a single-electron picture. In addition, we can rely on Bloch's theorem (Chap. 3, Sect. 3.2.3) which tells us how to express the complicated wavefunction as a product of a simple "free electron envelope" wavefunction and a complicated "fast-wiggling" wavefunction, the same in each cell. Thus, each electronic state is labeled by something which resembles the wavevector of a free electron, $\mathbf{k}$, and can be written as

$$\psi_{\mathbf{k}}(\mathbf{r}) = \frac{1}{\Omega^{1/2}}\, e^{i\mathbf{k}\cdot\mathbf{r}}\, u_{\mathbf{k}}(\mathbf{r}) , \tag{4.1}$$

called a *Bloch function* where we should recall that $u_{\mathbf{k}}$ has the periodicity of the lattice. This is just Eq. (3.12) with the normalization to the volume of the lattice, $\Omega$ explicitly introduced. Notice that, similar to the symmetries of the crystal in "real space" of all spatial coordinates $\mathbf{r}$ (as discussed in the previous chapter), there are also corresponding symmetries in the space of all wavevectors $\mathbf{k}$: For example, the state of an electron traveling along the $x$-axis of cubic crystal will be described by a wavefunction identical to that associated to an electron traveling along the $z$-axis, provided we replace $k_x$ with $k_z$. These symmetries allow us to consider only the first *Brillouin Zone* (BZ) of the crystal.



M. Fischetti, W.G. Vandenberghe, *Advanced Physics of Electron Transport in Semiconductors and Nanostructures*, Graduate Texts in Physics, DOI 10.1007/978-3-319-01101-1_4

It is enough to consider the Bloch wavefunctions with $\mathbf{k}$ inside the first BZ in order to obtain full knowledge of the electronic properties of the lattice. These are tremendous simplifications afforded by the symmetry of crystals.

Despite these powerful simplifications and approximations, the problem is still mathematically somewhat convoluted. So, having set the general theoretical scheme and having reached the point of computing the band structure of solids, at first we shall ignore quantitative aspects and discuss instead the band structure of solids only from a qualitative point of view.

We shall first present and discuss the full Hamiltonian of the crystal (without any simplification), consisting of kinetic and potential energy terms for both the nuclei and the electrons. We shall then decouple the motion of the ions (much heavier and slower) from that of the electrons (the so-called *adiabatic approximation*). The motion of the ions will be considered as semiclassical (since the ions are heavier, their wavelength will be very short, making classical mechanics a good approximation in describing their motion) and close to equilibrium. The description of the ionic motion so "linearized" around their equilibrium position (lattice vibrations or *phonons*) will be briefly considered. It will be revisited later in the context of the quantum theory of many-body systems. We shall then consider the electron Hamiltonian so decoupled from the ionic motion. We shall deal with the issue of how to approximate the complicated many-electron system with the picture of a single electron in the presence of the potential of the ions and all other electrons. In the next chapter, we shall move to the problem of how to solve the single-electron Schrödinger equation and obtain the band structure of solids. In this chapter, instead, we shall conclude by discussing only qualitative features of the band structure.

## 4.2 The Hamiltonian of a Crystal

The crystal can be described by the coordinates $\mathbf{r}_i$ and $\mathbf{R}_{l,\alpha}$ of the electrons and nuclei, respectively, and by their conjugate momenta $\mathbf{p}_i$ and $\mathbf{P}_{l,\alpha}$, respectively. Thus, the total Hamiltonian of the crystal will be

$$\widehat{H} = \sum_i \frac{\hat{p}_i^2}{2m_{\text{el}}} + \frac{1}{2}\sum_{i,j} \frac{e^2}{4\pi\epsilon_0|\mathbf{r}_i - \mathbf{r}_j|} + \sum_{l,\alpha} \frac{\widehat{P}_{l,\alpha}^2}{2M_\alpha} + \frac{1}{2}\sum_{l,\alpha,l',\alpha'} U(\mathbf{R}_{l,\alpha} - \mathbf{R}_{l',\alpha'}) + \sum_{i,\alpha} V^{(\alpha)}(\mathbf{r}_i - \mathbf{R}_{l,\alpha}) \,. \tag{4.2}$$

The first 2 terms describe the kinetic energy of the electrons and their potential energy due to the Coulomb interactions among all electron pairs (the factor 1/2 in front due to the fact that the sum $\sum_{i,j}$ would count each pair twice). The next two terms represent the kinetic and potential energy of the nuclei interacting via a potential $U$. Finally, the last term represents the potential energy of an electron in the presence of all the ions with charge $eZ_\alpha$ located at $\mathbf{R}_{l,\alpha} = \mathbf{R}_l + \tau_\alpha$. It is useful to recall that, in our notation, $\mathbf{R}_l$ identifies the lattice point (or, equivalently, cell coordinate or translation vector) $l$ and $\tau_\alpha$ the coordinate of ion $\alpha$ in the cell.

We have assumed that all electrons (core and valence/conduction) appear in the Hamiltonian above. Actually, since the state of the core electrons does not differ significantly when considering free ions and ions in the crystal (their binding energy being so large), it is customary to assume that the "nuclear" potentials $V^{(\alpha)}$ and $U$ are actually ionic potentials, so that only outer-shell electrons (valence for insulators and semiconductors, conduction for metals) appear in Eq. (4.2). For the time being, we shall ignore this difference and discuss the implications of each choice below. So, the terms "nuclear" and "ionic" will be interchangeable.

Note that the complexity of the problem is enormous. First, we must deal with the sheer size of the problem: For Si, for example, assuming pure nuclear potentials, we have $N$ nuclei and $14N$ electrons. Thus, the full wavefunction will depend on $15N \times 3$ variables. In addition, in order to account for the Fermionic nature of the electrons and for Pauli' principle, we must antisymmetrize the wavefunction under exchange of all pairs of electron coordinates. This is an obvious generalization of what we discussed when presenting qualitatively the He atom. Finally, whereas we know the form of the bare Coulomb potential due to an ion with charge $eZ_\alpha$ is $V^{(\alpha)}(\mathbf{r}) = Z_\alpha e^2/(4\pi\epsilon_0 r)$, the electron–electron and the inter-ionic potentials $U$ depend on the distribution of the electron charge, which is a function of the wavefunction (that is, of the unknown). This renders the problem *nonlinear*. The crystal symmetry will simplify the problem, obviously. But, equally obviously, we must look for reasonable ways to approximate and simplify the picture.

### 4.2.1 The Adiabatic Approximation

The first approximation we invoke relies on the observation that the nuclei are much heavier than the electrons, usually by a factor of $10^4$ or more. Therefore, the velocity of the nuclei will be $10^2$ times smaller than that of the electrons. It is reasonable to assume that, even if the ions move away from their equilibrium (lowest-energy) position, the electrons will adjust instantaneously to the new nuclear positions. In other words, the electrons can be assumed to follow the nuclear motion "adiabatically." Therefore, we can "decouple" the nuclear and electronic wavefunctions and express the full crystal wavefunction $\Xi(\mathbf{R},\mathbf{r})$ as:

$$\Xi(\mathbf{R},\mathbf{r}) \approx \Phi(\mathbf{R})\, \Psi_{\mathbf{R}}(\mathbf{r}) \,, \tag{4.3}$$

where $\mathbf{R}$ and $\mathbf{r}$ indicate the whole collection of nuclear and electronic coordinates. In other words, Eq. (4.3) says that the nuclear wavefunction, $\Phi(\mathbf{R})$, is not quantum mechanically correlated to the electronic coordinates (each nucleus carries its own electronic "cloud" with itself, as the electrons follow the nuclear motion instantaneously) and that the electronic wavefunction, $\Psi_{\mathbf{R}}(\mathbf{r})$, depends on the nuclear coordinate $\mathbf{R}$ as a "parameter," but not on the nuclear wavefunctions, instead adjusting instantaneously to the nuclear positions.

An equivalent way to understand the adiabatic approximation is to write explicitly the time dependence in Eq. (4.3):

$$\Xi(\mathbf{R},\mathbf{r},t) \approx \Phi(\mathbf{R},t)\, \Psi_{\mathbf{R}}(\mathbf{r},t) \,. \tag{4.4}$$

Since the time dependence of $\Psi$ is so much faster than that of $\Phi$, we can see that the electronic component $\Psi$ "sees" $\Phi$ as a constant as $\Psi$ itself oscillates at a high frequency. On the other hand, the nuclear component $\Phi$ will see a time-average of the electron density $|\Psi|^2$, since $\Psi$ will have undergone many ($\sim 10^2$) oscillations while $\Phi$ varies slowly. Therefore, it is appropriate to consider the ionic/nuclear and electronic motion as "decoupled" (or "disentangled") in the sense described above.

Having embraced this first approximation, the Schrödinger equation for the crystal, using the crystal Hamiltonian Eq. (4.2), and the adiabatic approximation, can be written as

$$\Psi_{\mathbf{R}}(\mathbf{r})\, \widehat{H}_{\mathrm{L}}\, \Phi(\mathbf{R}) + \Phi(\mathbf{R})\, \widehat{H}_{\mathrm{el}}\, \Psi_{\mathbf{R}}(\mathbf{r}) + \widehat{H}'\, \Phi(\mathbf{R})\, \Psi_{\mathbf{R}}(\mathbf{r}) = E\, \Phi(\mathbf{R})\, \Psi_{\mathbf{R}}(\mathbf{r}) \,, \tag{4.5}$$

where:

$$\widehat{H}_{\mathrm{L}} = \sum_{l,\alpha} \frac{\widehat{P}^2_{l,\alpha}}{2M_\alpha} + \frac{1}{2} \sum_{l,\alpha,l',\alpha'} U(\mathbf{R}_{l,\alpha} - \mathbf{R}_{l',\alpha'}) \tag{4.6}$$

is the lattice Hamiltonian,

$$\widehat{H}_{\mathrm{el}} = \sum_i \frac{\hat{p}_i^2}{2m_{\mathrm{el}}} + \sum_{i,l,\alpha} V^{(\alpha)}(\mathbf{r}_i - \mathbf{R}_{l,\alpha}) + \frac{1}{2}\sum_{i,j} \frac{e^2}{4\pi\epsilon_0|\mathbf{r}_i - \mathbf{r}_j|} , \tag{4.7}$$

is the electronic Hamiltonian with the nuclear coordinates $\mathbf{R}$ appearing as parameters, and the term

$$\begin{aligned} \widehat{H}' \,\Phi(\mathbf{R})\, \Psi_{\mathbf{R}}(\mathbf{r}) &= \widehat{H}_{\mathrm{L}}\, \Psi_{\mathbf{R}}(\mathbf{r})\, \Phi(\mathbf{R}) - \Psi_{\mathbf{R}}(\mathbf{r})\, \widehat{H}_{\mathrm{L}}\, \Phi(\mathbf{R}) \\ &= \Phi(\mathbf{R})\, \widehat{H}_{\mathrm{L}}\, \Psi_{\mathbf{R}}(\mathbf{r}) \sim -\Phi(\mathbf{R}) \sum_\alpha \frac{\hbar^2}{2M_\alpha} \nabla^2 \Psi_{\mathbf{R}}(\mathbf{r}) \end{aligned} \tag{4.8}$$

is the term which we neglect in the adiabatic approximation: it is of relative weight $m/M$, as mentioned above, since $\nabla^2\Psi_{\mathbf{R}}(\mathbf{r}) \sim \sum_i k_i^2 \sim m_{\mathrm{el}} N_{\mathrm{el}}$.

The adiabatic approximation is often referred to as the *Born–Oppenheimer approximation*. Of course, the fact that the ions may move away from their equilibrium positions does have an effect on the potential energy of the electrons. At very high temperatures these deviations may be very large: after all, at some sufficiently high temperature the crystal will melt! But at room temperature, in most solids these deviations are quite small and can be treated as small perturbations. In Chap. 13 we shall consider this small coupling between the electrons and the ionic motions. Since, as we shall see in Chap. 9, Sect. 9.4, the ionic motion is quantized into eigenmodes call *phonons*, this coupling is called the *electron–phonon* interaction.

### 4.2.2 The Ionic Hamiltonian (Phonons)

The nuclear (or ionic) Hamiltonian $\widehat{H}_{\mathrm{L}}$ can be simplified by assuming, as anticipated above, that the nuclei do not move appreciable away from their equilibrium position. The general problem of finding the equilibrium positions of the nuclei is straightforward in principle, but daunting in practice, even within the adiabatic approximation: We should assume a given set of ionic coordinates $\mathbf{R}$, solve the associated electronic Schrödinger equation, such as Eq. (4.11) below, find the total energy of the system, repeat the process assuming a new set of ionic coordinates, and look for the ionic configuration which gives us the lowest total energy. Newton-like schemes—amounting to the calculation of the forces acting on the nuclei—can be employed, but the procedure is still enormously complex. Recently (in the last 20 years) significant progress has been made, still using the adiabatic approximation, but using algorithms coupling "simulated annealing" with a fictitious Newtonian dynamics for the ions to improve the convergence to the lowest-energy ionic configuration. These techniques—developed by Roberto Car and Michele Parrinello [1]—now allow us to deal with systems of thousands of atoms. But we shall not go into the details of these sophisticated approaches. For now, we shall assume that we know "somehow" (experimentally, from total-energy calculations, from Car–Parrinello techniques) the lattice structure, and assume that the ions are at rest (or, at least, that their displacement away from their equilibrium positions is small enough to be considered a small perturbation).

So, having made the approximation that the nuclei are close to equilibrium, we can write $\mathbf{R}_\alpha = \mathbf{R}_\alpha^{(0)} + \mathbf{u}_\alpha$, where $\mathbf{u}_\alpha$ is the small displacement away from the equilibrium position $\mathbf{R}_\alpha^{(0)}$. We can then expand the inter-ionic (or inter-nuclear) potential $U(\mathbf{R}_{l,\alpha} - \mathbf{R}_{l',\alpha'})$ around equilibrium:

$$U(\mathbf{R}_{l,\alpha} - \mathbf{R}_{l',\alpha'}) \approx U^{(0)} + \frac{1}{2}\nabla_{l,\alpha}\nabla_{l',\alpha'}\, U^{(0)}\, \mathbf{u}_{l,\alpha}\, \mathbf{u}_{l',\alpha'} , \tag{4.9}$$

where the subscript 0 means that the quantity is evaluated at the equilibrium position ($U^{(0)} = U(\mathbf{R}^{(0)}_{l,\alpha} - \mathbf{R}^{(0)}_{l',\alpha'})$, for example). The notation $\nabla_{l,\alpha}$ stands for the vector $(\partial/\partial R_{l,\alpha x}, \partial/\partial R_{l,\alpha y}, \partial/\partial R_{l,\alpha z})$. Note that the first-order term (involving the gradient of $U$) vanishes, since we are at equilibrium, so that all forces acting on the ions/nuclei cancel. The "matrix" $D_{\alpha,\beta} = \frac{1}{4}\nabla_\alpha \nabla_\beta U$ (which is evaluated at the equilibrium position) therefore represents the force-matrix (or *dynamical matrix*) for the ions/nuclei. Using this expansion of the inter-ionic potential and setting $U^{(0)} = 0$, since it is just a rescaling of the zero-reference for energies, the full ionic Hamiltonian becomes

$$\widehat{H}_{\rm L} = \sum_{l,\alpha} \left[ \frac{\widehat{P}^2_{l,\alpha}}{2M_\alpha} + \sum_{l',\alpha'} D_{l,\alpha,l',\alpha'} \, \hat{\mathbf{u}}_{l,\alpha} \, \hat{\mathbf{u}}_{l',\alpha'} \right] , \tag{4.10}$$

which is a collection of coupled harmonic oscillators with conjugate variables $\widehat{\mathbf{P}}_{l,\alpha}$ and $\hat{\mathbf{u}}_{l,\alpha}$. These variables can be reduced to normal modes by diagonalizing the dynamical matrix, so that Eq. (4.10) takes the form of a sum over Hamiltonians of harmonic oscillators. This will give rise to acoustic and optical ionic waves. Upon quantization (as we shall see later in Chap. 9, Sect. 9.4), they will be called *phonons*. They are plane waves, weakly interacting with each other via anharmonic terms, that is, the higher-order terms which we have neglected in Eq. (4.9).

### 4.2.3 *The One-Electron Approximation: Hartree and Hartree–Fock*

Considering now the electronic Hamiltonian $\widehat{H}_{\rm el}$, the associated Schrödinger equation is

$$\left[ \sum_i \frac{\hat{p}_i^2}{2m_{\rm el}} + \sum_{i,l,\alpha} V^{(\alpha)}(\mathbf{r}_i - \mathbf{R}^{(0)}_{l,\alpha}) + \frac{1}{2}\sum_{i,j} \frac{e^2}{4\pi\epsilon_0 |\mathbf{r}_i - \mathbf{r}_j|} \right] \Psi(\mathbf{r}_1, \mathbf{r}_2, \ldots, \mathbf{r}_{N_{\rm el}})$$
$$= E\,\Psi(\mathbf{r}_1, \mathbf{r}_2, \ldots, \mathbf{r}_{N_{\rm el}}) , \tag{4.11}$$

where we have expressed explicitly the dependence on the electron coordinates, have indicated with $N_{\rm el}$ the total number of electrons in the system, and we have omitted the subscript $\mathbf{R}$, since we consider the nuclei in their equilibrium positions.

The problem, though simplified, is still unsolvable. Recall, for example, that the electronic wavefunction $\Psi(\mathbf{r}_1, \mathbf{r}_2, \ldots, \mathbf{r}_{N_{\rm el}})$ must be antisymmetrized under all exchanges of any two electron indices in order to obey Pauli's principle.

* *To see this, ignore spin and consider a 2-electron system with a normalized solution $\psi(\mathbf{r}_1, \mathbf{r}_2)$ of the associated Schrödinger equation. The linear combination $(1/2^{1/2})[\psi(\mathbf{r}_1, \mathbf{r}_2) - \psi(\mathbf{r}_2, \mathbf{r}_1)]$ is the required antisymmetric wavefunction. We see that if we try to place both electrons in the same state—so that swapping electrons results in the same state, that is $\psi(\mathbf{r}_1, \mathbf{r}_2) = \psi(\mathbf{r}_2, \mathbf{r}_1)$—the wavefunction vanishes, as required by Pauli's principle. If we want to account for spin, the situation is a little more complicated. The total wavefunction is now the product of the spatial and the spin wavefunctions. So, we can either have a symmetric spatial component and an antisymmetric spin component, or vice versa: In the first case, the electrons can form a "singlet" state of vanishing total spin (anti-parallel spins). Now we can form an antisymmetric spin state, $[|+\rangle|-\rangle - |-\rangle|+\rangle]/2^{1/2}$, and a symmetric spatial wavefunction, $(1/2^{1/2})[\psi(\mathbf{r}_1, \mathbf{r}_2) + \psi(\mathbf{r}_2, \mathbf{r}_1)]$. In the second case, the electrons can be in any of three spins states of total spin $s = 1$ and $s_z = -1, 0, +1$ (a symmetric triplet of spin states), so that the spatial component of their wavefunction must be antisymmetric.*

The first step towards a further simplification of the problem is to use one of the most common tools in solid-state physics: The *self-consistent field approximation*: We shall consider only one electron at the time, assuming that the effect of all other electrons can be lumped into an average field. This field will have to be solved self-consistently, since it depends on the density (and so wavefunctions) of all other electrons, and on the requirement that Pauli's principle is obeyed.

The easier way to reformulate the problem under this approximation is to follow Hartree [2] ignoring spin (and so antisymmetrization issues) and assuming that we can write the total electronic Hamiltonian as the sum of single-electron Hamiltonians, $\widehat{H}_{tot} = \widehat{H}_1 + \widehat{H}_2 + \cdots + \widehat{H}_{N_{\rm el}}$ accounting for the self-consistent field, obtaining the single-electron "Hartree equations":

$$\left[-\frac{\hbar^2}{2m_{\rm el}}\nabla^2 + V^{(\rm lat)}(\mathbf{r}) + \sum_{j\neq i} \frac{e^2}{4\pi\epsilon_0} \int \frac{\psi_j^*(\mathbf{r}')\psi_j(\mathbf{r}')}{|\mathbf{r}-\mathbf{r}'|}\mathrm{d}\mathbf{r}'\right] \psi_i(\mathbf{r}) = w_i \psi_i(\mathbf{r}) \,. \tag{4.12}$$

The sum extends over all occupied states, but not over $i$ itself. We have defined the "lattice potential" $V^{(\rm lat)}(\mathbf{r}) = \sum_{l,\alpha} V^{(\alpha)}(\mathbf{r} - \mathbf{R}_{l,\alpha}^{(0)})$. The integral term represents the electrostatic potential due to the charge density $e|\psi_j(\mathbf{r})|^2$ of the $j$-th electron, so that the sum constitutes the "mean field" due to all other electrons. Consistently with this approximation, the full wavefunction is expressed as the product of single-electron wavefunctions, $\Psi = \psi_1\psi_2\ldots\psi_{N_{\rm el}}$. Hartree suggested a variational solution for Eq. (4.12): We "guess" a functional form for all $\psi_j$'s expressed in terms of some parameters. We then vary these parameters in order to minimize the energy $\langle\Psi|\widehat{H}_H|\Psi\rangle$. The interpretation of the parameter $w_i$ is a little tricky. It may be tempting to call it the "single electron energy." However, we can never specify uniquely a single-particle energy in an interacting system: We can always add energy to one electron and subtract it from another without changing the total energy. Moreover, if we remove a single particle from the system, we do not change the total energy by simply subtracting the total energy of the particle we have removed. On the contrary, we also modify the total energy of the system by removing the interaction between all other particles and the one we have removed. As a result, if we compute the total energy of the system we do not find $\sum_i w_i$, as expected if the $w_i$'s were single-electron energies, but, instead, we find

$$\frac{\langle\Psi|\widehat{H}_H|\Psi\rangle}{\langle\Psi|\Psi\rangle} = \sum_i w_i - \frac{1}{2}\sum_{i,j} \frac{e^2}{4\pi\epsilon_0} \int \frac{\psi_j^*(\mathbf{r}')\psi_j(\mathbf{r}')\psi_i^*(\mathbf{r})\psi_i(\mathbf{r})}{|\mathbf{r}-\mathbf{r}'|}\,\mathrm{d}\mathbf{r}\mathrm{d}\mathbf{r}' \,, \tag{4.13}$$

because $\sum_i w_i$ accounts twice for the mutual interaction between pairs. The last term in Eq. (4.13) expresses this correction. More about the meaning of $w_i$ below.

An improved approximation consists in accounting for the required wavefunction antisymmetrization. The wavefunction must be formed from single-electron wavefunctions $\psi_i(\mathbf{r}_i)$ via the Slater determinant:

$$\Psi(\mathbf{r}_1, \mathbf{r}_2, \ldots, \mathbf{r}_{N_{\rm el}}) = \frac{1}{\sqrt{N_{\rm el}!}} \begin{vmatrix} \psi_1(\mathbf{r}_1) & \psi_1(\mathbf{r}_2) & \ldots & \psi_1(\mathbf{r}_{N_{\rm el}}) \\ \psi_2(\mathbf{r}_1) & \psi_2(\mathbf{r}_2) & \ldots & \ldots \\ \ldots & \ldots & \ldots & \ldots \\ \psi_{N_{\rm el}}(\mathbf{r}_1) & \ldots & \ldots & \ldots \end{vmatrix} \,. \tag{4.14}$$

Using this form of the wavefunction to calculate the expectation value of the Hamiltonian given by Eq. (4.11) and employing a variational method to minimize this expectation value, one obtains finally the "Hartree–Fock equations":

$$\begin{aligned} &\left[-\frac{\hbar^2}{2m_{\rm el}}\nabla^2 + V^{(\rm lat)}(\mathbf{r}) + \sum_{j\neq i} \frac{e^2}{4\pi\epsilon_0} \int \frac{\psi_j^*(\mathbf{r}')\psi_j(\mathbf{r}')}{|\mathbf{r}-\mathbf{r}'|}\,\mathrm{d}\mathbf{r}'\right] \psi_i(\mathbf{r}) \\ &-\sum_{j\neq i} \frac{e^2}{4\pi\epsilon_0}\,\psi_j(\mathbf{r}) \int \frac{\psi_j^*(\mathbf{r}')\psi_i(\mathbf{r}')}{|\mathbf{r}-\mathbf{r}'|}\,\mathrm{d}\mathbf{r}' = w_i\psi_i(\mathbf{r}) \,. \end{aligned} \tag{4.15}$$

These equations are still only approximations, since accounting for the full Slater determinant would lead to an infinite series of equations. But the essence of the Hartree–Fock approximation is the correction due to the last term on the left-hand side of Eq. (4.15). This is the *exchange energy* resulting from having exchanged the indices $i$ and $j$ of one single-electron wavefunction in the previous term. Physically, the exchange energy results from the fact that electrons in antisymmetrized spatial wavefunctions tend to "repel" each other—in order not to occupy the same "state"—and so the Coulomb repulsion (positive potential energy) is lowered. This explains the sign of the exchange term. The total energy is indeed lowered by an amount:

$$-\frac{1}{2}\sum_{i,j,i\neq j}\frac{e^2}{4\pi\epsilon_0}\int\frac{\psi_i^*(\mathbf{r}')\psi_j(\mathbf{r}')\psi_j^*(\mathbf{r})\psi_i(\mathbf{r})}{|\mathbf{r}-\mathbf{r}'|}\,d\mathbf{r}\,d\mathbf{r}' . \tag{4.16}$$

Note the indices swapped with respect to Eq. (4.13) above. Recall that we are still dealing with approximations. Customarily, the difference between the "exact" energy and the Hartree–Fock result is called *correlation energy*. Quantum calculations have been performed for the homogeneous electron gas. This last component of the total energy can loosely be attributed to the fact that electrons tend to have a correlated spatial distribution, further reducing their Coulomb repulsion. Finally, Koopmans has shown that the parameters $w_i$ in the Hartree–Fock single-particle equation, Eq. (4.15), has the magnitude (but opposite sign) of the ionization energy for that state in the crystal [3].

However, in most of the following we shall bypass the difficulty of accounting for them and we shall assume that the lattice potential, $V^{(\mathrm{lat})}$, is the *self-consistent lattice potential* which includes the (screening) repulsive Coulomb effect of all other electrons in the self-consistent mean-field approximation. This self-consistent potential, thanks to the symmetry of the problem, will retain its periodicity in the lattice structure. However, it is useful to mention that in self-consistent band-structure calculation methods, the exchange (or exchange-correlation) energy has been approximated with a "functional" of the electron density (and, so, of all single-electron wavefunctions). Its $1/r$ dependence implies a form $\sim \rho(\mathbf{r})^{1/3}$ for this functional, usually called the *Kohn–Sham functional* [4]. It is usually calculated assuming a free-electron model, as we shall see in the next chapter. Thus, the single-electron Schrödinger equation often used has the form

$$\left\{-\frac{\hbar^2}{2m_{\mathrm{el}}}\nabla^2+V^{(\mathrm{lat})}(\mathbf{r})+V^{(\mathrm{xc})}[\rho(\mathbf{r})]\right\}\psi(\mathbf{r})=w_i\,\psi(\mathbf{r}) , \tag{4.17}$$

where the exchange-correlation term has the form $V^{(\mathrm{xc})}(\rho)\propto\rho^{1/3}$ and, once more, we should recall that $V^{(\mathrm{lat})}$ includes the screening of all other electrons via the Hartree term, as we shall see in the next chapter. Note that, since the Kohn–Sham functional is a functional of the electron density, this theoretical approach is also known as *Density Functional Theory* (DFT).

## 4.3 Energy Bands: A Bird's Eye View

We are now ready to derive the band structure of solids by solving the single-electron Schrödinger equation in the Hartree approximation. Before delving into technical details, let's consider the qualitative picture of the band structure which emerges when we bring many atoms together into a lattice.

### *4.3.1 Coupling Atoms*

Consider two Si atoms, widely separated, as in Fig. 4.1 which shows the Coulomb potential around each nucleus ($\propto 1/r$, where $r$ is the distance from the nuclei), the energy of the 3 occupied levels, and their electronic configuration. Let's recall here the important fact that there are eight states available in the $n = 3$ energy level, two $3s$-states and six $3p$-states. However, Si has only 4 electrons in this shell, so that the outermost (valence) shell is only half-occupied.

Now let's bring the nuclei much closer, so that the potential around each nucleus is affected significantly by the other atom. This is illustrated schematically in Fig. 4.2. The "core" levels ($n = 1$ and $n = 2$) do not significantly feel the perturbation caused by the other nucleus, but the highest-energy occupied level $n = 3$ does feel a dramatic change. The $n = 3$ orbitals now spread over both nuclei. Clearly, their energy will change, because the potential itself has changed. Qualitatively, what happens is that, starting from the original hybrid $sp^3$ orbitals of a single atoms, we will have a "distortion" of each of them, caused by the new potential. In addition, we may form two superpositions of the orbitals, much like we had formed the hybrid $sp^3$. One superposition corresponds to orbitals with a large electron probability density between the nuclei. This is a "bonding orbital," since these "shared electrons" constitute a covalent bond. Since the electrons in this orbital sit close to the attractive nuclei, the energy of this orbital is low. The other linear combination of orbitals we may form corresponds to orbitals having a small electron density between the nuclei. These orbitals are "antibonding," since they do not contribute to the covalent bond between the two Si atoms. Their energy will be higher, since the electrons, on average, sit farther away from the nuclei.

The net effect is that the fourfold degenerate energy levels corresponding to the two $sp^3$ orbitals of the two separate atoms, each containing one (of a possible maximum of 2) electrons, have split into two lower-energy bonding states (each with a slightly different energy) and two higher-energy (also each at a slightly different energy) antibonding states. The electrons will occupy the lowest-energy states to form the bond. The antibonding states will remain unoccupied.

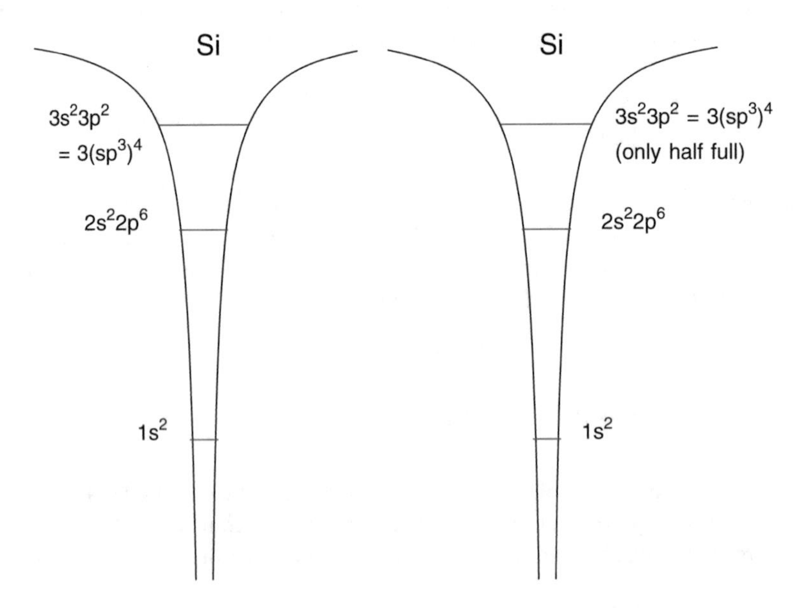


**Fig. 4.1** Schematic illustration of the ionic potentials of two isolated Si atoms. The *horizontal lines* show qualitatively the energy levels of the $2s$ and $2p$ "core" stets and of the $3s$ and $3p$ valence orbitals

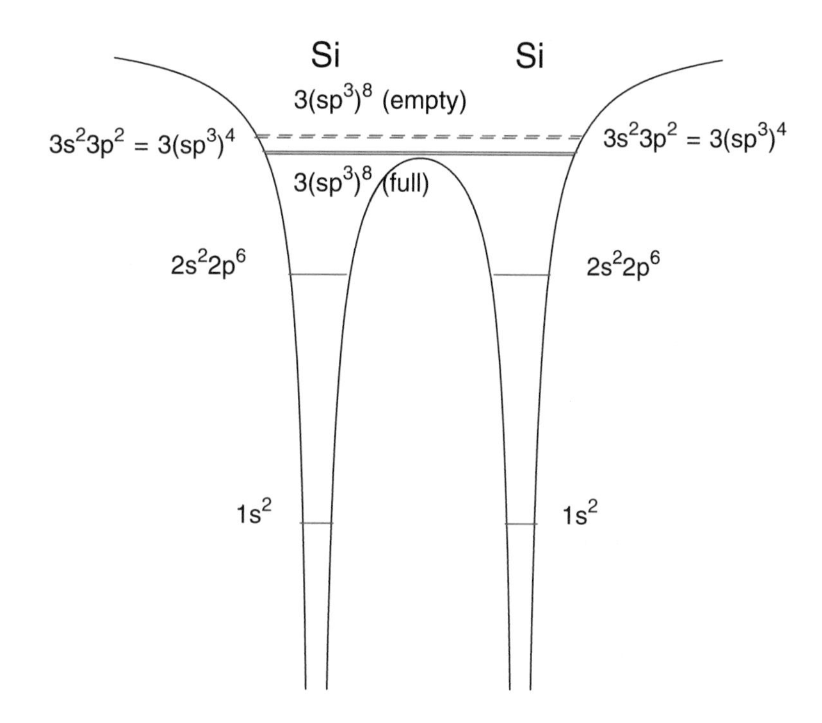


**Fig. 4.2** Schematic illustration of the ionic potentials of the two Si atoms shown in the previous figure, but now assumed to be in close proximity, as in the Si crystal. Whereas the core levels—indicated by the *horizontal lines*—are not affected by the proximity of the other atom, the valence states split into (occupied) bonding states and (unoccupied) antibonding states at a higher-energy

Let's now imagine what happens when we bring together three Si atoms: They will bond together forming a set of three occupied bonding states (at three different but very similar energies) and three antibonding unoccupied states.

## 4.3.2 Energy Bands

Now let's consider a Si crystal formed by, roughly, $N = 10^{22}$ atoms. What happens now is very similar to what we have seen in the case of two and three atoms. The $N$ atoms will contribute a total of $8N$ possible $sp^3$ orbitals, originally all at the same energy. They will split into $4N$ occupied orbitals contributing to the bond and $4N$ empty antibonding orbitals. The energy of each of the $4N$ bonding orbitals will be very close to that of any other bonding state. Indeed, they will spread over an energy interval of a few tens of an eV and we will have roughly $10^{22}$ energy levels in this small energy range. They are so many and so close together, that it makes sense to look at them as a "continuum" of infinitely many states populating an *energy band.* Since this is the band of bonding states made up using the valence electrons, it is called the *valence band* of the crystal. Similarly, the set of $4N$ antibonding orbitals will also form a band of unoccupied states. These states are spread away from the atoms throughout the whole crystal. Assuming we found a way to occupy one of these states (later on we shall see how to do this), the electron in this state will not be "forced" to bind the nuclei, but will be free to move around the crystal. If we apply an electric field, the electron will be free to move according to the external force, thus contributing to the conduction of current. Therefore, the band of antibonding states is called the *conduction band* of the crystal.

The valence and conduction bands are separated by an *energy gap* whose existence stems from considerations similar to those which caused the presence of forbidden energies in the isolated Si atom.

### *4.3.3 Metals, Insulators, Semiconductors*

We have just seen that the 4*N* valence electrons in a Si crystal occupy the entire valence band. On the other hand, the conduction band, which could contain also 4*N* electrons, is completely empty.

Suppose that we apply an electric field to a Si crystal at a very low temperature (by placing in contact with two metallic plates, for example, at the opposite ends of the crystal). In principle, the electrons would "like" to respond to the field and move away from the negatively charged plate towards the positively charged plate. But "motion" requires gaining kinetic energy. Therefore, the electrons in the valence band should acquire a bit of extra energy in order to move. But since the valence band is full and we are assuming an empty conduction band (low temperature), they may gain energy only by "jumping" to the next available empty states in the conduction band. But this jump requires a large amount of energy (about 1.1 eV at room temperature in Si). Therefore, the electrons cannot move, so no current will flow in the crystal, despite the electric field we have applied. In other words, a "perfect" Si crystal at a very low temperature (we'll see below why we stress the word "perfect") behaves like an insulator.

Not all crystals are like Si. For example, if we take a crystal of *N* Na ions, the valence band will be constituted by 2*N* states emerging from the *s* orbitals of each Na ion, but only *N* electrons populate the band, the "energy gap" now lying at a much larger energy. Therefore, the valence band will be only partially occupied, in this case half full, and electrons can acquire an arbitrarily small amount of energy from the applied electric field. Therefore, current will flow in Na: Sodium is a conductor. Other metals, instead, have "overlapping bands." The absence of a gap makes good electrical conductors, much like Na.

Glass (silicon dioxide, $SiO_2$, in its amorphous form) is another insulator, like Si, but its band gap is about one order of magnitude larger than that of Si (for Si we have seen that $E_g \approx 1.1\,\text{eV}$, while for $SiO_2$ we have $E_g \approx 9.2\,\text{eV}$). This gap is now much larger than the thermal energy and it is very difficult to place electrons in the conduction band. Deep ultraviolet light or the application of extremely high electric field, larger than $10^8$-to-$10^9$ V/m, are two ways one can achieve this, but these are extreme conditions that we ignore here, although they are concerns in ultra-small electronic devices. Therefore, glass will behave like an insulator even at elevated temperatures.

Figure 4.3 shows qualitatively the difference between semiconductors, insulators, and semiconductors. The difference between semiconductors and insulators is only quantitative: If the forbidden energy gap is very large, even the presence of impurities or a high temperature cannot modify the insulating properties of the material. For semiconductors, instead, thermal excitation of electrons across the gap or the addition of a small number of selected foreign atoms (impurities or "dopants") can alter dramatically their properties. We shall see how this process (*doping*) is intentionally done.

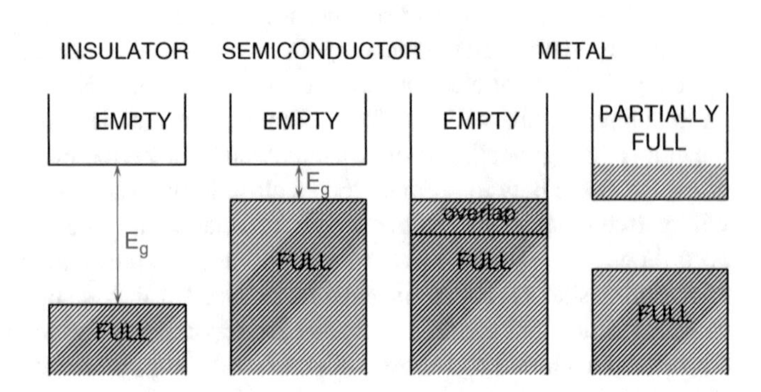


**Fig. 4.3** Schematic illustration of the energy bands in insulators, semiconductors, and metals

### 4.3.4 The Kronig–Penney Model

Before moving to methods used to calculate energy bands in solids, let's translate into mathematical terms the intuitive approach followed above. Let's consider a simple, well-known example, called the Kronig–Penney model. It can be solved almost analytically and it exhibits most of the features we will find in more realistic models. Consider a periodic potential consisting in many wells of width $a$ and energetic depth $U_0$ separated by potential barriers of width $b$. This is a one-dimensional "crystal" with a unit cell of size $a+b$. Let's choose $x=-b$ and $x=a$ as cell boundaries. We assume that the well occupies the region $0<x<a$, the barrier occupies the region $-b<x<0$ (Fig. 4.4).

The Schrödinger equation we must solve takes the form

$$\frac{\mathrm{d}^2\psi_a}{\mathrm{d}x^2}+\alpha^2\psi_a=0 \quad \text{for } 0<x<a$$

and

$$\frac{\mathrm{d}^2\psi_b}{\mathrm{d}x^2}+\beta^2\psi_b=0 \quad \text{for } -b<x<0$$

with $\alpha=(2m_{\mathrm{el}}E)^{1/2}/\hbar$ and

$$\beta=\begin{cases} i\beta_-;\ \beta_-=[2m_{\mathrm{el}}(U_0-E)]^{1/2} & 0<E<U_0 \\ \beta_+;\ \beta_+=[2m_{\mathrm{el}}(E-U_0)]^{1/2} & E>U_0 \end{cases}.$$

It is convenient to write the solutions as:

$$\psi_a(x)=A_a\ \sin(\alpha x)+B_a\ \cos(\alpha x)$$
$$\psi_b(x)=A_b\ \sin(\beta x)+B_b\ \cos(\beta x)$$

(later replacing sin and cos with sinh and cosh if necessary).

In order to find the coefficients $A_a$, $B_a$, $A_b$, and $B_b$, we must consider four boundary conditions: The first two express continuity of the wavefunction at the potential "steps":

$$\psi_a(0)=\psi_b(0)$$
$$\left.\frac{\mathrm{d}\psi_a}{\mathrm{d}x}\right|_0=\left.\frac{\mathrm{d}\psi_b}{\mathrm{d}x}\right|_0,$$

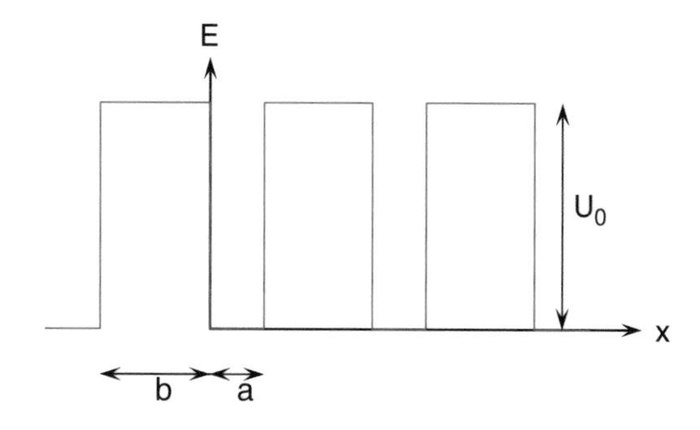


**Fig. 4.4** One-dimensional periodic potential used in the Kronig–Penney model

and two more express periodicity:

$$\psi_a(a) = \mathrm{e}^{\mathrm{i}k(a+b)} \; \psi_b(-b)$$
$$\left.\frac{\mathrm{d}\psi_a}{\mathrm{d}x}\right|_a = \mathrm{e}^{\mathrm{i}k(a+b)} \left.\frac{\mathrm{d}\psi_b}{\mathrm{d}x}\right|_{-b} .$$

These conditions imply

$$B_a = B_b$$
$$\alpha A_a = \beta A_b$$
$$A_a \sin(\alpha a) + B_a \cos(\alpha a) = \mathrm{e}^{\mathrm{i}k(a+b)} \left[-A_b \sin(\beta b) + B_b \cos(\beta b)\right]$$
$$\alpha A_a \cos(\alpha a) - \alpha B_a \sin(\alpha a) = \mathrm{e}^{\mathrm{i}k(a+b)} \left[\beta A_b \cos(\beta b) + \beta B_b \sin(\beta b)\right]$$

Using the first two equations, we can eliminate $A_b$ and $B_b$ from the last two:

$$A_a[\sin(\alpha a) + (\alpha/\beta)\mathrm{e}^{\mathrm{i}k(a+b)} \sin(\beta b)] + B_a[\cos(\alpha a) - \mathrm{e}^{\mathrm{i}k(a+b)} \cos(\beta b)] = 0$$
$$A_a[\alpha \cos(\alpha a) - \alpha \mathrm{e}^{\mathrm{i}k(a+b)} \cos(\beta b)] + B_a[-\alpha \sin(\alpha a) - \beta \; \mathrm{e}^{\mathrm{i}k(a+b)} \sin(\beta b)] = 0$$

We have reduced the problem to the solution of a linear homogeneous system in two unknowns. A nonzero solution exists only if the determinant of the matrix above vanishes. This simplifies to:

$$-\frac{\alpha^2 + \beta^2}{2\alpha\beta} \sin(\alpha a)\sin(\beta b) + \cos(\alpha a)\cos(\beta b) = \cos[k(a+b)]$$

This equation provides implicitly the “dispersion” $E(k)$. “Energy gaps” arise whenever the magnitude of the lhs exceeds unity. The dispersion can be analyzed inside the BZ (i.e., for $-\pi/(a+b) < k \leq \pi/(a+b)$ since $a+b$ is the fundamental translation “vector” and $G_n = 2\pi n/(a+b)$ are the reciprocal lattice “vectors.” Alternatively, one can plot $E(k)$ in an “extended zone” picture. In doing so, one can see that the dispersion (i.e., the energy-wavevector relation) is quasi-parabolic near the zone center (electron- or hole-like), it is “flat” (zero slope) at the zone-edge, and it approaches the parabolic dispersion of a free-electron at high energies.

## Problems

**4.1. (Si Molecules)**

(a) Why doesn’t the $Si_2$ molecule exist?
(b) Why, on the other hand, does the $SiH_4$ molecule exists? It is called “silane.”

**4.2. (An Alternative Form of the Kronig–Penney Model)** An alternative way to see how energy bands arise consists in solving a problem very similar to the Kronig–Penney model. Consider a one-dimensional periodic potential energy

$$V(x) == eV_0 \sum_{n=-\infty}^{\infty} \delta(x - na) . \tag{4.18}$$

This defines a potential energy that vanishes everywhere except at the lattice positions $x_n = na$ where it exhibits "spike" of amplitude $eV_0$. Therefore, $a$ is the periodicity of this lattice. Find the form of the Bloch waves, keeping in mind that at the locations $x_n$ the first derivative of the wavefunction is discontinuous. Find also a graphic procedure to obtain the dispersion $E(k)$.

## References

1. R. Car, M. Parrinello, Unified approach for molecular dynamics and density functional theory. Phys. Rev. Lett. **55**, 2471 (1995)
2. D.R. Hartree, The wave mechanics of an atom with a non-coulomb central field. Part I - theory and methods. Proc. Camb. Philol. Soc. **24**, 89–110 (1928)
3. T. Koopmans, Über die Zuordnung von Wellenfunktionen und Eigenwerten zu den einzelnen Elektronen eines Atoms. Physica **1**, 104 (1934)
4. W. Kohn, L.J. Sham, Self-consistent equations including exchange and correlation effects. Phys. Rev. **140**, A1133 (1965)

# Chapter 5
# The Electronic Structure of Crystals: Computational Methods

## 5.1 Overview

In the previous chapter we have outlined the general theoretical framework required to formulate the problem of finding the electronic properties of solids as a single-electron problem. Here we discuss in detail the most significant methods that can be used to solve the problem quantitatively. We start by describing methods based on the expansion of the unknown (Bloch) wavefunction into plane waves, explain why a "brute force" solution would not work, and introduce the concepts of orthogonalized plane waves (OPW) and model potentials that lead to *pseudopotentials*. We discuss their use in their "empirical" form, but, in their self-consistent implementation, they lead to density functional theory (DFT). This will be discussed separately in the next chapter, given the important role that it plays in solid-state physics, materials science, and nanoelectronics. We also present other methods not based on plane-wave expansions, namely tight-binding and $\mathbf{k} \cdot \mathbf{p}$ perturbation theory. For each method, we shall present some significant examples.

Since the topic treated here is so vast, we shall unavoidably skip some important "details." Texts exclusively devoted to the study of the electronic spectrum of crystals abound and, if additional details are needed, these details can be found there. In particular, Harrison presents an overview of the pseudopotential and tight-binding approximations [1]; the concept of pseudopotential, introduced by Hellman in 1935 [2], is reviewed, both as empirical model potential and ab initio, by Austin et al. [3], Harrison [4], and Heine [5]. Its most recent ab initio formulations are discussed, for example, by Kaxiras [6] and Pisani [7]. The use of linear combinations of atomic orbitals (also known as the *tight-binding* approximation) was introduced and discussed by Slater [8]. More details about the use of $\mathbf{k} \cdot \mathbf{p}$ perturbation theory are given by Luttinger and Kohn [9] and Kane [10], authors who first introduced it. A discussion about its application to the study of the valence bands of Si and Ge can be found in Kittel's text [11]. Finally, the development of the all-important exchange-correlation functional can be followed historically in the original articles [12, 13].

## 5.2 Plane Waves

Consider the single-electron Schrödinger equation (4.17) above. For the time being, we shall ignore exchange-correlation effects. We make use of Bloch's theorem, so that, by expanding the general solution over products of Bloch functions, we shall arrive at a matrix form for the Hamiltonian.



M. Fischetti, W.G. Vandenberghe, *Advanced Physics of Electron Transport in Semiconductors and Nanostructures*, Graduate Texts in Physics, DOI 10.1007/978-3-319-01101-1_5

In principle, the associated eigenvalue problem can be solved, of course. However, we shall see that, in practice, unless we do something about the lattice potential, it will be the size of the problem which will make it computationally impossible to solve.

We wish to solve the Schrödinger equation

$$-\frac{\hbar^2\nabla^2}{2m_{\text{el}}}\psi(\mathbf{r})+V^{(\text{lat})}(\mathbf{r})\psi(\mathbf{r})=E\,\psi(\mathbf{r})\,. \tag{5.1}$$

Since the lattice potential $V^{(\text{lat})}(\mathbf{r})$ is periodic, we use Bloch's theorem and consider the wavefunction

$$\psi(\mathbf{k},\mathbf{r})=\frac{1}{\Omega^{1/2}}\,e^{i\mathbf{k}\cdot\mathbf{r}}\sum_{\mathbf{G}}u_{\mathbf{kG}}\,e^{i\mathbf{G}\cdot\mathbf{r}}\,.$$

Inserting this into the Schrödinger equation, we obtain

$$\frac{\hbar^2}{2m_{\text{el}}}\sum_{\mathbf{G}}|\mathbf{k}+\mathbf{G}|^2u_{\mathbf{kG}}e^{i(\mathbf{k}+\mathbf{G})\cdot\mathbf{r}}+V^{(\text{lat})}(\mathbf{r})\sum_{\mathbf{G}}u_{\mathbf{kG}}\,e^{i(\mathbf{k}+\mathbf{G})\cdot\mathbf{r}}$$
$$=E(\mathbf{k})\sum_{\mathbf{G}}u_{\mathbf{kG}}\,e^{i(\mathbf{k}+\mathbf{G})\cdot\mathbf{r}}\,.$$

Now we can use a "standard technique" to solve equations in Fourier-transformed form: Multiplying it by $e^{-i(\mathbf{k}+\mathbf{G}')\cdot\mathbf{r}}$ and integrating over the whole volume of the crystal, this equation leads to

$$\left[\frac{\hbar^2}{2m_{\text{el}}}|\mathbf{k}+\mathbf{G}|^2-E(\mathbf{k})\right]u_{\mathbf{kG}}+\sum_{\mathbf{G}'}V_{\mathbf{GG}'}\,u_{\mathbf{kG}'}=0\,, \tag{5.2}$$

where $V_{\mathbf{GG}'}=V^{(\text{lat})}_{\mathbf{G}-\mathbf{G}'}$ is the Fourier transform of the lattice potential, $V^{(\text{lat})}(\mathbf{r})$. This linear homogeneous problem has nontrivial solutions only if the determinant of the coefficients vanishes, that is:

$$\text{Det}\left|\sum_{\mathbf{G}'}\left\{\left[\frac{\hbar^2}{2m_{\text{el}}}|\mathbf{k}+\mathbf{G}|^2-E(\mathbf{k})\right]\delta_{\mathbf{GG}'}+V_{\mathbf{GG}'}\right\}\right|=0\,. \tag{5.3}$$

The term $V_{\mathbf{GG}'}$ can be simplified using its periodicity:

$$V_{\mathbf{GG}'}=\frac{1}{\Omega}\int_{\Omega}e^{-i\mathbf{G}\cdot\mathbf{r}}\,V^{(\text{lat})}(\mathbf{r})\,e^{i\mathbf{G}'\cdot\mathbf{r}}\,d\mathbf{r}\,,$$

assuming that the wavefunctions have been properly normalized to the volume $\Omega$ of the crystal.

Now recall that $V^{(\text{lat})}$ is the sum of the ionic potentials in the WS cell. Since we must deal with indices for both cells and ions in each cell, we shall use, as usual, indices $l, m, \ldots$ for the cells, reserving the indices $\alpha$, $\beta$, etc., for the $N_{\text{ions}}$ ions in each cell. If the ions in the WS cell are at positions $\tau_\alpha$ ($\alpha=1,N_{\text{ions}}$) (relative to the origin in each cell), then:

$$V^{(\text{lat})}(\mathbf{r})=\sum_{l,\alpha}V^{(\alpha)}(\mathbf{r}-\mathbf{R}_l-\tau_\alpha)\,,$$

where $V^{(\alpha)}$ is the potential of ion $\alpha$. Then, setting $\mathbf{G}'' = \mathbf{G} - \mathbf{G}'$:

$$V_{\mathbf{G}\mathbf{G}'} = \frac{1}{\Omega} \sum_{l,\alpha} \int_{\Omega} \mathrm{e}^{-\mathrm{i}\mathbf{G}''\cdot\mathbf{r}}\, V^{(\alpha)}(\mathbf{r} - \mathbf{R}_l - \tau_\alpha)\, \mathrm{d}\mathbf{r}$$
$$= \frac{1}{\Omega} \sum_{l,\alpha} \mathrm{e}^{-\mathrm{i}\mathbf{G}''\cdot\mathbf{R}_l} \int_{\Omega} V^{(\alpha)}(\mathbf{r} - \mathbf{R}_l - \tau_\alpha)\, \mathrm{e}^{-\mathrm{i}\mathbf{G}''\cdot(\mathbf{r}-\mathbf{R}_l)}\, \mathrm{d}\mathbf{r}\,. \tag{5.4}$$

Setting $\mathbf{r}' = \mathbf{r} - \mathbf{R}_l - \tau_\alpha$ as dummy integration variable, we can rewrite this equation as:

$$V_{\mathbf{G}\mathbf{G}'} = \frac{1}{\Omega} \sum_{l,\alpha} \mathrm{e}^{-\mathrm{i}\mathbf{G}''\cdot\mathbf{R}_l} \int_{\Omega} V^{(\alpha)}(\mathbf{r}')\, \mathrm{e}^{-\mathrm{i}\mathbf{G}''\cdot(\mathbf{r}'+\tau_\alpha)}\, \mathrm{d}\mathbf{r}'$$
$$= \sum_{\alpha} \mathrm{e}^{-\mathrm{i}\mathbf{G}''\cdot\tau_\alpha} \frac{1}{N_{\mathrm{cell}}\Omega_{\mathrm{c}}} \sum_{l} \mathrm{e}^{-\mathrm{i}\mathbf{G}''\cdot\mathbf{R}_l} \int_{\Omega} V^{(\alpha)}(\mathbf{r}')\, \mathrm{e}^{-\mathrm{i}\mathbf{G}''\cdot\mathbf{r}'}\, \mathrm{d}\mathbf{r}'\,, \tag{5.5}$$

where $\Omega_{\mathrm{c}}$ is the volume of the cell. Since the ionic potentials $V^{(\alpha)}(\mathbf{r})$ are short-range, being "screened" by the other valence electrons, they decay very quickly at large distances. Thus, we can neglect the contribution to the integral coming from points $\mathbf{r}'$ outside the WS cell and assume that:

$$\int_{\Omega} V^{(\alpha)}(\mathbf{r}')\, \mathrm{e}^{-\mathrm{i}\mathbf{G}''\cdot\mathbf{r}'}\, \mathrm{d}\mathbf{r}' \approx \int_{\Omega_{\mathrm{c}}} V^{(\alpha)}(\mathbf{r}')\, \mathrm{e}^{-\mathrm{i}\mathbf{G}''\cdot\mathbf{r}'}\, \mathrm{d}\mathbf{r}'\,. \tag{5.6}$$

Since $\sum_l \mathrm{e}^{-\mathrm{i}\mathbf{G}''\cdot\mathbf{R}_l} = N_{\mathrm{cell}}$, the number of cells in the volume $\Omega$, we finally obtain

$$V_{\mathbf{G}\mathbf{G}'} = \sum_{\alpha} \mathrm{e}^{-\mathrm{i}(\mathbf{G}-\mathbf{G}')\cdot\tau_\alpha} \frac{1}{\Omega_{\mathrm{c}}} \int_{\Omega_{\mathrm{c}}} V^{(\alpha)}(\mathbf{r})\, \mathrm{e}^{-\mathrm{i}(\mathbf{G}-\mathbf{G}')\cdot\mathbf{r}}\, \mathrm{d}\mathbf{r}$$
$$= \sum_{\alpha} \mathrm{e}^{-\mathrm{i}(\mathbf{G}-\mathbf{G}')\cdot\tau_\alpha}\, V^{(\alpha)}_{\mathbf{G}-\mathbf{G}'}\,. \tag{5.7}$$

In the simple case of mono-atomic lattices (such diamond, Si, or Ge), all ionic potentials in each cell are the same, so, dropping the unnecessary index $\alpha$, we can rewrite this final expression in the simpler form:

$$V_{\mathbf{G}\mathbf{G}'} = \sum_{\alpha} \mathrm{e}^{-\mathrm{i}(\mathbf{G}-\mathbf{G}')\cdot\tau_\alpha} \frac{1}{\Omega_{\mathrm{c}}} \int_{\Omega_{\mathrm{c}}} V(\mathbf{r})\, \mathrm{e}^{-\mathrm{i}(\mathbf{G}-\mathbf{G}')\cdot\mathbf{r}}\, \mathrm{d}\mathbf{r}$$
$$= S(\mathbf{G}-\mathbf{G}')\, V_{\mathbf{G}-\mathbf{G}'}\,. \tag{5.8}$$

The factor $S(\mathbf{G})$ depends only on the location of the ions within the WS cell and is called the *structure factor*. The factor $V_{\mathbf{G}}$ is the Fourier transform of the atomic potential within the WS cell and is called *the form factor*.

A numerical solution of this problem can be obtained by considering a large-enough number, $N_{\mathbf{G}}$, of $\mathbf{G}$-vectors. The $N_{\mathbf{G}}$ eigenvalues, $E_n(\mathbf{k})$, give the dispersion in $N_{\mathbf{G}}$ bands, the $N_{\mathbf{G}}$ eigenvectors give the wavefunctions. The problem arises when we consider the number of $\mathbf{G}$-vectors which we should retain. If we take for $V^{(\alpha)}$ the bare ionic potential, $V^{(\alpha)}(\mathbf{r}-\mathbf{R}_\alpha) = -e^2/(4\pi\epsilon_0|\mathbf{r}-\mathbf{R}_\alpha|)$, we must account for the extremely short wavelengths that core electrons will have near the "singularity" $\mathbf{r} = \mathbf{R}_\alpha$. In other words: the Fourier components $V^{(\alpha)}_{\mathbf{G}}$ decay all too slowly with increasing magnitude of $\mathbf{G}$. For a typical crystal, the rank of the eigenvalue problem we would have to solve would approach or even exceed $10^6$. Even with the computing power available today, this is a nontrivial difficulty. It was an impossible task in the early days of semiconductor physics. Some approximation is definitely required.

### 5.2.1 *Orthogonalized Plane Waves*

The reason why the simple plane-wave method requires so many plane-waves stems, as we saw, from the necessity to solve also for the core states. Yet, these do not deviate significantly, or even at all, from their free-atom form. Moreover, the valence and conduction states should be orthogonal to these states. Therefore, it appears that the many high-frequency plane waves which create troubles are not even necessary. Thus, from these considerations, Herring [14] and later Callaway [15], had the idea of expanding the unknown single-electron wavefunctions over the subset of plane waves which are orthogonal to the core states. This is called the *orthogonalized plane-wave (OPW)* method.

One starts by defining normalized core eigenstates centered at individual ion positions (assumed to be frozen in their equilibrium position):

$$|t,l,\alpha\rangle = \psi_t(\mathbf{r}-\mathbf{R}_{l,\alpha}) \ , \tag{5.9}$$

where the index $t$ (we are quickly running out of letters of any alphabet to label our many indices!) labels the core level. We assume that these states are known from calculations of the atomic structure. OPWs then may be written as:

$$|\mathrm{OPW}_{\mathbf{k}}\rangle = |\mathbf{k}\rangle - \sum_{t,l,\alpha} |t,l,\alpha\rangle\langle t,l,\alpha|\mathbf{k}\rangle \ , \tag{5.10}$$

where

$$\langle t,l,\alpha|\mathbf{k}\rangle = \frac{1}{\Omega}\int \psi_t^*(\mathbf{r}-\mathbf{R}_{l,\alpha})\ \mathrm{e}^{\mathrm{i}\mathbf{k}\cdot\mathbf{r}}\ \mathrm{d}\mathbf{r} \tag{5.11}$$

does not depend on $l$ because of the periodicity of the lattice. Although the OPWs are not orthogonal to each other, they are orthogonal to all core states and form a complete basis for the conduction/valence states, which is what matters. We can now follow exactly the same procedure used above for plane waves and we will obtain a secular equation similar to Eq. (5.3) but, now, the coefficients $\langle \mathrm{OPW}_{\mathbf{G}}|V|\mathrm{OPW}_{\mathbf{G}'}\rangle$ will decay much faster with increasing magnitudes $G$ and $G'$, so that the size of the problem will be largely reduced.

## 5.3 The Concept of Pseudopotential

The idea of OPW leads directly to another idea: Without going into the trouble of calculating the core wavefunctions in order to deal with OPW, is there a way to modify the ionic potentials themselves—let's call them *pseudopotentials*—so that we will obtain directly valence wavefunctions that are orthogonal to the core states? Roughly speaking, this has been done following two logically different, but computationally very similar, paths. One path, followed by Harrison [4], relies on a more formal reformulation of the problem to extract this "pseudopotential" directly from the definition of OPWs. The second path, more intuitive and proposed originally by Hellman, Animalu, Heine, and co-workers [2, 3, 5], could be called a search for "model potentials" that give us the "correct" results for the valence states, giving up intentionally any information about the core states: After all, they are not needed since they do not influence significantly (or at all) the bonding properties of the crystal and, so, its electronic structure. Let us consider them both, discussing first the more formal approach.

### 5.3.1 *From OPWs to Pseudopotentials*

Consider again the OPWs, Eq. (5.10) and introduce the *projection operator*

$$\hat{\Pi} = \sum_{t,l,\alpha} |t,l,\alpha\rangle\langle t,l,\alpha| \,, \tag{5.12}$$

so that Eq. (5.10) can be written as:

$$|\mathrm{OPW}_{\mathbf{k}}\rangle = (1-\hat{\Pi})|\mathbf{k}\rangle \,. \tag{5.13}$$

The "real" wavefunction can be expanded as:

$$|\psi\rangle = (1-\hat{\Pi}) \sum_{\mathbf{k}} a_{\mathbf{k}}|\mathbf{k}\rangle \,. \tag{5.14}$$

Substitution of this expansion into the initial Schrödinger equation, Eq. (5.1), will give us back the OPW method. Instead, consider the equation above and rewrite it in terms of the "pseudo-wavefunction" $|\phi\rangle$:

$$|\psi\rangle = (1-\hat{\Pi})|\phi\rangle \,. \tag{5.15}$$

Note that $\phi(\mathbf{r})$ is identical to the real wavefunction $\psi(\mathbf{r})$ outside the cores, since the projector $\Pi$ vanishes there, but it is not necessarily orthogonal to the core states inside the cores. Hopefully, $\phi$ will be smooth inside the cores.

Let us now derive an equation for $\phi(\mathbf{r}) = \langle \mathbf{r}|\phi\rangle$ by substituting Eq. (5.15) into Eq. (5.1):

$$\begin{aligned} &-\frac{\hbar^2}{2m_{\mathrm{el}}}\nabla^2\phi(\mathbf{r}) + V^{(\mathrm{lat})}(\mathbf{r})\phi(\mathbf{r}) - \left[-\frac{\hbar^2}{2m_{\mathrm{el}}}\nabla^2 + V^{(\mathrm{lat})}(\mathbf{r})\right]\hat{\Pi}\phi(\mathbf{r}) \\ &+E\hat{\Pi}\phi(\mathbf{r}) = E\phi(\mathbf{r}) \,. \end{aligned} \tag{5.16}$$

Now notice that, since

$$-\frac{\hbar^2}{2m_{\mathrm{el}}}\nabla^2\psi_{t,\alpha}(\mathbf{r}) + V^{(\mathrm{lat})}(\mathbf{r})\ \psi_{t,\alpha}(\mathbf{r}) = E_{t,\alpha}\ \psi_{t,\alpha}(\mathbf{r})$$

(an equation expressing the fact that the core states are, after all, eigenstates of the original Hamiltonian) and from the definition of $\hat{\Pi}$, Eq. (5.12), we have

$$\left[-\frac{\hbar^2}{2m_{\mathrm{el}}}\nabla^2 + V^{(\mathrm{lat})}(\mathbf{r})\right]\hat{\Pi} = \sum_{t,\alpha} E_{t,\alpha}\ |t,\alpha\rangle\langle t,\alpha| \,. \tag{5.17}$$

Thus, we can rewrite Eq. (5.16) as:

$$-\frac{\hbar^2}{2m_{\rm el}}\nabla^2\phi(\mathbf{r})+[V^{(\rm pseudo)}\phi](\mathbf{r})=E\phi(\mathbf{r})\ , \tag{5.18}$$

where the pseudopotential has been defined as:

$$V^{(\rm pseudo)}=V^{(\rm lat)}(\mathbf{r})+\sum_{t,l,\alpha}(E-E_{t,\alpha})\ |t,l,\alpha\rangle\langle t,l,\alpha|\ . \tag{5.19}$$

Note that, transformed back into real space, this pseudopotential will be "nonlocal", since its action (as an operator) on the unknown wavefunction would involve an integration of the unknown with the core wavefunction $\langle t,\alpha|\mathbf{r}\rangle$. Equation (5.18) is called the *pseudopotential equation*. It can be rewritten explicitly as:

$$\left[-\frac{\hbar^2}{2m_{\rm el}}\nabla^2+V^{(\rm lat)}(\mathbf{r})\right]\phi(\mathbf{r})$$
$$+\sum_{t,\alpha}(E-E_{t,\alpha})\ \psi_{t,\alpha}(\mathbf{r}-\mathbf{R}_{l,\alpha})\int\psi^*_{t,\alpha}(\mathbf{r}')\phi(\mathbf{r}')\ d\mathbf{r}'=E\phi(\mathbf{r}). \tag{5.20}$$

Note that the eigenvalue $E$ is identical to the eigenvalue corresponding to the exact eigenfunctions. Indeed, let's assume that in Eq. (5.20) above we had a different eigenvalue $E'$. Then, multiplying this equation by the "exact" wavefunction $\psi^*$ and integrating, we would get:

$$E\int\psi^*\phi\ d\mathbf{r}=E'\int\psi^*\phi\ d\mathbf{r}\ , \tag{5.21}$$

since the second term in Eq. (5.20) vanishes because of the orthogonality between $\psi$ and the core states. Equation (5.21) shows that either the exact wavefunction is orthogonal to the pseudo-wavefunction or, when not, they have the same energy.

### 5.3.2 Model Potentials

An alternative way to reach the concept of pseudopotential relies on the idea of looking for a *model potential* that gives us wavefunctions that are "correct" (that is: identical to the "real" wavefunctions) outside the cores, no matter how inaccurate these "pseudo-wavefunctions" will be inside the cores. If we manage to do this, the binding properties of the crystal will be represented correctly. Similarly, we shall obtain the correct information about the energies of the valence and conduction states.

Figure 5.1 illustrates this idea: The dashed blue line describes (in cartoon fashion) the bare ionic potential. Large frequency components (that is, slowly decaying Fourier components $V^{(\alpha)}_{\mathbf{q}}$ as $q$ increases) still persist because of the fast variation of the ionic potential near the ion cores. The pseudopotentials are indicated by the red solid line and results from "chopping" the potential around the cores. The resulting Schrödinger equation will have the correct form outside the cores and will give us the correct wavefunctions in these regions *provided* we use the correct boundary conditions at the core boundaries (that is, we must somehow enforce that the pseudo-wavefunctions have the "correct" value and radial derivatives at the core boundaries).

It is easy to see the advantage of this formulation: we have eliminated all high-$q$ Fourier components of $V^{(\alpha)}_{\mathbf{q}}$, thus reducing significantly the size of our problem.

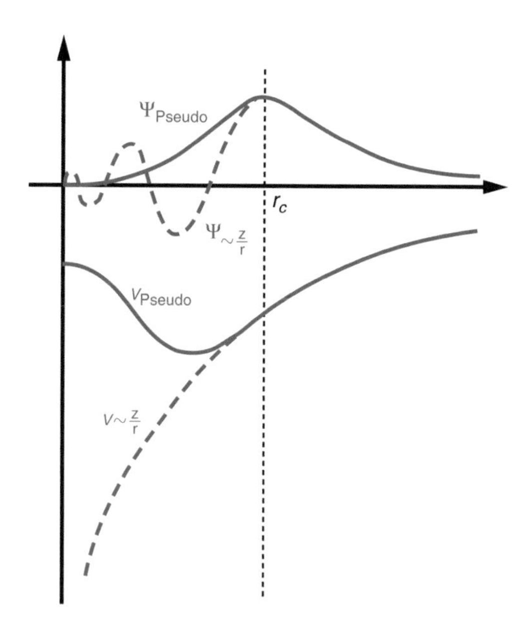


**Fig. 5.1** Schematic illustration of the concept of "pseudopotential": Within the "core" region the singularity of the all-electron, bare ionic potential is removed and replaced by a smoother function. The wavefunction inside the core is not physically meaningful, but as long as its value and the value of its radial derivative coincide with those of the "exact" wavefunction outside the core, the structure of the valence electrons, of the atomic bonds, and of the crystal structure will be correct. [Figure adapted from "Sketch Pseudopotentials" by Wolfram Quester. Original uploader was Wquester at en.wikipedia—Transferred from en.wikipedia; transferred to Commons by User:Leyo using CommonsHelper.self-made with inkscape. Licensed under Public Domain via Wikimedia Commons]

Note that the "model potential" by Animalu and Heine and Harrison's pseudopotential are essentially the same concept. Indeed it can be proven that in Eq. (5.19) above we may replace $E - E_{t,\alpha}$ with any function of $E$, $t$, and $\alpha$ and still reach the same result Eq. (5.21). This flexibility allows us to use the Animalu–Heine's model potential. Note also Harrison's pseudopotential is *nonlocal* [see the integral in Eq. (5.20)]. Indeed, nonlocal pseudo- and model-potentials are often more accurate. But local approximations may work and have been employed.

### 5.3.3 Density Functional Theory

Regardless of the perspective we take, we can construct our pseudo (or model) potential for any ion (say, Si, to fix the ideas) by solving self-consistently the Schrödinger equation for the Si atom using only the valence electrons. That is, we build a parametrized model potential which represents the bare nuclear Si potential ($-4e^2/(4\pi\epsilon_0 r)$) with the additional screening by the ten core electrons $1s^2 2s^2 2p^6$. We solve the Schrödinger equation for the four valence electrons (using the Kohn–Sham "density functional" in the Hartree–Fock approximation, for example, as we have seen in Eq. (4.17) in the previous chapter) by varying the parameters until we reproduce the known spectrum of the Si atom. This fixes the pseudopotential of the $Si^{+4}$ ion. We now insert this model potential into our Schrödinger equation Eq. (4.17) and solve it self-consistently with the charge of the valence electrons in the crystal. The pseudopotentials used in this DFT approaches are known as *self-consistent pseudopotentials*, since we require a solution of a nonlinear Schrödinger equation in which a portion of the potential is a functional of the unknowns (the electron density, as so the wavefunctions). While this scheme used to severely underestimate the energy gaps of most semiconductors (for example, the indirect gap of Si was underestimated by a factor of 2 in past decades), recent refinements of the model potentials and on the expression for the density functional, now render this method the algorithm of choice providing great accuracy. The self-consistency of this method allows its general applicability and portability, a virtue that model potentials (or the empirical pseudopotentials treated in the next section) do not possess. For example, unlike empirical pseudopotentials, the model potential for the Ga ion can be used to calculate the band structure of GaAs, GaN, GaP, etc. It also allows for calculations in strained materials, of defects in crystals, of ionic motion such as phonons (since the new electronic charge following the redistribution of the ions can be calculated). Given the increasingly important role that this method is playing in the communities of condensed matter physics, chemistry, biological sciences, materials science, and nanoelectronics, we shall discuss it in more depth in the next chapter.

### *5.3.4 Empirical Pseudopotentials (EP)*

For practical applications, a much simpler approximation is usually employed: Describing the Fourier transform (or series) of the pseudopotential with a parametrized algebraic expression (or simply by the values of its discrete components), the pseudopotential is treated in a fully empirical fashion, those parameters becoming simply quantities that are fitted to experimental results, such as band gaps at symmetry points in the first BZ, electron dispersion (when available), optical absorption, or electro-reflectance data.

One such empirical expression is given by Friedel et al. [16]:

$$V^{(\mathrm{Si})}(q) = \frac{b_1(q^2 - b_2)}{b_3\, \mathrm{e}^{b_4 q^2} - 1} . \tag{5.22}$$

The parameters $b_1$, $b_2$, $b_3$, and $b_4$ are determined by such a fitting procedure to experimental information. Local forms, such Eq. (5.22), have been used extensively and successfully. Examples, as well as alternative explicit forms for $V^{(\mathrm{pseudo})}(q)$ for various ions, parameters, and results, are discussed in a separate chapter and in Appendix B.

Further improvements to this local formulation of the pseudo- (or model) potential can be obtained by adding nonlocal terms, consistent with Harrison's result, as in Eq. (5.19). These terms essentially give us more freedom to extend the energy range over which we can expect to obtain a good match between the values of the wavefunction and its radial derivatives at the core boundary. Therefore, we can expect results that will be more reliable over a larger energy range, especially in the conduction band, after having fitted the parameters to experimental data obtained in a narrow energy range (usually, close to the band-gap or Fermi surface). Chelikowsky and Cohen [17] have expressed these corrections as additional terms—to be added to the local potential entering the one-electron Schrödinger equation—as terms that project the valence wavefunction in a subspace orthogonal to core states of angular momentum $l$. They express these terms in the energy-dependent form:

$$V_{\alpha,\mathrm{NL}}(\mathbf{r},E) = \sum_{l}^{\infty} A_l(E)\, f_l(r)\, \Pi_l , \tag{5.23}$$

where $A_l(E)$ represents the depth of the $l$-th angular momentum component of the ionic potential, $f_l(r)$ attempts to simulate the radial component of the core states to which the valence wavefunctions must be orthogonal, and $\Pi_l$ is the operator that projects onto the eigen-subspace of the angular momentum operator with quantum number $l$. In other words, this is just Eq. (5.19), having assumed that the core states lie in a subspace orthogonal to the projection identified by Eq. (5.23). For the semiconductors considered by Chelikowsky and Cohen, only terms with $l = 0$, 1, or 2 are required.

Empirical pseudopotentials are extremely powerful, since very few parameters are needed in order to obtain information about many properties of the crystals: Its band structure (that is, the electronic excitation spectrum or electron dispersion), its optical properties, and its dielectric properties, among others. If information about the vibrational eigenmodes of the crystal (phonons) is available, they can also provide the matrix elements for electron–phonon interactions.

When facing the problem of calculating the band structure of bulk crystals, very few parameters indeed are required: Since $V_{\mathbf{q}}^{(\mathrm{pseudo},\alpha)}$ is needed only for $\mathbf{q} = \mathbf{G}$ [see Eqs. (5.3) and (5.7)] and since $V_{\mathbf{G}}^{(\mathrm{pseudo},\alpha)}$ decays very quickly with increasing magnitude of $\mathbf{G}$, very few values of $V_{\mathbf{G}}^{(\mathrm{pseudo},\alpha)}$ are needed. For example, for cubic crystals the first $\mathbf{G}$ vectors are the eight vectors $(\pm1,\pm1,\pm1)$ of length $G^2 = 3$ (in units of $2\pi/a$), the vectors $(\pm2,0,0)$, $(0,\pm2,0)$, and $(0,0,\pm2)$ (called the "star" of vectors $(2,0,0)$), all of length $G^2 = 4$; the star of vectors $(2,2,0)$ of length $G^2 = 8$, the star of vectors $(1,1,3)$ of length $G^2 = 11$, etc. Since the pseudopotential, by symmetry, depends on the magnitude $G$ of $\mathbf{G}$, we must determine the first Fourier coefficients—the "form factors"— $V_{G^2=3}^{(\mathrm{pseudo},\alpha)}$, $V_{G^2=4}^{(\mathrm{pseudo},\alpha)}$, $V_{G^2=8}^{(\mathrm{pseudo},\alpha)}$, $V_{G^2=11}^{(\mathrm{pseudo},\alpha)}$,... Typically the empirical pseudopotential is truncated before reaching the star of the $(4,0,0)$ vectors of length $G^2 = 16$, so we have only four (complex) parameters that we are free to vary in order to fit experimental data. For Si and Ge, one can use only real form factors and, noticing that the structure factor, $S(\mathbf{G})$, vanishes on the star $G^2 = 4$ by symmetry, only three parameters are required to obtain a very accurate band structure.

Figure 5.2 shows one of the most commonly used forms of the local empirical pseudopotential of Si. Note the abrupt truncation for $q > 3.5 \times 2\pi/a$. Of course, the full expression for the ionic pseudopotential, $V^{(\mathrm{pseudo},\alpha)}(q)$, as given, for example, by Eq. (5.22), is needed when the lattice is distorted (when computing the change of the band structure caused by strain, for example), or when altering the periodicity of the crystal. This is the case when considering the "supercells" needed to study nanostructures, as discussed in Chap. 7. In these cases, the $\mathbf{G}$ vectors will change, so a knowledge of the empirical pseudopotential is needed for all values of $q$.

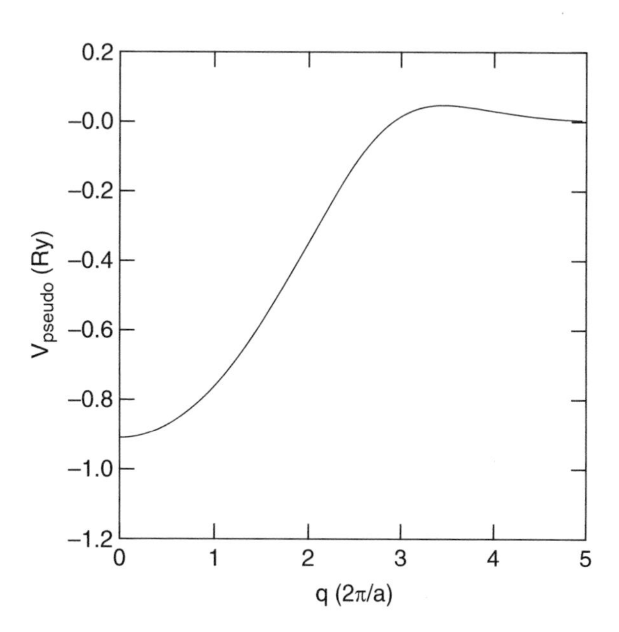


**Fig. 5.2** A commonly used local empirical pseudopotential for Si. This is the form proposed in [16]

We conclude this section with a brief discussion about the merits and shortcomings of empirical vs. self-consistent pseudopotentials.

Empirical pseudopotentials present the advantage of yielding the band structure of a crystal that matches experimental data. This, of course, is the very definition of the term "empirical." However, their "calibration" is performed only on a subset of known properties and additional properties, in general unknown, can be predicted. For example, having calibrated the empirical pseudopotential of Si to known band gaps and effective masses at the bottom of the conduction band, one can calculate effective masses in higher-energy local band minima. Similarly, the pseudo-wavefunctions and pseudopotentials can be used to calculate electron–phonon matrix elements, as we discuss in Chap. 13. These results can be used in high-field transport studies. Moreover, as long as the crystal structure is not distorted too much, studies can be reliably performed to assess the effect of small distortion (such a moderate strain) or to obtain the electronic structure in small structures, such as thin films or nanowires. This will be amply discussed in Chap. 7. But herein lies also their major drawback: The pseudopotentials are fitted to a specific structure. They embed, in their empirical nature, the spatial distribution of the valence electrons in this particular structure. This distribution determines how the real bare ionic potential is screened by valence charges. If the atomic configuration is altered, this distribution, and so also the shape of the screened ionic potential, will change. Therefore, empirical pseudopotentials can be trusted only when studying systems whose atomic configurations is not too different from the configuration to which they had been calibrated. A striking example is given by the fact that the empirical pseudopotential of Ga to be used in GaAs, for example, is not the same one that needs to be used to study GaSb, for example. The different size and electronic configuration of Ga in the GaAs lattice, bonded to As, is different from its electronic configuration in the GaSb lattice, now bonded to Sb.

The ab initio (or self-consistent) pseudopotentials used by DFT, instead, exhibit complementary advantages and disadvantages. They are "portable," in the sense that they are calibrated to isolated ions, so that the valence electrons are free to re-adjust to changing environment and atomic configurations. Indeed, DFT calculations are now extremely reliable in predicting the atomic configuration itself. On the other hand, since DFT is essentially a "ground state" theory [12, 13], there is no guarantee that the conduction bands will be treated correctly. This is indeed their major drawback, although recently significant progress has been made to solve this problem from both a theoretical and computational perspective.

## 5.4 Other Approximate Methods

So far, in this chapter we have considered only approaches to calculate the electronic properties of crystals based on plane-wave expansions. Alternative methods are available, based on different basis functions or on different approximations. Here we discuss two of the most common methods employed: The tight-binding method, based on expressing the unknown wavefunctions on the basis of localized atomic orbitals (LCAO is the term used mainly in chemistry), and the numerically convenient $\mathbf{k} \cdot \mathbf{p}$ perturbation theory.

### 5.4.1 *Tight-Binding*

The tight-binding approach is probably the simplest from a conceptual perspective. It's related to what's called *Linear Combination of Atomic Orbitals (LCAO)* in chemistry.

Suppose that we know the atomic orbitals of the atoms constituting the crystal, $\psi_t(\mathbf{r}-\mathbf{r}_\alpha)$, where $t$ denotes the set of quantum numbers that label the atomic orbitals and $\mathbf{r}_\alpha$ denotes, as usual, the coordinate of the ion $\alpha$. For a given $t$ we can always consider a linear combination of these orbitals (all having the same energy in each isolated ion). The idea indeed is to form such a combination and treat the "overlap" between the orbitals as the perturbation which will give rise to the band structure.

Thus, consider the linear combination:

$$\phi_t(\mathbf{r}) = \sum_{l,\alpha} c_{l,\alpha}\psi_t(\mathbf{r}-\mathbf{R}_{l,\alpha}) \,, \tag{5.24}$$

where, as usual, the ionic coordinates $\mathbf{R}_{l,\alpha}$ should be intended as the equilibrium positions. Of course, the wavefunction must still obey the requirements imposed by the periodicity of the lattice, and so it must be of Bloch form. This implies that the coefficients $c_{l,\alpha}$ must be equal to $e^{i\mathbf{k}\cdot\mathbf{R}_{l,\alpha}}$, so that

$$\phi_{\mathbf{k},t}(\mathbf{r}) = \frac{1}{N^{1/2}} \sum_{l,\alpha} e^{i\mathbf{k}\cdot\mathbf{R}_{l,\alpha}}\ \psi_t(\mathbf{r}-\mathbf{R}_{l,\alpha}) \,, \tag{5.25}$$

where $N$ is the number of orbitals we add. Note that if we assume that the orbitals $\psi_t$ are correctly normalized, $\phi_{\mathbf{k},t}$ will be also correctly normalized only if the overlap among the atomic orbitals vanishes, so that the normalization will be correct only at zero-th order. Since the states $\psi_t$ are eigenstates of the isolated ions, each with potential $v(\mathbf{r}-\mathbf{r}_\alpha)$, they satisfy the Schrödinger equation

$$\left[-\frac{\hbar^2}{2m_{\text{el}}}\nabla^2 + V^{(\alpha)}(\mathbf{r}-\mathbf{r}_\alpha)\right] \psi_t(\mathbf{r}-\mathbf{r}_\alpha) = E_t\ \psi_t(\mathbf{r}-\mathbf{r}_\alpha) \,. \tag{5.26}$$

The Schrödinger equation we wish to solve is

$$\left[-\frac{\hbar^2}{2m_{\text{el}}}\nabla^2 + V^{(\text{lat})}(\mathbf{r})\right] \phi_{\mathbf{k},t}(\mathbf{r}) = E_{\mathbf{k}}\ \phi_{\mathbf{k},t}(\mathbf{r}) \,, \tag{5.27}$$

where, as usual:

$$V^{(\text{lat})}(\mathbf{r}) = \sum_{l,\alpha} V^{(\alpha)}(\mathbf{r}-\mathbf{R}_{l,\alpha}) \,. \tag{5.28}$$

Inserting the linear combination Eq. (5.25) into the left-hand side of Eq. (5.27), we have

$$\begin{aligned}&\left[-\frac{\hbar^2}{2m_{\text{el}}}\nabla^2 + V^{(\text{lat})}(\mathbf{r})\right] \frac{1}{N^{1/2}} \sum_{l,\alpha} e^{i\mathbf{k}\cdot\mathbf{R}_{l,\alpha}}\ \psi_t(\mathbf{r}-\mathbf{r}_\alpha)\\ &= \sum_{l,\alpha}\left[E_t + \sum_{l];l'\beta\neq l\alpha} V^{(\alpha)}(\mathbf{r}-\mathbf{r}_{l',\beta})\right] e^{i\mathbf{k}\cdot\mathbf{r}_\alpha}\ \frac{1}{N^{1/2}}\ ,\psi_t(\mathbf{r}-\mathbf{R}_{l,\alpha}) \,. \end{aligned}\tag{5.29}$$

where $N = N_{\text{cell}}N_{\text{ion}}$ is the total number of ions in the crystal. In order to simplify the notation, for now we lump into the single index $\alpha$ the couple of indices $l,\alpha$, so now $\alpha$ runs over all ions in the crystal. With this simplified notation, the expectation value of the energy of the state $\phi_{\mathbf{k},t}$ can be written as:

$$\langle E_{\mathbf{k}} \rangle = \frac{\int \phi^*_{\mathbf{k},t}(\mathbf{r}) \, [-\hbar^2/(2m_{\rm el})\nabla^2 + V^{(\rm lat)}(\mathbf{r})] \, \phi_{\mathbf{k},t}(\mathbf{r}) \, d\mathbf{r}}{\int \phi^*_{\mathbf{k},t}(\mathbf{r}) \, \phi_{\mathbf{k},t}(\mathbf{r}) \, d\mathbf{r}}$$

$$= E_t + \frac{(1/N) \, \sum_{\alpha,\beta} \int \psi_t^*(\mathbf{r}-\mathbf{R}_\beta) \, \sum_{\gamma\neq\beta} V^{(\gamma)} \upsilon(\mathbf{r}-\mathbf{R}_\gamma) \, \psi_t(\mathbf{r}-\mathbf{R}_\alpha) \, \exp[i\mathbf{k}\cdot(\mathbf{R}_\alpha-\mathbf{R}_\beta)] d\mathbf{r}}{\int \phi^*_{\mathbf{k},t}(\mathbf{r}) \, \phi_{\mathbf{k},t}(\mathbf{r}) \, d\mathbf{r}} \, . \tag{5.30}$$

To leading order, we can assume that the normalization integral in the denominator is unity, as we noticed above. Also, keeping in mind that the overlap terms are assumed to be small, three-center integrals (that is, terms for which $\alpha \neq \beta \neq \gamma$ above) will be of at least one highest order in our perturbation expansion. Thus, we can ignore them and retain in the sum only terms for which either $\alpha = \beta$ or $\alpha = \gamma$.

The terms with $\alpha = \beta$ give a contribution of the form:

$$E_t + \frac{1}{N}\sum_\alpha \int \psi_t^*(\mathbf{r}-\mathbf{R}_\alpha) \, \sum_{\gamma\neq\alpha} V^{(\gamma)}(\mathbf{r}-\mathbf{R}_\gamma) \, \psi_t(\mathbf{r}-\mathbf{R}_\alpha) \, d\mathbf{r} \, . \tag{5.31}$$

Note that this term does not depend on $\mathbf{k}$ and it is simply the expectation value at each ion $\alpha$ of the potential due to all other ions $\gamma$. By symmetry, it does not depend on $\alpha$ and it can be written as:

$$E_t + \int \psi_t^*(\mathbf{r}-\mathbf{R}_\alpha) \, \sum_{\gamma\neq\alpha} V^{(\gamma)}(\mathbf{R}-\mathbf{R}_\gamma) \, \psi_t(\mathbf{r}-\mathbf{R}_\alpha) \, d\mathbf{r} \, . \tag{5.32}$$

Since the potential $V^{(\alpha)}$ of each ion is negative, this term is also negative and it expresses the binding energy of the crystal. It does not contribute to the band structure because of its independence of $\mathbf{k}$.

The terms for which $\alpha = \gamma$ are of the form:

$$E_t + \frac{1}{N} \sum_{\alpha,\beta\neq\alpha} e^{i\mathbf{k}\cdot(\mathbf{R}_\alpha-\mathbf{R}_\beta)} \int \psi_t^*(\mathbf{r}-\mathbf{R}_\beta) \, V^{(\alpha)}(\mathbf{r}-\mathbf{R}_\alpha) \, \psi_t(\mathbf{r}-\mathbf{R}_\alpha) \, d\mathbf{r} \, . \tag{5.33}$$

Again, by symmetry these terms do not depend on $\alpha$, so that once more the sum divided by $N$ can be taken as equal to any one term of the sum. Also, taking the origin at $\mathbf{R}_\beta$ this sum can be rewritten as:

$$E_t + \sum_{\alpha,\mathbf{R}_\alpha\neq\mathbf{0}} e^{i\mathbf{k}\cdot\mathbf{R}_\alpha} \int \psi_t^*(\mathbf{r}) \, V^{(\alpha)}(\mathbf{r}-\mathbf{R}_\alpha) \, \psi_t(\mathbf{r}-\mathbf{R}_\alpha) \, d\mathbf{r} \, . \tag{5.34}$$

This is the main result of the tight-binding approximation: The energy of the atomic orbital $E_t$ is modified by $\mathbf{k}$-dependent (band-structure) effects. Their strength depends on the amount of overlap among atomic orbitals. Thus, the method constitutes a good approximation if the overlap is not too large.

In general, several ($n$) atomic orbitals will be considered and the matrix

$$\langle t'|\widehat{H}_{\mathbf{k}}|t\rangle = \int \phi^*_{\mathbf{k},t'}(\mathbf{r}) \, [-\hbar^2/(2m_{\rm el})\nabla^2 + V^{(\rm lat)}(\mathbf{r})] \, \phi_{\mathbf{k},t}(\mathbf{r}) \, d\mathbf{r}$$

can be considered. Manipulating it as done above, one can obtain an effective tight-binding Hamiltonian of the form:

$$\langle t'|\widehat{H}_{\mathbf{k}}|t\rangle = E_t\delta_{t't} + T_{t't}(\mathbf{k}) \, .$$

where the "off diagonal" terms

$$T_{tt'}(\mathbf{k}) = \sum_{\mathbf{r}_\alpha \neq \mathbf{0}} e^{i\mathbf{k}\cdot\mathbf{R}_\alpha} \int \psi_{t'}^*(\mathbf{r})\; V^{(\alpha)}(\mathbf{r}-\mathbf{R}_\alpha)\; \psi_t(\mathbf{r}-\mathbf{R}_\alpha)\;,$$

involve overlap integrals of atomic orbitals. These terms can be calculated from known Hartree–Fock atomic orbitals or may even be the input for fully self-consistent calculations, known as *DFT tight-binding*. More commonly the "hopping integrals" appearing in the integrand of the expression above for $T_{tt'}(\mathbf{k})$ are treated as empirical (fitting) parameters (*empirical tight-binding*). The resulting matrix can be diagonalized, yielding the band structure. Of course, as for plane-wave methods, DFT can be used to calculate self-consistently the atomic orbitals and the overlap integrals. If a sufficiently large number of atomic orbitals is considered, the method becomes fully equivalent to the "conventional" plane-wave DFT method.

Common approximations made to deal with semiconductors include $s$ and $p^3$ states, often augmented by excited $s^*$ or $d^5$ orbitals to compensate for the neglected higher-energy orbitals (the so-called $sp^3s^*$ and $sp^3d^5$ models). The red curves in Figs. 5.4, 5.5, and 5.6 (pages 87–89) show the band structures of Si, Ge, and GaAs computed with the $sp^3s^*$ model of Peter Vogl and co-workers [18].

A simpler example we can consider here is the energy band originating from $s$ orbitals. Let's also assume an ideal cubic lattice. The integral we must consider is the term corresponding to Eq. (5.34) above:

$$\lambda = \int \psi_t^*(\mathbf{r})\; \upsilon(\mathbf{r}-\mathbf{r}_\alpha)\; \psi_t(\mathbf{r}-\mathbf{r}_\alpha)\; d\mathbf{r}\;, \tag{5.35}$$

which for $s$ states is the same for all neighbor ions. Then, the correction to the energy of the atomic orbital takes the form:

$$\lambda \sum_\alpha e^{-i\mathbf{k}\cdot\mathbf{r}_\alpha} = 2\lambda\,(\cos k_x a + \cos k_y a + \cos k_z a)\;, \tag{5.36}$$

where the sum extends only over the nearest neighbors. Since the potential $\upsilon$ is attractive, $\lambda < 0$. The band structure will be of the form:

$$E_{AO} - 2|\lambda|(\cos k_x a + \cos k_y a + \cos k_z a)\;. \tag{5.37}$$

Note that the energy-correction increases quadratically as we move away from the zone-center. Expanding Eq. (5.37) in powers of $k$ we have

$$E_{AO} - 2|\lambda| \left(3 - \frac{k^2 a^a}{2} + \cdots\right) = E_{AO} - 6|\lambda| - \frac{\hbar^2 k^2}{2m^*} + \cdots \tag{5.38}$$

where the effective mass is $m^* = \hbar^2/(2a^2|\lambda|)$.

### 5.4.2 $\mathbf{k}\cdot\mathbf{p}$ *Method and Effective Mass*

The $\mathbf{k}\cdot\mathbf{p}$ perturbation theory is of a nature different from the methods we have discussed so far. Yu and Cardona's text [19] presents a more comprehensive overview than what we provide here.

Consider again the Schrödinger equation we must solve, Eq. (5.1):

$$-\frac{\hbar^2\nabla^2}{2m_{\rm el}}\psi(\mathbf{r})+V^{(\rm lat)}(\mathbf{r})\psi(\mathbf{r})=E\,\psi(\mathbf{r})\,. \tag{5.39}$$

Using Bloch's theorem by setting $\psi(\mathbf{r})=\psi_{\mathbf{k}}(\mathbf{r})=u_{\mathbf{k}}(\mathbf{r})e^{i\mathbf{k}\cdot\mathbf{r}}$, inserting it into Eq. (5.39) and setting $\mathbf{p}=(\hbar/i)\nabla$, we obtain

$$\frac{1}{2m_{\rm el}}(\hbar\mathbf{k}+\hat{\mathbf{p}})^2u_{\mathbf{k}}+V^{(\rm lat)}(\mathbf{r})u_{\mathbf{k}}=E_{\mathbf{k}}u_{\mathbf{k}}\,. \tag{5.40}$$

The idea now is to assume that we know the solutions at a particular point $\mathbf{k}_0$ and consider the terms depending on $\mathbf{k}-\mathbf{k}_0$ as a perturbation. We can chose any point $\mathbf{k}_0$, but usually the point $\mathbf{k}_0=0$ is chosen. So, let's assume that we know how to solve

$$\frac{1}{2m_{\rm el}}\hat{p}^2u_0^{(n)}+V^{(\rm lat)}(\mathbf{r})u_0^{(n)}=E_0^{(n)}u_0^{(n)}\,, \tag{5.41}$$

where the index $n$ labels the band index. To order $k$ the leading perturbation term in Eq. (5.40) is

$$\widehat{H}_1=\frac{\hbar}{m_{\rm el}}\mathbf{k}\cdot\hat{\mathbf{p}}\,,$$

(hence the name of the method), and to second-order

$$\widehat{H}_2=\frac{\hbar^2k^2}{2m_{\rm el}}\widehat{I}\,.$$

Thus, the first-order correction to the eigenvalue $E_0^{(r)}$ will be

$$E_1^{(r)}=\frac{\hbar}{m_{\rm el}}\mathbf{k}\cdot\langle u_0^{(r)}|\hat{\mathbf{p}}|u_0^{(r)}\rangle\,. \tag{5.42}$$

If the crystal has a center of symmetry, the matrix element must vanish. So, we must consider second-order perturbation theory:

$$E_2^{(r)}=\frac{\hbar^2k^2}{2m_{\rm el}}+\frac{\hbar^2}{m_{\rm el}^2}\sum_s\frac{|\mathbf{k}\cdot\langle u_0^{(r)}|\hat{\mathbf{p}}|u_0^{(s)}\rangle|^2}{E_0^{(r)}-E_0^{(s)}}\,. \tag{5.43}$$

Let's define the "effective mass tensor" $m_{ij}$ (where $i$ and $j$ runs over the coordinate-indices $x$, $y$, $z$) via:

$$\frac{m_{\rm el}}{m_{ij}^*}=\delta_{ij}+\frac{2}{m_{\rm el}}\sum_s\frac{\langle u_0^{(r)}|\hat{p}_i|u_0^{(s)}\rangle\langle u_0^{(s)}|\hat{p}_j|u_0^{(r)}\rangle}{E_0^{(r)}-E_0^{(s)}}\,, \tag{5.44}$$

relation which is known as the *f-sum rule*. Then, the energy at the $\mathbf{k}$-point to second-order will be

$$E^{(r)}=E_0^{(r)}+\frac{\hbar^2}{2}\sum_{i,j}\frac{k_ik_j}{m_{ij}^*}\,. \tag{5.45}$$

Note that the dispersion of free electrons is modified by (energetically) "nearby" bands, since the effect of a band $s$ is seen from Eq. (5.43) or (5.44) to be inversely proportional to the energetic separation $E_0^{(r)} - E_0^{(s)}$. If only one band matters, then, since the momentum matrix element is of the order of $\hbar/a$ ($a$ is the lattice constant), setting $\Delta E = E_0^{(r)} - E_0^{(s)}$, we have

$$\frac{m_{\rm el}}{m^*} \sim 1 + \frac{2\hbar^2}{m_{\rm el} a^2 \Delta E} . \tag{5.46}$$

In the special case of the valence bands of cubic semiconductors, as we shall see, the three topmost bands are degenerate at the center of the Brillouin Zone, which is the symmetry point $\Gamma$. Therefore, we must use degenerate perturbation theory. This requires the full diagonalization of the perturbation Hamiltonian on the three-dimensional subspace spanned by the degenerate eigenvectors, that is, the matrix with elements:

$$\langle r|\widehat{H}_{\mathbf{k}\cdot\mathbf{p}}|s\rangle = \left( E_0^{(r)} + \frac{\hbar^2 k^2}{2m_{\rm el}} \right) \delta_{r,s} + \frac{\hbar^2}{m_{\rm el}^2} \sum_{i,j} k_i k_j \langle r|\hat{p}_i|s\rangle \langle s|\hat{p}_j|r\rangle , \tag{5.47}$$

having indicated with $|r\rangle$ the (degenerate) valence states $u_0^{(r)}(\mathbf{r})$. For Si and Ge, Dresselhaus, Kip, and Kittel [11], have proposed to use as unperturbed states $u_0$. These are the orbitals which transform like $yz$, $zx$, and $xy$ (which are a basis for the representation of the cubic point group at the symmetry point $\Gamma$); they have angular dependence of their wavefunctions of the form $\sin\theta\cos\theta\cos\phi$, $\sin\theta\cos\theta\cos\phi$, and $\sin^2\theta\cos\phi\sin\phi$. On symmetry arguments, they show that the matrix to be diagonalized (that is, whose eigenvectors give the structure of the three topmost valence bands) has the form:

$$\widehat{H}_{\mathbf{k}\cdot\mathbf{p}} = \begin{bmatrix} Lk_x^2 + M(k_y^2 + k_z^2) & Nk_xk_y & Nk_xk_z \\ Nk_xk_y & Lk_y^2 + M(k_x^2 + k_z^2) & Nk_yk_z \\ Nk_xk_z & Nk_yk_z & Lk_z^2 + M(k_x^2 + k_y^2) \end{bmatrix} , \tag{5.48}$$

where the coefficients $L$, $M$, and $N$ can either be fitted empirically or calculated from more complicated band-structure calculations or even tight-binding approximations using known atomic orbitals.

Figure 5.3 on page 86 shows the valence band structure of Si, also under biaxial tensile strain, computed with $\mathbf{k}\cdot\mathbf{p}$ perturbation theory, Eq. (5.48), (red dashed lines) compared with the results obtained using nonlocal empirical pseudopotentials.

## 5.5 A Computational Example

We conclude with a few computational examples that illustrate the results of the various methods just discussed to calculate the band structure of important semiconductors.

Figures 5.4, 5.5, and 5.6 show the band structure of Si, Ge, and GaAs calculated with nonlocal empirical pseudopotentials. Let's review briefly the procedure followed, but we shall consider only the simpler local approximation.

The starting point is the secular equation (5.3) with an expression for the Fourier coefficients of the (local) pseudopotential given by Eq. (5.7):

$$V_{\mathbf{G}} = V_{\mathbf{G}}^{(1)} \mathrm{e}^{-\mathrm{i}\mathbf{G}\cdot\tau_1} + V_{\mathbf{G}}^{(2)} \mathrm{e}^{-\mathrm{i}\mathbf{G}\cdot\tau_2} . \tag{5.49}$$

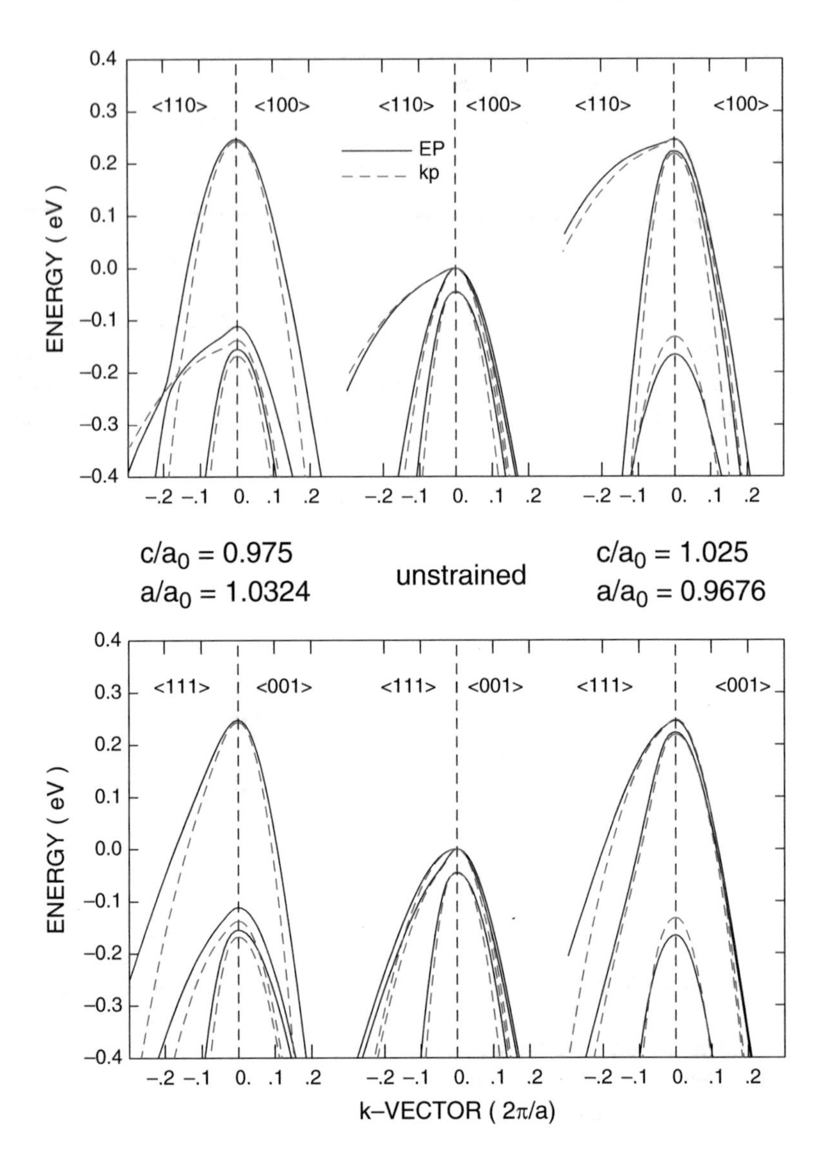


**Fig. 5.3** The valence band structure of Si under biaxial tensile strain, computed with $\mathbf{k} \cdot \mathbf{p}$ perturbation theory, Eq. (5.48), (*red dashed lines*) compared with the results obtained using nonlocal empirical pseudopotentials

In these cubic fcc crystals, such as Si and GaAs, the two ions in the primitive cell are at coordinates $\tau_1 = (0,0,0)$ and $\tau_2 = (a/4)(1,1,1)$, where $a$ (= 0.543 nm for Si) is the lattice constant. This is seen in the left frame of Fig. 3.1: The Ga and As ions are the "black" ion at the bottom left vertex and the "white" ion to which it is bonded. It is actually more convenient to shift the origin of our coordinates to the mid-point between the ions, so that $\tau_1 = \tau = (a/8)(1,1,1)$ and $\tau_2 = -\tau = -(a/8)(1,1,1)$. Then, Eq. (5.49) becomes

$$V_{\mathbf{G}} = V_{\mathbf{G}}^{(1)} e^{-i\mathbf{G}\cdot\tau} + V_{\mathbf{G}}^{(2)} e^{i\mathbf{G}\cdot\tau}$$
$$= [V_{\mathbf{G}}^{(1)} + V_{\mathbf{G}}^{(2)}] \cos(\mathbf{G}\cdot\tau) - i\,[V_{\mathbf{G}}^{(1)} - V_{\mathbf{G}}^{(2)}] \sin(\mathbf{G}\cdot\tau) \,. \quad (5.50)$$

This can be re-expressed in terms of a symmetric and an antisymmetric part:

$$V_{\mathbf{G}} = S^{(S)}(\mathbf{G}) V_G^{(S)} - i S^{(A)}(\mathbf{G}) V_A^{(S)} \,,$$

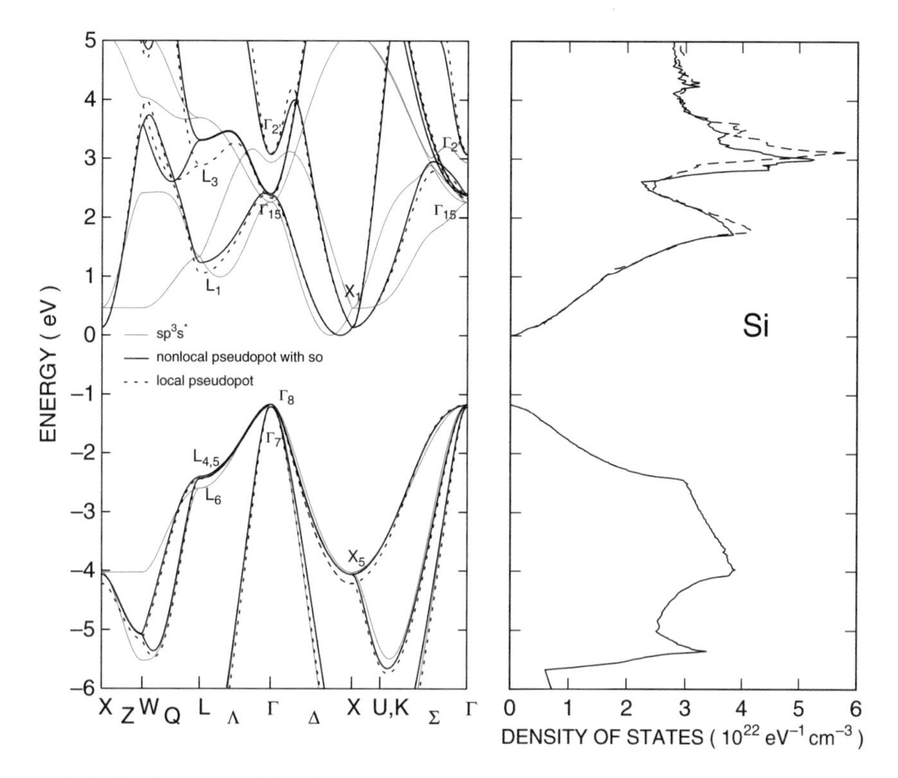


**Fig. 5.4** At *left*, the band structure of Si calculated using the nonlocal empirical pseudopotentials of Chelikowsky and Cohen [17] with spin–orbit corrections (*thick black lines*), using the local empirical pseudopotentials of Cohen and Bergstresser [20] (*dashed black lines*), and employing the $sp^3s^*$ tight-binding model of Vogl and co-workers [18] (*thin red lines*). At *right*, the density of states calculated from the nonlocal band structure at *left*

where

$$S^{(S)}(\mathbf{G}) = \cos(\mathbf{G}\cdot\tau) \,,$$

and

$$S^{(A)}(\mathbf{G}) = \sin(\mathbf{G}\cdot\tau) \,,$$

and, in terms of the atomic pseudopotentials $V_1$ and $V_2$ of the two ions in the cell,

$$V_G^{(S)} = \frac{1}{2}(V_G^{(1)} + V_G^{(2)}) \,,$$

$$V_G^{(A)} = \frac{1}{2}(V_G^{(1)} - V_G^{(2)}) \,,$$

where

$$V_{G,i} = \frac{2}{\Omega_{\mathrm{a}}} \int V_i(\mathbf{r})\, \mathrm{e}^{-\mathrm{i}\mathbf{G}\cdot\mathbf{r}} \mathbf{dr},$$

where $i = 1,2$ and $\Omega_{\mathrm{a}}$ is the atomic volume. We see immediately that for the diamond structure (Si and Ge), $V^{(A)} = 0$. Moreover, $S^{(S)}(2,0,0) = 0$, so that we do not need $V_G$ for $G^2 = 4$. Therefore, for

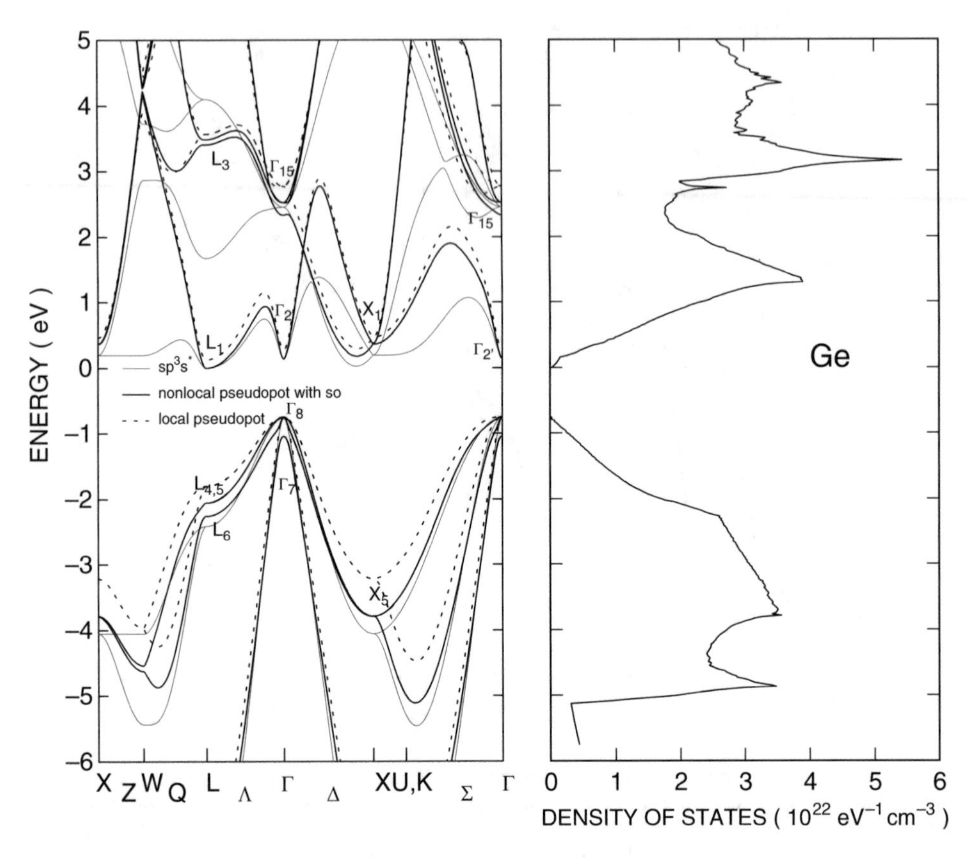


**Fig. 5.5** As in Fig. 5.4, but for Ge

Si and Ge we need only the symmetric form factors $V_{G^2=3}$, $V_{G^2=8}$ an $V_{G^2=11}$. For III–V compound semiconductors we need also the antisymmetric components, including $V_{G^2=4}$. In total, we need three real numbers for Ge and Si, three complex numbers (i.e., six real numbers) for III–V semiconductors.

The first such calculation—for cubic and zincblende type semiconductors of interest when dealing with semiconductor devices—was performed by Cohen and Bergestresser [20] for 14 semiconductors of the diamond and zincblende structure. For a given **k**, a set of **G**-vectors was employed, so that $\hbar^2|\mathbf{k}+\mathbf{G}|^2/(2m_{el}) < E_{cutoff}$. The proper choice of this energy "cut-off energy," $E_{cutoff}$, gave a maximum magnitude $G$ of about 7, resulting in about 20 **G**-vectors (so they employed "20 plane-waves"). This resulted in an eigenvalue problem, Eq. (5.2) of rank 20. For each **k**-vector, they obtained 20 eigenvalues. The first 4 give the 4 values $E_n(\mathbf{k})$ of the valence bands, with $n = 1,4$. The other eigenvalues correspond to the 16 conduction bands of lowest-energy. By moving **k** along symmetry directions in the first Brillouin Zone, they obtained plots similar to those shown in Figs. 5.4, 5.5, and 5.6.

Today, solving numerically large eigenvalue problems is routine, so that much larger matrices and typically much larger values for the cutoff energy $E_{cutoff}$ are chosen, resulting in 100–200 plane waves. In Chap. 7 we shall see that the study on nanostructures, handled by considering them periodic with a large period, requires including a number of plane waves of the order of $10^4$ to $10^6$. Algorithms based on fast Fourier transforms (FFT) avoid the need to store these huge matrices while "spectral folding methods" limit the calculations only to eigenvalues within a narrow energy range of interest, thus saving both memory and computation time. As we have already explained, the form factors are determined empirically, by fitting the obtained band structure to experimental information regarding

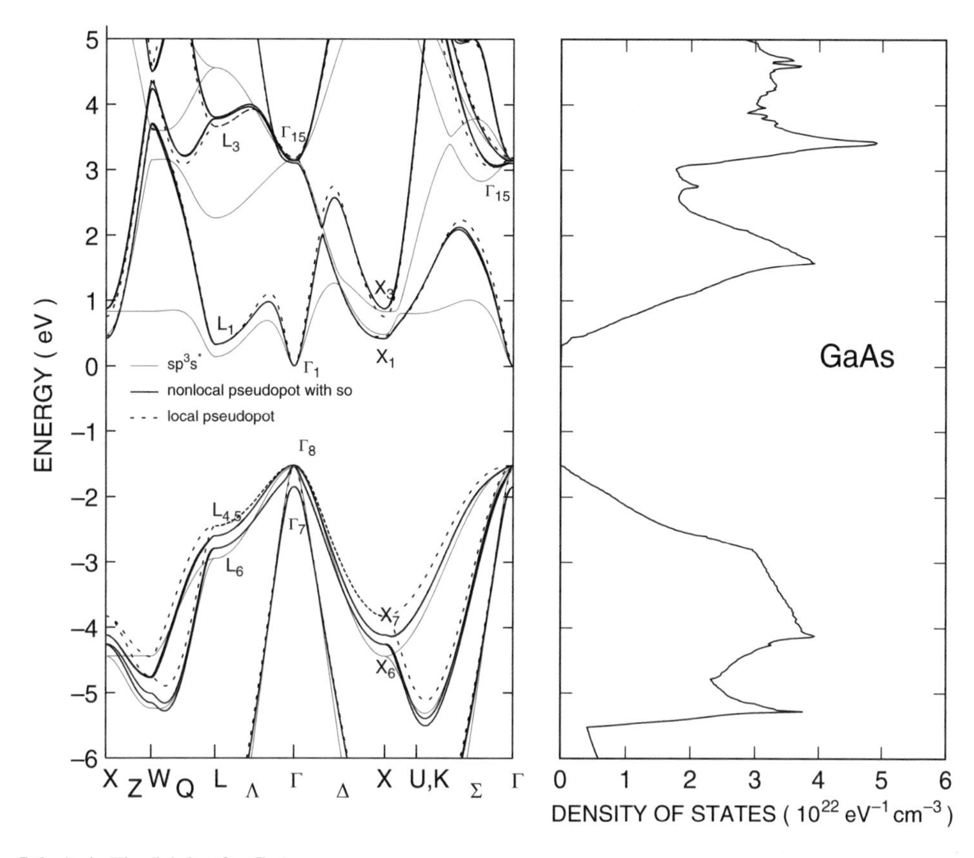


**Fig. 5.6** As in Fig. 5.4, but for GaAs

the density of states, the values of the energy splittings at symmetry points, and also about the effective masses (which we shall discuss below) at the top of the valence bands and the bottom of the conduction bands. The plots shown here have been obtained using about 120 plane waves and nonlocal empirical pseudopotentials of the form Eq. (5.19) and accounting also for spin–orbit interaction (discussed below), as given by Chelikowski and Cohen [17].

## 5.6 Qualitative Features of the Band Structure of Semiconductors

### *5.6.1 The Valence Bands in the Nearly-Free-Electron Model*

It is useful to consider the general features of the four (doubly degenerate) valence bands of fcc semiconductors. First of all, the first Brillouin Zone (BZ) of the fcc lattice is shown in Fig. 3.3 on page 44. It is the volume enclosed by the planes $k_x = \pm 1$, $k_y = \pm 1$, $k_z = \pm 1$, (the "little squares" parallel to the family of $\{1,0,0\}$ planes) and the $\{1,1,1\}$ planes $\pm k_x \pm k_y \pm k_z = 1.5$ (all **k**-vectors measured in units of $2\pi/a$). The principal symmetry points we shall consider are: (1) the zone-center, $\mathbf{k} = (0,0,0)$, called the $\Gamma$ point; (2) the 6 points $\mathbf{k} = (\pm 1,0,0)$, $(0,\pm 1,0$, and $(0,0,\pm 1)$, called the X points; the 8 points $\mathbf{k} = (1/2)(\pm 1,\pm 1,\pm 1)$, called the L points. Additional symmetry points shown are the K and U symmetry points (which are actually equivalent) and the W symmetry point. The 48-fold symmetry of the point group $O_h$ implies that a volume only 1/48 the size of the entire BZ is

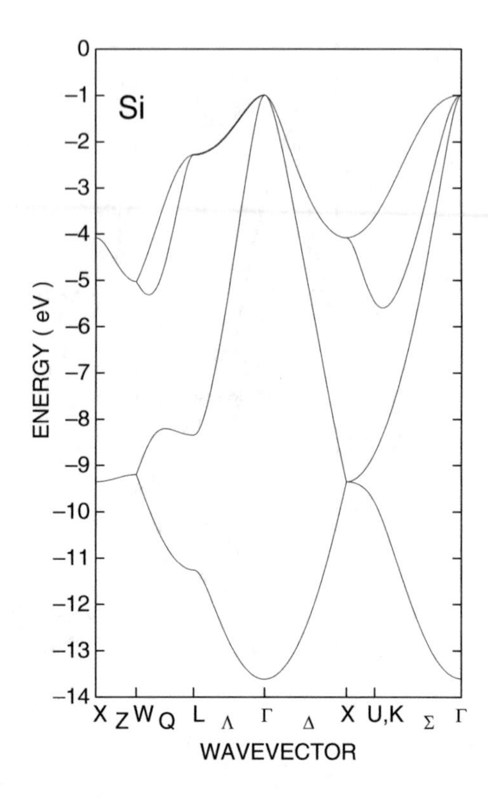


**Fig. 5.7** The structure of the valence bands of Si calculated using nonlocal pseudopotentials. Note the nearly free-electron dispersion of the lowest-energy band

sufficient to describe the band structure. This volume, defined by all points in the first octant ($k_x > 0$, $k_y > 0$, $k_z > 0$) such that $k_x + k_y + k_z < 1.5$, $1 \geq k_x \geq k_y \geq k_z$, is called the *irreducible wedge* of the first BZ.

Assume that the form factors $V_{\mathbf{G}}$ are small, so that, from Eq. (5.3), in the first band ($\mathbf{G}' = 0$) we have

$$E(\mathbf{k}) \approx \frac{\hbar^2 k^2}{2m_{\mathrm{el}}} , \tag{5.51}$$

as for free electrons. This "free electron" dispersion must be still be made compatible with the crystal structure, so it's actually called the *nearly-free electron model.* Equation (5.51) describes a parabolic band. Indeed in most semiconductors the first valence band is quite parabolic (see Figs. 5.3 or 5.7 on pages 86 and 90). We know that there are four valence bands, since diamond and zincblende semiconductors have four valence electrons and there are two atoms in each WS cell. Thus, the eight valence electrons will occupy four valence bands, two electrons in each band. Within this simple model, it is possible to calculate the Fermi energy, as it parallels the calculation in the case of the free-electron gas: At zero temperature, electrons will occupy a sphere of radius $k_{\mathrm{F}}$ such that:

$$\frac{2}{(2\pi)^3} \int_{k<k_{\mathrm{F}}} d\mathbf{k} = \frac{k_{\mathrm{F}}^3}{3\pi^2} = \frac{8}{\Omega_{\mathrm{WS}}} , \tag{5.52}$$

where $\Omega_{\mathrm{WS}} = a^3/4$ is the volume of the WS cell. Therefore, the Fermi energy will be

$$E_{\mathrm{F}} = \frac{\hbar^2 k_{\mathrm{F}}^2}{2m_{\mathrm{el}}} = \frac{\hbar^2}{2m_{\mathrm{el}}} \left( 3\pi^2 \frac{8}{\Omega_{\mathrm{WS}}} \right)^{2/3} \approx 12 \text{ eV} \tag{5.53}$$

for Si ($a = 0.543$ nm). As shown in the plot below, the width of the valence bands is indeed consistent with this free-electron estimate.

As $\mathbf{k}$ moves away from the zone-center at $\Gamma$ towards the edge of the zone, in general we would find a gap. Parenthetically, we note that Si has some strange properties at the zone-edge at X because of its time-reversal symmetry related to the fact that both ions in the cell are Si ions and there is an additional reflection symmetry. In the so-called *extended-zone scheme*, we would continue outside the first BZ with the parabolic dispersion. In the plot below, note how at the point X the first band "bounces back" into the BZ. This is the *reduced-zone scheme*: Instead of going into the second BZ, we remain within the first BZ, but set $\mathbf{G}' = (\pm 1, \pm 1, \pm 1)(2\pi/a)$ into Eq. (5.3), if we hit the zone-edge at L, set $\mathbf{G}' = (\pm 2, 0, 0)(2\pi/a)$ (and equivalent vectors) if we hit the zone-edge at X, etc. The second band continues to be parabolic, but the third and fourth valence bands are more complicated, because moving into the third and fourth BZs requires adding $\mathbf{G}$-vectors which are not simply along the direction in which we move. In general, these bands are much flatter, since the $\mathbf{k}$-vector does not change appreciably as it crosses the third and fourth zones along, say, the [100] direction. This is why these bands appear very different from free-electron bands.

### 5.6.2 The Origin of the Energy Gaps

Let's consider again Eq. (5.2) in a simplified context to understand the origin of the band gaps. Following Ridley [21], we rewrite that equation:

$$\left[\frac{\hbar^2}{2m_{\mathrm{el}}}|\mathbf{k}+\mathbf{G}|^2 - E(\mathbf{k})\right] u_{\mathbf{kG}} + \sum_{\mathbf{G}'} V_{\mathbf{GG}'}\, u_{\mathbf{kG}'} = 0\,. \tag{5.54}$$

We saw before that when the form factors are weak the solution of Eq. (5.54) is given approximately by

$$E(\mathbf{k}) \approx \frac{\hbar^2}{2m_{\mathrm{el}}}\, |\mathbf{k}+\mathbf{G}|^2\,, \tag{5.55}$$

provided $u_{\mathbf{kG}} \neq 0$. Let's consider first the nondegenerate case in which $u_{\mathbf{kG}} \neq 0$ only for one particular $\mathbf{G}$-vector, say $\mathbf{G}_a$, and set $E_a = \hbar^2|\mathbf{k}+\mathbf{G}_a|^2/(2m_{\mathrm{el}})$. We can also set $u_{\mathbf{kG}_a} = 1$ without loss of generality, since we are free to multiply the wavefunction by an arbitrary (normalization) constant. Then, a first-order correction to the solution Eq. (5.55) will be obtained by computing the first-order corrections to the coefficients $u_{\mathbf{kG}_b}$, with $\mathbf{G}_b \neq \mathbf{G}_a$, coefficients which were assumed to vanish at zero-th order:

$$u_{\mathbf{kG}_b} = \sum_{\mathbf{G}'} \frac{V_{\mathbf{G}_b-\mathbf{G}'}\, u_{\mathbf{kG}'}}{\hbar^2|\mathbf{k}+\mathbf{G}'|^2/(2m_{\mathrm{el}}) - E_b} \approx \frac{V_{\mathbf{G}_b-\mathbf{G}_a}\, u_{\mathbf{kG}_a}}{E_a - E_b}\,, \tag{5.56}$$

having set $E_b = \hbar^2|\mathbf{k}+\mathbf{G}_b|^2/(2m_{\mathrm{el}})$. Inserting this result into Eq. (5.54) we obtain

$$E(\mathbf{k}) \approx E_a + \sum_{\mathbf{G}'} \frac{|V_{\mathbf{G}'-\mathbf{G}_a}|^2}{E_a - E'}\,, \tag{5.57}$$

where $E' = \hbar^2|\mathbf{k}+\mathbf{G}'|^2/(2m_{\mathrm{el}})$. We see once more that bands "repel": The proximity of other bands (that is: other values of $\mathbf{G}$ in the reduced-zone scheme) at higher energies ($E' > E_a$) depresses the value of $E_a$. Bands at lower energies have the opposite effect. All bands push the $a$-band away, energetically.

Consider now the simple degenerate case in which we can obtain the same energy via Eq. (5.55) for two distinct vectors $\mathbf{G}_a$ and $\mathbf{G}_b$. We must consider the system

$$\begin{cases} [E_a - E(\mathbf{k})]u_{\mathbf{k}\mathbf{G}_a} + V_{\mathbf{G}_a-\mathbf{G}_b}\, u_{\mathbf{k}\mathbf{G}_b} = 0 \\ [E_b - E(\mathbf{k})]u_{\mathbf{k}\mathbf{G}_b} + V_{\mathbf{G}_b-\mathbf{G}_a}\, u_{\mathbf{k}\mathbf{G}_a} = 0 \end{cases} . \tag{5.58}$$

Setting $\Delta G = |\mathbf{G}_a - \mathbf{G}_b|$, the requirement that the determinant of the system of equations (5.58) vanishes yields

$$[E_a - E(\mathbf{k})][E_b - E(\mathbf{k})] - |V_{\Delta G}|^2 = 0 , \tag{5.59}$$

with solutions:

$$E(\mathbf{k}) = \frac{1}{2}(E_a + E_b) \pm \frac{1}{2}[(E_a + E_b)^2 - 4(E_a E_b - |V_{\Delta G}|)^2]^{1/2} . \tag{5.60}$$

Recall now our assumptions that we have two degenerate bands at zero-th order. Therefore, set $E_a = E_b$ above and

$$E(\mathbf{k}) = E_a \pm |V_{\Delta G}| , \tag{5.61}$$

which shows that we have a gap of magnitude $2|V_{\Delta G}|$ at this **k**-point. As we saw above when discussing the nearly-free-electron model, often degenerate free-electron values are obtained at the zone-edge. Indeed, these are exactly the points where we usually see the formation of gaps.

### 5.6.3 The Spin–Orbit Interaction

A final effect of significant importance in dealing with the structure of the bands of semiconductors (especially the valence bands near their maximum) is the so-called *spin–orbit* interaction. This is an effect which also matters in atoms. It originates from the fact that in the laboratory frame of reference, the charge of the nucleus (with core states) generates a pure electrostatic field, but the orbiting electron sees this field as a magnetic field, as if it were generated by the nucleus which appears to be a moving charge when "riding" with the electron around the nucleus. The electron also has a magnetic dipole generated by its spin. The spin–orbit interaction is just the interaction between the spin-induced magnetic dipole and the magnetic field seen by the electron. This interaction lifts the degeneracy between spin down and spin up electronic states in the absence of inversion symmetry.

The Hamiltonian for this interaction can be obtained classically: If $\mu$ is the magnetic moment associated with the electron spin and **B** is the magnetic field caused by the orbital motion of the electron, their interaction energy will be

$$H_{\text{so}} = -\mu \cdot \mathbf{B} . \tag{5.62}$$

Now, recall that the magnetic field seen by a particle moving with velocity **v** in the presence of an electric field **E** is

$$\mathbf{B} = -\frac{\mathbf{v}}{c} \times \mathbf{E} . \tag{5.63}$$

* *Although this derives from the relativistic transformation of the electromagnetic field, it can be derived simply using the requirement that in the frame in which the electron is at rest, only the Coulomb force of the charged nucleus acts on it. However, in the frame in which the electron moves with velocity* $\mathbf{v}$ *and the nucleus is at rest, only the Lorentz force acts on the electron. The two forces must be equal. From this, Eq. (5.63) follows.*

Writing $\mathbf{v} = \mathbf{p}/m_{\rm el}$ for the electron velocity in terms of its momentum, we have

$$H_{\rm so} = -\frac{1}{m_{\rm el}c}\,\mu\cdot\mathbf{E}\times\mathbf{p}\,. \tag{5.64}$$

Now recall that the electric field is just the gradient of the electrostatic potential $\phi$ and its radial component is the only non-vanishing component:

$$\mathbf{E} = -\frac{1}{r}\frac{\mathrm{d}\phi}{\mathrm{d}r}\,\mathbf{r}\,, \tag{5.65}$$

so that Eq. (5.64) becomes

$$H_{\rm so} = \frac{1}{2m_{\rm el}cr}\frac{\mathrm{d}\phi}{\mathrm{d}r}\,\mu\cdot\mathbf{r}\times\mathbf{p}\,, \tag{5.66}$$

with an additional factor of 1/2, which is discussed below. Now notice that $\mathbf{L} = \mathbf{r}\times\mathbf{p}$ is the orbital angular momentum and $\mu = e\mathbf{S}/(2m_{\rm el}c)$, where $\mathbf{S}$ is the spin angular momentum operator.

* *See Jackson's text [22], Sect. 5.6, to see that the magnetic moment* $\mathbf{M}$ *of a particle of charge* $q$*, mass* $m$*, and angular momentum* $\mathbf{L}$ *is given by* $\mathbf{M} = q\,\mathbf{L}/(2m_{\rm el}c)$*. This semiclassical result actually is violated in the present case: The magnetic moment caused by the electron spin is a little more than a factor of 2 larger than expected. The explanation of this* anomaly *is one of the most striking successes of the relativistic Dirac equation—which predicts a factor of 2—and of Quantum Electrodynamics (QED), which explains the remaining tiny discrepancy. However, this factor of almost 2 is almost exactly canceled by the so-called* Thomas factor*: We have expressed the interaction energy Eq. (5.64) in the frame in which the electron is at rest. Moving to the laboratory frame in which the nucleus is at rest, results in the appearance of an extra factor* $1/2$ *in the expression for the potential energy of the interaction. This is caused by the relativistic time dilatation between the electron and the nuclear rest frames. This time dilatation decreases the (Thomas) precession frequency of the electron, and so the interaction energy. Thus, except for a very small correction due to the fact that the gyro-magnetic ratio of the electron is 2.00232..., and not exactly 2, our result will be quantitatively correct.*

Thus, the Hamiltonian (now an operator) for the spin–orbit interaction is

$$\hat{H}_{\rm so} = \frac{e}{2m_{\rm el}^2c^2r}\frac{\mathrm{d}\phi}{\mathrm{d}r}\,\widehat{\mathbf{L}}\cdot\widehat{\mathbf{S}}\,. \tag{5.67}$$

We have briefly discussed before the properties of the angular momentum operators in Quantum Mechanics. Here we summarize just a few results applied to the three topmost valence states, which are $p$-like waves:

1. The eigenvalues and eigenstates $|l\rangle$ of the square of the general angular momentum operator $\widehat{L}^2$ are given in terms of an integer (orbital, spin of Bosons) or half-integer (spin of Fermions) index $l$:

$$\widehat{L}^2\,|lm\rangle = l(l+1)\,\hbar^2\,|lm\rangle\,.$$

2. The three components $\widehat{L}_x$, $\widehat{L}_y$, and $\widehat{L}_z$ of the angular momentum do not commute with each other but each of them commutes with $\widehat{L}^2$. Therefore, eigenstates of the total angular momentum are always degenerate, corresponding to different eigenstates of one of the components above, usually taken to be $\widehat{L}_z$. These eigenstates $|lm\rangle$ are given by the eigenvalue equation:

$$\widehat{L}_z\,|lm\rangle = m_{\mathrm{el}}\,\hbar\,|lm\rangle\;,$$

   $m$ being an integer or half-integer ranging from $-l$ to $+l$.
3. Angular momenta add vectorially, as in Classical Mechanics.

Now, in the simplest approximation the first valence band is made up of $s$-states with zero orbital angular momentum, so that the spin–orbit interaction does not alter its energy. On the contrary, the three highest-energy valence bands have orbital angular momentum $l = 1$ and, obviously, the electrons have spin angular momentum $s = \pm 1/2$. The total angular momentum $\mathbf{J}$ results from the vector sum of $\widehat{\mathbf{L}}$ and $\widehat{\mathbf{S}}$, so that

$$\widehat{J}^2 = \widehat{L}^2 + \widehat{S}^2 + 2\widehat{\mathbf{L}}\cdot\widehat{\mathbf{S}}\,.$$

Therefore, the expectation value $\langle\mathbf{L}\cdot\mathbf{S}\rangle$ can be expressed in terms of the known expectation values of $\widehat{\mathbf{J}}$, $\widehat{\mathbf{L}}$, and $\widehat{\mathbf{S}}$ as follows:

$$\langle\widehat{\mathbf{L}}\cdot\widehat{\mathbf{S}}\rangle = \frac{1}{2}\langle\widehat{J}^2 - \widehat{L}^2 - \widehat{S}^2\rangle = \frac{\hbar^2}{2}[j(j+1) - l(l+1) - s(s+1)]\,. \tag{5.68}$$

For $p$-states, we can have spin and orbital angular momentum parallel, so that we have a fourfold—two bands $\times$ two spins—degenerate state ($j = 3/2$ and $m = \pm 3/2, \pm 1/2$). Or spin and orbital angular momentum may be anti-parallel, so that we have a doubly degenerate—two spins only—state ($j = 1/2$ and $m = \pm 1/2$). Therefore, finally, from Eqs. (5.67) and (5.68) we deduce for the two states $j = 3/2$ an expectation value $\langle\widehat{\mathbf{L}}\cdot\widehat{\mathbf{S}}\rangle = \hbar^2/2$, whereas for the $j = 1/2$ state we have $\langle\widehat{\mathbf{L}}\cdot\widehat{\mathbf{S}}\rangle = -\hbar^2$, so that the two $j$-states are split by an energy:

$$\Delta_{\mathrm{so}} = \frac{3\hbar^2 e}{4m_{\mathrm{el}}^2 c^2 r}\,\frac{\mathrm{d}\phi}{\mathrm{d}r}\,. \tag{5.69}$$

The energy of the doubly degenerate $j = 3/2$ states is shifted up by an amount $\Delta_{\mathrm{so}}/3$ while that of the $j = 1/2$ state is shifted down by an amount $-2\Delta_{\mathrm{so}}/3$. The $j = 3/2$ states correspond to band usually called the *heavy-hole* and *light-hole* bands, the lone $j = 1/2$ state is called the *split-off band*, for obvious reasons. The magnitude of the spin–orbit energy $\Delta_{\mathrm{so}}$ varies from a negligibly small values in C, to a few meVs (44 meV in Si), to several tenths of one eV ($\sim$300 meV in Ge, and to the order of 1 eV or more in heavier elements, such as Sn or Pb). This is because the strength of the spin–orbit interaction obviously depends on the expectation value of the radius of the electronic orbit. Therefore, the closer the electrons to the nucleus, the stronger is the term $(\mathrm{d}\phi/\mathrm{d}r)/r$ in Eq. (5.69). This is why heavier elements usually have a larger spin–orbit coupling. Comparing the band structure of Si (Fig. 5.4 on page 87) with that of Ge (Fig. 5.5 on page 88), we can see the effect of the spin–orbit interaction: Note for Ge the large energy difference between the bands at the Γ-point for the three highest-energy valence bands. The splitting extends throughout the BZ. In the much lighter Si (Fig. 5.4 on page 87), the splitting is much smaller, barely visible in the energy scale of the plot.

### 5.6.4 The $\mathbf{k} \cdot \mathbf{p}$ Structure of the Bands of fcc Semiconductors Near Band-Extrema

We have already seen on page 83 how $\mathbf{k} \cdot \mathbf{p}$ perturbation theory can be used to calculate the effective masses in the neighborhood of the $\Gamma$ point. We also saw that degenerate perturbation theory is necessary. Kane has obtained the following expressions (which we shall not derive) using $4 \times 4$ matrices (three valence bands and the lowest-energy conduction band at $\Gamma$) [10]:

1. The conduction band at $\Gamma$ is spherical and (at the lowest order) parabolic with a dispersion approximately given by

$$E_{\rm c}(\mathbf{k}) = E_{{\rm gap},\Gamma} + \frac{\hbar^2 k^2}{2m_{\rm c}^*} , \quad (5.70)$$

where the conduction band effective mass is given by

$$\frac{m_{\rm el}}{m_{\rm c}^*} = 1 + \frac{2p_{{\rm c},v}^2}{3m} \left( \frac{2}{E_{{\rm gap},\Gamma}} + \frac{1}{E_{{\rm gap},\Gamma} + \Delta_{\rm so}} \right) , \quad (5.71)$$

where $E_{{\rm gap},\Gamma}$ is the direct gap at the $\Gamma$ point, and $p_{\rm c,v}$ is the momentum matrix element between the $s$ conduction band and the $p$ valence band(s). For Ge, $2p_{\rm v,c}^2/m_{\rm el} = 22.5$ eV; for GaAs, $2p_{\rm v,c}^2/m_{\rm el} = 21.5$ eV.

2. The "heavy hole" dispersion is given by

$$E_{\rm hh}(\mathbf{k}) = \frac{\hbar^2 k^2}{2m_{\rm hh}^*} , \quad (5.72)$$

with $m_{\rm hh}^* = 1$. Note the "wrong" sign of the dispersion: The effect of more remote bands is required to obtain the correct sign for the mass.

3. The "light hole" dispersion is given by

$$E_{\rm lh}(\mathbf{k}) = -\frac{\hbar^2 k^2}{2m_{\rm lh}^*} , \quad (5.73)$$

with

$$\frac{m_{\rm el}}{m_{\rm lh}^*} = \frac{4p_{\rm cv}^2}{3mE_{{\rm gap},\Gamma}} - 1 . \quad (5.74)$$

4. The "split-off hole" dispersion is given by

$$E_{\rm so}(\mathbf{k}) = -\Delta_{\rm so} - \frac{\hbar^2 k^2}{2m_{\rm so}^*} , \quad (5.75)$$

with

$$\frac{m_{\rm el}}{m_{\rm so}^*} = \frac{2p_{\rm cv}^2}{3m(\Delta_{\rm so} + E_{{\rm gap},\Gamma})} - 1 . \quad (5.76)$$

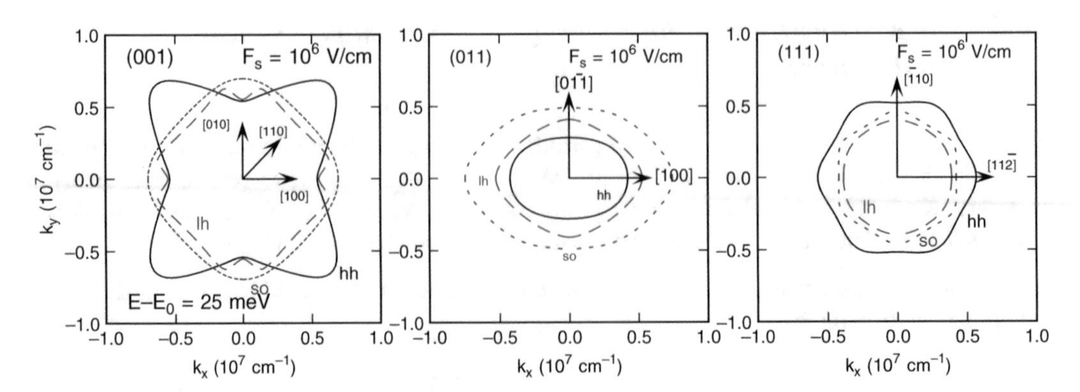


**Fig. 5.8** Equi-energy contours for the three highest-energy valence bands of a Si inversion layer subject to a confining perpendicular field of $10^6$ V/cm [23]. At *left*, we illustrate the *equi-energy lines* for a surface on the $(1,0,0)$ crystallographic plane; at *center* at *right* are the cases for the $(1,1,0)$ and for the $(1,1,1)$ surfaces, respectively, Although this plot illustrates the shape of the *equi-energy contour lines* in a more complicated situation, qualitatively it illustrates the strong anisotropy and warping of the valence bands for fcc group IV semiconductors or for III–V compound semiconductors. (Reprinted with permission from M.V. Fischetti, Z. Ren, P.M. Solomon, M. Yang, and K. Rim, J. Appl. Phys. vol. 94, p. 1079 (2003). Copyright 2003, AIP Publishing LLC)

5. *Warping*. Accounting for the effect of additional bands, the two topmost valence bands (hh and lh) are degenerate at $\mathbf{k} = 0$ and are "warped" with dispersion

$$E_{\rm hh/lh}(\mathbf{k}) = -\frac{\hbar^2}{2m_{\rm el}} \{Ak^2 \pm [B^2k^4 + C^2(k_x^2k_y^2 + k_y^2k_z^2 + k_z^2k_x^2)]^{1/2}\} , \tag{5.77}$$

where in Si $A = 4.0$, $B = 1.1$, and $C = 4.1$, which correspond to $m^*_{\rm hh} \approx 0.49m_{\rm el}$ and $m^*_{\rm lh} \approx 0.16m_{\rm el}$. For Ge, instead, $A = 13.1$, $B = 8.3$, and $C = 12.5$, which correspond to $m^*_{\rm hh} \approx 0.28m_{\rm el}$ and $m^*_{\rm lh} \approx 0.044m_{\rm el}$. Note, however, that the valence bands are so warped (as shown in Fig. 5.8) and nonparabolic that the use of the effective mass is always questionable. Figure 5.8 shows equi-energy surfaces for the Si valence bands, illustrating their strong anisotropy.

6. *Nonparabolicity*. As one moves away from the $\Gamma$ point, the dispersion becomes increasingly non-spherical and nonparabolic. The latter effect is often captured to the leading order by a *nonparabolicity parameter* $\alpha_{\rm np}$, such that, in terms of the parabolic dispersion $\gamma(\mathbf{k}) = \hbar^2k^2/(2m^*)$, the nonparabolic dispersion, $E(\mathbf{k})$, becomes

$$E(\mathbf{k}) \approx \gamma(\mathbf{k})[1 + \alpha_{\rm np}\gamma(\mathbf{k})] . \tag{5.78}$$

Since the parameter $\alpha_{\rm np}$ is almost always negative, we see that the dispersion tends to be "flattened" by nonparabolic corrections.

7. *Conduction dispersion at the L and X symmetry points.* The minima of the conduction bands at (or close to) the symmetry points L and X are described by prolate ellipsoidal equi-energy surfaces characterized by a longitudinal ($m_{\rm L}$) and a transverse ($m_{\rm T}$) effective mass:

$$E(\mathbf{k}) = \frac{\hbar^2}{2}\left(\frac{k_{\rm L}^2}{m_{\rm L}} + \frac{k_{\rm T1}^2 + k_{\rm T2}^2}{m_T}\right) , \tag{5.79}$$

where $k_{\rm L}$, $k_{\rm T1}$, and $k_{\rm T2}$ are the components of the $\mathbf{k}$-vector along the longitudinal and the two transverse axes of the ellipsoid. The minima of conduction bands of Si (near the X symmetry points) and of Ge (near the L symmetry points) are of this type.

Figure 5.8 illustrates these peculiar properties for the valence bands.

## Problems

**5.3. (Hole Effective Masses)** Using Eq. (5.78) and the values for the $\mathbf{k}\cdot\mathbf{p}$ parameters $A$, $B$, and $C$ listed there for Si and Ge, calculate the effective masses for the heavy- and light-hole bands along the $[100]$, $[110]$, and $[111]$ directions. Note that the values quoted here are averaged over all directions, so your results along specific directions do not have to be identical to these "average" values.

**5.4. (Group Velocity and Nonparabolicity)** Using the approximate $\mathbf{k}\cdot\mathbf{p}$ nonparabolic correction, Eq. (5.78), assuming $\gamma(\mathbf{k}) = \hbar^2k^2/(2m^*)$, derive the expression for the electron group velocity $\nabla_{\mathbf{k}}E(\mathbf{k})/\hbar$, as a function of $k$ to first-order in the nonparabolicity parameter $\alpha_{\rm np}$. The physical meaning of the group velocity will be discussed in Chap. 8, Sect. 8.1.3.

## References


1. W.A. Harrison, *Electronic Structure and the Properties of Solids* (Dover, New York, 1989)
2. H. Hellman, A new approximation method in the problem of many electrons. J. Chem. Phys. (Karpow?Institute for Physical Chemistry, Moscow) **3**, 61 (1935)
3. B.J. Austin, V. Heine, L.J. Sham, General theory of pseudopotentials. Phys. Rev. **127**, 276 (1962)
4. W.A. Harrison, *Pseudopotentials in the Theory of Metals*. Frontiers in Physics, vol. 25 (University of Virginia Press, New York, 1966)
5. V. Heine, The pseudopotential concept. Solid State Phys. (Acedemic Press) **24**, 1 (1970)
6. E. Kaxiras, *Atomic and Electronic Structure of Solids* (Cambridge University Press, Cambridge, 2003)
7. C. Pisani, *Quantum-Mechanical Ab-initio Calculation of the Properties of Crystalline Materials*. Lecture Notes in Chemistry (Springer, Berlin/Heidelberg, 1996)
8. J.C. Slater, G.F. Koster, Simplified LCAO method for the periodic potential problem. Phys. Rev. **94**, 1498 (1954)
9. J.M. Luttinger, W. Kohn, Motion of electrons and holes in perturbed periodic fields. Phys. Rev. **97**, 869 (1955)
10. E.O. Kane, Band structure of indium antimonide. J. Phys. Chem. Solids **1**, 249 (1957)
11. C. Kittel, *Quantum Theory of Solids*, 2nd edn. (Wiley, New York, 1987)
12. P. Hohenberg, W. Kohn, Inhomogeneous electron gas. Phys. Rev. **136**, B864 (1964)
13. W. Kohn, L.J. Sham, Self-consistent equations including exchange and correlation effects. Phys. Rev. **140**, A1133 (1965)
14. C. Herring, A new method for calculating wave functions in crystals. Phys. Rev. **57**, 1169 (1940)
15. J. Callaway, Orthogonalized plane wave method. Phys. Rev. **97**, 933 (1955)
16. P. Friedel, M.S. Hybertsen, M. Schlüter, Local empirical pseudopotential approach to the optical properties of Si/Ge superlattices. Phys. Rev. B **39**, 7974 (1989)
17. J.R. Chelikowsky, M.L. Cohen, Nonlocal pseudopotential calculations for the electronic structure of eleven diamond and zinc-blende semiconductors. Phys. Rev. **14**, 556 (1976)
18. P. Vogl, H.P. Hjalmarson, J.D. Dow, A semi-empirical tight binding theory of the electronic structure of semiconductors. J. Phys. Chem. Solids **44**, 365 (1983)
19. P. Yu, M. Cardona, *Fundamentals of Semiconductors: Physics and Materials Properties*, 3rd edn. (Springer, Berlin/Heidelberg, 2005)
20. M.L. Cohen, T.K. Bergstresser, Band structures and pseudopotential form factors for fourteen semiconductors of the diamond and zinc-blende structures. Phys. Rev. **141**, 789 (1966)
21. B.K. Ridley, *Quantum Processes in Semiconductors* (Oxford University Press, Oxford, 2000)
22. J.D. Jackson, *Classical Electrodynamics* (Wiley, New York, 1962)
23. M.V. Fischetti, Z. Ren, P.M. Solomon, M. Yang, K. Rim, Six-band $\mathbf{k}\cdot\mathbf{p}$ calculation of the hole mobility in silicon inversion layers: dependence on surface-orientation, strain, and silicon-thickness. J. Appl. Phys. **94**, 1079 (2003)

# Chapter 6
# Density Functional Theory

## 6.1 Overview

For many problems in physics, it is important to know the properties of the ground state. And while in principle it is possible to determine the exact ground state for a given many-body Hamiltonian, in practice it is impossible for almost any nontrivial system. Fortunately, in many systems a good approximation of the ground state properties can be determined computationally using Density Functional Theory (DFT).

In this chapter we present an introduction to DFT. The emphasis will be on its applications to electronic transport, but for those interested in a more in-depth treatment of DFT, we refer to specialized text books. Those already cited in the introduction of Chap. 3 represent a good example [1, 2]. In Sect. 6.2, we present the theoretical justification of DFT through the Hohenberg–Kohn theorem and the most popular way of evaluating the density functional through the Kohn–Sham formalism. In the subsequent sections, we show how DFT can be used to compute the equilibrium positions of the atoms and the band structure.

## 6.2 Theoretical Basis for DFT

The ground state of any system is the lowest-state eigenstate of its Hamiltonian. The ground state wavefunction $\Psi_0$ is the solution of the equation:

$$\widehat{H}\,\Psi_0(\mathbf{r}_1,\mathbf{r}_2,\ldots,\mathbf{r}_N) = E_0\,\Psi_0(\mathbf{r}_1,\mathbf{r}_2,\ldots,\mathbf{r}_N), \tag{6.1}$$

where $E_0$ is the smallest eigenvalue of the Hamiltonian $\widehat{H}$ that describes the electrons in the lattice using the Born–Oppenheimer approximation discussed in Chap. 4, Sect. 4.2.1:

$$\widehat{H} = \sum_i \left[ -\frac{\hbar^2}{2m_{\mathrm{el}}}\nabla_i^2 + V^{(\mathrm{lat})}(\mathbf{r}_i) + \sum_{j<i} \frac{e^2}{4\pi\epsilon_0|\mathbf{r}_i-\mathbf{r}_j|} \right], \tag{6.2}$$

where $V^{(\mathrm{lat})}(\mathbf{r})$ is, as usual, the potential energy due to the ions.

The problem, of course, lies in the fact that the potential $V^{(\mathrm{lat})}(\mathbf{r})$ is the sum of the ionic potentials screened by the electron charge. This charge density is a function of the electron wavefunctions, which are the unknowns of the problem. In addition, the Coulomb repulsion among the electrons (the last



M. Fischetti, W.G. Vandenberghe, *Advanced Physics of Electron Transport in Semiconductors and Nanostructures*, Graduate Texts in Physics, DOI 10.1007/978-3-319-01101-1_6

term in Eq. (6.2)) can be approximated as a Hartree term, plus exchange and correlation corrections that are difficult to express for an inhomogeneous electron gas. The major idea of DFT lies in the fact that, in the ground state of this system, these exchange and correlation effects depend only on the electron (charge) density of the system. This important result is part of what are known as the *Hohemberg–Kohn theorems* [3] and leads to an important but rigorous simplification of the problem: the full potential entering Eq. (6.2) can be expressed as a Hartree term (in principle straightforward to calculate via a solution of the Poisson equation) and an additional exchange-correlation energy that is a *functional*, $F[n(\mathbf{r})]$, of the electron density *only*. Hence the term "density functional."

To put these ideas in a more precise form, the ground state electron density is

$$n(\mathbf{r}) = N \int \ldots \int \mathrm{d}\mathbf{r}_2^3 \ldots \mathrm{d}\mathbf{r}_N^3 \; |\Psi(\mathbf{r},\mathbf{r}_2,\ldots,\mathbf{r}_N)|^2 \,, \tag{6.3}$$

with $N$ the number of particles in the ground state. In the ground state, this can now be found by minimizing the "density functional" $F[n(\mathbf{r})]$. For simplicity, we do not consider the effects of spin or magnetic fields, nor do we consider the case where the ground state is degenerate.

In the absence of degeneracy, for each potential due to the ions, $V^{(\mathrm{lat})}(\mathbf{r})$, there is one ground state wavefunction, $\Psi_0$. At the same time, if $\Psi_0$ is known, $V^{(\mathrm{lat})}(\mathbf{r})$ can be determined up to a constant through Eqs. (6.1)–(6.2). Furthermore, for each $\Psi_0$ there is a unique charge density, $n(\mathbf{r})$, through Eq. (6.3). Finally, as we have already mentioned, Hohenberg and Kohn have shown, using *reductio ad absurdum*, that each charge density is uniquely related to a potential and thus to a ground state [3]. Therefore, we can conclude that there is a one-to-one correspondence between the ground state charge density and the ground state wavefunctions, $n(\mathbf{r}) \leftrightarrow \Psi_0$, so that a wavefunction functional $\Psi_0[n(\mathbf{r})]$, and so also an energy functional $F[n(\mathbf{r})]$, exists. The remaining problem, of course, is to determine the form of this functional.

Remembering that $\Psi_0$ is the ground state, $\langle \Psi \mid \widehat{H} \mid \Psi \rangle$ reaches a minimum ($E_0$) for $\Psi = \Psi_0$. Therefore, minimizing the density functional

$$F[n(\mathbf{r})] = \langle \Psi_0[n(\mathbf{r})] \mid \widehat{H} \mid \Psi_0[n(\mathbf{r})] \rangle \tag{6.4}$$

with respect $n(\mathbf{r})$ results in the ground-state charge-density for the Hamiltonian $\widehat{H}$.

Equation (6.4) gives an exact expression for the density functional. However, evaluating the exact density functional is equally computationally prohibitive as solving for the ground state wavefunction. Luckily, approximate density functionals suitable for an efficient numerical evaluation have been developed. The results obtained using these functionals generally compare favorably to experimental results.

The most commonly used expression for the density functional $F[n(\mathbf{r})]$ is its Kohn–Sham form [4]. It relies on the fact that the density of the valence electrons in a crystal changes "slowly" when compared, for example, to the fast spatial variation of the density of the strongly confined core electrons. Therefore, one may rely approximately on the results obtained from an exact quantum-mechanical treatment of the homogeneous electron gas (as discussed, for example, in Kittel's [5], Mahan's [6], or Fetter and Walecka's [7] excellent texts): The functional can be approximated by a functional of the local density, $F[n(\mathbf{r})] \sim n(\mathbf{r})^{1/3}$, as in the Thomas–Fermi model originally formulated by Dirac [8]. We shall mention below more sophisticated approaches that go beyond this approximation that appropriately called the *local density approximation* (LDA).

As a result of this discussion, we reach the Kohn–Sham equation as a single-particle Schrödinger equation [4]:

$$\left( -\frac{\hbar^2}{2m_{\mathrm{el}}} \nabla^2 + v_{\mathrm{eff}}(\mathbf{r}) \right) \phi_i(\mathbf{r}) = w_i \; \phi_i(\mathbf{r}) \tag{6.5}$$

whose solutions $\phi_i(\mathbf{r})$ are known as the Kohn–Sham orbitals and where $v_{\mathrm{eff}}(\mathbf{r})$ is an effective "local" potential that captures the effects of the Coulomb interaction together with exchange and correlation effects. Recall that the eigenvalues $w_i$ are difficult to interpret physically, as we had already discussed in the context of the Hartree–Fock approximation (see Chap. 4, Sect. 4.2.3). Another tricky assumption we have to make at this point is to assume that the charge density of the non-interacting electron system described by Eq. (6.5) coincides with that of the interacting system [with the Hamiltonian from Eq. (6.2)]. In other words, that the Slater determinant, representing the correct form of the wavefunction of a system of non-interacting particles, represents the wavefunction also of the system of interacting electrons. With this assumption, the charge density can be expressed in a simple way:

$$n(\mathbf{r}) = \sum_i^N |\phi_i(\mathbf{r})|^2 \,. \tag{6.6}$$

Calculating the kinetic energy using the Kohn–Sham orbitals, the density functional can be written as:

$$F[n(\mathbf{r})] = \sum_i \int \mathrm{d}\mathbf{r}\; \phi_i^*(\mathbf{r})\nabla^2\phi_i(\mathbf{r}) + \int \mathrm{d}\mathbf{r}\; V^{(\mathrm{lat})}(\mathbf{r})\; n(\mathbf{r})$$
$$+ \frac{e^2}{8\pi\epsilon_0}\int \mathrm{d}\mathbf{r} \int \mathrm{d}\mathbf{r}'\; \frac{n(\mathbf{r})n(\mathbf{r}')}{|\mathbf{r}-\mathbf{r}'|} + E_{\mathrm{xc}}[n(\mathbf{r})] \,. \tag{6.7}$$

Minimizing the density functional amounts to setting:

$$v_{\mathrm{eff}}(\mathbf{r}) = V^{(\mathrm{lat})}(\mathbf{r}) + \frac{e^2}{4\pi\epsilon_0}\int dr\; \frac{n(\mathbf{r}')}{|\mathbf{r}-\mathbf{r}'|} + \frac{\delta E_{\mathrm{xc}}[n(\mathbf{r})]}{\delta n(\mathbf{r})} \tag{6.8}$$

and to solving Eq. (6.5) self-consistently.

Finally, using the LDA approximation for the exchange-correlation functional $E_{\mathrm{xc}}[n(\mathbf{r})]$ [8, 9]:

$$E_{\mathrm{xc,LDA}}[n(\mathbf{r})] = \int \mathrm{d}\,\mathbf{r}\; n(\mathbf{r})\; \epsilon_{\mathrm{xc}}[n(\mathbf{r})], \tag{6.9}$$

where $\epsilon_{\mathrm{xc}}(n)$ is the exchange-correlation energy of a homogeneous electron gas, $\sim n(\mathbf{r})^{1/3}$.

An improved approximation which can still be evaluated efficiently is the *generalized gradient approximation* (GGA) [10–12]. This approximation amounts to accounting (to the next order) for corrections to the exchange-correlation energy due to the "slow" spatial variations of the electron gas that, after all, is not strictly homogeneous in a crystal. The exchange-correlation functional in this next approximation is written in the following form that includes the simplest nonlocal correction one can think of, namely, the gradient of the density:

$$E_{\mathrm{xc,GGA}}[n(\mathbf{r})] = \int \mathrm{d}\mathbf{r}\; \epsilon_{\mathrm{xc}}[n(\mathbf{r}), \nabla n(\mathbf{r})]\; n(\mathbf{r}). \tag{6.10}$$

There are reasons why this is called the "generalized" gradient approximation. Indeed, the issue is more complicated than it seems: A naïve use of an expression that includes the gradient of the density, unfortunately, violates important conservation laws (namely: particle-number conservation). Therefore, an appropriate "generalization" is in order. We shall leave it at that, since the issue is too subtle and specialized for the purpose of this book. On the same "specialized" note, we observe that another, yet more involved, set of functionals are nowadays commonly used, the so-called *hybrid functionals* [13, 14]. In this "hybrid" form, $E_{\mathrm{xc,hybrid}}[n(\mathbf{r})]$ contains the exchange term taken to be a linear combination of the exchange of a conventional DFT functional, such as LDA or PBE [15] (see below), and the exchange energy obtained from the Hartree–Fock approximation.

## 6.3 * Relevance of DFT Calculations for Electronic Transport

Computer programs that use DFT to calculate atomic positions, electronic structure, phonon spectra (and more) have now become so sophisticated that, in a way, there is almost an "industry." DFT software packages are available that come in forms based on local atomic orbitals, all electrons (not based on pseudopotential), and, of course, plane waves. In addition, several packages are available to compute the atomic pseudopotentials themselves. The discussion of this chapter demands that we present some significant results, so that we may show what DFT can really do. Moreover, we wish to extend the discussion to subjects that are directly relevant to electron transport—namely and mainly, the calculation of the vibrational properties of a crystal, phonons, since they are the source of the main scattering processes for electrons moving in solids. In order to do so, we consider one such software package, the Vienna ab initio Simulation Package (VASP) [16–19]. VASP uses a plane-wave basis set for the Kohn–Sham orbitals. To reduce the size of the basis set, an extension of the pseudopotential method, the projector-augmented plane-wave method, is used [20]. Fast-Fourier Transforms (FFTs) are used to diagonalize efficiently the Kohn–Sham Hamiltonian. Moreover, optimized pseudopotentials for each atom are available from the extensive library included in the package. We shall not discuss these sophisticated choices. However, we must provide some detail, mainly for scientific "correctness" and for the sake of the occasional expert reader who may wonder about the exact procedure followed to obtain the results shown here. But, once more, from the perspective of this book, these are just "details" that the non-expert should not be concerned about.

### 6.3.1 *Performing a DFT Calculation*

In this somewhat "specialized section," we outline the process one must follow to use such a DFT package, VASP in our case, to perform a typical calculation.

As a first step, one must specify the atomic species, of course, and the following main system parameters:

- Unit cell basis vectors ($\mathbf{a}_1$, $\mathbf{a}_2$, $\mathbf{a}_3$)
- Basis atom types and positions ($\tau_i$)
- Density functional to be used (LDA vs GGA)
- Numerical approximations: K-point grid, energy cutoff/basis set, convergence threshold

VASP draws this information from four input files: POSCAR, POTCAR, KPOINTS, and INCAR. The file POSCAR contains the unit cell parameters. POTCAR contains the pseudopotential information, that is, which atoms one is considering. KPOINTS contains the size of the grids of the **k** points, either in the irreducible wedge or in the first BZ, that are required to store wavefunctions and to perform the necessary integrations. INCAR, finally, contains the remaining information: the cutoff energy used to limit the number of **G** vectors employed (and so, the size of the eigenvalue problem to be solved); information concerning additional ground state properties to be calculated (such as the forces on each atom); requests for additional output, such as the Kohn–Sham potential or the core charge, and computational parallelization parameters.

For example, when studying crystalline silicon the input will reflect the following properties and/or computational options:

- Silicon forms an fcc lattice with a lattice constant $a = 5.431$ and its basis vectors are $(a/2, a/2, 0)$, $(a/2, 0, a/2)$ and $(0, a/2, a/2)$.
- There are two silicon atoms in the unit cell: taking the first atom at $(0, 0, 0)$, the second atom can be found at $(a/4, a/4, a/4)$.

- One may chose the most popular density functional in solid-state physics, the Perdew–Burke–Ernzerhof (PBE) implementation of the GGA [15].
- The grid of **k** points is generated automatically through a method known as "Monckhorst–Pack." For example, one may request a $10 \times 10 \times 10$ Monckhorst–Pack K-point grid and the default cutoff energy and convergence threshold. This determines the numerical accuracy required to halt the self-consistent iteration required by the solution of the Kohn–Sham nonlinear eigenvalue problem.

As a result of this input, one obtains a ground state energy of $E = -10.8\,\text{eV}$. Note that this value depends on the kind of pseudopotential chosen and is only meaningful relative to other ground state energies. The total ground state density for a cubic cell consisting of four primitive cells, averaged along the [100] direction, is shown in Fig. 6.1 (left). The largest part of the electronic charge density can be found around the ions and will have no impact on the binding energy. When only considering the valence charge, i.e., without the charge contribution of the core electrons, shown in Fig. 6.1 (right), the charge density is still strongly peaked near the ions. Only a fraction of the valence charge is involved in the binding of neighboring atoms. To further smoothen the charge density and enable a reduction in the number of plane waves, plane-wave DFT codes base their calculations on pseudopotential-based Kohn–Sham orbitals. The resulting "pseudized" charge is shown in Fig. 6.2.

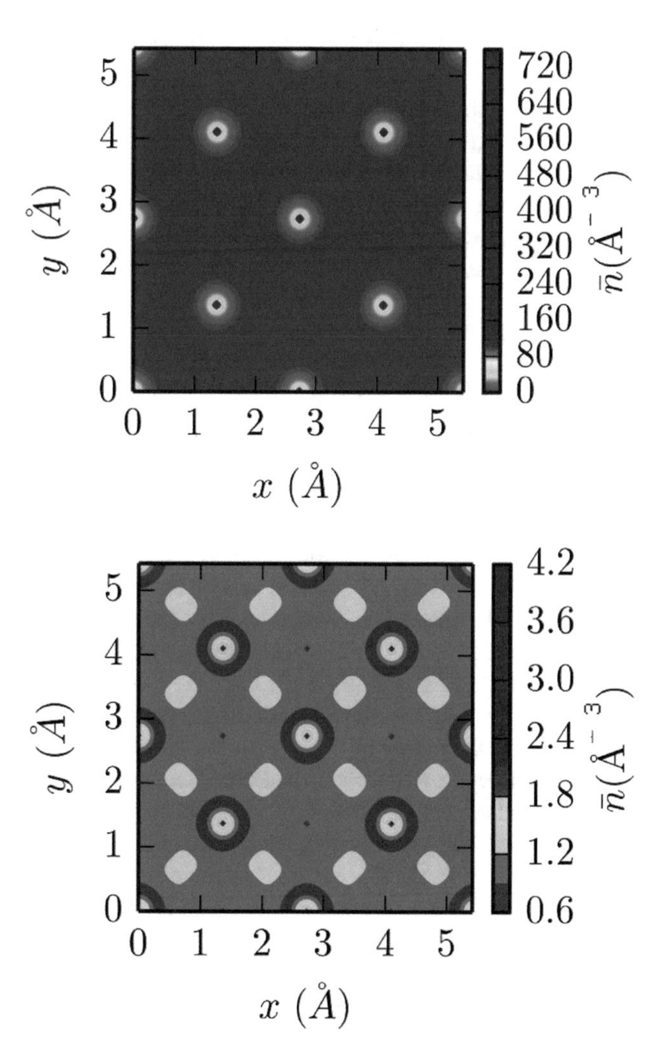


**Fig. 6.1** Total ground state density (*top*) and valence band ground state density (*bottom*) averaged along the [100] direction for a silicon unit cell containing eight atoms

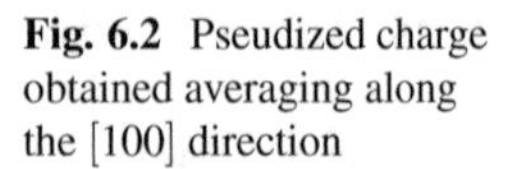
**Fig. 6.2** Pseudized charge obtained averaging along the [100] direction

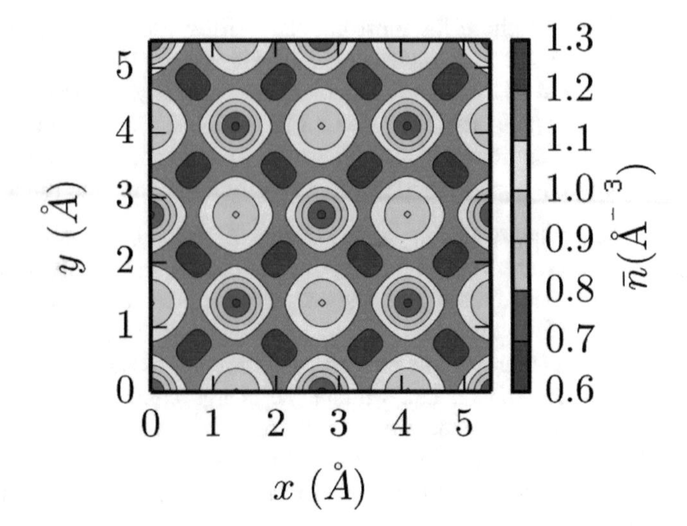


### 6.3.2 *Determining the Lattice Constant and Crystal Structure*

The ground state energy is only defined up to a constant and thus is not meaningful on its own. However, changing the lattice constant and recomputing the ground state energy for each lattice configuration enables us to determine the configuration with the smallest ground state energy. This process can be performed manually or one can also perform "atomic relaxation" studies. In this case, starting from an initial "guess" for the positions of the atoms (usually the experimental values, when known), the total ground state energy of the system is calculated for various configurations obtained by displacing atoms slightly from the initial positions. Calculating the derivatives of the total energy with respect to the ionic displacements, one can calculate the forces acting on each atom and, by methods such as steepest descent or the Car–Parrinello algorithm [21], one can pinpoint the equilibrium atomic configuration. However, caution should be exercised: There is no guarantee that a structure will relax to a minimal energy state, since the system can evolve towards a local minimum or a saddle point and remain "stuck" there. Giving an occasional stochastic "kick" out of a possible local minimum, as done in the "simulated annealing" algorithm [21], can minimize this danger.

As an example of a "system calibration" for bulk Si performed by varying the lattice constant manually, Fig. 6.3 shows the binding energy for 100 different values of the lattice constant, ranging from $a = 5$ to $a = 6$. The minimum ground state energy is found for $a = 5.43$, which is in excellent agreement with experimental data. This is a general feature of the DFT codes available at present: The value of the calculated band gaps remains somewhat underestimated. This is not surprising, since, as we saw, DFT is essentially a ground state theory (although progress is being made to extend its validity). However, the determination of the equilibrium atomic configuration is extremely reliable. Again, this is not surprising, exactly for the same reason.

In the following chapter we shall consider in detail how empirical pseudopotentials can be used to study what are called "low-dimensionality" or "nano" structures. These are thin slabs, nanowires, nanodots obtained by terminating the infinite lattice. Therefore, the periodicity (translation symmetry) is broken. Since the use of plane waves demands periodicity, this is recovered employing "supercells." This procedure is described in detail in Chap. 7, Sect. 7.1. Using the same procedure, two-dimensional materials and slabs can also be modeled using DFT, provided a supercell is constructed in the direction perpendicular to the surface. As an example, consider the interesting case of graphene. As explained in Sect. 3.3.4, this is a two-dimensional film of carbon atoms arranged in an hexagonal honeycomb structure:

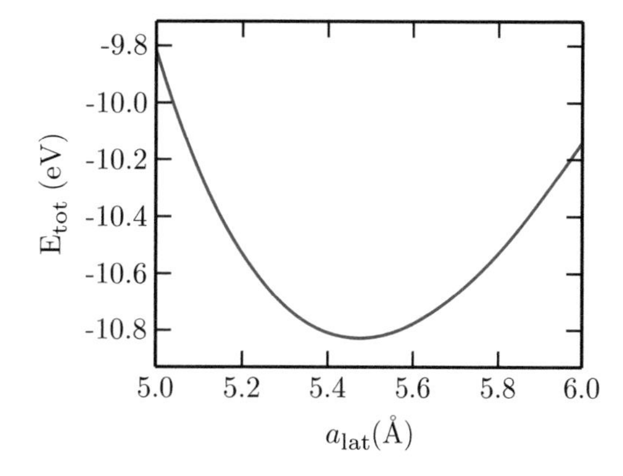


**Fig. 6.3** Ground state energy versus lattice constant for silicon

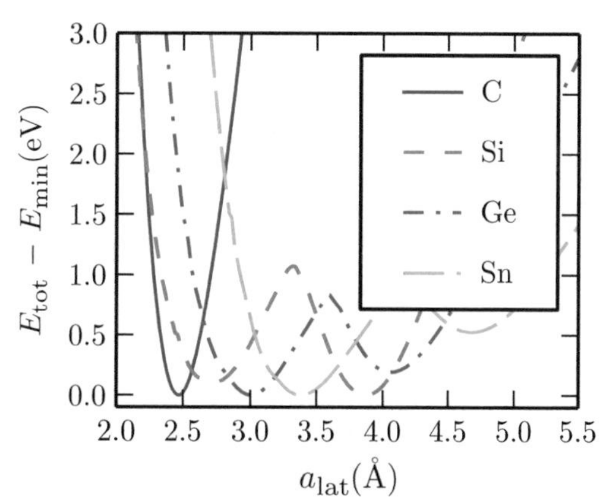


**Fig. 6.4** Energy vs buckling distance for optimal lattice constant

- Graphene's basis vectors are $(a/2, a\sqrt{3}/2, 0)$, $(a/2, -a\sqrt{3}/2, 0)$, and $(0, 0, c)$ where we take $c = 3a$
- The two carbon atoms can be found at $(0, 0, 0)$ and $(a/3, a/3, 0)$
- We opt for the PBE functional
- We use a $10 \times 10 \times 10$ Monckhorst–Pack K-point grid and use default energy cutoff and convergence threshold.

Taking $a = 2.46$, the ground state energy is $-18.45$ eV.

As an example of the relaxation process we have mentioned above, it is possible to determine the equilibrium structure of *buckled* hexagonal monolayer structures, like silicene and germanene. Such a structure is shown in Figs. 13.12 and 13.13, page 304. In a buckled hexagonal structure, the two atoms in the unit cell can be found at $(0, 0, 0)$ and at $(a/3, a/3, b)$ where $b$ is the buckling distance. DFT allows the simulation of hexagonal lattices with a lattice constant that varies from $a = 2$ up to $a = 5.5$. Employing a supercell with height $c = 5a$, the buckling distance is found by specifying an initial buckling distance $b_{\text{init}} = 0.01c$, which is subsequently relaxed by VASP to an equilibrium buckling distance $b_{\text{buckling}}$. In Fig. 6.4, the ground state energy for the relaxed buckled hexagonal lattices of C, Si, Ge, and Sn is shown. For carbon, a single minimum is observed which corresponds to the graphene lattice. For silicon, germanium, and tin, two minima can be distinguished. The phase corresponding to the first minimum is called the "high-buckled phase" and has a smaller lattice constant and a larger buckling distance compared to the second minimum, the "low-buckled phase." The high-buckled phase is metallic, while the band structure of the low-buckled phase resembles that of graphene. However, for germanium and tin the effects of spin–orbit coupling undoubtedly will play a significant role. This effect is not accounted for in these examples.

### *6.3.3 Calculating the Band Structure*

As we have already mentioned, despite the fact that, in principle, DFT only gives information about the ground state properties, it has become routine in the scientific community to assume that the Kohn–Sham orbitals give an approximation of the electronic band structure. The problem with this attitude is that the Kohn–Sham orbitals are not the real electronic states. As a not unexpected consequence, "DFT band structures" generally grossly underestimate the band gap in semiconductors and even predict small-gap semiconductors, such as InAs, to be metals.

Nevertheless, DFT band structures generally do paint a good qualitative picture for most materials. For example, silicon is correctly predicted to be an indirect-gap semiconductor with its conduction band minimum along the $\Delta$ direction. And often, DFT is the only (and therefore, automatically, the best) method to perform ab initio calculations of the band structure that is computationally feasible. Additionally, when experimental data are available, it is possible to use hybrid exchange-correlation functionals [13, 14] that use a Hartee–Fock/LDA mixing that can be tuned to obtain a "correct" band structure [13, 14, 22].

In Fig. 6.5, we show the silicon DFT band structure calculated using the PBE functional [15]. To obtain the band structure along high symmetry directions, first the ground state charge density is calculated by discretizing it on a $10 \times 10 \times 10$ grid. Subsequently, a "non-self-consistent" calculation is performed by taking the Kohn–Sham potential to compute the band structure at any chosen $\mathbf{k}$-point. Figure 6.5 shows the band structure from the L symmetry point to the $\Gamma$-point, the X-point, the U/K-point, and the back to the $\Gamma$-point again.

The maximum of the valence band occurs at the $\Gamma$-point, while the minimum of the conduction band occurs along the $\Gamma$-X axis, in agreement with the experimentally observed band structure. The DFT value of the band gap calculated using the PBE functional is 0.6 eV. This is a strong underestimation of the experimentally observed 1.12 eV band gap. A better optimization of hybrid functionals (called the "HSE03 functional measures"), one can obtain a value of the band gap of 1.15 eV, in much better agreement with experiments.

DFT also has the ability to deal with spin–orbit coupling of the electrons. This topic is discussed in Chap. 5, Sect. 5.6.3. Generally, the effects of spin–orbit coupling have only a small effect on the structural properties but they are important for the electronic band structure, especially for heavy elements.

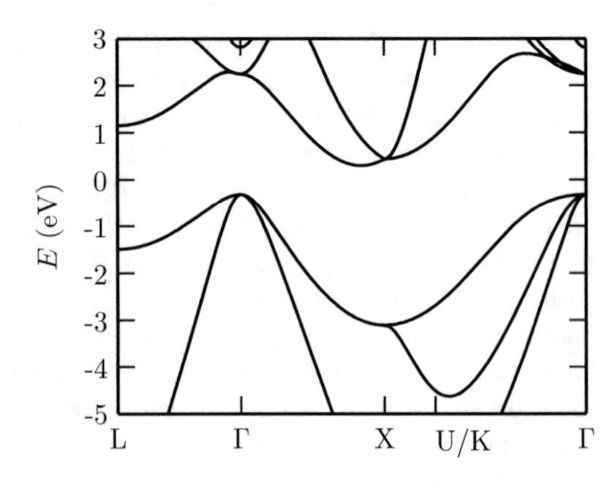


**Fig. 6.5** Silicon bandstructure calculated using the PBE exchange-correlation functional

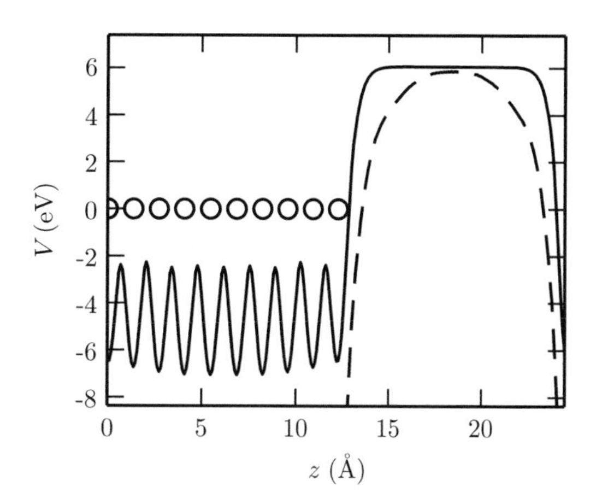


**Fig. 6.6** Hartree part (*solid*) and total Kohn–Sham (*dashed*) potential averaged on the $(x,y)$ plane for a silicon 10-atom slab along the [100] direction. *Circles* indicate the positions of the silicon atoms. The Hartree part can be seen to converge to the vacuum potential much more rapidly than the total potential

### 6.3.4 Calculating the Workfunction and Band Offsets

Other quantities of interest in semiconductor devices are the material workfunction, i.e., the difference between the vacuum level and the Fermi level, and the band offsets between two different materials, i.e., the energy difference between the top of the valence band or the bottom of conduction band of the two materials.

For metals, the Fermi level is readily available from DFT calculations. For semiconductors, instead, doping will determine the position of the Fermi level with respect to the valence and the conduction bands. The potential energy of vacuum is obtained by constructing a slab with a surface facing the vacuum. Averaging the Kohn–Sham potential along the direction of the slab surface on the $(x,y)$ plane, illustrated in Fig. 6.6, the vacuum potential energy is obtained. In practice, it is more convenient to average the Hartree term of the effective Kohn–Sham potential, since it converges towards the vacuum potential more rapidly when compared to the effective Kohn–Sham potential.

The workfunction is defined as the energy required to remove for an electron from the semiconductor the vacuum. In technical terms, the workfunction is the difference between the potential energy of an electron in vacuum and the Fermi level of the semiconductor. In Fig. 6.6, the vacuum energy is 6.07 eV, the Fermi level is 0.64 eV, and the resulting workfunction is 5.43 eV. However, since these results are relative to silicon, which is a semiconductor, the position of the Fermi level is not uniquely defined. The offsets with respect to the valence band (5.91 eV) and the conduction band (5.04 eV) are meaningful measures in semiconductors as long as the band gap is calculated correctly. Also, when considering experimental results, it is important to remember that the workfunction is a surface property and that the introduction of additional dipoles on the surface of any material will change its workfunction.

Similar to the workfunction, the offsets of the valence and conduction band between two semiconductors (s1 and s2) can be calculated using DFT. As a first step, a DFT calculation is performed on a supercell containing both semiconductors. Averaging the Kohn–Sham potential (or the Hartree part of the Kohn–Sham potential) over the $(x,y)$ plane and over the length of the original unit cell in the $z$ direction, the difference between the average potential $\Delta\tilde{V} = \tilde{V}_{s2} - \tilde{V}_{s1}$ in both semiconductors can be obtained. As a second step, a DFT band structure calculation is performed for the bulk cell for both semiconductors (with appropriate strain, if the compounds are not lattice-matched) and the difference of the conduction and valence with respect to the average potential, $\tilde{E}_{v/c1/2} = E_{v/c1/2} - \tilde{V}_{s1/2}$, is calculated. The band for the valence and conduction bands can now be calculated as $\Delta E_{v/c} = \tilde{E}_{v/c2} - \tilde{E}_{v/c1} - \Delta\tilde{V}$.

Figure 6.7 shows the potential, averaged on the $(x,y)$ plane, in a SiGe heterostructure. The difference between the average potential in the silicon and the germanium layer is $\Delta\tilde{V} = 2.32\,\text{eV}$.

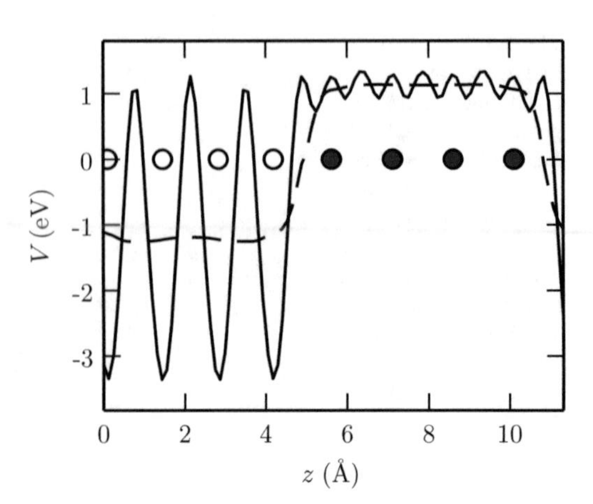


**Fig. 6.7** Hartree part of the Kohn–Sham potential for a SiGe heterostructure supercell averaged along the $x-y$ direction (*solid line*). *Open circles* indicate the position of silicon atoms while *filled circles* indicate the position of germanium atoms. The difference between the smoothened potential (*dashed*) in the silicon and the germanium region is $\Delta\tilde{V} = 2.32\,\mathrm{eV}$

Computing the silicon and the germanium band structures using hybrid functionals, the silicon and germanium band gaps are 1.18 and 0.24 eV, respectively. The germanium band gap is smaller than that of bulk germanium because of the strain required to lattice-match the germanium with the silicon. The computed band offsets are $\Delta E_v = 0.89\,\mathrm{eV}$ and $\Delta_c = -0.05\,\mathrm{eV}$ which is in line with experimental observations that the conduction bands of silicon and germanium are almost aligned.

### 6.3.5 Calculating the Phonon Spectrum

As we have mentioned before, DFT can also be used to calculate the phonon spectrum within the harmonic approximation. This is discussed in Chap. 9, Sect. 9.4. Moreover, the use of DFT to calculate the strength of the all-important electron–phonon interaction is discussed in Chap. 13. Thus, it is probably convenient for many readers to peruse those sections before returning to the present section for a second reading.

The calculation of the phonon spectrum requires the knowledge of the dynamical matrix. This, in turn, is calculated from the second-order force constants [see Eq. (9.93) at page 204]:

$$\tilde{D}_{ij\gamma\gamma'}(l'') = -\frac{\partial F_j(l'\gamma')}{\partial R_{l,\gamma,i}}, \tag{6.11}$$

where $l$ and $l'$ are the indices labeling the primitive cells, $\gamma$ and $\gamma'$ are the indices labeling the atoms in each cell, and $i,j$ are indices running over the three spatial dimensions. Therefore, if $\mathbf{R}_l$ is a lattice point and $\tau_\gamma$ is the position of atom $\gamma$ within the cell, $\mathbf{R}_{l,\gamma} = \mathbf{R}_l + \tau_\gamma$. Finally, in Eq. (6.11), the index $l''$ is such that $\mathbf{R}_{l''} = \mathbf{R}_l - \mathbf{R}_{l'}$. The quantity $\mathbf{F}(l'\gamma')$ is the force acting on the atom at $\mathbf{R}_{l'\gamma'}$.

* *Numerically, the force constants can be obtained in two equivalent ways: (1) from the forces calculated from the change of the total energy for structures with slightly displaced atomic positions, which is the method we have mentioned before and is known as the "small displacement method"; or (2) using density functional perturbation theory (DFPT) [23]. We shall not discuss this advanced method here. When using the small displacement method, the construction of a supercell consisting of repeated unit cells is required for small unit cells, so as to enable the description of non-nearest neighbor interactions. In the case of DFPT, there is no general real-space expression for the dynamical matrix and a new calculation is required at each* **k***-point of interest.*

A VASP post-processing package, called PHONOPY, permits the construction of a set of $3\times3\times3$ Si supercells with small atom displacements. DFT is used to obtain the forces acting on each atom

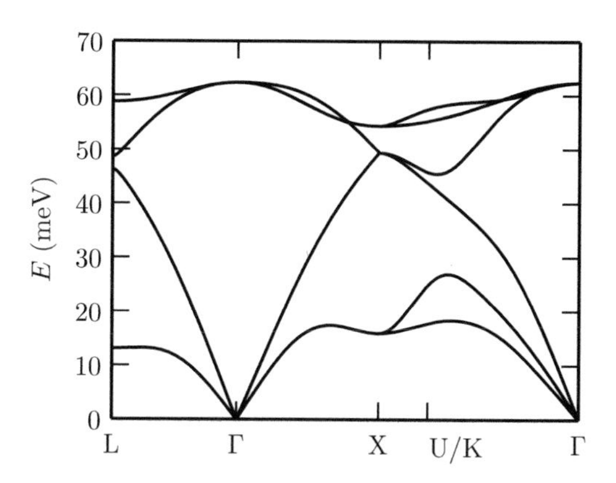


**Fig. 6.8** Silicon phonon spectra calculated by DFT using the small displacement method and a $3 \times 3 \times 3$ supercell

for each of these supercells, thus obtaining the force constants. Following a specific path in reciprocal space (going from the L-point $(\pi/a, \pi/a, \pi/a)$ to the Γ-point (0,0,0), the X-point $(2\pi/a, 0, 0)$, the $U/K$-point $(\pi/(2a), 2\pi/a, \pi/(2a))/(3\pi/(2a), 3\pi/(2a), 0)$ back to the Γ-point with 100-points along each line), PHONOPY can be used to construct and diagonalize the dynamical matrix for each of the k-points along this path. The resulting phonon spectrum is plotted in Fig. 6.8. This can be compared with the result obtained using the valence shell model [24–26], shown in Chap. 9, Fig. 9.1 on page 206.

## Problems

The following problems are appropriate for a class on Computational Physics and assume access to a DFT software package.

**6.1. (The Band Structure of Si)**

(a) Compute the silicon ground state charge density and energy using a DFT package (e.g., VASP, Quantum espresso, ...).
(b) Compute how the ground state energy of fcc Si changes when the lattice constant is increased from 5 AA to 6 AA.
(c) Compute band structure of Si from the results of (a) above.

**6.2. (Phonons in Si)** Compute the phonon spectrum of fcc Si using PHON or PHONOPY.

**6.3. (Graphenei)** Compute the band structure of graphene using DFT.

**6.4. (Silicene)** Silicene is the Si equivalent of graphene. However, it is not flat in an $sp^2$ bonding configuration, but it "buckles" into $sp^3$ bonding with unsaturated, dangling "out-of-plane" bonds. While we would prefer to use the suffix "*ene*" for unbuckled, $sp^2$-coordinated 2D lattices and "*ane*" for the buckled, $sp^3$-coordinated lattices, in the literature "silicane" denotes the buckled structure, whereas "silicane" denotes a buckled structure with the dangling bonds terminated by some functional group, usually by H. Having clarified this detail, construct the buckled hexagonal lattice of silicene with lattice constant 2.7 and determine the buckling distance by performing a relaxation.

**6.5. (A Thin Si Slab)**

(a) Construct a slab of Si, passivate it with hydrogen, and relax the atomic positions. See how the atoms at the surface relax.
(b) Using the slab from (a), determine the energy of the valence and conduction band of Si with respect to the potential energy in vacuum. This is the workfunction of Si.

## References


1. C. Pisani, *Quantum-Mechanical Ab-initio Calculation of the Properties of Crystalline Materials*. Lecture Notes in Chemistry (Springer, Berlin, Heidelberg, 1996)
2. E. Kaxiras, *Atomic and Electronic Structure of Solids* (Cambridge University Press, Cambridge, 2003)
3. P. Hohenberg, W. Kohn, Inhomogeneous electron gas. Phys. Rev. **136**, B864 (1964)
4. W. Kohn, L.J. Sham, Self-consistent equations including exchange and correlation effects. Phys. Rev. **140**, A1133 (1965)
5. C. Kittel, *Quantum Theory of Solids*, 2nd edn. (Wiley, New York, 1987)
6. G.D. Mahan, *Many-Particle Physics*. Physics of Solids and Liquids, 3rd edn. (Kluwer, Boston, 2000)
7. A.L. Fetter, J.D. Walecka, *Quantum Theory of Many-Particle Systems* (Dover, Mineola, New York, 2003)
8. P.A.M. Dirac, Note on exchange phenomena in the Thomas-Fermi atom. Proc. Cambridge Phil. Roy. Soc. **26**, 376 (1930)
9. R.G. Parr, W. Yang, *Density-Functional Theory of Atoms and Molecules* (Oxford University Press, Oxford, 1994)
10. D.C. Langreth, M.J. Mehl, Beyond the local-density approximation in calculations of ground-state electronic properties. Phys. Rev. B **28**, 1809 (1983)
11. J.P. Perdew, J.A. Chevary, S.H. Vosko, K.A. Jackson, M.R. Pederson, D.J. Singh, C. Fiolhais, Atoms, molecules, solids, and surfaces: applications of the generalized gradient approximation for exchange and correlation. Phys. Rev. B **46**, 6671 (1992)
12. A.D. Becke, Density-functional exchange-energy approximation with correct asymptotic behavior. Phys. Rev. A **38**, 3098 (1998)
13. A.D. Becke, A new mixing of Hartree–Fock and local density-functional theories. J. Chem. Phys. **98**, 1372 (1993)
14. J.P. Perdew, M. Ernzerhof, K. Burke, Rationale for mixing exact exchange with density functional approximations. J. Chem. Phys. **105**, 9982 (1996)
15. J.P. Perdew, K. Burke, M. Ernzerhof, Generalized gradient approximation made simple. Phys. Rev. Lett. **77**, 3865 (1996)
16. G. Kresse, J. Hafner, *Ab initio* molecular dynamics for liquid metals. Phys. Rev. B **47**, 558(R) (1993)
17. G. Kresse, Thesis, Technische Universität Wien, 1993
18. G. Kresse, J. Furthmüller, Efficiency of *ab initio* total energy calculations for metals and semiconductors using a plane-wave basis set. Comput. Mat. Sci. **6**, 15 (1996)
19. G. Kresse, J. Furthmüller, Efficient iterative schemes for *ab initio* total-energy calculations using a plane-wave basis set. Phys. Rev. B **54**, 11169 (1996)
20. P.E. Blöchl, Projector augmented-wave method. Phys. Rev. B **50**, 17953 (1994)
21. R. Car, M. Parrinello, Unified approach for molecular dynamics and density functional theory. Phys. Rev. Lett. **55**, 2471 (1995)
22. P.J. Stephens, F.J. Devlin, C.F. Chabalowski, M.J. Frisch, *Ab initio* calculation of vibrational absorption and circular dichroism spectra using density functional force fields. J. Phys. Chem. **98**, 11623 (1994)
23. S. Baroni, S. de Gironcoli, A. Dal Corso, P. Giannozzi, Phonons and related crystal properties from density-functional perturbation theory. Rev. Mod. Phys. **73**, 515 (2001)
24. B.G. Dick, A.W. Overhauser Jr., Theory of the dielectric constants of alkali halide crystals. Phys. Rev. **112**, 90 (1958)
25. K. Kunc, O.H. Nielsen, Phys. Commun. **17**, 47 (1979)
26. O.H. Nielsen, W. Weber, Phys. Commun. **18**, 101 (1979)

# Chapter 7
# * Electronic Structure of Low-Dimensionality Systems

The use of plane waves and empirical pseudopotentials—presented in Chap. 5, Sect. 5.3.4—to calculate the band structure of semiconductors (and other crystals) has proven useful to gain insight into the electronic excitation spectrum of solids. The "empirical' nature implies loss of strong predictive power and of "portability" of ionic (pseudo)potentials, but it results in a vast simplification of the numerical problem when compared to the ab initio methods presented in Chap. 6. Moreover, the small degree of "fitting" allowed by the technique affords, by definition, excellent agreement with experimental data. Since our focus is on electronic transport, not on structure calculations per se, empirical pseudopotentials represent the best choice to perform accurate calculations.

In this chapter we consider the use of empirical pseudopotentials to study the electronic structure (and, in later chapters, transport properties) of low-dimensional structures. These are regular assemblies of atoms that are arranged approximately in their bulk configurations, but the fact that they are "finite" (as opposite to the ideal infinite crystals we call "bulk"), thus presenting free surfaces (or surfaces interfacing dielectrics), gives rise to new physical properties that have proven to be extremely interesting from a scientific and technological perspective. For example, thin Si films are used as active channels in metal-oxide-semiconductor (MOS) field-effect transistors (FETs), as implemented in their Si-on-insulator (SOI) form. They are, in a way, "nanostructures," since these films are only a few nanometers thick. Electrons confined in these structures behave like particles in a box along the confinement direction. As a consequence, the band structure of these thin films, or "slabs," differs significantly from the band structure of bulk Si, since it has to reflect the discrete energy levels resulting from the confinement-induced quantization. Their electronic, transport, and even optical properties differ significantly from their bulk properties. For example, whereas bulk Si or Ge are indirect-gap semiconductors, when grown in the form of ultra-thin slabs, an effect called "band folding" (that is, the projection of the band structure along the crystallographic direction perpendicular to the film) renders them direct-gap materials. Electrons in these two-dimensional structures behave as a two-dimensional electron gas (2DEG), whose properties have been studied extensively [1]. More recently, technological fabrication and synthesis methods have made it possible to create films as thin as a single atomic layer. Graphene [2], silicene, germanene, stannanane, and transition-metal dichalcogenides (TMDs, such as $MoS_2$, for example) are extremely interesting for their peculiar physical and electronic properties (high carrier mobility, superconductivity, their nature as topological insulators, to name a few properties that we shall not discuss here) and also for possible use in electronic devices. Similarly, electronic states in semiconductor nanowires—that can be viewed as ultra-narrow slabs—will be affected by confinement-induced quantization along two dimensions. Electrons in these structures will behave like a one-dimensional electron gas (1DEG), whose electronic, transport, dielectric, and optical properties will also be significantly different from



M. Fischetti, W.G. Vandenberghe, *Advanced Physics of Electron Transport in Semiconductors and Nanostructures*, Graduate Texts in Physics, DOI 10.1007/978-3-319-01101-1_7

those of their bulk counterparts. Finally, "zero-dimensional" structures, such as small regular clusters of atoms ('quantum dots'), have raised interest for their possible applications in photovoltaic solar cells and solid-state injection lasers.

The fact that these structures do not exhibit the periodicity of bulk crystals demands some change in the theoretical approach used to study them, if we wish to use plane-wave methods. This is done in a simple (but computationally heavy) way, by rendering these structures artificially periodic, replicating them spatially along the directions of the broken translational symmetry. This is the concept of "supercell" that will be discussed next. Several structures that exhibit two-dimensional and one-dimensional electronic behavior will be studied next. In each case, we shall provide even minute details, such as the positions of the atoms, the lattice translation vectors, and the pseudopotential employed (in Appendix B), so that it will be possible to write an actual computer program using the algorithms presented in Chap. 5, Sect. 5.3.4. In this chapter we shall also provide a more complete and up-to-date list of references, more along the style of a technical article.

The bulk of this chapter, as well as Appendix B, is extracted from [3] and from a chapter of the book *"Nano-Electronic Devices: Semiclassical and Quantum Transport Modeling,"* edited by D. Vasileska and Stephen M. Goodnick (Springer, New York, 2011) [4]. Text, when extracted verbatim, and figures are here reproduced with permission from the publisher.

## 7.1 The Supercell Concept for Nanostructures

The "supercell" method is conceptually a trivial extension of the standard "bulk" plane-wave method, in which the primitive cell of the crystal (containing only two atoms in the fcc semiconductors of interest here) is considered. If, instead of bulk homogeneous solids, we are interested in studying finite structures, instead of employing the primitive lattice cell, a larger cell is considered, for example composed of many Si cells replicated $N$ times along the $z$-axis to form a Si layer of thickness $Na_0$ (where $a_0$ is the Si lattice constant) and $N_\mathrm{v}$ empty cells ("vacuum"), resulting in a cell of total extension $(N+N_\mathrm{v})a_0$ along the $z$-axis (see Fig. 7.1, left). This is the kind of supercell required to deal with inversion layers, thin semiconductor bodies, or quantum wells. Similarly, the cell may be extended along two directions (see Fig. 7.1, center), mimicking a quantum wire. A sufficiently "padding" consisting in a sufficiently large number of "vacuum cells" (just an isolation, several cells long, between adjacent structures) or insulating cells between adjacent layers or wires will guarantee avoiding artifacts due to the possible coupling between them (as in a superlattice).

It is important to note a useful feature of the empirical pseudopotential method: An external potential can be added without any significant additional numerical cost. This is particularly important, since these nanostructures are employed in situations that require an external applied bias to drive the electrons. As it is done for the ionic (pseudo)potentials, the external potential is also artificially rendered periodic with the period of the supercell. Therefore, we must consider its non-vanishing Fourier components that are of the form:

$$V_{\mathbf{G}}^{(\mathrm{ext})} = \frac{1}{\Omega_{\mathrm{sc}}} \int_{\Omega_{\mathrm{sc}}} \mathrm{d}\mathbf{r}\, V^{(\mathrm{ext})}(\mathbf{r})\, \mathrm{e}^{-\mathrm{i}\mathbf{G}\cdot\mathbf{r}} \,, \tag{7.1}$$

where $\Omega_{\mathrm{sc}}$ is the volume of the supercell. In the case of homogeneous 2DEGs, in which the artificial periodicity extends in one dimension (inversion layers, quantum wells, and thin bodies), $V^{(\mathrm{ext})}(\mathbf{r})$ and the external potential is assumed to depend only on the "confinement direction" $z$, Eq. (7.1) simplifies to

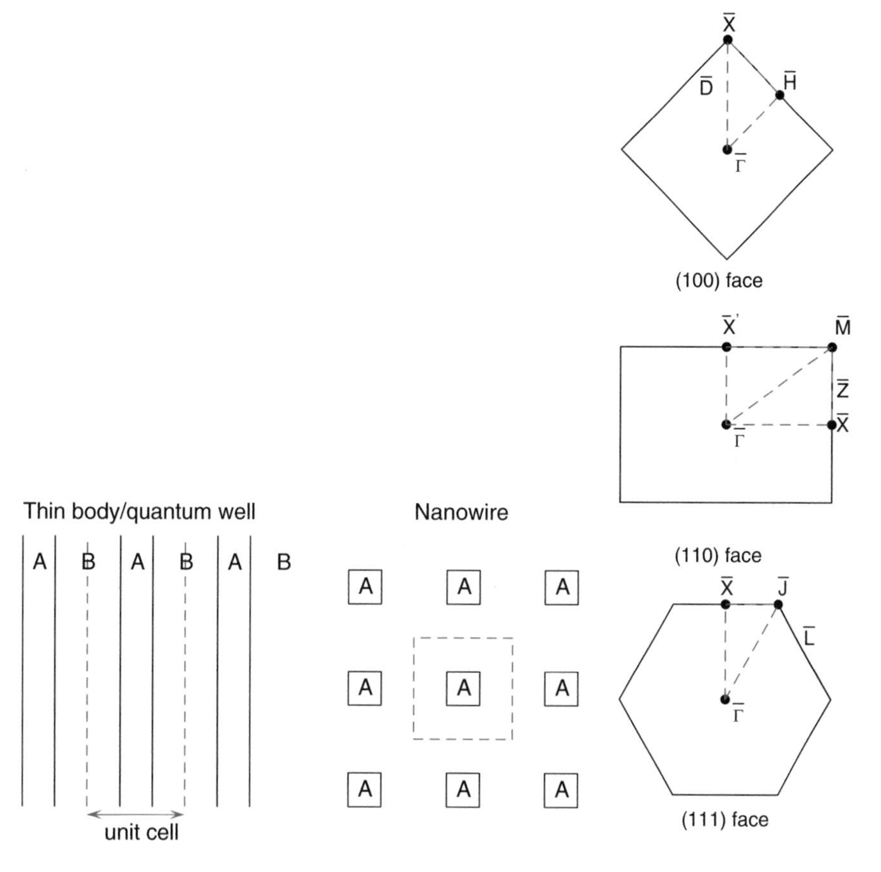


**Fig. 7.1** *Left*: Supercell scheme illustrating the choice of the supercell for an AB hetero-structure (A or B could be replaced by vacuum for free-standing films). *Center*: The same but for the case of a nanowire. *Right*: 2D Brillouin Zones for the (100), (110), and (111) fcc faces showing the conventional notation for symmetry points and symmetry lines and the "path" followed on the (100) face to plot the dispersion shown in the following figures

$$V_{\mathbf{G}}^{(\mathrm{ext})} = \delta_{\mathbf{G}_{\|},\mathbf{0}}\, V_{G_z^{(L)}}^{(\mathrm{ext})} = \delta_{\mathbf{G}_{\|},\mathbf{0}}\, \frac{1}{L_z} \int_0^{L_z} dz\, V^{(\mathrm{ext})}(z)\, \mathrm{e}^{-\mathrm{i}G_z^{(L)} z}\,, \tag{7.2}$$

where $L_z$ is the extension of the 1D supercell in the $z$ direction and $\mathbf{G}_{\|}$ is the projection of the reciprocal lattice wavevector $\mathbf{G}$ on the $(x,y)$ plane (the plane of the surface or interface). In the case of 1DEGs, such as graphene nanoribbons, nanowires, or carbon nanotubes, $V^{(\mathrm{ext})}(\mathbf{r})$ will depend only on the in-plane coordinates $\mathbf{R}$, so that

$$V_{\mathbf{G}}^{(\mathrm{ext})} = \delta_{G_z,0}\, V_{\mathbf{G}_{\|}}^{(\mathrm{ext})} = \delta_{G_z,0}\, \frac{1}{A} \int_A d\mathbf{R}\, V^{(\mathrm{ext})}(\mathbf{R})\, \mathrm{e}^{-\mathrm{i}\mathbf{G}_{\|}\cdot\mathbf{R}}\,, \tag{7.3}$$

where $A$ is the cross-sectional area of the 2D supercell.

Therefore, accounting also for the presence of an external potential, the electronic structure of the system will be obtained by solving the eigenvalue problem:

$$\sum_{\mathbf{G}'} \left[ \frac{\hbar^2}{2m} |\mathbf{k}+\mathbf{G}|^2 \delta_{\mathbf{G},\mathbf{G}'} + V_{\mathbf{G}-\mathbf{G}'}^{(\mathrm{lat})} + V_{\mathbf{G}-\mathbf{G}'}^{(\mathrm{ext})} \right] u_{\mathbf{G}',\mathbf{k}}^{(n)} = E_n(\mathbf{k})\, u_{\mathbf{G},\mathbf{k}}^{(n)}\,. \tag{7.4}$$

Note that we still use the full three-dimensional wavevector $\mathbf{k}$. However, we note that the dispersion $E_n(\mathbf{k})$ and the wavefunctions $u^{(n)}_{\mathbf{G},\mathbf{k}}$ do not depend "appreciably" on $k_z$ (the component perpendicular to the surface/interfaces) in the case of 2DEGs, and do not depend on $\mathbf{K}$ (the components on the cross-sectional plane) in the case of 1DEGs. By "appreciably" we mean to say that this is true as long as the electrons are confined, i.e., that they lie energetically below the energy of the vacuum. Therefore, under these circumstances, $\mathbf{k}$ may be considered the wavevector projected on the plane (or direction) perpendicular to the confinement direction (or plane) (see Fig. 7.1, right, illustrating the 2D Brillouin Zones of a face-centered cubic lattice, for example). As usual, we have indicated with $V^{(\mathrm{lat})}$ the lattice (pseudo)potential resulting from the sum over all ions $\alpha$ in the supercell of the ionic pseudopotentials $V^{(\mathrm{pseudo},\alpha)}_{\mathbf{G}}$, normalized to each atomic volume $\Omega_\alpha$ multiplied by the "structure factor" $\mathrm{e}^{-\mathrm{i}\mathbf{G}\cdot\tau_\alpha}$:

$$V^{(\mathrm{lat})}_{\mathbf{G}} = \frac{1}{\Omega_{\mathrm{sc}}} \sum_{\alpha} \mathrm{e}^{-\mathrm{i}\mathbf{G}\cdot\tau_\alpha}\, \Omega_\alpha\, V^{(\mathrm{pseudo},\alpha)}_{\mathbf{G}} , \tag{7.5}$$

where $\tau_\alpha$ is the position of ion $\alpha$ in the supercell. From now on, we shall drop the superscript "pseudo" and simply write $V_{\alpha,\mathbf{G}}$ for the ionic pseudopotentials. The wavefunction corresponding to the eigenvalue $E_n(\mathbf{k})$ is given by the Bloch expression:

$$\psi^{(n)}_{\mathbf{k}}(\mathbf{r}) = \frac{1}{\Omega_{\mathrm{sc}}}\, \mathrm{e}^{\mathrm{i}\mathbf{k}\cdot\mathbf{r}} \sum_{\mathbf{G}} u^{(n)}_{\mathbf{G},\mathbf{k}}\, \mathrm{e}^{\mathrm{i}\mathbf{G}\cdot\mathbf{r}} . \tag{7.6}$$

This method is "exact" (within the EP framework), but it can handle only closed systems. Thus, it is ideally suited to treat low-dimensionality confined cases, but it must be augmented by other techniques when we must deal with transport (i.e., open boundary conditions) problems. This will be discussed in Chap. 17.

In Appendix B we provide complete information about the form of the empirical pseudopotentials used in the examples we shall give below. We shall now discuss various examples of structures resulting in 2DEGs and 1DEGs.

## 7.2 Thin Layers (Thin Films and 2DEG, Graphene)

Electrons confined in two dimensions, the 2DEG, have been studied extensively, starting from the surface of liquid He, to electrons in the inversion layers of MOSFETs, to those in hetero-structures, and recently, in crystalline sheets one-atom-thick with the atoms arranged, usually, in a honeycomb hexagonal lattice. The reviews given by Ando et al. [1] and Geim and Novoselov [2] illustrate the physical and technological importance of these structures. Here we discuss how the use of empirical pseudopotentials can give us insight about the electronic properties of these systems.

### *7.2.1 Thin Si Layers*

Inversion layers of Si MOSFETs constitute, historically, the first example of a 2DEG that has given us results of high scientific and technological importance. Although the existence of a 2DEG had been predicted early on, it was first observed in 1964 [1]. The quantum confinement caused by a large electric field pushing the electrons against the Si–$SiO_2$ interface results in a quantization of the electron wavevector along the direction perpendicular to the interface. Therefore, the electron dispersion consists of two-dimensional "subbands," each describing electrons free to move on the plane of the interface, but with the component of their wavevector perpendicular to the interface constrained by quantization, much as the textbook case of a one-dimensional particle in a box. In Chap. 8, Sect. 8.2.1, we shall discuss this situation in detail within the context of the effective-mass approximation. As bulk Si is replaced by thin Si "bodies" (to use the jargon used in the Very Large Scale Integration—VLSI—community), additional confinement results from the finite thickness of the "slab." As confinement gains importance, electrons will be pushed to higher energies, beyond the validity of the effective-mass approximation. A correct description of the electronic structure of these systems renders mandatory the use of "atomistic" theoretical studies, such as DFT, empirical pseudopotentials, tight-binding, or $\mathbf{k}\cdot\mathbf{p}$ models. Here we see how empirical pseudopotentials can handle this problem.

#### 7.2.1.1 (100) Surfaces

Thin Si layers with $(100)$ surfaces, mimicking thin SOI or FinFET channels on $(100)$ wafers, are created by considering the usual fcc real-space primitive translation vectors

$$\mathbf{a}_1 = (a_0/2)\ (1,1,0)\ , \quad \mathbf{a}_2 = (a_0/2)\ (-1,1,0) \tag{7.7}$$

on the $(x,y)$ plane ($a_0 = 0.543$ nm being the Si lattice constant), but extending the cell in the $z$ direction defining the translation vector

$$\mathbf{a}_3 = a_0(0,0,N+N_{\mathrm{v}})\ , \tag{7.8}$$

where $N$ is an integer indicating the number of Si cells (of thickness $a_{\mathrm{Si}}$) considered and $N_{\mathrm{v}}$ is the number of vacuum cells used to isolate the periodically repeated Si layers. Si atoms are then placed in the supercell spanned by the translation vectors $\mathbf{a}_1$, $\mathbf{a}_2$, and $\mathbf{a}_3$ starting with the 4 atoms in the first (non-primitive) cell with coordinates:

$$\begin{cases} \tau_1 = (0,0,0) \\ \tau_2 = (a_0/4)\ (1,1,1) \\ \tau_3 = (a_0/4)\ (0,2,2) \\ \tau_4 = (a_0/4)\ (-1,1,3) \end{cases}, \tag{7.9}$$

and proceeding with $4N-4$ additional atoms labeled by the index $i = 4j+1$ (with $j = 1,N$) with coordinates:

$$\begin{cases} \tau_{4j+1} = \tau_1 + ja_0\hat{\mathbf{z}} \\ \tau_{4j+2} = \tau_2 + ja_0\hat{\mathbf{z}} \\ \tau_{4j+3} = \tau_3 + ja_0\hat{\mathbf{z}} \\ \tau_{4j+4} = \tau_4 + ja_0\hat{\mathbf{z}} \end{cases}, \tag{7.10}$$

where $\hat{\mathbf{z}}$ is the unit vector along $z$ and $j = 1,\ N-1$.

Clearly, at the "free" surfaces, there will be $sp^3$ orbitals that remain dangling, because of the missing atoms. These would cause the appearance of additional states mainly localized around the dangling orbitals (often called "dangling bonds") with an associated energy band that often falls in the energy gap. This is seen in nature, where indeed these "surface states" historically have created technological problems: For example, at the interface between Si and $SiO_2$, dangling bonds can trap and de-trap electrons, causing unwanted charging of the interface. This is also seen in calculations, and we shall show an example below. We should also remark that these dangling orbitals are not the lowest-energy configuration of a free surface. Under appropriate conditions, the free surface "reconstructs": For example, on a free Si (111) surface (with lone $sp^3$ orbitals pointing straight out of the surface), two surface atoms will move closer together to form a "bridging" covalent bond out of the two dangling orbitals. This results in a new periodic atomic arrangement at the surface: It is a reconstruction in which straight lines of Si–Si "dimers" form a new periodic structure, in this case called "$2 \times 1$": The un-reconstructed periodic structure consists of a 1-cell-by-1-cell ($1 \times 1$) pattern that now changes into 2 cells merging together via the dimers in a new "2-cells-by-one-cell" pattern. This also results in the "merging" of the bonds along the line of the dimers, forming an extended $\pi$ orbital. Other reconstructions exist, depending on how the surface is processed: $(2n+1) \times (2n+1)$ (the most common being the $7\times7$ reconstruction obtained at temperatures above 750 K); or a slightly different $2 \times 1$ reconstruction of the (100) surface. DFT can predict these reconstructions. Here, we assume that the surface, instead, remains unchanged in terms of its crystalline form. This is usually the morphology that is obtained experimentally when the free surface is "passivated" by an insulator.

The most convenient way to "passivate" our theoretical surface is to attach a H atom to each dangling orbital. This process is called "termination" of the surface. Empirical pseudopotentials for H that avoid the appearance of surface states are given in Appendix B. In our case of a (100) surface, we must terminate each surface cell with four hydrogen atoms at the following locations:

$$\begin{cases} \tau_{H,1} = \tau_1 + 0.158\, a_0(-1,1,-1) \\ \tau_{H,2} = \tau_1 + 0.158\, a_0(+1,-1,-1) \\ \tau_{H,3} = \tau_{4N} + 0.158\, a_0(-1,1,1) \\ \tau_{H,4} = \tau_{4N} + 0.158\, a_0(1,-1,1) \end{cases} . \tag{7.11}$$

The length of the Si–H bond, given here as $0.158\, a_0$, has been determined empirically, but it is not too different from the bond length observed experimentally. The positions of the H atoms are simply obtained by placing them along the direction of the dangling $sp^3$ orbital.

#### 7.2.1.2 (110) Surfaces

For the (110) surface, instead, the cell is described by the translation vectors:

$$\mathbf{a}_1 = a_0/\sqrt{2}(1,0,0)\,, \quad \mathbf{a}_2 = a_0(0,1,0) \tag{7.12}$$

on the $(x,y)$ plane and it is extended in the $z$ direction using the translation vector

$$\mathbf{a}_3 = a_0/\sqrt{2}(0,0,N+N_v)\,. \tag{7.13}$$

Atoms contained in the first cell have coordinates:

$$\begin{cases} \tau_1 = (0,0,0) \\ \tau_2 = (a_0/4)\,(0,1,2/\sqrt{2}) \\ \tau_3 = (a_0/2)\,(1/\sqrt{2},1,1/\sqrt{2}) \\ \tau_4 = (a_0/4)\,(2/\sqrt{2},3,0) \end{cases} . \tag{7.14}$$

The atomic coordinates for the remaining $4N - 4$ atoms in the supercell are obtained from the coordinates given above by translating them by an amount $ja_0/\sqrt{2}$ ($j$=1, $N$) along the $z$ direction, as in Eq. (7.10). Of course, we must terminate the surface dangling bonds with H atoms, as we have done for the (100) surface.

#### 7.2.1.3 (111) Surfaces

Finally, for the $(111)$ surface one may choose, among many possible equivalent selections of translation vectors:

$$\mathbf{a}_1 = (a_0/\sqrt{2})\ (1,0,0)\ , \quad \mathbf{a}_2 = (a_0/2)\ (1/\sqrt{2}, \sqrt{3/2}, 0) \tag{7.15}$$

on the $(x,y)$ plane. The cell is extended along the $z$ by defining the translation vector

$$\mathbf{a}_3 = (a_0\sqrt{3})\ (0,0,N+N_\mathrm{v})\ . \tag{7.16}$$

As we just said, this is not the only possible selection. For example, choosing $\mathbf{a}'_1 = a_0/2(1/\sqrt{2}, -\sqrt{3/2}, 0)$ instead of $\mathbf{a}_1$, one would obtain the same end result, since $\mathbf{a}'_1 = \mathbf{a}_1 - \mathbf{a}_2$. In this case one must consider six atoms in the first cell. Their coordinates are

$$\begin{cases} \tau_1 = (0,0,0) \\ \tau_2 = a_0\ (0,0,\sqrt{3}/4) \\ \tau_3 = a_0\ (0,1/\sqrt{6},1/\sqrt{3}) \\ \tau_4 = a_0\ (0,1/\sqrt{6},\sqrt{3}/4+1/\sqrt{3}) \\ \tau_5 = a_0\ (1/(2\sqrt{2}),1/(2\sqrt{6}),2/\sqrt{3}) \\ \tau_6 = a_0\ (1/(2\sqrt{2}),1/(2\sqrt{6}),2/\sqrt{3}+\sqrt{3}/4) \end{cases} . \tag{7.17}$$

As usual, atomic coordinates for the remaining $6N - 6$ atoms in the supercell are obtained by translating the coordinates above by an amount $ja_0\sqrt{3}$ ($j$=1, $N$) along the $z$ direction and dangling bonds at the free surfaces are terminated with H atoms.

#### 7.2.1.4 Band-Structure Calculations

Our interest is in the electron transport properties of the structure. The calculation of the band structure is, obviously, the first step. By itself, it shows immediately something extremely interesting (see Fig. 7.2): A thin Si film becomes a *direct-gap* structure. This is the result of projecting the bulk BZ onto the 2D BZ. Of the six minima of the conduction band, one each along the equivalent $\{100\}$ directions, those along the $[100]$ and the $[\bar{1}00]$ directions are projected at the center of the 2D BZ, the $\overline{\Gamma}$ symmetry point. The top of the valence band, already at the center of the 3D BZ, will also be projected on the center of the 2D BZ. Therefore, provided that the energy of the other four minima of the conduction band—that lie along the $[010]$, $[0\bar{1}0]$, $[001]$, and $[00\bar{1}]$ directions—does not exhibit a lower energy, the band gap will now be at $\overline{\Gamma}$. This is the result of what is called *band folding*. We should add that we actually expect the minima along the $[100]$ and $[\bar{1}00]$ directions to be at a lower energy. Indeed, thinking of particle in a box, the ground state energy is proportional to $1/m^*$, where $m^*$ is the effective mass. In our case, this mass will be the longitudinal mass, $m_\mathrm{L}$ for these minima, while it will be the much smaller ($\approx 0.19\ m_\mathrm{el}$ vs. $\approx 0.91\ m_\mathrm{el}$), for the minima at $\overline{\mathrm{X}}$, so the widening of the gap caused by the quantum confinement is expected to be larger for the minima at $\overline{\mathrm{X}}$ than for those at $\overline{\Gamma}$. (As a matter of terminology, the quantized states in the conduction band at $\overline{\Gamma}$ are called "unprimed" states, those at $\overline{rmX}$ "primed.") This is indeed seen in Fig. 7.2. Attempts to exploit this new property of thin Si in optical applications have been frequent. The crucial issue is the practical realization of a film thin enough to prevent decoherence of the electron wavefunctions (via collisions with phonons, defects, interface imperfections), so that quantization and band folding can be achieved.

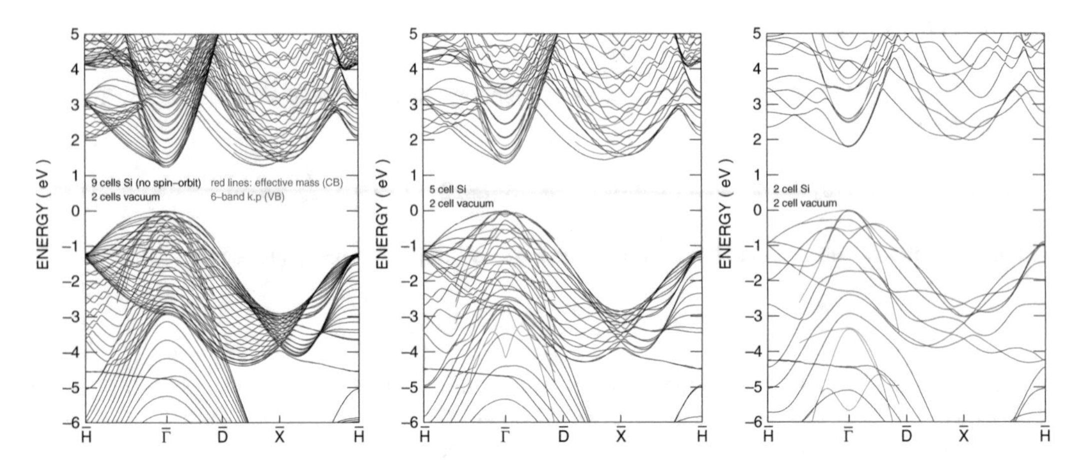


**Fig. 7.2** Band structure of a (100) Si slab in vacuum terminated by H atoms. The slab thickness is 9 Si cells (*left*), 5 cells (*center*), and 2 cells (*right*). Note the direct gap caused by band folding, the quantized subbands in the conduction and valence bands, the widening of the gap caused by the confinement, and, barely visible, the lifting of the twofold degeneracy of the unprimed states (known as "valley splitting") caused by the symmetry breaking due to the external potential. Note also at the $\overline{\text{X}}$-point the presence of two additional 2D valleys. The *red dashed lines* are parabolic bands (conduction) and $\mathbf{k}\cdot\mathbf{p}$ approximations to the problem (calculated assuming vanishing wavefunction at the Si–vacuum interface) illustrating the significant effect of the EP band structure. The spin–orbit interaction has been neglected to obtain these results to gain computational speed. Finally, the zero of the energy has been set arbitrarily at the top of the valence bands. [Used from M.V. Fischetti, Bo Fu, S. Narayanan, and J. Kim, in *Nano-Electronic Devices: Semiclassical and Quantum Transport Modeling*, Dragica Vasileska and Stephen M. Goodnick eds. (Springer, New York, 2011), pp. 183–247, with kind permission from Springer Science+Business Media]

A second quantity in which we are mostly interested is the *ballistic conductance*. This concept is defined in Chap. 17, Sect. 17.2.1. It represents the "quantum conductance," $2e^2/h$, multiplied by the number of conductive channels (bands and state in each band) available for an electron to travel through the structure [see Eq. (17.11)]. It can be viewed as a first semi-quantitative way to characterize the ability of the structure to carry current. Therefore, it helps us to identify the best structures for potential applications. In our case, comparing films of different thickness, of different materials, and with surfaces cleaved along different crystallographic planes is an excellent first step to achieve this goal.

Coming now to computational details, Eq. (7.4) can be solved conventionally using the Si and H pseudopotentials from the literature. As an example, here we show results obtained using the local empirical pseudopotentials from [5, 6] (see Appendix B, Table B.1).

The density of states is computed using the two-dimensional version of the Gilat–Raubenheimer algorithm described below (see the introduction of Chap. 12 and, in that chapter, Eq. (12.37) and the paragraph following this equation for the notation and meaning of the symbols):

$$\mathscr{D}_{2\text{D}}(E) = 2\sum_n \int \frac{d\mathbf{k}}{(2\pi)^2}\,\delta[E_n(\mathbf{k})-E] = \frac{1}{2\pi^2}\sum_{jn}{}' \frac{L(w_{jn})}{|\nabla_{2\text{D}}E_{jn}|}\,. \tag{7.18}$$

The ballistic conductance along the direction characterized by the unit vector $\hat{\mathbf{n}}$ can be computed in a similar way:

$$G_{2\text{D}}(E) = 2\,e^2\sum_n \int \frac{d}{(2\pi)^2}\,\upsilon_n(\mathbf{k})\cdot\hat{\mathbf{n}}\,\delta[E_n(\mathbf{k})-E] = \frac{e^2}{\pi h}\sum_{jn}{}' \frac{\nabla_{2\text{D}}E_{jn}\cdot\hat{\mathbf{n}}}{|\nabla_{2\text{D}}E_{jn}|}\,L(w_{jn})\,, \tag{7.19}$$

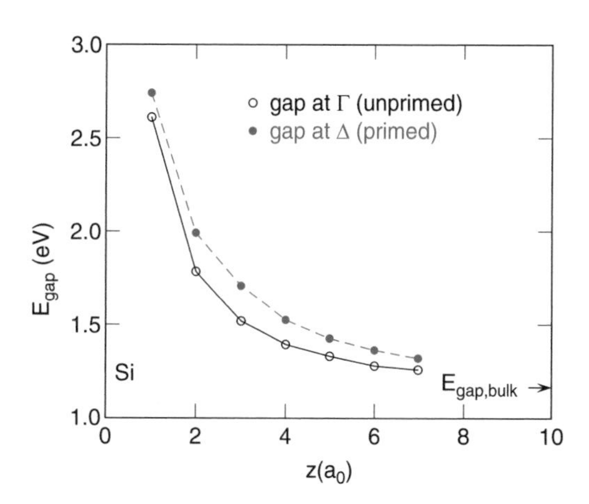


**Fig. 7.3** Variation of the Si band gap at the (100) and (001) minima as a function of body thickness

where $\upsilon_n(\mathbf{k})$ is the group velocity in band $n$ at the point $\mathbf{k}$ and the integration must be extended only over states whose group velocity along the direction $\hat{\mathbf{n}}$ is positive.

Figure 7.2 shows the band structure of (100) Si layers of thickness equal to $9a_{\rm Si}$, $5a_{\rm Si}$, and $2a_{\rm Si}$, all separated by a vacuum $2a_{\rm Si}$ "thick." The obvious results are, as already pointed out, the appearance of a direct gap and of many subbands. Several details, most appreciated by specialists and mentioned here for completeness, are also evident. For example, the appearance of an additional doubly degenerate conduction band at the $X$ point, a band already obtained by Esseni and Palestri [7] using a linear combination of bulk bands (LCBB) and denoted by them as M3, M4. Such a band is completely missed by simpler models based on the effective-mass approximation and has significant effects on electron transport, as we shall see in Chap. 13, Sect. 13.4.1. In Fig. 7.2 one can also see the expected widening of the gap as the film is made thinner. A quantitative summary of this observation is given by Fig. 7.3 that shows the dependence of the energy gap on the thickness of the film.

As stated before, an external potential $V^{(\rm ext)}(z)$ can be added in Eq. (7.4). This is particularly relevant to the practical case of channels in Si MOSFETs, since these are heavily biased by the gate contact that is used to modulate the electron density, and so the current in the transistor. It may be appropriate to return to this section after having read the discussion given in Sect. 8.2.1: There the same problem is tackled using the effective-mass approximation. In Figure 7.4 we show the squared amplitude of the lowest-lying unprimed/primed conduction-band and valence-band wavefunctions (averaged over the area of the cell on the $(x,y)$ plane) obtained in the presence of a "triangular well" potential whose Fourier components $V^{(\rm ext)}_{G_z}$ are given by $a_{\rm Si}F_{\rm s}(N+N_{\rm v})/2$ for $G_z=0$ and:

$$V^{(\rm ext)}_{G_z} = \frac{iF_{\rm s}}{G_z}\,, \tag{7.20}$$

where $F_{\rm s}$ is the surface field ($= 5\times10^5$ V/cm in Fig. 7.4). Note the oscillation of the fast-varying periodic component of the Bloch function and the expected slow-varying envelope. Figures 7.5, 7.6, 7.7, 7.8, 7.9, and 7.10 show details of the band structure and DOS near the gap for 2-cell, 3-cell, and 9-cell thin Si layers (the latter ones in the presence of a constant confining field) with surfaces of different orientations, while Fig. 7.11 compares the DOS obtained using the supercell method with what is obtained for parabolic bands for various surface orientations.

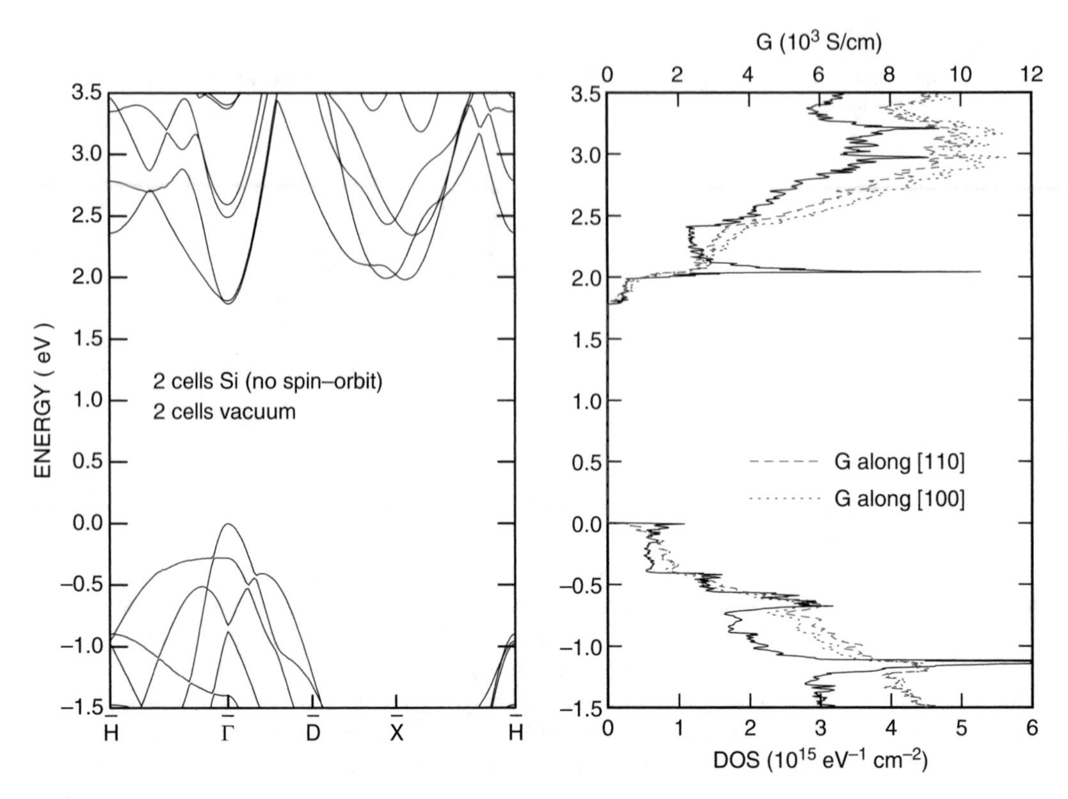


**Fig. 7.4** Squared amplitude—averaged over a cell on the plane of the slab—of the wavefunctions in a 9-cell-thick Si layer in vacuo with H termination and a triangular-well potential with a field of $5 \times 10^5$ V/cm. At *left* are shown the wavefunctions of the unprimed states, at *center* those of the primed electron states, and at *right* the hole states. [Used from M.V. Fischetti, Bo Fu, S. Narayanan, and J. Kim, in *Nano-Electronic Devices: Semiclassical and Quantum Transport Modeling*, Dragica Vasileska and Stephen M. Goodnick eds. (Springer, New York, 2011), pp. 183–247, with kind permission from Springer Science+Business Media]

### 7.2.2 III–V Hetero-Channels

Hetero-structures composed of thin layers of III–V compound semiconductors constitute the building blocks of most opto-electronic devices that we use daily, such as light-emitting diodes (LEDs), and injection lasers. The book by Bastard and collaborators [8] gives a review of the subject.

These hetero-structures are typically fabricated using a crystal-growth technique called Molecular Beam Epitaxy (MBE): In a chamber kept under very high vacuum, beams of the desired elements are aimed at a substrate. By keeping the desired ratio of elements according to the stoichiometry of the compound to be grown, atomic layers of the desired material can be deposited. This technique allows excellent control of the chemical composition of the layers, of their crystal structure, and of their thickness. Binary III–V compound semiconductors, such as GaAs or InAs, can be grown on a substrate, usually consisting of InP wafers that can be obtained commercially in relatively large diameters. Ternary alloys, such as $In_xGa_{1-x}As$, can also be grown by controlling the ratio of the In, Ga, and As beams. In typical applications, compounds that are "lattice matched" to the InP substrate are preferred. Indeed, the ternary alloy composed of a mole fraction $x$ of InAs and a mole fraction $(1-x)$ of GaAs, denoted by $In_xGa_{1-x}As$, in general has a lattice constant different from that of InP. Only when $x = 0.53$ does the ternary alloy have the same lattice constant. This is very desirable, since it eliminates concerns caused by strain. If strain is desired, or a non-lattice-matched alloy is desired

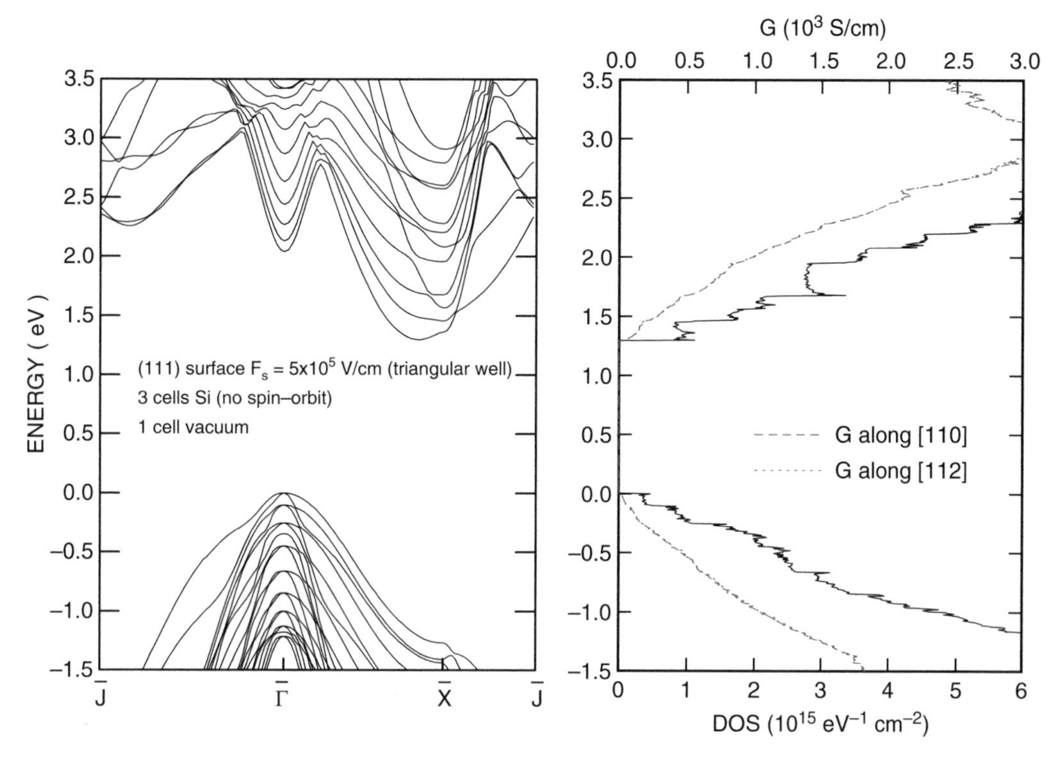


**Fig. 7.5** Band structure (*left*), density of states, and ballistic conductance along the [100] and [110] directions (*right*) of a (100) Si slab as in the right frame of Fig. 7.2. [Used from M.V. Fischetti, Bo Fu, S. Narayanan, and J. Kim, in *Nano-Electronic Devices: Semiclassical and Quantum Transport Modeling*, Dragica Vasileska and Stephen M. Goodnick eds. (Springer, New York, 2011), pp. 183–247, with kind permission from Springer Science+Business Media]

(for example, using a larger In mole fraction to control the electronic properties), strained layers can be grown without crystalline defects up to a critical thickness. Beyond this critical thickness, it will be energetically more favorable to relieve the stress by creating crystal defects, such as stacking faults.

A typical application is represented by quantum-well lasers. Layers of GaAs are grown first, then several nm of $In_{0.53}Ga_{0.47}As$ are grown afterwards, and GaAs is deposited again. Since the band alignment of these materials is such that the conduction-band bottom of $In_{0.53}Ga_{0.47}As$ sits at a lower energy than that of GaAs, a confining quantum well is formed. Similarly, a quantum well for holes is formed in the valence bands. When carriers are injected into the structure, electrons and holes will loose energy and be captured by the well. They will then recombine radiatively emitting light. Keeping the structure in an electromagnetic cavity, stimulated emission will occur, that is, the laser action.

Optical applications have made hetero-structure a thriving industry. However, the low effective mass and related high velocity of electrons in most III–V compound semiconductors have triggered interest by the VLSI industry, attracted by their potential use as active channels in fast transistors. The structure that we consider here is indeed a typical research application of this type.

Regarding the theoretical study of these structures, ternary alloys require a comment. We have seen how to use pseudopotentials, empirical or self-consistent, to study Si or GaAs. In order to study a ternary alloy, $In_xGa_{1-x}As$, for example, in principle we should consider a supercell large enough to contain a fraction $x/2$ of In atoms, a fraction $(1-x)/2$ of Ga atoms, and a fraction 1/2 As atoms, all in zincblende cells. Clearly, we would need to consider very large cells. Moreover, we would be able to study structures with a limited number of mole fractions $x$ (that must result in commensurate integer number of atoms). Most important, we would have to perform a very large number of calculations,

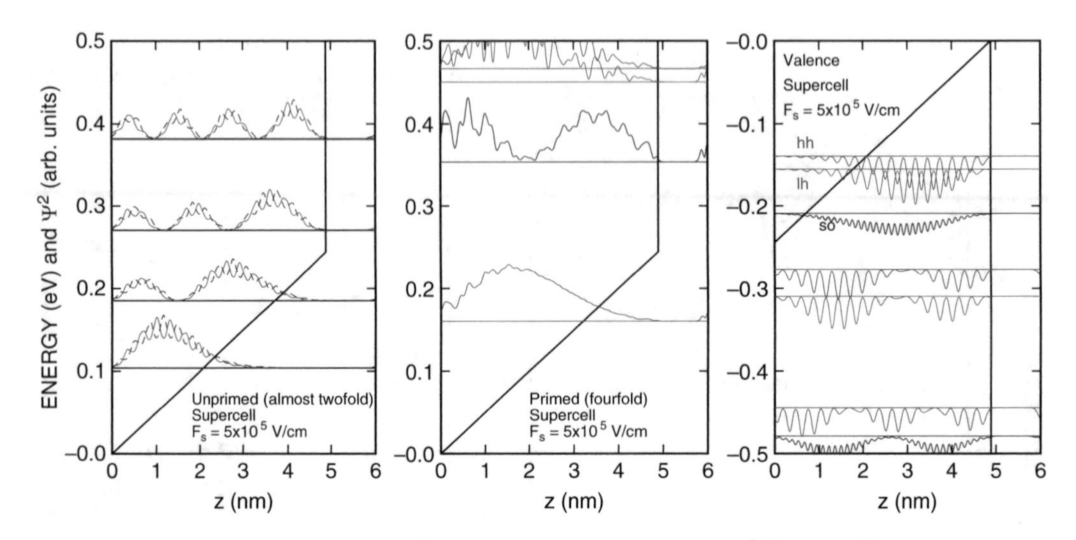


**Fig. 7.6** Band structure (*left*), density of states, and ballistic conductance along the $[1\bar{1}0]$ and $[11\bar{2}]$ directions (*right*) of a (111) Si slab. The film is 3-cell ($3\sqrt{3}a_0$) thick, H terminated and 1 cell of vacuum separates the periodically repeated films. A constant field of $5 \times 10^5$ V/cm is applied perpendicularly to the slab to mimic a triangular-well confining potential

each for a different random spatial distribution of In and Ga atoms. The common way to bypass this problem is to use what is normally called the *virtual crystal approximation* (VCA). We consider a zincblende lattice composed of As atoms and "virtual" In/Ga atoms obtained by interpolating linearly their properties, their pseudopotentials in our case. To be explicit, the pseudopotential $V_x^{(\mathrm{In/Ga})}$ of this virtual atom will be obtained from the pseudopotentials $V^{(\mathrm{In})}(q)$ and $V^{(\mathrm{Ga})}(q)$ of In and Ga in InAs and GaAs, respectively, as the linear combination:

$$V_x^{(\mathrm{In/Ga})}(q) = x\,\frac{\Omega_{\mathrm{In}}}{\Omega_{\mathrm{sc}}}\,V^{(\mathrm{In})}(q) + (1-x)\,\frac{\Omega_{\mathrm{Ga}}}{\Omega_{\mathrm{sc}}}\,V^{(\mathrm{Ga})}(q)\,, \tag{7.21}$$

where $\Omega_{\mathrm{In}}$ and $\Omega_{\mathrm{Ga}}$ are the (normalization) atomic volumes of In and Ga in InAs and GaAs, respectively, and $\Omega_{\mathrm{sc}}$ is the volume of the cell or supercell of the structure we intend to study. Clearly, this approximation lets us perform a much simpler calculation. The price we pay is some loss of accuracy, especially our inability to study fluctuations present in the system. For example, the band structure we will obtain for $\mathrm{In_{0.53}Ga_{0.47}As}$ will be "smooth," as in a binary alloy. In reality, free electrons will feel fluctuations going from one cell of InAs to the next of GaAs, resulting in *alloy scattering*. This effect will have to be treated in some other way. For completeness, regarding the VCA we also mention that the lattice constant of ternary alloys is usually estimated using *Vegard's law* [9], which consists in a linear interpolation of the lattice constants of the binary compounds. If $a_A$ is the lattice constant of the binary compound $A$ (think of $A$ as InAs), and $a_B$ is the lattice constant of the binary compound $B$ (think of $B$ as GaAs), then the lattice constant $a_{AB,x}$ of the ternary alloy $A_xB_{1-x}$ (think of $\mathrm{In}_x\mathrm{Ga}_{1-x}\mathrm{As}$) will be

$$a_{AB,x} = x\,a_A + (1-x)a_B + bx(1-x)\,, \tag{7.22}$$

where we have added a "bowing parameter" $b$ that reflects deviations from linearity. On a similar note, a rule-of-thumb to estimate the valence- and conduction-band discontinuities at interfaces is the so-called *Anderson's 60/40 rule* [10], which states that in a type-I band alignment (in which the band gap of the smaller-gap material is fully contained within the gap of the larger-gap material), 60 % of the

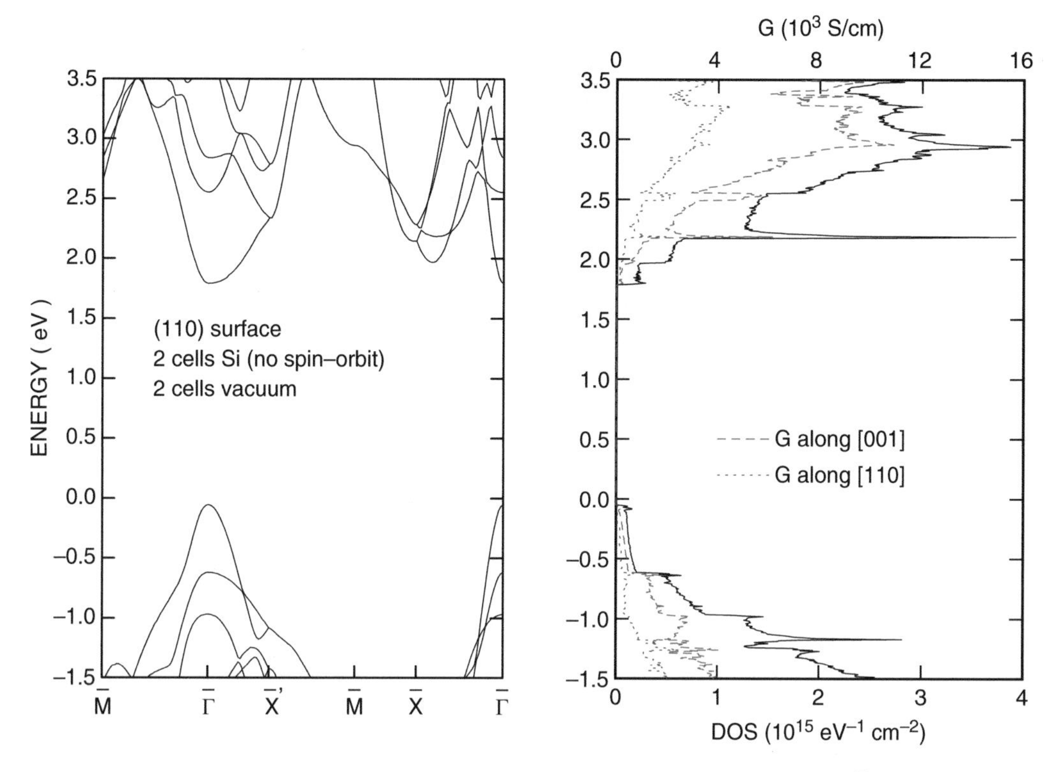


**Fig. 7.7** Band structure (*left*), density of states, and ballistic conductance along the [001] and $[1\bar{1}0]$ directions (*right*) of a (110) Si slab The slab is 2-cell $(2a_0/\sqrt{2})$ thick and H terminated. A similar thickness of vacuum padding has been employed. The twofold degenerate absolute minimum of the conduction band is at $\mathbf{K} = (0, 0.15)(2\pi/a_0)$, while a fourfold degenerate minimum is at $\mathbf{K} = (0.85/\sqrt{2}, 0)(2\pi/a_0)$. Note that the energetic ordering of these minima is opposite to what shown in Fig. 7.10 because the large nonparabolicity of the dispersion around the twofold minimum along the [110] direction weighs heavily at the high energies shown here for this very thin film. [Used from M.V. Fischetti, Bo Fu, S. Narayanan, and J. Kim, in *Nano-Electronic Devices: Semiclassical and Quantum Transport Modeling*, Dragica Vasileska and Stephen M. Goodnick eds. (Springer, New York, 2011), pp. 183–247, with kind permission from Springer Science+Business Media]

difference of the band gaps shows as the offset of the conduction bands, 40 % as valence-band offset. Both Vagard's law and Anderson's rule have found successful applications in a limited number of cases, but they should be considered as empirical rules with limited validity. For example, Anderson's rule has been applied to some III–V hetero-structures, but we saw in Chap. 6, Sect. 6.3.4 a counter-example to Anderson's rule for the Si–Ge system: In that system, almost all of the difference of the band gaps shows as a valence-band offset.

As it has been the case for Si inversion layers, the use of the effective-mass approximation has provided early insight. However, additional complications arise in the context of III–V hetero-structures. These are mainly due to the smaller potential barriers that confine the electrons. The Si–$SiO_2$ interface presents a barrier of about 3 eV to conduction electrons. For all practical purposes, this can be approximated by an infinite barrier (unless one wishes to worry about the tiny detail of the penetration of the wavefunction inside the oxide). But the band discontinuities present in hetero-structures are about one order of magnitude smaller. This presents the advantage of opening the path to the fabrication and use of *superlattices*: periodic repetition of a hetero-structure. However, the disadvantage from a theoretical viewpoint is that now electrons can easily move across interfaces (this is the whole point, after all!), but issues arise: How do we "match" the wavefunction across interfaces? And if transport occurs on the plane of the interface with electronic wavefunctions extending in both

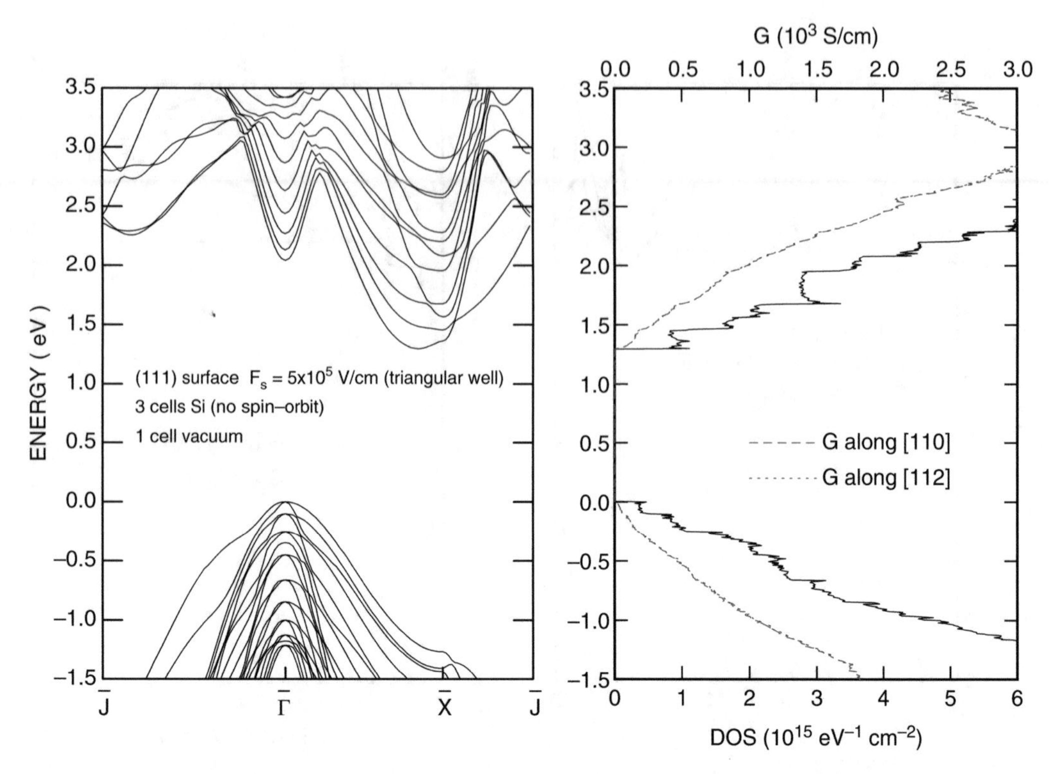


**Fig. 7.8** Band structure (*left*), density of states, and ballistic conductance along the $[1\bar{1}0]$ and $[11\bar{2}]$ directions (*right*) of a (111) Si slab. The film is 3-cell ($3\sqrt{3}a_0$) thick, H terminated and 1 cell of vacuum separates the periodically repeated films. A constant field of $5 \times 10^5$ V/cm is applied perpendicularly to the slab to mimic a triangular-well confining potential. [Used from M.V. Fischetti, Bo Fu, S. Narayanan, and J. Kim, in *Nano-Electronic Devices: Semiclassical and Quantum Transport Modeling*, Dragica Vasileska and Stephen M. Goodnick eds. (Springer, New York, 2011), pp. 183–247, with kind permission from Springer Science+Business Media]

the GaAs and the InGaAs layers, what in-plane effective mass do we use? Atomistic calculations can help. We consider here the specific case of a rather complicated hetero-structure and show how one can tackle these problems. Note that the use of the supercell method renders this structure a "superlattice," but we shall be interested on the effect that the results will have on electron transport "on the plane," as opposed to what is called "vertical transport," since we consider this structure as the channel of a III–V-based MOSFET.

The structure we consider consists a (100) lattice-matched (to InP, as we have explained before) $In_{0.53}Ga_{0.47}As/InP/Al_{48}In_{0.52}As$ hetero-channel that mimics a typical III–V MOSFET channel. The supercell used in the calculation consists of a composite $In_{0.53}Ga_{0.47}As/InP$ channel with a 4-cell-thick $In_{0.53}Ga_{0.47}As$ layer, an equally thick InP layer, and of an "insulating" 3-cell-thick $Al_{48}In_{0.52}As$ back layer. Figure 7.12 shows the band alignment and the electronic structure obtained using the empirical pseudopotentials from Zunger's group [11–13] with parameters listed in Table B.5 of Appendix B. These parameters have been calibrated to obtain the "correct" band alignment and, so, the confining wells. Since insulators such as $HfO_2$ or $Al_2O_3$ are used to passivate and insulate layers, it is interesting to see the effect of replacing the weakly insulating $Al_{0.48}In_{0.52}As$ back layer with an "ad-hoc" insulator mimicking $HfO_2$ (with empirical pseudopotentials reported in the literature). This is shown in the right panel of Fig. 7.12. This insulator behaves even too realistically: We can see the appearance of "interface states" (with wavefunctions localized at the InP/insulator interface) in this case. They give rise to an energy band within the gap, exactly as dangling bonds do in real life.

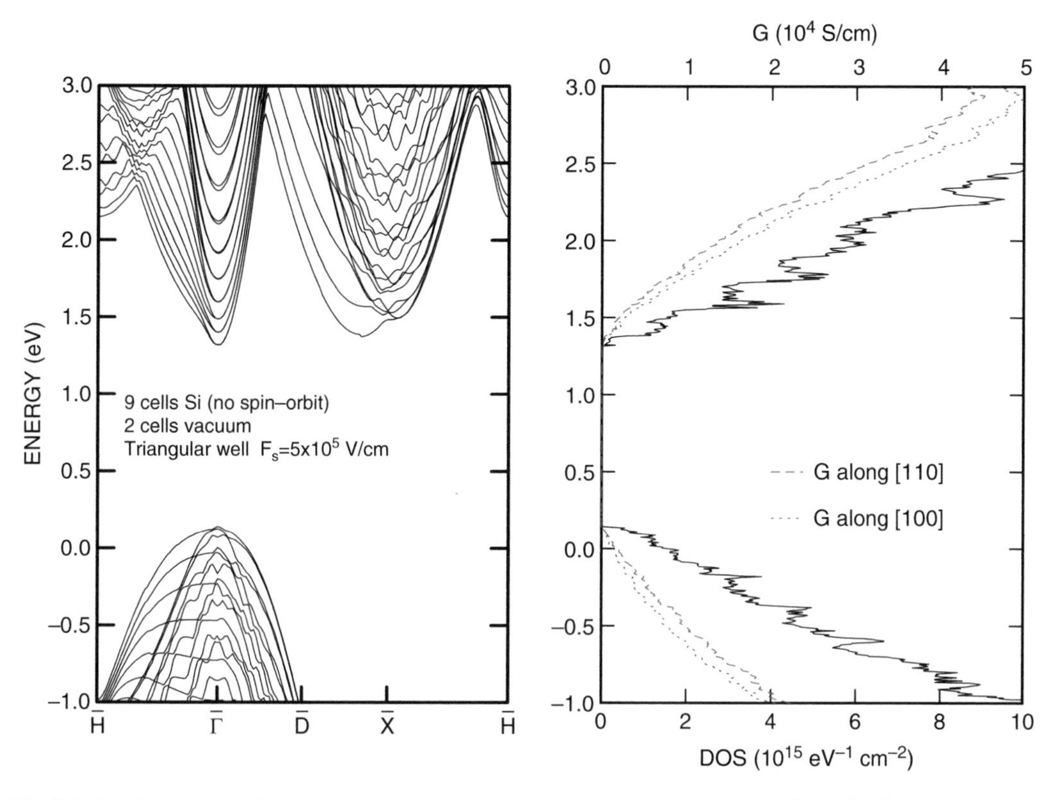


**Fig. 7.9** Band structure (*left*), density of states, and ballistic conductance along the [100] and [110] directions (*right*) of a (100) Si slab as in the left frame of Fig. 7.2, but with a triangular-well potential with a field of $5\times10^5$ V/cm. The zero for the energy has been set at the top of the valence bands in the absence of the applied field, as in the left frame of Fig. 7.2, to judge the shift of the subband energies in the presence of the external field. [Used from M.V. Fischetti, Bo Fu, S. Narayanan, and J. Kim, in *Nano-Electronic Devices: Semiclassical and Quantum Transport Modeling*, Dragica Vasileska and Stephen M. Goodnick eds. (Springer, New York, 2011), pp. 183–247, with kind permission from Springer Science+Business Media]

Figure 7.13 shows the variation of the energy gap as a function of the thickness of the composite channel, while Fig. 7.14 shows how the band structure is modified by the application of a parabolic potential (mimicking the potential of an inversion layer) of the form:

$$V^{(\mathrm{ext})}(z) = V_0 \left(1 - \frac{2z}{L} + \frac{z^2}{L^2}\right) , \tag{7.23}$$

(where $L$ is the extension of the supercell in the $z$ direction, $L = a_{InP}N$, with $N$ the total number of cells employed and $V_0 = F_\mathrm{s}L/2$ is the total voltage drop in the cell expressed in terms of the surface field $F_\mathrm{s}$) with Fourier components $F_\mathrm{s}L/3$ for $G_z = 0$ and

$$V_{G_z}^{(\mathrm{ext})} = -\frac{F_\mathrm{s}}{2}\left(\frac{2}{LG_z^2} - \frac{\mathrm{i}}{G_z}\right) \tag{7.24}$$

for $G_z \neq 0$. Figure 7.14 presents results obtained for $F_\mathrm{s} = \pm 5 \times 10^5$ V/cm, for electron and hole confinement, respectively. These are of interest in the study of $n$- and $p$-channel devices. Finally, the electron and hole wavefunctions in the $In_{0.53}Ga_{0.47}As$/InP/$Al_{48}In_{0.52}As$ hetero-layer for the cases of

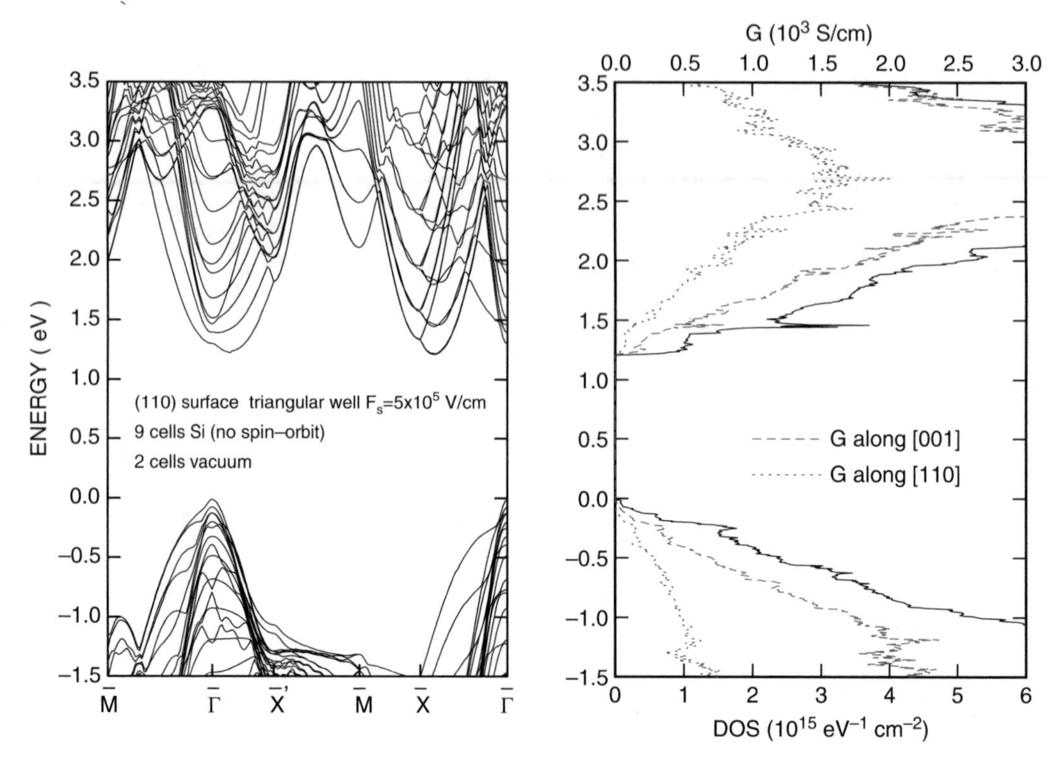


**Fig. 7.10** Band structure (*left*), density of states, and ballistic conductance along the [001] and [$1\bar{1}0$] directions (*right*) of a (110) Si slab 9-cell ($9a_0/\sqrt{2}$) thick with a triangular-well potential with a field of $5\times10^5$ V/cm. Comparing with Fig. 7.7, note that the twofold minimum $\mathbf{K} = (0,0.15)(2\pi/a_0)$ is now at an energy about 12 meV higher than the fourfold minimum at $\mathbf{K} = (0.85/\sqrt{2},0)(2\pi/a_0)$. From estimates based on conventional effective masses [1] we expect this same ordering, but an energy difference of about 28 meV. The difference is likely due to nonparabolic effects. [Used from M.V. Fischetti, Bo Fu, S. Narayanan, and J. Kim, in *Nano-Electronic Devices: Semiclassical and Quantum Transport Modeling*, Dragica Vasileska and Stephen M. Goodnick eds. (Springer, New York, 2011), pp. 183–247, with kind permission from Springer Science+Business Media]

flat-band, electron and hole confinement are shown in Fig. 7.15. Note how potentially intricate issues (such as matching the wavefunctions at interfaces, determining the effective mass to be used when the wavefunction extends over two materials, etc.) are bypassed by the supercell method. The shape of the envelope of the wavefunctions agrees with our naïve expectations based on the band discontinuities shown and on the "envelope" idea.

Figure 7.16 shows the large nonparabolicity of the InGaAs conduction band, an effect that obviously translates into significant effects on electron transport and on the gate capacitance of InGaAs-based devices. Finally, the right frame of Fig. 7.17 shows the 2D band structure of this "superlattice" in the left frame and the corresponding dispersion calculated keeping the 2D in-plane wavevector $\mathbf{K}$ at the $\bar{\Gamma}$-point while varying the $z$-component, $k_z$, along the direction perpendicular to the interfaces. Dispersionless curves (i.e., $k_z$-independent) are expected for purely two-dimensional, confined states. This is indeed the case at low energies both in the conduction and valence bands. At higher energies we can see the onset of a dispersion, a symptom of the delocalized nature of states (superlattice "minibands") whose energy exceeds the confinement energy caused by the band alignment at the $In_{0.53}Ga_{0.47}As/Al_{0.48}Ga_{0.52}As$ interface. Strikingly different is the case of the free-standing H-terminated Si layer of Fig. 7.5: The strong confinement due to the large vacuum workfunction causes an almost completely dispersionless behavior for energies as large as 5 eV in the conduction band, as clearly seen in Fig. 7.18.

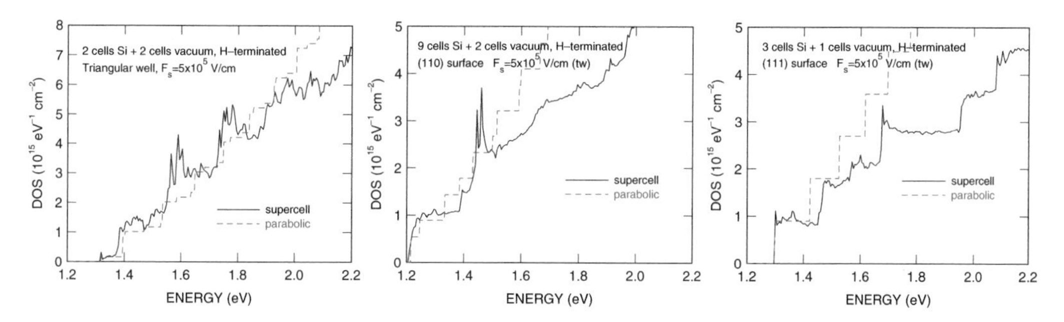


**Fig. 7.11** Density of states of the (100), (110), and (111) Si slabs of the previous figures (*solid black line*) compared to the DOS calculated using a two-ladder (for the (100) and (110) surfaces) or one-ladder (for the (111) surface) parabolic band structure with longitudinal and transverse masses of 0.19 and 0.91 m. For the (100) surface note that the energy of the ground state unprimed (primed) subband obtained using the supercell method is 0.103 eV (0.158 eV), compared to the "parabolic" value of 0.109 eV (0.184 eV). [Used from M.V. Fischetti, Bo Fu, S. Narayanan, and J. Kim, in *Nano-Electronic Devices: Semiclassical and Quantum Transport Modeling*, Dragica Vasileska and Stephen M. Goodnick eds. (Springer, New York, 2011), pp. 183–247, with kind permission from Springer Science+Business Media]

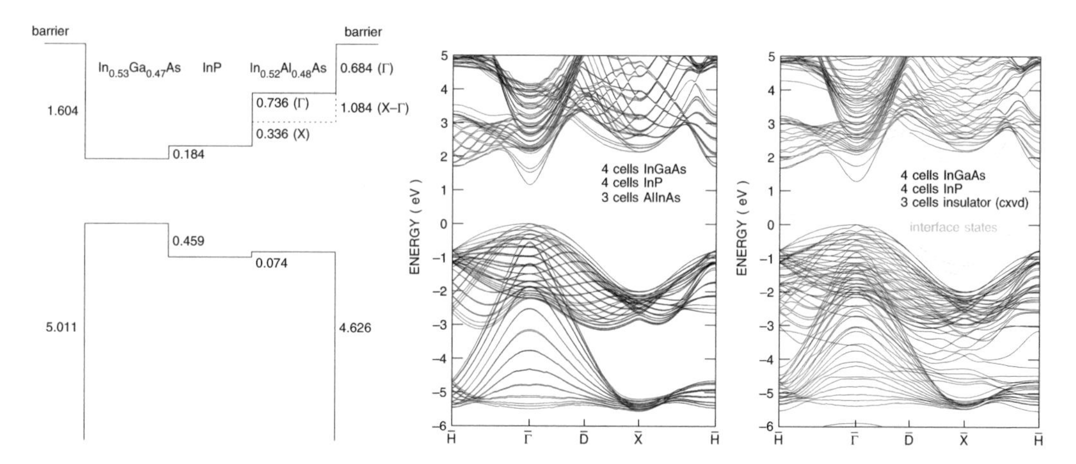


**Fig. 7.12** *Left*: Band diagram showing the discontinuities/alignments for a (100) lattice-matched (to InP) $In_{0.53}Ga_{0.47}As/InP/Al_{48}In_{0.52}As$ periodic hetero-structure resulting from the Zunger's atomic pseudopotentials and accounting for spin–orbit interaction. *Center*: Band structure for the system at left with 4-cells/4-cells/3-cells layer thickness (1 cell = 1 InP cell = 0.586 nm). *Right*: Band structure for a system as in the frames at *left*, but with the AlInAs "insulator" replaced by an ad-hoc polar insulator mimicking the band alignment of $HfO_2$. Note the emergence of interface states, caused by the presence of the insulator. As in the thin Si case, for convenience the spin–orbit interaction has been neglected here. [Used from M.V. Fischetti, Bo Fu, S. Narayanan, and J. Kim, in *Nano-Electronic Devices: Semiclassical and Quantum Transport Modeling*, Dragica Vasileska and Stephen M. Goodnick eds. (Springer, New York, 2011), pp. 183–247, with kind permission from Springer Science+Business Media]

### 7.2.3 Graphene

Graphene constitutes the first example of a 2DEG confined not by band discontinuities between two different materials, but by the ionic potentials of the atoms arranged in a two-dimensional lattice. Once more, we refer to the review paper by Geim and Novoselov [2], who first studied and produced graphene layers by "exfoliating" graphite (that is, by stripping a single layer with scotch tape) and were awarded the Nobel prices in Physics in 2010 for this work.

* *Actually, the existence of two-dimensional crystals was a surprise for many. In the mid-late 1960s, Mermin and Wagner [14, 15], and Hohenberg [16] proved that crystals in two (and one) dimension(s)*

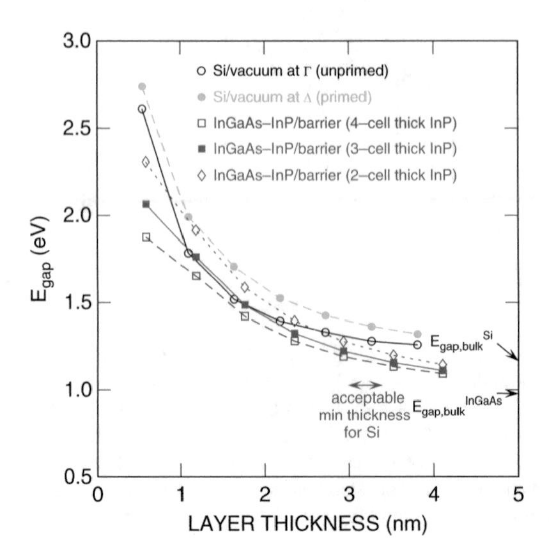


**Fig. 7.13** Variation of the band gap of the AlInAs/InGaAs/InP quantum-well structure as a function of InGaAs thickness for three different values of the InP thickness and a 3-cell-thick insulating barrier layer compared with the Si (100) and (001) band gaps. This plot may be used to estimate the sensitivity of the threshold voltage to thickness variations in ultra-thin-body Si and InP-InGaAs channel FETs. [Used from M.V. Fischetti, Bo Fu, S. Narayanan, and J. Kim, in *Nano-Electronic Devices: Semiclassical and Quantum Transport Modeling*, Dragica Vasileska and Stephen M. Goodnick eds. (Springer, New York, 2011), pp. 183–247, with kind permission from Springer Science+Business Media]

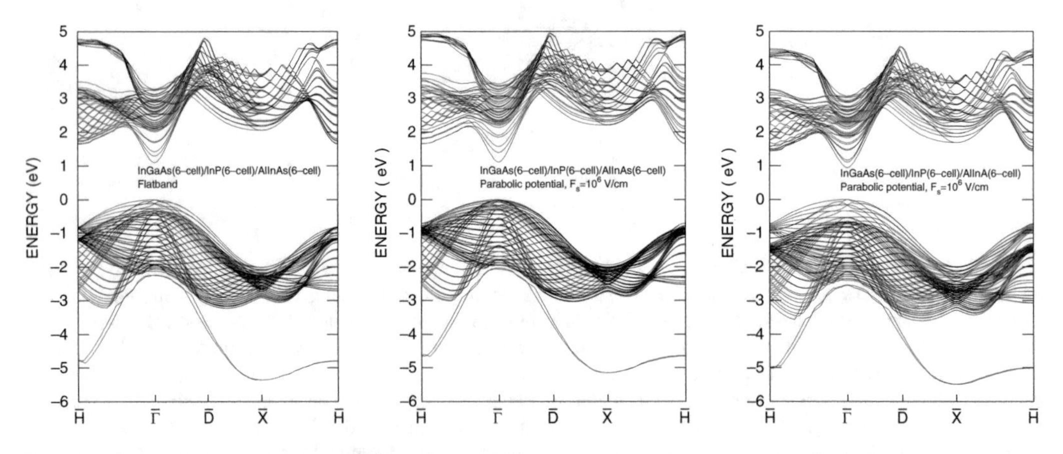


**Fig. 7.14** In-plane dispersion for an InGaAs/InP/AlInAs hetero-channel under flat-band conditions (*left*), with a parabolic potential with a surface electric field of $10^6$ V/cm (*center*, confinement for electrons) and $-10^6$ V/cm (*right*, confinement for holes). [Used from M.V. Fischetti, Bo Fu, S. Narayanan, and J. Kim, in *Nano-Electronic Devices: Semiclassical and Quantum Transport Modeling*, Dragica Vasileska and Stephen M. Goodnick eds. (Springer, New York, 2011), pp. 183–247, with kind permission from Springer Science+Business Media]

*cannot exist. The argument, often formulated in rather abstract terms of Quantum Field Theory, states that the spontaneous breaking of a continuous symmetry (in our case: the continuous rotational symmetry that must be broken in order to form a crystal that does not possess a* continuous *symmetry) in two or fewer dimensions cannot occur at a finite (nonzero) temperature, because the correlation of the associated massless Goldstone bosons would diverge... We can explain it in a simpler way. However, even this "simple" explanation may be understood after having assimilated the concepts presented in Chap. 9, Sect. 9.4, and in Chap. 10, Sect. 10.1.5.*

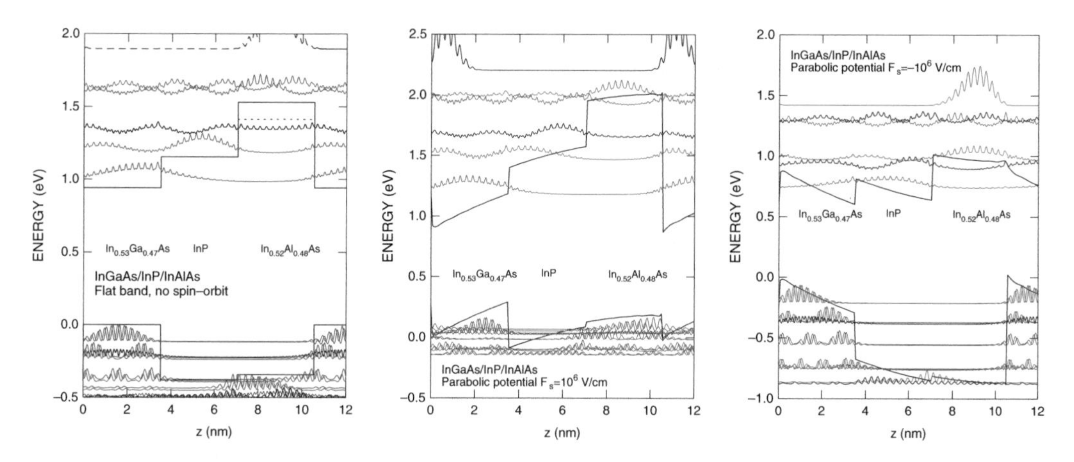


**Fig. 7.15** Conduction- and valence-band wavefunctions in an InGaAs/InP/AlInAs channel (each layer 6-cell thick) under flat-band condition (*left*) or in the presence of a parabolic potential with a surface electric field of $10^6$ V/cm (*center*, confinement for electrons) and $-10^6$ V/cm (*right*, confinement for holes), as in the previous figure. This plot emphasizes the main strength of the method: Complicated issues related to matching envelope wavefunctions at hetero-interfaces, nonparabolic effects, the value of the in-plane effective mass when electronic wavefunction spans several different materials, etc. are all bypassed in the "correct" way. [Used from M.V. Fischetti, Bo Fu, S. Narayanan, and J. Kim, in *Nano-Electronic Devices: Semiclassical and Quantum Transport Modeling*, Dragica Vasileska and Stephen M. Goodnick eds. (Springer, New York, 2011), pp. 183–247, with kind permission from Springer Science+Business Media]

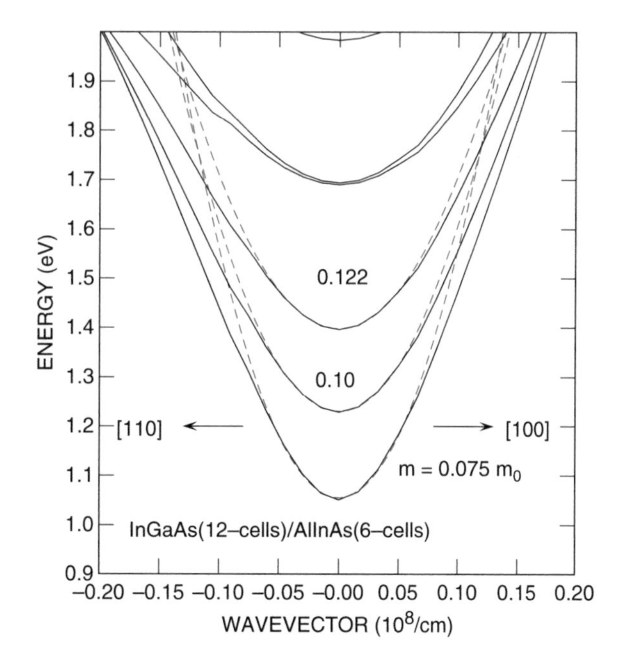


**Fig. 7.16** Band structure near the bottom of the conduction band for the superlattice shown in the previous figure. The anisotropy of the bands as well as the large nonparabolic effects is clearly visible. The *dashed lines* illustrate the parabolic dispersion obtained using the effective mass calculated from the curvature of the dispersion at the bottom of the conduction band

*Consider first a three-dimensional crystal. The number of quantized vibrational modes of a lattice (phonons) at temperature T is given by the Bose–Einstein distribution, Eq. (10.27), which we can rewrite in the form:*

$$\langle N_{\mathbf{q}} \rangle_{\text{th}} = \frac{1}{\exp\left(\frac{k_{\text{B}}T}{\hbar\omega_{\mathbf{q}}}\right) - 1}, \tag{7.25}$$

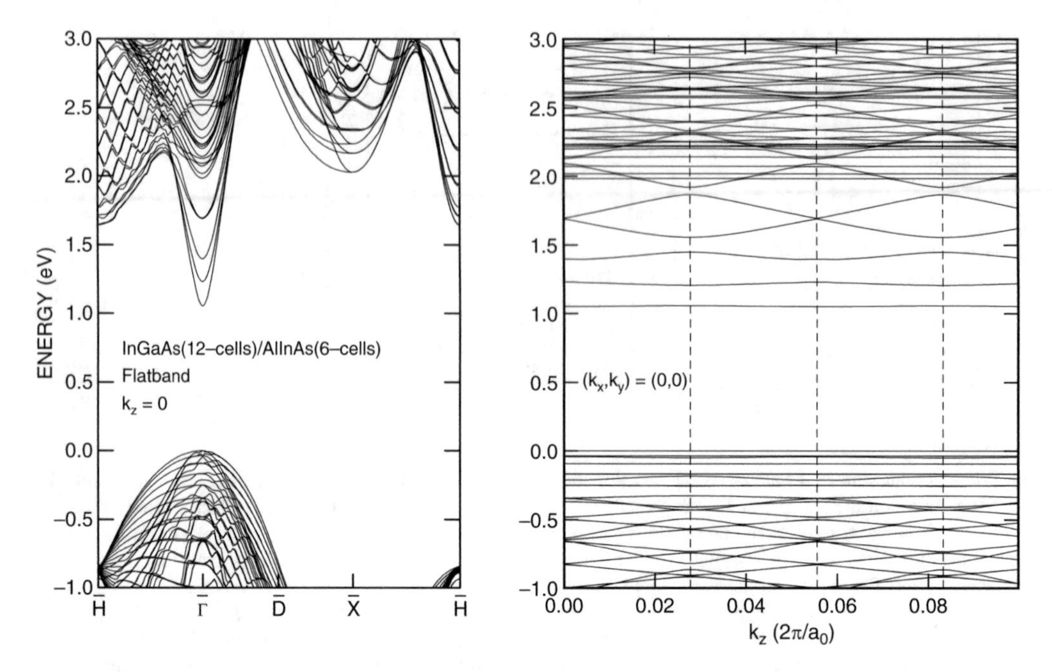


**Fig. 7.17** The 2D band structure of the hetero-layer of the previous figures (*left*) is shown against the dispersion along the "transverse" $k_z$ direction for $(k_x,k_y) = \mathbf{0}$ (the $\overline{\Gamma}$ point of the 2D BZ, *right*). The *vertical dashed lines* at *right* show the size of the BZs along the $k_z$ direction ($a_0 = 0.586$ nm) equal to 1/18 $(2\pi/a_0)$. For fully 2D states the energy is dispersionless, that is, independent of $k_z$. However, as the energy of the subband-bottoms exceeds the confinement energy caused by the $In_{0.53}Ga_{0.47}As/Al_{0.48}Ga_{0.52}As$ band alignment, the states become "extended 3D states" and acquire the "normal" bulk (superlattice miniband) dispersion. From this plot one can extract an $In_{0.53}Ga_{0.47}As/Al_{0.48}Ga_{0.52}As$ conduction-band discontinuity $\Delta E_c \approx 0.5$ eV and a valence-band discontinuity $\Delta E_v \approx 0.2$ eV. Plots of this type are useful in judging the correctness of the band alignment caused by the choice of the interpolation of the bulk pseudopotentials as $q \to 0$

*where* $k_B$ *is Boltzmann's constant and* $\omega_{\mathbf{q}}$ *is the frequency of the phonon. This is a wave of ions oscillating around their equilibrium positions, moving along the direction of the wavevector* $\mathbf{q}$ *with wavelength* $\lambda = q/(2\pi)$*. Those waves that represent essentially propagation of sound in the crystal, the so-called acoustic phonons, have a frequency that depends linearly on their wavevector for large wavelengths; that is, assuming an isotropic crystal (the same in all directions),* $\omega_q \approx c_s q$*, where* $c_s$ *is the velocity of sound. Therefore, for long wavelengths (i.e.,* $q \to 0$*) we can approximate* $\langle N_{\mathbf{q}} \rangle_{\text{th}} \approx k_B T/(\hbar c_s q)$*. The volume density,* $N/\Omega$*, of these "acoustic waves" with wavelength larger than some arbitrary wavelength* $\lambda_{\max}$ *will be obtained by integrating the number of phonons at a given wavelength over all wavelengths larger than* $\lambda_{\max}$*; or, equivalently, for all wavevectors smaller than* $q_{\max} = 2\pi/\lambda_{\max}$*:*

$$\frac{N}{\Omega} \approx \int_{q \le q_{\max}} \frac{\mathrm{d}\mathbf{q}}{(2\pi)^3} \frac{k_B T}{\hbar c_s q} = \frac{k_B T}{2\pi^2 \hbar c_s} \int_0^{q_{\max}} \mathrm{d}q\, q = \frac{k_B T}{\hbar c_s q_{\max}} \frac{q_{\max}^3}{4\pi^2} \,. \tag{7.26}$$

*Whatever this number might be, it is finite.*

*Now, instead of a three-dimensional crystal, consider a two-dimensional one. Looking at one example of phonons in a two-dimensional crystal, the phonon frequencies in graphene shown in Fig. 13.6 in Chap. 13, we see that the phonons labeled "ZA" have a frequency that, for* $Q \to 0$*, goes to zero as* $aQ^2$*, where a is some constant whose values do not matter in this discussion. (We denote by upper-case letters quantities in two dimensions.) These are waves associated with carbon atoms*

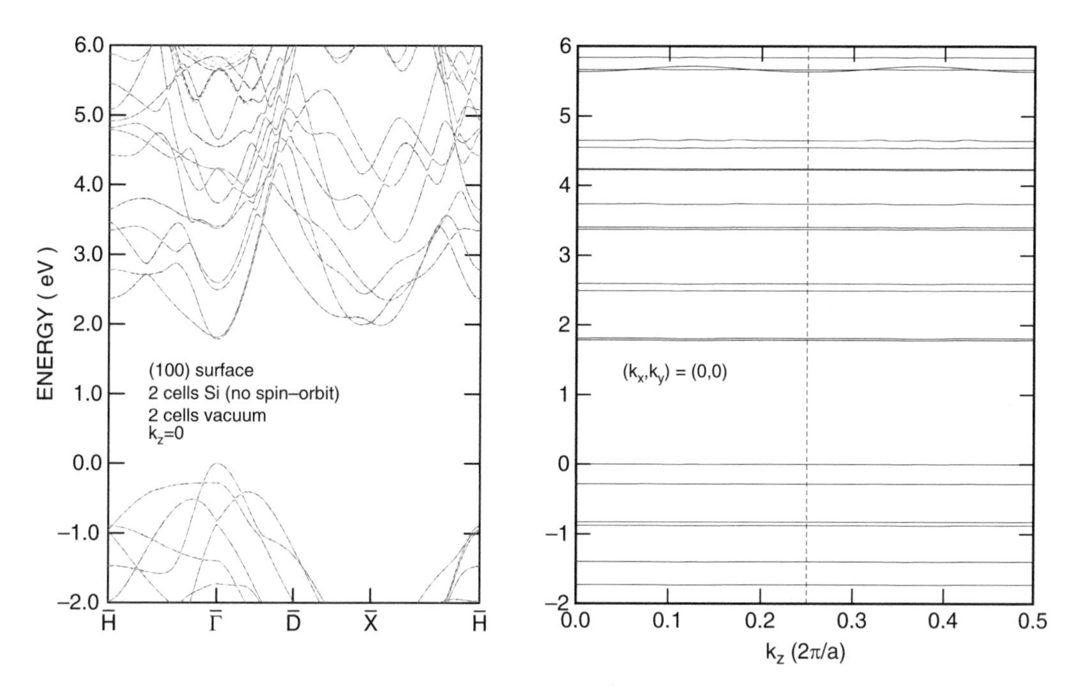


**Fig. 7.18** The 2D band structure of the H-terminated free-standing Si layer of Fig. 7.5 (*left frame*) is shown against the dispersion along the "transverse" $k_z$ direction for $(k_x, k_y) = \mathbf{0}$ (the $\overline{\Gamma}$ point of the 2D BZ, *right frame*), as in the previous figures. Note the absence of any appreciable dispersion up to an energy of about 5 eV, indicating the quantum confinement caused by the high vacuum workfunction. This is also emphasized at *left*: *Curves of different colors* correspond to calculations performed varying the in-plane components $\mathbf{K}$ of the wavevector on the 2D BZ and selecting values of 0.05 (*red dashed lines*), 0.10 (*blue dotted lines*), and 0.25 $2\pi/a_0$ (*green dot-dashed lines*) for the out-of-plane component $k_z$. The 2D band structure is once more independent of $k_z$ for energies up to 5 eV in the conduction band

*that move "up and down," out of the plane of the graphene sheet. The reason why now the frequency goes to zero faster at long wavelength is due to the fact that there are no atoms "above and below." Therefore, the restoring force that causes the atoms to oscillate above and below the plane is due to the "diagonal" pull of other atoms on the plane. This makes the out-of-plane "spring constant" smaller as the wavelength increases, as a simple geometrical argument shows. This is an intrinsic property of two-dimensional systems. Now, let's compute the areal density, $N_{2D}/A$, of these out-of-plane acoustic waves with wavelength larger than $\lambda_{max}$:*

$$\frac{N}{A} \approx \int_{Q \leq Q_{max}} \frac{\mathrm{d}\mathbf{Q}}{(2\pi)^2} \frac{k_B T}{\hbar a Q^2} = \frac{k_B T}{2\pi \hbar a} \int_0^{Q_{max}} \mathrm{d}Q \, \frac{1}{Q} \to \infty \,. \tag{7.27}$$

*The problem results from two effects arising from the reduced dimensionality of the crystal: First, comparing to the three-dimensional case, we have gained a factor of $Q$ in the denominator because of the different ZA phonon frequency. Moreover, we have also lost a factor of $Q$ in the numerator, because of the different phase-space in two dimensions. The important result is that the energy of these out-of-plane waves is so small that they can be excited in infinite numbers. The crystal would just disintegrate, oscillating out of the plane with increasing amplitude at longer wavelength.*

*So: Why does graphene exist? Simply because it comes in finite size and it deviates from full planarity. Graphene usually comes in the form of "flakes." Moreover, even in large areas, unavoidably it forms ripples. The length-scale set by the size of the flakes and by the ripples is what sets a nonzero lower limit for the integral in Eq. (7.27), so that the logarithmic singularity is avoided. We should*

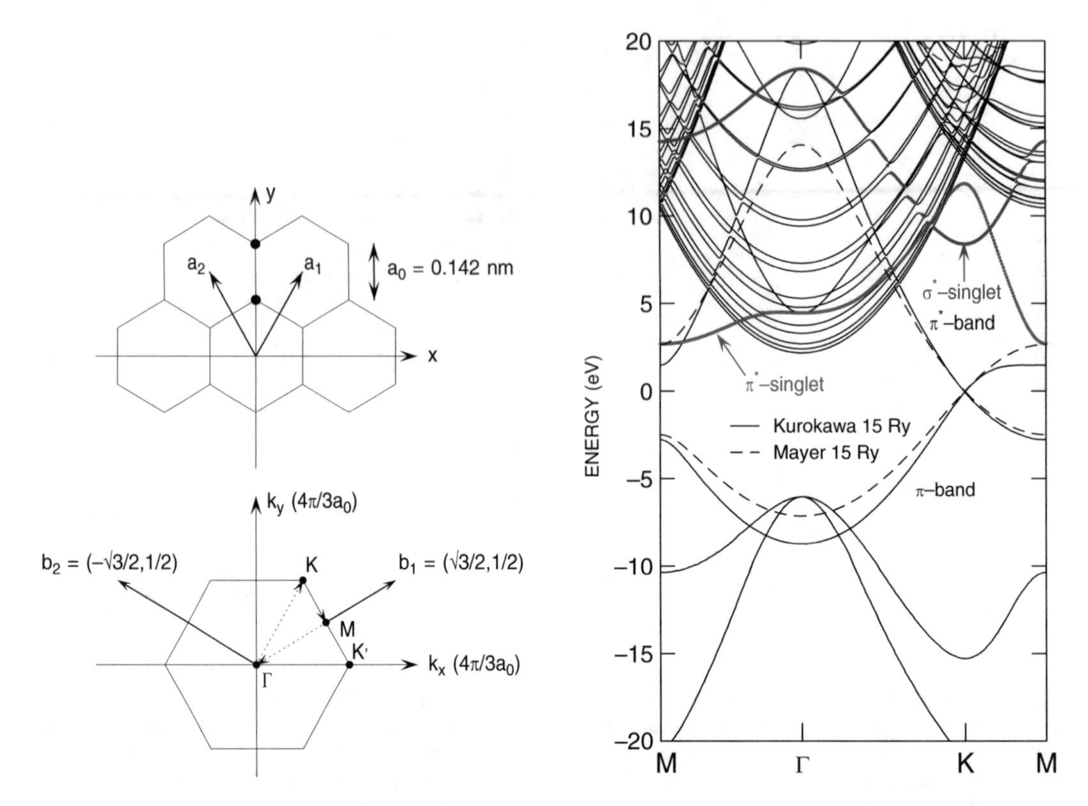


**Fig. 7.19** *Top left*: Primitive translation vectors and position of the atoms (*dots*) for graphene. *Bottom left*: Hexagonal 2D Brillouin Zone of graphene, main symmetry points, and trajectory in **k**-space employed in the plot at *right*. *Right*: Band structure for graphene obtained using the Kurokawa (*solid lines*) and Mayer (*dashed lines*) pseudopotentials. A separation of $10\sqrt{3}a_0/2$ along $z$ has been assumed between adjacent sheets. The bands highlighted in *color* and indicated by *arrows* are the $\sigma^*$ (*blue*) and $\pi^*$ (*red*) singlet bands whose interaction and hybridization result in an interesting and unexpected behavior of the band gap in single-wall zigzag $(n,0)$ carbon nanotubes of small diameter. The use of Kurokawa pseudopotentials, in particular, yields relatively small energies for the $\pi^*$-singlet band along the $M$-$\Gamma$ line ($\approx 2.5$ eV above the Fermi level, compared to energies three times as large obtained using the self-consistent LDA [18, 19]). (Reprinted with permission from M.V. Fischetti, J. Kim, S. Narayanan, Z.-Y. Ong, C. Sachs, D.K. Ferry, and S.J. Aboud, J. Phys.: Cond. Matter vol. 25, 473202 (2013). Copyright 2013, Institute of Physics)

*stress that Mermin himself commented that his results may nevertheless allow the existence of "two-dimensional systems of less than astronomical size to display crystalline order."*

The band structure of graphene presents extremely interesting features. Using a simple tight-binding model, its band structure shows the absence of a gap, with the valence and conduction bands crossing with an almost-linear dispersion at the symmetry point K (see Fig. 7.19). This linear dispersion and the degree of freedom associated with the two triangular sublattices, usually called A-sublattice and B-sublattice, make it possible to describe this dispersion with a Dirac equation [see Eq. (C.44) in Appendix C] for massless electrons with spin replaced by the A/B-sublattice "pseudospin". A full discussion of this connection between graphene and massless Dirac Fermions can be found in [17]. Now we should return to our main discussion, namely, the use of empirical pseudopotentials to calculate the band structure of graphene.

The band structure of an infinite graphene sheet can be calculated assuming the sheet is a supercell layer separated periodically by the neighbor sheets by a distance $N\frac{a_0\sqrt{3}}{2}$ [see Eq. (7.28) below], thus using 1D supercells as in the case of thin Si layers or hetero-layers discussed above. Examples of results available in the literature, with which one can compare the quality of results obtained by using empirical pseudopotentials, are those by Reich [18] and by Khoshnevisan [19] for graphene and also the (5,5) CNT (see below). These results have been obtained using ab initio calculations.

The real-space structure of graphene can be described by a two-atom 2D unit cell with basis vectors [17]:

$$\mathbf{a}_1 = \frac{a_0\sqrt{3}}{2}\,(1,\sqrt{3},0)\ , \quad \mathbf{a}_2 = \frac{a_0\sqrt{3}}{2}\,(-1,\sqrt{3},0)\ , \quad \mathbf{a}_3 = \frac{a_0\sqrt{3}}{2}\,(0,0,N)\ , \tag{7.28}$$

where $a_0$ is the bond length $\approx 0.142$ nm and $N$ (=10 in the results presented below) is the separation between sheets along the $z$-axis. The coordinates of the two C atoms in the cell are

$$\tau_1 = a_0\,(0,1,0)\ , \quad \tau_2 = a_0\,(0,2,0)\ , \tag{7.29}$$

while the basis vectors in reciprocal space are

$$\mathbf{b}_1 = \frac{4\pi}{3a_0}\left(\frac{\sqrt{3}}{2},\frac{1}{2},0\right)\ , \quad \mathbf{b}_2 = \frac{4\pi}{3a_0}\left(-\frac{\sqrt{3}}{2},\frac{1}{2},0\right)\ , \quad \mathbf{b}_3 = \frac{4\pi}{3a_0}\left(0,0,\frac{\sqrt{3}}{2N}\right)\ . \tag{7.30}$$

Figure 7.19 shows the basis geometry in real and reciprocal space, as well as the band structure obtained using the Kurokawa [20] and Mayer [21] empirical pseudopotentials described in Appendix B.

Compared to ab initio results [18, 19], employing the Kurokawa carbon pseudopotential results in the "correct" behavior. Of particular interest is the behavior of the extended $\pi$ (bonding) and $\pi^*$ (antibonding) orbitals that result from the $p_z$ atomic orbital of each C atom $sp^2$-bonded to three other C atoms on the plane. The energies of these $\pi$-$\pi^*$ bands are "correct" for energies close to the Fermi level. Most important, we see the correct band-crossing ("Dirac" point) at the $K$ symmetry point, arguably the most interesting feature of the graphene band structure. However, these results are not in complete agreement with the DFT results. In particular, a set of bands at $\Gamma$ appear at a lower energy than in the ab initio results. These bands compress the $\pi$-$\pi^*$-band energetic separation near $k = 0$. The Mayer pseudopotential, being an empirical one-electron model, by definition fails to account for the 2$s$ valence states and also misses many higher-energy states. Nevertheless, it reproduces satisfactorily the $\pi$-$\pi^*$-band energetic separation near the Fermi level. The Fermi velocity at the Dirac point is calculated to be about $\upsilon_F \approx 9.5 \times 10^7$ cm/s when using the Kurokawa pseudopotentials and $\approx 8.8 \times 10^7$ cm/s when using the Mayer pseudopotentials, both values in good agreement with DFT results. These are all about 15 % smaller than experimental data. Theoretical calculations that can reproduce this experimental value [22] must rely on a model (called $GW$, $G$ standing for the Green's function and $W$ for the screened electron–electron potential), that goes beyond DFT and that we shall not attempt to discuss here. Aryasetiawan and Gunnarsson [23] give a detailed description of this method that is known to yield values for the band gap that are in better agreement with experiments than those obtained using DFT.

## 7.3 Nanowires, Nanotubes, and Nanoribbons

Scaling VLSI devices to the nanometer scale presents several challenges. The main concern is the ever-growing power density. The complementary metal-oxide-semiconductor (CMOS) technology has addressed this issue in the past, by always placing an $n$-channel and a $p$-channel device in series between the electric ground and the power supply, so that, except during switching, there is no direct path for the current to flow from the power supply to ground. Switching is when the "real action" happens, but it is a temporary condition that happens only 10–15 % of the time ("duty cycle"). This all works fine, as long as the devices do not "leak" current when they are in their "off" state. As devices shrink, this leakage becomes larger, because the gate finds it more and more difficult to push all charge carriers outside the channel: The electric field lines originating from the gate now tend to be terminated at the source and drain contacts, that are ever closer, rather than at the substrate, as is desired. Therefore, having the gate increasingly closer to the active region of the channel becomes imperative if we wish to "turn off" the devices. The best scenario in this electrostatic battle is constituted by very thin and narrow channels surrounded by the gate. The concept of Si "fins" surrounded by a gate on three sides has led to FinFETs. But even better is the idea of having thin and narrow bodies surrounded by the gate from all sides. This amounts to considering nanowires. Hence, the great interest in one-dimensional structures, and 1DEG.

The term "one dimensional" is probably used a bit too liberally: One-dimensional conduction properly occurs only when the linear dimensions of the cross section of the wire are smaller than the electron coherence length. That is, when electrons do not scatter appreciably along the cross-sectional dimension, and a constructive interference pattern of standing waves can be sustained, thus leading to quantization of momentum on the cross-sectional plane. Here, we shall assume that this is the case. Small nanowires (NWs) and carbon nanotubes (CNTs) represent such structures. Clearly, their small cross-sectional dimensions will cause both a relaxation of the atoms at the edges of the structure and strong confinement on the cross-sectional plane. We shall discuss amply the latter issue. Regarding a possible, if not likely, atomic relaxation at the edges of these small structures, it is indeed an issue. Judicious use of DFT to obtain the relaxed atomic position and empirical pseudopotentials to obtain the band structure seems to be a good compromise between convenience and accuracy. We shall present below, dealing with circular cross-section Si NWs, an example of this problem.

## *7.3.1 Si Nanowires*

We have considered Si nanowires with their axis along the [100], [110], and [111] directions. For those who wish to compare the results obtained using empirical pseudopotentials with published results, the literature is rich. Covering a variety of models employed, we may mention that Nehari [24] and Neophytou [25] have employed a tight-binding model; Sacconi et al. [26] have used a technique called "linear combination of bulk bands" (LCBB); Scheel and collaborators [27] have employed empirical tight-binding (ETB). Finally, Lee and co-workers [28] have performed first-principles (DFT) calculations. Here we consider both rectangular cross-section nanowires, easier to study, and circular cross-section wires, perhaps of a geometry closer to what is obtained experimentally.

### 7.3.1.1 Rectangular Cross-Section Nanowires

For the rectangular cross-section nanowires, consider rectangular supercells with sides $Na_0$ long and H terminations using the Zunger pseudopotentials. For [100] wires with (110) sides, the supercell is square and constructed by considering the cell defined by the translation vectors $\mathbf{a}_1$, $\mathbf{a}_2$, and $\mathbf{a}_3$:

$$\begin{cases} \mathbf{a}_1 = a_0\,(1/2,1/2,0) \\ \mathbf{a}_2 = a_0\,(-1/2,1/2,0) \\ \mathbf{a}_3 = a_0\,(0,0,1) \end{cases} , \tag{7.31}$$

and atomic coordinates $\tau_\alpha$:

$$\begin{cases} \tau_1 = a_0\ (0,0,0) \\ \tau_2 = a_0\ (1/4,1/4,1/4) \\ \tau_3 = a_0\ (0,1/2,1/2) \\ \tau_4 = a_0\ (-1/4,1/4,3/4) \end{cases} . \tag{7.32}$$

For a square-section nanowire with $(110)$ sides, this cell is replicated $N_c$ times in the "diagonal" directions $x=y$ and $x=-y$. An additional layer of atoms is added to obtain a symmetric configuration (see Fig. 7.20, left) and $N_v$ cells of vacuum padding are added. The resulting supercell is now spanned by the translation vectors:

$$\begin{cases} \mathbf{b}_1 = a_0 N_t\ (1/2,1/2,0) \\ \mathbf{b}_2 = a_0 N_t\ (-1,1,0) \\ \mathbf{b}_3 = a_0\ (0,0,1) \end{cases} , \tag{7.33}$$

where $N_t = N_c + N_v$. The supercell has a cross section of size $\sqrt{2}N_t a_0 \times \sqrt{2}N_t a_0$ and contains $(4N_c(N_c+1)+1)$ Si atoms plus additional $(8N_c+4)$ H atoms terminating the dangling bonds at the surface of the wire. The Si–H distance was taken to be $0.158\ \sqrt{3}\ a_0$ in all cases, as we saw before discussing the termination of Si surfaces.

For $[110]$ wires, instead (using the same notation as above):

$$\begin{cases} \mathbf{a}_1 = (a_0/\sqrt{2})\ (1,0,0) \\ \mathbf{a}_2 = a_0\ (-0,1,0) \\ \mathbf{a}_3 = (a_0/\sqrt{2})\ (0,0,1) \end{cases} , \tag{7.34}$$

$$\begin{cases} \tau_1 = a_0\ (0,0,0) \\ \tau_2 = a_0\ (0,1/4,1/(2\sqrt{2})) \\ \tau_3 = a_0\ (1/(2\sqrt{2}),1/2,1/(2\sqrt{2})) \\ \tau_4 = a_0\ (1/(2\sqrt{2}),3/4,0) \end{cases} . \tag{7.35}$$

For a rectangular cross-section nanowire with $(001)$ and $(1\bar{1}0)$ sides, this cell is replicated $N_{xc}$ times along the $x$ direction and $N_{yc}$ times along the $y$ direction, adding, as before, an additional layer of atoms to obtain a symmetric configuration in each direction (see Fig. 7.20, center) and $N_v$ cells of vacuum padding are added. The resulting supercell is now defined by the translation vectors:

$$\begin{cases} \mathbf{b}_1 = a_0(N_{xc}+N_v)\sqrt{2}\ (1,0,0) \\ \mathbf{b}_2 = a_0(N_{yc}+N_v)\ (0,1,0) \\ \mathbf{b}_3 = a_0\sqrt{2}\ (0,0,1) \end{cases} . \tag{7.36}$$

The area of the rectangular cross section of the supercell is now $\sqrt{2}(N_{xc}+N_v)a_0 \times (N_{yc}+N_v)a_0$. This supercell contains $(4N_{xc}N_{yc}+2N_{yc}+N_{xc}+1)$ Si atoms and additional $(4N_{xc}+4N_{yc}+4)$ H atoms.

Nanowires with their axis along the $[111]$ directions are a bit more complicated and are computationally more demanding, since their period along the axis of the wire is much larger (unfortunately, the semiconductors of interest do not possess body-centered cubic symmetry!). First, construct a cell spanned by the translation vectors:

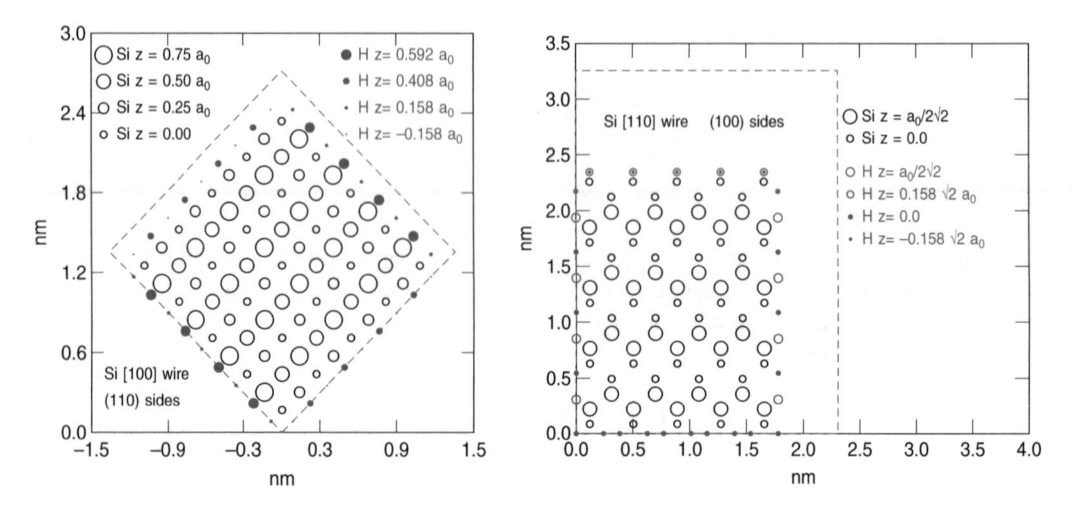


**Fig. 7.20** *Left:* Position of the Si atoms (and hydrogen terminations) for a 4-cell × 4-cell square cross-section nanowire with side 1.535 nm long, axis along the [100] direction, and sides on the (110) planes. The *blue dashed line* indicates the size of the supercell including one cell of vacuum—of thickness $a_0/\sqrt{2}$—separating adjacent wires. *Right:* Position of the Si and H atoms for a 4-cell × 4-cell rectangular cross-section nanowire of dimension 1.5384 × 2.172 nm with axis along the [110] direction and sides on the (001) and $(1\bar{1}0)$ planes. In this case the vacuum "padding" separating the wires to form the 2D supercell (indicated by the *dashed blue lines*) has a thickness of $\sqrt{2}a_0$ along the $x$ direction and $2a_0$ along $y$, including the space occupied by the H terminations

$$\begin{cases} \mathbf{a}'_1 = (a_0/\sqrt{2})\,(1,0,0) \\ \mathbf{a}'_2 = [a_0/(2\sqrt{2})]\,(1,\sqrt{3},0) \\ \mathbf{a}'_3 = a_0\,(0,0,\sqrt{3}) \end{cases}, \tag{7.37}$$

containing six atoms with coordinates:

$$\begin{cases} \tau_1 = a_0\,(0,0,0) \\ \tau_2 = a_0\,(0,0,\sqrt{3}/4) \\ \tau_3 = (a_0/\sqrt{3})\,(0,1/\sqrt{2},1) \\ \tau_4 = a_0\,(0,1/\sqrt{6},\sqrt{3}/4+1/\sqrt{3}) \\ \tau_5 = a_0\,(1/(2\sqrt{2}),1/(2\sqrt{6}),2/\sqrt{3}) \\ \tau_6 = a_0\,(1/(2\sqrt{2}),1/(2\sqrt{6}),2/\sqrt{3}+\sqrt{3}/4) \end{cases}. \tag{7.38}$$

Note in that in Eq. (7.37) the length of the vector $\mathbf{a}'_3$ (i.e., $a_0\sqrt{3}$) is much larger than in the previous cases of [100] and [110] wires, $a_0$ and $a_0/\sqrt{2}$, respectively. Replicating this cell on the cross-sectional $(x,y)$ plane along the directions determined by $\mathbf{a}'_1$ and $\mathbf{a}'_2$ results in a diamond-shaped cross section. Thus, by combining two cells, a cell is obtained with a rectangular cross section on the $(x,y)$ plane. The perpendicular translation vectors are

$$\begin{cases} \mathbf{a}_1 = a_0\,(1/\sqrt{2},0,0) \\ \mathbf{a}_2 = a_0\,(0,\sqrt{3/2},0) \\ \mathbf{a}_3 = a_0\,(0,0,\sqrt{3}) \end{cases}, \tag{7.39}$$

and the cell now contains 12 atoms with coordinates given by Eq. (7.38) above for $i = 1, 6$ and by $\tau_{i+6} = \tau_i + \mathbf{a}'_2$. Replicating this cell $N_{\mathrm{xc}}$ times along $x$ and $N_{\mathrm{yc}}$ times along $y$ yields a rectangular cross-section wire with sides on the $(1\bar{1}0)$ and $(11\bar{2})$ faces (see Fig. 7.21 for such a cell with additional vacuum padding). The supercell translation vectors are now:

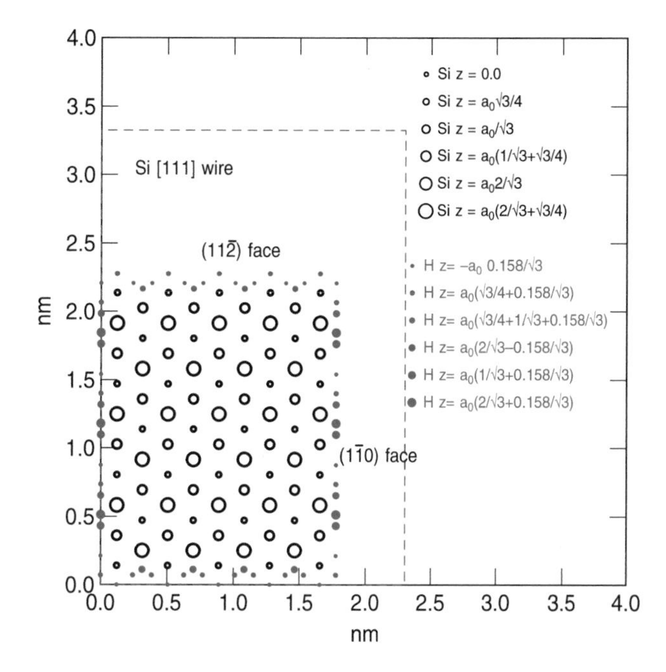


**Fig. 7.21** As in the previous figure, but for a Si 4-cell $\times$ 3-cell nanowire with axis along the $[111]$ direction and with $(1\bar{1}0)$ and $(11\bar{2})$ sides

$$\begin{cases} \mathbf{b}_1 = a_0(N_{\rm xc}+N_{\rm v})\ (1/\sqrt{2},0,0) \\ \mathbf{b}_2 = a_0(N_{\rm yc}+N_{\rm v})\ (0,\sqrt{3/2},0) \\ \mathbf{b}_3 = a_0\sqrt{3}\ (0,0,1) \end{cases} . \tag{7.40}$$

The resulting supercell has a cross section of area $(N_{\rm xc}+N_{\rm v})a_0/\sqrt{2}\times(N_{\rm yc}+N_{\rm v})a_0\sqrt{3/2}$ and contains $(12N_{\rm xc}N_{\rm yc}+2N_{\rm xc}+6N_{\rm yc}+2)$ Si atoms and additional $(8N_{\rm xc}+12N_{\rm yc}+6)$ H atoms.

In the case of nanowires, our interests are similar to those that motivate us in the case of thin films and hetero-layers. The band structure and the ballistic conductance are the main results that give us information about the electron transport properties of these structures before worrying about scattering, which, of course, will be the focus of the following chapters.

Figures 7.22 and 7.23 shows the band structure for square-section Si NWs with sides 2–5 cells long separated by 1 cell of "vacuum." Note the energy gap increasing with decreasing wire size (as expected). The density of states of all subbands $n$ is calculated from its definition:

$$\mathscr{D}_{\rm 1D}(E) = 2\sum_{n,i}\int\frac{{\rm d}E'}{2\pi}\left|\frac{{\rm d}E_n(k_{z,n,i})}{{\rm d}k_z}\right|^{-1}\delta(E'-E)\ , \tag{7.41}$$

where the index $i$ labels the $p_n \geq 0$ solutions $k_{z,n,i}$ such that $E(k_{z,n,i}) = E'$ and $E_n(k_z)$ is the dispersion in (sub)band $n$. The ballistic conductance, $G_{1D}(E)$, along the wire axis at energy $E$ is given by

$$\begin{aligned} G_{\rm 1D}(E) &= 2e^2\frac{1}{2}\sum_n\int\frac{{\rm d}k_z}{2\pi}\ \upsilon_n(k_z)\ \delta\left[E_n(k_z)-E\right] \\ &= 2e^2\frac{1}{2}\sum_{n,i}\int\frac{{\rm d}E'}{2\pi}\ \upsilon_n(k_{z,n,i})\left|\frac{{\rm d}E_n(k_{z,n,i})}{{\rm d}k_z}\right|^{-1}\delta(E'-E) = \frac{2e^2}{h}\frac{1}{2}\sum_n p_n\ , \end{aligned} \tag{7.42}$$

where $\upsilon_n(k_{z,n,i})$ is the group velocity $(1/\hbar){\rm d}E_n(k_{z,n,i})/{\rm d}k_z$ at the $k_z$-point $k_{z,n,i}$, and the factor of $1/2$ in the equation above reflects the fact that the sum should be performed only over $k_z$-points

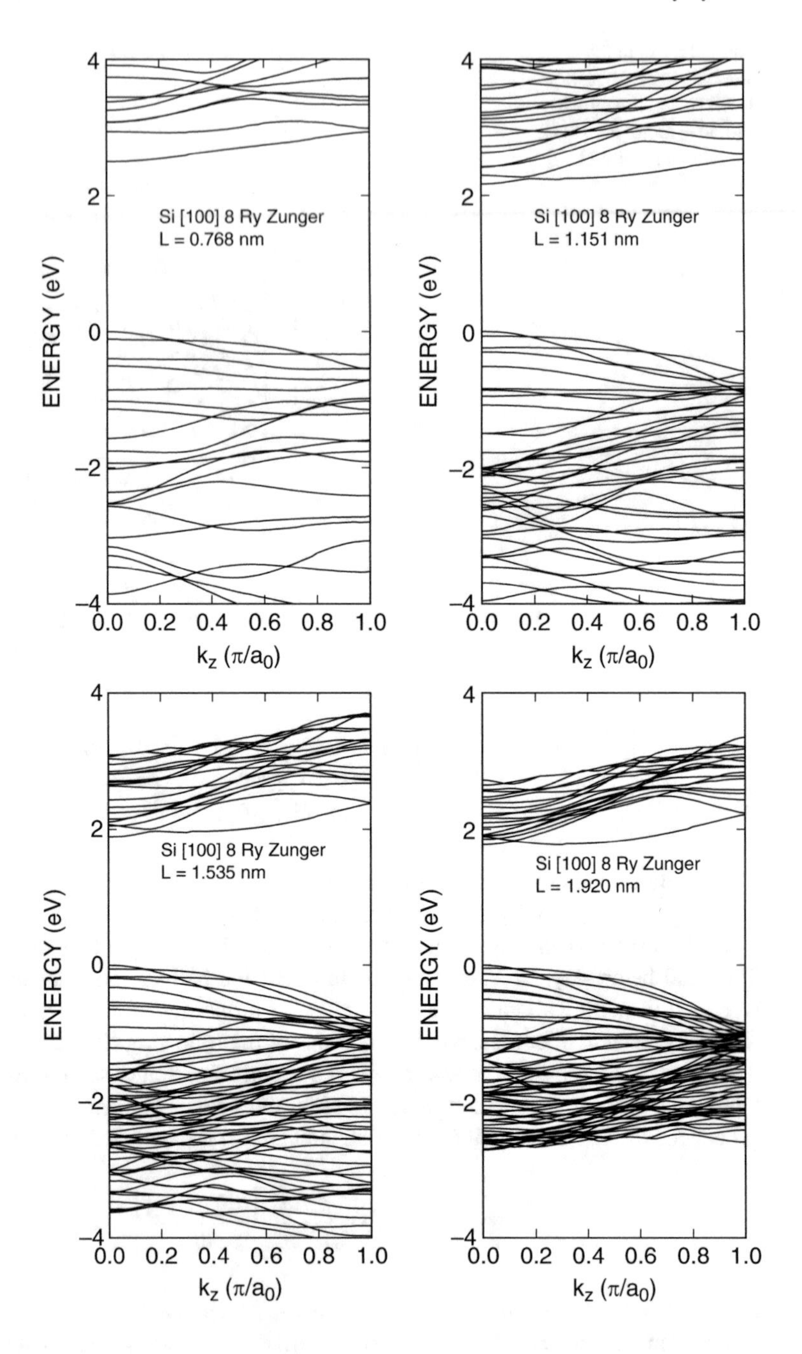


**Fig. 7.22** Band structure of four square cross-section H-terminated [100] Si nanowires with sides of four different lengths. The wires are separated by a one-cell thickness of vacuum. The results have been obtained using a cutoff energy of 8 Ry and empirical pseudopotentials from [5]. [Used from M.V. Fischetti, Bo Fu, S. Narayanan, and J. Kim, in *Nano-Electronic Devices: Semiclassical and Quantum Transport Modeling*, Dragica Vasileska and Stephen M. Goodnick eds. (Springer, New York, 2011), pp. 183–247, with kind permission from Springer Science+Business Media]

corresponding to a positive group velocity $v_n(k_{z,n,i})$, and so, by symmetry, over 1/2 of the entire 1D BZ. The ballistic conductance of free-standing [100], [110], and [111] nanowires of similar rectangular cross sections is shown in Fig. 7.24. Note the larger conductance for both electrons and holes in the [100] wire and the smaller conductance of the [111] nanowire whose many bands are "flat" and exhibit few crossings. Finally, Fig. 7.25 shows the squared wavefunctions for the six lowest-energy conduction-band states for the [100] wire with sides 2 cell-long separated by 1 vacuum cell.

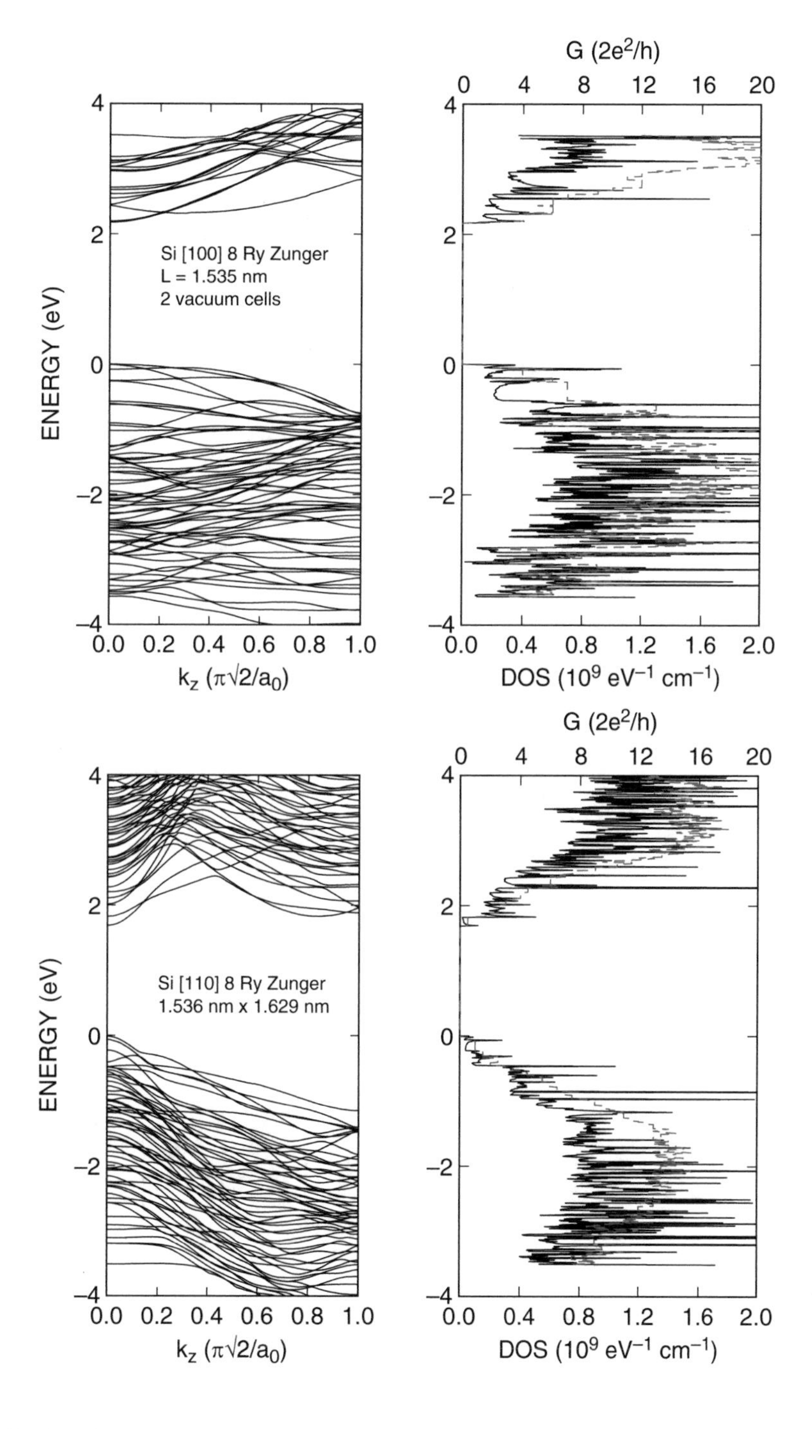


**Fig. 7.23** Band structure, density of states, and conductance of free-standing H-terminated [100] (*top*) and [110] (*bottom*) Si nanowires with an ("almost" for the [110] wire) square cross section of the indicated dimensions with 2 cells of vacuum padding. [Used from M.V. Fischetti, Bo Fu, S. Narayanan, and J. Kim, in *Nano-Electronic Devices: Semiclassical and Quantum Transport Modeling*, Dragica Vasileska and Stephen M. Goodnick eds. (Springer, New York, 2011), pp. 183–247, with kind permission from Springer Science+Business Media]

#### 7.3.1.2 Circular Cross-Section Nanowires

Although nanowires with square or rectangular cross section can be fabricated (for example, etching narrow and short Si "fins"), more conventional processing yields NWs with rather irregular cross sections, resembling circles or ellipses rather than squares. Here we consider ideal Si NWs with a

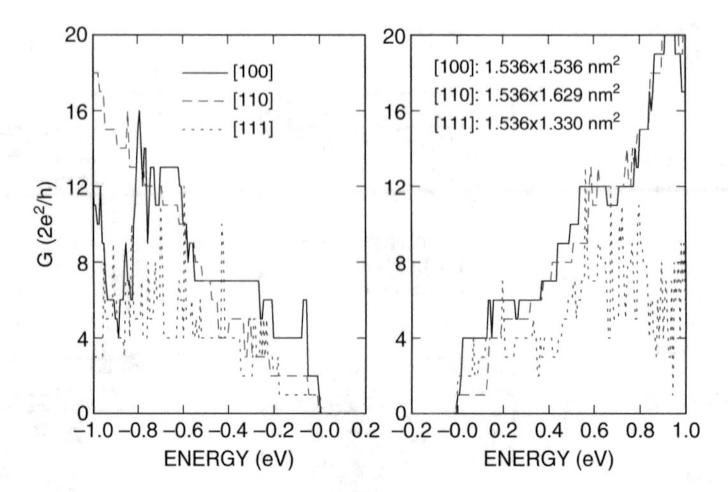


**Fig. 7.24** Valence-band (*left*) and conduction-band (*right*) ballistic conductance of the [100], [110], and [111] Si nanowires of the previous figures. The zero-energy has been set at the band-edge in both plots. [Used from M.V. Fischetti, Bo Fu, S. Narayanan, and J. Kim, in *Nano-Electronic Devices: Semiclassical and Quantum Transport Modeling*, Dragica Vasileska and Stephen M. Goodnick eds. (Springer, New York, 2011), pp. 183–247, with kind permission from Springer Science+Business Media]

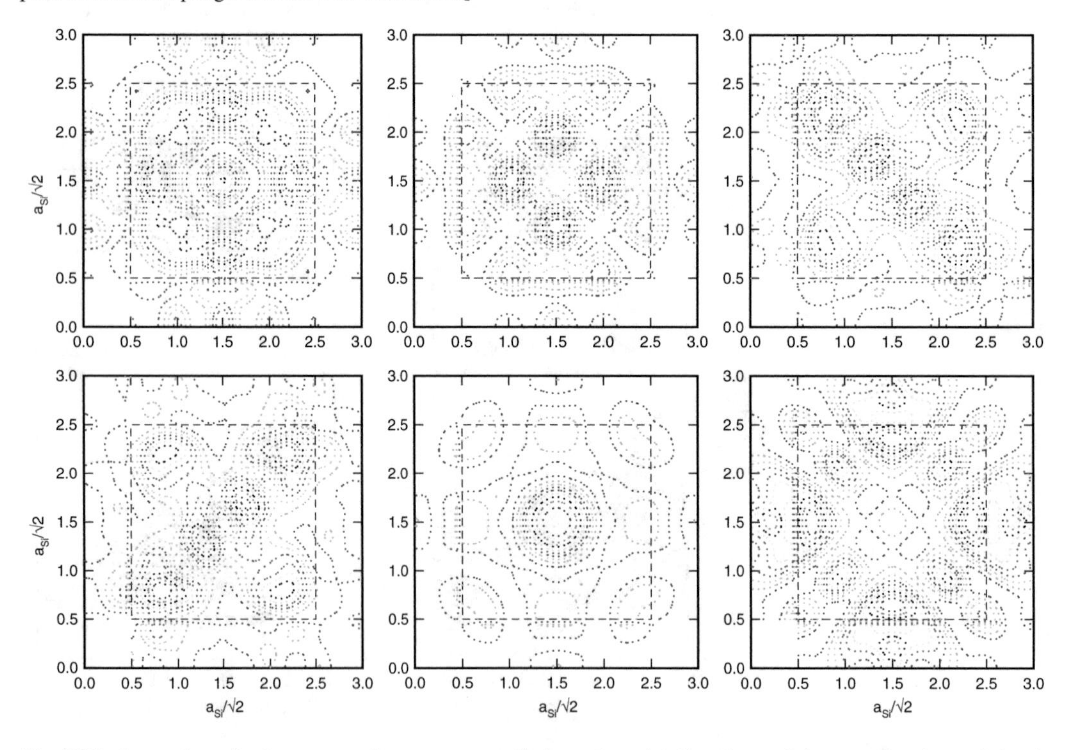


**Fig. 7.25** Squared amplitude—averaged over a supercell along the axial direction—of the wavefunctions of the six lowest-energy conduction-band states in the smallest [100] square cross-section Si nanowire of the previous figure. The *contour lines* are drawn at intervals of $0.1\ |\psi|^2_{\text{max}}$, where $|\psi|^2_{\text{max}}$ is the maximum of the squared amplitude of the wavefunction. The *square* with sides $2a_{\text{Si}}/\sqrt{2} = 0.768$ nm long indicated by the *dashed lines* shows the "nominal" size of the Si square, while the surrounding area is vacuum, 1 cell thick. Note the wavefunctions "spilling" into the vacuum especially at higher energy. In particular, note in the last frame how the highest-energy state exhibits tunneling into the neighboring supercell, indicating that a thicker vacuum is necessary in order to "isolate" the wire

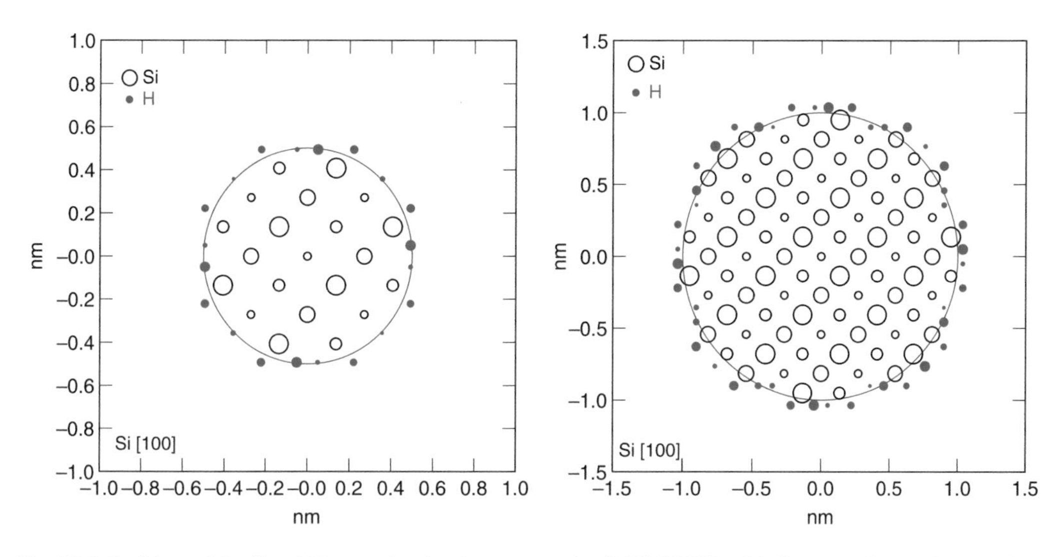


**Fig. 7.26** Positions of the Si and H atoms for circular cross-section [100] Si NWs with diameter of 1 nm (*left*) and 2 nm (*right*)

circular cross section. Note that when the diameter of the NW shrinks, the word "circular" becomes ill-defined, since, ultimately, the wires are made of cubic cells.

For circular cross-section nanowires we have considered similar cells, but truncated on the cross-sectional plane in order to obtain cross sections as circular as possible. Figure 7.26 shows the atomic positions for H-terminated [100] Si NWs with diameters of 1 and 2 nm. The band structure, density of states, and ballistic conductance for 1-to-2.5 nm Si nanowires are shown in Fig. 7.27. The squared amplitudes of the wavefunctions of the lowest-energy conduction-band states for the 2 nm diameter NW are illustrated in Fig. 7.28.

We have mentioned before the possibility, or even likelihood, that atoms may relax at the edges of these structures. Figure 7.29 shows the positions of the Si atoms in a circular cross-section NW with a diameter of 1 nm, as taken as the positions in the bulk, or by allowing them to relax, using DFT. The differences in the resulting band structure are not dramatic, but these small differences may matter in transport studies.

Note that in these small wires the ground state wavefunctions have polar symmetry (being characterized by an angular momentum quantum number $l = 0$), but the first excited state has quadrupole symmetry ($l = 2$). This is due to the fact that the states at $\Gamma$ originate from a superposition of states in the four ellipsoidal equi-energy surfaces with transverse and longitudinal masses along the principal $x$ and $y$ directions. A smaller energy is thus obtained for states whose "lobes" probe the larger longitudinal mass by extending along the positive and negative $x$ and $y$ directions, resulting in a fourfold symmetry. On the contrary, the dipole-like ($l = 1$) states result from states whose lobes probe the smaller transverse mass, thus yielding a larger kinetic energy. Finally, the dependence of the band gap as a function of nanowire diameter is shown in Fig. 7.30. The expected $D^{-2}$-dependence is seen, deviations at the smallest diameters being caused by the finite confinement potential.

### 7.3.2 *Graphene Nanoribbons*

Graphene, of course, has attracted so much interest because of its structural and electronic properties. In particular its high carrier mobility has excited engineers, its Dirac-like dispersion, pseudospin, and analogy with concepts derived from Quantum Field Theory have excited physicists. However, in

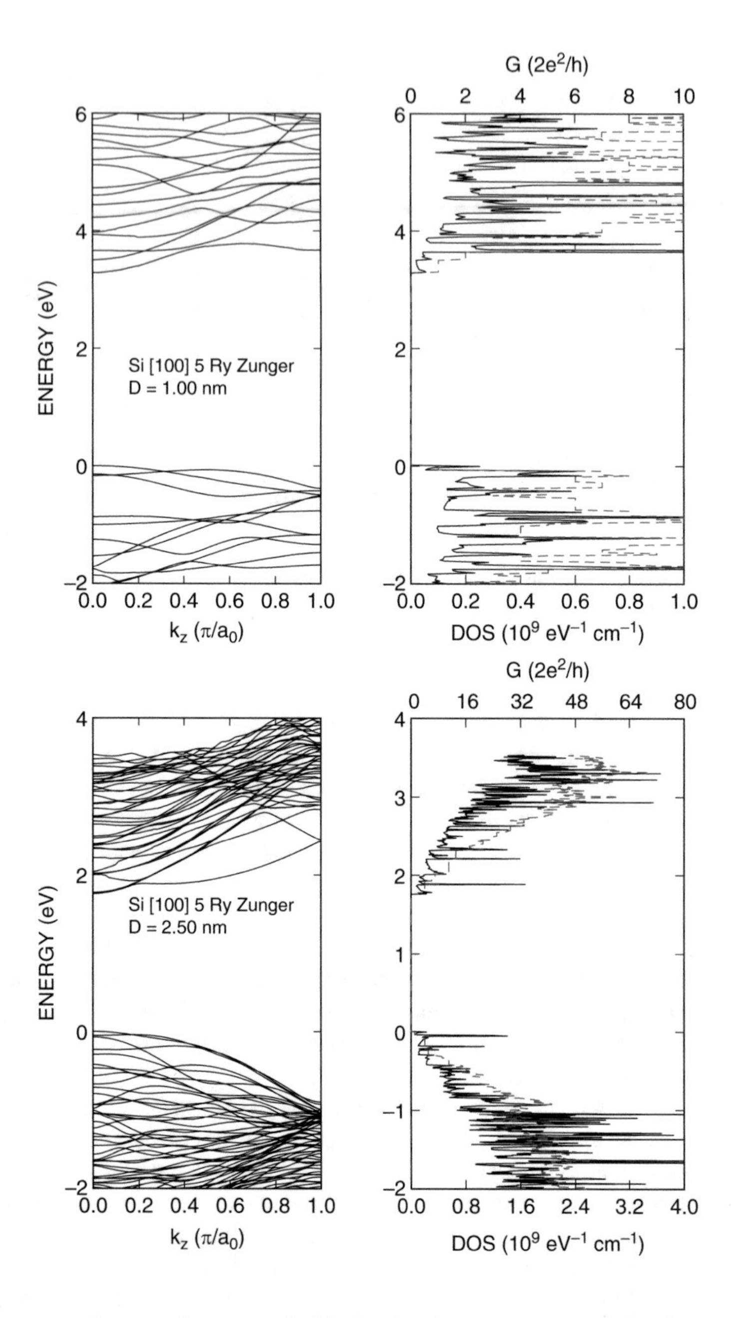


**Fig. 7.27** Band structure, DOS, and ballistic conductance for two circular cross-section H-terminated Si nanowires with diameter of 1 nm (*top*) and 2.5 nm (*bottom*)

practical terms, the absence of a gap renders graphene unsuitable for (or, in any event, very hard to apply to) nanoelectronics applications. Graphene nanoribbons, on the other hand, do possess a gap in some of their many possible atomic arrangements. Therefore, we consider here two types of graphene nanoribbons (GNRs): Those obtained by "cutting" a strip out of a graphene sheet with cuts along "armchair" lines, resulting in GNRs with armchair edges (AGNRs), or along zigzag lines, obtaining zigzag-edge GNRs (ZGNRs). So, looking at the top-left picture in Fig. 7.19, AGNRs are obtained by cutting an infinite graphene sheet along two parallel lines along the vertical axis, while ZGNRs are obtained with cuts along the direction of the primitive vector $\mathbf{a}_1$.

GNRs can be described by their chirality and their width characterized by the number $N_a$ of atomic lines. For AGNRs, the dependence of the energy gap on their width follows three types of

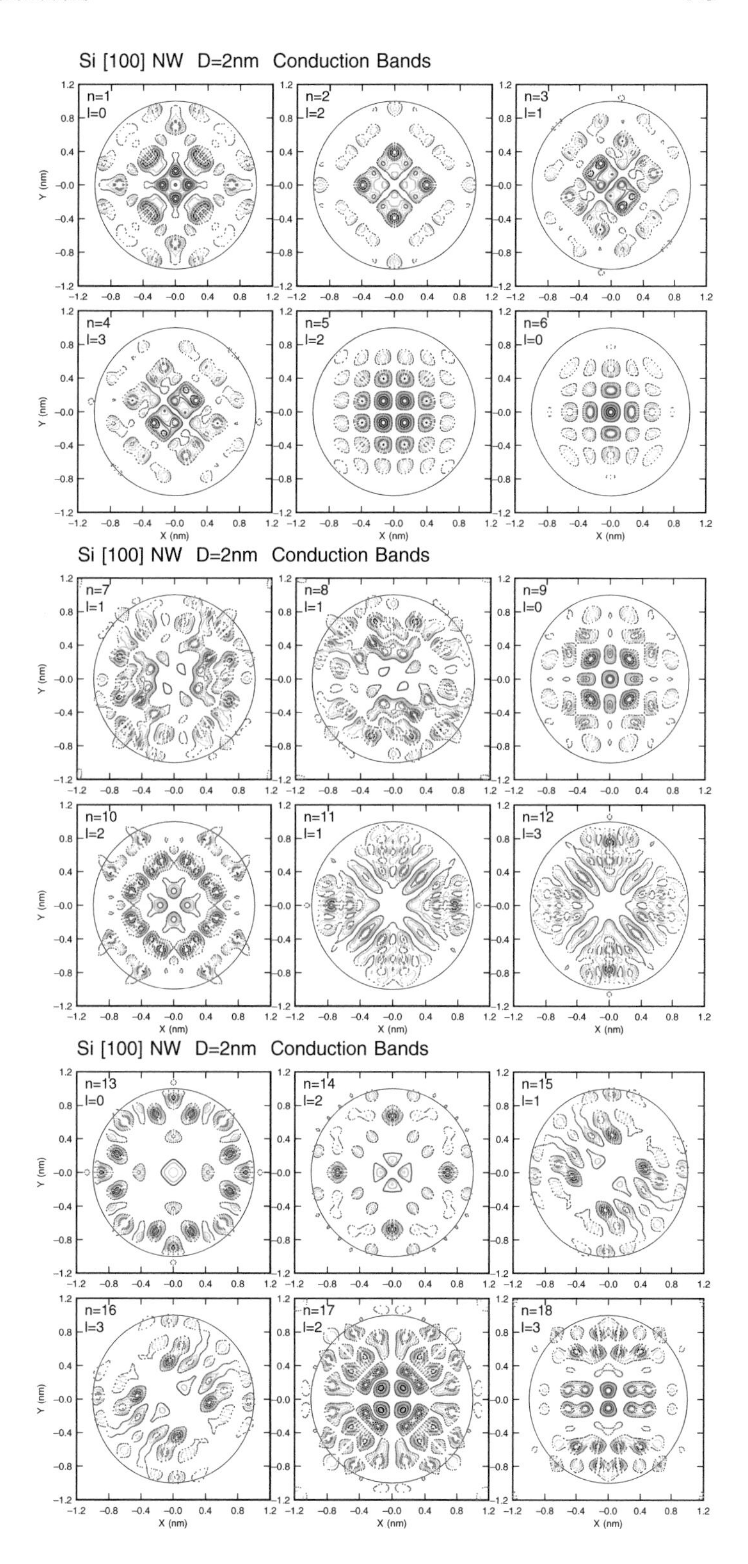


**Fig. 7.28** Square amplitude of the wavefunctions of the 18 lowest-energy conduction-band states for the 2 nm diameter H-terminated [100] Si NWs of the previous figures. The angular momentum quantum number $l$ indicates the most significant lowest-$l$ component of each wavefunction

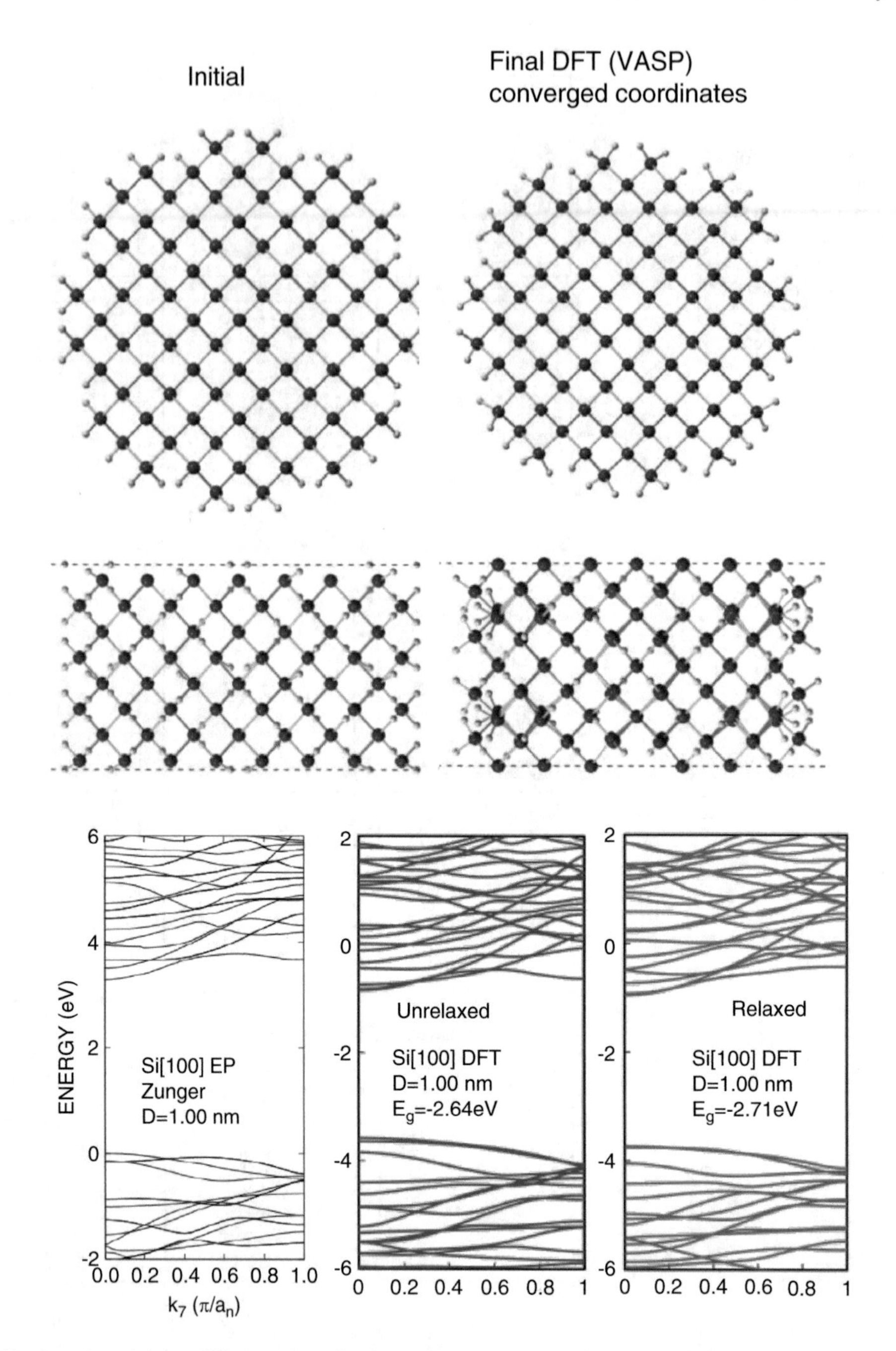


**Fig. 7.29** Atomic positions of Si atoms in a circular cross-section nanowire as taken from the bulk (*top left*) and after having relaxed the structure using DFT (*top right*). The *two bottom frames* show the band structure calculated using empirical pseudopotentials and DFT before relaxation, and DFT on the fully relaxed structure. (Figures kindly provided by Dr. Shela Aboud)

trends, depending on whether $N_a = 3p$, $3p+1$, or $3p+2$ [see Fig.7.31 for the geometry of such an armchair (AGNR)]. Tight-binding calculations [29–31] and calculations based on the massless Dirac equation [32, 33] predict $E_{g,3p} \geq E_{g,3p+1} > E_{g,3p+2} = 0$, so $3p+2$-type GNRs are predicted to be semimetallic. On the contrary, ab initio DFT (LDA and GW) calculations [34–36] predict

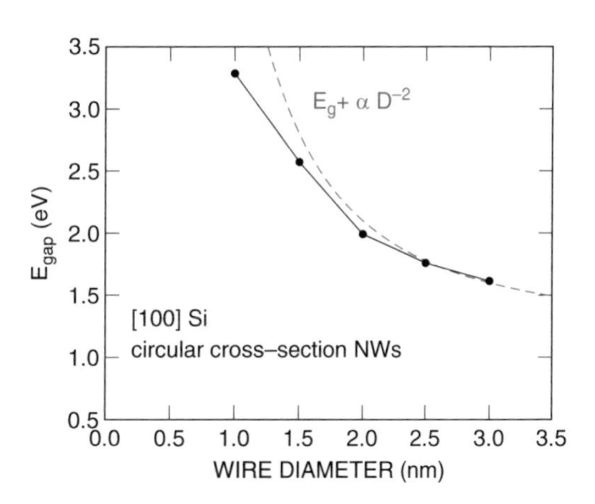


**Fig. 7.30** Diameter dependence of the band gap of H-terminated circular cross-section Si nanowires. The expected dependence on inverse of the square of the diameter is seen with deviations at the smallest diameters due to the finite height of the Si–vacuum confining potential barrier

$E_{g,3p+1} \geq E_{g,3p} > E_{g,3p+2} \neq 0$. Thus, all GNRs should be semiconducting, the difference between tight-binding and ab initio results originating mainly, according to Son et al. [34], from the change of the C–C bond length along the edges. However, as we shall see, the use of empirical pseudopotentials results in the same ab-initio behavior without accounting for this effect, hinting, instead, at some inherent inadequacy of the tight-binding method. The cause of this extremely interesting behavior of the energy gap of AGNRs is intrinsically related to the aromatic properties of these chains. We shall discuss it shortly.

It should be also noted that there is no experimental information about the band gap of AGNRs of a given width. It is still almost impossible to control the width of GNRs down to a single atomic line. As a result, experiments usually give information about the behavior of system of GNRs with different widths, therefore washing out interesting "fluctuations." At present, we must trust theoretical results. On this note, we should note that calculation performed using the GW method we mentioned before yields values for the band gaps that are much larger than those obtained using empirical pseudopotentials or DFT. For example, one may compare the results reported in [34] with those given in [36].

The important point is that the empirical pseudopotential approach is able to reproduce the behavior found using first-principle calculations even without accounting for the edge-bonds distortion. The use of Kurokawa pseudopotential also accounts for the correct behavior of bare-edge states. The main problems with these local pseudopotentials stem from their empirical non-self-consistent nature leading to their inability to predict the correct semiconducting behavior of zigzag-edge nanoribbons. In these ribbons, spin polarization plays a major role and accounting for this effect requires self-consistent methods including exchange-correlation (actually, mainly exchange). In addition, and possibly unrelated to this, is the problem that, when applied to carbon nanotubes, they predict an excessively low energy of the $\pi^*$ singlet in CNTs of some chirality (as in the $(n,0)$ CNTs with $n \leq 10$ discussed below), in disagreement with first-principle results. We must wait for experimental information before blaming DFT, GW, or empirical pseudopotentials.

#### 7.3.2.1 Armchair-Edge GNRs

We have promised to revisit in some detail the extremely peculiar dependence of the band gap of AGNRs on ribbon-width. A brilliant explanation has been given by Dr. Aboud in [3]. Here we quote her verbatim.

> However this feature, required from a practical perspective, exhibits a qualitative behavior significantly different from the more customary particle-in-a-box situation in which the zero-point energy (and so the gap) increases monotonically with increasing confinement. Instead, one observes the emergence of three families of AGNRs characterized by the number of atomic layers $N_a$ along the width of the AGNR (see Fig. 7.31), the energy gap oscillating as $N_a$ takes the values $3p$, $3p+1$, or $3p+2$, where $p$ is an integer. This "peculiar" behavior is due to

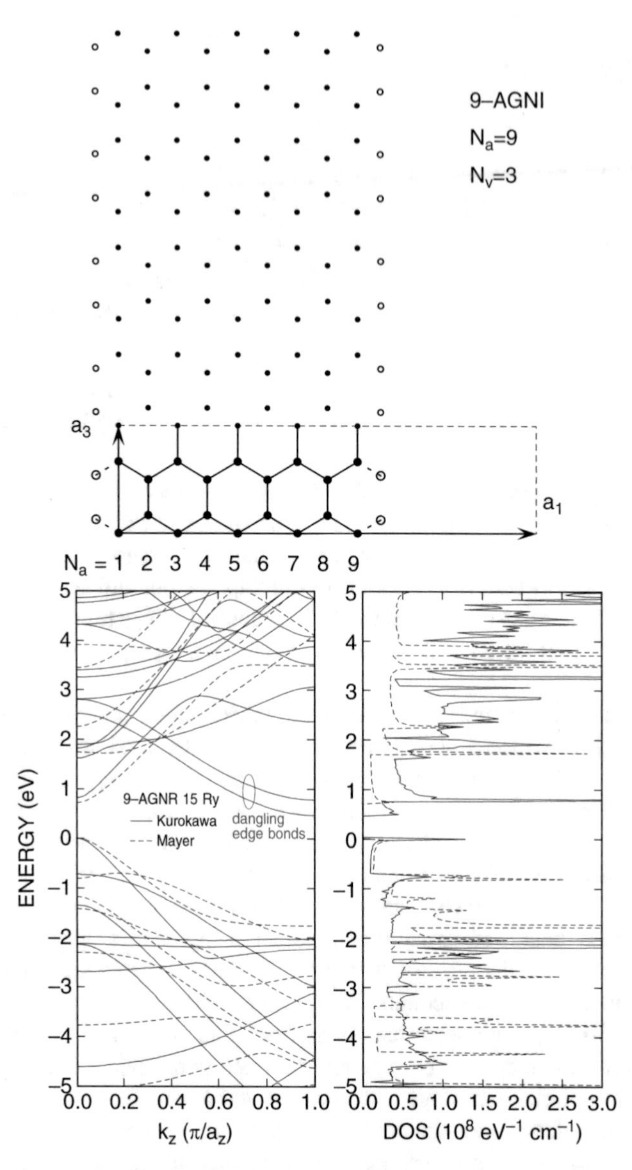


**Fig. 7.31** *Left*: Projection on the $(x,z)$ plane of the supercell used to treat an $N_a = 9$ armchair-edges graphene nanoribbon. The *dashed lines* define the supercell—including the 3 vacuum "cells" separating adjacent ribbons—while the *thick dots* show the C atoms in the supercell, the *thinner dots* simply showing the periodically replicated structure of the ribbon. Hydrogen terminations are shown as *circles*. A distance of $4\sqrt{3}a_0$ between planes has been assumed in the calculations by defining a primitive translation vector of such a length along the direction perpendicular to the plane of the ribbon. *Right*: Band structure and density of states for the bare-edge 9-AGNR illustrated at *right*. Results obtained using Kurokawa (*solid lines*) and Mayer (*dashed lines*) pseudopotentials are shown. Note the bands associated with edge states resulting from the edge-C dangling bonds. Here and in the following graphene ribbons are separated by $N_v\sqrt{3}a_0$, with $N_v = 4$ (unlike the choice of $N_v = 3$ made to sketch the ribbon in the *left panel* of this figure), along the plane of the ribbon and by $N_y\sqrt{3}a_0$, with $N_y = 3$, along the direction perpendicular to the sheets. The energy has been set to zero at the top of the valence band. (Reprinted with permission from M.V. Fischetti, J. Kim, S. Narayanan, Z.-Y. Ong, C. Sachs, D.K. Ferry, and S.J. Aboud, J. Phys.: Cond. Matter vol. 25, 473202 (2013). Copyright 2013, Institute of Physics)

the spatial distribution of the Clar resonance structures (*i.e.*, Clar sextets) [37]. This is a well-characterized effect in the chemistry community, most notably for explaining the behavior of poly-aromatic hydrocarbons [38], and it has been recognized by Ezawa [30, 31] in the electronics community. We shall refer to this property as the "claromatic" behavior (or "claromaticity") of AGNRs [39].

In simple terms, in aromatic compounds C atoms are bonded with $sp^2$-coordination, the remaining ($p_z$) bonds forming the $\pi$ band. However, one alternative possible way to visualize the formation of the $\pi$ band is to consider these bonds resonating among different C atoms as in a benzene ring. In graphene and AGNRs we can draw the hexagonal lattice and connect neighboring C atoms with double bonds wherever possible. In so doing, we obtain a structure in which some of the hexagonal rings will have three double-bonded C atoms. These rings are labeled 'Clar sextets'. Note that the presence of two adjacent Clar sextets is not allowed, as this would imply the presence of C atoms with 5 bonds. In graphene, clearly there is no unique configuration (pattern) of Clar sextets: We can consider many equivalent patterns, the structure resonating among all of these possible patterns, thus forming the $\pi$ band. Each possible configuration of Clar sextets is called a 'Kékule pattern'. In AGNRs, when the edge bonds are terminated by a single H atom, the edge-rings will be necessarily Clar sextets because the terminating H atoms force single edge-bonds. This will happen at both edges. Therefore, depending on the width of the ribbon, we may have three possible cases: 1. two possible resonating Kékule patterns exist. The $\pi$-electrons will be delocalized throughout the ribbon creating resonant paths in the center region of the ribbon, they will contribute to the binding of the structure, and this will give rise to a large gap/ionization energy. 2. Only one resonant Kékule pattern is possible. The $\pi$-electrons are localized at the edges of the ribbon leading to a less stable structure and a small gap. 3. Many possible Kékule patterns can be formed, some delocalizing and some localizing the electrons, yielding an intermediate gap. These three different possibilities occur respectively for ribbons with width given by a number of atomic lines $N_a = 3p+1$ (two Kékule patterns), $N_a = 3p$ (one Kékule pattern), and $N_a = 3p+2$ (many Kékule patterns) giving rise to the band gap scaling $E_{g,3p+1} > E_{g,3p} > E_{g,3p+2}$. If we terminate the edges with, say, $H_2$ instead, the opposite will happen, since the edge rings will be prevented from being Clar sextets by the double terminating edge bonds and the appearance of Kékule patterns (claromaticity) as a function of ribbon-width will be different. We note parenthetically that this behavior is absent in zigzag-edge GNRs (ZGNRs). Also, the lack of a gap in these structures makes them less interesting for nanoelectronics applications; as a consequence electron transport in ZGNRs has not been studied as extensively as in AGNRs and we shall ignore them here. Nevertheless, a discussion of their [local empirical pseudopotential]-based band-structure is given below.

The Kékule patterns can be visualized through computationally generated Scanning Tunneling Microscope (STM) images using the wavefunctions (which can be obtained from DFT or EP) through the Tersoff-Hamann approximation [40, 41] of the STM current:

$$I(x,y,z,U) = \sum_n |\Psi_n(x,y,z)|^2 \left[f(E_F - E_n) - f(E_F + eU - E_n)\right], \tag{7.43}$$

where $E_F$ is the Fermi energy, $U$ is the applied potential, and $E_n$ is the energy in band $n$ corresponding to the wavefunction $\Psi_n$. A plot of the calculated STM images from DFT simulations using the "Vienna Ab Initio Simulation Package" (VASP) [42–44] is shown in Figs. 7.32 and 7.33 for $N_a = 9$, 10 and 11 AGNRs. These VASP simulations have been performed using the projector-augmented wave method (PAW). Exchange-correlation is represented with the revised Perdew-Burke-Ernzerhof (PBE) model of the generalized gradient approximation (GGA). A plane wave cutoff of 550 eV was applied with a $\Gamma$-centered Monkhorst-Pack mesh of $11\times1\times1$ in $k$-space. A Methfessel-Paxton Gaussian smearing of order 1 with a width of 0.05 was used to accelerate convergence. Geometric optimization was performed with the conjugate-gradient algorithm until the absolute value of the forces was less than 0.3 eV/Å. Optimization of the graphene unit cell resulted in a C–C bond length of 1.426 Å. The wavefunctions are sampled at a distance $z = 0.2$ nm above the AGNRs and a value of $eU = -0.5$ eV (which captures the highest two valence bands) was chosen for the applied potential energy.

In comparison, empirical tight binding (ETB) calculations [29–31] and calculations based on the Weyl (massless Dirac) equation [32, 33] have found a different width-dependence compared to DFT and EP and predict $E_{g,3p} > E_{g,3p+1} > E_{g,3p+2} = 0$, so $3p+2$ GNRs are predicted to be semimetallic. This is due to the inability of nearest-neighbours-only models to capture the basic physics of the resonant Clar structures, as noted by Zhao et al. [45]. On the contrary, ab initio DFT (LDA and with GW corrections) calculations [34–36] also predict $E_{g,3p+1} \geq E_{g,3p} > E_{g,3p+2} \neq 0$, which implies that all AGNRs are semiconducting. This reveals a big shortcoming of ETB (in its nearest-neighbor implementation) compared to ab initio results. While Son et al. [34] have emphasized the importance of the relaxation of the edge C–C bonds, we find that this is a secondary contribution to the width-dependence of the AGNR band gap. Indeed we find the same ab initio behavior ourselves even without accounting for this effect. Note also that the value of the calculated band gap increases dramatically when performing GW calculations (compare the results of [34] with those of [36], for example)."

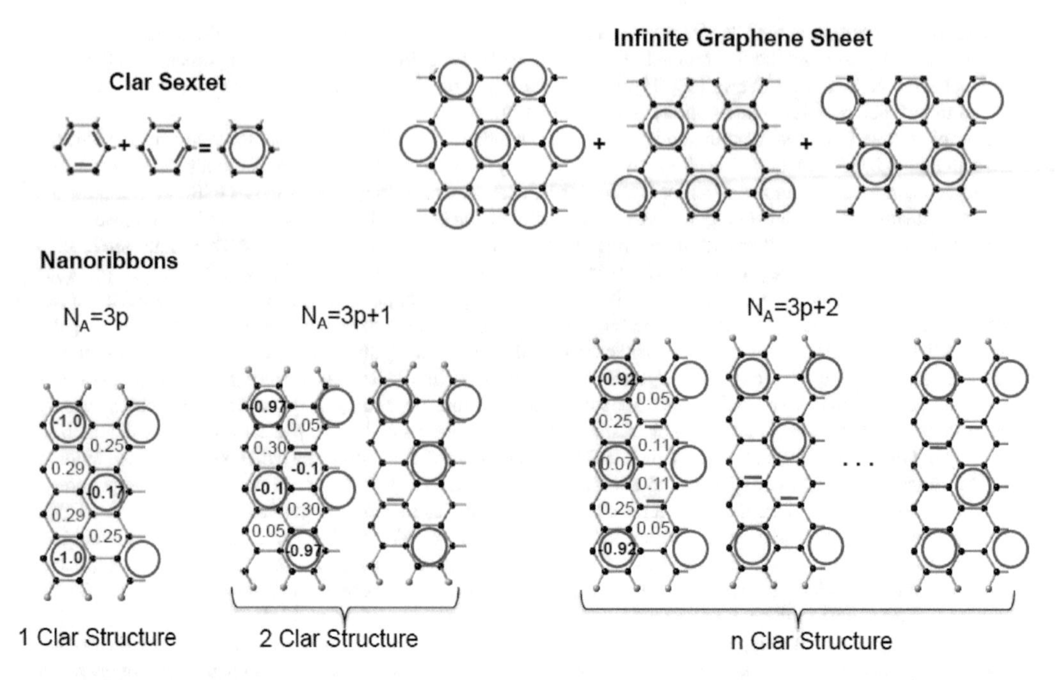


**Fig. 7.32** Illustration of the possible Clar resonances in graphene and armchair-edge graphene nanoribbons. (Figure kindly provided by Dr. Shela Aboud)

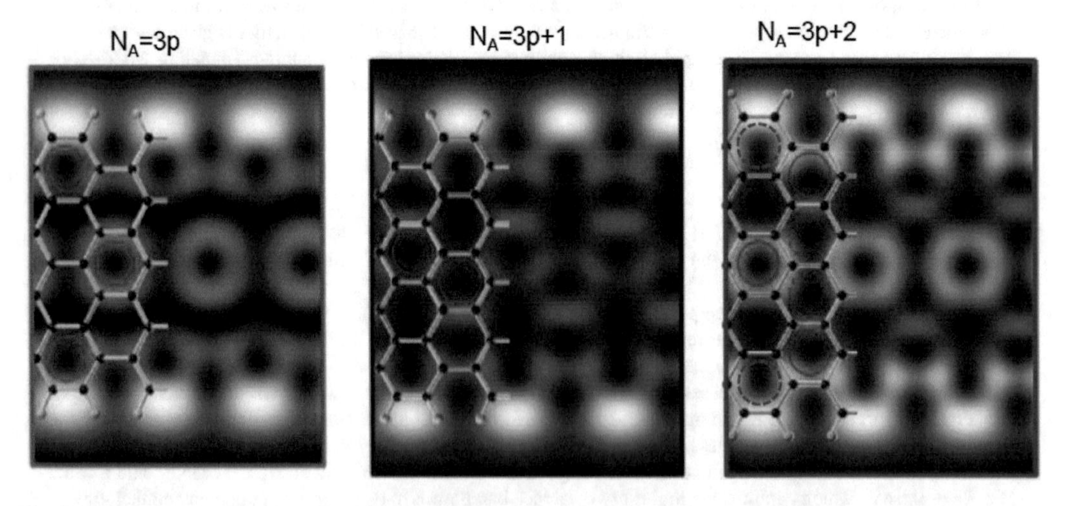


**Fig. 7.33** Illustration of the possible Clar resonances in armchair-edge graphene nanoribbons superimposed to DFT calculations of the electron charge density in the highest-energy conduction band. Figures kindly provided by Dr. Shela Aboud

After this rather long preamble about the band gap of ANGRs, we can discuss now how this behavior is captured by empirical pseudopotentials. The supercell employed to treat an AGNR is described by the primitive translation vectors in real space

$$\mathbf{a}_1 = ((N_c + N_v)\sqrt{3}a_0, 0, 0)\ , \quad \mathbf{a}_2 = (0, N_y\sqrt{3}a_0, 0)\ , \quad \mathbf{a}_3 = (0, 0, 3a_0)\ , \tag{7.44}$$

where $a_0 \approx 0.1422$ nm is the "standard" C–C bond length and $N_c$ is the number of cells between the edges (i.e., along the $x$ direction). $N_c$ is related to the width parameter of the ribbon [34], $N_a$, via

$N_a = 2N_c$ (for even $N_a$) or $N_a = 2N_c + 1$ (for odd $N_a$). Also, $N_v$ is the number of cells separating the ribbons along the $x$ (transverse) direction, and $N_y$ is the number of cells of extension $\sqrt{3}a_0$ separating the planes on which adjacent ribbons lie (in the "vertical" $y$ direction). All results presented below have been obtained using $N_v = 4$ and $N_y = 3$. In each supercell there will be $4N_c + 2$ (odd $N_a$) or $4N_c$ (even $N_a$) atoms. The four atoms in the cell adjacent to the "left" edge will have coordinates:

$$\tau_1 = (0,0,0) \ , \quad \tau_2 = a_0 \left( \frac{\sqrt{3}}{2}, 0, \frac{1}{2} \right) \ , \quad \tau_3 = a_0 \left( \frac{\sqrt{3}}{2}, 0, \frac{3}{2} \right) \ , \quad \tau_4 = a_0 \ (0,0,2) \ . \tag{7.45}$$

The remaining atom coordinates can be obtained by translating these coordinates $N_c - 1$ times by an amount $j\sqrt{3}a_0$ (with $j = 1, N_c - 1$) along the positive $x$ direction:

$$\begin{aligned} \tau_{4j+1} &= \tau_1 + j\sqrt{3}\,a_0\hat{x} \ , \quad \tau_{4j+2} = \tau_2 + j\sqrt{3}\,a_0\hat{x} \ , \\ \tau_{4j+3} &= \tau_3 + j\sqrt{3}\,a_0\hat{x} \ , \quad \tau_{4j+4} = \tau_4 + j\sqrt{3}\,a_0\hat{x} \ . \end{aligned} \tag{7.46}$$

For odd $N_a$ two more C atoms must be added at the "right" edge by translating atoms 1 and 4 in Eq. (7.45) by an amount $\sqrt{3}a_0N_c$:

$$\tau_{4N_c+1} = \tau_1 + N_c\sqrt{3}\,a_0\hat{x} \ , \quad \tau_{4N_c+2} = \tau_4 + N_c\sqrt{3}\,a_0\hat{x} \ . \tag{7.47}$$

The dangling $sp^2$ $\sigma$ orbitals of the edge C atoms can be saturated by adding H atoms along the direction of the missing C–C bonds assuming a C–H bond length equal to that of methane ($CH_4$), 0.10919 nm.

The case of zigzag-edge C nanoribbons (ZGNRs) is similarly treated by defining the primitive translation vectors:

$$\mathbf{a}_1 = ((N_c + N_v)\ 3\,a_0, 0, 0) \ , \quad \mathbf{a}_2 = (0, N_y\,a_0, 0) \ , \quad \mathbf{a}_3 = (0, 0, \sqrt{3}\,a_0) \ , \tag{7.48}$$

and the coordinates of the four atoms in the cell adjacent to the "left" edge:

$$\begin{aligned} \tau_1 &= a_0 \left( 0, 0, \frac{\sqrt{3}}{2} \right) \ , \quad \tau_2 = a_0 \left( \frac{1}{2}, 0, 0 \right) \ , \\ \tau_3 &= a_0 \left( \frac{3}{2}, 0, 0 \right) \ , \quad \tau_4 = a_0 \left( 2, 0, \frac{\sqrt{3}}{2} \right) \ . \end{aligned} \tag{7.49}$$

The remaining atomic coordinates are obtained by replicating $N_c - 1$ times the coordinates of the first four C atoms by translating them by an amount $j3a_0$ (with $j = 1, N_c - 1$) along the positive $x$ direction:

$$\begin{aligned} \tau_{4j+1} &= \tau_1 + j3a_0\hat{x} \ , \quad \tau_{4j+2} = \tau_2 + j3a_0\hat{x} \ , \\ \tau_{4j+3} &= \tau_3 + j3a_0\hat{x} \ , \quad \tau_{4j+4} = \tau_4 + j3a_0\hat{x} \ . \end{aligned} \tag{7.50}$$

Termination of the edge-C dangling bonds can be done by adding two H atoms bonded to the first and last C atom in the supercell.

Figure 7.31 (right) shows the band structure of a "bare" (as opposite to H-terminated) 9-AGNR. Note that the use of the Mayer pseudopotentials yields a reasonable energy gap (when compared to first-principle results [34]) and also a reasonable dispersion for the topmost valence band and

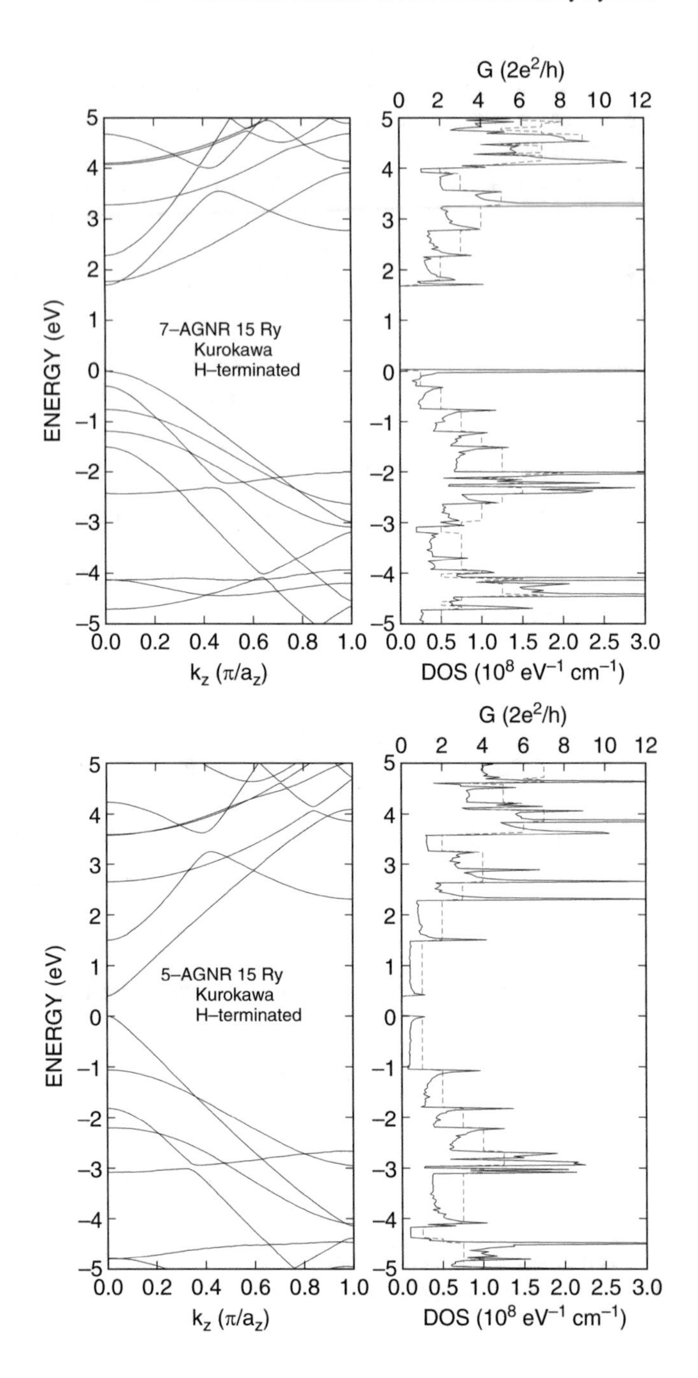


**Fig. 7.34** Band structure and density of states for an $N_a = 7$ (*left*) and $N_a = 5$ (*right*) armchair-edge graphene nanoribbon with H termination of the edge C atoms. Note that while tight-binding models predict a semimetallic (no gap) behavior for the 5-AGNR, empirical pseudopotentials yield semiconducting behavior even in the absence of the distortion of the edge C–C bonds found by first-principle calculations [34]. [Used from M.V. Fischetti, Bo Fu, S. Narayanan, and J. Kim, in *Nano-Electronic Devices: Semiclassical and Quantum Transport Modeling*, Dragica Vasileska and Stephen M. Goodnick eds. (Springer, New York, 2011), pp. 183–247, with kind permission from Springer Science+Business Media]

lowest-energy conduction bands, while missing by definition other valence bands, several highest-energy conduction bands, and also the well-known edge states which enter the $\pi$-$\pi^*$ gap. These states are clearly noticeable in the results obtained by using the Kamakura pseudopotentials: These pseudopotentials account for all bands, yield a reasonable gap at $k = 0$, as well as the edge states which can be removed by H termination [46]. This is shown in Fig. 7.34. Note that using the Kurokawa pseudopotentials, the gaps are in agreement with the first-principles, non-GW-corrected results of [34], as shown in Fig. 7.35. Only for the smallest-width ribbon (3-AGNR) the gap is noticeably smaller than what obtained from first-principle calculations, presumably because of the growing importance

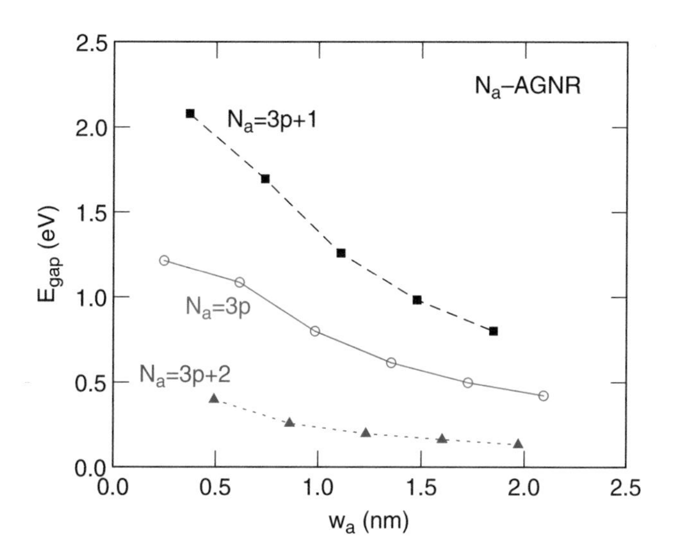


**Fig. 7.35** The three ladders of the band gap at $k = 0$ as a function of ribbon-width for armchair-edge graphene nanoribbons. These results have been obtained using the Kurokawa's pseudopotentials with H-terminated $sp^2$ $\sigma$ C edge bonds and are in excellent qualitative agreement and good quantitative agreement with the non-GW-corrected DFT+LDA calculation of Son et al. [34]. GW corrections yield much large gaps for the quasiparticle energy [36], while tight-binding models predict qualitatively incorrect gaps and an incorrect ordering of the three ladders [34]

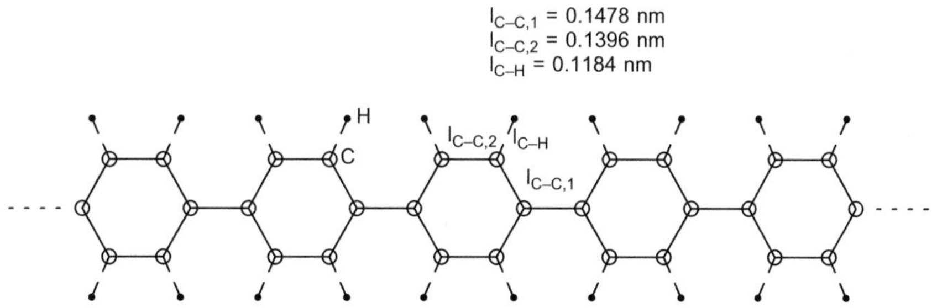

**Fig. 7.36** Atomic structure of polyparaphenylene (PPP). Note the similarity with a 3-AGNR from which it differs only for the different C–C bond lengths (assumed there to be 0.1422 nm for all C–C bonds) and also for the C–H bond lengths (assumed to be 0.10919 nm for the 3-AGNR)

of edge-bond distortion noticed by Son et al. [34]. It should be remarked that experiments [47] have not confirmed this chirality dependence of the band gap. Querlioz and co-workers have attributed this to disorder of the chirality (or, equivalently, line-edge roughness) [48]. Tseng et al. [49] have similarly argued that the clustering of the gap around values corresponding to the largest $3p+1$ gap is the result of line-edge roughness (LER or simply ER) which allows electron transmission only at the largest "local" gap.

A structure similar to that of a 3-AGNR is given by the polyparaphenylene (PPP) chain, which is identical to an H-terminated 3-AGNR, but with different C–C and C–H bond lengths, as seen in Fig. 7.36. Figure 7.37 shows the band structure, DOS, and ballistic conductance of PPP. As for the case of the 3-AGNR, the energy gap obtained using the Kurokawa pseudopotentials ($\approx$ 1.2 eV) is lower than what is measure experimentally [50] ($\approx$ 2.4 eV), in agreement with LDA calculations [34]. It appears that the small gap is due to the appearance of an "additional" band in the gap, as it will be discussed below when dealing with $(n, 0)$ nanotubes.

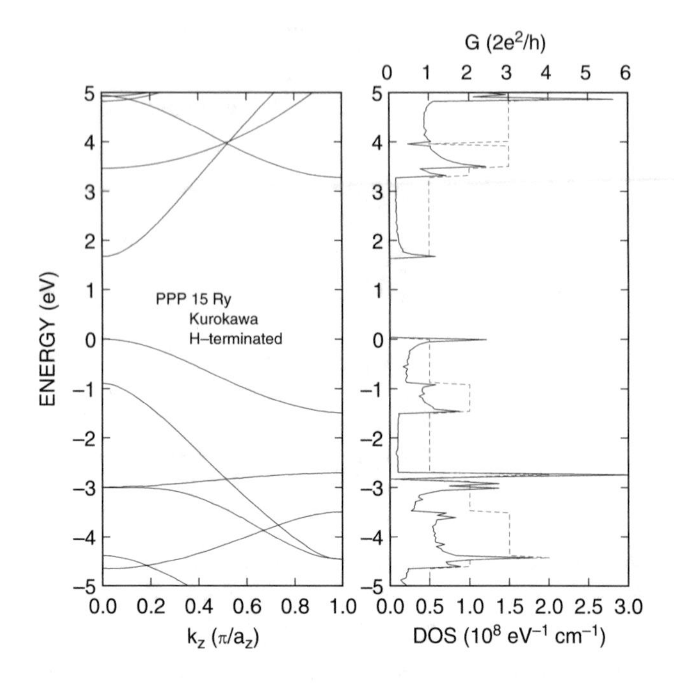


**Fig. 7.37** Band structure (*left*), DOS, and ballistic conductance (*right*) of PPP calculated using the Kurokawa pseudopotentials. Note that the gap between the bottom CB and second highest-energy VB is similar to the experimental value, but an "additional" band enters the gap

#### 7.3.2.2 Zigzag-Edge GNRs

In Fig. 7.38 we show the band structure of an $N_a$=4 zigzag-edge graphene nanoribbon (4-ZGNR) and of an 8-ZGNR obtained using the Mayer pseudopotentials with dangling bonds for the edge C atoms and using the Kurokawa pseudopotentials with H-terminated C edge bonds. Note that in this case both choices of pseudopotentials result in semimetallic behavior, the $\pi$ and $\pi^*$ bands overlapping slightly, in agreement with the results obtained by Ezawa [31] (who, however, also predicts metallic behavior for $N_a = 2p + 2$ armchair nanoribbons) and with the LDA results by Pisani et al. [51] for mono-hydrogenated non-magnetic nanoribbons. Although the shape of the bands appears qualitatively in agreement with the LDA results of [36], the semimetallic behavior of this ZGNR emerges from the fact that we have not accounted for spin polarization effects [52].

One can speculate about other possible shortcomings of the empirical pseudopotential approach. The obvious first concern stems from a possible inaccuracy of the empirical C pseudopotentials themselves. However, while Kurokawa's C pseudopotentials were calibrated to the diamond structure, the H pseudopotentials had been fit to the electronic structure of trans-polyacetylene, which resembles very closely the hydrogenated edges of ZGNRs. The neglect of the spin–orbit interaction may constitute another possible cause of concern. On the one hand, Kan and co-workers [53] have shown that the ZGNR bands are not spin-degenerate. However, spin polarization of edge states is known to emerge not from the spin–orbit interaction which is very small in C [54], but from another type of interaction said to be *Hubbard-like*. We shall not delve into this issue here. There is more: Kan et al. have also shown that ZGNRs may be metallic or semiconducting depending on the functional groups (H, $NH_2$, $CH_3$, and $NO_2$) used to terminate the $sp^2$ $\sigma$ orbitals of the edge C atoms. This would point at some possibly wrong assumptions made in terminating these orbitals with the Kurokawa's H pseudopotential and assuming the $CH_4$ C–H bond length. A final possible source of concerns may be the distortion of the edge C–C bonds emphasized by Son et al. [34] or of the C–H edge bonds.

Despite these doubts and concerns, the overall picture that emerges from calculations based on empirical pseudopotentials is qualitatively in agreement with more sophisticated approaches and captures the essential physical ingredients, especially the Claromatic behavior of GNRs.

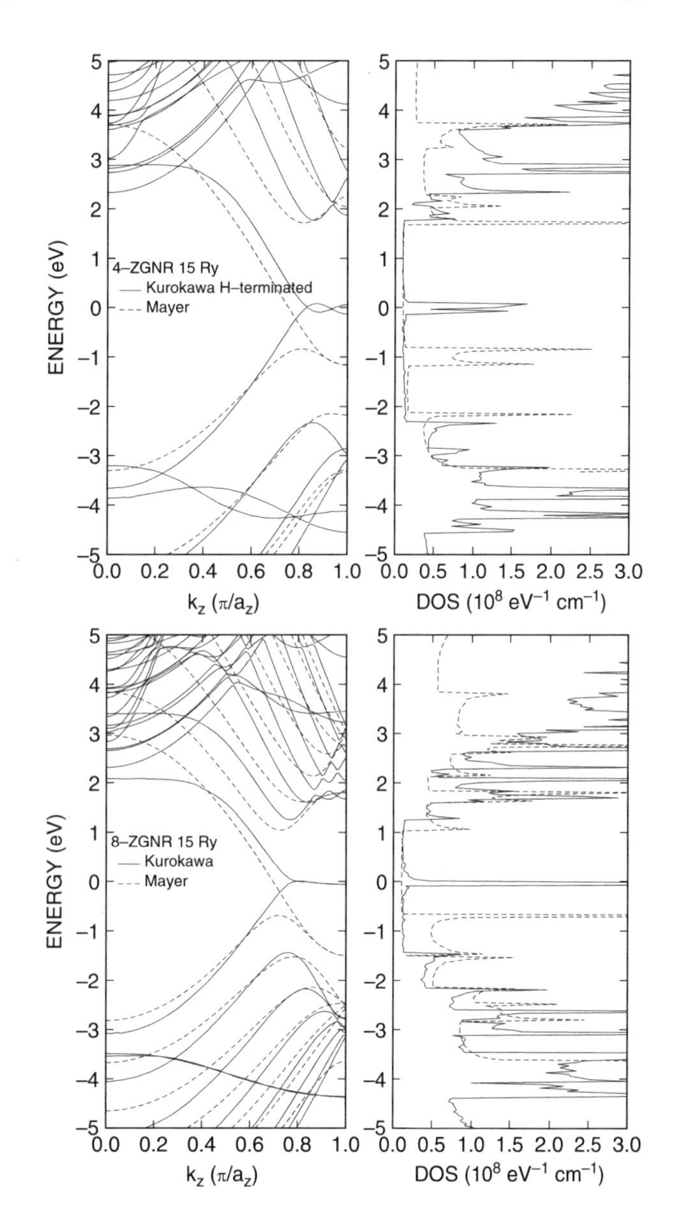


**Fig. 7.38** Band structure and density of states for an $N_a = 4$ graphene nanoribbon (4-ZGNR, *top*), and of an 8-ZGNR (*bottom*) with zigzag edges obtained using the Mayer pseudopotentials with bare edges (*dashed lines*) and the Kurokawa pseudopotentials with H termination of the edge C atoms (*solid lines*). In both cases the ZGNRs exhibit semimetallic behavior, in agreement with Ezawa's tight-binding results [31] and Pisani's "non-magnetic" LDA calculations [51]. A gap is expected to open when accounting for spin polarization effects, as shown by first-principle self-consistent LSDA calculations [34, 51]. [Used from M.V. Fischetti, Bo Fu, S. Narayanan, and J. Kim, in *Nano-Electronic Devices: Semiclassical and Quantum Transport Modeling*, Dragica Vasileska and Stephen M. Goodnick eds. (Springer, New York, 2011), pp. 183–247, with kind permission from Springer Science+Business Media]

### 7.3.3 Carbon Nanotubes

In the previous section, we have discussed two types of GNRs: Those with armchair edges and those with zigzag edges. We had defined them, looking at the top-left picture in Fig. 7.19, as obtained by cutting an infinite graphene sheet along two parallel lines along the vertical axis (armchair) or along direction of the primitive vector $\mathbf{a}_1$ (zigzag). These are just two particular crystallographic directions. In general, we can define a *chiral vector*

$$\mathbf{C}_{m,n} = m\mathbf{a}_1 + n\mathbf{a}_2 \, . \tag{7.51}$$

If we denote the direction of the vector $\mathbf{C}_{m,n}$ with $(m,n)$, then zigzag direction corresponds to $(n,0)$, the armchair direction to $(n,n)$. Figure 7.39 illustrates the geometry. Suppose now that we cut a ribbon along the $(n,0)$ direction and roll it so that the two armchair edges are brought together. This gives

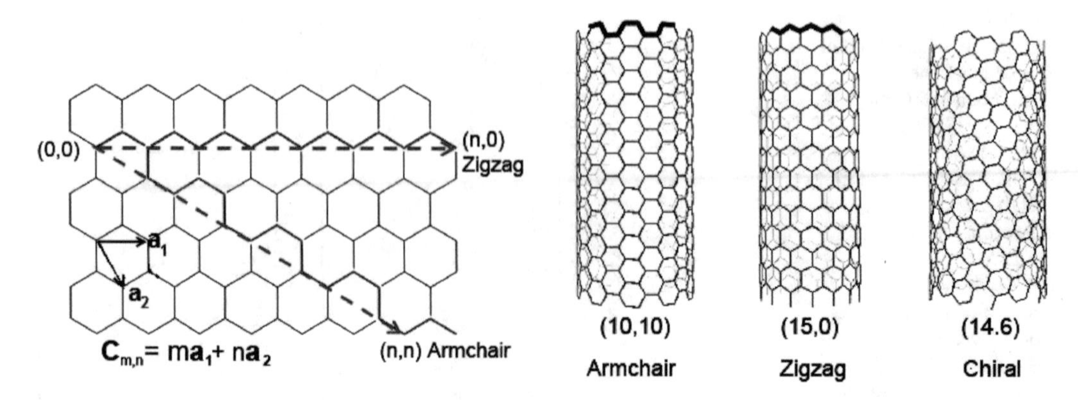


**Fig. 7.39** The chiral directions (*left*) and a schematic representation of armchair, zigzag, and chiral CNTs

us a "tube" whose direction along the circumference of a cross section is zigzag. We call this an $(n,0)$ zigzag carbon nanotube, or $(n,0)$ CNT. Similarly, cutting an $(n,n)$ ribbon and rolling it over to bring the two zigzag edges together, we obtain a tube with an armchair cross section, an $(n,n)$ CNT. This is illustrated in the right frame of Fig. 7.39. CNTs can also be obtained by cutting along arbitrary directions. In this case, the crystallographic lines appear to be twisted along the CNT, as shown by the rightmost CNT in the figure. These CNTs are called *chiral*. The chirality of CNTs affects enormously their electronic properties. We have seen that in GNRs the structure of the Clar resonances causes zigzag-edge GNRs to be metallic, armchair-edge GNRs to be semiconducting. These properties carry over to CNTs for exactly the same reasons, their claromaticity, just generalized to arbitrary chiral directions $m,n$: CNTs are metallic if $|n-m|$ is a multiple of 3, a different manifestation of the oscillations of the band gap of AGNRs. Thus, all armchair CNTs, $(n,n)$, are metallic. After all, they can be thought as a rolled-up ZGNR, which is metallic. Parenthetically, note the small size of CNTs: Their diameter $D$ is determined by the chirality via simple geometric considerations, as $D=(m^2+n^2+nm)^{1/2}a/\pi$, where $a \approx 0.246$ is the lattice constant of graphene.

We can finally perform the calculation of these interesting electronic properties of CNTs. The atomic positions required as input to set-up the empirical pseudopotential Hamiltonian can be either painfully calculated "by hand" or, more conveniently, using the on-line Java tool TubGen v3.3 [55]. Comparison can be made with the qualitative analysis by Ajiki and Ando [17], by Reich [18], Gulseren [56], Sharma [57], and Miyake and Saito [58, 59] for the diameter dependence of zigzag semiconducting nanotubes, and by Mayer [21] and Khoshnevisan [19] for the band structure of (5,5) and (10,0) CNTs. Figure 7.40 shows the position of the C atoms in a (5,5) armchair nanotube and in a (13,0) zigzag CNT, while in Fig. 7.41 we show the band structure and density of states for these CNTs. These data have been obtained using a supercell with square cross section of sides 1 and 1.4 nm long for the (5,5) and (13,0) CNTs and both the Kurokawa and Mayer pseudopotentials with a cutoff energy of 15 Ry. (The use of a larger cutoff, 25 Ry, does not show appreciable effects on the calculated band structure, as far as the few bands close to the Fermi level are concerned.) Results using a real-space approach [60] are also shown in the case of the (5,5) nanotube. The small difference obtained by using real-space or **k**-space methods can be attributed to the truncation at high spatial frequencies (large **G**-vectors) by the latter more than by the proximity of the "neighbor" nanotube implied by the supercell periodicity. Indeed, increasing the size of the supercell beyond twice the diameter of the nanotube does not cause any appreciable difference in the results of the supercell **k**-space method, as shown in Fig.7.42. In all cases the Mayer pseudopotentials yield the correct behavior of the $\pi$ and $\pi^*$ bands near the Fermi level, but miss by construction the deeper $2s$ and $2p$ valence states as well as many additional higher-energy states. By contrast, the pseudopotentials proposed by Kurokawa yield results much closer to ab initio results in the cases of (5,5) and large-

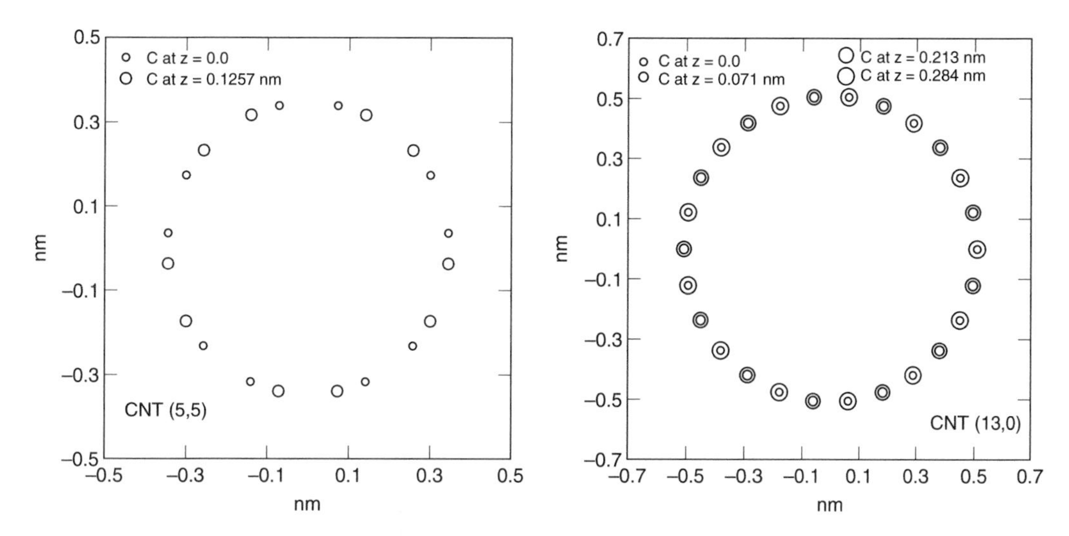


**Fig. 7.40** Positions of the C atoms in an armchair (5,5) (*left*) and in a zigzag (13,0) (*right*) CNT

diameter $(n,0)$ CNTs. However, in the latter cases [see, for example, the case of (10,0) nanotubes shown in Fig. 7.42] the $\pi^*$ singlet is pushed within the $\pi$-$\pi^*$ doublet gap resulting in an energy gap at $\Gamma$ much lower than expected from ab initio calculations [18, 56]. Blase [61] and later Gulseren et al. [56] have attributed this to $\sigma^* - \pi^*$ hybridization due to the high curvature of small-radius CNTs, an effect which seems to be overestimated by the Kurokawa pseudopotentials (see the $\sigma^*$ and $\pi^*$ singlet bands for graphene already mentioned as responsible for this effect in the caption of Fig. 7.19). First-principles calculations predict metallic behavior for *all* $(n,0)$ CNTs for $n \leq 6$, the Kurokawa empirical pseudopotentials predict this behavior for $n \leq 9$, while employing the Mayer one-electron pseudopotentials results in metallic behavior only for $n = 3p$ ($p$ = integer) for any $n$, since these pseudopotentials do not account for the $\sigma^*$ and $\pi^*$ singlet bands.

Looking at Fig. 7.43 we should also note that for zigzag nanotubes with a diameter smaller than about 2 nm (that is, for $(n,0)$ CNTs with $n \leq 5$), the Kurokawa pseudopotentials fail quite dramatically, always yielding semimetallic behavior and unexpected dispersion.

Once more, we should spend a few words of caution regarding the "correctness" (or lack thereof) of results obtained using the Mayer or Kurokawa pseudopotentials. We have compared these results with "first-principles" calculations (usually DFT+LDA and the occasional GGA or GW correction) [18, 19, 56]. The energy gaps and dispersion found in these papers are not always consistent among themselves. For example, the MP3 (Quantum Chemistry) approach followed by Bulusheva et al. [62] yields significantly different gaps, while the quality of experimental data on the gap dependence on tube diameter, all of them from the same Harvard group [63–65], is hard to assess, given the daunting practical difficulty of isolating CNTs of the same chirality (and, so, diameter). Also, the small density of states associated with the $\pi^*$-singlet band could render it hard to detect optically and electrically.

Finally, Figs. 7.44 and 7.45 show the contour plots of the squared amplitude of the wavefunctions of the three lowest-energy and three highest-energy conduction- and valence-band states for the (5,5) and (13,0) nanotubes, respectively.

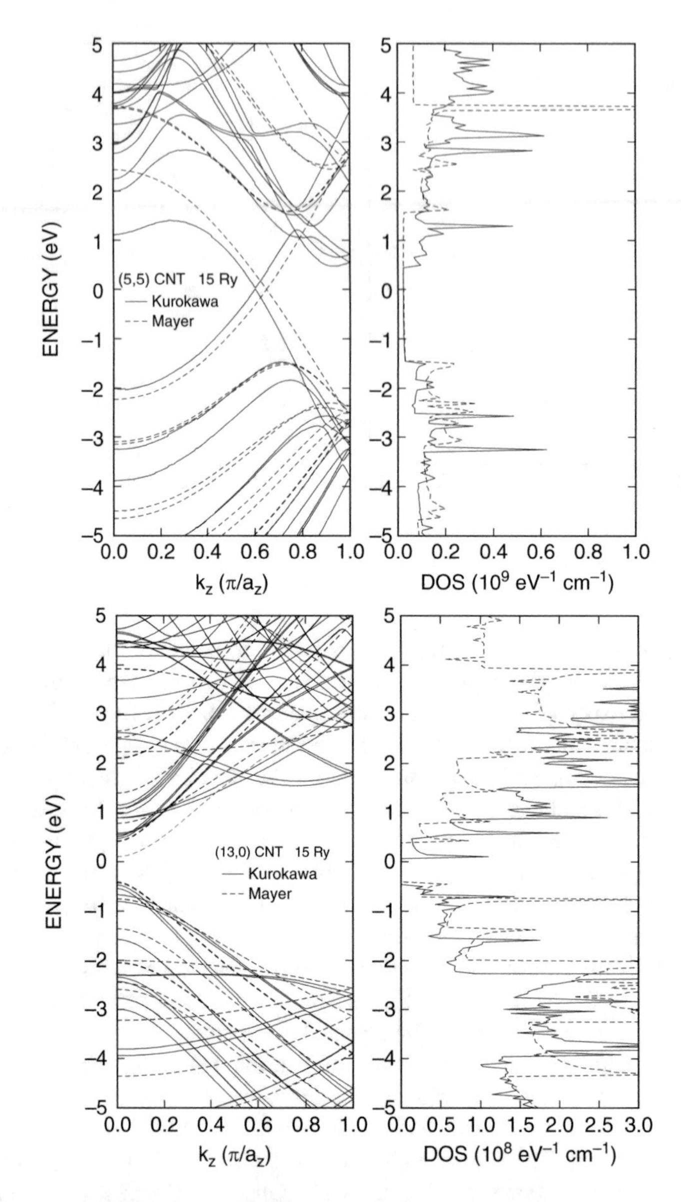


**Fig. 7.41** Band structure and density of states (DOS) of the metallic armchair (5,5) CNT (*left*) and of the semiconducting (13,0) zigzag CNT (*right*). The energy has been set to zero at mid-gap or band crossing, which is approximately equal to the Fermi level. The dispersion has been obtained using the "bulk C" local empirical pseudopotentials of Kurokawa et al. [20] (*solid lines*), and those of Mayer [21] (*dashed lines*), which should provide better results since they have been calibrated to graphene. However, the value of the band gap obtained using Kurokawa's pseudopotentials (0.574 eV) agrees with the values obtained using LDA [57] and CGA [56] (yielding, respectively, 0.669 and 0.625 eV) much better than the result (0.817 eV) obtained using Mayer's pseudopotentials. For the (5,5) CNT results obtained by Zhang and Polizzi [60] using a real-space approach with Mayer's pseudopotentials are also shown (*circles*). [Used from M.V. Fischetti, Bo Fu, S. Narayanan, and J. Kim, in *Nano-Electronic Devices: Semiclassical and Quantum Transport Modeling*, Dragica Vasileska and Stephen M. Goodnick eds. (Springer, New York, 2011), pp. 183–247, with kind permission from Springer Science+Business Media]

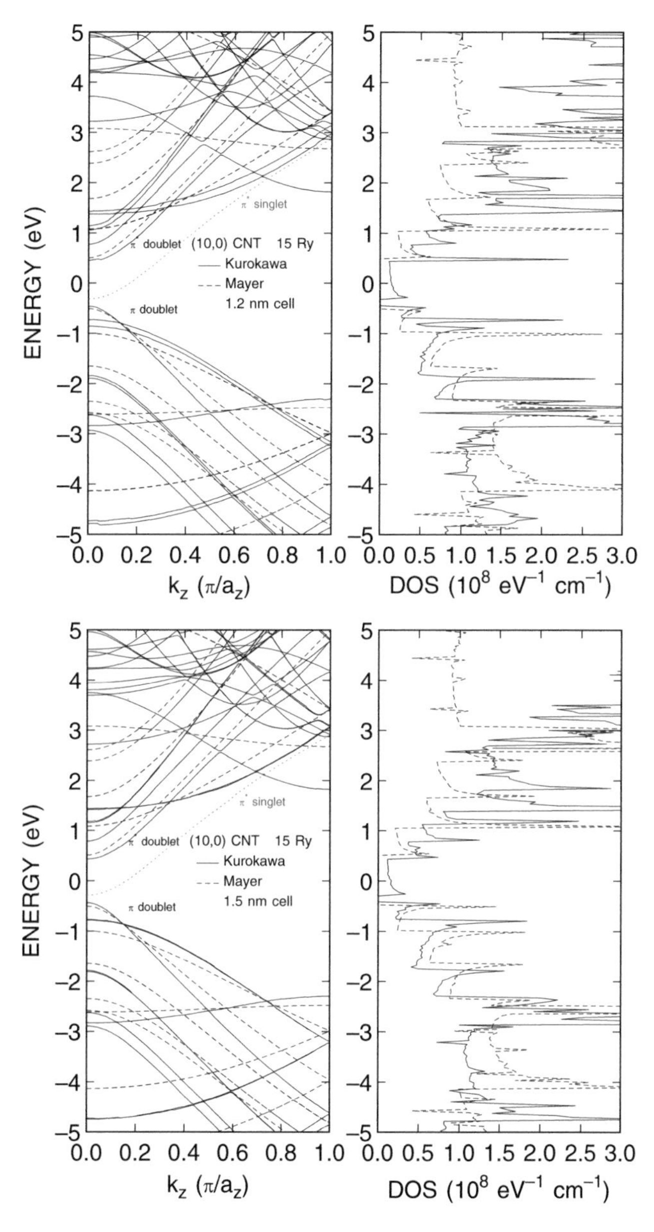


**Fig. 7.42** Band structure and density of states (DOS) of the semiconducting (10,0) zigzag CNT obtained, as in Fig. 7.41, using the "bulk C" local empirical pseudopotentials of Kurokawa et al. [20] (*solid lines*), and those of Mayer [21] (*dashed lines*). The energy has been set to zero at mid-gap or band crossing, which is approximately equal to the Fermi level. These results are obtained using a supercell with square cross section of 1.2 nm. The use of a slightly larger supercell (1.5 nm) yields very similar results, with the bands shifting by no more than 20 meV, as shown in the bottom frames. Note that the value of the band gap obtained using Kurokawa's pseudopotentials (0.1443 eV) is significantly smaller than the values obtained using LDA [57] and GGA [56] (0.764 eV in both cases) because of the presence of the $\pi^*$ singlet band (shown by a *red dotted line*) which the choice of Kurokawa pseudopotentials pushes to low energies inside the $\pi$-$\pi^*$ doublet gap. However, the magnitude of the $\pi$-$\pi^*$ doublet gap (0.8695 eV) is much closer to the expected value. Results obtained using Mayer's pseudopotentials yield a much larger gap, 1.0141 eV. [Used from M.V. Fischetti, Bo Fu, S. Narayanan, and J. Kim, in *Nano-Electronic Devices: Semiclassical and Quantum Transport Modeling*, Dragica Vasileska and Stephen M. Goodnick eds. (Springer, New York, 2011), pp. 183–247, with kind permission from Springer Science+Business Media]

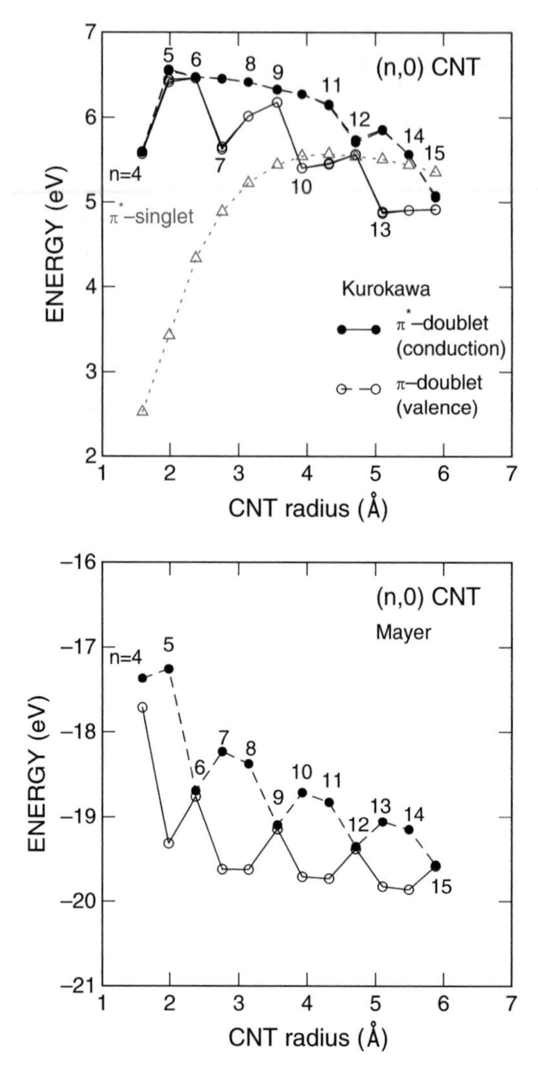


**Fig. 7.43** *Top*: Maximum (minimum) energy of the conduction $\pi^*$-bands (valence $\pi$-bands) as a function of diameter of $(n,0)$ CNTs obtained using the Kurokawa pseudopotentials. Note the quasi-periodic oscillations of the $\pi - \pi^*$ gap as $n$ varies between $3p$ (with $p$ an integer), corresponding to a very small gap which would vanish in absence of curvature effects, $3p+2$ and $3p+1$, the latter case yielding the largest gap in analogy with the situation observed for graphene nanoribbons in Fig. 7.35. Note, however, that the hybridization of the $\sigma^*$ and $\pi^*$ orbitals caused by the increasing curvature of the CNTs at small diameters pushes the energy of the $\pi^*$-singlet states within the $\pi - \pi^*$ gap for $n$ smaller than about 15 and ultimately closes the gap for $n < 10$. First-principle results predict this "gap closing" for $n < 7$, instead [56]. Results for $n < 5$ are suspect because of the very large curvature effects which have to be treated using first-principle methods. *Bottom*: As in the top frame, but showing results obtained using the Mayer pseudopotentials. Note the correct periodic oscillations of the gap with chiral number $n$ with period 3. However, the inability of the Mayer pseudopotentials to yield the singlet $\pi^*$ and $\sigma^*$ bands results in the prediction of semiconducting behavior also for small-diameter nanotubes, notably, the (4,0) and (5,0) CNTs. [Used from M.V. Fischetti, Bo Fu, S. Narayanan, and J. Kim, in *Nano-Electronic Devices: Semiclassical and Quantum Transport Modeling*, Dragica Vasileska and Stephen M. Goodnick eds. (Springer, New York, 2011), pp. 183–247, with kind permission from Springer Science+Business Media]

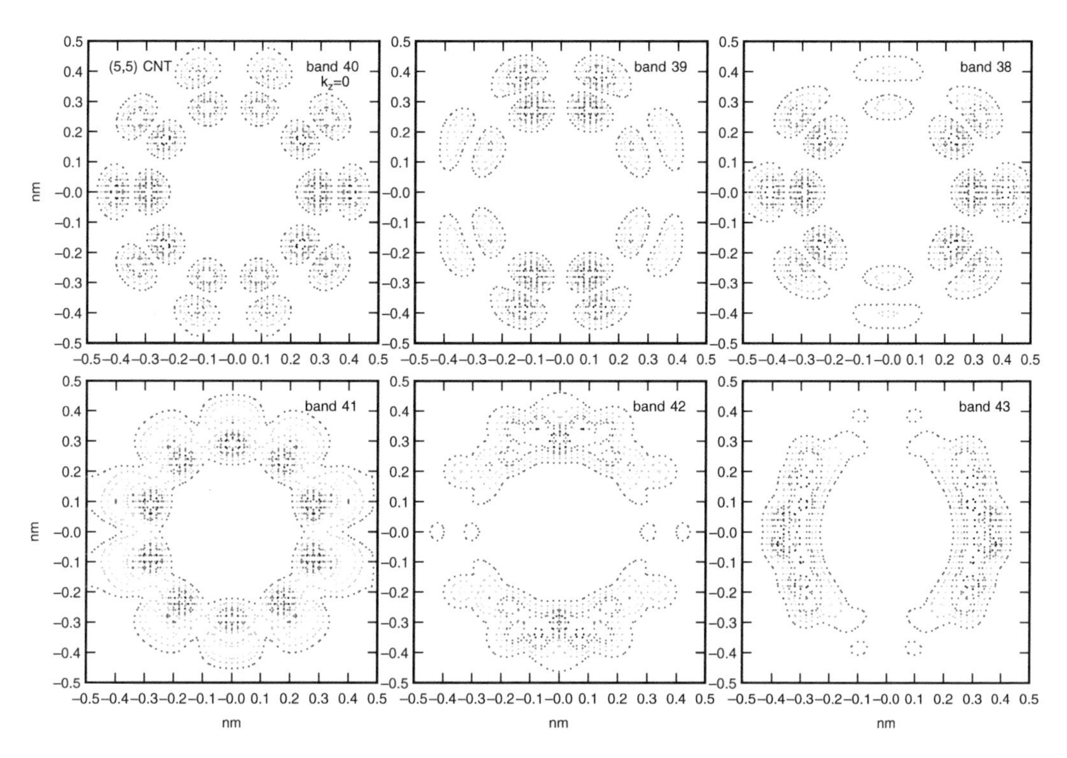


**Fig. 7.44** Contour plots of the squared amplitude of the wavefunctions—averaged over a cell along the axial direction—of the three highest-energy bands below the Fermi level (*top*) and the lowest-energy bands above the Fermi level (*bottom*) at the point $k_z$=0 for a (5,5) metallic CNT. Clearly evident are the $p_z$ orbitals. The Kurokawa pseudopotentials have been used to obtain the results presented in this figure

# References


1. T. Ando, A.B. Fowler, F. Stern, Electronic properties of two-dimensional systems. Rev. Mod. Phys. **54**, 437 (1982)
2. A.K. Geim, K.S. Novoselov, The rise of graphene. Nat. Mater. **6**, 183 (2007)
3. M.V. Fischetti, J. Kim, S. Narayanan, Z.-Y. Ong, C. Sachs, D.K. Ferry, S.J. Aboud, Pseudopotential-based studies of electron transport in graphene and graphene nanoribbons. J. Phys. Condens. Matter **25**, 473202 (2013)
4. M.V. Fischetti, B. Fu, S. Narayanan, J. Kim, Semiclassical and quantum electronic transport in nanometer-scale structures: empirical pseudopotential band structure, Monte Carlo simulations and Pauli master equation, in *Nano-Electronic Devices: Semiclassical and Quantum Transport Modeling*, ed. by D. Vasileska, S.M. Goodnick (Springer, New York, 2011), pp. 183–247
5. S.B. Zhang, C.-Y. Yeh, A. Zunger, Electronic structure of semiconductor quantum film. Phys. Rev. B **48**, 11204 (1993)
6. L.-W. Wang, A. Zunger, Electronic structure pseudopotential calculations of large (apprx. 1000 atoms) Si quantum dots. J. Phys. Chem. **98**, 2158 (1994)
7. D. Esseni, P. Palestri, Linear combination of bulk bands method for investigating the low-dimensional electron gas in nanostructured devices. Phys. Rev. B **72**, 165342 (2005)
8. G. Bastard, J.A. Brum, R. Ferreira, Electronic states in semiconductor heterostructures, in *Solid state physics: Semiconductor Heterostructures and Nanostructures*, ed. by H. Ehrenreich, D. Turnbull (Academic, New York, 1991), p. 259
9. L. Vegard, Die Konstitution der Mischkristalle und die Raumfüllung der Atome. Z. Phys. **5**, 17 (1921)
10. R.L. Anderson, Germanium-gallium arsenide heterojunction. IBM J. Res. Dev. **4**, 283 (1960)
11. K. Mäder, A. Zunger, Empirical atomic pseudopotentials for AlAs/GaAs superlattices, alloys, and nanostructures. Phys. Rev. B **50**, 17393 (1994)
12. L. Bellaiche, S.-H. Wei, A. Zunger, Localization and percolation in semiconductor alloys: GaAsN vs GaAsP. Phys. Rev. B **54**, 17568 (1996)

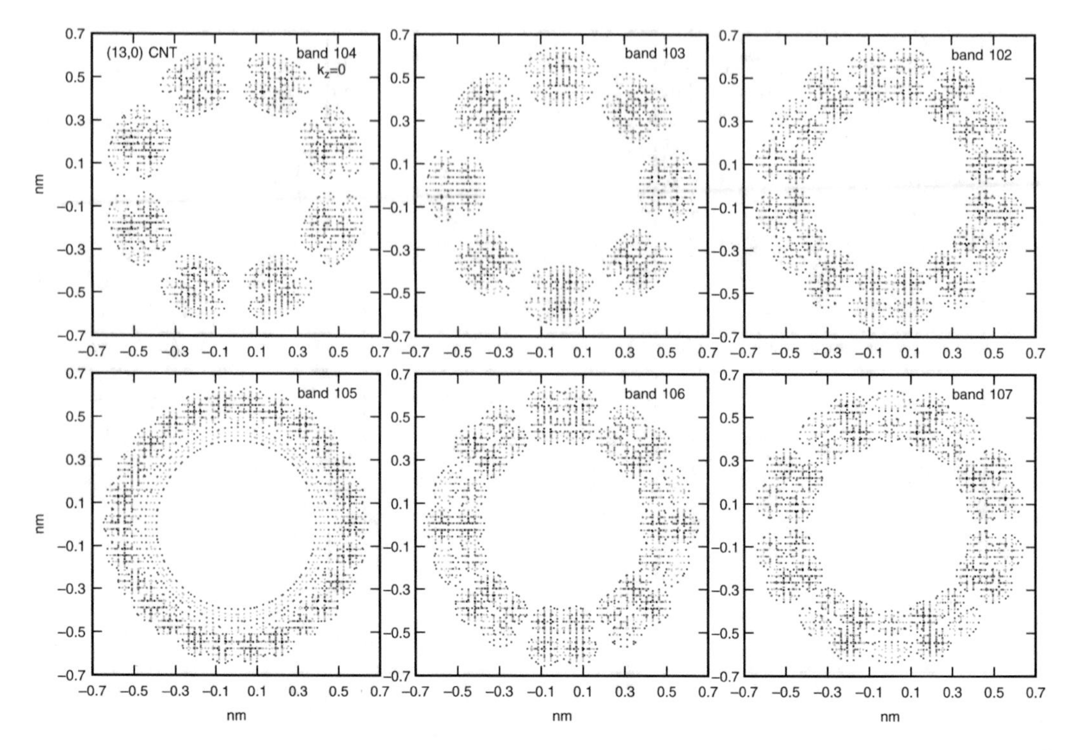


**Fig. 7.45** Contour plots of the squared amplitude of the three highest-energy valence bands (*top*) and the lowest-energy conduction bands (*bottom*) at the point $k_z = 0$ for a (13,0) semiconducting CNT, as in the previous figure


13. L. Bellaiche, L.-W. Wang, S.-H. Wei, A. Zunger, Resonant hole localization and anomalous optical bowing in InGaN alloys. Appl. Phys. Lett. **74**, 1842 (1999)
14. N.D. Mermin, H. Wagner, Absence of ferromagnetism or antiferromagnetism in one- or two-dimensional isotropic Heisenberg models. Phys. Rev. Lett. **17**, 1133 (1966)
15. N.D. Mermin, Crystalline order in two dimensions. Phys. Rev. **176**, 250 (1968)
16. P.C. Hohenberg, Existence of long-range order in one and two dimensions. Phys. Rev. **158**, 383 (1967)
17. H. Ajiki, T. Ando, Electronic states of carbon nanotubes. Jpn. J. Appl. Phys. **62**, 1255 (1993)
18. S. Reich, C. Thomsen, Electronic band structure of isolated and bundled carbon nanotubes. Phys. Rev. B **65**, 155411 (2002)
19. B. Khoshnevisan, Z.S. Tabatabaean, Effect of different types of pseudopotentials on study of electronic dispersion for graphene and a (5,5) SWCNT. Appl. Phys. A **92**, 371 (2008)
20. Y. Kurokawa, S. Nomura, T. Takemori, Y. Aoyagi, Large-scale calculation of optical dielectric functions of diamond nanocrystallites. Phys. Rev. B **61**, 12616 (2000)
21. A. Mayer, Band structure and transport properties of carbon nanotubes using a local pseudopotential and a transfer-matrix technique. Carbon **42**, 2057 (2004)
22. P.E. Trevisanutto, C. Giorgetti, L. Reining, M. Ladisa, V. Olevano, *Ab initio* GW many-body effects in graphene. Phys. Rev. Lett. **101**, 226405 (2008)
23. F. Aryasetiawan, O. Gunnarsson, The GW method. Rep. Prog. Phys. **61**, 237 (1998)
24. K. Nehari, N. Cavassilas, J.L. Autran, M. Bescond, D. Munteanu, M. Lannoo, Influence of band structure on electron ballistic transport in silicon nanowire MOSFET's: an atomistic study. Solid State Electron. **50**, 716 (2006)
25. N. Neophytou, A. Paul, M.S. Lundstrom, G. Klimeck, Bandstructure effects in silicon nanowire electron transport. IEEE Trans. Electron Dev. **55**, 1286 (2008)
26. F. Sacconi, M.P. Persson, M. Povolotsky, L. Latessa, A. Pecchia, A. Gagliardi, A. Balint, T. Fraunheim, A. Di Carlo, Electronic and transport properties of silicon nanowires. J. Comput. Electron. **6**, 329 (2007)
27. H. Scheel, S. Reich, C. Thomsen, Electronic band structure of high-index silicon nanowires. Phys. Status Solidi B **242**, 2474 (2005)
28. Y. Lee, T. Nagata, K. Kakushima, K. Shiraishi, H. Iwai, A study on electronic structure of silicon nanowires with diverse diameters and orientations for high performance FET, in *Proc. International Workshop on Density Functional Theory*, Tokyo (2008), p. 83

29. M. Fujita, K. Wakabayashi, K. Nakada, K. Kusakabe, Peculiar localized state at zigzag graphite edge. J. Phys. Soc. Jpn. **65**, 1920 (1996)
30. M. Ezawa, Peculiar width dependence of the electronic properties of carbon nanoribbons. Phys. Rev. B **73**, 045432 (2006)
31. M. Ezawa, Peculiar bandgap structure of graphene nanoribbons. Phys. Status Solidi C **4**, 489 (2007)
32. L. Brey, H.A. Fertig, Electronic states of graphene nanoribbons studied with the Dirac equation. Phys. Rev. B **73**, 235411 (2006)
33. K.-I. Sasaki, S. Murakami, R. Saito, Gauge field for edge state in graphene. J. Phys. Soc. Jpn. **75**, 074713 (2006)
34. Y.-W. Son, M.L. Cohen, S.G. Louie, Energy gaps in graphene nanoribbons. Phys. Rev. Lett. **97**, 216803 (2006)
35. V. Barone, O. Hod, G. Scuseria, Electronic structure and stability of semiconducting graphene nanoribbons. Nano Lett. **6**, 2748 (2006)
36. L. Yang, C.-H. Park, Y.-W. Son, M.L. Cohen, S.G. Louie, Quasiparticle energies and bandgaps in graphene nanoribbons. Phys. Rev. Lett. **99**, 186801 (2007)
37. E. Clar, *The Aromatic Sextet* (Wiley, New York, 1972)
38. T. Wassman, A.P. Seitsonen, A.M. Saitta, M. Lazzeri, F. Mauri, Clar's theory, $\pi$-electron distribution, and geometry of graphene nanoribbons. J. Am. Chem. Soc. **132**, 3440 (2010)
39. A. Balaban, D.J. Klein, Claromatic carbon nanostructures. J. Phys. Chem. C **113**, 19123 (2009)
40. J. Tersoff, D.R. Hamann, Theory of the scanning tunneling microscope. Phys. Rev. Lett. **50**, 1998 (1983)
41. J. Tersoff, D.R. Hamann, Theory of the scanning tunneling microscope. Phys. Rev. B **31**, 805 (1985)
42. G. Kresse, J. Hafner, *Ab initio* molecular dynamics for liquid metals. Phys. Rev. B **47**, 558(R) (1993)
43. G. Kresse, J. Furthmüller, Efficiency of *ab initio* total energy calculations for metals and semiconductors using a plane-wave basis set. Comput. Mater. Sci. **6**, 15 (1996)
44. G. Kresse, J. Furthmüller, Efficient iterative schemes for *ab initio* total-energy calculations using a plane-wave basis set. Phys. Rev. B **54**, 11169 (1996)
45. P. Zhao, J. Guo, Modeling edge effects in graphene nanoribbon field-effect transistors with real and mode space methods. J. Appl. Phys. **105**, 034503 (2009)
46. H. Sevincli, M. Topsakal, S. Ciraci, Superlattice structures of graphene-based armchair nanoribbons. Phys. Rev. B **78**, 245402 (2008)
47. X. Li, X. Wang, L. Zhang, S.W. Lee, H.J. Dai, Chemically derived, ultrasmooth graphene nanoribbon semiconductors. Science **319**, 1229 (2008)
48. D. Querlioz, Y. Apertet, A. Valenkin, K. Huet, A. Bournel, S. Galdin-Retailleau, P. Dollfus, Suppression of the orientation effects on bandgap in graphene nanoribbons in the presence of edge disorder. Appl. Phys. Lett. **92**, 042108 (2008)
49. F. Tseng, D. Unluer, K. Holcomb, M.R. Stan, A.W. Ghosh, Diluted chirality dependence in edge rough graphene nanoribbon field-effect transistors. Appl. Phys. Lett. **94**, 223112 (2009)
50. H. Eckhardt, L.W. Shaklette, K.Y. Jen, R.L. Elsenbaumer, The electronic and electrochemical properties of poly(phenylene vinylenes) and poly(thienylene vinylenes): an experimental and theoretical study. J. Chem. Phys. **91**, 1303 (1989)
51. L. Pisani, J.A. Chan, B. Montanari, N.M. Harrison, Electronic structure and magnetic properties of graphitic ribbon. Phys. Rev. B **75**, 064418 (2007)
52. D. Gunlycke, P.A. Areshkin, J. Li, J.M. Mintmire, C.T. White, Graphene nanostrip digital memory device. Nano Lett. **7**, 3608 (2007)
53. E. Kan, Z. Li, J. Yang, J.G. Hou, Half-metallicity in edge-modified zigzag graphene nanoribbons. J. Am. Chem. Soc. **130**, 4224 (2008)
54. G. Dresselhaus, M. Dresselhaus, J.G. Madrovies, Spin-orbit interaction in graphite. Carbon **4**, 433 (1966)
55. J.T. Frey, D.J. Doren, TubGen 3.3 Web Interface, http://turin.nss.udel.edu/research/tubegenonline.html (2015)
56. O. Gulseren, T. Yildirim, S. Caraci, Systematic *ab initio* study of curvature effects in carbon nanotubes. Phys. Rev. B **65**, 153405 (2002)
57. M. Sharma, A. Tiwari, U.S. Sharma, *Ab initio* study of electronic band structure of zigzag single wall carbon nanotubes, in *Proc. International Workshop on New Trends in Science and Technology*, Ankara, 3–4 November 2008, http://ntst08.cankaya.edu/proceedings/proceedings/Manoj/SharmaPaper.doc
58. T. Miyake, S. Saito, Quasiparticle band structure of carbon nanotubes. Phys. Rev. B **68**, 155424 (2003)
59. T. Miyake, S. Saito, Band-gap formation in (n,0) single-walled carbon nanotubes (n=9,12,15,18): a first-principles study. Phys. Rev. B **72**, 073404 (2005)
60. D. Zhang, E. Polizzi, Efficient modeling techniques for atomistic-based electronic density calculations. J. Comput. Electron. **7**, 427 (2008)
61. X. Blase, L.X. Benedict, E.L. Sherly, S.G. Louie, Hybridization effects and metallicity in small radius carbon nanotubes. Phys. Rev. Lett. **72**, 1878 (1994)
62. L.G. Bulusheva, A.V. Okotrub, D.A. Romanov, D. Tomanek, Electronic structure of (n, 0) zigzag carbon nanotubes: cluster and crystal approach. J. Phys. Chem. A **102**, 975 (1998)

63. T.W. Odom, J. Huang, P. Kim, C.M. Lieber, Atomic structure and electronic properties of single-walled carbon nanotubes. Nature (London) **391**, 62 (1998)
64. T.W. Odom, J. Huang, P. Kim, C.M. Lieber, Structure and electronic properties of carbon nanotubes. J. Phys. Chem. B **104**, 2794 (2000)
65. M. Ouyang, J. Huang, C.L. Cheung, C.M. Lieber, Energy gaps in 'metallic' single-walled carbon nanotubes. Science **292**, 702 (2001)

# Chapter 8
# Single-Electron Dynamics in Crystals

## 8.1 Equations of Motion for Electrons in Crystals

Having discussed the rather complicated band structure of semiconductors and nanostructures, we must now consider how single electrons will behave in these complicated energy bands under the action of an external force.

We shall first consider a remarkable simplification of the very complicated Schrödinger equation which results from adding a potential $V^{(\mathrm{ext})}(\mathbf{r},t)$ to the lattice potential itself. Under some restrictive conditions, we shall obtain an equation very similar to the Schrödinger equation for an electron in free space, the major modification being the dispersion (described often by an effective mass) and the meaning of the "wavefunction" itself, which will now be the "envelope" of the periodic Bloch part. This is the so-called *envelope* or *effective mass* approximation that has been originally derived by Luttinger and Kohn [1]. We shall then consider what are known as the *acceleration* and *Feynman* theorems, which will allow us to treat the electron moving under an external field $\mathbf{E}$ with equations similar to those valid for electrons in free space.

### 8.1.1 The Envelope Approximation

From Bloch's theorem, we know that the general solution of the Schrödinger equation in the lattice is labeled by a wavevector $\mathbf{k}$ and a band index $n$ and has the Bloch form:

$$\psi_{n\mathbf{k}}(\mathbf{r}) = \frac{1}{\Omega^{1/2}}\, \mathrm{e}^{i\mathbf{k}\cdot\mathbf{r}}\, u_{\mathbf{k}}^{(n)}(\mathbf{r}) = \frac{1}{\Omega^{1/2}}\, \mathrm{e}^{\mathrm{i}\mathbf{k}\cdot\mathbf{r}} \sum_{\mathbf{G}} u_{\mathbf{k}+\mathbf{G}}^{(n)}\, \mathrm{e}^{\mathrm{i}\mathbf{G}\cdot\mathbf{r}}\ . \tag{8.1}$$

Consider now adding an external potential $V^{(\mathrm{ext})}(\mathbf{r})$ to the lattice potential. We assume that the external potential is weak enough so that it does not cause transitions to other bands, so that we may consider only one band, labeled by $m$, with dispersion $E_m(\mathbf{k})$. We also assume that this weak potential does not cause changes of the Bloch component $u_{\mathbf{k}}^{(m)}(\mathbf{r}) = \sum_{\mathbf{G}} u_{\mathbf{k}+\mathbf{G}}^{(m)}\, \mathrm{e}^{\mathrm{i}\mathbf{G}\cdot\mathbf{r}}$. Under these assumptions we wish to show that the wavefunction in a small-enough neighborhood of $\mathbf{k}_0$ can be approximated by:

$$\psi_m(\mathbf{r}) = \phi(\mathbf{r})\, u_{\mathbf{k}_0}^{(m)}(\mathbf{r}) = \phi(\mathbf{r}) \sum_{\mathbf{G}} u_{\mathbf{k}_0+\mathbf{G}}^{(m)}\, \mathrm{e}^{\mathrm{i}\mathbf{G}\cdot\mathbf{r}}$$



M. Fischetti, W.G. Vandenberghe, *Advanced Physics of Electron Transport in Semiconductors and Nanostructures*, Graduate Texts in Physics, DOI 10.1007/978-3-319-01101-1_8

and the "envelope" $\phi(\mathbf{r})$ satisfies the wave equation:

$$[E_m(-\mathrm{i}\nabla) + eV^{(\mathrm{ext})}(\mathbf{r})]\ \phi(\mathbf{r}) = E\ \phi(\mathbf{r})\ . \tag{8.2}$$

Let's prove this statement.

The full Schrödinger equation including explicitly the lattice potential is

$$\left[-\frac{\hbar^2}{2m}\nabla^2 + V^{(\mathrm{lat})}(\mathbf{r}) + V^{(\mathrm{ext})}(\mathbf{r})\right]\ \psi(\mathbf{r}) = E\ \psi(\mathbf{r})\ . \tag{8.3}$$

Let's expand the full wavefunction $\psi$ into a series of Bloch functions (8.1) as follows:

$$\psi(\mathbf{r}) = \sum_{n,\mathbf{k}} c_{n\mathbf{k}}\ \psi_{n\mathbf{k}}(\mathbf{r}) = \frac{1}{\Omega^{1/2}} \sum_{n,\mathbf{k}} c_{n\mathbf{k}}\ \mathrm{e}^{\mathrm{i}\mathbf{k}\cdot\mathbf{r}} \sum_{\mathbf{G}} u^{(n)}_{\mathbf{k}+\mathbf{G}}\ \mathrm{e}^{\mathrm{i}\mathbf{G}\cdot\mathbf{r}}\ . \tag{8.4}$$

Since each Bloch function $\psi_{n\mathbf{k}}$ is an eigenvector of the Eq. (8.3) when $V^{(\mathrm{ext})}$=0, inserting the expansion (8.4) into Eq. (8.3) we get

$$\sum_{n,\mathbf{k}} \left[E_{n\mathbf{k}}\ c_{n\mathbf{k}}\ \psi_{n\mathbf{k}}(\mathbf{r}) + V^{(\mathrm{ext})}(\mathbf{r})\ c_{n\mathbf{k}}\ \psi_{n\mathbf{k}}(\mathbf{r})\right] = E \sum_{n,\mathbf{k}} c_{n\mathbf{k}}\ \psi_{n\mathbf{k}}(\mathbf{r})\ . \tag{8.5}$$

Now let's multiply both sides of this equation by $\psi^*_{m\mathbf{k}'}$ and integrate over the entire volume. Note that:

$$\sum_{n,\mathbf{k}} c_{n\mathbf{k}} \int \psi^*_{m\mathbf{k}'}(\mathbf{r})\ \psi_{n\mathbf{k}}(\mathbf{r})\ \mathrm{d}\mathbf{r} = \frac{1}{\Omega} \sum_{n,\mathbf{k}} \sum_{\mathbf{G},\mathbf{G}'} c_{n\mathbf{k}}\ u^{(m)*}_{\mathbf{k}'+\mathbf{G}'}\ u^{(n)}_{\mathbf{k}+\mathbf{G}} \int \mathrm{e}^{\mathrm{i}(\mathbf{k}+\mathbf{G}-\mathbf{k}'-\mathbf{G}')\cdot\mathbf{r}}\ \mathrm{d}\mathbf{r}$$

and the integral on the right hand side will be nonzero only if $\mathbf{k} - \mathbf{k}' + \mathbf{G} - \mathbf{G}' = 0$. If we restrict $\mathbf{k}$ and $\mathbf{k}'$ within the first BZ (which we can always do, from the definition (8.1) of Bloch waves), this condition can be met only when $\mathbf{k} - \mathbf{k}' = 0$ and $\mathbf{G} - \mathbf{G}' = 0$ separately. Thus the expression above becomes

$$\sum_{n} c_{n\mathbf{k}'} \sum_{\mathbf{G}'} u^{(m)*}_{\mathbf{k}'+\mathbf{G}'}\ u^{(n)}_{\mathbf{k}'+\mathbf{G}'} = \sum_{n} c_{n\mathbf{k}'}\ \delta_{nm} = c_{m\mathbf{k}'}, \tag{8.6}$$

since Bloch waves in different bands at the same $\mathbf{k}$-point are eigenvectors with different eigenvalues and so are orthogonal. Using Eq. (8.6), we obtain from Eq. (8.5) (switching the roles of $\mathbf{k}$ and $\mathbf{k}'$):

$$E_{m\mathbf{k}}\ c_{m\mathbf{k}} + \sum_{n\mathbf{k}'} c_{n\mathbf{k}'} \int \psi^*_{m\mathbf{k}}\ V^{(\mathrm{ext})}(\mathbf{r})\ \psi_{n\mathbf{k}'}(\mathbf{r}) = E\ c_{m\mathbf{k}}\ . \tag{8.7}$$

So far we have simply recast the original Schrödinger equation (8.3) into a matrix form, but we have accomplished nothing. Let's now consider the matrix element of the external potential,

$$\langle m\mathbf{k}|V^{(\mathrm{ext})}|n\mathbf{k}'\rangle = \int \psi^*_{m\mathbf{k}}(\mathbf{r})\ V^{(\mathrm{ext})}(\mathbf{r})\ \psi_{n\mathbf{k}'}(\mathbf{r})\ \mathrm{d}\mathbf{r}\ . \tag{8.8}$$

Let's write the potential in terms of its Fourier components:

$$V^{(\mathrm{ext})}(\mathbf{r}) = \sum_{\mathbf{q}} V_{\mathbf{q}}^{(\mathrm{ext})} \, \mathrm{e}^{\mathrm{i}\mathbf{q}\cdot\mathbf{r}} , \tag{8.9}$$

and insert this into Eq. (8.8):

$$\langle m\mathbf{k}|V^{(\mathrm{ext})}|n\mathbf{k}'\rangle = \frac{1}{\Omega} \sum_{\mathbf{q}} V_{\mathbf{q}}^{(\mathrm{ext})} \int u_{\mathbf{k}}^{(m)*}(\mathbf{r}) \, u_{\mathbf{k}'}^{(n)}(\mathbf{r}) \, \mathrm{e}^{-\mathrm{i}(\mathbf{k}'-\mathbf{k}-\mathbf{q})\cdot\mathbf{r}} \, \mathrm{d}\mathbf{r} . \tag{8.10}$$

Now express the integration over the entire volume $\Omega$ into the sum of integrations over cells $l$ by writing $\mathbf{r} = \mathbf{R}_l + \rho$:

$$\begin{aligned}\langle m\mathbf{k}|V^{(\mathrm{ext})}|n\mathbf{k}'\rangle &= \frac{1}{\Omega} \sum_{\mathbf{q}} V_{\mathbf{q}}^{(\mathrm{ext})} \sum_{l} \mathrm{e}^{-\mathrm{i}(\mathbf{k}'-\mathbf{k}-\mathbf{q})\cdot\mathbf{R}_l} \\ &\times \int_{\Omega} u_{\mathbf{k}}^{(m)*}(\rho) \, u_{\mathbf{k}'}^{(n)}(\rho) \, \mathrm{e}^{-\mathrm{i}(\mathbf{k}'-\mathbf{k}+\mathbf{q})\cdot\rho} \, \mathrm{d}\rho . \end{aligned} \tag{8.11}$$

The term $\sum_l \mathrm{e}^{-\mathrm{i}(\mathbf{k}'-\mathbf{k}-\mathbf{q})\cdot\mathbf{R}_l}$ will be nonzero only when $\mathbf{k}' - \mathbf{k} - \mathbf{q} = \mathbf{G}''$, where $\mathbf{G}''$ is some vector of the reciprocal lattice, while the integrals of the Bloch components over the cell volume $\Omega_{\mathrm{c}}$ will be equal in all cells. Therefore:

$$\begin{aligned}\langle m\mathbf{k}|V^{(\mathrm{ext})}|n\mathbf{k}'\rangle &= \frac{1}{\Omega} \sum_{\mathbf{q}\mathbf{G}''} N_{\mathrm{cells}} \, V_{\mathbf{q}}^{(\mathrm{ext})} \, \delta_{\mathbf{k}'-\mathbf{k}-\mathbf{q},\mathbf{G}''} \int_{\Omega_{\mathrm{c}}} u_{\mathbf{k}}^{(m)*}(\rho) \, u_{\mathbf{k}'}^{(n)}(\rho) \, \mathrm{e}^{-\mathrm{i}\mathbf{G}''\cdot\rho} \, \mathrm{d}\rho \\ &= \sum_{\mathbf{G}''} V_{\mathbf{k}'-\mathbf{k}-\mathbf{G}''}^{(\mathrm{ext})} \frac{1}{\Omega_{\mathrm{c}}} \int_{\Omega_{\mathrm{c}}} u_{\mathbf{k}}^{(m)*}(\mathbf{r}) \, u_{\mathbf{k}'}^{(n)}(\mathbf{r}) \, \mathrm{e}^{-\mathrm{i}\mathbf{G}''\cdot\rho} \, \mathrm{d}\rho , \end{aligned} \tag{8.12}$$

since $\Omega = \Omega_{\mathrm{c}} N_{\mathrm{cells}}$. Now notice that, looking at Eq. (8.1), $u_{\mathbf{k}+\mathbf{q}+\mathbf{G}''}^{(n)}(\rho)\mathrm{e}^{-\mathrm{i}\mathbf{G}''\cdot\rho} = u_{\mathbf{k}+\mathbf{q}}^{(n)}(\rho)$. Therefore:

$$\langle m\mathbf{k}|V^{(\mathrm{ext})}|n\mathbf{k}'\rangle = \sum_{\mathbf{G}''} V_{\mathbf{k}'-\mathbf{k}-\mathbf{G}''}^{(\mathrm{ext})} \frac{1}{\Omega_{\mathrm{c}}} \int_{\Omega_{\mathrm{c}}} u_{\mathbf{k}}^{(m)*}(\rho) \, u_{\mathbf{k}'}^{(n)}(\rho) \, \mathrm{d}\rho . \tag{8.13}$$

This new form of the matrix element (8.8) now allows us to make two additional approximations to further simplify the problem.

1. If the external potential is varying slowly enough to have nonzero Fourier components $V_{\mathbf{q}}$ only for small values of $q$, we can retain only the term with $\mathbf{G}''$=0 above, that is, $V_{\mathbf{k}'-\mathbf{k}}^{(\mathrm{ext})}$.
2. Let's assume that the wave packet we have considered originally, Eq. (8.4), is a spatially broad packet (that is, extending over several lattice cells), so it is narrow in $\mathbf{k}$-space, and only a small range of $\mathbf{k}$-vectors centered around a value $\mathbf{k}_0$ is involved.
3. Since only small values of $q = |\mathbf{k} - \mathbf{k}'|$ are retained thanks to assumption 1 above, if we assume that the Bloch components do not vary too quickly with $\mathbf{k}$, then it reasonable to assume also that $u_{\mathbf{k}'}^{(n)}(\rho) \approx u_{\mathbf{k}}^{(n)}(\rho)$ and

$$\int_{\Omega_{\mathrm{c}}} u_{\mathbf{k}}^{(m)*}(\rho) \, u_{\mathbf{k}'}^{(n)}(\rho) \, \mathrm{d}\rho \approx \int_{\Omega_{\mathrm{c}}} u_{\mathbf{k}}^{(m)*}(\rho) \, u_{\mathbf{k}}^{(n)}(\rho) \, \mathrm{d}\rho \approx \Omega_{\mathrm{c}} \, \delta_{m,n} ,$$

as one can see following the same procedure followed to arrive at Eq. (8.6).

Then, finally:

$$\langle m\mathbf{k}|V^{(\mathrm{ext})}|n\mathbf{k}'\rangle \approx V_{\mathbf{k}'-\mathbf{k}}\,\delta_{m,n}\ . \tag{8.14}$$

Our starting Schrödinger equation (8.7) in matrix form now has the simpler appearance:

$$E_{m\mathbf{k}}\,c_{m\mathbf{k}} + \sum_{\mathbf{k}'} V^{(\mathrm{ext})}_{\mathbf{k}'-\mathbf{k}}\,c_{m\mathbf{k}'} = E\,c_{m\mathbf{k}}\ , \tag{8.15}$$

which shows that the potential is weak enough to prevent interband transitions. We are thus able to restrict Eq. (8.3) to a single band.

Let's now define the *envelope wavefunction* in band $m$ via:

$$\phi(\mathbf{r}) = \frac{1}{\Omega^{1/2}} \sum_{\mathbf{k}} c_{m\mathbf{k}}\,\mathrm{e}^{\mathrm{i}\mathbf{k}\cdot\mathbf{r}}\ , \tag{8.16}$$

that is, the same expression we used to define the full, "exact" wavefunction given by Eq. (8.4), but having replaced the Bloch functions $\psi_{n\mathbf{k}}$ with plane waves $\mathrm{e}^{\mathrm{i}\mathbf{k}\cdot\mathbf{r}}$. Thus, from the definition (8.4), recalling that only a few $\mathbf{k}$-vectors around $\mathbf{k}_0$ are involved

$$\begin{aligned}\psi(\mathbf{r}) &= \frac{1}{\Omega^{1/2}} \sum_{\mathbf{k}} c_{m\mathbf{k}}\,\mathrm{e}^{\mathrm{i}\mathbf{k}\cdot\mathbf{r}} \sum_{\mathbf{G}} u^{(m)}_{\mathbf{k}+\mathbf{G}}\,\mathrm{e}^{-\mathrm{i}\mathbf{G}\cdot\mathbf{r}} \\ &\approx \frac{1}{\Omega^{1/2}} \sum_{\mathbf{k}} c_{m\mathbf{k}}\,\mathrm{e}^{\mathrm{i}\mathbf{k}\cdot\mathbf{r}} \sum_{\mathbf{G}} u^{(m)}_{\mathbf{k}_0+\mathbf{G}}\,\mathrm{e}^{-\mathrm{i}\mathbf{G}\cdot\mathbf{r}} \\ &= \phi(\mathbf{r})\,u^{(m)}_{\mathbf{k}_0}(\mathbf{r})\ ,\end{aligned} \tag{8.17}$$

which is the same expression for a Bloch wave with the envelope $\phi(\mathbf{r})$ replacing the plane-wave modulation $\mathrm{e}^{\mathrm{i}\mathbf{k}\cdot\mathbf{r}}$. Thus, in the presence of a weakly varying external potential, the function $\phi$ "modulates" the periodic component $u^{(m)}_{\mathbf{k}_0}$, much like plane waves modulate them in the absence of a perturbation. The analogy with free electrons is immediate: A plane wave is the solution in the absence of any potential, the wavefunction $\phi$ is the solution in the presence of the potential. In a lattice, $\mathrm{e}^{\mathrm{i}\mathbf{k}\cdot\mathbf{r}}u^{(m)}_{\mathbf{k}_0}$ is the solution in the presence of the lattice potential only, $\phi(\mathbf{r})u^{(m)}_{\mathbf{k}_0}$ is the solution in the presence of an additional potential. Indeed, it is easy to see now that the envelope $\phi$ obeys the equation:

$$[E_m(-\mathrm{i}\nabla) + V^{(\mathrm{ext})}(\mathbf{r})]\,\phi(\mathbf{r}) = E\,\phi(\mathbf{r})\ , \tag{8.18}$$

where $E_m(\mathbf{k})$ is the dispersion in band $m$. This can be easily proved by inserting Eq. (8.17) into (8.18) and reaching Eq. (8.15).

Equation (8.18) is known under many names. When we use the effective mass concept and approximate $E(\mathbf{k}) \approx \hbar^2 k^2/(2m^*)$, Eq. (8.18) becomes

$$\left[-\frac{\hbar^2}{2m^*}\nabla^2 + V^{(\mathrm{ext})}(\mathbf{r})\right]\phi(\mathbf{r}) = E\,\phi(\mathbf{r})\ , \tag{8.19}$$

which is identical to the Schrödinger equation. Therefore, it takes the name of "effective mass" equation (or approximation). Probably, in its more general form, it should take the name of "envelope" wave equation. It was first derived by Slater and by Kohn and Luttinger, and it is sometimes called the *Kohn–Luttinger* equation.

### 8.1.2 Effective Mass

We have already introduced the concept of "effective mass" before, dealing with the $\mathbf{k}\cdot\mathbf{p}$ perturbation theory [see Eq. (5.44)]. The usefulness of this concept can be understood as follows. Semiconductors are characterized by full valence bands, empty conduction bands, and a relatively small energy gap between the valence and conduction bands. "Doping" results in the introduction of excess electrons occupying the bottom of the conduction band, or the elimination of some electrons from the valence bands (which is viewed as the introduction of holes occupying the top of the valence band). Thus, the "action" happens at the top of the valence band and/or at the bottom of the conduction band. At these extrema of the bands we can expand the dispersion as:

$$E(\mathbf{k}) = E(\mathbf{k}_0) + \frac{1}{2}(\mathbf{k}-\mathbf{k}_0)\cdot\nabla^2_{\mathbf{k}}E(\mathbf{k})\,|_{\mathbf{k}_0}\cdot(\mathbf{k}-\mathbf{k}_0) + \cdots$$

(the first-order term in $\nabla E$ vanishes since we are at an extremum of the band). Thus, replacing the "momentum" (measured from the band extremum $\mathbf{k}_0$)

$$\mathbf{k}-\mathbf{k}_0 \rightarrow -\mathrm{i}\nabla$$

thanks to the envelope approximation, we have

$$\begin{aligned} E(-\mathrm{i}\nabla) &= E(\mathbf{k}_0) + \frac{1}{2}(-\mathrm{i}\nabla)\cdot\nabla^2_{\mathbf{k}}E(\mathbf{k})\,|_{\mathbf{k}_0}\cdot(-\mathrm{i}\nabla) + \cdots \\ &= E(\mathbf{k}_0) - \frac{1}{2}\sum_{ij}\frac{\partial E}{\partial k_\mathrm{i}\partial k_j}\,|_{\mathbf{k}_0}\,\frac{\partial}{\partial x_\mathrm{i}}\frac{\partial}{\partial x_j} + \cdots . \end{aligned} \quad (8.20)$$

Let's now set $E(\mathbf{k}_0) = 0$ by a suitable choice of the reference energy and let's measure $\mathbf{k}$ from $\mathbf{k}_0$. Let's ignore higher-order terms in the expansion above. Let's define the inverse "effective mass" tensor $1/m^*_{ij}$ via:

$$\frac{1}{m^*_{ij}} = \frac{1}{\hbar^2}\frac{\partial E}{\partial k_\mathrm{i}\partial k_j}\,|_{\mathbf{k}_0} .$$

Then

$$E(\mathbf{k}) \approx \sum_{ij}\frac{\hbar^2 k_\mathrm{i} k_j}{2m^*_{ij}}$$

and the envelope wave equation, Eq. (8.18), becomes

$$\left[-\sum_{ij}\frac{\hbar^2}{2m^*_{ij}}\frac{\partial}{\partial x_\mathrm{i}}\frac{\partial}{\partial x_j} + eV^{(\mathrm{ext})}(\mathbf{r})\right]\,\psi(\mathbf{r}) = E\,\psi(\mathbf{r}) ,$$

which, as we have already noticed, looks like a "conventional" Schrödinger equation.

For GaAs, the effective mass in the CB is actually a scalar (since the first CB is formed by $s$-type orbitals), $m^* \approx 0.063\ m_0$. For Si, the first CB has ellipsoidal equi-energy surfaces, so that the effective mass tensor is diagonal along the $[100]$ crystal direction:

$$m = \begin{pmatrix} m_{\rm L} & 0 & 0 \\ 0 & m_{\rm T} & 0 \\ 0 & 0 & m_{\rm T} \end{pmatrix} ,$$

where $m_{\rm L} \approx 0.91\, m_0$ and $m_{\rm T} \approx 0.19\, m_0$.

### 8.1.3 Group Velocity and Acceleration Theorems

We have established so far that—under the approximations listed on pages 163–166 and with the expansion (8.20)—we can eliminate the complicated periodic potential $V^{(\rm lat)}$ from our problem, "hide" the periodic Bloch component $u_{\mathbf{k}}^{(m)}$ (with the aforementioned exceptions which we shall encounter) and work only with the envelope wavefunction $\phi$, solution of a quasi-free Schrödinger equation in the presence of the weakly varying external potential $V^{(\rm ext)}$. In other words, we can treat electrons as they were in free space, not in the crystal, provided we replace their mass. How far can we extend the analogy? Are we also able to describe the motion of the electron wavepacket (8.4) as if we were in free space? In particular, how does the electron momentum evolve with time in the presence of an external electric field? And how does the electron velocity evolve?

We shall answer these questions "rigorously" below. But, first, let's give a compact intuitive answer: We know from the "correspondence principle" of Quantum Mechanics that the center of motion of a wave packet can be obtained from classical mechanics by considering the quantum Hamiltonian and rendering it "classical" by replacing $-i\hbar\nabla$ with the momentum $\hbar\mathbf{k}$. Thus, we can consider the equivalent classical Hamiltonian stemming from Eq. (8.18):

$$H_{\rm cl}(\mathbf{p},\mathbf{r}) = E(\mathbf{p}/\hbar) + V^{(\rm ext)}(\mathbf{r}).$$

Then, from the Hamilton equations of motion (see page 7) and recall that $\mathbf{p} = \hbar\mathbf{k}$):

$$\mathbf{v} = \frac{\mathrm{d}\mathbf{r}}{\mathrm{d}t} = \nabla_{\mathbf{p}} E(\mathbf{p}/\hbar) = \frac{1}{\hbar}\,\nabla_{\mathbf{k}} E(\mathbf{k}) ,$$

and

$$\frac{\mathrm{d}\mathbf{k}}{\mathrm{d}t} = \frac{1}{\hbar}\frac{\mathrm{d}\mathbf{p}}{\mathrm{d}t} = -\frac{1}{\hbar}\,\nabla V^{(\rm ext)}(\mathbf{r}) = \frac{\mathbf{F}}{\hbar} = -\frac{e}{\hbar}\,\mathbf{E} ,$$

where $\mathbf{F}$ is the (conservative) force from the potential $V$ and $\mathbf{E}$ is the electric field.

This shows that the electron velocity is simply the group velocity of a packet, as for electromagnetic waves with dispersion $\omega(\mathbf{k}) = E(\mathbf{k})/\hbar$ and the electron "crystal momentum" $\hbar\mathbf{k}$ (which is *not* the "real momentum" since adding any $\mathbf{G}$-vector to it does not alter anything, simply shifting the electron to another equivalent BZ) evolves with time as if it were the "real" momentum. Clearly, this nice picture must be justified more rigorously, since the assumption that we can use the equivalent classical Hamiltonian $H_{\rm cl}$ is, so far, unjustified.

We shall now prove the following two "acceleration theorems":

**Theorem 1.** *If a spatially and temporally constant electric field* $\mathbf{E}$ *is applied to the crystal, then*

$$\frac{\mathrm{d}\mathbf{k}}{\mathrm{d}t} = -\frac{e}{\hbar}\mathbf{E} , \tag{8.21}$$

*and the electron remains in the same band.*

**Theorem 2.** *If $\langle \mathbf{v} \rangle$ is the expectation value of the electron velocity in a state $|n\mathbf{k}\rangle$, then:*

$$\langle \mathbf{v} \rangle = \frac{\mathrm{i}}{\hbar} \langle [\widehat{H}, \hat{\mathbf{r}}] \rangle = \frac{1}{\hbar} \nabla_{\mathbf{k}} E_n(\mathbf{k}) , \tag{8.22}$$

*in the absence of magnetic fields.*

Consider first Theorem 2, since its validity will be used to prove Theorem 1.

**Proof of Theorem 2.** Consider the matrix element

$$\langle n\mathbf{k}|[\widehat{H}, \hat{\mathbf{r}}]|n\mathbf{k}\rangle = \int u_{\mathbf{k}}^{(n)*}(\mathbf{r})\mathrm{e}^{-\mathrm{i}\mathbf{k}\cdot\mathbf{r}} [\widehat{H}, \hat{\mathbf{r}}] u_{\mathbf{k}}^{(n)}(\mathbf{r})\mathrm{e}^{\mathrm{i}\mathbf{k}\cdot\mathbf{r}} \, \mathrm{d}\mathbf{r} . \tag{8.23}$$

Now,

$$\nabla_{\mathbf{k}}(\mathrm{e}^{-\mathrm{i}\mathbf{k}\cdot\mathbf{r}} \widehat{H} \mathrm{e}^{\mathrm{i}\mathbf{k}\cdot\mathbf{r}}) = -\mathrm{i}\, \mathrm{e}^{-\mathrm{i}\mathbf{k}\cdot\mathbf{r}} \mathbf{r} \widehat{H} \mathrm{e}^{\mathrm{i}\mathbf{k}\cdot\mathbf{r}} + \mathrm{i}\, \mathrm{e}^{-\mathrm{i}\mathbf{k}\cdot\mathbf{r}} \widehat{H} \mathbf{r} \mathrm{e}^{\mathrm{i}\mathbf{k}\cdot\mathbf{r}} = \mathrm{i}\, \mathrm{e}^{-\mathrm{i}\mathbf{k}\cdot\mathbf{r}} [\widehat{H}, \mathbf{r}] \mathrm{e}^{\mathrm{i}\mathbf{k}\cdot\mathbf{r}} . \tag{8.24}$$

Now note that

$$H(\mathbf{p}, \mathbf{r}) \, \mathrm{e}^{\mathrm{i}\mathbf{k}\cdot\mathbf{r}} = \mathrm{e}^{\mathrm{i}\mathbf{k}\cdot\mathbf{r}} H(\mathbf{p}+\mathbf{k}, \mathbf{r}) .$$

Thus:

$$\begin{aligned} \langle n\mathbf{k}|[\widehat{H}, \hat{\mathbf{r}}]|n\mathbf{k}\rangle &= -\mathrm{i} \int u_{\mathbf{k}}^{(n)*}(\mathbf{r}) \, \nabla_{\mathbf{k}}(\mathrm{e}^{-\mathrm{i}\mathbf{k}\cdot\mathbf{r}} \widehat{H} \mathrm{e}^{\mathrm{i}\mathbf{k}\cdot\mathbf{r}}) \, u_{\mathbf{k}}^{(n)}(\mathbf{r}) \, \mathrm{d}\mathbf{r} \\ &= -\mathrm{i} \int u_{\mathbf{k}}^{(n)*}(\mathbf{r}) \, \nabla_{\mathbf{k}}[\widehat{H}(\mathbf{p}+\mathbf{k}, \mathbf{r})] \, u_{\mathbf{k}}^{(n)}(\mathbf{r}) \, \mathrm{d}\mathbf{r} . \end{aligned} \tag{8.25}$$

Now we use what's known as the *Feynman* or *Hellmann–Feynman theorem*: If the Hamiltonian $\widehat{H}$ depends on a parameter $\lambda$, then:

$$\frac{\partial}{\partial \lambda} \langle n\mathbf{k}|\widehat{H}|n\mathbf{k}\rangle = \left\langle n\mathbf{k} \left| \frac{\partial \widehat{H}}{\partial \lambda} \right| n\mathbf{k} \right\rangle . \tag{8.26}$$

Thus Eq. (8.25) becomes $-\mathrm{i}\nabla_{\mathbf{k}} E_n(\mathbf{k})$ and we get the final result:

$$\langle \mathbf{v} \rangle = \frac{\mathrm{i}}{\hbar} \langle n\mathbf{k}|[\widehat{H}, \hat{\mathbf{r}}]|n\mathbf{k}\rangle = \frac{1}{\hbar} \nabla_{\mathbf{k}} E_n(\mathbf{k}) . \tag{8.27}$$

**Proof of Theorem 1.** If the external force due to the field is $-e\mathbf{E}$, the rate at which the electron energy (call it $W$ to avoid confusion with the magnitude $E$ of the electric field) will be

$$\frac{\mathrm{d}W}{\mathrm{d}t} = -e\, \mathbf{E} \cdot \langle \mathbf{v}(\mathbf{k}) \rangle = -\frac{e}{\hbar} \mathbf{E} \cdot \nabla_{\mathbf{k}} E_n(\mathbf{k}) . \tag{8.28}$$

Since $E_n(\mathbf{k})$ does not depend on time, being the dispersion of the crystal, any time dependence of the electron energy must be associated with the time variation of $\mathbf{k}$. Thus we must also have

$$\frac{\mathrm{d}W}{\mathrm{d}t} = \frac{\mathrm{d}\mathbf{k}}{\mathrm{d}t} \cdot \nabla_{\mathbf{k}} E_n(\mathbf{k}) . \tag{8.29}$$

So, from Eqs. (8.28) and (8.29):

$$\frac{d\mathbf{k}}{dt} = -\frac{e}{\hbar}\,\mathbf{E}\,, \tag{8.30}$$

or, writing $\mathbf{F} = -e\mathbf{E}$ for the force,

$$\frac{d(\hbar\mathbf{k})}{dt} = \mathbf{F}\,. \tag{8.31}$$

This result may be derived more rigorously using time-dependent perturbation theory using $-e\mathbf{E}\cdot\mathbf{r}$ as perturbing Hamiltonian (see Ridley's text [2], pages 48 and 49 or Kittel's book [3]) or expressing the electric field in terms of a vector potential and "Houston accelerated waves" (See Kittel's book again, same chapter).

* *Let's reproduce here Kittel's first proof: When using perturbation theory, it would be tempting to account for the presence of the electric field via a scalar potential $\phi = -e\mathbf{r}\cdot\mathbf{F}$. But the fact that this potential is unbound would create some mathematical problems. Instead, let's express the electric field in terms of a vector potential*

$$\mathbf{A} = -\mathbf{E}\,t\,, \tag{8.32}$$

*which yields the electric field via the general expression:*

$$\mathbf{E} = -\nabla\phi - \frac{\partial\mathbf{A}}{\partial t}\,. \tag{8.33}$$

*Then the Hamiltonian can be written as:*

$$\widehat{H} = \frac{1}{2m}(\mathbf{p}+e\mathbf{A})^2 + V^{(\mathrm{lat})}(\mathbf{r}) = \frac{1}{2m}(\mathbf{p}+e\mathbf{E}t)^2 + V^{(\mathrm{lat})}(\mathbf{r})\,. \tag{8.34}$$

*In writing Eq. (8.34) we have employed what is known as the* minimal coupling*: In the presence of an electromagnetic field, the Schrödinger equation for a single electron can be written by simply replacing the momentum operator $\hat{\mathbf{p}}$ with $\hat{\mathbf{p}}+e\mathbf{A}$. This ultimately originates from considerations based on the theory of special relativity, as briefly discussed in Appendix C.*

*The Hamiltonian given by Eq. (8.34) retains the periodicity of the lattice, thanks to the fact that we have used the vector potential. Therefore, the solutions must have the Bloch form*

$$\phi_{\mathbf{k}}(\mathbf{r},\mathbf{E},t) = e^{i\mathbf{k}\cdot\mathbf{r}}\,u_{\mathbf{k}}^{(n)}(\mathbf{r},\mathbf{E},t)\,, \tag{8.35}$$

*where $t$ should be viewed as a parameter. Note that in this formulation $\mathbf{k}$ remains a good quantum number, unaffected by the field (this is indeed the scheme often used in the context of "accelerated" or "Houston" waves). Now notice that the kinetic energy term in the Hamiltonian acting on such a Bloch wave will be $(\mathbf{p}+e\mathbf{E}t+\hbar\mathbf{k})^2/(2m)$. If we now compare the kinetic energy terms for the two Bloch waves $u_{\mathbf{k}}^{(n)}(\mathbf{r},\mathbf{E},t)$ and $u_{\mathbf{k}'}^{(n)}(\mathbf{r},\mathbf{E},t')$, we see that the Hamiltonians will be equal only if*

$$e\mathbf{E}t + \hbar\mathbf{k} = e\mathbf{E}t' + \hbar\mathbf{k}'\,, \tag{8.36}$$

*or*

$$\frac{d\mathbf{k}}{dt} = -\frac{e}{\hbar}\,\mathbf{E} \tag{8.37}$$

*so that states at* $\mathbf{k}, t$ *are identical to those at* $\mathbf{k}', t'$. *Thus, an electron which is in the state* $\mathbf{k}$ *at t=0, at a later time t will still be in the same state* $\mathbf{k}$, *but all of its other properties, including its kinetic energy, will be those corresponding to the state which was originally* $\mathbf{k} - \mathbf{E}t/\hbar$.

*Finally, an "adiabatic" argument due to Kohn and Shockley shows that the electron remains in the same band if the change of the Hamiltonian during a period* $1/\omega_{nm}$ *(where* $\hbar\omega_{nm} = E_n(\mathbf{k}) - E_m(\mathbf{k})$ *is the "interband" energy) is smaller than* $\hbar\omega_{nm}$ *itself,*

$$\frac{\partial \widehat{H}}{\partial t} \frac{1}{\omega_{nm}} << \hbar\omega_{nm} , \tag{8.38}$$

*condition which is met very easily. Note, however, that interband tunneling is the most notable exception to this "rule."*

For electrons in nondegenerate bands the result (8.30) may be generalized to the case of weak magnetic fields $\mathbf{B}$:

$$\frac{\mathrm{d}\mathbf{k}}{\mathrm{d}t} = -\frac{e}{\hbar}(\mathbf{E} + \mathbf{v} \times \mathbf{B}) . \tag{8.39}$$

Recall that we have called "crystal momentum" the quantity $\hbar\mathbf{k}$. It is the momentum associated with the envelope, ignoring the (much larger) momentum associated with the cell-periodic Bloch component $u_{\mathbf{k}}$. If we assume parabolic bands, $E(\mathbf{k}) = \hbar^2 k^2/(2m^*)$, we will have $\mathbf{v}(\mathbf{k}) = \hbar\mathbf{k}/m^*$, or $\hbar\mathbf{k} = m^*\mathbf{v}$ (having indicated simply with $\mathbf{v}$ the group velocity. The total momentum, instead, will be $\mathbf{p} = m\mathbf{v}$, so that the ratio between the crystal and total momentum will simply be $m^*/m$.

Note that for holes the picture is the same: The top of the VB has a "negative" effective mass. However, talking about "missing electrons" ("holes") is more convenient and this is accomplished by considering these fictitious particles as positively charged (that is, moving in the opposite directions as electrons under an external field), and by "flipping" all signs in the equations of motion, thus recovering a positive mass. Unfortunately, the structure of the valence bands is more complicated: The parabolic approximation breaks down very early, the effective mass is highly anisotropic, and three bands are energetically very close (degenerate at the center of the BZ, spin–orbit coupling reducing the degeneracy to two bands). Thus, one needs to consider always at two bands (the so-called heavy and light hole bands), often three (the "split-off" band).

## 8.2 Some Important Applications

We now discuss several important applications of the simplified description of the electron dynamics we have just outlined.

### *8.2.1 The Two-Dimensional Electron Gas*

Chapter 7 has discussed in detail the band structure of low-dimensionality structures. Sections 7.2.1 and 7.2.2, in particular, have dealt with thin layers of Si and III–V compound semiconductors also in the presence of a confining vertical field. The envelope approximation we have just considered actually permits vastly simpler calculations if we are content with limiting our attention to low-energy states (that is, electronic states whose energy is only a few thermal energies, $k_B T$, away from the minimum of the conduction band or the maximum of the valence band(s)). This is an interesting range of energies, since electronic transport in semiconductor structures involves, *usually*, only electrons or holes close to equilibrium, so with a small kinetic energy.

Consider, for example, a III–V hetero-structure like the complicated multi-layer shown in Fig. 7.12. If we are interested only in low-energy states, it is enough to consider the InGaAs layer. Indeed, in any device (transistor or injection laser) based on such a structure, these low-energy states will be the most occupied and considering only the active role of this layer will provide a sufficient physical picture in *most* cases. If the potential barriers (the conduction band offset) confining the electrons in this layer are sufficiently larger than the electron energies of interest, we can approximate the InGaAs quantum well (QW) with the prototypical "box" of elementary texts of Quantum Mechanics, assuming infinite potential "walls" at both sides.

Now, consider Eq. (8.18). The minimum of the conduction band of InGaAs occurs at the $\Gamma$ point. The high symmetry of this point implies that, at low energies, a single isotropic, effective mass $m^*$ will suffice to approximate the electron dispersion. Therefore, in the form given by Eq. (8.19), and having separated the variables in the usual way, the envelope equation Eq. (8.18) describes free electrons moving on the plane of the interfaces (the 2DEG indeed) using a one-dimensional Schrödinger equation for a free electron in a "box," (the only difference being a smaller mass, $m^*$). This causes the quantum confinement that we had painfully obtained in Sect. 7.2.2. In the limited range of validity implied by the validity Eq. (8.18) and of the approximations we have just made to simplify the hetero-structure, the problem has been converted from a numerically laborious supercell/empirical pseudopotential problem to the solution of the Schrödinger equation for an electron in a box, an easy homework problem for an undergraduate course of Quantum Mechanics.

Indeed, this simple approach has guided the development of electronic devices based on III–V hetero-structures. Clearly, subtle problems abound: How do we match the wavefunctions at the interfaces between different materials (say, between InGaAs and InP in our case), if we wish to account for the finite height of the confining barriers? How do we account for nonparabolic corrections, extremely strong in small-mass, small-gap semiconductors, if the well is very narrow and even the ground state is at an energy high enough to require nonparabolic corrections? The literature on the subject is extremely vast. Ultimately, for the very narrow nanostructures of current interest, accurate "atomistic" calculations based on empirical pseudopotentials or tight-binding should guide our study. But Eq. (8.18) is a tool that can be used to obtain qualitatively (and, sometimes, quantitatively) correct information with little effort.

Another important example has been considered in Sect. 7.2.1: The case of electrons in a Si inversion layer. As shown in Fig. 8.1, the potential energy at the Si–$SiO_2$ interface in a MOSFET exhibits a "notch." If we consider the wavelength of an electron at equilibrium (called its "thermal wavelength"), this is given by the de Broglie relation $\lambda_{th} = h/p_{th} = 2\pi/k_{th}$, where $p_{th} = \hbar k_{th}$ is the average momentum of electrons at thermal equilibrium. For the average carrier energy at thermal equilibrium we have $E_{th} = (3/2)k_B T = \hbar^2 k_{th}^2/(2m^*)$, so that, at 300 K,

$$\lambda_{th} = \frac{2\pi\hbar}{(3m^* k_B T)^{1/2}} \approx 6.2 \left(\frac{m_0}{m^*}\right)^{1/2} \text{nm}. \tag{8.40}$$

Consider the Si–$SiO_2$ interface in inversion/strong inversion. The electron density $n_s$ (the "sheet density" per unit area) is of the order of $10^{11}$-to-$10^{13}$ cm$^{-2}$, corresponding to an interface field $F_s = en_s/\epsilon_s$ of the order of $10^4$-to-$10^6$ V/cm as it can be easily derived from Gauss law. Thus, an electron of thermal energy ($(3/2)k_B T \approx 40$ meV) will be "squeezed" by the field against the interface over a confining distance $\Delta z \approx 3k_B T/(2F_s)$, very much like the particle in a box considered a few paragraphs above. This distance is of the order of 40-to-0.4 nm, comparable or even smaller than the electron thermal wavelength. We are not allowed to ignore the wave-like nature of electrons when we confine them so tightly: We expect that discrete energy levels will emerge from the confinement. If we confine a particle in a region of width $\Delta z$, by Heisenberg's principle the particle momentum

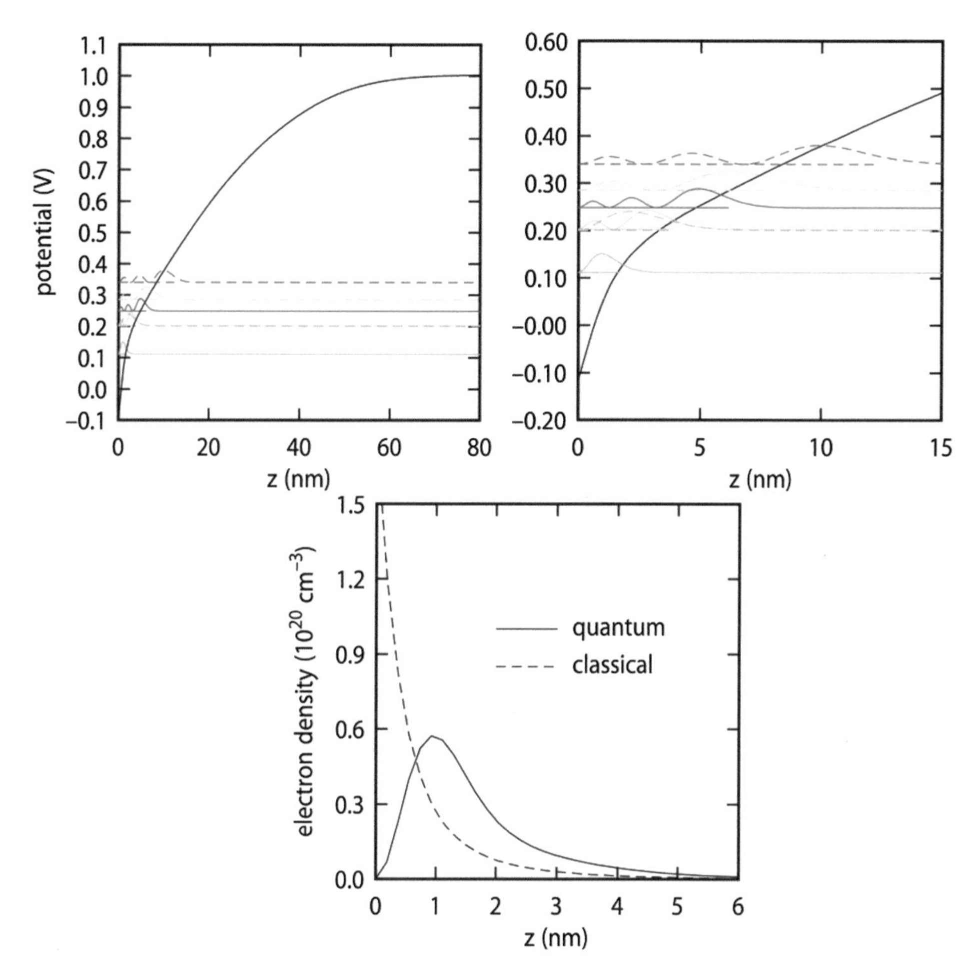


**Fig. 8.1** *Top*: Self-consistent potential energy (*black solid line*) and electron wavefunctions $\zeta_\mu(z)$ (*color lines*; *solid* for the "unprimed ladder," *dashed* for the "primed" ladder) for the three lowest-energy eigenstates in a Si inversion layer. The electron sheet density is $6 \times 10^{12}$ cm$^{-2}$. *Middle*: An enlarged version of the plot at the *top*. *Bottom*: Comparison of the electron density distribution calculated classically (that is, ignoring quantization) and quantum mechanically

will suffer an uncertainty $\Delta k \sim 1/\Delta z$, so that the confined particle will have a minimum energy $E_0 \sim \hbar^2 \Delta k^2/(2m^*) \sim \hbar^2/(2\Delta z^2 m^*)$, called the "zero-point energy." In strong inversion this energy may be comparable to (or even larger than) the thermal energy, and quantum effects due to the confinement should not be ignored. The major changes to the classical picture that these quantum-mechanical correction provides are

1. The electron charge in inversion layers is removed somewhat (typically by a length $\Delta t$ of the order of 1.0 nm) from the interface. This causes a reduction of the gate capacitance in accumulation, $C_{\text{ox}}$, since the "effective" thickness of the oxide is increased by the amount $\epsilon_{\text{ox}}/\epsilon_{\text{Si}}\Delta t$.
2. The threshold voltage shifts to higher values, since we must move the Fermi level in the inversion layer to an energy higher by an amount $\sim E_0$ (the "zero-point" energy mentioned above).
3. The properties of electron transport (and so, in particular, the electron mobility) are modified, since both the "shape" of the wavefunctions and the density of states are modified.

* *The rigorous treatment of this problem is somewhat laborious, since it requires the solution of a coupled, nonlinear problem that requires an iterative approach. It is convenient to proceeds as*

*follows: The envelope equation, Eq. (8.18), contains an external potential energy term, $V^{(\mathrm{ext})}(\mathbf{r}) = -e\phi^{(\mathrm{ext})}$ ($\phi^{(\mathrm{ext})}$ is the electrostatic potential), which must be obtained from a solution of Poisson equation*

$$\nabla \cdot [\epsilon \nabla V^{(\mathrm{ext})}(\mathbf{r})] = e^2 \left[p(\mathbf{r}) - N_A(\mathbf{r}) - n(\mathbf{r})\right] , \tag{8.41}$$

*where $N_A$, $n$, and $p$ are the volume densities of the acceptors (p-type doping) impurities, and of the free electrons, and holes, respectively. The acceptor density is assumed as constant, whereas $p$ can be treated using a classical model. We review it here for convenience.*

$$\begin{aligned} p &= \frac{1}{2\pi^2} \left(\frac{2m_\mathrm{h}}{\hbar^2}\right)^{3/2} \int_0^\infty \mathrm{d}E\, E^{1/2} f[E(\mathbf{k})] \\ &= \frac{1}{4} \left(\frac{2m^* k_\mathrm{B} T}{\pi\hbar^2}\right)^{3/2} \frac{2}{\sqrt{\pi}} \int_0^\infty \frac{x^{1/2}}{1+\mathrm{e}^{x-\eta}}\, \mathrm{d}x , \end{aligned} \tag{8.42}$$

*where to reach the last expression we have changed the integration variable to $x = E/(k_\mathrm{B}T)$, we have set $\eta = E_\mathrm{F}/(k_\mathrm{B}T)$. Also, $m_\mathrm{h}$ is some approximation for the hole effective mass, $E_\mathrm{F}$ is the Fermi level measured from the top of the valence band (so $E_\mathrm{F}(x) = V(x) - E_\mathrm{gap}$) and we have multiplied and divided by $2/\sqrt{\pi}$ to recover the expression for the pre-factor usually employed in the literature, labeled $N_\mathrm{v}$ below, the "effective density of states" in the valence band. This integral cannot be solved in closed form, but it is a well-known integral (appropriately called Fermi–Dirac integral of order 1/2 and labeled by $\mathscr{F}_{1/2}(\eta)$). So, the hole density is usually written in terms of this integral as:*

$$p = \frac{1}{4} \left(\frac{2m_\mathrm{h} k_\mathrm{B} T}{\pi\hbar^2}\right)^{3/2} \mathscr{F}_{1/2}(\eta) = N_\mathrm{v}\, \mathscr{F}_{1/2}(\eta) , \tag{8.43}$$

*having defined $N_\mathrm{v} = (1/4)[(2m_\mathrm{h}k_\mathrm{B}T)/(\pi\hbar^2)]^{3/2}$. In the limit of nondegenerate statistics (which we can assume here for $N_A < 10^{18}\,\mathrm{cm}^{-3}$ or so), $\eta$ is very large and negative so that:*

$$\frac{2}{\sqrt{\pi}} \int_0^\infty \frac{x^{1/2}}{1+\mathrm{e}^{x-\eta}}\, \mathrm{d}x \approx \frac{2}{\sqrt{\pi}} \int_0^\infty x^{1/2}\, \mathrm{e}^{\eta - x}\, \mathrm{d}x = \mathrm{e}^{\eta} , \tag{8.44}$$

*and:*

$$p(z) \approx N_\mathrm{v}\, \mathrm{e}^{[V^{(\mathrm{ext})}(z) - E_\mathrm{gap}]/(k_\mathrm{B}T)} , \tag{8.45}$$

*when measuring energies from the Fermi level.*

*Now let us return to the problem at hand. Because of the symmetry of the problem, assuming that the $z$ axis is perpendicular to the plane of the Si–$SiO_2$ interface, assuming to be on the technological important $\{1,0,0\}$ plane, we see that all dependent variables of the problem depend only on $z$. Therefore, Eq. (8.41) simplifies to:*

$$\frac{\mathrm{d}}{\mathrm{d}z}\left[\epsilon\, \frac{\mathrm{d}}{\mathrm{d}z} V^{(\mathrm{ext})}(z)\right] = e^2 \left[p(z) - N_A - n(z)\right] . \tag{8.46}$$

*We can also separate variables in Eq. (8.18), writing the wavefunctions as:*

$$\psi(\mathbf{r}) = \frac{1}{A^{1/2}}\, \mathrm{e}^{\mathrm{i}\mathbf{K}\cdot\mathbf{R}}\, \zeta(z) . \tag{8.47}$$

*Here A is the normalization area on the $(x,y)$ plane of the interface, on which electrons move freely as a 2DEG, and* **K** *and* **R** *denote the two-dimensional wavevector and coordinate on this plane. The function $\zeta(z)$ is a solution, yet to be determined, of the equation:*

$$-\frac{\hbar^2}{2m_z}\frac{\mathrm{d}^2}{\mathrm{d}z^2}\zeta(z)+V^{(\mathrm{ext})}(z)\zeta(z)=E\,\zeta(z)\,. \tag{8.48}$$

*The quantity $m_z$ requires a brief discussion. For the $\{1,0,0\}$ interface we have assumed, Si exhibits six equivalent minima of the conduction band, four with the light transverse effective mass, $m_\mathrm{T}$, along z, and two with the heavy longitudinal effective mass, $m_\mathrm{L}$, along that direction. Therefore, we must solve two different Eq. (8.48), one for each value, $m_\mathrm{T}$ or $m_\mathrm{L}$, for the mass $m_z$. From each of these eigenvalue problems we will obtain two "ladders" of eigenstates, usually termed the "unprimed" and "primed" ladders, depending on whether $m_z=m_\mathrm{T}$ or $m_z=m_\mathrm{L}$. The calculation of the electron charge density is a bit complicated. Having solved the two Eq. (8.48), we must calculate which states are occupied at equilibrium using the Fermi–Dirac distribution. This, regrettably, requires another long digression.*

*Let's proceed as follows. Consider a fully closed system with a discrete spectrum (such as a QW or inversion layer) with eigenvalues and eigenfunctions $E_\mu$ and $\zeta_\mu$, respectively. Let's also assume that we are at thermal equilibrium at temperature T. Then the occupation of each state will be determined by the Fermi–Dirac distribution. In 1D, the occupation of a state at energy $E_\mu$ will be simply given by $f_\mathrm{FD}(E_\mu)=\{1+\exp[(E_\mu-E_\mathrm{F})/(k_\mathrm{B}T)]\}^{-1}$ where $E_\mathrm{F}$ is the Fermi level and $k_\mathrm{B}$ is Boltzmann constant. However, this is not true for a 3D system: In this case, there are two more degrees of freedom labeled by a continuous "index," the in-plane wavevector* **K**. *Thus, we must first calculate the density of states of our system for a given total energy E, find the Fermi energy for a given electron density, and finally, calculate the spatial distribution of the electric charge.*

*The density of states (DoS) at a given energy E is defined as:*

$$\mathscr{D}(E)=2\sum_\mu\int\frac{\mathrm{d}\mathbf{K}}{(2\pi)^2}\,\delta[E_\mu(\mathbf{K})-E]\,, \tag{8.49}$$

*where $E_\mu(\mathbf{K})=E_\mu-\hbar^2K_x^2/(2m_x)-\hbar^2K_y^2/(2m_y)$, the first factor of 2 is due to spin degeneracy, and $m_x$ and $m_y$ are the appropriate masses on the $(x,y)$-plane. Here the energy $E_\mu$ is an eigenvalue of the 1D Schrödinger problem discussed above, the additional term being the kinetic energy in the "other" two dimensions in which the 2D electron gas (2DEG) is confined. A simple free-electron-like dispersion has been assumed with a mass m which we may regard as an "effective mass." The integration of Eq. (8.49) is easily done:*

$$\begin{aligned}\mathscr{D}(E)&=\frac{2}{(2\pi)^2}\sum_\mu\int_0^\infty \mathrm{d}KK\int_0^{2\pi}\mathrm{d}\phi\;\delta[E_\mu(K)-E]\\&=\frac{1}{\pi}\sum_\mu\int_0^\infty \mathrm{d}E'\;\delta[E_\mu+E'-E]=\frac{m_\mathrm{d}}{\pi\hbar^2}\sum_\mu\theta(E-E_\mu)\,,\end{aligned} \tag{8.50}$$

*where $\theta(x)$ is the step function equal to zero if $x<0$, to 1 otherwise. The quantity $m_\mathrm{d}=\sqrt{m_xm_y}$ is the DOS mass on the $(x,y)$ plane. The relation between the electron density and the Fermi level can be obtained in a similar way:*

$$n=2\sum_\mu\int\frac{\mathrm{d}\mathbf{K}}{(2\pi)^2}f_\mathrm{FD}[E_\mu(\mathbf{K})]\,, \tag{8.51}$$

*which can be integrated as follows:*

$$n = \frac{2}{(2\pi)^2} \sum_\mu \int_0^\infty \mathrm{d}KK \int_0^{2\pi} \mathrm{d}\phi\, f_{\mathrm{FD}}[E_\mu(K)]$$
$$= \frac{m_\mathrm{d} k_\mathrm{B} T}{\pi \hbar^2} \sum_\mu \int_{(E_\mu - E_\mathrm{F})/(k_\mathrm{B}T)}^\infty \mathrm{d}\eta\, \frac{1}{1+\mathrm{e}^\eta}\,, \tag{8.52}$$

*having introduced the integration variable $\eta = (E_\mu + E(K) - E_\mathrm{F})/(k_\mathrm{B}T)$. Thus, multiplying the numerator and denominator of the integrand in the last expression by $\mathrm{e}^{-\eta}$:*

$$n = \frac{m_\mathrm{d} k_\mathrm{B} T}{\pi \hbar^2} \sum_\mu \int_{(E_\mu - E_\mathrm{F})/(k_\mathrm{B}T)}^\infty \mathrm{d}\eta\, \frac{\mathrm{e}^{-\eta}}{1+\mathrm{e}^{-\eta}}$$
$$= \frac{m_\mathrm{d} k_\mathrm{B} T}{\pi \hbar^2} \sum_\mu \ln\left[1 + \mathrm{e}^{(E_\mathrm{F} - E_\mu)/(k_\mathrm{B}T)}\right]\,. \tag{8.53}$$

*This expression allows us to determine the Fermi level $E_\mathrm{F}$ for a given electron density $n$. Newton's method, the bisection method, or a combination of these two methods (i.e., bisection at first, then a switch to Newton's method once we are getting close to the solution) can be used. Once the Fermi energy has been obtained, the density of electrons in an eigenlevel $\mu$ can be calculated from:*

$$n_\mu = \frac{m_\mathrm{d} k_\mathrm{B} T}{\pi \hbar^2} \ln\left[1 + \mathrm{e}^{(E_\mathrm{F} - E_\mu)/(k_\mathrm{B}T)}\right]\,, \tag{8.54}$$

*so the electron density distribution will be:*

$$n(z) = \frac{m_\mathrm{d} k_\mathrm{B} T}{\pi \hbar^2} \sum_\mu \ln\left[1 + \mathrm{e}^{(E_\mathrm{F} - E_\mu)/(k_\mathrm{B}T)}\right] |\zeta_\mu(z)|^2\,, \tag{8.55}$$

*since any state in the level $\mu$ (that is, regardless of the value of the in-plane wavevector* **K***) will contribute with the same probability distribution $|\zeta_\mu(z)|^2$.*

*We are finally close to the end! We start from some initial guess for the external potential $V^{(\mathrm{ext})}(z)$ taking, for example, from the depletion approximation [4]. We solve Eq. (8.48) twice, once for each value of the mass $m_z$. We insert the eigenvalues $E_\mu$ and eigen-wavefunction, $\zeta_\mu(z)$ just obtained into Eq. (8.55) to obtain the electron density $n(z)$ by summing over the twofold and fourfold degenerate valleys (one per mass $m_z$). The DOS mass entering this equation, $m_\mathrm{d}$, is just $m_\mathrm{T}$ for the ellipsoids with effective mass $m_\mathrm{L}$, and $(m_l m_\mathrm{T})^{1/2}$ for the ellipsoids with effective mass $m_\mathrm{L}$ (see Problem 8.1.). We then use Eq. (8.44) [or Eq. (8.45), if appropriate] to calculate $p(z)$ and insert it, together with $n(z)$, into the Poisson equation, Eq. (8.46). The "new" potential energy, $V^{(\mathrm{ext})}(z)$, is inserted back into the Schrödinger equation Eq. (8.48) and this iterative procedure continues until some criterion of convergence is met. Details are given in the review of 2DEG by Ando and co-workers [5]. Here we show the result in Fig. 8.1. This figures shows not only the potential energy obtained from this self-consistent procedure, but also comparison between the electron distribution obtained quantum mechanically and classically. Quantum effects result in a shift of the electron charge away from the interface, an example of the common "quantum repulsion from a barrier," that has important practical effects.*

### 8.2.2 Zener–Bloch Oscillations and the Stark Ladder

Considering a crystal with a homogeneous electric field, Eq. (8.21) can be used to evaluate the time dependence of the electron wavevector:

$$\mathbf{k}(t) = -\frac{e}{\hbar}\mathbf{E}\,t\,. \tag{8.56}$$

But we know that periodicity holds in reciprocal space also. Under the constant acceleration the wavevector will eventually exit the first BZ. If we adopt the reduced-zone scheme, it will return inside the BZ. Thus, the electron velocity will reverse sign and the electron will undergo spatial oscillations. These are called *Zener–Bloch oscillations*. Assume for simplicity that we apply a field along the $x$-axis, so that only $k_x$ changes with time with periodicity $2\pi/a$. The oscillation will have a period $T$ such that

$$\frac{eE}{\hbar}T = \frac{2\pi}{a}\,, \tag{8.57}$$

corresponding to a frequency

$$\omega_{\mathrm{ZB}} = \frac{aeE}{\hbar}\,. \tag{8.58}$$

This implies that electrons are "trapped" into quantized levels of spatial extent $L_{\mathrm{ZB}}$ such that $2eEL_{\mathrm{ZB}} = E_{\mathrm{band}}$, where $E_{\mathrm{band}}$ is the width of the band, thus:

$$L_{\mathrm{ZB}} = \frac{E_{\mathrm{band}}}{2eE}\,. \tag{8.59}$$

Figure 8.2 illustrates the origin of these "levels," effect usually referred to as the "Stark ladder," for the obvious similarity to the Stark effect in atoms. For a typical semiconductor $a \approx 0.5$ nm, so from Eq. (8.58) for an electric field of strength $\approx 10^5$ V/cm, we have $\omega_{\mathrm{ZB}} \approx 10^{13}$/s. Note that in this picture the application of an electric field does not result in charge transport! The electrons simply oscillate with spatial amplitude given by $L_{\mathrm{ZB}}$ above but, on average, they do not move. These oscillations also provide the potential for excellent generators of THz radiation. The reason why Zener–Bloch oscillations have not been observed experimentally (despite strong efforts) and why crystals do carry current is obviously due to the existence of scattering. In order for Zener–Bloch oscillations to occur, the electron must be accelerated without undergoing collisions for times $1/\omega_{\mathrm{ZB}}$ longer than the average collision time $\tau$, or

$$\omega_{\mathrm{ZB}}\tau >> 1. \tag{8.60}$$

Usually $\tau$ is shorter than $10^{-12}$ s, often as short as $10^{-14}$ s, so that ZB oscillations are prevented from occurring.

### 8.2.3 Landau Levels

Another important application of the equation describing the electron dynamics is the study of their behavior in the presence of a magnetic field. From classical electrodynamics we know that a charged

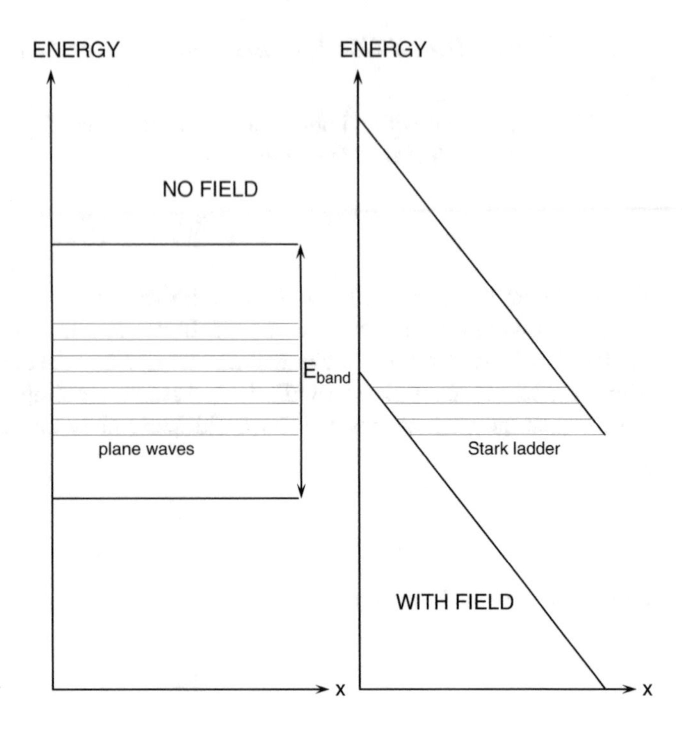


**Fig. 8.2** Schematic illustration of the Stark ladder

particle moving on the $(x, y)$ plane, in the presence of a uniform and constant magnetic field along the $z$ axis, moves along circular orbits (cyclotron orbits) on the plane. In particular, in a finite sample, the Lorentz force acting on the electrons results in the "Hall effect": A voltage arises along the $y$ direction when an electric field is applied along the $x$ direction. This corresponds to what is called the "Hall conductance" $\sigma_{xy}$. A quantum-mechanical treatment of this system yields different and interesting results. The orbital energy is quantized into what are called "Landau levels" [6] that are degenerate: Many electrons can occupy the same orbit. The number of states corresponding to the same cyclotron radius (or Landau level) increases linearly with the strength of the magnetic field. It is this quantization of the orbits that results in the quantization of the Hall conductance of a 2DEG: Changing the density of the 2D electrons by applying a "gate" bias, the Hall conductance $\sigma_{xy}$ exhibits plateaus in steps of the "universal conductance" $e^2/h$ ($h$ is the Planck's constant). This is called the "integer Quantum Hall effect" (IQHE) [7]. It was observed for the first time by von Klitzing in 1978 (and interpreted in 1980). For this discovery he was awarded the Nobel Prize for Physics 5 years later.

If we consider the effect of a magnetic field, we can employ Eq. (8.39) to look at the time evolution of the electron wavevector. Let's assume that the magnetic field $B$ is a uniform field along the $z$ direction. Then, the equations of motion read (in SI units):

$$\begin{cases} \hbar \dfrac{\mathrm{d}k_x}{\mathrm{d}t} = -eBv_y \\ \hbar \dfrac{\mathrm{d}k_y}{\mathrm{d}t} = eBv_x \\ \hbar \dfrac{\mathrm{d}k_z}{\mathrm{d}t} = 0. \end{cases} \tag{8.61}$$

Assuming parabolic bands ($E = \hbar^2k^2/2m^*$), the equations for the $x$ and $y$ components of the wavevector become

$$\begin{cases} \frac{\mathrm{d}k_x}{\mathrm{d}t} = -\frac{eB}{m^*}k_y \\ \frac{\mathrm{d}k_y}{\mathrm{d}t} = \frac{eB}{m^*}k_x . \end{cases} \quad (8.62)$$

Taking the time derivative of one equation and inserting into the other we get

$$\begin{cases} \frac{\mathrm{d}^2k_x}{\mathrm{d}t^2} = -\,\omega_c^2 k_x \\ \frac{\mathrm{d}^2k_y}{\mathrm{d}t^2} = -\,\omega_c^2 k_y , \end{cases} \quad (8.63)$$

where $\omega_c = eB/m^*$ is the cyclotron frequency. We see that electrons move along closed orbits $k_x^2 + k_y^2 =$ constant in the $x, y$-plane, while translating along the $z$-axis as free particles. So far this is exactly the same picture which describes the motion of a classical electron. To account for quantum effects we must consider the envelope wave equation [see Eq. (8.19)]:

$$\frac{1}{2m^*}[\hat{\mathbf{p}} + e\mathbf{A}]^2\phi(\mathbf{r}) = E\,\phi(\mathbf{r}) , \quad (8.64)$$

where $\mathbf{A}$ is the vector potential

$$\mathbf{A} = \left(-y\frac{B}{2}, x\frac{B}{2}, 0\right) . \quad (8.65)$$

Actually, Eq. (8.19) was derived under a different set of circumstances: Here we consider a perturbation of the kinetic energy, while Eq. (8.19) was derived considering a perturbation of the potential energy. So, it is not at all obvious that we may use the effective mass approximation. Nevertheless, perturbation theory may be used to show that Eq. (8.64) is approximately correct (see Ridley's text for a proof). Then, inserting the expression (8.65) into Eq. (8.64), we have

$$\widehat{H} = \frac{1}{2m^*}[\hat{\mathbf{p}} + e\mathbf{A}]^2 = \frac{\hat{p}^2}{2m^*} + \frac{eB\hbar}{4m^*}(-yk_x + xk_y) + \frac{e^2B^2}{8m^*}(x^2 + y^2) . \quad (8.66)$$

With a few manipulations of this expression, Eq. (8.64) can be rewritten as:

$$\left\{\frac{\hat{p}^2}{2m^*} + \frac{1}{2m^*}\left[\left(\frac{1}{2}m^*\omega_c x + \hbar k_y\right)^2 + \left(\frac{1}{2}m^*\omega_c y - \hbar k_x\right)^2\right]\right\}\phi(\mathbf{r})$$
$$= \left[E + \frac{\hbar^2}{2m^*}(k_x^2 + k_y^2)\right]\,\phi(\mathbf{r}) . \quad (8.67)$$

As remarked by Ridley [2], apart for the term $\hat{p}^2/(2m^*)$, this is the equation for a classical particle orbiting with frequency $\omega_c$ on the $x, y$-plane. Thus, we can rewrite this equation as:

$$\left[\frac{\hat{p}^2}{2m^*} + \frac{1}{2}m^*\omega_c R^2\right]\,\phi(\mathbf{r}) = E'\,\phi(\mathbf{r}) , \quad (8.68)$$

where $E' = E + \hbar^2 k_z^2/(2m^*)$, $R$ is the radius of the classical orbit, and we know that $k_x^2 + k_y^2 = k_R^2$ is a constant. The energy $E'$ will be the energy of the uniform translation along the $z$ axis, $\hbar^2 k_z^2/(2m^*)$, plus the energy—dependent on a particular value of $k_R$—due to the orbital motion. We can now reformulate Eq. (8.68) in order to show more clearly the separation of these two motions:

$$\left[\frac{\hat{p}_z^2}{2m^*}+\frac{\widehat{P}^2}{2m^*}+\frac{1}{2}m^*\omega_c R^2\right]\phi(\mathbf{r})=\left(E+\frac{\hbar^2k_z^2}{2m^*}\right)\phi(\mathbf{r}) , \tag{8.69}$$

where $\widehat{\mathbf{P}}$ is the momentum on the $x,y$-plane. If we separate the orbital $(x,y)$ variables from the $z$ variable, we can write $\phi(\mathbf{r})=\zeta(z)X(\mathbf{R})$ and we see that the equation for $\zeta$ describes simply the uniform motion along $z$, while the equation for $X$ is the equation of a harmonic oscillator. Thus the total energy will be the sum of the two terms:

$$E'=\frac{\hbar^2k_z^2}{2m^*}+\left(n+\frac{1}{2}\right)\hbar\omega_c , \tag{8.70}$$

and

$$\phi(\mathbf{r})=C\mathrm{e}^{\mathrm{i}k_zz}\,H_n\left[\left(\frac{m^*\omega}{\hbar}R\right)^{1/2}\right]\,\mathrm{e}^{-m^*\omega_cR^2/(2\hbar)} , \tag{8.71}$$

where $H_n$ is the Hermite polynomial of $n$th degree.

The energy levels labeled by the quantum number $n$ are called "Landau levels." They are clearly degenerate, since we can select $k_x$ and $k_y$ in many ways while obtaining the same "orbital" energy. We can view these states as the classical states, each $k_R$ giving a different energy, now "lumped" together by quantum effects. In order to estimate the resulting degeneracy, $g_B$, note that the number of states in a "shell" of width $\Delta k_R$ in the $k_x,k_y$-plane will be, as usual:

$$g_B=\frac{1}{(2\pi)^2}\,2\pi k_R\Delta k_R=\frac{k_R\Delta k_R}{2\pi} . \tag{8.72}$$

On the other hand, the total change of the "orbital" energy $\hbar^2k_R^2/(2m^*)$ as we vary $k_R$ over this shell must be equal to the energy "jump" $\hbar\omega_c$:

$$\frac{\mathrm{d}}{\mathrm{d}k_R}\left(\frac{\hbar^2k_R^2}{2m^*}\right)\Delta k_R=\hbar\omega_c , \tag{8.73}$$

which implies

$$k_R\Delta k_R=\frac{m^*}{\hbar}\omega_c . \tag{8.74}$$

From Eqs. (8.72) and (8.74) we get

$$g_B=\frac{1}{2\pi}\frac{m^*}{\hbar}\omega_c=\frac{eB}{2\pi\hbar} . \tag{8.75}$$

As in the case of Zener–Bloch oscillations, Landau levels may be observed only if the electron manages to orbit at least once before scattering:

$$\omega_c\tau>>1 \quad \text{or} \quad \mu B>>1 , \tag{8.76}$$

where $\mu=e\tau/m^*$ is the electron mobility (as we shall see later). This condition is met relatively easily at low temperatures and with magnetic fields of a few Tesla.

### 8.2.4 Plasma Oscillations

Let's consider a semiconductor doped $n$-type with a density $N_D$ donors per $cm^3$. By charge neutrality obviously we will have a density $n_0 = N_D$ of electrons in the conduction band. In this average picture nothing interesting happens. However, if we bring the semiconductor at a nonzero temperature, the thermal, random motion of the electrons in the background of positive charges (we assume here the "jellium" model, in which the dopants are taken to be a uniform "jell-O" background of charge) will cause an occasional deviation of the electron density, $\delta n = n - n_0$, in some regions of the crystal. How do these deviations $\delta n$ vary with time?

Let's write the equations which control the dynamics of the electrons. First, we have the equation of motion

$$\frac{\mathrm{d}\mathbf{k}}{\mathrm{d}t} = -\frac{e}{\hbar}\mathbf{E} , \tag{8.77}$$

(here $\mathbf{E}$ is the electric field). Second, we have Gauss law,

$$\nabla \cdot \mathbf{E} = -\frac{e}{\epsilon_s}\, \delta n , \tag{8.78}$$

(where $\epsilon_s$ is the dielectric constant of the semiconductor) which relates the divergence of the electric field to the charge density, and, finally, we have charge conservation expressed by the continuity equation

$$\frac{\mathrm{d}n}{\mathrm{d}t} = -\nabla \cdot (\mathbf{v}n) . \tag{8.79}$$

Let's assume parabolic bands, so that the electron velocity is simply $\mathbf{v} = \hbar\mathbf{k}/m^*$. Let's take the time derivative of Eq. (8.79):

$$\frac{\mathrm{d}^2 \delta n}{\mathrm{d}t^2} = -\frac{\mathrm{d}}{\mathrm{d}t}\nabla \cdot (\mathbf{v}n) = -\nabla \cdot \frac{\mathrm{d}}{\mathrm{d}t}(\mathbf{v}n) . \tag{8.80}$$

Let's ignore second-order effects, so that we can ignore the derivatives of $n$ on the rhs of the last equation. Then:

$$\frac{\mathrm{d}^2 \delta n}{\mathrm{d}t^2} = -\nabla \cdot \left( \frac{\mathrm{d}\mathbf{v}}{\mathrm{d}t} n \right) \approx -\frac{\hbar}{m^*}\, n_0\, \nabla \cdot \left( \frac{\mathrm{d}\mathbf{k}}{\mathrm{d}t} \right) . \tag{8.81}$$

Let's now use Eqs. (8.77) and (8.78) on the rhs of this equation:

$$\frac{\mathrm{d}^2 \delta n}{\mathrm{d}t^2} = \frac{e n_0}{m^*}\, \nabla \cdot \mathbf{E} = -\frac{e^2 n_0}{\epsilon_s m^*} \delta n . \tag{8.82}$$

Introducing the "Plasma frequency" $\omega_p^2 = e n_0/(\epsilon_s m^*)$ we can write this equation as:

$$\frac{\mathrm{d}^2 \delta n}{\mathrm{d}t^2} = -\omega_p^2\, \delta n, \tag{8.83}$$

with solution $\delta n \propto \mathrm{e}^{\mathrm{i}\omega_p t}$ which shows that the electron density $\delta n$ oscillates with frequency $\omega_p$. We should view this as an "eigenmode" of the gas of conduction electrons. When excited by thermal effects or other external perturbations, the electron gas exhibits a "collective behavior": Density-waves of electrons travel through the crystal, all at a frequency $\omega_p$. This is the first example we meet of an *elementary excitation* of the solid. We shall revisit this concept, as well as the particular examples of plasma waves, in the following.

### 8.2.5 Shallow-Impurity States

The last example deals with the use of the "envelope" approximation, Eq. (8.18). It is to study this particular problem that Luttinger and Kohn first derived this equation. In doped semiconductors ionized impurities are introduced so that one obtains a controllable density of electrons in the conduction band (or holes in the valence bands, for $p$-type doping). How can we describe the electronic states associated with each impurity? How can we estimate the ionization energy of the dopant atoms, so that we can understand under which conditions the impurities will be ionized? This is actually the problem which Kohn and Luttinger studied to derive Eq. (8.18) itself. This is indeed the prototypical problem for which the approximations made earlier to derive the envelope equation hold true: The potential of the impurity can be approximately described as

$$V_{\text{imp}}(r) \approx -\frac{e}{4\pi\epsilon_s r} , \tag{8.84}$$

which is exactly the potential of the hydrogen atom, with one major difference: The dielectric constant of semiconductors is about a factor of 10 (often more) larger than the permittivity of vacuum. Thus, the potential will be a factor of 10 (often more) weaker. The radius of the ground state of the hydrogen atom (the Bohr radius) is given by

$$a_{\text{B}} = \frac{4\pi\epsilon_0\hbar^2}{e^2 m} \approx 0.05\,\text{nm}, \tag{8.85}$$

while the energy of the ground state (the ionization energy) is the Rydberg constant:

$$Ry = \frac{1}{2}\left(\frac{e^2}{4\pi\epsilon_0}\right)^2 \frac{m}{\hbar^2} \approx 13.6\,\text{eV}. \tag{8.86}$$

We see that in a semiconductor the radius of the "narrowest" orbit will be a factor of $(m/m^*)(\epsilon_s/\epsilon_0) \sim 10^2$ larger, since the effective mass is often a factor of 10 smaller than the electron mass. This gives orbital radii of tens of nm, several cells wide. Conversely, the "effective Rydberg," $Ry^*$, will be a factor of $(m/m^*)(\epsilon_s/\epsilon_0)^2 \sim 10^3$ smaller, and so of the order of a few tens of meV.

These estimates should be viewed as "sanity" checks to justify the use of the hydrogen-like potential (8.84): The orbits are large enough to allow the use of the macroscopic dielectric constant and ignore anisotropic effects which are surely going to matter at small distances (remnants of the $sp^3$ tetragonal bond-charge). The potential at these large distances will be weak and slowly varying, so that the use of the envelope equation (8.18) is fully justified. The small ionization energies are consistent with our naïve expectation that the dopants will be ionized at all but the lowest ($<$20 K or so) temperatures.

## Problems

**8.1. (DOS Effective Mass in Si)** Show that the two-dimensional density-of-states effective mass to be used in Eq. (8.55) is just $m_{\mathrm{T}}$ for the ellipsoids with effective mass $m_z = m_{\mathrm{L}}$, it is $(m_{\mathrm{L}} m_{\mathrm{T}})^{1/2}$ for the ellipsoids with effective mass $m_z = m_{\mathrm{T}}$, as stated in the text.

## References

1. J.M. Luttinger, W. Kohn, Motion of electrons and holes in perturbed periodic fields. Phys. Rev. **97**, 869 (1955)
2. B.K. Ridley, *Quantum Processes in Semiconductors* (Oxford University Press, Oxford, 2000)
3. C. Kittel, *Quantum Theory of Solids*, 2nd edn. (Wiley, New York, 1987)
4. S. Sze, M.K. Lee, *Semiconductor Devices: Physics and Technology* (Wiley, New York, 2012)
5. T. Ando, A.B. Fowler, F. Stern, Rev. Mod. Phys. **54**, 437 (1982)
6. L.D. Landau, E.M. Lifschitz, *Quantum Mechanics: Non-relativistic Theory. Course of Theoretical Physics*, vol. 3, 3rd edn. (Pergamon, London, 1977)
7. Z.F. Ezawa, *Quantum Hall Effects: Recent Theoretical and Experimental Developments*, 3rd edn. (World Scientific, New York, 2013)

# Part III
# Elementary Excitations, Statistical Mechanics, and Dielectric Response

# Chapter 9
# Elementary Excitations in Solids

## 9.1 Lagrangian and Hamiltonian Formulation for Fields

Second quantization is necessary when we want to study many-particle effects. Historically, it was born out of the relativistic description of particles, since ordinary Quantum Mechanics (QM) could not handle it: The number of particles is fixed in conventional QM, but a relativistic description must allow for creation and annihilation of particles. It soon became apparent that this new description was extremely useful in describing solid-state phenomena and statistical mechanics, since second quantization, dealing with fields, is intrinsically a description of systems with infinitely many (... well, very many indeed in solid-state systems, albeit not infinitely many...) degrees of freedom. It is thus very well suited to describe collective phenomena as elementary excitations above the ground state: Elementary excitations—such as phonons, plasmons, magnons, etc.—can be created (emissions) and destroyed (absorption), similarly to relativistic particles. In addition, for small perturbations away from equilibrium, the Hamiltonian can often be linearized and algebraically manipulated so that it takes the form of the Hamiltonian of a collection of harmonic oscillators (as we shall see for free electrons, plasmons, and phonons below). These manipulations often take the name of Bogoliubov transformations. [These have been first introduced in 1947 by Nicolai Bogoliubov in the context of superfluidity [1]. Since his original paper is in Russian, the interested reader may find the subject amply discussed, in English, in the text by Blaizot and Ripka [2]]. So, the whole machinery of simple harmonic oscillators can be used to handle the much more complicated collective excitations.

In "first quantization" (i.e., conventional QM), we start by describing the system classically in the Lagrangian form. We then translate the description into the Hamiltonian form, so that the equations of motion are expressed in terms of Poisson brackets between the Hamiltonian, the dynamical variables (observables), and the canonical coordinates (generalized coordinates and conjugate momenta). Quantization is finally performed by promoting the dynamic variables to operators on a suitable Hilbert space and converting the Poisson brackets to commutators. In second quantization, we follow a similar route: Think of a system as described by the generalized coordinates $\phi_{\mathrm{i}}$, where i runs over the degrees of freedom of the system. As the number of degrees of freedom approaches infinity, we replace the discrete index i with a continuum of indices $\mathbf{r}$ and our generalized coordinates become a field $\phi(\mathbf{r})$. Thus, we start from a Lagrangian description of the system in terms of Lagrangian fields (as done, classically, in elasticity theory), convert it to a Hamiltonian description, and convert the fields to operators on some (more complicated) Hilbert space, promoting the Poisson brackets involving fields to commutators (or anticommutators, as discussed below).



M. Fischetti, W.G. Vandenberghe, *Advanced Physics of Electron Transport in Semiconductors and Nanostructures*, Graduate Texts in Physics, DOI 10.1007/978-3-319-01101-1_9

The use of the term "second quantization" is due to the fact that the first fields to be quantized were solutions of the Dirac equation. Thus, the procedure appeared to be a quantization of wavefunctions, already quantized in conventional QM, and so it seemed that we had quantized the system twice. But this is not the right way of looking at it: What we have done is to have dropped altogether the concept of "particle" as the primary concept and replaced it with fields, from which particles (or "quasi-particles", in solid-state physics) will emerge as excitations of the field away from its ground state.

In this chapter we shall first review the Lagrangian and Hamiltonian formulations of fields and describe the canonical quantization procedure in a general case. We shall then apply this procedure first to the Schrödinger field. Finally, we shall discuss two examples of great relevance to charge transport in solids: plasmons and phonons. We will conclude with the quantization of the electromagnetic field. Although this is not an elementary excitation of a solid, the formalism is similar and its application to condensed-matter physics are many and important.

The Lagrangian and Hamiltonian formulation of fields and their quantization can be found in many texts of Quantum Field Theory (QFT). Lurié's is an old-time favorite [3]. A concise but exceptionally clear presentation in the context of non-relativistic fields can be found in Schiff's text [4]. His rigorous presentation represents one of very few examples that can be found in a text not exclusively dedicated to many-body problems or to QFT. In the context of many-body problems in condensed matter theory, the texts by Fetter and Walecka [5] and Mahan [6] are the most comprehensive and clear.

### 9.1.1 Classical Lagrangian Density

We discuss how, knowing how to deal with a system with a finite number of degrees of freedom, we can describe in Lagrangian and Hamiltonian terms a system with infinitely many (actually, a continuum) of degrees of freedom. A typical example we could consider is the displacement of a violin string. At each location $x$ along the string, the displacement of the string from its equilibrium position is $\phi(x)$. We would like to find a Lagrangian and Hamiltonian description of the "plucked" string. Clearly, we have way too many degrees of freedom, one for each location $x$.

Therefore, instead of defining a Lagrangian, we define a *Lagrangian density* at each location $x$:

$$\mathscr{L}\left(\phi, \frac{\partial \phi}{\partial x}, \frac{\partial \phi}{\partial t}, x, t\right), \tag{9.1}$$

so that the Lagrangian is obtained by integrating over the length $D$ of the string:

$$L(t) = \int_0^D \mathscr{L}\left(\phi, \frac{\partial \phi}{\partial x}, \frac{\partial \phi}{\partial t}, x, t\right) \mathrm{d}x. \tag{9.2}$$

The equations of motion can be obtained from a generalized Hamilton principle requiring

$$\delta \mathscr{I} = \delta \int_{t_1}^{t_2} L(t)\, \mathrm{d}t = \delta \int_{t_1}^{t_2} \int_0^D \mathscr{L}\left(\phi, \frac{\partial \phi}{\partial x}, \frac{\partial \phi}{\partial t}, x, t\right) \mathrm{d}x\, \mathrm{d}t = 0. \tag{9.3}$$

Following a generalized version of the procedure followed before (pages 5–6), this variational principle leads to Euler–Lagrange equations of the form

$$\frac{\mathrm{d}}{\mathrm{d}t}\frac{\partial \mathscr{L}}{\partial \dot{\phi}} + \frac{\mathrm{d}}{\mathrm{d}x}\left(\frac{\partial \mathscr{L}}{\partial(\partial \phi/\partial x)}\right) - \frac{\partial \mathscr{L}}{\partial \phi} = 0. \tag{9.4}$$

In three dimensions and in the case of several "fields" $\phi_i$, the Euler–Lagrange equations become

$$\frac{\mathrm{d}}{\mathrm{d}t}\frac{\partial\mathcal{L}}{\partial\dot{\phi}_i}+\sum_{k=1}^{3}\frac{\mathrm{d}}{\mathrm{d}x_k}\left(\frac{\partial\mathcal{L}}{\partial(\partial\phi_i/\partial x_k)}\right)-\frac{\partial\mathcal{L}}{\partial\phi_i}=0 \quad \text{for } j=1,2,\ldots \tag{9.5}$$

It is convenient to introduce the *functional derivative of $\mathcal{L}$ with respect to $\phi_i$* as:

$$\frac{\delta\mathcal{L}}{\delta\phi_i}=\frac{\partial\mathcal{L}}{\partial\phi_i}-\sum_{k=1}^{3}\frac{\partial}{\partial x_k}\left(\frac{\partial\mathcal{L}}{\partial(\partial\phi_i/\partial x_k)}\right) \tag{9.6}$$

This notation has the advantage of allowing us to ignore the dependence of the Lagrangian density on the spatial derivatives of $\phi$. Similarly,

$$\frac{\delta\mathcal{L}}{\delta\dot{\phi}_i}=\frac{\partial\mathcal{L}}{\partial\dot{\phi}_i} \tag{9.7}$$

which is simpler looking, since $\mathcal{L}$ does not depend on the spatial gradients of $\dot{\phi}_i$. Thus, Eq. (9.5) takes the simpler-looking form:

$$\frac{\mathrm{d}}{\mathrm{d}t}\frac{\delta\mathcal{L}}{\delta\dot{\phi}_i}-\frac{\delta\mathcal{L}}{\delta\phi_i}=0, \tag{9.8}$$

which is formally similar to the "usual" Euler–Lagrange equations.

### 9.1.2 Conjugate Momenta and Classical Hamiltonian Density

We can push the analogy with the usual Hamiltonian theory even further: Let's define the conjugate momenta (which are fields now!):

$$\pi_i=\frac{\delta\mathcal{L}}{\delta\dot{\phi}_i}, \tag{9.9}$$

and the Hamiltonian:

$$H=\int\mathcal{H}\,\mathrm{d}\mathbf{r}, \tag{9.10}$$

where $\mathcal{H}$ is the Hamiltonian density:

$$\mathcal{H}=\sum_k\pi_k\dot{\phi}_k-\mathcal{L}. \tag{9.11}$$

### 9.1.3 Hamilton's Equations

We can show that the Euler–Lagrange equations (9.8) become the Hamilton equations:

$$\frac{\delta H}{\delta \phi_k} = -\dot{\pi}_k, \qquad \frac{\delta H}{\delta \pi_k} = \dot{\phi}_k. \tag{9.12}$$

Finally, given any function $A$ which can be represented as the integral of a density $\mathscr{A}$,

$$A = \int \mathscr{A} \, d\mathbf{r}, \tag{9.13}$$

one can show that its time derivative is given by:

$$\frac{dA}{dt} = \{A, H\} + \frac{\partial A}{\partial t}, \tag{9.14}$$

where the symbol $\{\ldots,\ldots\}$ is the obvious generalization of the Poisson brackets:

$$\{A,B\} = \int \sum_k \left( \frac{\delta A}{\delta \phi_k} \frac{\delta B}{\delta \pi_k} - \frac{\delta A}{\delta \pi_k} \frac{\delta B}{\delta \phi_k} \right) d\mathbf{r} = \int \{A((\mathbf{r}), B(\mathbf{r})\} \, d\mathbf{r}. \tag{9.15}$$

where the Poisson bracket at a given point has been defined as:

$$\{A(\mathbf{r}), B(\mathbf{r}')\} = \sum_k \left[ \frac{\delta A(\mathbf{r})}{\delta \phi_k} \frac{\delta B(\mathbf{r}')}{\delta \pi_k} - \frac{\delta A(\mathbf{r})}{\delta \pi_k} \frac{\delta B(\mathbf{r}')}{\delta \phi_k} \right]. \tag{9.16}$$

Note that the following relations hold for the Poisson brackets between canonical variables (see the analogous relations Eq. (1.30) for systems with a finite number of degrees of freedom):

$$\{\phi_i(\mathbf{r}), \pi_j(\mathbf{r}')\} = \delta_{ij}\delta(\mathbf{r}-\mathbf{r}'), \qquad \{\phi_i(\mathbf{r}), \phi_j(\mathbf{r}')\} = 0, \qquad \{\pi_i(\mathbf{r}), \pi_j(\mathbf{r}')\} = 0, \tag{9.17}$$

### 9.1.4 Canonical Quantization of Fields

The procedure to be followed in order to quantize the field $\phi$ is strictly parallel to the canonical quantization scheme outlined on pages 12–14, although several issues are significantly more complicated:

1. We define/select/guess the Hilbert space associated with the system. As we shall see below, this task is not as straightforward as in the case of "conventional" QM. Let's defer to a later section a short discussion on this issue. For the moment let's just say that we shall work on what's called "Fock space" $\mathscr{F}$.
2. We re-express the field equations (Maxwell equations, the wave equation for the strings of our "quantum violin," Dirac equation, or even Schrödinger equation, as we shall see below) in their Hamiltonian formulations.

3. We promote the fields $\phi$ and $\pi$ to operators acting on the Hilbert space $\mathscr{F}$. These operators will satisfy the *canonical commutation rules* (CCRs):

$$[\hat{\phi}_i(\mathbf{r}), \hat{\pi}_j(\mathbf{r}')] = i\hbar\, \delta_{ij}\delta(\mathbf{r}-\mathbf{r}') \; ; \qquad [\hat{\phi}_i(\mathbf{r}), \hat{\phi}_j(\mathbf{r}')] = 0 \; ; \qquad [\hat{\pi}_i(\mathbf{r}), \hat{\pi}_j(\mathbf{r}')] = 0. \tag{9.18}$$

4. The equation of motion will follow from the "usual" commutation rules with the Hamiltonian (in the Heisenberg representation).

## 9.2 Quantization of the Schrödinger Field: Electrons

Let's consider, as a first example, the quantization of the Schrödinger Field. This example—in addition to being relevant since it constitutes the basic framework needed to describe also electrons in a crystal—shows how the Hamiltonian of free (that is: non-interacting among themselves) particles can be written as the sum of Hamiltonians of harmonic oscillators, each oscillator corresponding to one particular single-electron state, each excited state of each oscillator representing the presence of one particle in that particular state. The main new ingredient we shall encounter is the anticommutation rules (instead of "commutation" rules) the electron operators must obey in order to satisfy Pauli's exclusion principle.

### 9.2.1 *The Lagrangian and Hamiltonian Densities*

As we said, the main idea is to consider the "usual" Schrödinger equation

$$i\hbar\frac{\partial\psi}{\partial t} = -\frac{\hbar^2}{2m}\nabla^2\psi + V(\mathbf{r},t)\;\psi, \tag{9.19}$$

but consider the function $\psi(\mathbf{r})$ as a "classical field" (much like the electric field of Maxwell's equation, or the displacement field of our violin string), rather than a "wavefunction." "Classical" here means "not yet quantized." The next step consists in finding a Lagrangian density (a functional of the electronic wavefunction $\psi$) which gives us back the Schrödinger equation. It can be shown that:

$$\mathscr{L}(\psi,\dot{\psi},t) = i\hbar\psi^*(\mathbf{r},t)\,\frac{\partial\psi(\mathbf{r},t)}{\partial t} - \frac{\hbar^2}{2m}\nabla\psi^*(\mathbf{r},t)\cdot\nabla\psi(\mathbf{r},t) - V(\mathbf{r},t)\;\psi^*(\mathbf{r},t)\;\psi(\mathbf{r},t). \tag{9.20}$$

Indeed, using the usual variational procedure which yields the Euler equations, one finds that the field $\psi(\mathbf{r},t)$ which corresponds to an extremum of the Lagrangian L(t) = $\int d\mathbf{r}\mathscr{L}(\psi,\dot{\psi},t)$ is a solution of the equation

$$i\hbar\frac{\partial\psi}{\partial t} = -\frac{\hbar^2}{2m}\nabla^2\psi + V(\mathbf{r},t)\;\psi, \tag{9.21}$$

i.e., Schrödinger equation, which is what we wanted.

*Actually, using Eq. (9.8) one would obtain the complex conjugate of Eq. (9.21). Equivalently, instead of varying $\psi$ to derive Eq. (9.8) from Eq. (9.3), one may vary the complex conjugate of Eq. (9.20) over $\psi^*$ and obtain the Lagrange equation*

$$\frac{\mathrm{d}}{\mathrm{d}t}\frac{\delta\mathscr{L}^*}{\delta\dot{\psi^*}_i}-\frac{\delta\mathscr{L}^*}{\delta\psi_i^*}=0, \tag{9.22}$$

*from which Eq. (9.21) follows.*

Now we define the canonical momentum density

$$\pi(\mathbf{r},t)=\frac{\delta\mathscr{L}}{\delta(\partial\psi/\partial t)}=\mathrm{i}\,\hbar\,\psi^*(\mathbf{r},t), \tag{9.23}$$

and the Hamiltonian density:

$$\mathscr{H}(\psi,\pi,t)=\pi(\mathbf{r},t)\,\frac{\partial\psi(\mathbf{r},t)}{\partial t}-\mathscr{L}(\mathbf{r},t)=-\frac{\mathrm{i}\hbar}{2m}\,\nabla\pi\cdot\nabla\psi-\frac{\mathrm{i}}{\hbar}V\,\pi\psi. \tag{9.24}$$

We are now ready to quantize the field $\psi$.

### 9.2.2 Canonical Quantization and Commutation Rules

Second quantization is now performed by promoting the functions $\psi$ and $\pi$ to operators satisfying the equal-time CCRs, which are a generalization of the canonical Poisson brackets between coordinates and momenta in the classical description of systems with a finite number of degrees of freedom:

$$[\hat{\psi}(\mathbf{r},t),\hat{\pi}(\mathbf{r}',t)]=\mathrm{i}\hbar\delta^{(3)}(\mathbf{r}-\mathbf{r}')\;;\quad[\hat{\psi}(\mathbf{r},t),\hat{\psi}(\mathbf{r}',t)]=0\;;\quad[\hat{\pi}(\mathbf{r},t),\hat{\pi}(\mathbf{r}',t)]=0, \tag{9.25}$$

which, in turn, imply

$$[\hat{\psi}(\mathbf{r},t),\hat{\psi}^\dagger(\mathbf{r}',t)]=\delta^{(3)}(\mathbf{r}-\mathbf{r}')\;;\quad[\hat{\psi}(\mathbf{r},t),\hat{\psi}(\mathbf{r}',t)]=0\;;\quad[\hat{\psi}^\dagger(\mathbf{r},t),\hat{\psi}^\dagger(\mathbf{r}',t)]=0. \tag{9.26}$$

Note that Hamilton's equations become

$$\dot{\hat{\psi}}=\frac{\delta\widehat{\mathscr{H}}}{\delta\hat{\pi}}=\frac{\partial\widehat{\mathscr{H}}}{\partial\hat{\pi}}-\sum_k\frac{\partial\widehat{\mathscr{H}}}{\partial(\partial\hat{\pi}/\partial x_k)}=-\frac{\mathrm{i}}{\hbar}V\hat{\psi}+\frac{\mathrm{i}\hbar}{2m}\nabla^2\hat{\psi}, \tag{9.27}$$

which is the original Schrödinger equation, while the other equation,

$$\dot{\hat{\pi}}=-\frac{\delta\widehat{\mathscr{H}}}{\delta\hat{\psi}}=-\frac{\partial\widehat{\mathscr{H}}}{\partial\hat{\psi}}+\sum_k\frac{\partial\widehat{\mathscr{H}}}{\partial(\partial\hat{\psi}/\partial x_k)}=\frac{\mathrm{i}}{\hbar}V\hat{\pi}-\frac{\mathrm{i}\hbar}{2m}\nabla^2\hat{\pi}, \tag{9.28}$$

is simply the complex-conjugate equation, as can be seen using Eq. (9.23).

### 9.2.3 The Hamiltonian as a Collection of Harmonic Oscillators

At this point, a crucial step consists in expanding the field as:

$$\hat{\psi}(\mathbf{r},t) = \sum_{\mu} \hat{c}_{\mu}(t)\, u_{\mu}(\mathbf{r}), \tag{9.29}$$

$$\hat{\psi}^{\dagger}(\mathbf{r},t) = \sum_{\mu} \hat{c}^{\dagger}_{\mu}(t)\, u^{*}_{\mu}(\mathbf{r}), \tag{9.30}$$

where the (numeric) functions $u_{\mu}$ are the stationary eigenfunctions of Eq. (9.21) with eigenvalues $E_{\mu}$. The label $\mu$ will actually be a set of three labels (in addition to the degree of freedom associated to spin, ignored for the time being), possibly belonging to a continuous spectrum, for electrons in three dimensions. It may be just the three components of the **k**-vector when $V = 0$, it may be the set of integers $(n,l,m)$ for atomic orbitals, or any other set of "quantum numbers." Now, the crucial step: Some algebra shows that the Hamiltonian (the spatial integral of the Hamiltonian density) can be written as:

$$\widehat{H}(t) = \sum_{\mu} E_{\mu}\, \hat{c}^{\dagger}_{\mu}(t)\, \hat{c}_{\mu}(t). \tag{9.31}$$

* *In order to derive this expression we need to derive the (anti)commutation rules for the operators* $\hat{c}_{\mu}$ *and* $\hat{c}^{\dagger}_{\mu}$*, which follow from the (anti)commutation rules for the field, Eqs. (9.25) and (9.26): As we will discuss below, the only non-vanishing (anti)commutators are*

$$[\hat{c}_{\mu}, \hat{c}^{\dagger}_{\mu'}] = \delta_{\mu,\mu'}, \tag{9.32}$$

*all other (anti)commutators vanishing. We also need to remark that the expression given by Eq. (9.24) for the Hamiltonian density is ambiguous when interpreted in the operator sense, since the product AB of two numbers can be written as AB or BA. But if these quantities are operators such that* $[\widehat{A},\widehat{B}] \neq 0$*, we must decide which of the two expressions we should use. Commonly the "symmetrized" product* $(\widehat{A}\widehat{B}+\widehat{B}\widehat{A})/2$ *is employed. In so doing, we reach the result:*

$$\widehat{H}(t) = \sum_{\mu} \left[ E_{\mu}\, \hat{c}^{\dagger}_{\mu}(t)\, \hat{c}_{\mu}(t) + \frac{1}{2} \right], \tag{9.33}$$

*instead of Eq. (9.31). The term* $1/2$ *will be discussed below, but we ignore it for the moment.*

*It is convenient here to review the algebraic method can be used to quantize a simple harmonic oscillator in one dimension. Its Hamiltonian is*

$$\widehat{H} = \frac{\hat{p}^2}{2M} + \frac{K\hat{x}^2}{2}. \tag{9.34}$$

*Define the operators:*

$$\hat{a}_{\pm} = \frac{1}{(2M)^{1/2}}\, \hat{p} \pm \mathrm{i} \left(\frac{K}{2}\right)^{1/2} \hat{x}. \tag{9.35}$$

*Note that* $\hat{a}_+^\dagger = \hat{a}_-$ *and that, using the commutation rule* $[\hat{p},\hat{x}] = -\mathrm{i}\hbar$:

$$\widehat{H} = \hat{a}_+\hat{a}_- + \frac{1}{2}\hbar\omega = \hat{a}_-\hat{a}_+ - \frac{1}{2}\hbar\omega, \tag{9.36}$$

*where* $\omega = (K/M)^{1/2}$. *Note also that*

$$[\hat{a}_+,\hat{a}_-] = -\hbar\omega, \qquad [\widehat{H},\hat{a}_\pm] = \pm\hbar\omega\, a_\pm. \tag{9.37}$$

*Using these algebraic relations, we can construct the entire spectrum of the Hamiltonian (9.34) using the ground state and the operators* $a_\pm$. *Indeed, suppose that we know the ground state* $|0\rangle$, *such that:*

$$\widehat{H}|0\rangle = E_0|0\rangle. \tag{9.38}$$

*Using the first of Eq. (9.37),*

$$\widehat{H}\hat{a}_+|0\rangle = (E_0 + \hbar\omega)\,\hat{a}_+|0\rangle, \tag{9.39}$$

*which shows that starting from the ground state* $|0\rangle$ *we can "create" excited states by applying repeatedly the "creation" (or "raising") operator* $\hat{a}_+$. *What is left to do in order to find the eigenvalues corresponding to these states is to find the ground state energy* $E_0$. *Note that*

$$\widehat{H}\hat{a}_-|0\rangle = (E_0 - \hbar\omega)\hat{a}_-|0\rangle. \tag{9.40}$$

*Since we have assumed that* $E_0$ *is the lowest possible energy, this equation implies that* $\hat{a}_-|0\rangle$ *must not exists, that is,* $\hat{a}_-|0\rangle = 0$. *But from the first part of Eq. (9.36),*

$$\hat{a}_+\hat{a}_-|0\rangle = 0 \quad \rightarrow \quad \left(\widehat{H} - \frac{1}{2}\hbar\omega\right)|0\rangle = 0. \tag{9.41}$$

*This has the form of the eigenvalue equation for the ground state, so that* $E_0 = \hbar\omega/2$. *Therefore, the eigenstates of* $\widehat{H}$ *are the states* $|n\rangle = \hat{a}_+^n|0\rangle$ *with eigenvalues* $E_n = (n+1/2)\hbar\omega$.
*The "raising" operator* $\hat{a}_+$ *corresponds to the creation operator* $\hat{c}_\mu^\dagger$, *the 'lowering' operator* $\hat{a}_-$ *to the annihilation operator* $\hat{c}_\mu$ *defined above. They satisfy exactly the same algebraic and commutation relations.*

Note now the similarity of Eq. (9.31) with the Hamiltonian of the harmonic oscillator. We are going to use this fact to draw an algebraic analogy with the "ladder operator" formalism of the quantum harmonic oscillator to discover the eigenstates of the Hamiltonian. Let's define the "position" and "momentum" operators:

$$\hat{q}_\mu = \left(\frac{E_\mu}{2}\right)^{1/2}(\hat{c}_\mu + \hat{c}_\mu^\dagger), \qquad \hat{p}_\mu = -\mathrm{i}\left(\frac{E_\mu}{2}\right)^{1/2}(\hat{c}_\mu - \hat{c}_\mu^\dagger). \tag{9.42}$$

Then, we see that the Hamiltonian can be rewritten as

$$\widehat{H}(t) = \sum_\mu \frac{1}{2}\left[(\hat{p}_\mu^2 + \hat{q}_\mu^2) - E_\mu\right], \tag{9.43}$$

which is indeed the Hamiltonian of a collection of harmonic oscillators if we identify the operators $\hat{c}_\mu$ and $\hat{c}_\mu^\dagger$ with the "usual" lowering (annihilator) and raising (creator) operators for each oscillator $\mu$.

The term "$-E_\mu$" inside the square brackets in Eq. (9.43) corrects for the zero-point energy of the infinitely many oscillators $\mu$. It is clearly a troublesome diverging term. The reasons for its appearance are very deep and are briefly discussed below. It is usually referred to as the "energy of the vacuum." It is just a constant (albeit an infinite one!) and if we decide to measure all energies from this value, we can ignore it.

### 9.2.4 Anticommutation Rules

We have performed canonical quantization by employing the commutation rules Eqs. (9.25) and (9.26) above. However, these do not guarantee that Pauli's exclusion principle will be obeyed. Jordan and Wigner have shown that we can fix this problem by employing canonical *anti*commutation rules (CAR)

$$\begin{aligned} [\hat{\psi}(\mathbf{r},t),\hat{\pi}(\mathbf{r}',t)]_+ &= i\hbar\delta^{(3)}(\mathbf{r}-\mathbf{r}') \\ [\hat{\psi}(\mathbf{r},t),\hat{\psi}(\mathbf{r}',t)]_+ &= 0 \\ [\hat{\pi}(\mathbf{r},t),\hat{\pi}(\mathbf{r}',t)]_+ &= 0, \end{aligned} \tag{9.44}$$

(where $[\widehat{A},\widehat{B}]_+ = \widehat{A}\widehat{B}+\widehat{B}\widehat{A}$) instead of Eq. (9.25) above. These imply also:

$$\begin{aligned} [\hat{\psi}(\mathbf{r},t),\hat{\psi}^\dagger(\mathbf{r}',t)]_+ &= \delta^{(3)}(\mathbf{r}-\mathbf{r}') \\ [\hat{\psi}(\mathbf{r},t),\hat{\psi}(\mathbf{r}',t)]_+ &= 0 \\ [\hat{\psi}^\dagger(\mathbf{r},t),\hat{\psi}^\dagger(\mathbf{r}',t)]_+ &= 0. \end{aligned} \tag{9.45}$$

The reason behind the selection of *anti*-commutation rules is very deep. Streater and Whightman, in their book entitled *"PCT, Spin, Statistics, and all that"* [7] have shown that free particles of half-integer spin **must** be associated with a quantum field $\hat{\psi}(\mathbf{r})$ such that the CAR, Eq. (9.45) are satisfied, or one would violate at least one of several "axioms" (relativistic invariance, causality, locality, etc.) which constitute the foundation of QFT. Similarly, for integer-spin particles, the field must be such that the "CCRs" are valid:

$$[\hat{\psi}(\mathbf{r},t),\hat{\psi}^\dagger(\mathbf{r}',t)]_- = \delta^{(3)}(\mathbf{r}-\mathbf{r}'), \tag{9.46}$$

all other commutators vanishing, as we saw before. An intuitive way to understand the spin-statistics theorem relies on the following argument. The need to symmetrize (under exchange of two particles) many-particle wavefunctions of integer-spin particles and antisymmetrize those of half-integer-spin particles stems from the fact that rotations by an angle $\theta$ introduce a phase $e^{is\theta}$ for particles with spin $s$. Exchanging two particles amounts to rotating the vector connecting the two particles by an angle $\theta = \pi$. If the spin is a half-integer, this results in a change of sign. Unfortunately, this argument is not rigorously valid. The "correct" proof must rely on relativity and on the invariance of the theory under Lorentz transformations. The argument is involved and interested readers could find it presented in excruciating detail in the text by Streater and Wightman we have mentioned before.

One should add that the Schrödinger equation does not deal with spin, so we may, at this stage, choose commutators instead. Indeed, the same Schrödinger equation applies to all particles, Fermions or not. But in choosing commutators, we would run into troubles, because we would not get Fermi–Dirac statistics. And since we will have to account for spin later, we might as well start with anticommutators.

### 9.2.5 Fock Space for Fermions

In this framework, the states of a many-electron system belong to a complicated functional space built as follows. Let's consider only free electrons, so that the index $\mu$ is the 3D-wavevector $\mathbf{k}$ and the functions $u_\mu = u_\mathbf{k}$ are plane waves, and let's start with the "vacuum" state $|0\rangle$, that is, the ground state of the system. This is the state in which each oscillator $\mathbf{k}$ is in its ground state. Its energy diverges, because of the zero-point term 1/2 in Eq. (9.33) above. But we should think as subtracting this energy from the zero of our energy scale and carry on ignoring this problem. As we have just seen, the operator $\hat{c}_\mathbf{k}$ and its conjugate $\hat{c}_\mathbf{k}^\dagger$ can be viewed as lowering and raising operators, respectively. The latter can be thought as creating an electron in the eigenstate $\mathbf{k}$ of the momentum operator when acting on the vacuum state. Thus, the wavefunction corresponding to this state will simply be:

$$u_\mathbf{k}(\mathbf{r}) = \langle \mathbf{r}|\hat{c}_\mathbf{k}^\dagger|0\rangle = \frac{1}{(2\pi)^{3/2}}\, e^{-i\mathbf{k}\cdot\mathbf{r}} \tag{9.47}$$

that is, just the corresponding one-particle plane wave in the real-space representation. The operator $\hat{c}_\mathbf{k}^\dagger$ is called a "creation" operator. Let's call $|\mathbf{k}\rangle = \hat{c}_\mathbf{k}^\dagger|0\rangle$ the state in which there is only one electron in the system and this electron is in an eigenstate $\mathbf{k}$. This is an eigenstate of the Hamiltonian with eigenvalue $E_\mathbf{k}$. The conjugate operator $\hat{c}_\mathbf{k}$ destroys an electron in the state $\mathbf{k}$. Thus $\hat{c}_\mathbf{k}|\mathbf{k}\rangle = |0\rangle$. The fact that we deal with fermions (and not bosons) translates into the property that the operators $\hat{c}_\mathbf{k}$ and $\hat{c}_{\mathbf{k}'}^\dagger$ satisfy the canonical *anti*commutation rules:

$$[\hat{c}_\mathbf{k}, \hat{c}_{\mathbf{k}'}^\dagger]_+ = \delta_{\mathbf{k},\mathbf{k}'}, \quad [\hat{c}_\mathbf{k}, \hat{c}_{\mathbf{k}'}]_+ = 0, \quad [\hat{c}_\mathbf{k}^\dagger, \hat{c}_{\mathbf{k}'}^\dagger]_+ = 0. \tag{9.48}$$

Note now that:

$$\hat{c}_\mathbf{k}^\dagger \hat{c}_\mathbf{k}|0\rangle = (1 - \hat{c}_\mathbf{k}\hat{c}_\mathbf{k}^\dagger)|0\rangle = |0\rangle - |0\rangle = 0, \tag{9.49}$$

that is, the "zero" vector (not to be confused with the vacuum state!). Similarly:

$$\hat{c}_\mathbf{k}^\dagger \hat{c}_\mathbf{k}|\mathbf{k}\rangle = \hat{c}_\mathbf{k}^\dagger|0\rangle = |\mathbf{k}\rangle, \tag{9.50}$$

so that the operator $n_\mathbf{k} = \hat{c}_\mathbf{k}^\dagger \hat{c}_\mathbf{k}$ is a sort of "number" operator, which counts the number of electrons in the state $\mathbf{k}$.

Note that if we try to add another $\mathbf{k}$-electron to the state $\mathbf{k}$ we get

$$\hat{c}_\mathbf{k}^\dagger|\mathbf{k}\rangle = \hat{c}_\mathbf{k}^\dagger \hat{c}_\mathbf{k}^\dagger|0\rangle = -\hat{c}_\mathbf{k}^\dagger \hat{c}_\mathbf{k}^\dagger|0\rangle, \tag{9.51}$$

the last step relying on the CAR, so we must have $\hat{c}_\mathbf{k}^\dagger|\mathbf{k}\rangle = 0$ for this equation to be valid. Indeed, this is just Pauli's principle: We can't put two electrons in the same state. This follows directly from having used *anti*commutation rules for the creation and annihilation operators.

Let's also note the following:

$$\widehat{H}\,|\mathbf{k}\rangle = \sum_{\mathbf{k}'} E_{\mathbf{k}'}\,\hat{c}^{\dagger}_{\mathbf{k}'}\hat{c}_{\mathbf{k}'}\hat{c}^{\dagger}_{\mathbf{k}}\,|0\rangle. \tag{9.52}$$

Let's now use the anticommutation rule $\hat{c}_{\mathbf{k}'}\hat{c}^{\dagger}_{\mathbf{k}} = \delta_{\mathbf{k},\mathbf{k}'} - \hat{c}^{\dagger}_{\mathbf{k}}\hat{c}_{\mathbf{k}'}$:

$$\widehat{H}\,|\mathbf{k}\rangle = \sum_{\mathbf{k}'} E_{\mathbf{k}'}\hat{c}^{\dagger}_{\mathbf{k}'}(\delta_{\mathbf{k}',\mathbf{k}} - \hat{c}^{\dagger}_{\mathbf{k}}\hat{c}_{\mathbf{k}'})|0\rangle. \tag{9.53}$$

Note that the second term inside the parentheses, $\hat{c}^{\dagger}_{\mathbf{k}'}\hat{c}^{\dagger}_{\mathbf{k}}\hat{c}_{\mathbf{k}'}|0\rangle$, vanishes, since the first operator on the right annihilates an electron $\mathbf{k}'$ in the vacuum state, which yields zero. Thus,

$$\widehat{H}|\mathbf{k}\rangle = E_{\mathbf{k}}\hat{c}^{\dagger}_{\mathbf{k}}|0\rangle = E_{\mathbf{k}}\,|\mathbf{k}\rangle. \tag{9.54}$$

This shows that indeed the state $|\mathbf{k}\rangle = \hat{c}^{\dagger}_{\mathbf{k}}|0\rangle$ is an eigenstate of the Hamiltonian with eigenvalue $E_{\mathbf{k}}$.

From the discussion above, it follows that an arbitrary state of our many-electron system can be written as:

$$|\{n_\lambda\}\rangle = |n_1 n_2 \ldots n_N\rangle = (\hat{c}^{\dagger}_1)^{n_1}(\hat{c}^{\dagger}_2)^{n_2}\ldots(\hat{c}^{\dagger}_N)^{n_N}|0\rangle, \tag{9.55}$$

where the integers $n_i$ are restricted to taking only the values 0 or 1 and the indices $1,2,\ldots N$ label wavevectors $\mathbf{k}_1$, $\mathbf{k}_2,\ldots\mathbf{k}_N$ (and also spin indices, if necessary). The notation $|\{n_\lambda\}\rangle$ identifies all possible combinations $\lambda$ of $N$-electron states. Note that these states are already antisymmetric. For example, the wavefunction of a 2-electron state will be

$$\langle\mathbf{r}',\mathbf{r}|\mathbf{k}',\mathbf{k}\rangle = \langle\mathbf{r}',\mathbf{r}|\hat{c}^{\dagger}_{\mathbf{k}'}\hat{c}^{\dagger}_{\mathbf{k}}|0\rangle = \frac{1}{\sqrt{2}}\,\frac{1}{(2\pi)^3}\,(\,e^{\mathrm{i}\mathbf{k}'\cdot\mathbf{r}'}\,e^{\mathrm{i}\mathbf{k}\cdot\mathbf{r}} - e^{\mathrm{i}\mathbf{k}\cdot\mathbf{r}'}\,e^{\mathrm{i}\mathbf{k}'\cdot\mathbf{r}}\,). \tag{9.56}$$

The general state $|n_1 n_2 \ldots n_N\rangle$ is obviously an eigenstate of the Hamiltonian with eigenvalue $n_1E_1 + n_2E_2 + \ldots$, as one can show following the derivation we have used above to reach Eq. (9.54).

So far we have considered a Schrödinger field, which does not account for spin. If we add it to the picture and label with $\sigma = \pm 1/2$ the spin variable, we have the new CAR:

$$[\hat{c}^{\sigma}_{\mathbf{k}}, \hat{c}^{\sigma'\dagger}_{\mathbf{k}'}]_{+} = \delta_{\sigma,\sigma'}\,\delta_{\mathbf{k},\mathbf{k}'}, \tag{9.57}$$

all other anticommutators vanishing. Everything carries on as before, the only complication being related to the fact that for many-electron states the antisymmetrization of the wavefunctions must account for the quantum-mechanical rules of adding angular momenta. Thus, we will have four types of antisymmetrized 2-electron states: One spin-singlet state, with a symmetric spatial component of the wavefunction, and three spin-triplet states, with an antisymmetric spatial component.

Starting from the "usual" single-particle Hilbert space $\mathcal{H}$, $L^2(\mathbb{R}^3)\otimes\mathcal{S}$, (where $\otimes$ means 'tensor product' of vector spaces and $\mathcal{S}$ is the 1/2-spin-space, the space associated with the two-dimensional irreducible representation of the special unitary group SU(2)), the total functional space associated with the many-electron system can be formally defined as:

$$\mathcal{H}\;\oplus\;\mathcal{H}\otimes\mathcal{H}\;\oplus\;\mathcal{H}\otimes\mathcal{H}\otimes\mathcal{H}\ldots = \sum_{N=1}^{\infty}\oplus\prod_{j=1}^{N}\otimes\mathcal{H}, \tag{9.58}$$

which—after proper topological closure and compactification—is called *Fock space*.

* *Out of curiosity, it is interesting to note here a major difference between first and second quantization, a difference which is the deep source of the infinite zero-point energy we saw above and of many other infinities which affect QFT: In ordinary QM (first quantization), there is only one Hilbert space—up to unitary transformations. That's the reason Heisenberg's matrix-mechanics and Schrödinger's wave equation are actually the same thing, just viewed in two different, but equivalent, representations. Not so in many-body theory or QFT: For every Hamiltonian, there is only one representation which is "correct." Unfortunately, Fock-space is the representation of free particles. As soon as we turn on any interactions, problems arise because we are in the wrong representation. Renormalization is a typical problem. This sad state of things is known as "Haag's theorem" [8, 9], which is known to state that the interaction picture exists only if there is no interaction. Ultimately, this depends on the following consideration: From axiom 9 of the Copenhagen interpretation (Chap. 1, page 16), the Hilbert space associated with two isolated systems brought together is the tensor product of the Hilbert spaces associated with the two systems. When the two systems are strongly coupled, any arbitrary state of the coupled system cannot necessarily be expressed as the tensor product of two states of the isolated systems. However, the set of these tensor-product states constitutes a complete basis in the larger Hilbert space and we can always express any arbitrary state of the strongly coupled systems as a linear superposition of these basis states. This is true for an arbitrary large, but* finite, *number of systems. So, for example, it is true for a system of N interacting electrons, as long as N is finite. But this property fails when $N \to \infty$, which is the definition of Fock space. In other words, an arbitrary state of the interacting electrons does not belong to Fock space. "Constructive QFT" is the branch of theoretical physics which attempts to "construct" a proper representation for every interaction, thus bypassing renormalization problems.*

## 9.3 Quantization of the Charge Density: Plasmons

As a second example, let's consider the plasma oscillations of an electron gas that we saw already on page 181. This is a very interesting example which shows clearly how "elementary excitations" emerge: We know how to solve the problem of electrons in a crystal when the system is in its ground state: The ions are "frozen" in their equilibrium positions and the electrons are all in their lowest-energy state, filling all states up to the Fermi level. In tackling the problem of what happens when electrons acquire additional energy (via an external perturbation, typically) while interacting among themselves via long-range Coulomb interactions is quite a complicated task. So, we "linearize" the problem by assuming that only first-order deviations from the ground state occur. In retaining only the leading terms of the interaction Hamiltonian (in our case, the Coulomb potential energy) we retain only terms quadratic in the deviation from the ground state. By decomposing the system into normal modes (by taking Fourier transforms in the example below), the Hamiltonian will be the recast into the form of the sum of many Hamiltonians, each corresponding to a different wavelength, and each being the sum of a kinetic energy term (quadratic in the velocity or canonically conjugate momenta) and a potential energy term (quadratic in the displacement coordinates, which are the canonical coordinates). Thus, for each wavelength there will be a corresponding Hamiltonian of a single harmonic oscillator. We can now "quantize" these oscillators using the Dirac algebraic method of raising (creation) and lowering (annihilation) operators, as we saw for the Schrödinger field. These are just the creation and annihilation operators for each wavelength. So, for each wavelength there may be no corresponding excitation at all (the oscillator is in its ground state), one excitation (the oscillator is in its first excited state), etc. Looking back at the Schrödinger field, there each excitation was interpreted as one electron "created" at a given wavelength. Here, in complete analogy, we shall call these excitations "quasi particles." The suffix "on" attached to the name given to these excitations (plasm**ons**, phon**ons**, magn**ons**, etc.) reflects this analogy.

So, let's tackle the problem of the plasma excitations of a homogeneous, bulk electron gas. It will be instructive, at the end, to compare the resulting normalization of the plasmon field with the result of simpler semiclassical energetic arguments.

### 9.3.1 Lagrangian Density

The classical Lagrangian density of the electron gas can be written as:

$$\mathscr{L}(\dot{\mathbf{u}},\phi) = \frac{1}{2}\, n\, m\, \dot{\mathbf{u}}(\mathbf{r})\cdot\dot{\mathbf{u}}(\mathbf{r}) - \frac{1}{2}\delta\rho(\mathbf{r})\,\phi(\mathbf{r}), \tag{9.59}$$

where $en = \rho$ is the uniform charge density, $m$ is the electron mass, $\phi(\mathbf{r})$ is the potential, and $\mathbf{u}(\mathbf{r})$ is the displacement of the electron gas at position $\mathbf{r}$. This displacement is related to the charge fluctuations $\delta\rho(\mathbf{r})$ via the continuity equation:

$$\frac{\delta\rho(\mathbf{r})}{\rho} = -\nabla\cdot\mathbf{u}(\mathbf{r}). \tag{9.60}$$

Thus, Poisson equation can be written as:

$$\nabla^2\phi(\mathbf{r}) = \frac{\delta\rho(\mathbf{r})}{\epsilon_s} = -\frac{\rho}{\epsilon_s}\nabla\cdot\mathbf{u}(\mathbf{r}), \tag{9.61}$$

where $\epsilon_s$ is the dielectric constant of the solid.

### 9.3.2 Going to Fourier Transforms

We consider, as usual, their Fourier decomposition:

$$\phi(\mathbf{r}) = \sum_{\mathbf{q}} \phi_{\mathbf{q}}\, e^{i\mathbf{q}\cdot\mathbf{r}}, \tag{9.62}$$

and

$$\mathbf{u}(\mathbf{r}) = \sum_{\mathbf{q}} \mathbf{u}_{\mathbf{q}}\, e^{i\mathbf{q}\cdot\mathbf{r}}. \tag{9.63}$$

Equation (9.61) implies

$$-\sum_{\mathbf{q}} q^2\phi_{\mathbf{q}} e^{i\mathbf{q}\cdot\mathbf{r}} = -i\,\frac{\rho}{\epsilon_s}\sum_{\mathbf{q}} \mathbf{q}\cdot\mathbf{u}_{\mathbf{q}}\, e^{i\mathbf{q}\cdot\mathbf{r}}. \tag{9.64}$$

Since plasmons are longitudinal oscillations, $\mathbf{q}\cdot\mathbf{u}_{\mathbf{q}} = qu_q$, so that $\phi_q = i\rho\, u_q/(\epsilon_s q)$. Keeping in mind that the total energy must be real, we should replace the terms $\dot{\mathbf{u}}(\mathbf{r})\cdot\dot{\mathbf{u}}(\mathbf{r})$ and $\delta\rho(\mathbf{r})\,\phi(\mathbf{r})$ in the expression for the Lagrangian density Eq. (9.59) with $\dot{\mathbf{u}}(\mathbf{r})^*\cdot\dot{\mathbf{u}}(\mathbf{r})$ and $\delta\rho(\mathbf{r})^*\,\phi(\mathbf{r})$, respectively. Then, the kinetic energy can be written as:

$$T = \frac{1}{2} \int \mathrm{d}\mathbf{r}\, n\, m\, \dot{\mathbf{u}}(\mathbf{r})^* \cdot \dot{\mathbf{u}}(\mathbf{r}) = \frac{1}{2}\, n\, m \sum_{\mathbf{q}} \dot{\mathbf{u}}^*_{-\mathbf{q}} \cdot \dot{\mathbf{u}}_{\mathbf{q}}, \tag{9.65}$$

where the dots indicate derivatives with respect to time. The potential energy can be calculated ignoring second-order corrections to the charge density, that is, by treating $\rho(\mathbf{r}) = en$ as a constant in Eq. (9.60), as we had done before when dealing with classical plasma oscillations. This is where the "linearization" approximation comes into play. Then:

$$V = \frac{1}{2} \int \mathrm{d}\mathbf{r}\, \delta\rho(\mathbf{r})^* \phi(\mathbf{r}) = +\frac{1}{2} \frac{e^2 n^2}{\epsilon_s} \sum_{\mathbf{q}} \mathbf{u}^*_{-\mathbf{q}} \cdot \mathbf{u}_{\mathbf{q}}. \tag{9.66}$$

Thus, from Eqs. (9.65) and (9.66), the Lagrangian for the system can be written as:

$$L = T - V = \frac{1}{2}\, n\, m \sum_{\mathbf{q}} \dot{\mathbf{u}}^*_{-\mathbf{q}} \cdot \dot{\mathbf{u}}_{\mathbf{q}} - \frac{1}{2} \frac{e^2 n^2}{\epsilon_s} \sum_{\mathbf{q}} \mathbf{u}^*_{-\mathbf{q}} \cdot \mathbf{u}_{\mathbf{q}}. \tag{9.67}$$

### 9.3.3 The Hamiltonian

Let's now define the Hamiltonian. First, we define the canonical conjugate momentum

$$\pi_{\mathbf{q}} = \frac{\partial L}{\partial \dot{\mathbf{u}}_{\mathbf{q}}} = \frac{\partial T}{\partial \dot{\mathbf{u}}_{\mathbf{q}}} = n\, m\, \dot{\mathbf{u}}^*_{\mathbf{q}}. \tag{9.68}$$

Then, we apply the usual Legendre transform to get the Hamiltonian $H = T + V$:

$$H = \frac{1}{2} \sum_{\mathbf{q}} \left\{ \frac{1}{nm}\, \pi^*_{-\mathbf{q}} \cdot \pi_{\mathbf{q}} + \frac{e^2 n^2}{\epsilon_s}\, \mathbf{u}^*_{-\mathbf{q}} \cdot \mathbf{u}_{\mathbf{q}} \right\}. \tag{9.69}$$

### 9.3.4 Canonical Quantization

Quantization can be now performed imposing the CCRs

$$[\hat{u}_{i,\mathbf{q}}, \hat{\pi}_{j,\mathbf{q}'}] = \mathrm{i}\hbar\, \delta_{ij}\, \delta_{\mathbf{q}\mathbf{q}'}. \tag{9.70}$$

Following a standard procedure, we rewrite the Hamiltonian as

$$\widehat{H} = \frac{1}{2} \sum_{\mathbf{q}} [\, \widehat{P}^\dagger_{-\mathbf{q}} \widehat{P}_{\mathbf{q}} + \widehat{Q}^\dagger_{-\mathbf{q}} \widehat{Q}_{\mathbf{q}} \,], \tag{9.71}$$

having set:

$$\widehat{P}_{\mathbf{q}} = \frac{\hat{\pi}_{\mathbf{q}}}{(nm)^{1/2}}, \tag{9.72}$$

$$\widehat{Q}_{\mathbf{q}} = \left(\frac{e^2 n^2}{\epsilon_s}\right)^{1/2} \hat{u}_{\mathbf{q}}. \tag{9.73}$$

Let's now define, as usual, the creation and annihilation operators via the relations:

$$\widehat{Q}_{\mathbf{q}} = \left(\frac{\hbar\omega_P}{2}\right)^{1/2} (\hat{b}_{\mathbf{q}} + \hat{b}^{\dagger}_{-\mathbf{q}}), \tag{9.74}$$

$$\widehat{P}_{\mathbf{q}} = \mathrm{i}\left(\frac{\hbar\omega_P}{2}\right)^{1/2} (\hat{b}_{\mathbf{q}} - \hat{b}^{\dagger}_{-\mathbf{q}}), \tag{9.75}$$

where we have introduced the plasma frequency $\omega_P = [(e^2 n)/(\epsilon_s m)]^{1/2}$. Thus, finally, the Hamiltonian Eq. (9.69) becomes

$$\widehat{H} = \frac{\hbar\omega_P}{2} \sum_{\mathbf{q}} \left\{\hat{b}^{\dagger}_{\mathbf{q}}\hat{b}_{\mathbf{q}} + \hat{b}_{\mathbf{q}}\hat{b}^{\dagger}_{\mathbf{q}}\right\}. \tag{9.76}$$

Using the definitions (9.72)–(9.75) and the CCR (9.70), the only non-vanishing commutator for the operators $\hat{b}^{\dagger}_{\mathbf{q}}$ and $\hat{b}_{\mathbf{q}}$ is

$$[\hat{b}^{\dagger}_{\mathbf{q}}, \hat{b}_{\mathbf{q}}] = -\frac{1}{\omega_P}\left(\frac{en}{\epsilon_s m}\right)^{1/2} = -1. \tag{9.77}$$

This shows that the operators $\hat{b}^{\dagger}_{\mathbf{q}}$ and $\hat{b}_{\mathbf{q}}$ behave like raising (creation) and lowering (annihilation) operators, so that the Hamiltonian Eq. (9.76) becomes a collection of harmonic oscillators

$$\widehat{H} = \hbar\omega_P \sum_{\mathbf{q}} \left\{\hat{b}^{\dagger}_{\mathbf{q}}\hat{b}_{\mathbf{q}} + \frac{1}{2}\right\}, \tag{9.78}$$

while the displacement field and the potential can be expressed in terms of the creation and annihilation operators

$$\hat{u}_{\mathbf{q}} = \left(\frac{\hbar\omega_P\epsilon_s}{2e^2 n^2}\right)^{1/2} (\hat{b}^{\dagger}_{-\mathbf{q}} + \hat{b}_{\mathbf{q}}), \tag{9.79}$$

$$\hat{\phi}_{\mathbf{q}} = -\frac{\mathrm{i}}{q}\left(\frac{\hbar\omega_P}{2\epsilon_s}\right)^{1/2} (\hat{b}^{\dagger}_{-\mathbf{q}} + \hat{b}_{\mathbf{q}}). \tag{9.80}$$

As usual, the diverging contribution in Eq. (9.78) will be ignored and we shall write

$$\widehat{H} = \hbar\omega_P \sum_{\mathbf{q}} \hat{b}^{\dagger}_{\mathbf{q}}\hat{b}_{\mathbf{q}}. \tag{9.81}$$

### 9.3.5 *Fock Space for Bosons*

The Hilbert-like space associated with the system of bosons (as plasmons carry no spin) can be constructed essentially in the same way we saw above for the case of Schrödinger Fermions. The obvious difference is that, since we use commutation instead of anticommutation rules, we can put as many particles as we wish in each quantum state. Thus, the most general state of Fock space will have the form

$$|\{N_\lambda\}\rangle = |N_1 N_2 \ldots N_N\rangle = (\hat{b}_1^\dagger)^{N_1} (\hat{b}_2^\dagger)^{N_2} \ldots (\hat{b}_N^\dagger)^{N_N} |0\rangle, \tag{9.82}$$

where now states with $N_i > 1$ will also exist. Note that the squared norm of a state $|N_{\mathbf{q}}\rangle$ will be $N_{\mathbf{q}}!$, so that states of the form (9.82) must be normalized by dividing by $\sqrt{N_{\mathbf{q}}!}$.

### 9.3.6 *A Calculation: Thermal Expectation Value of the Plasmon Field*

Let's assume that we are at thermal equilibrium. Since plasmons have zero spin, their population will obey the Bose–Einstein statistics and the thermal average $\langle N_{\mathbf{q}}\rangle_{\mathrm{th}}$ of the number $N_{\mathbf{q}}$ of plasmons of momentum $\mathbf{q}$ present at temperature $T$ will be given by:

$$\langle N_{\mathbf{q}}\rangle_{\mathrm{th}} = \left[\exp\left(\frac{\hbar\omega_{\mathbf{q}}}{k_{\mathrm{B}}T}\right) - 1\right]^{-1}. \tag{9.83}$$

If we want to calculate the thermally averaged root-mean-square (rms) amplitude of the field associated with the plasmons, we can first calculate the expectation value of the (squared) amplitude on the Fock state $|\, N_{\mathbf{q}}\rangle = (N_{\mathbf{q}}!\,)^{-1/2}\, (\hat{b}_{\mathbf{q}}^\dagger)^{N_{\mathbf{q}}} \,|\, 0\rangle$ (where $|\, 0\rangle$ is the vacuum) containing $N_{\mathbf{q}}$ plasmons of wavevector $\mathbf{q}$. This is

$$\begin{aligned}\langle N_{\mathbf{q}} \,|\, \hat{\phi}_{\mathbf{q}}^\dagger \hat{\phi}_{\mathbf{q}} \,|\, N_{\mathbf{q}}\rangle &= \frac{\hbar\omega_{\mathrm{P}}}{2\epsilon_{\mathrm{s}}} \frac{1}{q^2} \langle 0|(\hat{b}_{\mathbf{q}})^{N_{\mathbf{q}}} (\hat{b}_{[\mathbf{q}} + \hat{b}_{\mathbf{q}}^\dagger)(\hat{b}_{\mathbf{q}}^\dagger + \hat{b}_{\mathbf{q}})(\hat{b}_{\mathbf{q}}^{-\dagger})^{N_{\mathbf{q}}}|0\rangle \\ &= \frac{\hbar\omega_{\mathrm{P}}}{2\epsilon_{\mathrm{s}}} \frac{1}{q^2} (1 + N_{\mathbf{q}} + N_{-\mathbf{q}}).\end{aligned} \tag{9.84}$$

We can then take the thermal average of this expression, so that:

$$\langle \hat{\phi}_{\mathbf{q}}\rangle_{\mathrm{th}} = \left(\frac{\hbar\omega_{\mathrm{P}}}{2\epsilon_{\mathrm{s}}} \frac{1}{q^2}\right)^{1/2} (1 + 2\langle N_{\mathbf{q}}\rangle_{\mathrm{th}})^{1/2}. \tag{9.85}$$

Setting $N_{\mathbf{q}} = 0$ above, it shows that the plasmon zero-point motion at wavevector $\mathbf{q}$ is associated with a field of rms amplitude

$$|\, \phi_{\mathbf{q}} \,| = \left(\frac{\hbar\omega_{\mathrm{P}}}{2\epsilon_{\mathrm{s}}}\right)^{1/2} \frac{1}{q}. \tag{9.86}$$

### 9.3.7 A Semiclassical Analogy

Can we obtain this result without going through the procedure of canonical quantization? Stern and Ferrel [10] have provided the following semiclassical argument, based on energetic considerations: Let us write the magnitude of the plasmon field at wavevector $\mathbf{q}$ as:

$$E_{q,\omega}(\mathbf{r},t) = a_q \cos(\mathbf{q}\cdot\mathbf{r} - \omega_P t), \tag{9.87}$$

so that the electrostatic potential is

$$\phi_{q,\omega}(\mathbf{r},t) = -\frac{a_q}{q} \sin(\mathbf{q}\cdot\mathbf{r} - \omega_P t). \tag{9.88}$$

For harmonic oscillators such as plasmons, the time-averaged kinetic energy, $\langle T \rangle$, is equal to the time-averaged potential energy, $\langle U \rangle_{\text{time}}$. Thus, the total energy, $W$, associated with the zero-point plasmon field over a volume $\Omega$, *including* its own self-energy (which explains the absence of the factor 1/2 in front of the integral below), will be

$$W = \langle T \rangle_{\text{time}} + \langle U \rangle_{\text{time}} = 2\langle U \rangle_{\text{time}} = 2 \left\langle \int_\Omega d\mathbf{r}\ \phi^*_{q,\omega}(\mathbf{r})\ \rho_{q,\omega}(\mathbf{r}) \right\rangle_{\text{time}} = \epsilon_s \mid a_q \mid^2 \Omega, \tag{9.89}$$

where $\epsilon_s$ must be understood to be the high-frequency dielectric constant $\epsilon^\infty$ in our case, since $\rho_{q,\omega}(\mathbf{r})$ is just a microscopic polarization charge, so that only the background (i.e., valence) dielectric response should be considered. Quantum mechanically, this energy must represent the zero-point plasmon energy within the volume $\Omega$, so that $W = \frac{1}{2}\ \hbar\omega_P\Omega$, which implies

$$\mid \phi_{\mathbf{q}} \mid = \frac{\mid a_q \mid}{q} = \left(\frac{\hbar\omega_P}{2\epsilon_s}\right)^{1/2} \frac{1}{q}, \tag{9.90}$$

in agreement with Eq. (9.86).

## 9.4 Quantization of the Vibrational Degrees of Freedom: Phonons

In the first part of this text we had a quick look at the ionic Hamiltonian (see pages 60–61), but we left it there, promising to come back and revisit this term. Here we shall do just that as usual rewriting it in such a way that we will end up with a collection of harmonic oscillators. A very useful reading that may fill the many gaps we are going to leave open is the well-known text by Ziman [11], a book that does not show its age.

### 9.4.1 Canonical Quantization

Let $\mathbf{R}^{(0)}_{l\gamma}$ be the equilibrium position of the ion (labeled by the index $\gamma$ inside the unit cell $l$) and let $\delta\mathbf{R}_{l\gamma}$ be its displacement from $\mathbf{R}^{(0)}_{l\gamma}$. Let also $\mathbf{R}_l$ be the coordinate-origin of the cell $l$. In terms of the momentum coordinate of ion $\gamma$ in cell $l$, $\mathbf{P}_{l\gamma} = M_\gamma \delta\dot{\mathbf{R}}_{l\gamma}$, where $M_\gamma$ is the mass of the ion, the ionic Hamiltonian is

$$H_{\rm ion} = \frac{1}{2}\sum_{l\gamma}\left[\frac{P_{l\gamma}^2}{M_\gamma} + \sum_{l'\neq l,\gamma'}\sum_{ij}\frac{\partial^2 U(\mathbf{R}_{l\gamma}^{(0)} - \mathbf{R}_{l'\gamma'}^{(0)})}{\partial R_{l\gamma i}\,\partial R_{l'\gamma' j}}\,\delta R_{l\gamma i}\,\delta R_{l'\gamma' j}\right], \tag{9.91}$$

where $U$ is the total potential energy of the lattice. Note that here the indices $i$ and $j$ run over the spatial-coordinate indices $x, y, z$. Using Bloch's theorem we can now expand $\delta\mathbf{R}_{l\gamma}$ in its spatial Fourier coefficients at wavevector $\mathbf{q}$ and assume a time dependence $e^{i\omega t}$:

$$\delta\mathbf{R}_{l\gamma} = \frac{1}{(N_{\rm cell}M_\gamma)^{1/2}}\sum_{\mathbf{q}}\xi_{\mathbf{q}\gamma}\,e^{i\mathbf{q}\cdot\mathbf{R}_l}\,e^{i\omega t}, \tag{9.92}$$

where $N_{\rm cell}$ is the number of cells in the crystal. It should be noted that the periodicity of the lattice implies that the 'polarization vector' $\xi_{\mathbf{q}\gamma}$ does not depend on the cell index $l$, the entire dependence on the cell being absorbed by the phase $e^{i\mathbf{q}\cdot\mathbf{R}_l}$. We also define the "lattice Fourier transform" of the dynamical matrix as

$$D_{ij\gamma\gamma'}(\mathbf{q}) = \sum_{l''}\tilde{D}_{ij\,\gamma\gamma'}(l'')\,e^{i\mathbf{q}\cdot(\mathbf{R}_l - \mathbf{R}_{l'})}, \tag{9.93}$$

where

$$\tilde{D}_{ij\gamma\gamma'}(l'') = -\frac{\partial^2 U(\mathbf{R}_{l\gamma}^{(0)} - \mathbf{R}_{l'\gamma'}^{(0)})}{\partial R_{l\gamma i}\,\partial R_{l'\gamma' j}}, \tag{9.94}$$

where $\mathbf{R}_{l''} = \mathbf{R}_l - \mathbf{R}_{l'}$. Inserting the expansion (9.92) into the equations of motion determined by the Hamiltonian (9.91), for each $\mathbf{q}$ we obtain the homogeneous linear system

$$\sum_{j\gamma\gamma'}\left[\tilde{D}_{ij\gamma\gamma'}(\mathbf{q}) + \omega^2\,\delta_{ij}\,\delta_{\gamma\gamma'}\right]\xi_{\mathbf{q}\gamma' j} = 0, \tag{9.95}$$

where $\tilde{D}_{ij\gamma\gamma'}(\mathbf{q}) = D_{ij\gamma\gamma'}(\mathbf{q})/(M_\gamma M_{\gamma'})^{1/2}$. The associated secular equation determines the eigenfrequencies $\omega_{\mathbf{q}\eta}$ of oscillations labeled by a "branch index" $\eta$ running over 3 (one LA and 2 TA) acoustic modes and over $3(N_{\rm ions} - 1)$ optical modes, $N_{\rm ion}$ being the number of ions in the primitive cell.

Formally, the eigenvalue problem given by Eq. (9.95) sets the framework that allows us to quantize the vibrational modes of the lattice. In practice, if we need to obtain a quantitatively accurate phonon spectrum, we need to calculate somehow the dynamical matrix. This requires the knowledge of the forces acting on the ions of the crystal. We have seen in Chap. 6, Sect. 6.3.5, how DFT can be used for this purpose. Alternative semi-empirical models have also been proposed in the past, when DFT was not computationally viable as it is today. Without going into details, we may mention here the Born-von Kármán model [12], the "Keating's valence force field" model [13], or the "valence shell model" [14–16]. These models are largely empirical, relying on the lattice symmetry, on the nature of the bond and ion–ion potential, and employ parameters related, of course, to the elastic properties of the crystal (Fig. 9.1).

Figure 9.2 on page 208 shows the dispersion for the three optical and three acoustic phonons in Si obtained using the 'valence shell model' mentioned above. These spectra can be compared to the results shown in Fig. 6.8 in Chap. 6, Sect. 6.3.5, page 109, obtained using the *ab initio* method of DFT.

Returning to the general formulation of the problem, we can now write the general solution of the equations of motion as follows: Let's normalize the polarization vectors $\xi_{\mathbf{q}\gamma}$ for each eigenmode $\eta$ and call $\mathbf{e}_{\mathbf{q}\gamma}^{(\eta)}$ the unit vectors so obtained. We can now express the general solution as a linear combination of these normalized eigenmodes, Eq. (9.92), with coefficients $Q_{\mathbf{q}\eta}$ which will play the role of canonical coordinates. Note that these coefficients are determined up to an overall multiplicative constant, which will be set by the CCR conditions below. We can view this as a simple rotation of the basis vectors. Thus, we can write for the "displacement field" $\delta\mathbf{R}_{l\gamma}^{(\eta)}$ for each mode $\eta$:

$$\delta\mathbf{R}_{l\gamma}^{(\eta)} = \frac{1}{(N_{\text{cell}}M)^{1/2}} \sum_{\mathbf{q}} \mathbf{e}_{\mathbf{q}\gamma}^{(\eta)} \, Q_{\mathbf{q}\eta} \, e^{\mathrm{i}\mathbf{q}\cdot\mathbf{R}_l}. \tag{9.96}$$

We can now define a conjugate momentum

$$\mathbf{P}_{l\gamma}^{(\eta)} = \delta\dot{\mathbf{R}}_{l\gamma}^{(\eta)} = \left(\frac{M}{N_{\text{cell}}}\right)^{1/2} \sum_{\mathbf{q}} \mathbf{e}_{\mathbf{q}\gamma}^{(\eta)} \, P_{\mathbf{q}\eta} \, e^{\mathrm{i}\mathbf{q}\cdot\mathbf{R}_l}, \tag{9.97}$$

and quantize the system by promoting the canonical variables $P_{\mathbf{q}\eta}$ and $Q_{\mathbf{q}\eta}$ to operators on Fock space and imposing the CCR (as phonons do not carry spin)

$$[\widehat{Q}_{\mathbf{q}\eta}, \widehat{P}_{\mathbf{q}'\eta'}] = \mathrm{i}\hbar \, \delta_{\mathbf{q}\mathbf{q}'} \, \delta_{\eta\eta'}, \quad [\widehat{P}_{\mathbf{q}\eta}, \widehat{P}_{\mathbf{q}'\eta'}] = 0, \quad [\widehat{Q}_{\mathbf{q}\eta}, \widehat{Q}_{\mathbf{q}'\eta'}] = 0. \tag{9.98}$$

Then, using Eqs. (9.95)–(9.97), the ion Hamiltonian (9.91) becomes

$$\widehat{H}_{\text{ion}} = \frac{1}{2} \sum_{\mathbf{q}\eta} [\, \widehat{P}_{-\mathbf{q}\eta}^{\dagger} \widehat{P}_{\mathbf{q}\eta} + \omega_{\mathbf{q}\eta}^2 \, \widehat{Q}_{-\mathbf{q}\eta}^{\dagger} \widehat{Q}_{\mathbf{q}\eta} \,]. \tag{9.99}$$

Once more, this is the Hamiltonian of a collection of harmonic oscillators (phonons). Defining the annihilations and creation operators (going to the infinite-volume normalization):

$$\widehat{Q}_{\mathbf{q}\eta} = \left(\frac{\hbar}{2\omega_{\mathbf{q}\eta}}\right)^{1/2} (\hat{b}_{\mathbf{q}\eta} + \hat{b}_{-\mathbf{q}\eta}^{\dagger}), \tag{9.100}$$

$$\widehat{P}_{\mathbf{q}\eta} = \mathrm{i} \left(\frac{\hbar\omega_{\mathbf{q}\eta}}{2}\right)^{1/2} (\hat{b}_{\mathbf{q}\eta} - \hat{b}_{-\mathbf{q}\eta}^{\dagger}), \tag{9.101}$$

finally the phonon Hamiltonian can be rewritten in terms of the creation and annihilation operators as:

$$\widehat{H}_{\text{ion}} = \sum_{\mathbf{q}\eta} \hbar\omega_{\mathbf{q}\eta} \, \hat{b}_{\mathbf{q}\eta}^{\dagger} \, \hat{b}_{\mathbf{q}\eta} \tag{9.102}$$

(ignoring as usual the diverging zero-point contribution) in analogy with Eqs. (9.31) and (9.81). Note that now, since phonons carry no spin, the creations and annihilation operators $\hat{b}_{\mathbf{q}\eta}^{\dagger}$ and $\hat{b}_{\mathbf{q}\eta}$ satisfy commutation rules like Eq. (9.48) for the electron operators, but with commutators in place of anticommutators.

### 9.4.2 A Simple Example: A Linear Chain

Let's consider the simpler case of a 1D chain of atoms of mass $M$. We can drop the indices $i,j$ (since we are in 1D) as well as the indices $\gamma$ (since we consider only one atom in each 1D cell). We may remain for the time being at the classical level, since quantization does not affect the phonon spectrum. The ionic Hamiltonian becomes

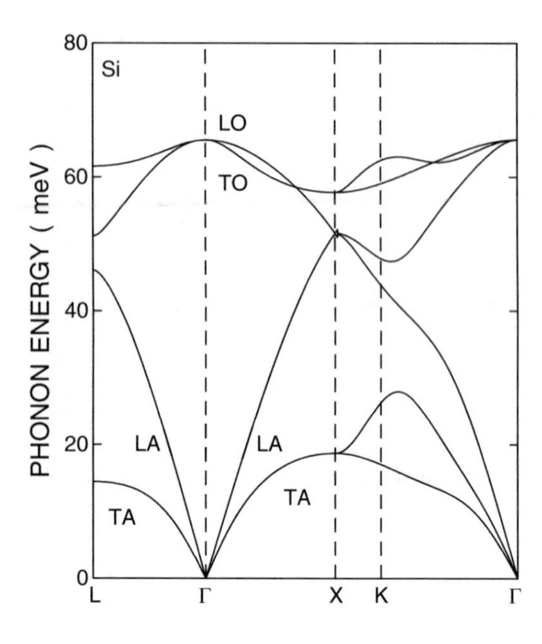


**Fig. 9.1** Silicon phonon spectra calculated using a model known as "valence shell model"

$$H_{\text{ion}} = \frac{1}{2}\sum_{l}\left[\frac{P_l^2}{2M} + \sum_{l'} D_{ll'}\delta R_l \delta R_{l'}\right].$$

Note that:

$$D_{l,l'} = D_{l-l',0} \text{ by translation symmetry}$$

and

$$\sum_{l} D_{0,l} = 0 \text{ since for equal shifts } \delta R_l \text{ the crystal energy does not change.}$$

Put it differently, since $D_{0,l}$ is the force acting on the ion 0 due to ion $l$, the equation above states that at equilibrium the total force due to all other ions acting on the ion 0 must vanish. The Hamilton equation for the ionic momentum will be

$$\dot{P}_l = -\frac{\partial H_{\text{ion}}}{\partial \delta R_l} = -\sum_{l'} D_{ll'}\delta R_{l'},$$

or

$$M\frac{\mathrm{d}^2\delta R_l}{\mathrm{d}t^2} = -\sum_{l'} D_{ll'}\delta R_{l'},$$

Let's also make the simplifying assumption that each ion interacts only with its nearest neighbor (so that we can solve the problem "by hand" and not numerically). Then:

$$M\frac{\mathrm{d}^2\delta R_l}{\mathrm{d}t^2} = D_{l,l}\delta R_l + D_{l,l+1}\delta R_{l+1} + D_{l,l-1}\delta R_{l-1}.$$

By translation symmetry $D_{l,l} = D_{0,0}$. Without loss of generality we can consider the case $l = 0$. So we can consider only the elements $D_{0,0}$, $D_{0,\pm 1}$, and $D_{\pm 1,0}$. Using again the translation symmetry we have

$$D_{0,1} = D_{-1,0},$$

but since $D_{ll'} = D_{l'l}$ (it's only a switch in the order of the derivatives), we also have

$$D_{0,1} = D_{1,0},$$

so that $D_{0,1} = D_{1,0} = D_{-1,0} = D_{0,-1}$.
By the property $\sum_l D_{0,l} = 0$ above we have

$$D_{0,0} + D_{0,1} + D_{0,-1} = 0 \rightarrow D_{0,0} = -2\ D_{1,0}.$$

Let's simply call $D = D_{1,0}$ (the "spring" constant of the inter-ionic force). Then the Hamilton equation of motion becomes simply:

$$M \frac{\mathrm{d}^2 \delta R_l}{\mathrm{d}t^2} = D\left[-2\ \delta R_l + \delta R_{l+1} + \delta R_{l-1}\ \right].$$

Now let's set

$$\delta R_l = \mathbf{e}_q\ e^{\mathrm{i}qla}\ e^{\mathrm{i}\omega t},$$

where $a$ is the equilibrium distance between the ions in the chain (our "lattice constant"). Then:

$$-M\ \omega^2 = D\left[\ 2\ \cos(qa) - 2\ \right],$$

becomes our trivial "secular equation" from which we obtain the dispersion

$$\omega = 2\ \left|\sin\left(\frac{qa}{2}\right)\right|\ \left(\frac{D}{M}\right)^{1/2}.$$

Note that the periodicity in $2\pi/a$ for $q$ defines the BZ of the 1D chain.

### 9.4.3 *Types of Phonons*

As noted, in a 3D cell having $N_{\mathrm{ions}}$ ions, there will be three acoustic phonons (no relative motion between ions, the whole cell vibrates) and three $(N_{\mathrm{ions}} - 3)$ optical phonons (characterized by relative displacements among ions in the same cell). The former are purely acoustic waves (like density waves) which affect electrons by modifying (linearly) the band structure with microscopic stress (they act like waves distorting the lattice constant as they travel). This effect is called "deformation potential" interaction after Bardeen and Shockley. Optical phonons in polar materials (such as GaAs) additionally carry an associated dipole field which scatters electrons in an obvious matter.
Acoustic and optical phonons can be "transverse" or "longitudinal,", depending on whether the displacement $\delta R$ (also called "polarization") is normal or parallel to the direction of travel $\mathbf{q}$.

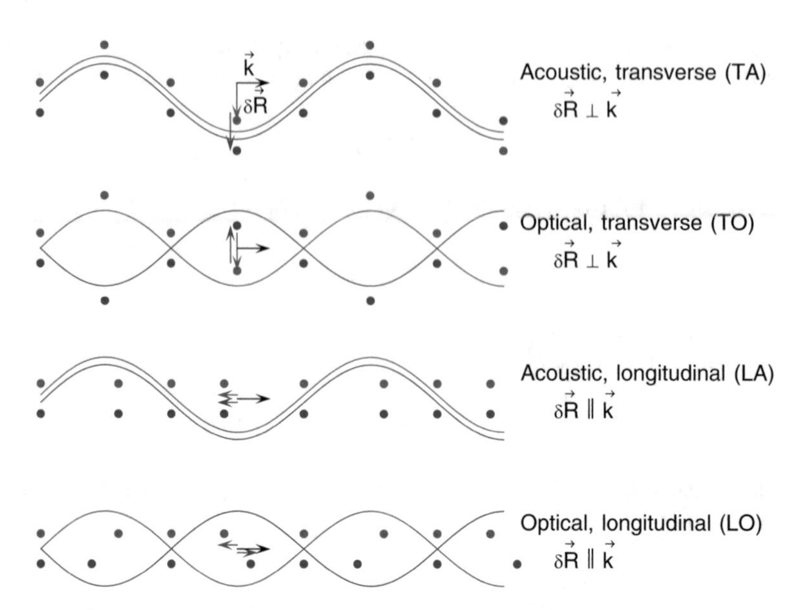


**Fig. 9.2** Schematic illustration of the atomic displacement associated with acoustic and optical, transverse and longitudinal phonons in a diatomic linear chain. Acoustic modes (*top and third lines*) involve an in-phase displacement of all ions in a cell, here denoted by *red and blue dots*. Optical modes (*second and fourth lines*) are associated with ions in a given cell moving out-of-phase. Transverse modes (*top two lines*) are associated with an ionic displacement along the normal to the direction of propagation of the wave (to the right in this diagram), whereas for longitudinal excitations (*two bottom lines*) the displacement is parallel to the direction of propagation (or wavevector)

## 9.5 Quantization of the Electromagnetic Fields: Photons

So far we have considered "excitations" of the ions + electrons system. By this we mean that in dealing with phonons, for example, we have identified the "vacuum" state $|0\rangle$ with the ground state of the ionic Hamiltonian (that is: ions at rest in their equilibrium positions). Excited states are realized by applying the creation operator to the ground state, thus obtaining states such as $|\mathbf{q}\eta\rangle = \hat{b}^{\dagger}_{\mathbf{q}\eta}|0\rangle$ describing the excitation of a phonon of polarization $\eta$ and wavevector $\mathbf{q}$. Similarly, plasmons are "collective" excitations of the electronic Hamiltonian. And, in a way, even electronic states $|\mathbf{k}\rangle = \hat{c}^{\dagger}_{\mathbf{k}}|0\rangle$ can be viewed as electrons excited from the valence band into the conduction band, if we view the state $|0\rangle$ as the state in which all electrons populate the valence band (up to the Fermi level), which is the ground state of a semiconductor crystal at zero temperature. We could take as index $\mu$ the crystal momentum $\mathbf{k}$ and a band index $n$, so that the functions $u_{\mathbf{k}n}(\mathbf{r})$ in Eq. (9.30), on page 193, would be Bloch functions in band $n$, and $|\mathbf{k}n\rangle = \hat{c}^{\dagger}_{\mathbf{k}n}|0\rangle$ would correspond to the excitation of an electron into the conduction band $n$.
In this section we temporarily abandon the environment of a crystal and consider instead the electromagnetic field in free space. The formalism will be the same, but now the ground state $|0\rangle$ will be electromagnetic vacuum. Considering the quantization of the electromagnetic field in vacuum will allow us to introduce "photons" as particles (treated on the same footing as electrons before) and will allow us to later calculate the Einstein's coefficients appearing in the matrix elements of radiative processes. Indeed, this is the first and foremost example of a field requiring "second quantization" in order to reconcile the particle/wave nature of photons.

The procedure we follow here intentionally avoids a relativistic formulation of the electromagnetic field as a gauge theory. Essentially, we follow Heitler's book [17]: Originally published by Clarendon Press in 1936, in many ways it remains useful "for all practical purposes."

### *9.5.1 Maxwell's Equations*

Consider Maxwell's equation in SI units:

$$\nabla \cdot \mathbf{D} = \rho, \tag{9.103}$$

$$\nabla \cdot \mathbf{B} = 0, \tag{9.104}$$

$$\nabla \times \mathbf{E} = -\frac{\partial \mathbf{B}}{\partial t}, \tag{9.105}$$

$$\nabla \times \mathbf{H} = \mathbf{J} + \frac{\partial \mathbf{D}}{\partial t}. \tag{9.106}$$

In vacuum and in the absence of external sources ($\mathbf{J} = 0$, $\rho = 0$), recalling that $\mathbf{D} = \epsilon \mathbf{E}$, $\mathbf{B} = \mu \mathbf{H}$, and that $\epsilon\mu = 1/c^2$, we have

$$\nabla \cdot \mathbf{E} = 0, \tag{9.107}$$

$$\nabla \cdot \mathbf{B} = 0, \tag{9.108}$$

$$\nabla \times \mathbf{E} = -\frac{\partial \mathbf{B}}{\partial t}, \tag{9.109}$$

$$\nabla \times \mathbf{B} = \frac{1}{c^2}\frac{\partial \mathbf{E}}{\partial t}. \tag{9.110}$$

### *9.5.2 Lagrangian Density*

Recalling the expression for the energy of the electromagnetic field, we could immediately write down the Hamiltonian, express it in terms of Fourier components of the fields or potentials (that is, electromagnetic plane waves) and "quantize" the result. However, let's follow the canonical procedure, starting from the definition of the Lagrangian density:

$$\mathscr{L} = \frac{\epsilon}{2}E^2 - \frac{1}{2\mu}B^2 = \frac{\epsilon}{2}(\dot{\mathbf{A}} + \nabla\phi)^2 - \frac{1}{2\mu}(\nabla \times A)^2, \tag{9.111}$$

having introduced the usual scalar and vector potentials $\phi$ and $\mathbf{A}$ such that

$$\mathbf{E} = -\dot{\mathbf{A}} - \nabla\phi, \qquad \mathbf{B} = \nabla \times \mathbf{A}. \tag{9.112}$$

Note that Eqs. (9.108) and (9.109) are automatically satisfied thanks to the definitions (9.112). The remaining Maxwell's equations, Eqs. (9.107) and (9.110), must be derived from the Euler–Lagrange equations. The first one,

$$\frac{\partial}{\partial t}\left(\frac{\partial \mathscr{L}}{\partial \dot{\phi}}\right) + \sum_j \frac{\partial}{\partial x_j}\left(\frac{\partial \mathscr{L}}{\partial(\partial\phi/\partial x_j)}\right) - \frac{\partial \mathscr{L}}{\partial \phi} = 0 \tag{9.113}$$

results in

$$\nabla^2\phi = 0, \tag{9.114}$$

which is Eq. (9.107). The second Lagrange equation, which we can write in a somewhat "sloppy" vector notation as

$$\frac{\partial}{\partial t}\left(\frac{\partial \mathscr{L}}{\partial \dot{\mathbf{A}}}\right)+\sum_j \frac{\partial}{\partial x_j}\left(\frac{\partial \mathscr{L}}{\partial(\partial \mathbf{A}/\partial x_j)}\right)-\frac{\partial \mathscr{L}}{\partial \mathbf{A}}=0, \tag{9.115}$$

results in the equation:

$$\epsilon\,\frac{\partial}{\partial t}(\dot{\mathbf{A}}+\nabla\phi)-\frac{1}{\mu}\nabla\times\nabla\times\mathbf{A}=0, \tag{9.116}$$

which is Eq. (9.110).

### 9.5.3 Hamiltonian Density

Let's now define the canonical conjugate momentum density:

$$\mathbf{P}=\frac{\partial \mathscr{L}}{\partial \dot{\mathbf{A}}}=\epsilon(\dot{\mathbf{A}}+\nabla\phi), \tag{9.117}$$

so that the Hamiltonian density will be

$$\begin{aligned}\mathscr{H}=\mathbf{P}\cdot\dot{\mathbf{A}}-\mathscr{L}&=\frac{\epsilon}{2}(\dot{\mathbf{A}}+\nabla\phi)^2+\frac{1}{2\mu}(\nabla\times\mathbf{A})^2-\epsilon(\dot{\mathbf{A}}+\nabla\phi)\cdot\nabla\phi\\&=\frac{1}{2}\left(\frac{\mathbf{P}^2}{\epsilon}+\frac{(\nabla\times\mathbf{A})^2}{\mu}\right)-\epsilon(\dot{\mathbf{A}}+\nabla\phi)\cdot\nabla\phi.\end{aligned} \tag{9.118}$$

When integrated over the entire space—in order to get the Hamiltonian—the last term in the expression above vanishes. Indeed integrating it by parts:

$$\epsilon\int(\dot{\mathbf{A}}+\nabla\phi)\cdot\nabla\phi\;d\mathbf{r}=\int\mathbf{P}\cdot\nabla\phi\;d\mathbf{r}=-\int\phi\;\nabla\cdot\mathbf{P}\,d\mathbf{r}=0, \tag{9.119}$$

by Green's (or "divergence" or "Gauss") theorem, assuming that $\mathbf{P}$ vanishes at infinity or satisfies periodic boundary conditions at the edge of the normalization volume. Therefore:

$$H=\int\mathscr{H}\;d\mathbf{r}=\frac{1}{2}\int\left[\frac{\mathbf{P}^2}{\epsilon}+\frac{(\nabla\times\mathbf{A})^2}{\mu}\right]\;d\mathbf{r}. \tag{9.120}$$

Let's now represent the potentials as plane waves of polarization $\lambda$ and wavevector $\mathbf{k}$:

$$\mathbf{A}(\mathbf{r},t)=\frac{1}{V^{1/2}}\sum_{\mathbf{k}\lambda}q_{\mathbf{k}\lambda}(t)\;\mathbf{e}_{\mathbf{k}\lambda}\;e^{i\mathbf{k}\cdot\mathbf{r}}+\mathrm{cc}, \tag{9.121}$$

and

$$\mathbf{P}(\mathbf{r},t)=\frac{1}{V^{1/2}}\sum_{\mathbf{k}\lambda}p_{\mathbf{k}\lambda}(t)\;\mathbf{e}_{\mathbf{k}\lambda}\;e^{i\mathbf{k}\cdot\mathbf{r}}+\mathrm{cc}, \tag{9.122}$$

where $V$ is the normalization volume and $\mathbf{e}_{\mathbf{k}\lambda}$ is the polarization vector. Then, using the orthonormalization of the polarization vectors, the Hamiltonian can be written in terms of the 'Fourier coefficients' $q_{\mathbf{k}\lambda}$ and $p_{\mathbf{k}\lambda}$ as follows:

$$H = \sum_{\mathbf{k}\lambda} \left( \frac{1}{\epsilon} p^*_{-\mathbf{k}\lambda} p_{\mathbf{k}\lambda} + \frac{k^2}{\mu} q^*_{-\mathbf{k}\lambda} q_{\mathbf{k}\lambda} \right). \tag{9.123}$$

Equation (9.123)—as one could have seen also earlier in Eq. (9.120)—is taking the desired shape of the Hamiltonian of a collection of harmonic oscillators. Let's now recast it in the usual form involving creation and annihilation operators.

### *9.5.4 Equations of Motion*

In order to do that, let's see what type of time dependence the quantities $q_{\mathbf{k}\lambda}$ and $p_{\mathbf{k}\lambda}$ will exhibit, after having been promoted to operators following the usual canonical quantization. The equations of motion for them can be derived by noticing that they obey the CCR:

$$[\hat{q}_{\mathbf{k}\lambda}, \hat{p}^\dagger_{\mathbf{k}'\lambda'}] = [\hat{q}^\dagger_{\mathbf{k}\lambda}, \hat{p}_{\mathbf{k}'\lambda'}] = \mathrm{i}\hbar\, \delta_{\mathbf{k}'\mathbf{k}}\, \delta_{\lambda'\lambda}. \tag{9.124}$$

Then:

$$\dot{\hat{q}}_{\mathbf{k}\lambda} = \frac{1}{\mathrm{i}\hbar} [\hat{q}_{\mathbf{k}\lambda}, H] = \frac{1}{\mathrm{i}\hbar} \sum_{\mathbf{k}'\lambda'} \frac{1}{\epsilon} [\hat{q}_{\mathbf{k}\lambda}, \hat{p}^\dagger_{\mathbf{k}'\lambda'}]\, \hat{p}_{\mathbf{k}'\lambda'} = \frac{1}{\epsilon} \sum_{\mathbf{k}'\lambda'} \delta_{\mathbf{k}'\mathbf{k}}\, \delta_{\lambda'\lambda}\, \hat{p}_{\mathbf{k}'\lambda'} = \frac{\hat{p}_{\mathbf{k}\lambda}}{\epsilon}. \tag{9.125}$$

Similarly,

$$\dot{\hat{p}}_{\mathbf{k}\lambda} = \frac{1}{\mathrm{i}\hbar} [\hat{p}_{\mathbf{k}\lambda}, H] = -\frac{k^2}{\mu}\, \hat{q}_{\mathbf{k}\lambda}. \tag{9.126}$$

Taking the time derivative of Eq. (9.125) and using Eq. (9.126), we have

$$\ddot{\hat{q}}_{\mathbf{k}\lambda} = \frac{\dot{\hat{p}}_{\mathbf{k}\lambda}}{\epsilon} = -\frac{k^2}{\epsilon\mu}\, \hat{q}_{\mathbf{k}\lambda}. \tag{9.127}$$

This has the general solution:

$$\hat{q}_{\mathbf{k}\lambda}(t) = \widehat{A}_{\mathbf{k}\lambda}\, e^{ikct} + \widehat{A}^\dagger_{\mathbf{k}\lambda}\, e^{-ikct}. \tag{9.128}$$

Inserting this result into Eq. (9.125), we obtain

$$\hat{p}_{\mathbf{k}\lambda}(t) = \mathrm{i}c\epsilon k \widehat{A}_{\mathbf{k}\lambda}\, e^{ikct} - \mathrm{i}c\epsilon k \widehat{A}^\dagger_{\mathbf{k}\lambda}\, e^{-ikct}. \tag{9.129}$$

The "coefficients" $\widehat{A}_{\mathbf{k}\lambda}$ and $\widehat{A}^\dagger_{\mathbf{k}\lambda}$ exhibit all of the properties of annihilation and creation operators except for an overall normalization. Indeed, from the CCRs (9.124) and from the definitions (9.128) and (9.129) we have

$$[\widehat{A}_{\mathbf{k}\lambda}, \widehat{A}^\dagger_{\mathbf{k}'\lambda'}] = -\frac{\hbar}{2kc\epsilon}\, \delta_{\mathbf{k}'\mathbf{k}} \delta_{\lambda\lambda'}. \tag{9.130}$$

Therefore, it is convenient to define instead:

$$\hat{q}_{\mathbf{k}\lambda}(t) = \left(\frac{\hbar}{2kc\epsilon}\right)^{1/2} [\hat{a}_{\mathbf{k}\lambda}\, e^{\mathrm{i}kct} + \hat{a}^{\dagger}_{-\mathbf{k}\lambda}\, e^{-\mathrm{i}kct}], \tag{9.131}$$

$$\hat{p}_{\mathbf{k}\lambda}(t) = \mathrm{i}k\left(\frac{\epsilon}{\mu}\right)^{1/2}\left(\frac{\hbar}{2kc\epsilon}\right)^{1/2} [\hat{a}_{\mathbf{k}\lambda}\, e^{\mathrm{i}kct} - \hat{a}^{\dagger}_{-\mathbf{k}\lambda}\, e^{-\mathrm{i}kct}], \tag{9.132}$$

or, inverting these relations:

$$\hat{a}_{\mathbf{k}\lambda} = e^{-\mathrm{i}kct}\left(\frac{kc\epsilon}{2\hbar}\right)^{1/2}\left[\hat{q}_{\mathbf{k}\lambda} - \frac{\mathrm{i}}{k}\left(\frac{\mu}{\epsilon}\right)^{1/2}\hat{p}_{-\mathbf{k}\lambda}\right], \tag{9.133}$$

and

$$\hat{a}^{\dagger}_{\mathbf{k}\lambda} = e^{\mathrm{i}kct}\left(\frac{kc\epsilon}{2\hbar}\right)^{1/2}\left[\hat{q}_{\mathbf{k}\lambda} + \frac{\mathrm{i}}{k}\left(\frac{\mu}{\epsilon}\right)^{1/2}\hat{p}_{-\mathbf{k}\lambda}\right], \tag{9.134}$$

Then, the CCRs become

$$[\hat{a}_{\mathbf{k}\lambda}, \hat{a}^{\dagger}_{\mathbf{k}'\lambda'}] = \delta_{\mathbf{k}'\mathbf{k}}\delta_{\lambda\lambda'}, \tag{9.135}$$

and the Hamiltonian (9.123) becomes

$$\widehat{H} = \sum_{\mathbf{k}\lambda} \hbar ck\left[\hat{a}^{\dagger}_{\mathbf{k}\lambda}\hat{a}_{\mathbf{k}\lambda} + \frac{1}{2}\right] \rightarrow \sum_{\mathbf{k}\lambda} \hbar\omega_{\mathbf{k}}\, \hat{a}^{\dagger}_{\mathbf{k}\lambda}\hat{a}_{\mathbf{k}\lambda}, \tag{9.136}$$

having ignored the "usual" diverging zero-point term and having used the fact that $kc = \omega_{\mathbf{k}}$ is the energy of a photon of frequency $\nu = \omega_{\mathbf{k}}/(2\pi)$.

### 9.5.5 The Quantum Field and the Importance of Second Quantization

In addition to the many reasons why second quantization is necessary, we wish to point out another aspect that is not commonly emphasized. From Eq. (9.131) we see how the "position" operator $\hat{q}_{\mathbf{k}\lambda}$ can be expressed in terms of the single-photon operators $\hat{a}_{\mathbf{k}\lambda}$ and $\hat{a}^{\dagger}_{\mathbf{k}\lambda}$. Similarly, Eq. (9.132) relates the "momentum" operator $\hat{p}_{\mathbf{k}\lambda}$ to the same quantities. In turn, Eqs. (9.121) and (9.122) tell us how to express the vector and scalar potentials (via the momentum operator $\mathbf{P}_{\mathbf{k}\lambda}$) in terms of the operators $\hat{q}_{\mathbf{k}\lambda}$ and $\hat{p}_{\mathbf{k}\lambda}$. As a result, we can express the vector and scalar potentials in terms of $\hat{a}_{\mathbf{k}\lambda}$ and $\hat{a}^{\dagger}_{\mathbf{k}\lambda}$. For example:

$$\widehat{\mathbf{A}}(\mathbf{r},t) = \frac{1}{V}\sum_{\mathbf{k}\lambda}\left(\frac{\hbar}{2kc\epsilon}\right)^{1/2} [\hat{a}_{\mathbf{k}\lambda}\, e^{\mathrm{i}kct} + \hat{a}^{\dagger}_{-\mathbf{k}\lambda}]\ \mathbf{e}_{\mathbf{k}\lambda}\, e^{-\mathrm{i}kct}e^{\mathrm{i}\mathbf{k}\cdot\mathbf{r}}. \tag{9.137}$$

From this—and the corresponding equation for $\dot{\widehat{\mathbf{A}}} + \nabla\hat{\phi}$—we can derive the amplitude of the electric and magnetic fields:

$$\widehat{\mathbf{B}}(\mathbf{r},t) = \nabla \times \widehat{\mathbf{A}}(\mathbf{r},t) = \frac{\mathrm{i}}{V} \sum_{\mathbf{k}\lambda} \left( \frac{\hbar}{2kc\epsilon} \right)^{1/2} [\hat{a}_{\mathbf{k}\lambda}\, e^{ikct} + \hat{a}^{\dagger}_{-\mathbf{k}\lambda}\, e^{-ikct}]\; (\mathbf{k} \times \mathbf{e}_{\mathbf{k}\lambda})\, e^{\mathrm{i}\mathbf{k}\cdot\mathbf{r}}, \tag{9.138}$$

and

$$\widehat{\mathbf{E}}(\mathbf{r},t) = \frac{\widehat{\mathbf{P}}(\mathbf{r},t)}{\epsilon} = \frac{\mathrm{i}}{V} \sum_{\mathbf{k}\lambda} \frac{k}{\epsilon} \left( \frac{\hbar}{2kc\mu} \right)^{1/2} [\hat{a}_{\mathbf{k}\lambda}\, e^{ikct} - \hat{a}^{\dagger}_{-\mathbf{k}\lambda}\, e^{-ikct}]\; \mathbf{e}_{\mathbf{k}\lambda}\, e^{\mathrm{i}\mathbf{k}\cdot\mathbf{r}}. \tag{9.139}$$

*This is the basic result of second quantization: It tells us the amplitude of the field associated to a single quantum of excitation.* Without second quantization, how could we know the strength of the electric field associated to a photon? Similar relations apply to electrons: Eq. (9.80) gives us the amplitude of the (scalar) potential associated to a single quantum of plasma excitations. Similarly, Eq. (9.100), together with Eq. (9.96), allows us to calculate the displacement of each ion in terms of the phonon creation and annihilation operators:

$$\delta\widehat{\mathbf{R}}^{(\eta)}_{l\gamma} = \frac{1}{(NM_{\gamma})^{1/2}} \sum_{\mathbf{q}\eta} \mathbf{e}^{(\eta)}_{\mathbf{q}\gamma} \left( \frac{\hbar}{2\omega_{\mathbf{q}\eta}} \right)^{1/2} (\hat{b}_{\mathbf{q}\eta} + \hat{b}^{\dagger}_{-\mathbf{q}\eta})\, e^{\mathrm{i}\mathbf{q}\cdot\mathbf{R}_l}. \tag{9.140}$$

The algebraic rules allowing us to evaluate easily expectation values of the creation and annihilation operators on the ground state $|0\rangle$ will ultimately allow us to calculate the strength of interactions between electrons and these excitations (phonons, plasmons, photons), a result which we could not have accomplished without the information, given by second quantization, regarding the strength of the field carried by a single excitation.

## Problems

**9.1.** Consider spinless particles in free space described by the Schrödinger field (so that the creation and annihilation operators satisfy the canonical *commutation* rules). Consider the total number operator

$$\widehat{N} = \sum_{\mathbf{k}} \hat{n}_{\mathbf{k}} = \sum_{\mathbf{k}} \hat{c}^{\dagger}_{\mathbf{k}} \hat{c}_{\mathbf{k}}.$$

Show that $\widehat{N}$ is a constant of motion, that is, that

$$\mathrm{i}\hbar\dot{\widehat{N}} = [\widehat{N}, \widehat{H}] = 0.$$

For $\widehat{H}$ you should use Eq. (9.31) and you must also show first that Eq. (9.31) follows from the spatial integration of the Hamiltonian density, Eq. (9.24).

**9.2.** A vector of Fock space representing $n_{\mathbf{k}}$ bosons with wavevector $\mathbf{k}$ can be written as

$$|N_{\mathbf{k}}\rangle = (\hat{c}^{\dagger}_{\mathbf{k}})^{N_{\mathbf{k}}} |0\rangle.$$

**(a)** Show that the norm $\alpha$ of $|N_{\mathbf{k}}\rangle$ is given by $\alpha = N_{\mathbf{k}}!^{1/2}$.

**(b)** Consider the "position" and "momentum" operators $\hat{q}_{\mathbf{k}}$ and $\hat{p}_{\mathbf{k}}$ given by Eq. (9.42) (with $\mu = \mathbf{k}$). Show that, with the proper normalization:

$$\langle\langle N_{\mathbf{k}}|\hat{q}_{\mathbf{k}}|N_{\mathbf{k}}+1\rangle\rangle_{\mathrm{th}} = \langle\langle N_{\mathbf{k}}+1|\hat{q}_{\mathbf{k}}|N_{\mathbf{k}}\rangle\rangle_{\mathrm{th}} = \left(\frac{E_{\mathbf{k}}}{2}\right)^{1/2} (\langle N_{\mathbf{k}}\rangle_{\mathrm{th}}+1)^{1/2},$$

and that

$$\langle\langle N_{\mathbf{k}}|\hat{p}_{\mathbf{k}}|N_{\mathbf{k}}+1\rangle\rangle_{\mathrm{th}} = -\ \langle\langle N_{\mathbf{k}}+1|\hat{p}_{\mathbf{k}}|N_{\mathbf{k}}\rangle\rangle_{\mathrm{th}} = -\mathrm{i}\left(\frac{E_{\mathbf{k}}}{2}\right)^{1/2} (\langle N_{\mathbf{k}}\rangle_{\mathrm{th}}+1)^{1/2},$$

all other matrix element vanishing.

**9.3.** Why does the energy $\hbar\omega_{\mathbf{q}\eta}$ of the acoustic branch of phonons goes to zero as $q \to 0$? What is the physical reason?

**9.4.** Dealing with the quantization of the electromagnetic field, we have considered "free space"; that is, we have assumed no charge or current, $\rho = 0$ and $\mathbf{j} = 0$. Now show that in the presence of charges and currents, the Lagrangian density has the form

$$\mathscr{L} = \frac{\epsilon E^2}{2} + \frac{B^2}{2\mu} + \alpha\,\rho\phi + \beta\,\mathbf{j}\cdot\mathbf{A}.$$

Determine the constants $\alpha$ and $\beta$ so that the Euler–Lagrange equations yield Maxwell's equations, Eqs. (9.103) and (9.106).

## References

1. N. Bogoliubov, On the theory of superfluidity. J. Phys. (USSR) **11**, 23 (1947)
2. J.-P. Blaizot, G. Ripka, *Quantum Theory of Finite Systems* (MIT Press, Cambridge, MA, 1985)
3. D. Lurié, *Particles and Fields* (Wiley, New York, 1968)
4. L.I. Schiff, *Quantum Mechanics* (McGraw-Hill, New York, 1968)
5. A.L. Fetter, J.D. Walecka, *Quantum Theory of Many-Particle Systems* (Dover, Mineola, NY, 2003)
6. G.D. Mahan, *Many-Particle Physics (Physics of Solids and Liquids)*, 3rd edn. (Kluwer, Boston, 2000)
7. R.F. Streater, A.S. Wightman, *Spin and Statistics, and All That* (Princeton University Press, Princeton, NJ, 2000)
8. R. Haag, On quantum field theories. Matematisk-fysiske Meddelelser **29**, 12 (1955)
9. D. Hall, A.S. Wightman, A theorem on invariant analytic functions with applications to relativistic quantum field theory. Mat. Fys. Medd. Dan. Vid. **31**, 1 (1957)
10. E.A. Stern, R.A. Ferrel, Surface plasma oscillations of a degenerate electron gas. Phys. Rev. **120**, 130 (1960)
11. J. Ziman, *Electrons and Phonons: The Theory of Transport Phenomena in Solids* (Oxford University Press, Oxford, 1960)
12. R.B. Leighton, The vibrational spectrum and specific heat of a face-centered cubic crystal. Rev. Mod. Phys. **20**, 165 (1948)
13. P.N. Keating, Effect of invariance requirements on the elastic strain energy of crystals with application to the diamond structure. Phys. Rev. **145**, 637 (1966)
14. B.G. Dick Jr., A.W. Overhauser, Theory of the dielectric constants of alkali halide crystals. Phys. Rev. **112**, 90 (1958)
15. K. Kunc, O.H. Nielsen, Phys. Commun. **17**, 47 (1979)
16. O.H. Nielsen, W. Weber, Phys. Commun. **18**, 101 (1979)
17. W. Heitler, *The Quantum Theory of Radiation*, 3rd edn. (Dover, Mineola, NY, 2010)

# Chapter 10
# Elements of Quantum Statistical Mechanics

## 10.1 Quantum Statistical Mechanics

With a few exceptions, up to this point we have considered quantum systems in *pure* states. That is, we have considered the system to be in a particular state associated with either a wavefunction $\psi$ or a particular state in Fock space, $|n_{\mathbf{k}}\rangle$. In practice, such a characterization applies only to systems simple enough to give us a realistic hope of measuring accurately in which state the system finds itself. The electronic state of an isolated hydrogen atom or the spin of a single electron may be examples of such simple systems. However, as soon as the system becomes sufficiently large (that is, it has many degrees of freedom), such a description becomes "unrealistic."

On the one hand we can never hope to measure accurately the position and velocity (or, quantum mechanically, the wavefunctions) of all molecules in a macroscopic volume of gas, or of all electrons and ions in a semiconductor chip. This "admission of ignorance" forces us to describe such large systems in a statistical—rather than deterministic—fashion. On the other hand, while one may consider such an "admission of ignorance" a practical consideration which, in principle, should not affect the deep nature of how we describe the system, it is probably correct to regard such "ignorance" an intrinsic property of large, macroscopic systems [1]. It is such ignorance, after all, which forces us to lose information about the system, thus introducing irreversibility into our description of the physical world. And the irreversible flowing of time can hardly be considered a "practical consideration" which, in principle, could be ignored when describing the world!

In order to describe macroscopic systems, "Statistical Mechanics" has been developed starting from Maxwell, Gibbs, Boltzmann, and moving into "Quantum Statistical Mechanics" (Einstein's early "quantum" work). We cannot discuss it here as accurately and deeply as the subject deserves. However, in order to understand the deep problems surrounding the issue of charge transport in solids (an irreversible process: your laptop gets hot and heat cannot be fully recovered into more ordered forms of energy, so the second law of Thermodynamics teaches us), we must at least mention some concepts of Quantum Statistical Mechanics. Texts such as Jancel's [2] and the "encyclopedic" volumes by Zubarev and collaborators [3, 4] should be consulted by interested readers.



M. Fischetti, W.G. Vandenberghe, *Advanced Physics of Electron Transport in Semiconductors and Nanostructures*, Graduate Texts in Physics, DOI 10.1007/978-3-319-01101-1_10

### 10.1.1 Statistical Ensembles

The basic idea needed to account for our "ignorance" consists in introducing the concept of a "statistical ensemble" (originally devised by Gibbs). Rather than considering a single chunk of semiconductor, we consider an ideal collection of $N$ *identically prepared* copies of our original chunk. In conventional (i.e., deterministic) mechanics, classical or quantum, we describe with infinite accuracy the state of the system by specifying, for example, in which of the possible wavefunctions $\Psi_\alpha(\mathbf{r}_1, \mathbf{r}_2, \ldots, \mathbf{r}_{N_p})$ (states labeled by an index $\alpha$—which can vary from 1 to $N_{\text{states}}$—and functions of the coordinates $\mathbf{r}_i$ $(i = 1, N_p)$ of the particles constituting the system) we shall find the system. Thus, we specify a particular $\alpha$.

In Statistical Mechanics, instead, we consider a large number of systems $N_{\text{systems}}$ and subsequently specify the number of systems in state $\alpha$, $n_\alpha$. Equivalently, we shall talk of one system and the probability $P_\alpha = n_\alpha / N_{\text{systems}}$ of finding the system in the state $\alpha$.

Most commonly, three ensembles are considered:

- The microcanonical ensemble describing a system with fixed volume in which all possible states with a given energy $E$ and number of particles $N$ have equal probability of being occupied.
- The canonical ensemble describing a system with fixed volume and fixed number of particles $N$ but with a given *expectation value* for the energy $\langle E \rangle$ of the ensemble determined by the temperature ($T$).
- The grand canonical ensemble describing a system with fixed volume but with a given *expectation value* for the number of particles $\langle N \rangle$ and the energy of the ensemble determined by the chemical potential ($\mu$) and the temperature, respectively.

### 10.1.2 Density Matrix

The properties of an ensemble are determined by the states in the system $\Psi_\alpha$ together with the probability $P_\alpha = n_\alpha / N_{\text{systems}}$ that the system is in this state. However, it is convenient to combine the probability and the wavefunction into a single entity "the density matrix" which is defined as

$$\rho(\mathbf{r}, \mathbf{r}') = \sum_\alpha \Psi^*_\alpha(\mathbf{r}') P_\alpha \Psi_\alpha(\mathbf{r}) \tag{10.1}$$

in real-space form in first quantization or

$$\hat{\rho} = \sum_\alpha \mid \alpha \rangle P_\alpha \langle \alpha \mid \tag{10.2}$$

in first or second quantization in bra-ket notation.

The density matrix behaves as an operator yielding the probability that a system is in a given state as

$$\hat{\rho} \mid \alpha \rangle = P_\alpha \mid \alpha \rangle . \tag{10.3}$$

All probabilities, which are the eigenvalues of the density matrix, have to be numbers between 0 and 1 and the sum of all probabilities has to equal one, which results in the property $\text{Tr}(\hat{\rho}) = 1$. The overall expectation value of an observable $\widehat{O}$ for an ensemble is given by

$$\langle \widehat{O} \rangle = \sum_{\alpha} P_{\alpha} \langle \alpha \mid \widehat{O} \mid \alpha \rangle = \sum_{\alpha\alpha'} P_{\alpha} \langle \alpha \mid \alpha' \rangle \langle \alpha' \mid \widehat{O} \mid \alpha \rangle \tag{10.4}$$

$$= \sum_{\alpha} \langle \alpha \mid \hat{\rho} \widehat{O} \mid \alpha \rangle \tag{10.5}$$

$$= \mathrm{Tr}(\hat{\rho} \widehat{O}) \; . \tag{10.6}$$

where we have used the orthogonality $P_{\alpha/\alpha'} \langle \alpha \mid \alpha' \rangle = P_{\alpha} \delta_{\alpha\alpha'}$ and the completeness of the basis $\alpha$: $\sum_{\alpha} \mid \alpha \rangle \langle \alpha \mid = 1$.

Equation (10.6) can be applied in the single-particle picture where $\hat{\rho}$ and $\widehat{O}$ are conventional matrices and the trace is the sum of the diagonal elements. But Eq. (10.6) applies equally well in the second quantized picture where $\hat{\rho}$ and $\widehat{O}$ are operators acting on Fock space and the trace is defined in Eq. (10.5) and the sum runs over a complete set of states spanning Fock space.

The amount of information contained in a quantum ensemble is given by its von Neumann entropy [5]:

$$S = -k_{\mathrm{B}} \mathrm{Tr}(\hat{\rho} \log(\hat{\rho})) \tag{10.7}$$

and using the definition of the density matrix, the entropy can be rewritten using the probabilities for each state to occur

$$S = -k_{\mathrm{B}} \sum_{\alpha} P_{\alpha} \log P_{\alpha} \; . \tag{10.8}$$

When $k_{\mathrm{B}}$ is taken to be unity, Eq. (10.8) is Shannon's entropy from information theory. Equation (10.8) is also Gibbs' entropy from Classical Mechanics when $P_{\alpha}$ is taken as the probability of a state occurring in a classical ensemble.

If a system is in a pure state $\Psi_{\alpha}$, the system is in an extremely ordered state and there is no real need to invoke an ensemble. Nevertheless, an ensemble describing a pure state has $P_{\alpha'} = \delta_{\alpha\alpha'}$, $P_{\alpha'} \log P_{\alpha'} = 0$ and an entropy $S = 0$. On the other hand, if the system is in a completely random state $P_{\alpha} = 1/N_{\text{systems}}$ and the entropy reaches a maximum $S = \log(N_{\text{systems}})$ where $N_{\text{systems}}$ measures the number of degrees of freedom in the system. In Sect. 10.1.4, we will derive the grand canonical ensemble as an ensemble with only two constraints: a fixed expectation value of its energy and a fixed expectation value of the number of particles in the system. All other degrees of freedom are unrestricted and any state which satisfies both constraints is equally probable. Mathematically, the grand canonical ensemble is the ensemble with *maximizing the entropy* whilst satisfying both constraints.

### *10.1.3 Examples of Density Matrices*

To familiarize ourselves with the density matrix, we consider the spin of an electron. For the electron, there are two are spin states available, $|+\rangle$ and $|-\rangle$, spin up and spin down along the $z$-axis. The density matrix

$$\hat{\rho} = \begin{bmatrix} 1/2 & 0 \\ 0 & 1/2 \end{bmatrix} , \tag{10.9}$$

corresponds to an ensemble of totally unpolarized electrons ($P_{+} = P_{-} = 0.5$). The ensemble spin expectation value will be

$$\langle \sigma_z \rangle = \mathrm{Tr}[\hat{\sigma}_z \hat{\rho}] = \frac{1}{2} \mathrm{Tr}[\hat{\sigma}_z] = 0 \; , \tag{10.10}$$

as well as $\langle \sigma_{x,y} \rangle = 0$ recalling that

$$\hat{\sigma}_x = \begin{bmatrix} 0 & 1 \\ 1 & 0 \end{bmatrix} \quad , \quad \hat{\sigma}_y = \begin{bmatrix} 0 & \mathrm{i} \\ -\mathrm{i} & 0 \end{bmatrix} \quad , \quad \hat{\sigma}_z = \begin{bmatrix} 1 & 0 \\ 0 & -1 \end{bmatrix} . \tag{10.11}$$

On the other hand, the density matrix

$$\hat{\rho} = \begin{bmatrix} 1 & 0 \\ 0 & 0 \end{bmatrix} , \tag{10.12}$$

corresponds to an ensemble of completely polarized, spin-up electrons. The ensemble average of the spin will be

$$\langle \sigma_z \rangle = \mathrm{Tr}[\hat{\sigma}_z \hat{\rho}] = \sigma_{z,11} = 1 . \tag{10.13}$$

Thus, the electrons are all in pure spin-up states.

Now let's consider our electrons as being in a (coherent) superposition of spin-up and spin-down states. The state vector corresponding to this situation is

$$|\mathbf{s}\rangle = \frac{1}{\sqrt{2}} \left[(1,0) + (0,1)\right] = \frac{1}{\sqrt{2}} \, (1,1) . \tag{10.14}$$

The density matrix built out of this state will be

$$\hat{\rho} = \mathbf{s}^* \mathbf{s} = \frac{1}{2} \begin{bmatrix} 1 & 1 \\ 1 & 1 \end{bmatrix} , \tag{10.15}$$

$\langle \sigma_x \rangle = 1$ and $\langle \sigma_y \rangle = \langle \sigma_z \rangle = 0$.

Note that, unlike in Eqs. (10.9) and (10.12) considered above, the off-diagonal elements are nonzero in Eq. (10.15). The off-diagonal elements represent the fact that the system is in a coherent superposition of quantum states. This has no classical analogue and is a feature of Quantum Mechanics.

### *10.1.4 The Grand Canonical Ensemble*

Without any knowledge about a system, there is no preference with regard to the system being in one state compared to another and each state is assumed to be occupied by a particle with equal probability: $p_\alpha = p_{\alpha'}$. However, some knowledge is available: an isolated system is subject to the conservation of energy. Even if our description of the system and its Hamiltonian is only approximate, the total energy of the system must remain unchanged. The ensemble describing a closed system of which the total energy is known is called the microcanonical ensemble and has $p_\alpha = \delta_{E_\alpha,E}/N_E$ where $E$ is the energy of the microcanonical ensemble, $E_\alpha$ is the energy of state $\alpha$, $\delta_{E_\alpha,E}$ equals 1 when $E = E_\alpha$ and 0 otherwise, and $N_\alpha = \sum_\alpha \delta_{E_\alpha,E}$ the total number of states with energy $E$.

Considering an electrical contact, however, the assumption that our contact is an isolated system is no longer valid because there is exchange of energy, e.g., with phonons, and exchange of particles, e.g., with another contact. If our contact is connected to—or is a part of—a large reservoir, energy and particles can flow back and forth between the contact and the reservoir. Such a system is described by the grand canonical ensemble.

If the contact and the reservoir are in equilibrium, the expectation value of the energy and the number of particles flowing out of the contact into the reservoir vanish ($\mathrm{d}\langle E\rangle/\mathrm{d}t = \mathrm{d}\langle N\rangle/\mathrm{d}t = 0$). The microcanonical requirement that the total energy of the system $E$ is known is now replaced by the requirement that the expectation value of the total energy $\langle E\rangle$ is known. For the microcanonical ensemble we implicitly assumed the number of particles remained unchanged but for our contact described by the grand canonical ensemble, the expectation value on the number of particles $\langle N\rangle$ is taken to be known.

An expression for the grand canonical ensemble ($\rho_{\mathrm{GC}}$) can be obtained by finding the ensemble which maximizes the entropy while satisfying $\mathrm{Tr}(\widehat{H}\hat{\rho}) = \langle E\rangle$ and $\mathrm{Tr}(\widehat{N}\hat{\rho}) = \langle N\rangle$. Any density matrix further has the property $\mathrm{Tr}(\hat{\rho}) = 1$.

Mathematically, the entropy maximum can be found by imposing $\delta S = 0$ while introducing Lagrange multipliers ($\beta_0$, $\beta_E$, and $\beta_N$) for the different constraints

$$\delta \mathrm{Tr}\left(\beta_0\hat{\rho} + \beta_E\widehat{H}\hat{\rho} + \beta_N\widehat{N}\hat{\rho} - k_{\mathrm{B}}\hat{\rho}\log(\hat{\rho})\right) = 0, \text{or}, \tag{10.16}$$

$$\mathrm{Tr}\left(\left(\beta_0\widehat{I} + \beta_E\widehat{H} + \beta_N\widehat{N} - k_{\mathrm{B}}\log(\hat{\rho}) - k_{\mathrm{B}}\widehat{I}\right)\delta\hat{\rho}\right) = 0 \tag{10.17}$$

for any $\delta\hat{\rho}$, which requires

$$\beta_0\widehat{I} + \beta_E\widehat{H} + \beta_N\widehat{N} - k_{\mathrm{B}}\log(\hat{\rho}) - k_{\mathrm{B}}\widehat{I} = 0. \tag{10.18}$$

Solving for $\hat{\rho}$ and relabeling the Lagrange multipliers gives the grand canonical ensemble in its usual form

$$\hat{\rho} = \exp\left(-\beta(\widehat{H} - \mu\widehat{N})\right)/\mathscr{Z} \tag{10.19}$$

The Lagrange multipliers have been relabeled: $\beta = \beta_E/k_{\mathrm{B}} = 1/k_{\mathrm{B}}T$, $\mu = \beta_N/\beta_E$, and $\beta_0$ ensure that $\mathrm{Tr}(\hat{\rho}) = 1$ which is accomplished by introducing the partition function

$$\mathscr{Z} = \mathrm{Tr}\left(\exp\left(-\beta(\widehat{H} - \mu\widehat{N})\right)\right). \tag{10.20}$$

If the Hamiltonian is diagonalized by the operators $\hat{c}_{\mathbf{k}}$, the grand canonical ensemble is given by

$$\hat{\rho} = 1/\mathscr{Z}\prod_{\mathbf{k}}\exp\left(-\beta(E_{\mathbf{k}} - \mu)\hat{c}_{\mathbf{k}}^{\dagger}\hat{c}_{\mathbf{k}}\right). \tag{10.21}$$

### 10.1.5 Distribution Functions

The probability that a single-particle state $\mathbf{k}$ is occupied in a many-particle system described by an ensemble $\hat{\rho}$ is called the distribution function and is defined as

$$f(\mathbf{k}) = \mathrm{Tr}(\hat{c}_{\mathbf{k}}^{\dagger}\hat{c}_{\mathbf{k}}\hat{\rho}). \tag{10.22}$$

Equation (10.22) can be evaluated for the grand canonical ensemble using some algebra. A first relation valid for any ensemble follows from the commutation/anticommutation properties of bosons/fermions

$$\mathrm{Tr}(\hat{c}_{\mathbf{k}}\hat{c}_{\mathbf{k}}^{\dagger}\hat{\rho}) \mp \mathrm{Tr}(\hat{c}_{\mathbf{k}}^{\dagger}\hat{c}_{\mathbf{k}}\hat{\rho}) = 1. \tag{10.23}$$

For Fermions, Eq. (10.23) is an expression of Pauli's principle as it states that the sum of the probabilities of finding an electron and not finding an electron (or finding a hole) $\mathrm{Tr}(\hat{c}_{\mathbf{k}}\hat{c}^\dagger_{\mathbf{k}}\hat{\rho})$ equals unity. Invoking the invariance of the trace with respect to cyclic permutation gives $\mathrm{Tr}(\hat{c}_{\mathbf{k}}\hat{c}^\dagger_{\mathbf{k}}\hat{\rho}) = \mathrm{Tr}(\hat{c}^\dagger_{\mathbf{k}}\hat{\rho}\hat{c}_{\mathbf{k}})$ and we proceed by writing $\mathrm{Tr}(\hat{c}^\dagger_{\mathbf{k}}\hat{\rho}\hat{c}_{\mathbf{k}})$ as a function of $\mathrm{Tr}(\hat{c}^\dagger_{\mathbf{k}}\hat{c}_{\mathbf{k}}\hat{\rho})$.

Keeping the grand canonical ensemble in mind, the commutation/anticommutation relations can be applied inside the exponential to yield a second relation

$$\exp\left(-\beta(E_{\mathbf{k}}-\mu)\hat{c}_{\mathbf{k}}\hat{c}^\dagger_{\mathbf{k}}\right) = \exp\left(-\beta(E_{\mathbf{k}}-\mu)\right)\exp\left(\mp\beta(E_{\mathbf{k}}-\mu)\hat{c}^\dagger_{\mathbf{k}}\hat{c}_{\mathbf{k}}\right). \tag{10.24}$$

From the series expansion of the exponential $\exp(\hat{c}\hat{c}^\dagger) = \hat{c}\hat{c}^\dagger + \hat{c}\hat{c}^\dagger\hat{c}\hat{c}^\dagger/2 + \cdots$, it follows that $\exp(\hat{c}\hat{c}^\dagger)\hat{c} = \hat{c}\exp(\hat{c}^\dagger\hat{c})$ and furthermore for fermions $\hat{c}\hat{c} = 0$ and thus $\exp(\pm\hat{c}^\dagger\hat{c})\hat{c} = \hat{c}$. Using the diagonal representation of the grand canonical ensemble [Eq. (10.21)], knowing that $\hat{c}_{\mathbf{k}}$ commutes with $\hat{c}_{\mathbf{k}'}$ for $\mathbf{k} \neq \mathbf{k}'$ and using Eq. (10.24)

$$\hat{\rho}\hat{c}_{\mathbf{k}} = \exp(\beta(E_{\mathbf{k}}-\mu))\hat{c}_{\mathbf{k}}\hat{\rho}. \tag{10.25}$$

resulting in the relation

$$\mathrm{Tr}(\hat{c}_{\mathbf{k}}\hat{c}^\dagger_{\mathbf{k}}\hat{\rho}) = \exp(\beta(E_{\mathbf{k}}-\mu))\mathrm{Tr}(\hat{c}^\dagger_{\mathbf{k}}\hat{c}_{\mathbf{k}}\hat{\rho}). \tag{10.26}$$

Substituting Eq. (10.26) into Eq. (10.23) results in the Bose–Einstein distribution for Bosons and the Fermi–Dirac distribution for Fermions

$$f(\mathbf{k}) = \frac{1}{e^{\beta(E_{\mathbf{k}}-\mu)} \mp 1}. \tag{10.27}$$

Note that the distribution is only a function of the energy and it is more common to write $f(E)$ compared to $f(\mathbf{k})$.

In practice, formally using the ensemble of a many-particle system is a tedious process and the distribution function provides a much more convenient entity. Moreover, the distribution function contains sufficient information to evaluate the expectation value of all single-particle operators. Considering a single-particle operator

$$\widehat{O} = \sum_{\mathbf{k}\mathbf{k}'}\langle\mathbf{k} \mid \widehat{O} \mid \mathbf{k}'\rangle\hat{c}^\dagger_{\mathbf{k}}\hat{c}_{\mathbf{k}'}, \tag{10.28}$$

the expectation value of the ensemble is

$$\langle\widehat{O}\rangle = \mathrm{Tr}(\hat{\rho}\widehat{O}) = \sum_{\mathbf{k}\mathbf{k}'}\langle\mathbf{k} \mid \widehat{O} \mid \mathbf{k}'\rangle\mathrm{Tr}(\hat{\rho}\hat{c}_{\mathbf{k}}\hat{c}'_{\mathbf{k}}) \tag{10.29}$$

$$= \sum_{\mathbf{k}} f(\mathbf{k})\langle\mathbf{k} \mid \widehat{O} \mid \mathbf{k}\rangle. \tag{10.30}$$

The last step is a consequence of $\rho$ being diagonalized by the $\mathbf{k}$ basis. For example, in real-space notation the charge density can be computed as

$$\rho(\mathbf{r}) = \sum_{\mathbf{k}} f(\mathbf{k})|\psi_{\mathbf{k}}(\mathbf{r})|^2 \tag{10.31}$$

where $\psi_{\mathbf{k}}(\mathbf{r})$ is the solution of the single-particle Schrödinger equation.

### 10.1.6 Equation of Motion for the Density Matrix

When the eigenstates of the ensemble are not eigenstates of the Hamiltonian of the system, the ensemble will evolve over time. The evolution is determined by the time-dependent Schrödinger equation

$$\mathrm{i}\hbar\frac{\partial\Psi}{\partial t} = \widehat{H}\,\Psi\,. \tag{10.32}$$

The time dependence of the density matrix is easily derived:

$$\begin{aligned}\mathrm{i}\hbar\frac{\partial\hat{\rho}}{\partial t} &= \mathrm{i}\hbar\sum_{\alpha}\left(\frac{\partial|\alpha\rangle}{\partial t}p_{\alpha}\langle\alpha| + |\alpha\rangle p_{\alpha}\frac{\partial\langle\alpha|}{\partial t}\right)\\ &= (\widehat{H}|\alpha\rangle p_{\alpha}\langle\alpha| - |\alpha\rangle p_{\alpha}\langle\alpha|\widehat{H} = [\widehat{H},\hat{\rho}]\end{aligned} \tag{10.33}$$

where, as usual, $[\widehat{A},\widehat{B}] = \widehat{A}\widehat{B} - \widehat{B}\widehat{A}$ the commutator.

This equation has the same form of a famous equation in classical Statistical Mechanics, the "Liouville" equation, in which the commutator is obviously replaced by a Poisson bracket and the density matrix is replaced by the classical distribution function. This quantum mechanical analogue is known as the *Liouville–von Neumann* equation. Note two main considerations which differentiate the time evolution of the density matrix $\hat{\rho}$ from that of an observable:

1. The time evolution of an observable $\widehat{Q}$ is given by Heisenberg's equation

$$[\widehat{Q},\widehat{H}] = i\hbar\frac{\partial\widehat{Q}}{\partial t}\,. \tag{10.34}$$

   Note the sign difference comparing with Eq. (10.33).
2. Observables $\widehat{Q}$ evolve in time in the Heisenberg representation, but are constant in the Schrödinger representation. The density matrix $\hat{\rho}$ behaves in the opposite way: It is constant in the Heisenberg representation, it evolves in time in the Schrödinger representation.

For these two reasons it is customary to call $\hat{\rho}$ a "matrix" rather than an "operator." It is, of course, an operator mathematically speaking. Yet, it does not behave exactly as an observable.

## References


1. I. Prigogine, *From Being to Becoming* (Freeman, New York, 1980)
2. R. Jancel, D. ter Haar, *Foundations of Classical and Quantum Statistical Mechanics* (Elsevier, Amsterdam, 1963)
3. D.N. Zubarev, V. Morozov, G. Ropke, *Statistical Mechanics of Nonequilibrium Processes: Basic Concepts, Kinetic Theory* (Wiley, New York, 1996)
4. D.N. Zubarev, V. Morozov, G. Ropke, *Statistical Mechanics of Nonequilibrium Processes: Relaxation and Hydrodynamic Processes* (Wiley, New York, 1997)
5. L.E. Reichl, *A Modern Course in Statistical Physics* (Wiley-VCH Verlag GmbH & Co. KGaA, Weinheim, 2009)

# Chapter 11
# Dielectric Properties of Semiconductors

## 11.1 Outline

In this section we discuss the response of the semiconductor to a perturbing potential. The practical reason which forces us to discuss this subject is, of course, our need to handle the problem of carriers which scatter with the external potential. Unfortunately, any external perturbation causes not only the "single-particle" scattering process we are interested in, but also a "many-body" rearrangement of all electrons, valence and conduction, as well as of the ions (especially if the crystal is ionic). Thus, we must consider both the electronic and the ionic dielectric properties. This rearrangement modifies the scattering potential, so that the scattering event depends on these many-body effects as well.

In introductory textbooks of electromagnetic theory, the effect is "lumped" into a dielectric constant of the medium: If **E** is the "external" electric field applied to a dielectric medium, this field will induce a redistribution of the charges in the medium which will give rise to an additional polarization field **P**. Assuming a linear relation between the perturbation **E** and the "response" **P** (a typical example of what is indeed known as "linear response theory"), we write

$$\mathbf{P} = \chi\,\mathbf{E}\,, \tag{11.1}$$

where $\chi$ (the Greek letter "chi") is called the "dielectric susceptibility." Therefore, the total electric displacement field **D** resulting from *both* the applied field **E** *and* the response **P** of the medium will be

$$\mathbf{D} = \epsilon_0(\mathbf{E}+\mathbf{P}) = \epsilon_0(\mathbf{E}+\chi\,\mathbf{E}) = \epsilon_0(1+\chi)\mathbf{E} = \epsilon\,\mathbf{E}\,, \tag{11.2}$$

where $\epsilon$, the dielectric constant, absorbs the effect of the polarization charges.

Such a "macroscopic averaging" may be too much of a simplification in our case. In general, the response of the system depends on both the wavelength and the frequency of the external perturbation. In addition, semiconductors are not isotropic media. Thus a single dielectric "constant" is actually a frequency and wavevector dependent matrix. We shall derive it in a few cases, assuming for the spatial and temporal dependence of the perturbation a simple harmonic form of the type $e^{i\mathbf{q}\cdot\mathbf{r}}e^{i\omega t}$. Assuming a "linear response," we can always Fourier-decompose any external perturbation, function of space and time, analyze each Fourier component, one at the time, and derive the response of the system for a particular value of **q** and $\omega$. Adding the response at all wavelengths and frequencies at the end, we may recover the response under the arbitrary perturbation.



M. Fischetti, W.G. Vandenberghe, *Advanced Physics of Electron Transport in Semiconductors and Nanostructures*, Graduate Texts in Physics, DOI 10.1007/978-3-319-01101-1_11

## 11.2 Ionic Response

Let's consider first the response of the ions, which we shall treat in rather simple terms.

### 11.2.1 *Optical Phonons in a Linear Chain*

Let's go back to our example of a linear chain of ions (page 205) and modify the problem slightly, so that we now have two ions per "cell," as illustrated Fig. 11.1. The ions are connected by springs with elastic constant $D$, as before. The ions have alternating masses, $M$ and $m$, as shown, with $M > m$. The equilibrium distance between two adjacent atoms is $a$.

The equations of motion can be derived by writing two equations, one for the displacement $u_1$ of the atoms of mass $M$ and another one for the displacement $u_2$ of the atoms of mass $m$. So, denoting by $u_{1,j}$ the displacement of the ion of mass $M$ at location $ja$ and by $u_{2,j}$ the displacement of the ion of mass $m$ at location $(j+1)a$, the equations of motion are

$$M\frac{\mathrm{d}^2 u_{1,j}}{\mathrm{d}t^2} = D(u_{2,j} - u_{1,j}) + D(u_{2,j-1} - u_{1,j}) \,. \tag{11.3}$$

$$m\frac{\mathrm{d}^2 u_{2,j}}{\mathrm{d}t^2} = D(u_{1,j+1} - u_{2,j}) + D(u_{1,j} - u_{2,j}) \,. \tag{11.4}$$

Defining

$$u_{1,j} = \upsilon_1 \, \mathrm{e}^{\mathrm{i}(2jqa - \omega t)} \,, \tag{11.5}$$

$$u_{2,j} = \upsilon_2 \, \mathrm{e}^{\mathrm{i}(2jqa - \omega t)} \,, \tag{11.6}$$

the equations of motion become

$$-M\,\omega^2\,\upsilon_1 = D\upsilon_2 - D\upsilon_1 + D\upsilon_2 \mathrm{e}^{-2iqa} - D\upsilon_1 \,, \tag{11.7}$$

$$-m\,\omega^2\,\upsilon_2 = D\upsilon_1 \mathrm{e}^{2iqa} - D\upsilon_2 + D\upsilon_1 - D\upsilon_2 \,, \tag{11.8}$$

or:

$$(M\,\omega^2 - 2D)\upsilon_1 + D(1 + \mathrm{e}^{-2iqa})\upsilon_2 = 0 \,, \tag{11.9}$$

$$D(1 + \mathrm{e}^{2iqa})\upsilon_1 + (m\,\omega^2 - 2D)\upsilon_2 = 0 \,. \tag{11.10}$$

This $2 \times 2$ linear homogeneous system (which is exactly the system given by Eq. (9.95) in the more general case considered above) admits nontrivial solutions only if the determinant of the coefficients vanishes, that is

$$(M\,\omega^2 - 2D)(m\,\omega^2 - 2D) - D^2(1 + \mathrm{e}^{2iqa})(1 + \mathrm{e}^{-2iqa}) = 0 \,. \tag{11.11}$$

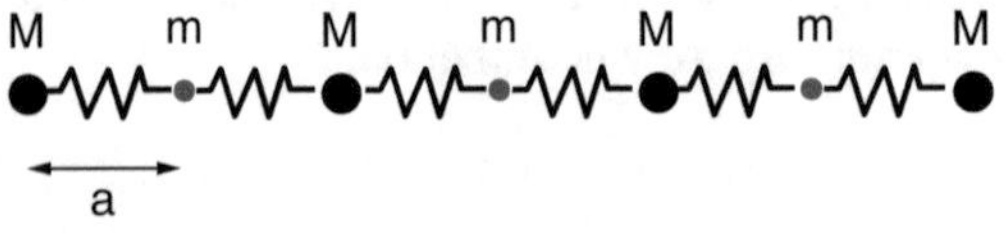

**Fig. 11.1** Schematic illustration of a linear diatomic chain of atoms with different masses

The two solutions of these quadratic equations are

$$\omega_{\pm}^2 = \frac{D}{mM} \left\{ (m+M) \pm \left[ (m+M)^2 - 4mM \sin^2(qa) \right]^{1/2} \right\} . \tag{11.12}$$

The choice of the "plus" sign leads to an "optical" branch that at low wavelengths ($q \to 0$) and at short wavelengths, approaching the zone boundary ($q \to \pi/(2a)$), behaves as

$$\omega_{+}^2 \to \begin{cases} 2D\left(\frac{1}{m} + \frac{1}{M}\right) & \text{for } q \to 0 \\ \frac{2D}{m} & \text{for } q \to \frac{\pi}{2a} . \end{cases} \tag{11.13}$$

The choice of the "minus" sign yields an acoustic mode:

$$\omega_{-}^2 \to \begin{cases} \frac{D}{M+m} a^2 q^2 \to c_s^2 q^2 & \text{for } q \to 0 \\ \frac{2D}{M} & \text{for } q \to \frac{\pi}{2a} . \end{cases} \tag{11.14}$$

The optical solution corresponds to a mode in which the two different ions oscillate out of phase, the acoustic solution corresponds to a mode in which the different ions oscillate in phase.

### *11.2.2 Response in the Infrared*

Now consider Eqs. (11.7) and (11.8) and focus on the long-wavelength (infrared) limit $qa \to 0$ when adding the perturbation of an electric field $Ee^{-i\omega t}$: Assuming a positive charge $e$ for the ion with mass $M$ and negative charge $-e$ for the ion with mass $m$, those equations become

$$-M\,\omega^2\,\upsilon_1 = 2D(\upsilon_2 - \upsilon_1) + eE , \tag{11.15}$$

$$-m\,\omega^2\,\upsilon_2 = 2D(\upsilon_1 - \upsilon_2) - eE . \tag{11.16}$$

Dividing the first equation by $M$, the second one by $m$, and subtracting the second from the first, we obtain

$$\upsilon_2 - \upsilon_1 = \frac{eE/\mu}{(2D/\mu) - \omega^2} , \tag{11.17}$$

where $\mu$ is the "reduced mass" $1/(m^{-1} + M^{-1})$. From Eq. (11.13) we recognize that $2D/\mu = \omega_{\mathrm{TO}}^2$ is the squared frequency of the optical mode in the diatomic linear chain.

Now, the application of the external field $E$ has caused a polarization of the pair, the negative ion moving in the opposite direction of the positive ion. Thus, this polarization results in an additional electric field

$$P_{\mathrm{ion}} = \frac{1}{\epsilon(\infty)}\, eN(\upsilon_2 - \upsilon_1) = \frac{1}{\epsilon(\infty)}\, \frac{e^2N/\mu}{\omega_{\mathrm{TO}}^2 - \omega^2}\, E , \tag{11.18}$$

where $N$ is the number of ion pairs per unit volume and $\epsilon(\infty)$ is the dielectric constant in absence of the ionic response (so, at a frequency large enough so that the ions cannot respond). Note the resonant behavior at the frequency $\omega_{\mathrm{TO}}$. In general, this resonant behavior of Eq. (11.18) translates into a frequency dependence of the form:

$$\epsilon_{\mathrm{ion}}(\omega) = \epsilon(\infty) + [\epsilon(0) - \epsilon(\infty)]\, \frac{\omega_{\mathrm{TO}}^2}{\omega_{\mathrm{TO}}^2 - \omega^2} , \tag{11.19}$$

where $\omega_{TO}$ in the case of a 3D crystal (more general than our simple 1D chain) is the frequency of the transverse optical modes at long wavelengths. What Eq. (11.19) tells us is that the dielectric "constant" approaches a smaller value $\epsilon(\infty)$ at frequencies so large that the ions do not respond to the external perturbation. Thus $\epsilon(\infty) = \epsilon_s$, the "static" electronic dielectric constant of the semiconductor [see below, Eq. (11.42)]. As the frequency decreases, the ions begin to respond and, after having gone through the resonance at $\omega_{TO}$, their response results in a larger dielectric "constant" $\epsilon(0)$. The factor $\epsilon(0) - \epsilon(\infty) = e^2 N/(\mu \omega_{TO}^2)$ represents the contribution of ions to the dielectric response and it is also known as the "oscillator strength" or "$f$-strength." Note that Eq. (11.19) can also be written as:

$$\epsilon_{ion}(\omega) = \epsilon(\infty) \frac{\omega_{LO}^2 - \omega^2}{\omega_{TO}^2 - \omega^2} , \tag{11.20}$$

where $\omega_{LO}$ is such that $\epsilon_{ion}(\omega_{LO}) = 0$. This shows that the dielectric constant is negative for $\omega_{TO} < \omega < \omega_{LO}$, which means that electromagnetic waves do not propagate through the crystal: they will be damped, as the polarization of the crystal is in the opposite direction of the external field, thus canceling it. To see this mathematically, recall that in a dielectric medium the relation between wavevector and frequency of an electromagnetic wave is

$$k^2 = \frac{\epsilon(\omega)}{\epsilon_0 c^2} \omega^2 . \tag{11.21}$$

Whenever $\epsilon < 0$, $k$ will acquire an imaginary part which tells us that the wave becomes damped.

We have yet to understand the physical meaning of $\omega_{LO}$. Since

$$\nabla \cdot \mathbf{D} = \epsilon(\omega) \nabla \cdot \mathbf{E} = 0 , \tag{11.22}$$

assuming $\mathbf{E} = \mathbf{E}_0 e^{i\mathbf{k}\cdot\mathbf{r}}$, for transverse modes we will have $\mathbf{k} \cdot \mathbf{E}_0 = 0$, so that Eq. (11.22) is trivially satisfied when $\epsilon(\omega) \neq 0$. However, when we deal with longitudinal modes, we satisfy Eq. (11.22) with $\epsilon(\omega) = 0$. We see that *longitudinal modes correspond to the zeros of the dielectric function, transverse modes correspond to its poles*. We also see from Eq. (11.19) and the property $\epsilon(\omega_{LO}) = 0$ that

$$\omega_{LO} = \left[ \frac{\epsilon(0)}{\epsilon(\infty)} \right]^{1/2} \omega_{TO} . \tag{11.23}$$

This expression is known as the Lyddane–Sachs–Teller relation.

Note, finally, that since $\mathbf{D}(\omega) = \epsilon(\omega)\mathbf{E}(\omega)$, and since in the absence of external charges $\mathbf{D}(\omega) = 0$, transverse modes must have a vanishing microscopic electric field $\mathbf{E}(\omega)$. Instead, longitudinal modes can have $\mathbf{E}(\omega) \neq 0$ and still have $\mathbf{D}(\omega) = 0$, since $\epsilon(\omega_{LO}) = 0$. Thus, *electrons will interact via polar interactions only with longitudinal modes*.

## 11.3 Electronic Response

Let's now consider the response of the electrons under an external perturbing potential. We shall not discuss the problem in its full generality: Even for the case of a homogeneous and isotropic electron gas the problem has a long and complicated history. Instead, we shall rely on two main simplifications.

1. First, we shall restrict our discussion to the so-called longitudinal response. We shall assume that our medium is isotropic (in some case thanks to a suitable transformation, such as the Herring–Vogt transformation of **k**-space coordinates usually employed to "isotropize" the Si equi-energy ellipsoids) [1], or we shall only consider the longitudinal polarization field, that is, the microscopic field (generated by the redistribution of the charges under the external field) along the direction of the external field itself. Transverse components, such as those which may arise from transverse vector potentials in dynamic situations, will be ignored. A notable example of a situation where "local transverse fields" may be important is discussed when dealing with a microscopic theory of the electron–phonon interaction
2. The second approximation we shall embrace consists in assuming a "quasi-equilibrium" carrier distribution. This approximation, dictated exclusively by practical computing difficulties, will be discussed below.

The textbooks by Fetter and Walecka [2] and Mahan [3], as usual, provide a precious source of additional details and clarifications.

### *11.3.1 Random Phase Approximation*

Let's consider a "free" semiconductor, i.e. a semiconductor described by the Schrödinger equation Eq. (5.1). Let's denote by $\widehat{H}_0$ the free Hamiltonian at the left-hand side of Eq. (5.1), by $\psi_{\mathbf{k},\mu}(\mathbf{r}) = \langle \mathbf{r}|\mathbf{k},\mu\rangle$ the (Bloch) plane waves which are the eigenfunctions of the lattice Hamiltonian $\widehat{H}_0$ with eigenvalues $E_\mu(\mathbf{k})$ for each band $\mu$.

If we apply an external perturbation described by the potential $\varphi^{(\mathrm{ex})}(\mathbf{r},t)$, the charges in the semiconductor will rearrange themselves, causing an additional "polarization potential" (or "screening potential") $\varphi^{(\mathrm{pol})}(\mathbf{r},t)$ which modifies the original external potential. Thus, the "actual" potential in the material will be the sum of the external perturbation and of the polarization potential,

$$\varphi(\mathbf{r},t) = \varphi^{(\mathrm{ex})}(\mathbf{r},t) + \varphi^{(\mathrm{pol})}(\mathbf{r},t) . \tag{11.24}$$

We are interested in describing the total potential $\varphi$, since this is the effective perturbation acting on the material. We may think of $\varphi^{(\mathrm{ex})}$ as the "bare" potential of a positively charged donor impurity in n-type Si, described by the $1/r$ Coulomb potential, while $\varphi$ will be the screened potential felt by the conduction electrons, resulting from a single positive charge of the impurity ion and from the negative cloud of free conduction electrons attracted by the impurity. Two approximations are very commonly made to reduce the complexity of the problem: First, different Fourier components of the perturbation and of the response of the system are considered independently. Cross-terms mixing different "wavelengths" are ignored, on the grounds that their phases will vary wildly, resulting in their cancelation. This is the so-called Random Phase Approximation (RPA). We thus transform this many-body problem to a single-particle problem, thanks to this "mean-field" approximation. Secondly, we assume that both the external and the polarization potentials are weak perturbations of the "free" Hamiltonian $\widehat{H}_0$. This is the so-called "linear screening approximation" which allows us to express the polarization charge in simple first-order perturbation theory.

Thanks to the RPA, we can state the problem more simply: Let's expand the external perturbation in plane waves:

$$\varphi^{(\mathrm{ex})}(\mathbf{r},t) = \sum_{\mathbf{q}} \varphi_{\mathbf{q}}^{(\mathrm{ex})} \, \mathrm{e}^{-\mathrm{i}\mathbf{q}\cdot\mathbf{r}-\mathrm{i}\omega t} , \tag{11.25}$$

having assumed a simple harmonic time dependence. Then, we are interested in finding the "relative dielectric" function $\epsilon^{(\mathrm{r})}(\mathbf{q},\omega) = \epsilon(\mathbf{q},\omega)/\epsilon_0$, where $\epsilon_0$ is the vacuum permittivity, such that

$$\varphi_{\mathbf{q}}^{(\mathrm{ex})} = \varphi_{\mathbf{q}} - \varphi_{\mathbf{q}}^{(\mathrm{pol})} = \epsilon^{(\mathrm{r})}(\mathbf{q},\omega)\ \varphi_{\mathbf{q}}\ , \tag{11.26}$$

or $\varphi_{\mathbf{q}} = \varphi_{\mathbf{q}}^{(\mathrm{ex})}/\epsilon(\mathbf{q},\omega)$ and so $\epsilon(\mathbf{q},\omega)/\epsilon_0 = 1 - \varphi_{\mathbf{q}}^{(\mathrm{pol})}/\varphi_{\mathbf{q}}$. The usefulness of this expression lies in the fact that, once we know the dielectric function $\epsilon(\mathbf{q},\omega)$, the knowledge of the simple external potential is sufficient to determine the net response of the system, without having to re-evaluate the self-consistent redistribution of the charges internally to the system.

We shall proceed as follows: First, we shall employ the approximation of linear response to determine the "new" wavefunctions of the system under the perturbation of the net potential $\varphi$, assuming that it is known. From the perturbed wavefunctions we shall derive the "polarization" charge,

$$\rho^{(\mathrm{pol})}(\mathbf{r},t) = \sum_{\mathbf{q}} \delta\rho_{\mathbf{q}}\ \mathrm{e}^{-\mathrm{i}\mathbf{q}\cdot\mathbf{r}-\mathrm{i}\omega t}\ , \tag{11.27}$$

and, from this, the polarization potential. Summing the external and polarization potentials, we shall finally solve for the net potential self-consistently.

Now we express the wavefunctions, $\phi_{\mathbf{k}}^{(\mu)}(\mathbf{r},t)$, of the perturbed system (lattice Hamiltonian plus perturbing potential) in terms of the unperturbed wavefunctions $\psi_{\mathbf{k}}^{(\mu)}(\mathbf{r},t) = \langle \mathbf{r},t|\mathbf{k},\mu\rangle$, using first-order perturbation theory (see pages 437–443):

$$\begin{aligned}\phi_{\mathbf{k},\mu}(\mathbf{r},t) &= \psi_{\mathbf{k}}^{(\mu)}(\mathbf{r},t) + \delta\psi_{\mathbf{k},\mu}(\mathbf{r},t) \\ &= \psi_{\mathbf{k}}^{(\mu)}(\mathbf{r},t) + \sum_{\mathbf{k}',\mu'} \frac{e\langle \mathbf{k}',\mu'|\varphi|\mathbf{k},\mu\rangle}{E_\mu(\mathbf{k}) - E_{\mu'}(\mathbf{k}') + \hbar\omega + \mathrm{i}\hbar s}\ \psi_{\mathbf{k}'}^{(\mu')}(\mathbf{r},t)\end{aligned} \tag{11.28}$$

and

$$\begin{aligned}\phi_{\mathbf{k},\mu}^{*}(\mathbf{r},t) &= \psi_{\mathbf{k}}^{(\mu)*}(\mathbf{r},t) + \delta\psi_{\mathbf{k}}^{(\mu)*}(\mathbf{r},t) \\ &= \psi_{\mathbf{k}}^{(\mu)*}(\mathbf{r},t) + \sum_{\mathbf{k}',\mu'} \frac{e\langle \mathbf{k},\mu|\varphi|\mathbf{k}',\mu'\rangle}{E_\mu(\mathbf{k}) - E_{\mu'}(\mathbf{k}') - \hbar\omega - \mathrm{i}\hbar s}\ \psi_{\mathbf{k}'}^{(\mu')*}(\mathbf{r},t)\ .\end{aligned} \tag{11.29}$$

Parenthetically, here we use the Greek letter $\mu$ to indicate the band index, instead of the "usual" index $n$, in order to avoid confusion with the electron density.

* *These expressions are likely to present a surprise, as they contain a "strange" imaginary term $\pm\mathrm{i}\hbar s$ in the denominator. Its origin lies in the fact that when we have a continuous spectrum, we must "regularize" the behavior of the coefficient appearing in the equation above when $\mathbf{k}' \to \mathbf{k}$. The standard "trick" is to add a small imaginary part to the denominator, do the calculations, and at the end take the limit $s \to 0$. This is also known as taking the "principal part" of the integral. We shall discuss it at length below, understanding what it means physically. For the time being, we may ignore it.*

From Eq. (11.25) and with some manipulations with Bloch waves, we have

$$\langle \mathbf{k}',\mu'|\varphi|\mathbf{k},\mu\rangle = \varphi_{\mathbf{k}-\mathbf{k}'} \sum_{\mathbf{G}} \langle \mathbf{k}'+\mathbf{G},\mu'|\mathrm{e}^{\mathrm{i}(\mathbf{k}-\mathbf{k}')\cdot\mathbf{r}}|\mathbf{k},\mu\rangle\ . \tag{11.30}$$

The "new" charge density is obtained by summing the individual contributions $e|\phi_{\mathbf{k},\mu}|^2$ over all occupied states $\phi_{\mathbf{k},\mu}$. Denoting by $p(\mathbf{k},\mu)$ the occupation of each state—not necessarily the equilibrium occupation number—we have

$$\begin{aligned}
\rho(\mathbf{r},t) &= e\sum_{\mathbf{k},\mu} p(\mathbf{k},\mu)\ |\phi_{\mathbf{k}}^{(\mu)}(\mathbf{r},t)|^2 \\
&= e\sum_{\mathbf{k},\mu} p(\mathbf{k},\mu)\ |\psi_{\mathbf{k}}^{(\mu)}(\mathbf{r},t)\ +\ \delta\psi_{\mathbf{k}}^{(\mu)}(\mathbf{r},t)|^2 \\
&\simeq \rho_0(\mathbf{r},t)\ +\ e^2\sum_{\mathbf{k},\mu} p(\mathbf{k},\mu)\ \psi_{\mathbf{k}}^{(\mu)*}(\mathbf{r},t) \\
&\times \sum_{\mathbf{G},\mathbf{k}',\mu'} \frac{\varphi_{\mathbf{k}-\mathbf{k}'}\ \langle \mathbf{k}'+\mathbf{G},\mu'|\mathrm{e}^{\mathrm{i}(\mathbf{k}-\mathbf{k}')\cdot\mathbf{r}}|\mathbf{k},\mu\rangle}{E_\mu(\mathbf{k})-E_{\mu'}(\mathbf{k}')+\hbar\omega+\mathrm{i}\hbar s}\ \psi_{\mathbf{k}'}^{(\mu')}(\mathbf{r},t)\ +\ \mathrm{cc}\ ,
\end{aligned} \tag{11.31}$$

where cc means "complex conjugate." Now notice that under the sums we may swap the dummy variables $\mathbf{k}$ with $\mathbf{k}'$ and $\mu$ with $\mu'$ in the cc term. Finally, expressing the product $\psi_{\mathbf{k}}^{(\mu)*}(\mathbf{r},t)\psi_{\mathbf{k}'}^{(\mu')}(\mathbf{r},t)$ in terms of its Fourier transform,

$$\begin{aligned}
\psi_{\mathbf{k}}^{(\mu)*}(\mathbf{r},t)\ \psi_{\mathbf{k}'}^{(\mu')}(\mathbf{r},t) &= \sum_{\mathbf{q}} \langle \mathbf{k},\mu|\mathrm{e}^{\mathrm{i}\mathbf{q}\cdot\mathbf{r}}|\mathbf{k}',\mu'\rangle\ \mathrm{e}^{-\mathrm{i}\mathbf{q}\cdot\mathbf{r}} \\
&= \sum_{\mathbf{G}'} \langle \mathbf{k},\mu|\mathrm{e}^{\mathrm{i}(\mathbf{k}'-\mathbf{k})\cdot\mathbf{r}}|\mathbf{k}'+\mathbf{G}',\mu'\rangle\ \mathrm{e}^{-\mathrm{i}(\mathbf{k}'-\mathbf{k})\cdot\mathbf{r}}
\end{aligned} \tag{11.32}$$

and setting $\mathbf{q}=\mathbf{k}'-\mathbf{k}$, we obtain

$$\begin{aligned}
\delta\rho(\mathbf{r},t) = e^2\sum_{\mathbf{k},\mu}\sum_{\mathbf{G},\mathbf{q},\mu'} &\frac{[p(\mathbf{k},\mu)-p(\mathbf{k}+\mathbf{q},\mu')]|\langle \mathbf{k}+\mathbf{q}+\mathbf{G},\mu'|\mathrm{e}^{\mathrm{i}\mathbf{q}\cdot\mathbf{r}}|\mathbf{k},\mu\rangle|^2}{E_\mu(\mathbf{k})-E_{\mu'}(\mathbf{k}+\mathbf{q})+\hbar\omega+\mathrm{i}\hbar s} \\
&\times \varphi_{\mathbf{q}}\ \mathrm{e}^{-\mathrm{i}\mathbf{q}\cdot\mathbf{r}}\ .
\end{aligned} \tag{11.33}$$

Comparing Eq. (11.27) with Eq. (11.32), we see that

$$\delta\rho_{\mathbf{q}} = e^2\sum_{\mathbf{G},\mathbf{k},\mu,\mu'} \frac{[p(\mathbf{k},\mu)-p(\mathbf{k}+\mathbf{q},\mu')]|\langle \mathbf{k}+\mathbf{q}+\mathbf{G},\mu'|\mathrm{e}^{\mathrm{i}\mathbf{q}\cdot\mathbf{r}}|\mathbf{k},\mu\rangle|^2}{E_\mu(\mathbf{k})-E_{\mu'}(\mathbf{k}+\mathbf{q})+\hbar\omega+\mathrm{i}\hbar s}\ \varphi_{\mathbf{q}}\ . \tag{11.34}$$

Using Poisson equation, $\nabla^2\varphi^{(\mathrm{pol})} = -\delta\rho/\epsilon_0$, we can easily express the Fourier components of the polarization potential in terms of the Fourier components of the charge variation $\delta\rho_{\mathbf{q}}$:

$$\varphi_{\mathbf{q}}^{(\mathrm{pol})} = \frac{\delta\rho_{\mathbf{q}}}{q^2\epsilon_0}\ . \tag{11.35}$$

Combining Eqs. (11.26), (11.34), and (11.35), we can finally identify the dielectric function with

$$\epsilon^{(\mathrm{r})}(\mathbf{q},\omega) = 1\ -\ \frac{e^2}{q^2\epsilon_0}\sum_{\mathbf{G},\mathbf{k},\mu,\mu'} \frac{[p(\mathbf{k},\mu)-p(\mathbf{k}+\mathbf{q},\mu')]|\langle \mathbf{k}+\mathbf{q}+\mathbf{G},\mu'|\mathrm{e}^{\mathrm{i}\mathbf{q}\cdot\mathbf{r}}|\mathbf{k},\mu\rangle|^2}{E_\mu(\mathbf{k})-E_{\mu'}(\mathbf{k}+\mathbf{q})+\hbar\omega+\mathrm{i}\hbar s}\ . \tag{11.36}$$

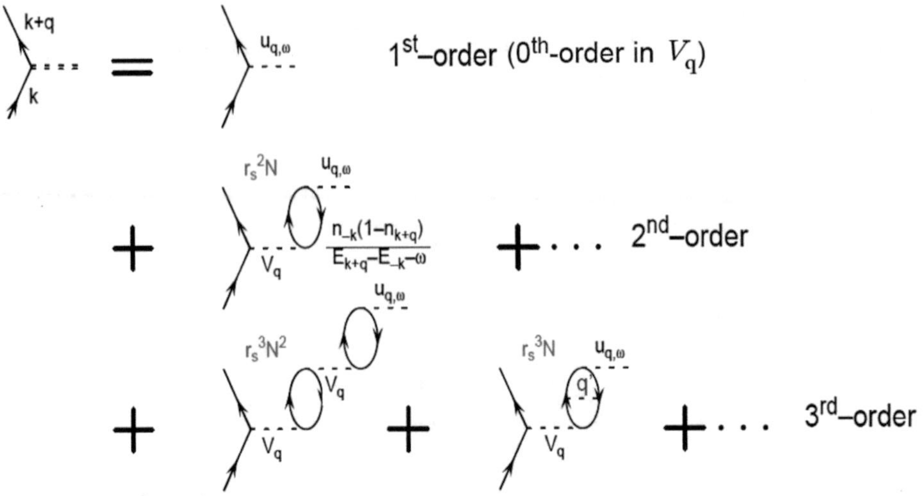

**Fig. 11.2** Diagrams explaining the meaning of the Random Phase Approximation. These diagrams are called *Goldstone diagrams*. *Vertical arrows* represent electrons traveling forward in time; *horizontal dashed lines* represent interactions via a bare potential $V_{\mathbf{q}}$, $u_{\mathbf{q},\omega}$ represents the screened interaction. Vertices count the number of times the interaction enters as a factor, i.e., the perturbation order of the interaction. The RPA amounts to assuming only diagrams containing "Fermion loops" dominate. This is true at large electron densities

*The validity of the RPA. *The RPA can be formulated simply as follows: Suppose that $V_{\mathbf{q}}$ is the Fourier component of the potential through which electrons interact. We want to calculate the screened potential $u_{\mathbf{q}\omega}$. The frequency $\omega$ here appears because we wish to account also for the dynamic response of the other electrons that screen $V_{\mathbf{q}}$. We proceed by considering the electrons interacting with another particle by considering a perturbation series in which the first term (zero-th order) represents the bare interaction; the first-order term represents the electron interacting with the mediation of another electron in the surrounding electron gas; and the* N*th-order term represents the interaction mediated by n surrounding electrons. Graphically, we may represent the electron in its initial state $|\mathbf{k}\rangle$ as an arrow propagating forward in time, an arrow pointing upwards in the diagram shown in Fig. 11.2. A horizontal dashed line represents an interaction that terminates at the position of whatever the electron is interacting with (an impurity, another electron, and such). At zero-th order (that is, for the unscreened potential), the electron interacts via $V_{\mathbf{q}}$. This is the topmost diagram. At the next order in our perturbation series, the only process that appears is a process in which the original electron interacts with another electron, exchanging momentum* **q**, *and the interaction now terminates via this electron. From first-order perturbation theory, this term contributes a factor*

$$V_{\mathbf{q}} \frac{n_{-\mathbf{k}}(1 - n_{\mathbf{k}+\mathbf{q}})}{E_{\mathbf{k}+\mathbf{q}} - E_{-\mathbf{k}} + \hbar\omega} \, . \tag{11.37}$$

*There are no other terms at first order. There will be n processes of this type, if n is the total number of electrons in the gas. At the next order, there will be processes that simply add another intermediate electron (the left-most diagram in the bottom line of diagrams). There will now be $n^2$ of them, since each one of the "loops" shown in the figure can be chosen in n different ways, one per electron in the gas. If we add all terms of the sum, we have a geometric series of powers of terms given by Eq. (11.37). Equations (11.36) and (11.45) represent exactly the sum of this series. Indeed, consider again Eq. (11.26) and rewrite it in terms of a geometric series of a susceptibility (or polarizability) $\chi(\mathbf{q},\omega)$ as*

$$\varphi_{\mathbf{q}} = \varphi_{\mathbf{q}}^{(\text{ext})} + \chi(\mathbf{q},\omega)\varphi_{\mathbf{q}} = \varphi_{\mathbf{q}}^{(\text{ext})} + \chi(\mathbf{q},\omega)\varphi_{\mathbf{q}}^{(\text{ext})} + \chi(\mathbf{q},\omega)^2\varphi_{\mathbf{q}}^{(\text{ext})} + \cdots$$
$$= \frac{\varphi_{\mathbf{q}}^{(\text{ext})}}{1 - \chi(\mathbf{q},\omega)} = \frac{\varphi_{\mathbf{q}}^{(\text{ext})}}{\epsilon(\mathbf{q},\omega)}\,, \tag{11.38}$$

*where we have set $\epsilon_0 = 1$ for convenience. This is exactly the sum of the RPA perturbation series at all orders in the interaction.*

*This method was first proposed by Bohm and Pines in the early 1950s [4–7], but the question arose about the validity of this approximation. Indeed, at a given order $N \geq 2$, there will be other diagrams that do not contain loops: For example, the last second-order diagram shown in Fig. 11.2 that contains a self-interaction of the intermediate electron, or other diagrams consisting of a lower-order diagram and a disconnected loop. All of these diagrams are ignored by the RPA, making it a somewhat dubious approximation. Only a few years later, in 1957, Gell–Mann and Bruckner [8] showed that, at any given order, the number of these other diagrams is of the order of at most $n^{N-1}$, whereas, as we just saw, there are $\sim n^N$ diagrams containing loops. (For example, the second second-order diagram shown in the figure appears only n times, because the same electron interacts at four vertices. On the contrary, there are $n^2$ two-loop diagrams.) These are exactly those processes that involve a different wavevector, $q'$, thus mixing the various wavelengths. Therefore, at large electron densities, the RPA is justified.*

## 11.3.2 Special Limits (Long-Wavelength, Static, Response of 2DEGs and 1DEGs)

### 11.3.2.1 Static Dielectric Constant

We are interested in situations in which the occupation of the states, $p(\mathbf{k},\mu)$, is the occupation of the perfect crystal at zero temperature, *plus* a small perturbation representing the free carriers. Thus,

$$p(\mathbf{k},\mu) = p_0(\mathbf{k},\mu) + f(\mathbf{k},\mu)\,, \tag{11.39}$$

where $p_0$ is unity for all states in the valence bands, zero otherwise, and $f$ is the distribution of free electrons (holes) in the conduction (valence) bands. It is convenient to separate the contributions of the two different populations, since the population $p_0$ is largely unaffected by the free carriers, and the free carriers themselves matter almost exclusively in our context.

Considering the dielectric response of the valence electrons, we note that almost all of the perturbing potentials $\varphi$ we have to deal with are characterized by frequencies much smaller than the gap of the semiconductor. The Coulomb potential responsible for interband impact ionization represents a notable exception. Since contributions to the sum in Eq. (11.36) can be nonzero only when $p(\mathbf{k},\mu) = 1$ and $p(\mathbf{k}+\mathbf{q},\mu') = 0$, or vice versa, the largest contributions will come from occupied (empty) states $|\mathbf{k},\mu\rangle$ and empty (occupied) states $|\mathbf{k}+\mathbf{q},\mu'\rangle$ whose energy difference is the smallest. In other words, only states across the valence–conduction gap ought to be considered. Ignoring this special case, we can perform the sum over the topmost valence band and the lowest-lying conduction band, $\mu = \upsilon$ and $\mu' = \mathrm{c}$. We can use the "sum rule" (which we shall not prove):

$$\sum_f (E_f - E_i)|\langle f|\mathrm{e}^{\mathrm{i}\mathbf{q}\cdot\mathbf{r}}|i\rangle|^2 = \frac{\hbar^2 q^2}{2m_{\text{el}}}\,, \tag{11.40}$$

which is an extension of the sum rule for oscillator strengths. Assuming $E_c(\mathbf{k}+\mathbf{q}) - E_v(\mathbf{k}) \simeq E_{\mathrm{gap}}$, which holds approximately for wavelengths $1/q$ long enough to result in almost "vertical" transitions between $\mathbf{k}$ and $\mathbf{k}+\mathbf{q}$, and assuming also, as mentioned above, that $\hbar\omega \ll E_{\mathrm{gap}}$, we have

$$\sum_{\mathbf{G}} |\langle \mathbf{k}+\mathbf{q}+\mathbf{G}, c | \mathrm{e}^{\mathrm{i}\mathbf{q}\cdot\mathbf{r}} | \mathbf{k}, \upsilon \rangle|^2 = \frac{\hbar^2 q^2}{2m_{\mathrm{el}} E_{\mathrm{gap}}} . \tag{11.41}$$

Here $E_{\mathrm{gap}}$ is the separation between the valence and conduction band, averaged over the entire BZ. In practice, since the largest contributions to the sum in Eq. (11.36) come from regions of high DOS, the direct gap at the symmetry point $X$ represents a good approximation to the quantity $E_{\mathrm{gap}}$ considered here. Thus, from Eqs. (11.36) and (11.41), accounting for a factor of 2 resulting from the cases $p(\mathbf{k},c) = 1$, $p(\mathbf{k}+\mathbf{q},\upsilon) = 0$, and $p(\mathbf{k},\upsilon) = 0$, $p(\mathbf{k}+\mathbf{q},c) = 1$, we have

$$\epsilon^{(\mathrm{r})}(\mathbf{q},0) \simeq 1 + \left(\frac{\hbar\omega_{\mathrm{P}}}{E_{\mathrm{gap}}}\right)^2 \equiv \epsilon_{\mathrm{s}}^{(\mathrm{r})} , \tag{11.42}$$

where the bulk, valence-band plasma frequency is given by

$$\omega_{\mathrm{P}} = \left(\frac{e^2 n_v}{\epsilon_0 m}\right)^{1/2} . \tag{11.43}$$

This is the static dielectric constant of the semiconductor, which approaches a constant at the long wavelength we have considered.

#### 11.3.2.2 Response of the Free Carriers

Considering now the response of the free carriers, we restrict ourselves to carriers in a single band, since the corresponding denominator is often (but not always!) the smallest. We shall also consider here only perturbations such that $q \ll G$, so that the matrix element of $\mathrm{e}^{\mathrm{i}\mathbf{q}\cdot\mathbf{r}}$ between states in the same band becomes unity, that is,

$$\sum_{\mathbf{G}} |\langle \mathbf{k}+\mathbf{q}+\mathbf{G}, \mu | \mathrm{e}^{\mathrm{i}\mathbf{q}\cdot\mathbf{r}} | \mathbf{k}, \mu \rangle|^2 \simeq \sum_{\mathbf{G}} |\langle \mathbf{k}+\mathbf{q}+\mathbf{G}, \mu | \mathbf{k}, \mu \rangle|^2 \simeq 1 . \tag{11.44}$$

A notable case for which this approximation fails is the case of zone-edge phonons, whose screening must be treated by retaining the matrix element above. Then, dropping for now the band indices, we can now rewrite Eq. (11.36) as follows

$$\frac{\epsilon(\mathbf{q},\omega)}{\epsilon_{\mathrm{s}}} = 1 - \frac{e^2}{q^2 \epsilon_{\mathrm{s}}} \sum_{\mathbf{k}} \frac{f(\mathbf{k}) - f(\mathbf{k}+\mathbf{q})}{E(\mathbf{k}) - E(\mathbf{k}+\mathbf{q}) + \hbar\omega + \mathrm{i}\hbar s} . \tag{11.45}$$

Except for the static dielectric constant $\epsilon_{\mathrm{s}}$ in place of the vacuum value, $\epsilon_0$, this is exactly the so-called Lindhard's expression valid for a free electron gas. As a matter of convenience, we introduce a "screening wavevector" $\beta(\mathbf{q},\omega)$ and recast Eq. (11.45) in the form:

$$\epsilon(\mathbf{q},\omega) = \epsilon_{\mathrm{s}} \left[1 + \frac{\beta^2(\mathbf{q},\omega)}{q^2}\right] . \tag{11.46}$$

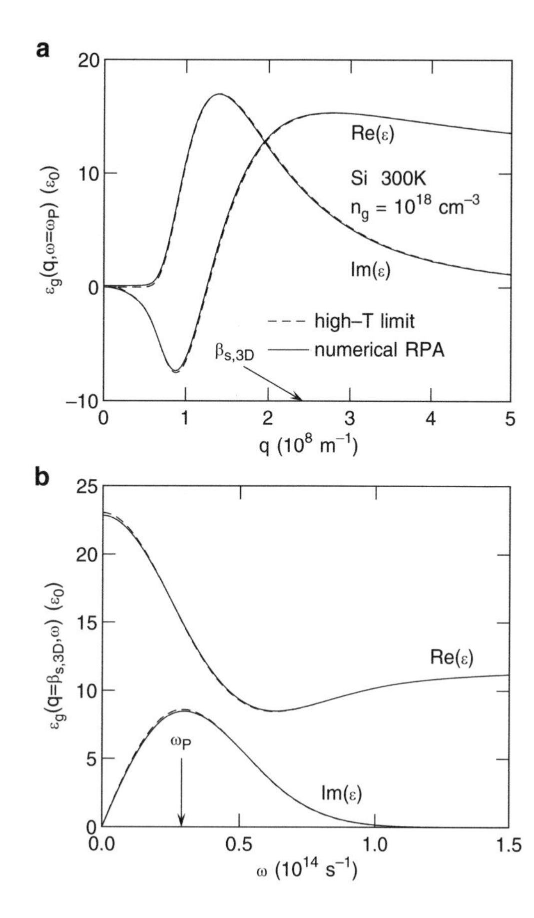


**Fig. 11.3** Real and imaginary parts of the bulk Si dielectric function at a nondegenerate electron concentration of $10^{18}$ cm$^{-3}$ computed using either the numerical-RPA expression (*solid lines*) or the nondegenerate high-T expression (*dashed lines*) as a function of $Q$ at the plasma frequency (**a**) or at the Thomas–Fermi screening wavelength $\beta_{s,3D}$ (**b**). (Reprinted with permission from M.V. Fischetti, J. Appl. Phys. 89, 1232 (2001). Copyright 2001, AIP Publishing LLC)

Thus, the final answer to our initial problem ("What is the 'net' potential $\varphi(\mathbf{r},t)$, once the external perturbation $\varphi^{(\mathrm{ex})}(\mathbf{r},t)$ is known?") can be expressed as

$$\varphi(\mathbf{r},t) = \sum_{\mathbf{q}} \frac{\varphi_{\mathbf{q}}^{(\mathrm{ex})}}{\epsilon^{(\mathrm{r})}(\mathbf{q},\omega)}\, \mathrm{e}^{\mathrm{i}\mathbf{q}\cdot\mathbf{r}} \mathrm{e}^{\mathrm{i}\omega t}\,. \tag{11.47}$$

The evaluation of the electronic component of the dielectric function requires the knowledge of the electron distribution function $p(\mathbf{k})$. In transport problems, this may actually be the main unknown, so that self-consistent schemes are required. But even when $p$ is known (as in thermodynamic equilibrium), the evaluation of the RPA expression Eq. (11.36) or even (11.45) is, in general, quite complicated. Therefore, we shall now consider various cases in which this expression, ultimately needed to screen the scattering potentials, can be evaluated in closed form. This can be done by employing a simple effective mass, parabolic band approximations for the conduction band, and assuming that the distribution function is well approximated by its thermal equilibrium value. Indeed, screening is particularly important in high density regions of the devices. In these regions the large density results in a small resistivity and, as a consequence, in small electric fields that prevent significant deviations from equilibrium (small "carrier heating"). Thus, we may set $f(\mathbf{k}) \simeq f_0(\mathbf{k})$, the equilibrium distribution function.

We shall now derive well-known expressions for the dielectric constant in the static- and long-wavelength limits, obtaining the Debye–Hückel/Thomas–Fermi expression and the semiclassical

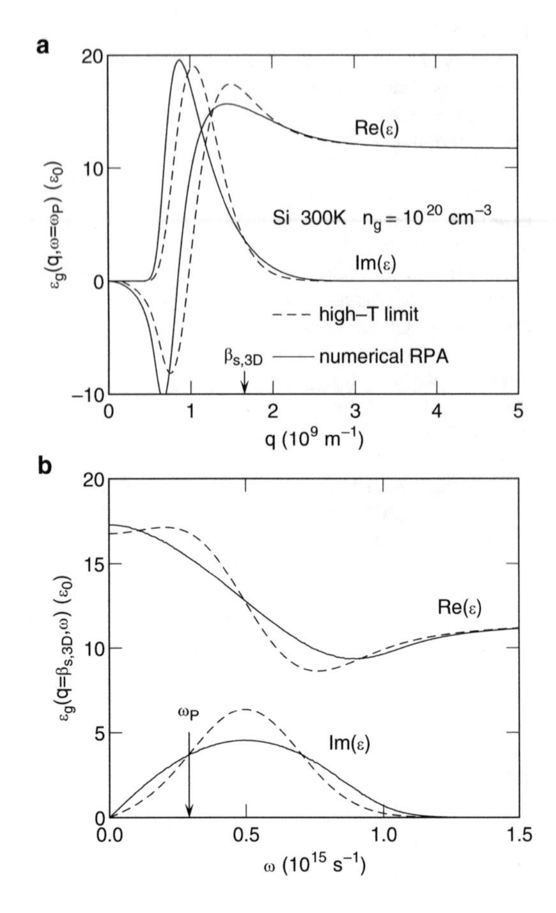


**Fig. 11.4** As in Fig. 11.3, but for the degenerate case corresponding to an electron concentration of $10^{20}$ cm$^{-3}$. (Reprinted with permission from M.V. Fischetti, J. Appl. Phys. 89, 1232 (2001). Copyright 2001, AIP Publishing LLC)

plasma frequency, respectively. Afterwards, we shall derive more general expressions—valid at all wavelengths and frequencies—in the two limits of a nondegenerate, high-temperature electron gas, and of a degenerate gas at zero temperature. Figures 11.3 and 11.4 (from Ref. [9]) give an example of the dielectric response of free carriers in Si, calculated using the RPA.

*Static Debye–Hückel/Thomas–Fermi Screening.* For static perturbations (e.g., impurity scattering), we set $\hbar\omega = 0$ and employ the following approximations valid at long wavelengths:

$$E(\mathbf{k}+\mathbf{q}) - E(\mathbf{k}) \simeq \mathbf{q}\cdot\nabla_{\mathbf{k}}E(\mathbf{k}) , \tag{11.48}$$

$$f_0(\mathbf{k}+\mathbf{q}) - f_0(\mathbf{k}) \simeq \mathbf{q}\cdot\frac{\partial f_0}{\partial E}\nabla_{\mathbf{k}}E(\mathbf{k}) , \tag{11.49}$$

so that

$$\epsilon(\mathbf{q},0) \simeq \epsilon_{\mathrm{s}} - \frac{e^2}{q^2}\int \mathrm{d}E\;\mathscr{D}(E)\left(-\frac{\partial f_0}{\partial E}\right) = \epsilon_{\mathrm{s}} + \frac{e^2}{q^2}\int \mathrm{d}E\;\mathscr{D}(E)\left(-\frac{\partial f_0}{\partial E_{\mathrm{F}}}\right) , \tag{11.50}$$

where $\mathscr{D}(E)$ is the density of states at energy $E$. Thus, we obtain the static screening parameter at long wavelengths:

$$\beta_{\mathrm{s}}^2 = \frac{e^2}{\epsilon_{\mathrm{s}}}\int \mathrm{d}E\;\mathscr{D}(E)\left(\frac{\partial f_0}{\partial E_{\mathrm{F}}}\right) = \frac{e^2}{\epsilon_{\mathrm{s}}}\frac{\partial n}{\partial E_{\mathrm{F}}}. \tag{11.51}$$

In the nondegenerate case (i.e., at very high temperatures), $\partial n/\partial E_{\rm F} \simeq n/(k_{\rm B}T)$ and we have the Debye–Hückel expression:

$$\beta_{\rm DH}^2 \simeq \frac{e^2 n}{\epsilon_{\rm s} k_{\rm B} T}. \tag{11.52}$$

In the opposite case of a degenerate gas (low temperature), $\partial f/\partial E_{\rm F} \simeq \delta(E - E_{\rm F})$ and we have the Thomas–Fermi expression:

$$\beta_{\rm TF}^2 \simeq \frac{e^2}{\epsilon_{\rm s}} \mathscr{D}(E_{\rm F}) = \frac{e^2}{\epsilon_{\rm s}} \frac{3n}{2E_{\rm F}}. \tag{11.53}$$

*Plasma Oscillations.* We can consider now the opposite limit of dynamic effects at long wavelengths, i.e., the limit $q \to 0$. Let's recast Eq. (11.45) as follows (ignoring the imaginary part):

$$\frac{\epsilon(\mathbf{q},\omega)}{\epsilon_{\rm s}} = 1 - \frac{e^2}{\epsilon_{\rm s} q^2} \sum_{\mathbf{k}} \left\{ \frac{f(\mathbf{k})}{E(\mathbf{k}) - E(\mathbf{k}+\mathbf{q}) + \hbar\omega} - \frac{f(\mathbf{k}+\mathbf{q})}{E(\mathbf{k}) - E(\mathbf{k}+\mathbf{q}) + \hbar\omega} \right\}. \tag{11.54}$$

Inside the second sum let's set $\mathbf{k}' = -\mathbf{k} - \mathbf{q}$:

$$\frac{\epsilon(\mathbf{q},\omega)}{\epsilon_{\rm s}} = 1 - \frac{e^2}{\epsilon_{\rm s} q^2} \left\{ \sum_{\mathbf{k}} \frac{f(\mathbf{k})}{E(\mathbf{k}) - E(\mathbf{k}+\mathbf{q}) + \hbar\omega} - \sum_{\mathbf{k}'} \frac{f(-\mathbf{k}')}{E(-\mathbf{k}'-\mathbf{q}) - E(-\mathbf{k}') + \hbar\omega} \right\}. \tag{11.55}$$

Since $E(-\mathbf{k}') = E(\mathbf{k}')$ and $f(-\mathbf{k}') = f(\mathbf{k}')$:

$$\frac{\epsilon(\mathbf{q},\omega)}{\epsilon_{\rm s}} = 1 - \frac{e^2}{\epsilon_{\rm s} q^2} \left\{ \sum_{\mathbf{k}} \frac{f(\mathbf{k})}{E(\mathbf{k}) - E(\mathbf{k}+\mathbf{q}) + \hbar\omega} - \sum_{\mathbf{k}'} \frac{f(\mathbf{k}')}{E(\mathbf{k}'+\mathbf{q}) - E(\mathbf{k}') + \hbar\omega} \right\}. \tag{11.56}$$

Relabeling $\mathbf{k}$ the dummy variable $\mathbf{k}'$ inside the second term:

$$\begin{aligned} \frac{\epsilon(\mathbf{q},\omega)}{\epsilon_{\rm s}} &= 1 - \frac{e^2}{\epsilon_{\rm s} q^2} \sum_{\mathbf{k}} f(\mathbf{k}) \left\{ \frac{1}{E(\mathbf{k}) - E(\mathbf{k}+\mathbf{q}) + \hbar\omega} - \frac{1}{E(\mathbf{k}+\mathbf{q}) - E(\mathbf{k}) + \hbar\omega} \right\} \\ &= 1 - \frac{e^2}{\epsilon_{\rm s} q^2} \sum_{\mathbf{k}} f(\mathbf{k}) \left\{ \frac{1}{E(\mathbf{k}) - E(\mathbf{k}+\mathbf{q}) + \hbar\omega} + \frac{1}{E(\mathbf{k}) - E(\mathbf{k}+\mathbf{q}) - \hbar\omega} \right\}, \end{aligned} \tag{11.57}$$

which we can write as

$$\frac{\epsilon(\mathbf{q},\omega)}{\epsilon_{\rm s}} = 1 + \frac{e^2}{\epsilon_{\rm s} q^2} \sum_{\mathbf{k}} \frac{2f(\mathbf{k})[E(\mathbf{k}) - E(\mathbf{k}+\mathbf{q})]}{(\hbar\omega)^2 - [E(\mathbf{k}) - E(\mathbf{k}+\mathbf{q})]^2}. \tag{11.58}$$

Now let's expand for small deviation $\mathbf{q}$ from $\mathbf{k}$ as we had done above in Eq. (11.49), but retaining also terms of order $q^2$ (for reasons we shall understand soon enough):

$$E(\mathbf{k}) - E(\mathbf{k}+\mathbf{q}) \approx -\nabla_{\mathbf{k}} E(\mathbf{k}) \cdot \mathbf{q} - \frac{1}{2} \mathbf{q} \cdot \nabla^2 E(\mathbf{k}) \cdot \mathbf{q}. \tag{11.59}$$

Inserting this into Eq. (11.58) we see that the term linear in $q$ integrates out to zero, since it is odd in $\mathbf{q}$. Thus the first non-vanishing term will be $(1/2)\mathbf{q}\cdot\nabla^2 E(\mathbf{k})\cdot\mathbf{q} = (q^2/2)\partial^2 E/\partial q^2 = \hbar^2 q^2/(2m^*)$. Thus:

$$\frac{\epsilon(\mathbf{q},\omega)}{\epsilon_{\mathrm{S}}} = 1 - \frac{e^2}{\epsilon_{\mathrm{S}} q^2} \sum_{\mathbf{k}} \frac{f(\mathbf{k})\hbar^2 q^2}{m^*(\hbar\omega)^2} \ . \tag{11.60}$$

Since we have ignored spin,

$$\sum_{\mathbf{k}} f(\mathbf{k}) = n \ . \tag{11.61}$$

Thus, Eq. (11.60) becomes

$$\frac{\epsilon(\mathbf{q},\omega)}{\epsilon_{\mathrm{S}}} = 1 - \frac{e^2 m}{\epsilon_{\mathrm{S}}\omega^2} = 1 - \frac{\omega_{\mathrm{P}}^2}{\omega^2} \ , \tag{11.62}$$

where $\omega_{\mathrm{P}}$ is the plasma frequency associated with the free carriers. It is given by the expression after Eq. (9.75), with the density of electrons in the valence bands replaced by the (much smaller) density of free carriers. The singularity of the dielectric function identifies the frequency of the plasma modes. By expanding the numerator to higher orders in $q$ and considering also the imaginary part of $\epsilon$, one could draw a very close analogy between this (quantum) treatment of the eigenmodes of the electron gas and the classical picture we drew before when dealing with the root-mean-square (rms) amplitude of the field associated with plasmons.

*Nondegenerate Screening at High Temperature*. In the nondegenerate limit (i.e., $f_0(\mathbf{k}) \simeq \exp[-E(\mathbf{k})/(k_{\mathrm{B}}T)]$), we can do better, by finding an expression which describes a general dynamical situation for arbitrary wavelengths, and which reduces to the Debye–Hückel expression in the static- and long-wavelength limits. Following Fetter and Walecka [2], we can convert the sum in Eq. (11.45) into an integral over $\mathbf{k}$. We obtain (the derivation is outlined on pages 237–242):

$$\mathrm{Re}[\beta^2(\mathbf{q},\omega)] \simeq \beta_{\mathrm{DH}}^2 \left(\frac{mk_{\mathrm{B}}T}{2\hbar^2 q^2}\right)^{1/2} \left\{ \Phi\left[\left(\frac{m}{2k_{\mathrm{B}}T}\right)^{1/2}\left(\frac{\omega}{q}+\frac{\hbar q}{2m_{\mathrm{el}}}\right)\right] - \Phi\left[\left(\frac{m}{2k_{\mathrm{B}}T}\right)^{1/2}\left(\frac{\omega}{q}-\frac{\hbar q}{2m_{\mathrm{el}}}\right)\right]\right\} \tag{11.63}$$

$$\mathrm{Im}[\beta^2(\mathbf{q},\omega)] \simeq \beta_{\mathrm{DH}}^2 \left(\frac{\pi m\omega^2}{2k_{\mathrm{B}}Tq^2}\right)^{1/2} \exp\left(-\frac{m\omega^2}{2k_{\mathrm{B}}Tq^2} - \frac{\hbar^2 q^2}{8mk_{\mathrm{B}}T}\right) \frac{\sinh[\hbar\omega/(2k_{\mathrm{B}}T)]}{\hbar\omega/(2k_{\mathrm{B}}T)} \ , \tag{11.64}$$

where

$$\Phi(x) = \frac{1}{\pi^{1/2}} \mathscr{P} \int_{-\infty}^{+\infty} \mathrm{d}y \, \frac{\mathrm{e}^{-y^2}}{x-y} \tag{11.65}$$

is the real part of the *plasma dispersion function* [10] and $m$ is replaced by the effective mass of the lowest-lying valley or conduction band (or valence, for holes). Useful asymptotic and series expansions for the function $\Phi$ are

$$\Phi(x) \simeq \begin{cases} \frac{1}{x}\left(1+\frac{1}{2}x^{-2}+\cdots\right) x \gg 1 \\ 2x\left(1-\frac{2}{3}x^2+\cdots\right) x \ll 1, \end{cases}$$

so that

$$g_1(x) \simeq \begin{cases} \frac{8\pi}{x^2}\left(1+\frac{8\pi}{x^2}+\cdots\right) x \gg 1 \\ 1 - x^{\frac{2}{24\pi}} + \cdots x \ll 1. \end{cases}$$

In the static limit, $\omega \to 0$, we have

$$\beta(\mathbf{q},\omega \to 0)^2 \to \beta_{\mathrm{DH}}^2\, g_1(q\ell)\,, \tag{11.66}$$

where

$$\ell = \left(\frac{2\pi\hbar^2}{mk_{\mathrm{B}}T}\right)^{1/2}$$

is the thermal wavelength and

$$g_1(x) = \frac{2\pi^{1/2}}{x}\Phi\left(\frac{x}{4\pi^{1/2}}\right).$$

At long wavelengths, we recover the static Debye–Hückel expression, $\beta(\mathbf{q} \to \mathbf{0}, 0) \to \beta_{\mathrm{DH}}$, while at short distances screening effects disappear as $1/q^2$. The imaginary part of the dielectric function is related to the lifetime of the electron response: As a function of $\omega$ it peaks at the plasma frequency, while the full-width at half-maximum corresponds to the lifetime plasmons have before decaying into single-particle excitations (electrons "surfing" the waves), a process called "Landau damping."

* Derivation of Eqs. (11.63) and (11.64). *First of all, we must prove the following identity:*

$$\frac{1}{x - x_0 \pm \mathrm{i}\eta} = \mathscr{P} \mp \mathrm{i}\pi\delta(x - x_0)\,, \tag{11.67}$$

*where* $\mathscr{P}$ *means that we should take the "Cauchy principal value" (or "part") when integrating. This identity goes under various names, such as* Sokhotski–Weierstrass *or* Sokhotski–Plemelj theorem *("Sochocki" is the Polish spelling, transliterations vary).*

*In order to prove this, we must recall Cauchy's theorem and the "residue" theorem: If* $f(z)$ *is a function of the complex variable* $z$ *with poles at* $z_j$ *(that is, singularities of the "benign" form* $f(z) \sim a_j/(z - z_j)$ *where* $a_j = Res[f(z_j)]$ *is called the "residue" of* $f$ *at* $z_j$*), then:*

$$\oint f(z)\,\mathrm{d}z = 2\pi\mathrm{i}\sum_j Res[f(z_j)]\,, \tag{11.68}$$

*where the integration contour is a closed loop, the integration path runs counterclockwise around the contour, and the sum extends over all residues enclosed by the contour. This is the "residue" theorem. In particular, if* $f$ *is analytic (i.e., does not have singularities) inside the contour, then the integral along the closed loop will vanish (Cauchy's theorem).*

*Now let's consider the integral:*

$$\oint_C \frac{f(z)}{z-x_0}\,\mathrm{d}z\,, \tag{11.69}$$

*where $f(z)$ is a function which is analytic everywhere (which means $f$ has infinitely many derivatives on the entire complex plane) and the closed contour $C$ is illustrated below in Fig. 11.5.*

*Since we do not enclose the singularity within the contour,*

$$\oint_C \frac{f(z)}{z-x_0}\,\mathrm{d}z = 0\,. \tag{11.70}$$

*Note that the residue of the integrand at $z = x_0$ is $f(x_0)$ and, so, integrating along the full little circle around the pole $x_0$ we get $2\pi\mathrm{i}f(x_0)$. If we integrate clockwise around the upper half of the circle—as indicated in the Fig. 11.5—we get $-\mathrm{i}\pi f(x_0)$. Therefore, breaking the integration contour into its components:*

$$\begin{aligned}\oint_C \frac{f(z)}{z-x_0}\,\mathrm{d}z &= \lim_{\eta\to 0}\int_{\infty}^{x_0-\eta}\frac{f(x)}{x-x_0}\,\mathrm{d}z + \int_{C_0}\frac{f(z)}{z-x_0}\,\mathrm{d}z \\ &\quad + \lim_{\eta\to 0}\int_{x_0+\eta}^{\infty}\frac{f(x)}{x-x_0}\,\mathrm{d}z + \int_{C_\infty}\frac{f(z)}{z-x_0}\,\mathrm{d}z \\ &= \mathscr{P}\int_{-\infty}^{\infty}\frac{f(x)}{x-x_0}\,\mathrm{d}z - \mathrm{i}\pi f(x_0) + \int_{C_\infty}\frac{f(z)}{z-x_0}\,\mathrm{d}z = 0,\end{aligned} \tag{11.71}$$

*having defined the Cauchy principal value:*

$$\mathscr{P}\int_{-\infty}^{\infty}\frac{f(x)}{x-x_0}\,\mathrm{d}z = \lim_{\eta\to 0}\left(\int_{\infty}^{x_0-\eta}\frac{f(x)}{x-x_0}\,\mathrm{d}z + \int_{x_0+\eta}^{\infty}\frac{f(x)}{x-x_0}\,\mathrm{d}z\right)\,. \tag{11.72}$$

*But the integration along the entire contour can be written as the integration over the upper semi-circle at infinity and the integration along the real axis:*

$$\oint_C \frac{f(z)}{z-x_0}\,\mathrm{d}z = \int_{-\infty}^{\infty}\frac{f(x)}{x-x_0}\,\mathrm{d}z + \int_{C_\infty}\frac{f(z)}{z-x_0}\,\mathrm{d}z\,, \tag{11.73}$$

*and so, from Eqs. (11.71) and (11.72):*

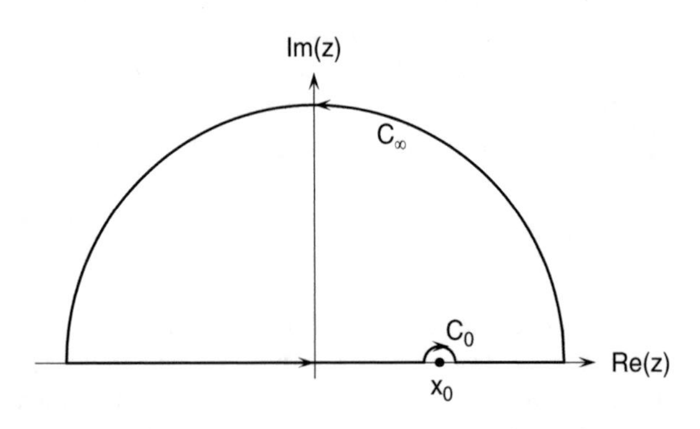


**Fig. 11.5** Integration contour used in Eq. (11.68)

$$\int_{-\infty}^{\infty} \frac{f(x)}{x-x_0} \, dz = \mathscr{P} \int_{-\infty}^{\infty} \frac{f(z)}{z-x_0} \, dz \; - \; i\pi f(x_0) \; . \tag{11.74}$$

*This may seem like a roundabout way of proving our statement, since we could have simply rewritten the integral over the real axis in the desired form without having to deal with the integration over the closed contour. However, having integrated over the entire contour C, we can now note that if we had chosen to integrate counterclockwise around the lower-half little circle* $C_0$ *around* $x_0$ *we would have obtained a contribution of* $+i\pi f(x_0)$*, but we would have enclosed the pole inside the big contour, so that we would have had on the left-hand side of Eq. (11.71) the term* $2i\pi f(x_0)$ *instead of* 0*. Bringing the term to the left-hand side, it would add up with* $+i\pi f(x_0)$ *giving* $-i\pi f(x_0)$*, as before. This shows that we can go around the singularity either clockwise or counterclockwise and still obtain the same result.*

*Yet, one arbitrary choice still remains. So far we have chosen to close the contour by integrating over the path* $C_\infty$ *in the upper half of the complex plane. This can be rephrased mathematically by moving the singularity at* $z = x_0$ *slightly below the real axis, thus setting* $x_0 \to x_0 - i\eta$*, so that closing the loop along* $c_\infty$ *would not give us any residue. Thus, it is often convenient to rewrite Eq. (11.74) as*

$$\int_{-\infty}^{\infty} \frac{f(x)}{x-x_0+i\eta} \, dz = \mathscr{P} \int_{-\infty}^{\infty} \frac{f(x)}{x-x_0} \, dz \; - \; i\pi f(x_0) \; , \tag{11.75}$$

*which is equivalent to Eq. (11.67) after integration.*

*On the contrary, we might as well have chosen to close the loop by integrating over the lower half of the complex plane by setting* $x_0 \to x_0 + i\eta$*. We would have obtained the same result, but the sign of the* $\delta$*-function term would have been opposite:*

$$\int_{-\infty}^{\infty} \frac{f(x)}{x-x_0-i\eta} \, dz = \mathscr{P} \int_{-\infty}^{\infty} \frac{f(x)}{x-x_0} \, dz \; + \; i\pi f(x_0) \; . \tag{11.76}$$

*What does the choice of the sign mean physically? To understand this, we must go back to time-dependent perturbation theory: From pages 439–443, assuming that the perturbation was present at* $t \to -\infty$*, the fifth equation may be rewritten as*

$$a_k(t) = -\frac{1}{\hbar} \int_{-\infty}^{t} H_{kj}(t') \, e^{-i(E_k/\hbar - E_j/\hbar + \omega)t'} \, dt' \; . \tag{11.77}$$

*Replacing* $k \to \mathbf{k}$*,* $j \to \mathbf{k}+\mathbf{q}$*, and* $H_{kj} \to \varphi_{\mathbf{q}}$ *in our continuum case, we see that the denominator* $E(\mathbf{k}) - E(\mathbf{k}+\mathbf{q}) + \hbar\omega$ *in Eq. (11.45) originates from the integration above, having assumed that the perturbation* $\varphi^{(ex)}(\mathbf{r},t)$ *has the harmonic time dependence* $e^{-i\omega t}$ *shown in Eq. (11.25). Therefore, adding a term* $\pm i\hbar s$ *to* $E(\mathbf{k}) - E(\mathbf{k}+\mathbf{q}) + \hbar\omega$ *is equivalent to assuming that the time dependence of the perturbation is of the form* $e^{\pm st}e^{-i\omega t}$*. Physically this is called an "adiabatic turn-on" of the perturbation: If we select* $E(\mathbf{k}) - E(\mathbf{k}+\mathbf{q}) + \hbar\omega + i\hbar s$*, the perturbation acquires an exponential damping backwards in time, which means that the perturbation was turned on slowly in the past (at* $t \to \infty$*) and the perturbed wavefunction responds* after *the perturbation has been turned on. This corresponds to the choice of a* retarded *wavefunction. Conversely, the opposite choice for the sign would correspond to an* advanced *wavefunction. In general, it is the particular boundary or initial condition of our problem which fixes the sign-choice for the "regularizing" term* $e^{\pm st}$*. For example, If we were dealing with emission and absorption of phonons, we could consider the retarded wavefunction in the case of absorption (the phonon present at* $t \to -\infty$ *results in a perturbed wavefunction at later times), the advanced wavefunction in the case of emission (the wavefunction causes the presence of the perturbing phonon at later times). In both cases, the turn-on and turn-off are called "adiabatic" because the limit* $s \to 0$ *is always implied and this corresponds to a turn-on* *(or turn-off) which is infinitely slow. In our case, the choice of the retarded wavefunction leads us to the sign-choice we have made so far.*

*Let's now consider Eq. (11.45) and convert the sum over* **k** *into an integral:*

$$\frac{\epsilon(\mathbf{q},\omega)}{\epsilon_{\rm s}} = 1 - \frac{e^2}{q^2\epsilon_{\rm s}} \int \frac{d\mathbf{k}}{(2\pi)^3} \frac{f(\mathbf{k})-f(\mathbf{k}+\mathbf{q})}{E(\mathbf{k})-E(\mathbf{k}+\mathbf{q})+\hbar\omega+i\hbar s} . \tag{11.78}$$

*Let's consider the integral*

$$\mathscr{I} = \int \frac{d\mathbf{k}}{(2\pi)^3} \frac{f(\mathbf{k})-f(\mathbf{k}+\mathbf{q})}{E(\mathbf{k})-E(\mathbf{k}+\mathbf{q})+\hbar\omega+i\hbar s} . \tag{11.79}$$

*Let's define*

$$\mathbf{k}' = \mathbf{k} + \frac{1}{2}\mathbf{q} , \tag{11.80}$$

*so that*

$$\mathbf{k} = \mathbf{k}' - \frac{1}{2}\mathbf{q} , \quad \mathbf{k}+\mathbf{q} = \mathbf{k}' + \frac{1}{2}\mathbf{q} , \quad d\mathbf{k} = d\mathbf{k}' . \tag{11.81}$$

*Then:*

$$\mathscr{I} = \int \frac{d\mathbf{k}'}{(2\pi)^3} \frac{f(\mathbf{k}-\mathbf{q}/2)-f(\mathbf{k}'+\mathbf{q}/2)}{E(\mathbf{k}'-\mathbf{q}/2)-E(\mathbf{k}'+\mathbf{q}/2)+\hbar\omega+i\hbar s} . \tag{11.82}$$

*Since we integrate over the entire* **k***-space, we can let* $\mathbf{k}' \to -\mathbf{k}'$ *inside the first integral. Since* $f(\mathbf{k}) = f(-\mathbf{k})$*, we can rewrite this integral as*

$$\mathscr{I} = \int \frac{d\mathbf{k}'}{(2\pi)^3} f(\mathbf{k}+\mathbf{q}/2) \left\{ \frac{1}{E(\mathbf{k}'+\mathbf{q}/2)-E(\mathbf{k}'-\mathbf{q}/2)+\hbar\omega+i\hbar s} - \frac{1}{E(\mathbf{k}'-\mathbf{q}/2)-E(\mathbf{k}'+\mathbf{q}/2)+\hbar\omega+i\hbar s} \right\} . \tag{11.83}$$

*Now:*

$$E(\mathbf{k}'+\mathbf{q}/2) = \frac{\hbar^2}{2m_{\rm el}} \left( k^2 + \frac{1}{4}q^2 + \mathbf{k}\cdot\mathbf{q} \right) , \tag{11.84}$$

$$E(\mathbf{k}'-\mathbf{q}/2) = \frac{\hbar^2}{2m_{\rm el}} \left( k^2 + \frac{1}{4}q^2 - \mathbf{k}\cdot\mathbf{q} \right) , \tag{11.85}$$

*so:*

$$\mathscr{I} = \int \frac{d\mathbf{k}'}{(2\pi)^3} f(\mathbf{k}+\mathbf{q}/2) \left\{ \frac{1}{\hbar\omega+i\hbar s+(\hbar^2/m)\mathbf{k}\cdot\mathbf{q}} - \frac{1}{\hbar\omega+i\hbar s-(\hbar^2/m)\mathbf{k}\cdot\mathbf{q}} \right\} . \tag{11.86}$$

*Let's now use Boltzmann statistics with Fermi level* $E_{\rm F}$ *(indeed we are within the high-temperature, nondegenerate approximation) and introduce cylindrical coordinates with the z-axis along the direction of* $\mathbf{q}$*:*

$$\mathbf{k} = \mathbf{k}_{\perp} + \hat{\mathbf{q}} k_z , \tag{11.87}$$

*where* $\hat{\mathbf{q}}$ *is the unit vector along the direction of* $\mathbf{q}$. *Then:*

$$\mathscr{I} = e^{\beta E_F} \int \frac{d\mathbf{k}_{\perp}}{(2\pi)^2} \exp\left(-\beta \frac{\hbar^2 k_{\perp}^2}{2m_{el}}\right) \int \frac{dk_z}{2\pi} \exp\left(-\beta \frac{\hbar^2 (k_z + q/2)^2}{2m_{el}}\right)$$
$$\times \left\{ \frac{1}{\hbar\omega + i\hbar s + (\hbar^2/m) k_z q} - \frac{1}{\hbar\omega + i\hbar s - (\hbar^2/m) k_z q} \right\} , \tag{11.88}$$

*where, as usual,* $\beta = 1/(k_B T)$. *The integration over* $\mathbf{k}_{\perp}$ *can be done easily: Going to polar coordinates and setting* $E_{\perp} = \hbar^2 k_{\perp}^2/(2m_{el})$, *so that* $k_{\perp} dk_{\perp} = (m/\hbar^2) dE_{\perp}$, *we have*

$$\int \frac{d\mathbf{k}_{\perp}}{(2\pi)^2} \exp\left(-\beta \frac{\hbar^2 k_{\perp}^2}{2m_{el}}\right) = \int_0^{2\pi} d\phi \int_0^{\infty} \frac{dk_{\perp}}{(2\pi)^2} k_{\perp} e^{-\beta\hbar^2 k_{\perp}^2/(2m_{el})}$$
$$= \frac{1}{2\pi} \frac{m}{\hbar^2} \int_0^{\infty} dE_{\perp} e^{-\beta E_{\perp}} = \frac{m}{2\pi\hbar^2\beta} . \tag{11.89}$$

*In order to handle the remaining integral over* $k_z$, *let's set the new integration variable (writing now* $k$ *for* $k_z$*):*

$$\begin{aligned} x^2 &= \frac{\beta\hbar^2}{2m_{el}} \left(k + \frac{q}{2}\right)^2 , \\ k + \frac{q}{2} &= \left(\frac{2m_{el}}{\beta\hbar^2}\right)^{1/2} x , \\ dk &= \left(\frac{2m_{el}}{\beta\hbar^2}\right)^{1/2} dx , \\ \frac{\hbar^2 kq}{m} &= \left(\frac{2\hbar^2}{m\beta}\right)^{1/2} qx - E_q , \end{aligned} \tag{11.90}$$

*where* $E_q = \hbar^2 q^2/(2m_{el})$. *Thus we have*

$$\mathscr{I} = \frac{m e^{\beta E_F}}{2\pi\beta\hbar^2} \left(\frac{2m_{el}}{\beta\hbar^2}\right)^{1/2} \int_{-\infty}^{\infty} dx\, e^{-x^2} \left\{ \frac{1}{\hbar\omega + i\hbar s + E_q - [2\hbar^2/(m\beta)]^{1/2} qx} \right.$$
$$\left. - \frac{1}{\hbar\omega + i\hbar s - E_q + [2\hbar^2/(m\beta)]^{1/2} qx} \right\} , \tag{11.91}$$

*or, setting*

$$x_{\pm} = \left(\frac{m\beta}{2\hbar^2}\right)^{1/2} \frac{1}{q} \left[\hbar\omega \pm \frac{\hbar^2 q^2}{2m_{el}} + i\hbar s\right] , \tag{11.92}$$

*we now have*

$$\mathscr{I} = \frac{m\mathrm{e}^{\beta E_\mathrm{F}}}{2\pi\beta\hbar^2} \frac{m}{2\pi\hbar^2 q} \int_{-\infty}^{\infty} \mathrm{d}x \left\{ \frac{\mathrm{e}^{-x2}}{x_- + x} - \frac{\mathrm{e}^{-x2}}{x_+ + x} \right\} . \tag{11.93}$$

*Now we make use of Eq. (11.67). Considering first the real part, defining the "plasma dispersion function"*

$$\Phi(y) = \frac{1}{\pi^{1/2}} \mathscr{P} \int_{-\infty}^{\infty} \mathrm{d}x \, \frac{\mathrm{e}^{-x^2}}{x - y} , \tag{11.94}$$

*after a little algebra we get*

$$Re(\mathscr{I}) = \frac{n}{\hbar q} \left( \frac{1}{2} m\beta \right)^{1/2} \left[ \Phi(y_-) - \Phi(y_+) \right] , \tag{11.95}$$

*where*

$$y_\pm = \left( \frac{\beta m}{2} \right)^{1/2} \left( \frac{\omega}{q} \pm \frac{\hbar q}{2m_\mathrm{el}} \right) . \tag{11.96}$$

*This leads to Eq. (11.63). The imaginary part, Eq. (11.64), should follow quite easily now from the imaginary part of Eq. (11.93) using (11.67) as well.*

*Degenerate Screening at Low Temperature.* In the opposite limit of a degenerate electron gas at low temperature, we can approximate the Fermi function $f_0(\mathbf{k})$ with a step function at the Fermi level. Then, the screening wavevector becomes

$$\beta(\mathbf{q},\omega)^2 \simeq \beta_{TF}^2 \left\{ \frac{1}{2} + \frac{1}{8\eta^3} \left[ [(\gamma+\eta^2)^2 - \eta^2] \ln \left| \frac{\eta+\eta^2+\gamma}{\eta-\eta^2-\gamma} \right| \right. \right.$$
$$\left. \left. + (\gamma-\eta^2)^2 - \eta^2] \ln \left| \frac{\eta+\eta^2-\gamma}{\eta-\eta^2+\gamma} \right| \right] \right\} , \tag{11.97}$$

where $\eta = q/(2k_\mathrm{F})$ and $\gamma = (\hbar\omega + is)/(4E_\mathrm{F})$. In the limit $q \to 0$ we find

$$\frac{\epsilon(q,\omega)}{\epsilon_\mathrm{s}} = 1 - \frac{\omega_p^2}{\omega^2} \left( 1 + \frac{3}{5} \frac{\upsilon_\mathrm{F} q^2}{\omega^2} \right) , \tag{11.98}$$

where $\upsilon_\mathrm{F} = \hbar k_\mathrm{F}/m$ is the Fermi velocity, which shows how higher-order corrections to Eq. (11.60) introduce a (weak) dispersion for the plasma modes. In the static limit, Eq. (11.97) becomes

$$\beta(\mathbf{q},0)^2 \simeq \beta_{TF}^2 \left\{ \frac{1}{2} + \frac{4k_\mathrm{F}^2 - q^2}{8k_\mathrm{F} q} \ln \left| \frac{2k_\mathrm{F} + q}{2k_\mathrm{F} - q} \right| \right\} . \tag{11.99}$$

As $q \to 0$ this expression tends to the Thomas–Fermi expression Eq. (11.53), while screening effects become weaker at short distances, similarly to what is implied by the high-temperature limit, Eq. (11.66). These results have been obtained by Stern [11].

#### 11.3.2.3 *Screened Impurity and Friedel's Sum Rule

A particularly important application is given by the static potential of an ionized impurity, such as dopant impurities in semiconductors. Let's assume that the impurity has a net charge $eZ$, and let's keep the discussion as simple as possible. From Eq. (11.45), the full expression Eq. (11.63), the static limit Eq. (11.66), and from the approximate expressions for the plasma dispersion function, we have

$$\varphi(\mathbf{r}) = \frac{eZ}{\epsilon_s}\sum_{\mathbf{q}}\frac{\mathrm{e}^{\mathrm{i}\mathbf{q}\cdot\mathbf{r}}}{q^2+\beta(\mathbf{q},0)^2} = \frac{eZ}{\epsilon_s r}\frac{-\mathrm{i}}{4\pi^2}\int_{-\infty}^{\infty}\mathrm{d}q\,q\,\frac{\mathrm{e}^{iqr}}{q^2+\beta(q,0)^2}\,, \tag{11.100}$$

having assumed that the screening wavevector depends only on the magnitude of $\mathbf{q}$, since the impurity potential does not depend on time. At large distances away from the impurity ($r >> 1/\beta$), we can approximate $\beta$ with its long-wavelength limit, Eq. (11.51), obtaining the well-known Yukawa-type potential:

$$\varphi(\mathbf{r}) = \frac{eZ}{4\pi\epsilon_s r}\mathrm{e}^{-\beta_s r}\,. \tag{11.101}$$

As we have seen, the particular form of the screening parameter $\beta(q,0)$ may be quite complicated, depending on the approximation embraced. Thus, expression (11.101) above will fail at short distances, when the $q$-dependence of $\beta(q,0)$, as given in Eq. (11.66), is important. Therefore, the scattering cross section obtained from Eq. (11.101) above may be inaccurate. However, we can follow a completely different path to derive a particular condition on $\beta$, valid generally for arbitrary potentials. We shall consider the simpler case of zero temperature, but the general formulation at arbitrary temperature can be obtained by modifying all sums up to the Fermi level as sums over all states weighted by their thermal occupation.

In the absence of the impurity, the electron gas consists of states occupied to the Fermi level. Instead of considering plane waves in a box, as usually done, we can consider a spherical volume and express the plane waves in spherical harmonics.

* *We write the Laplace operator in polar coordinates and write Schrödinger equation as*

$$-\frac{\hbar^2}{2m^*}\frac{1}{r^2\sin\theta}\left[\sin\theta\frac{\partial}{\partial r}\left(r^2\frac{\partial\psi}{\partial r}\right)+\frac{\partial}{\partial\theta}\left(\sin\theta\frac{\partial\psi}{\partial\theta}\right)+\frac{1}{\sin\theta}\frac{\partial^2\psi}{\partial\phi^2}\right]$$
$$+V(r)\psi = E\psi\,. \tag{11.102}$$

*Separating variables, we find that the azimuthal component $\Phi(\phi)$ must satisfy the equation*

$$\frac{1}{\Phi(\phi)}\frac{\partial^2\Phi}{\partial\phi^2} = constant\,. \tag{11.103}$$

*Since we require single-valued solutions, the constant must be equal to $-m^2$, where $m$ is an integer, so that $\Phi(\phi) = \mathrm{e}^{\pm im\phi}$. The polar component $\Theta(\theta)$ must satisfy the equation:*

$$\frac{1}{\sin\theta}\frac{\partial}{\partial\theta}\left(\sin\theta\frac{\partial\Theta}{\partial\theta}\right) - \frac{m^2}{\sin^2\theta}\Theta = -\,l(l+1)\Theta\,, \tag{11.104}$$

*where the "separation constant" has been set to the total angular momentum quantum number $l(l+1)$. Finally, the radial component of the wavefunction must satisfy the equation:*

$$-\frac{\hbar^2}{2m^*}\frac{1}{r^2}\frac{\mathrm{d}}{\mathrm{d}r}\left(r^2\frac{\mathrm{d}R}{\mathrm{d}r}\right)+\left[\frac{\hbar^2 l(l+1)}{2m^* r^2}+V(r)\right]R(r)=E\,R(r)\,. \tag{11.105}$$

*At large distances $r$ the potential $V$ becomes negligible and the solution of Eq. (11.105) is the spherical Bessel function $j_{kl}(r)$ (where as usual $k=(2m^*E)^{1/2}/\hbar$). Its asymptotic behavior at large $r$ is $\sim \sin(kr-\pi l/2)/r$.*

Each state is then labeled by the magnitude of the **k**-vector and by its angular momentum quantum number $l$:

$$\psi_{k,l}(r,\theta)\ \sim\ \frac{1}{r}\sin\left(kr-\frac{1}{2}l\pi\right)\,P_l(\cos\theta)\,, \tag{11.106}$$

(where $P_l(x)$ is the Legendre polynomial of order $l$) at large distances. We assume that the wavefunctions vanish on some boundary $r=R$, where $R$ is sufficiently large so that the (screened) potential vanishes. Therefore, the allowed values of $k$ are labeled by an integer $n$ such that $k_nR=n\pi+\pi l/2$. The number of electrons, $N_R$, in the volume can be obtained by counting all states up to the Fermi level, that is

$$N_R\sim 2\sum_{n<n_\mathrm{F},\ l\leq n}(2l+1)\,, \tag{11.107}$$

where a factor 2 comes from spin, the factor $(2l+1)$ represents the number of all possible angular momentum states for each $l$, and the maximum "allowed" radial quantum number $n_\mathrm{F}$ is set by the Fermi level, i.e., $n_\mathrm{F}=k_\mathrm{F}R/\pi-l/2$. If we now introduce the impurity potential, the radial component of the new wavefunction obeys the equation:

$$\left[-\frac{1}{r^2}\frac{\mathrm{d}}{\mathrm{d}r}\left(r^2\frac{\mathrm{d}}{\mathrm{d}r}\right)+\frac{l(l+1)}{r^2}+\frac{2me}{\hbar^2}\varphi(r)-k^2\right]\,R_{l,k}(r)=0\,. \tag{11.108}$$

Now, thanks to the spherical symmetry of the scattering potential, we know that at large distances the angular dependence of the wavefunction must remain unaltered. All the external potential can do is to shift the phases of the partial radial waves. This is known as the *partial-wave method* [12]. Therefore, at large distances the wavefunctions can be approximated by

$$\phi_{k,l}(r,\theta)\ \sim\ \frac{1}{r}\,\sin\left[kr-\frac{1}{2}l\pi+\eta_l(k)\right]\,P_l(\cos\theta)\,, \tag{11.109}$$

where $\eta_l(k)$ is the phase shift of the partial wave, given by

$$\sin[\eta_l(k)]=-\frac{2m_\mathrm{el}ek}{\hbar^2}\int_0^\infty \mathrm{d}r\ r^2 j_l(kr)R_{l,k}(r)\varphi(r). \tag{11.110}$$

Note that in the Born approximation, we assume that all shifts are small and we substitute $R_{l,k}(r)$ with its "free particle" expression $R_{l,k}(r)\simeq j_l(kr)$, so that

$$\eta_l^\mathrm{Born}(k)\simeq -\frac{2m_\mathrm{el}ek}{\hbar^2}\int_0^\infty \mathrm{d}r\ r^2 j_l^2(kr)\varphi(r). \tag{11.111}$$

It is easy to see that the phase shift results in additional electrons being placed in the volume: For every $l$-state, the maximum index $n_F$ is increased by $\eta_l(k_F)/\pi$. Therefore, the change in the number of electrons in the volume is given by

$$\delta N_R = \frac{2}{\pi}\sum_l (2l+1)\eta_l(k_F). \tag{11.112}$$

This extra charge must be accounted for by the electrons required to screen the perturbing potential. Therefore, since precisely $Z$ electrons are needed to neutralize the charge of the impurity at large distances, we must have

$$Z = \frac{2}{\pi}\sum_l (2l+1)\eta_l(k_F). \tag{11.113}$$

This is *Friedel's sum rule* [13]. This result depends only on the assumption that the Fermi level is unchanged far away from the impurity and that the perturbing potential must be neutralized by excess carriers at some finite distance.

In the presence of a general potential, we can compute the scattering cross section as follows. We start with a screened potential of the Yukawa form (11.101), where $\beta$ is now treated as a parameter to be determined. We compute the phase shifts and fix $\beta$ so that Friedel's sum rule, Eq. (11.113), is satisfied. Finally, we obtain the cross section from the well-known formula

$$\sigma(k,\theta) = \frac{1}{k^2}\left|\sum_l (2l+1)\sin[\eta_l(k)]e^{i\eta_l(k)}P_l(\cos\theta)\right|^2 . \tag{11.114}$$

This result is usually more accurate than the use of the matrix element $\varphi_{\mathbf{q}}$, as in the Born approximation. In particular, the difference between attractive and repulsive Coulomb scattering (which vanishes in the Born approximation), the charge oscillations associated with either the wavefunctions (11.109) or the screening parameter (11.66), and the general dependence of $\beta$ on $q$ will be accounted for.

#### 11.3.2.4 *Two-Dimensional Screening

The response of a two-dimensional electron (or hole) gas to an external perturbation can be analyzed in the same way. There are, however, slight complications caused mainly by the fact that carriers can occupy many subbands. Therefore, it is useful to derive here a general expression for screening caused by carriers in inversion layers or quantum wells. A complete discussion can be found in the Ando–Fowler–Stern review [14], in Ridley's text [15], but a specific discussion has been given by Dahl and Sham [16]. Here we shall simplify somewhat the picture. In particular, we shall consider here only longitudinal screening. Intersubband excitations, considered by Dahl and Sham, can play a non-negligible role, particularly in (resonant) dynamic situations and give rise to a transverse component of the polarizability, since intersubband transitions cause a redistribution of the charge density in the direction normal to plane of quantization.

Assuming that the external perturbation is isotropic in the plane of confinement, it is convenient to express it in terms of its Fourier–Bessel expansion:

$$\varphi^{(ex)}(\mathbf{R},z,t) = 2\pi\int dQ\, Q\, J_0(QR)\, \varphi_Q^{(ex)}(z)e^{i\omega t}, \tag{11.115}$$

where $J_0$ is the spherical Bessel function of order 0 and the variables $\mathbf{R}$ and $\mathbf{Q}$ denote the coordinate and wavevector in the quantization plane and $z$ is the coordinate normal to the plane. The electron wavefunctions in each subband labeled by $\mu$ can be written as

$$\psi_{\mathbf{K}}^{(\mu)}(\mathbf{R},z) = \mathrm{e}^{\mathrm{i}\mathbf{K}\cdot\mathbf{R}}\zeta^{(\mu)}(z)\sum_{\mathbf{G}}\mathrm{e}^{\mathrm{i}\mathbf{G}\cdot\mathbf{r}}u_{\mathbf{G}}(\mathbf{k},0) \equiv |\mathbf{K},\mu\rangle. \quad (11.116)$$

Under the *net* perturbation $\varphi = \varphi^{(\mathrm{ex})} + \varphi^{(\mathrm{pol})}$, the wavefunctions are modified, similarly to Eq. (11.28), as follows

$$\psi_{\mathbf{K}}^{(\mu)}(\mathbf{R},z) \to \phi_{\mathbf{K}}^{(\mu)}(\mathbf{R},z) = \psi_{\mathbf{K}}^{(\mu)}(\mathbf{R},z) + \sum_{\mathbf{K}',\mu'}\frac{e\langle\mathbf{K}',\mu'|\varphi|\mathbf{K},\mu\rangle}{E_\mu(\mathbf{K}) - E_{\mu'}(\mathbf{K}') + \hbar\omega + \mathrm{i}\hbar s}\psi_{\mathbf{K}'}^{(\mu')}(\mathbf{R},z). \quad (11.117)$$

We have noticed before that the matrix element $\langle\mathbf{K}',\mu'|\varphi|\mathbf{K},\mu\rangle$ is, in principle, a quite complicated object, since it involves the Bloch wavefunctions themselves. However, assuming as we did before that the external perturbation is of wavelength much longer than any nonzero $\mathbf{G}$-vector, the overlap term over Bloch factors yields unity and

$$\langle\mathbf{K}',\mu'|\varphi|\mathbf{K},\mu\rangle \simeq \varphi_{\mathbf{K}-\mathbf{K}',\mu\mu'} \equiv \int \mathrm{d}z\; \zeta^{(\mu)}(z)\; \varphi_{\mathbf{K}-\mathbf{K}'}(z)\; \zeta^{(\mu')}(z). \quad (11.118)$$

In complete analogy with the derivation of Eq. (11.34), we have for the Fourier–Bessel components of the induced charge density:

$$\delta\rho_Q(z) = 2e^2\sum_{\mathbf{K},\mu,\mu'}\frac{[g_\mu f^{(2\mathrm{D})}(\mathbf{K},\mu) - g_{\mu'}f^{(2\mathrm{D})}(\mathbf{K}+\mathbf{Q},\mu')]}{E_\mu(\mathbf{K}) - E_{\mu'}(\mathbf{K}+\mathbf{Q}) + \hbar\omega + \mathrm{i}\hbar s}\varphi_{Q,\mu\mu'}\zeta_\mu(z)\zeta\mu'(z), \quad (11.119)$$

where we have indicated by $g_\mu$ the degeneracy of the $\mu$th subband and with $f^{(2\mathrm{D})}$ the Fermi–Dirac distribution for the two-dimensional dispersion. The form of the Poisson equation replacing Eq. (11.35) in a two-dimensional context is less transparent. Indeed, both in the case of inversion layers as well as of quantum wells, the discontinuity of the permittivity $\epsilon_\mathrm{s}$ across interfaces, such as the Si–$SiO_2$ or GaAs–$Al_xGa_{1-x}As$ interfaces, should be handled by the appropriate boundary conditions, reflecting the presence of image charges. In order to keep the discussion at a general level, we need to consider the Green's function $G_Q(z,z')$ such that

$$\left(\frac{\mathrm{d}^2}{\mathrm{d}z^2} - Q^2\right)G_Q(z,z') = \delta(z-z'), \quad (11.120)$$

subject to the boundary conditions appropriate to the geometry at hand. A notable example is the geometry of two semi-infinite media (namely, Si and $SiO_2$ characterized by the dielectric constants $\epsilon_\mathrm{s}$ and $\epsilon_\mathrm{ox}$), for which we have (see Ando–Fowler–Stern [14]):

$$G_Q(z,z') = \frac{1}{2Q}\left[\mathrm{e}^{-Q|z-z'|} + \frac{\epsilon_\mathrm{s} - \epsilon_\mathrm{ox}}{\epsilon_\mathrm{s} + \epsilon_\mathrm{ox}}\mathrm{e}^{-Q(z+z')}\right]. \quad (11.121)$$

Expressions for more complicated geometries have been given by Dahl and Sham [16]. Then, from Eq. (11.120) we can obtain the self-consistent equation for the net potential:

$$\varphi_Q(z) = \varphi_Q^{(\mathrm{ex})}(z) + \int \mathrm{d}z'\; G_Q(z,z')\frac{\delta\rho_Q(z')}{\epsilon_\mathrm{s}}. \quad (11.122)$$

Only the matrix elements over the envelope wavefunctions $\zeta^{(\mu)}(z)$ are required in order to evaluate the scattering rates caused by the perturbation $\varphi^{(\text{ex})}$. From Eqs. (11.118), (11.119), and (11.122) we have

$$\varphi_{Q,\mu\mu'} = \varphi^{(\text{ex})}_{Q,\mu\mu'} - \sum_{\lambda,\lambda'} \frac{\beta^{(2\text{D})}_{\lambda\lambda'}(\mathbf{Q},\omega)}{Q} \mathcal{G}_{Q,\mu\mu',\lambda\lambda'}\ \varphi_{Q,\lambda\lambda'}, \tag{11.123}$$

where the "form factor" $\mathcal{G}$ is given by

$$\mathcal{G}_{Q,\mu\mu',\lambda\lambda'} = 2\int \mathrm{d}z\mathrm{d}z'\ QG_Q(z,z')\zeta^{(\mu)}(z)\zeta^{(\mu')}(z)\zeta^{(\lambda)}(z')\zeta^{(\lambda')}(z'), \tag{11.124}$$

and the two-dimensional screening wavevector is given by

$$\beta^{(2\text{D})}_{\mu\mu'}(\mathbf{Q},\omega) = \frac{e^2}{2\epsilon_\text{s}} \sum_{\mathbf{K}} \frac{g_\mu f^{(2\text{D})}(\mathbf{K},\mu) - g_{\mu'} f^{(2\text{D})}(\mathbf{K}+\mathbf{Q},\mu')}{E_{\mu'}(\mathbf{K}+\mathbf{Q}) - E_\mu(\mathbf{K}) - \hbar\omega - \mathrm{i}\hbar s}. \tag{11.125}$$

The self-consistent Eq. (11.123) can be recast in terms of a dielectric matrix:

$$\varphi_{Q,\mu\mu'} = \sum_{\lambda\lambda'} [\epsilon_{\mu\mu',\lambda\lambda'}(\mathbf{Q},\omega)]^{-1} \varphi^{(\text{ex})}_{Q,\lambda\lambda'}, \tag{11.126}$$

where

$$\epsilon_{\mu\mu',\lambda\lambda'}(\mathbf{Q},\omega) = \delta_{\mu\mu'}\delta_{\lambda\lambda'} - \frac{\beta^{(2\text{D})}_{\lambda\lambda'}(\mathbf{Q},\omega)}{Q} \mathcal{G}_{Q,\mu\mu',\lambda\lambda'}. \tag{11.127}$$

The problem of finding the net potential from the external perturbation is now translated to the problem of inverting this dielectric matrix.

As described by Eq. (11.126), the problem is a complicated one. A relative simplification can be obtained by noting that terms of $\mathcal{G}_{Q,\mu,\mu',\lambda\lambda'}$ with $\lambda'$ different from $\lambda$ do not contribute to a net change of the charge in the layer, but simply to a polarization of the charge in the direction normal to the plane of quantization. This is seen by integrating Eq. (11.119) along $z$: for $\lambda \neq \lambda'$ we have $\delta\rho_Q = 0$. Thus we can consider only the "longitudinal" dielectric function, by retaining only "diagonal" terms $\mathcal{G}_{Q,\mu\mu',\lambda\lambda}$ and setting $\beta^{(2\text{D})}_{\lambda\lambda} \equiv \beta^{(2\text{D})}_{\lambda}$. Thus, the change of the charge density becomes proportional to the expectation value of the potential in each subband, which is the customary assumption. The problem given by Eq. (11.126) can now be solved in two steps. First, the "diagonal" problem $\mu = \mu'$ is solved by inverting the matrix $\epsilon_{\mu\mu,\lambda\lambda}(\mathbf{Q},\omega)$. Once the diagonal matrix elements $\varphi_{Q,\mu\mu,\lambda\lambda}$ are known, the off-diagonal terms can be evaluated directly from Eq. (11.126).

Equation (11.125) is very similar to the Lindhard expression, Eq. (11.45), and we can obtain a closed-form expression for it in the same limits considered above. We shall consider only the diagonal (in the subband indices) screening wavevector $\beta^{(2\text{D})}_{Q,\mu\mu}$.

*Static Screening, Long-Wavelength Limit.* For static perturbations, we proceed along the path followed to derive Eq. (11.48) obtaining

$$\beta^{(2\text{D})}_{s,\mu} = \frac{e^2}{2\epsilon_\text{s}} \int \mathrm{d}E\ \mathcal{D}_\mu(E) \left[\frac{\partial f^{(2\text{D})}(\mathbf{k},\mu)}{\partial E_\text{F}}\right] = \frac{e^2}{2\epsilon_\text{s}} \frac{\partial n_\mu}{\partial E_\text{F}}, \tag{11.128}$$

where $\mathscr{D}_\mu$ and $n_\mu$ refer to the DOS and density in the $\mu$th subband, respectively. In the nondegenerate case (i.e., at high temperatures), $\partial n_\mu/\partial E_F \simeq n_\mu/(k_B T)$ and we obtain the two-dimensional equivalent of the Debye–Hückel limit:

$$\beta^{(2D)}_{DH,\mu} \simeq \frac{e^2 n_\mu}{2\epsilon_s k_B T}. \tag{11.129}$$

In the opposite case of a degenerate gas (low temperature), $\partial f/\partial E_F \simeq \delta(E-E_F)$ and we have the equivalent of the Thomas–Fermi expression:

$$\beta^{(2D)}_{TF,\mu} \simeq \frac{e^2}{2\epsilon_s}\mathscr{D}_\mu(E_F) = \frac{e^2}{2\epsilon_s}\frac{g_\mu m_{d,\mu}}{\pi\hbar^2}, \tag{11.130}$$

for occupied subbands, $m_{d,\mu}$ being the DOS effective mass in the $\mu$th subband. Note that this expression implies, somewhat counter-intuitively, that the screening wavevector is independent of the density of free carriers. This depends on our perhaps oversimplified assumption of ignoring higher-order terms in $\mathbf{Q}$ in order to derive Eq. (11.128). We shall see below [Eq. (11.139)] that a more accurate derivation overcomes this difficulty, at least partially. However, it is generally true that screening in the two-dimensional quantum limit depends weakly on the density of free carriers, as compared to the bulk case.

*Two-Dimensional Plasma Oscillations.* Proceeding in analogy with Eqs. (11.60) and (11.62), we obtain in the limit of long wavelengths and high frequencies, $\hbar\omega \gg E_\mu(\mathbf{K}+\mathbf{Q}) - E_\mu(\mathbf{K})$ for all states of interest:

$$\beta^{(2D)}_\mu(\mathbf{Q},\omega) \simeq \frac{e^2 g_\mu n_\mu}{2\epsilon_s m_\mu \omega^2} Q^2. \tag{11.131}$$

This expression, when substituted into the dielectric matrix, Eq. (11.127), is still quite complicated. In the simple case of only one subband ($\mu = 0$) occupied, and in the isotropic case described by a single effective mass $m_\mu$, we have

$$\frac{\epsilon_0(\mathbf{Q},\omega)}{\epsilon_s} \simeq 1 - \frac{\omega^{(2D)}_P(Q)^2}{\omega^2}, \tag{11.132}$$

where the "plasma frequency" now exhibits a strong dispersion, unlike the almost dispersionless behavior of the bulk plasmons:

$$\omega^{(2D)}_P(Q)^2 = \frac{e^2 n_0}{2\epsilon_s m_0}\, Q\, \mathscr{G}_{Q,00,00} \simeq \frac{e^2 n_0}{2\bar{\epsilon} m_0} Q, \tag{11.133}$$

where in deriving the last expression we have ignored the thickness of the two-dimensional layer (i.e., $\zeta^{(0)}(z) \simeq \delta(z)$) to evaluate explicitly the form factor $\mathscr{G}$, we have considered the geometry of two semi-infinite media [i.e., the Green function is given by Eq. (11.121)], and we have set $\bar{\epsilon} = (\epsilon_s + \epsilon_{ox})/2$. The presence of boundaries will alter the dispersion of the plasma modes, because of the $Q$-dependence of the form factor $\mathscr{G}$. A notable case is the presence of a gate insulator of finite thickness: In the limit of a very thin insulator, the dependence of the plasma frequency on $Q$ will actually become linear [14].

*Nondegenerate Screening at High Temperature.* In the high-temperature limit, we can follow once more Fetter and Walecka as we have done before in the case of three dimensions, the only difference being an overall factor $2/\ell_\mu$. Thus, in analogy with Eqs. (11.63) and (11.64), we have

$$\mathrm{Re}[\beta_\mu^{(2D)}(\mathbf{Q},\omega)] \simeq \beta_{\mathrm{DH},\mu}^{(2D)} \frac{\pi^{1/2}}{Q\ell_\mu} \left\{ \Phi\left[\left(\frac{m_\mu}{2k_\mathrm{B}T}\right)^{1/2}\left(\frac{\omega}{Q}+\frac{\hbar Q}{2m_\mu}\right)\right] - \Phi\left[\left(\frac{m_\mu}{2k_\mathrm{B}T}\right)^{1/2}\left(\frac{\omega}{Q}-\frac{\hbar Q}{2m_\mu}\right)\right]\right\} \quad (11.134)$$

$$\mathrm{Im}[\beta_\mu^{(2D)}(\mathbf{Q},\omega)] \simeq \beta_{\mathrm{DH},\mu}^{(2D)} \frac{\pi\hbar\omega}{Q\ell_\mu k_\mathrm{B}T} \exp\left(-\frac{m_\mu\omega^2}{2k_\mathrm{B}TQ^2}-\frac{\hbar^2Q^2}{8m_\mu k_\mathrm{B}T}\right)\frac{\sinh[\hbar\omega/(2k_\mathrm{B}T)]}{\hbar\omega/(2k_\mathrm{B}T)}. \quad (11.135)$$

In the long-wavelength limit we have

$$\beta_\mu^{(2D)}(Q,\omega\to 0) \simeq \beta_{\mathrm{DH},\mu}^{(2D)} g_1(Q\ell_\mu). \quad (11.136)$$

*Screening in the Quantum Limit.* In the so-called extreme quantum limit of zero temperature, the two-dimensional electron gas is fully degenerate and we assume that only the ground state subband is occupied. Under these conditions Stern has obtained an analytic expression for the dielectric function [11]:

$$\beta^{(2D)}(Q,\omega) = \frac{e^2 m_0 n_0}{\hbar^2 k_\mathrm{F} Q}\left\{\left[C_+(A_+^2-1)^{1/2}+C_-(A_-^2-1)^{1/2}-\frac{Q}{2k_\mathrm{F}}\right] + \mathrm{i}\left[D_-(1-A_+^2)^{1/2}+D_-(1-A_-^2)^{1/2}\right]\right\}, \quad (11.137)$$

where

$$A_\pm = \frac{Q}{2k_\mathrm{F}} \pm \frac{\omega}{v_\mathrm{F}Q},$$

and

$$\begin{cases} C_\pm = \mathrm{sgn}(A_\pm) \\ D_\pm = 0 \end{cases} \quad |A_\pm| > 1,$$

$$\begin{cases} C_\pm = 0 \\ D_\pm = 1 \end{cases} \quad |A_\pm| < 1.$$

In the long-wavelength limit, $Q \to 0$, the expression above provides higher-order (in $Q$) corrections to the dispersion of the plasma modes, exactly as we found in three dimensions:

$$\frac{\epsilon^{(2D)}(Q,\omega)}{\epsilon_\mathrm{s}} \to 1 - \left[\omega_p^{(2D)}(Q)^2 + \frac{3\epsilon_\mathrm{s}}{4\bar{\epsilon}}Q^2 v_\mathrm{F}^2\right]\frac{1}{\omega^2}, \quad (11.138)$$

while in the static limit, $\omega \to 0$, we find

$$\beta^{(2D)}(Q,\omega) \to \begin{cases} \beta_\mathrm{TF}^{(2D)} & Q < 2k_\mathrm{F} \\ \beta_\mathrm{TF}^{(2D)}\{1-[1-(2k_\mathrm{F}/Q)^2]^{1/2}\} & Q > 2k_\mathrm{F}, \end{cases} \quad (11.139)$$

thus overcoming the counter-intuitive results of Eq. (11.130) that the screening length does not depend on density in the degenerate limit.

Finally, we consider the two-dimensional version of the screened impurity potential and Friedel's sum rule: By arguments similar to those used to derive Eq. (11.113) we have, in the limit of zero temperature and only one subband occupied:

$$Z = \frac{2}{\pi} g_0 \sum_{l=-\infty}^{\infty} \eta_l(k_F), \tag{11.140}$$

where $g_0$ is the degeneracy of the ground state subband, while the cross section derived from the partial-waves analysis is

$$\sigma(k, \theta) = \frac{2}{\pi k} \sum_{l=-\infty}^{\infty} \left| e^{i[l\theta + \eta_l(k)]} \sin[\eta_l(k)] \right|^2 . \tag{11.141}$$

## Problems

**11.1. (Interface Plasma Oscillations).** Consider a bulk doped semiconductor with an electron concentration $n$. Assume that the semiconductor is terminated with a flat surface covered by an insulator with static dielectric constant $\epsilon_{ins}$. So, we have the semiconductor in the half-space $z \leq 0$ and the insulator in the half-space $z > 0$. Assume the long-wavelength limit, so that the dielectric function of the semiconductor is given by Eq. (11.62). At the semiconductor/insulator interface there will be a polarization charge that is responsible for the discontinuity of the $z$-component of an electric field, so that $\epsilon(\mathbf{q}, \omega) E_z(z = 0_-) = \epsilon_{ins} E_z(z = 0^+)$. Assume that this electric field has the form $\mathbf{E}_0 \exp(i\mathbf{q} \cdot \mathbf{r} + i\omega t)$. Show that in this system there will be an interface plasma oscillation with frequency $\omega_{SP} = \omega_P / \sqrt{1 + (\epsilon_{ins}/\epsilon_s)}$.

**11.2. (Plasma Oscillations of a 2DEG).** Equation (11.133) shows that the frequency of plasma oscillations of a 2DEG, $\omega_P(Q)$, vanishes as their wavelength grows to infinity. On the contrary, in the same limit the plasma frequency of a 3DEG remains constant and nonzero. In simple physical terms, explain why this must be so.

**11.3. (Screened Coulomb Potential).** Show that using the static Debye–Hückel expression, Eq. (11.52), to express the dielectric function, Eq. (11.47), of a bulk semiconductor, the screened Coulomb potential due to a charge $e$ takes the Yukawa form:

$$V^{(screened)}(\mathbf{r}) = \frac{e^2}{4\pi\epsilon_s} \frac{e^{-\beta_{DH} r}}{r} . \tag{11.142}$$

## References

1. C. Herring, E. Vogt, Transport and deformation-potential theory for many-valley semiconductors with anisotropic scattering. Phys. Rev. **101**, 944 (1957)
2. A.L. Fetter, J.D. Walecka, *Quantum Theory of Many-Particle Systems* (Dover, Mineola, New York, 2003)
3. G.D. Mahan, *Many-Particle Physics*. Physics of Solids and Liquids, 3rd edn. (Kluwer, Boston, 2000)
4. D. Bohm, D. Pines, A collective description of electron interactions. I. Magnetic Interactions. Phys. Rev. **82**, 625 (1951)
5. D. Bohm, D. Pines, A collective description of electron interactions: II. Collective vs individual particle aspects of the interactions. Phys. Rev. **85**, 338 (1952)

6. D. Bohm, D. Pines, A collective description of electron interactions: III. Coulomb interactions in a degenerate electron gas. Phys. Rev. **92**, 609 (1953)
7. D. Pines, A collective description of electron interactions: IV. Electron interaction in metals. Phys. Rev. **92**, 626 (1953)
8. M. Gell-Mann, K. Brueckner, Correlation energy of an electron gas at high density. Phys. Rev. **106**, 364 (1957)
9. M.V. Fischetti, Long-range Coulomb interactions in small silicon devices. Part II: effective electron mobility in thin-oxide structures, J. Appl. Phys. **89**, 1232 (2001)
10. B.D. Fried, S.D. Conte, *The Plasma Dispersion Function - The Hilbert Transform of the Gaussian* (Academic Press, New York, 1961)
11. F. Stern, Polarizability of a two-dimensional electron gas. Phys. Rev. Lett. **18**, 546 (1967)
12. A. Messiah, *Quantum Mechanics* (Dover, New York, 2014)
13. J. Friedel, The distribution of electrons round impurities in monovalent metals. Philos. Mag. **43**, 153 (1952)
14. T. Ando, A.B. Fowler, F. Stern, Electronic properties of two-dimensional systems. Rev. Mod. Phys. **54**, 437 (1982)
15. B.K. Ridley, *Quantum Processes in Semiconductors* (Oxford University Press, Oxford, 2000)
16. D.A. Dahl, L.J. Sham, Electrodynamics of quasi-two-dimensional electrons. Phys. Rev. B **16**, 651 (1977)

# Part IV
# Electron Scattering in Solids

# Chapter 12
# Generalities About Scattering in Semiconductors

## 12.1 Scattering Theory

So far, we have considered only ideal crystals at equilibrium and the elementary excitations associated with the ionic and electronic degrees of freedom, phonons and plasmons, respectively. However, we have not considered the interaction between these excitations. This was indeed our original plan: In Chap. 4 we wrote the total crystal Hamiltonian (see pages 58–63), and, thanks to the Born–Oppenheimer approximation, we considered the electronic states assuming that the ions are frozen in their equilibrium positions. Later, in Chap. 9, Sect. 9.4, we have considered the ionic excitations away from equilibrium, but we have not considered the effects of their displacement on the structure of the electronic states. Using first-order perturbation theory (or, equivalently, a linear-response approximation), we shall describe these effects as "collisions" between electrons and phonons.

Collisions are indeed crucial phenomena which allow us to understand most of the electrical properties of crystals, since they control their conductivity. In addition to electron–phonon collisions, we shall also consider the electron collisions in the screened Coulomb field of an ionized impurity (such as a dopant), the "transitions" (more than scattering or collisions) induced by electromagnetic fields (electron–photon or "radiative" processes), and electron–electron collisions which result from fluctuations of the (Hartree or Hartree–Fock) mean field. The use of perturbation theory implies that Fermi's Golden Rule and the first Born approximation will be our main tools to evaluate the frequency at which these processes occur.

We shall first consider in general terms the form these scattering rates take in the case of electrons moving in bulk crystals in three dimensions. However, we have seen that electrons (and holes) can be confined in two dimensions (2DEG) in inversion layers of MOS field-effect transistors, in the quantum wells present in hetero-structures, and in the mono-atomic planes of graphene, silicene, or transition-metal dichalcogenides (TMDs). They can also be confined in one dimension (1DEG), as in nanowires or carbon nanotubes (CNTs). Therefore, we shall also consider the scattering rates for these confined electrons.

Chapters 13, 14, and 15 will discuss in detail how scattering rates for electron/phonon, electron/ionized-impurity, and electron/electron interactions can be evaluated numerically and in closed form within the effective-mass approximation and in other specific cases. In this chapter, instead, we shall provide practical "recipes" to calculate the collision rates numerically, using the band structure of the crystal obtained using one of the computational schemes discussed in Chaps. 5 and 6. We shall also discuss the use of these algorithms for a two-dimensional BZ, resulting actually in a much simpler formulation [see Eq. (12.37) below]. The one-dimensional case is simpler yet, as we shall see.



M. Fischetti, W.G. Vandenberghe, *Advanced Physics of Electron Transport in Semiconductors and Nanostructures*, Graduate Texts in Physics, DOI 10.1007/978-3-319-01101-1_12

## 12.2 Scattering in Three Dimensions

### 12.2.1 Scattering Rates

We consider the case of an infinite crystal without any external potential applied (such as from the bias applied at contacts). In this case, we know that the electronic states—unperturbed by whatever "scattering" perturbation we shall consider shortly—are described by the Bloch wavefunctions:

$$\psi_{\mathbf{k}}^{(n)}(\mathbf{r}) = \frac{1}{\Omega^{1/2}} \, \mathrm{e}^{\mathrm{i}\mathbf{k}\cdot\mathbf{r}} \sum_{\mathbf{G}} u_{\mathbf{G},\mathbf{k}}^{(n)} \, \mathrm{e}^{\mathrm{i}\mathbf{G}\cdot\mathbf{r}} \,, \tag{12.1}$$

where $\Omega$, as usual, is the normalization volume, $\mathbf{k}$ is the wavevector, and $n$ is the index labeling the band. This is an over-idealized situation and one may wonder why it should be considered at all. Indeed, we want to study how electrons move in the crystal and this motion requires some electric field that drives the electrons. We also wish to understand how they scatter with perturbations.

The reason why Bloch functions matter at all is explained in Chap. 18. To anticipate and summarize that discussion in rough terms, assume that indeed an electric field is applied externally. Assume also that the length-scale over which the potential energy associated with the external field changes significantly is much larger than the distance between two collisions suffered *on average* by the electron, the so-called *mean free path* or *coherence length*. By "changes significantly" we mean a change comparable to the electron kinetic energy or to the thermal energy $k_{\mathrm{B}}T$.

* *As a side comment, we may also add the effect of a spatially dependent doping profile as a potential source of variations of the potential energy. In this case, mean free path and coherence length are different concepts, since loss of coherence requires collisions with scatterers that possess internal degrees of freedom, such as phonons or other electrons, so that "memory" of the phase of a Bloch wave is indeed lost after the collision. We shall discuss this issue in Chap. 18.*

In these conditions, the electron will have "lost memory" of its quantum state before the last collision by the time its potential energy has changed significantly. This corresponds to situations in which all spatial variations induced by the external field are slow enough. In this case, one can ignore the spatial dependence of the electron wavefunction caused by the changing external potential, since "coherent" transport, characterized by a pure wavefunction, will not occur. Moreover, things do not change appreciably, as seen by the electron, between two collisions. Then, we can assume that locally, over the distance between successive collisions, the crystal appears almost field-free and approximate satisfactorily the electron wavefunction with the Bloch wave of the ideal crystal. This is the regime of *semiclassical transport*, since, besides the obvious role played by Quantum Mechanics in determining the band structure and Bloch wavefunction, scattering occurs "locally," ignoring interference effects between the wavefunctions scattered by other collisions that have occurred previously, and ignoring also the effect that the variation of the potential over the mean free path has on the wavefunction itself.

As a result, scattering processes handled in this ideal, field-free, uniform crystal do play a major role when we wish to study electron transport in systems whose size exceeds the electron coherence length. Most textbooks that deal with electron transport in solids make use of this approximation. Only when we move to the study of electron transport over small distances, such as in nanostructures, do quantum effects play a major role in transport. We are indeed interested in these situations and in the next sections of this chapter we will take a more general viewpoint and deal with the problem of treating collisions in these cases.

According to the assumptions just discussed, here we consider a crystal without any external potential (besides the perturbations we wish to study, of course) and an electron in band $n$ with a 3D wavevector $\mathbf{k}$ scattering with a perturbing potential

$$V^{(\eta)}(\mathbf{r},t) = \sum_{\mathbf{q}} V_{\mathbf{q}}^{(\eta)} \, \mathrm{e}^{\mathrm{i}\mathbf{q}\cdot\mathbf{r}} \, \mathrm{e}^{\mathrm{i}\omega_{\mathbf{q}}^{(\eta)}t} \,. \tag{12.2}$$

The index $\eta$ here labels a general perturbation. However, in writing the perturbation in this form, we are considering a perturbation that varies spatially as a plane wave of wavelength $\lambda = 2\pi/q$ and that oscillates in time with a radian frequency $\omega_{\mathbf{q}}^{\eta}$ that may in general depend on the wavevector $\mathbf{q}$. Such a perturbation will generally correspond to some excitation of the lattice with energy $\hbar\omega_{\mathbf{q}}^{\eta}$. Phonons are a typical example of this type, with the "index" $\eta$ now identifying the branch, TA, LA, TO, or LO. This is the most general case of a perturbation. In fact, if a perturbation does not exhibit a harmonic dependence on space or time, the linearity of perturbation theory allows us to decompose it into its Fourier components, study each component separately, and "reassemble" the full response at the end. Perturbations that do not depend on time, such as the potential of an impurity or defect in the crystal, can be treated in a similar way. However, in this case and strictly speaking, we should use the first Born approximation. Since its result usually coincides with time-dependent perturbation theory, as expressed by Fermi's Golden Rule (see Appendix A), for simplicity we shall limit our discussion to this time-dependent case, unless explicitly noted.

$^*$ A comment regarding the notation. *Since we conduct our discussion in the context of second quantization, the states we consider are states in Fock space. However, as we shall see in detail in the next chapter, Chap. 13, page 274, all the perturbation we shall consider, with the exception of the long-range Coulomb interactions among electrons, will have nonzero matrix elements only when involving an initial and final state that differ only in the occupation of one electronic state. Therefore, we shall use a simplified notation to express the matrix element, writing only the quantum numbers (wavevector, band, spin) describing the initial and final states of the electron affected by the perturbation. The notation will be identical to the notation used in ordinary Quantum Mechanics, but additional effects will appear (Pauli blocking, stimulated and spontaneous emissions, for example), as explained by the discussion starting on page 274. So, we should keep in mind that we are still in a second-quantization formulation of the problem, even though the simplified notation misleadingly may suggest a single-electron picture. Furthermore, the elementary excitations with which electrons interact—we shall consider only phonons here—are usually assumed to remain at thermal equilibrium, since we are considering weak coupling between these excitations and the electron gas. Therefore, their occupation, in principle affected by scattering, will appear only via their Bose–Einstein thermal occupation.*

So, from Fermi's Golden Rule, the frequency at which such a perturbation scatters an electron can be expressed as

$$\frac{1}{\tau_{\eta}(\mathbf{k},n)} = \frac{2\pi}{\hbar} \sum_{\mathbf{k}'n'\mathbf{q}} |\langle \mathbf{k}'n'|V_{\mathbf{q}}^{(\eta)}|\mathbf{k}n\rangle|^2 \, \delta[E_n(\mathbf{k}) - E_{n'}(\mathbf{k}') \pm \hbar\omega_{\mathbf{q}}^{(\eta)}] \, . \tag{12.3}$$

As we have just discussed, we compute the matrix element for this process using the bulk Bloch wavefunctions. Therefore:

$$\langle \mathbf{k}'n'|V_{\mathbf{q}}^{(\eta)}|\mathbf{k}n\rangle = \frac{1}{\Omega} \, V_{\mathbf{q}}^{(\eta)} \int \mathrm{d}\mathbf{r} \; \mathrm{e}^{\mathrm{i}(\mathbf{k}-\mathbf{k}')\cdot\mathbf{r}} \sum_{\mathbf{G}\mathbf{G}'} u_{\mathbf{G}',\mathbf{k}'}^{(n')*} \, u_{\mathbf{G},\mathbf{k}}^{(n)} \, \mathrm{e}^{\mathrm{i}(\mathbf{G}-\mathbf{G}'+\mathbf{q})\cdot\mathbf{r}} \, . \tag{12.4}$$

We can proceed in the conventional way by expressing $\mathbf{r}$ as the sum of a lattice vector $\mathbf{R}_l$ and a vector $\rho$ spanning a single cell. Then:

$$\langle \mathbf{k}'n'|V_{\mathbf{q}}^{(\eta)}|\mathbf{k}n\rangle = \frac{1}{\Omega} V_{\mathbf{q}}^{(\eta)} \sum_{l} \sum_{\mathbf{G}\mathbf{G}'} \mathrm{e}^{\mathrm{i}(\mathbf{k}-\mathbf{k}'+\mathbf{G}-\mathbf{G}'+\mathbf{q})\cdot\mathbf{R}_l}$$

$$\times \int_{\Omega_{\mathrm{c}}} \mathrm{d}\rho \; u_{\mathbf{G}',\mathbf{k}'}^{(n')*} \, u_{\mathbf{G},\mathbf{k}}^{(n)} \, \mathrm{e}^{\mathrm{i}(\mathbf{k}-\mathbf{k}'+\mathbf{G}-\mathbf{G}'+\mathbf{q})\cdot\rho} \, . \tag{12.5}$$

The sum over lattice sites yields $N_{\mathrm{cell}} \sum_{\mathbf{G}''} \delta(\mathbf{k}-\mathbf{k}'+\mathbf{q}-\mathbf{G}'')$, i.e., the argument of the complex exponential must be some reciprocal-lattice vector $\mathbf{G}'' = \mathbf{G}-\mathbf{G}'$ in order to obtain a non-vanishing contribution. Thus, with some renaming of the dummy $\mathbf{G}$-vectors, we can write

$$\langle \mathbf{k}'n'|V_{\mathbf{q}}^{(\eta)}|\mathbf{k}n\rangle = \sum_{\mathbf{G}} \delta(\mathbf{q}+\mathbf{k}-\mathbf{k}'+\mathbf{G})\, V_{\mathbf{k}-\mathbf{k}'+\mathbf{G}}^{(\eta)} \frac{1}{\Omega_{\mathrm{c}}} \times \int_{\Omega_{\mathrm{c}}} \mathrm{d}\rho \sum_{\mathbf{G}'\mathbf{G}''} u_{\mathbf{G}'',\mathbf{k}'}^{(n')*}\, u_{\mathbf{G}',\mathbf{k}}^{(n)}\, \mathrm{e}^{\mathrm{i}(\mathbf{G}+\mathbf{G}'-\mathbf{G}'')\cdot\rho} \,. \tag{12.6}$$

We have intentionally left this expression in a form that can be simplified. Indeed, the last factor is the "overlap" factor,

$$\begin{aligned} \mathscr{I}_{\mathbf{k},\mathbf{k}',n,n';\mathbf{G}}^{(3\mathrm{D})} &= \frac{1}{\Omega_{\mathrm{c}}} \int_{\Omega_{\mathrm{c}}} \mathrm{d}\rho \sum_{\mathbf{G}'\mathbf{G}''} u_{\mathbf{G}'',\mathbf{k}'}^{(n')*}\, u_{\mathbf{G}',\mathbf{k}}^{(n)}\, \mathrm{e}^{\mathrm{i}(\mathbf{G}+\mathbf{G}'-\mathbf{G}'')\cdot\rho} \\ &= \frac{1}{\Omega_{\mathrm{c}}} \int_{\Omega_{\mathrm{c}}} \mathrm{d}\rho\, u_{\mathbf{k}'}^{(n')*}(\rho)\, \mathrm{e}^{\mathrm{i}\mathbf{G}\cdot\rho}\, u_{\mathbf{k}}^{(n)}(\rho) \,, \end{aligned} \tag{12.7}$$

expressed also in terms of the Bloch waves $u_{\mathbf{k}}^{(n)}(\mathbf{r}) = \sum_{\mathbf{G}} u_{\mathbf{G},\mathbf{k}}^{(n)} \mathrm{e}^{\mathrm{i}\mathbf{G}\cdot\mathbf{r}}$. Now note that

$$\int_{\Omega_{\mathrm{c}}} \mathrm{d}\rho\, \mathrm{e}^{\mathrm{i}\mathbf{G}\cdot\rho} = \Omega_{\mathrm{c}}\, \delta_{\mathbf{G},\mathbf{0}} \,, \tag{12.8}$$

so that Eq. (12.7) can also be written in a much simpler and more transparent form:

$$\mathscr{I}_{\mathbf{k},\mathbf{k}',n,n';\mathbf{G}}^{(3\mathrm{D})} = \sum_{\mathbf{G}'} u_{\mathbf{G}'-\mathbf{G},\mathbf{k}'}^{(n')*}\, u_{\mathbf{G}',\mathbf{k}}^{(n)} \,. \tag{12.9}$$

Thus, noticing that in Eq. (12.3) the sum over $\mathbf{q}$ now reduces to a discrete sum over $\mathbf{G}$s, we reach the final expression:

$$\frac{1}{\tau_\eta(\mathbf{k},n)} = \frac{2\pi}{\hbar} \sum_{\mathbf{k}'n'} \sum_{\mathbf{G}} |V_{\mathbf{k}-\mathbf{k}'+\mathbf{G}}^{(\eta)}\, \mathscr{I}_{\mathbf{k},\mathbf{k}',n,n';\mathbf{G}}^{(3\mathrm{D})}|^2\, \delta[E_n(\mathbf{k}) - E_{n'}(\mathbf{k}') \pm \hbar\omega_{\mathbf{k}-\mathbf{k}'+\mathbf{G}}^{(\eta)}] \,. \tag{12.10}$$

We should remark that, although we have explicitly written $\omega_{\mathbf{k}-\mathbf{k}'+\mathbf{G}}^{(\eta)}$ as a $\mathbf{G}$-dependent quantity, crystal symmetry requires the dispersion to be periodic in $\mathbf{k}$-space, so that the $\mathbf{G}$ dependence is trivial. The same observation applies to several equations below.

Now notice that contributions to the sum over $\mathbf{G}$ resulting from vectors of large magnitude $G$ decrease quickly as $G$ grows. This is due to the fact that the exponential inside the overlap factor (12.7) will oscillate fast as $G$ grows beyond the spectral range of the Bloch functions. Moreover, the magnitude of the matrix element $V_{\mathbf{q}}^{(\eta)}$ often decreases quickly at large $q$ (as in Coulomb interactions, polar electron–phonon coupling,...). In these circumstances large-$G$ contributions will be small. Therefore, it is acceptable in most instances to consider only the smallest $G$ (i.e., $\mathbf{G}$=0, corresponding to $N$ or *Normal* processes) in the sum above:

$$\frac{1}{\tau_\eta(\mathbf{k},n)} = \frac{2\pi}{\hbar} \sum_{\mathbf{k}'n'} |V_{\mathbf{k}-\mathbf{k}'}^{(\eta)}\, \mathscr{I}_{\mathbf{k},\mathbf{k}',n,n'}^{(3\mathrm{D})}|^2\, \delta[E_n(\mathbf{k}) - E_{n'}(\mathbf{k}') \pm \hbar\omega_{\mathbf{k}-\mathbf{k}'}^{(\eta)}] \,, \tag{12.11}$$

with the understanding that the transfer wavevector $\mathbf{k}-\mathbf{k}'$ must be mapped inside the first Brillouin Zone (BZ).

### 12.2.2 *Numerical Evaluation

The numerical evaluation of the scattering rates when the electron band structure is known numerically from, say, DFT or empirical pseudopotential calculations could be the topic of an entire textbook, so we shall not render the presentation heavier than necessary by discussing numerical issues in depth. Interested readers may find numerical details in articles authored by one of the authors of this book [1, 2].

Briefly: it is assumed that the band structure has been calculated numerically and tabulated over a discrete mesh of **k**-points—a set of points labeled $\{\mathbf{k}_i\}$—throughout the BZ, so that interpolation may be used to calculate the dispersion $E_n(\mathbf{k})$ and its gradient $\nabla_{\mathbf{k}} E_n(\mathbf{k})$ for an arbitrary **k**-point. Algorithms to calculate the density of states have been devised by Gilat and Raubenheimer [3, 4] and these can be used to calculate the scattering rates. These algorithms assume that the dispersion in each 3D mesh element ("cube"), labeled by an index $j$ and with "central energy" $E_{jn}$, can be linearized around the center of the cube. Therefore, one can estimate the maximum and minimum energy spanned by each cube $j$. This permits the identification of all cubes that are allowed by energy conservation to contain the final state after a collision. Moreover, the density of states in each cube at a given energy $E$ can be calculated from the area, $A(w_{jn})$, of the surface obtained by cutting the cube with a plane perpendicular to the gradient $\nabla E_{jn}$ at the center of the cube and displaced from the center by a length $w_{jn}$ (in reciprocal space) $(E-E_{jn})/|\nabla E_{jn}|$ along the direction of $\nabla E_{jn}$. This method, originally applied to calculate the phonon DOS in crystals, yields results much more accurate than those obtained using naïve histograms to approximate numerically the delta-function. As we have mentioned in the introduction of this chapter, we shall use an analogous scheme also to deal with scattering in two dimensions [see Eq. (12.37) below]. Again: The one-dimensional case is much simpler.

Therefore, we assume that we have discretized the BZ with "cubes" centered around points $\mathbf{k}_j$, and some method for band-structure calculations has been used to calculate the energy $E_{jn}$ and energy gradient $\nabla E_{jn}$ at the "center point" $\mathbf{k}_j$ for band $n$. Using the Gilat–Raubenheimer scheme we have already outlined [3, 4], the scattering rate is calculated numerically from the following expression:

$$\frac{1}{\tau_\eta(\mathbf{k},n)} \approx \frac{2\pi}{\hbar} \sum_{jn'}{}' |V^{(\eta)}_{\mathbf{k}-\mathbf{k}_j} \mathscr{I}^{(3D)}_{\mathbf{k},\mathbf{k}_j,n,n'}|^2 \frac{1}{(2\pi)^3} \frac{A(w_{jn'})}{|\nabla E_{jn'}|}, \tag{12.12}$$

where $[1/(2\pi)^3]A(w_{jn'})/|\nabla E_{jn'}|$ is the Gilat–Raubenheimer density of states in the $j$th cube in band $n'$, and the primed-sum symbol indicates that the sum extends only over "energy-conserving" cubes.

## 12.3 Scattering in Two Dimensions

### 12.3.1 Scattering Rates

Let's now consider the case of scattering in two-dimensional transport, such as for electrons of the 2DEG present in inversion layers, quantum wells, thin films, and 2D systems such as graphene. However, we must note that here we shall consider a "bulk" perturbation given by Eq. (12.2). This form exhibits a spatial dependence also on the $z$ direction, perpendicular to the $(x,y)$ plane of the 2DEG. This is the case, for example, of scattering with bulk phonons in an inversion layer, or scattering with ionized impurities. On the contrary, in pure 2D structures, such as ultra-thin films or 2D materials, like graphene, some perturbations will also be purely two-dimensional, most notably "confined" phonons (or "slab modes"). In this case, the perturbation will not depend on $z$ and will be of the form

$$V^{(\eta)}(\mathbf{R},t) = \sum_{\mathbf{Q}} V_{\mathbf{Q}}^{(\eta)} \, \mathrm{e}^{\mathrm{i}\mathbf{Q}\cdot\mathbf{R}} \, \mathrm{e}^{\mathrm{i}\omega_{\mathbf{Q}}^{(\eta)} t} \, . \tag{12.13}$$

We shall consider such a "true" two-dimensional case in Sect. 13.4.2 when we will discuss electron–phonon scattering in graphene. Here, instead, we shall retain the $z$-dependent, bulk form of Eq. (12.2). This applies to the case of electrons in a Si inversion layer interacting with bulk Si phonons and ionized impurities. However, the difference is not a major one: The case of a pure 2D perturbation can be treated following the discussion we give here, and simply setting equal to 1 the many factors of $e^{iq_z z}$ appearing in the equations below.

In this case, using the supercell approach to describe the system, the electron wavefunction has the same form of Eq. (12.1), although the number of **G**-vectors required is much larger, as a consequence of the larger (super)cell. For reasons that will become clear shortly, we rewrite Eq. (12.1) in a very slightly modified notation:

$$\psi_{\mathbf{k}}^{(n)}(\mathbf{r}) = \frac{1}{\Omega_{(2\mathrm{D})}^{1/2} L_{\mathrm{c}}^{1/2}} \, \mathrm{e}^{\mathrm{i}\mathbf{K}\cdot\mathbf{R}} \, \mathrm{e}^{\mathrm{i}k_z z} \sum_{\mathbf{G}} u_{\mathbf{G},\mathbf{k}}^{(n)} \, \mathrm{e}^{\mathrm{i}\mathbf{G}\cdot\mathbf{r}} \, . \tag{12.14}$$

We have simply rewritten the normalization volume $\Omega$ as $\Omega_{2\mathrm{D}} L_{\mathrm{c}}$, where $\Omega_{2\mathrm{D}}$ is the normalization area of the 2D system and $L_{\mathrm{c}}$ is the periodicity of the supercell along the direction perpendicular to the 2D system. Moreover, we have split the "envelope" $\mathrm{e}^{\mathrm{i}\mathbf{k}\cdot\mathbf{r}}$ into the product of an on-plane 2D plane wave, $\mathrm{e}^{\mathrm{i}\mathbf{K}\cdot\mathbf{R}}$, and a plane wave, $\mathrm{e}^{\mathrm{i}k_z z}$, along the direction normal to the plane. Moreover, **R** and **K** are the on-plane spatial coordinates and wavevectors, and $z$ and $k_z$ are the corresponding out-of-plane variables. We shall comment shortly on the obvious fact that the electron dispersion in this situation is not expected to depend on $k_z$. What we must emphasize now is that, as we saw in the previous section, scattering rates calculated using this wavefunction will reflect the situation of a homogeneous 2D system, free from externally applied variation of the potential energy. As such, they will be useful to treat "semiclassical 2D transport."

However, there is a more general way to view the content of this section on "scattering in 2D." Suppose that we have to study a 3D structure, uniform on the $(x,y)$ plane, but now with a strong variation of the potential along the $z$ direction. Think of a very short $p-n$ diode, infinite and uniform on the $(x,y)$ plane, but with contacts at $z=0$ and $z=L$ and a strong bias applied between these two contacts. The study of electron transport in this system using empirical pseudopotentials, for example, requires solving a Schrödinger equation with "open" boundary conditions at $z=0$ and $z=L$; that is, with boundary conditions that account for the possibility of injecting waves from the contacts. This will be discussed in Chap. 17. The fact that the crystal symmetry is preserved on the $(x,y)$ plane still makes it possible to use plane waves on this plane. But the loss of periodicity along the $z$ direction requires care. We can assume that spatial variations along $z$, caused by the varying doping profile and by the external bias, are slow over a unit cell. This is a condition much less restrictive than assuming a slow variation over an electron mean free path or coherence length, since these usually are much larger. In this case, we can step beyond "semiclassical transport," since we can consider the full "exact" wavefunction as it varies along $z$ without assuming it is a plane wave. Indeed, the slow variation of the inhomogeneities over a unit cell allows us to consider a generalization of Eq. (12.14) as follows

$$\psi_{\mathbf{k}}^{(n)}(\mathbf{r}) = \frac{1}{\Omega_{2\mathrm{D}}^{1/2} L_{\mathrm{c}}^{1/2}} \, \mathrm{e}^{\mathrm{i}\mathbf{K}\cdot\mathbf{R}} \sum_{\mathbf{G}} \phi_{\mathbf{G},\mathbf{k}}^{(n)}(z) \, \mathrm{e}^{\mathrm{i}\mathbf{G}\cdot\mathbf{r}} \, . \tag{12.15}$$

Note that we have replaced the plane-wave envelope $\mathrm{e}^{\mathrm{i}k_z z}$ of the ideal, homogeneous crystal, with a more general dependence on $z$ expressed by the "Bloch factors" $\phi_{\mathbf{G},\mathbf{k}}^{(n)}(z)$, now $z$-dependent. We shall not worry here about the differential equation that these terms, $\phi^{(n)}_{\mathbf{G},\mathbf{k}}(z)$, must obey. The interested readers may satisfy their curiosity looking at the gory details given in a technical paper [5]. We should note that, taking this perspective, the "band index" $n$ now takes an altogether different meaning: In addition to labeling bands, it is also a quantum number that labels the continuum spectrum of waves injected by the contacts. We shall retain the notation $n$, but we should keep in mind that this variable may actually be continuous. In this case, the area $\Omega_{2D}$ and length $L_c$ take a different meaning: $\Omega_{2D}$ becomes the area of the supercell and $L_c$ the length of the device.

The conclusion we draw from taking this more general perspective is that the formalism presented in this section can be viewed in two completely different ways: Scattering in "real" 2D systems with electrons confined on the $(x,y)$ plane, or, instead, scattering in 3D, but accounting for strong inhomogeneities along the $z$ direction.

If we are interested in studying transport of electrons confined in 2D, we should recall that we are describing the system within the supercell approach. This means that we could use, in principle, a full 3D description of transport and express the scattering rates with the expression we have derived in the previous sub-section. This is implied by the form of Eq. (12.14), which, after all, is identical to the "bulk" form of Eq. (12.1). This would be a way to describe transport in a mixed bulk/confined situation. Two important such cases are those of 1. electrons in an inversion layer which are confined in two dimensions at low energy, but become fully delocalized bulk electrons at energies larger than the potential energy deep in the substrate and 2. electrons undergoing "vertical transport" in a superlattice and undergoing "trapping" into 2D quantum-well states. This mixed "bulk/2D" problem has been investigated in the past and we see that the supercell approach can provide—at least in principle—a viable and correct solution.

Yet, ignoring for simplicity these cases, it is convenient to note that whenever we deal only with fully confined states on the $(x,y)$ plane, their energy $E_n(\mathbf{k}) = E_n(\mathbf{K},k_z)$ and Bloch coefficients $u^{(n)}_{\mathbf{G},\mathbf{k}} = u^{(n)}_{\mathbf{G},\mathbf{K},k_z}$ do not depend on the "out-of-plane" component $k_z$ of their wavevector, as seen in Figs. 7.17 and 7.18. Therefore, in Eq. (12.3) the sum (or integral) over $k_z$ is trivial.

If, instead, we view the problem as 3D transport in a system open along $z$, then the $k_z$-dependence disappears altogether in Eq. (12.14), since $k_z$ is not a good "quantum number" anymore, since it is an eigenvalue of the translation operator along $z$ (and so, also of the momentum' operator $\widehat{p}_z$) that does not commute with the Hamiltonian because of the loss of homogeneity along $z$.

In either case, it is convenient to make use of this independence from $k_z$ and reduce the numerical complexity of the problem by writing the scattering rate of an electron in band (or subband) $n$ and in-plane wavevector $\mathbf{K}$ due to a perturbation potential $V^{(\eta)}_{\mathbf{q}}$ as a sum only over 2D states as follows

$$\frac{1}{\tau^{\eta}(\mathbf{K},n)} = \frac{2\pi}{\hbar} \sum_{\mathbf{K}'n'\mathbf{q}} |\langle \mathbf{K}'n'|V^{(\eta)}_{\mathbf{q}}|\mathbf{K}n\rangle|^2 \, \delta[E_n(\mathbf{K}) - E_{n'}(\mathbf{K}') \pm \hbar\omega^{(\eta)}_{\mathbf{q}}] \, . \tag{12.16}$$

The matrix element $\langle \mathbf{K}'n'|V^{(\eta)}_{\mathbf{q}}|\mathbf{K}n\rangle$ is given by

$$\langle \mathbf{K}'n'|V^{(\eta)}_{\mathbf{q}}|\mathbf{K}n\rangle = \frac{1}{AL_c} \int \mathrm{d}z \int \mathrm{d}\mathbf{R} \; \psi^*_{\mathbf{K}'n'}(\mathbf{R},z) \; V^{(\eta)}_{\mathbf{q}} \mathrm{e}^{\mathrm{i}q_z z} \, \mathrm{e}^{\mathrm{i}\mathbf{Q}\cdot\mathbf{R}} \; \psi_{\mathbf{K}n}(\mathbf{R},z) \, , \tag{12.17}$$

where the wavefunctions can be expressed as

$$\psi_{\mathbf{K}n}(\mathbf{r}) = \frac{\mathrm{e}^{\mathrm{i}\mathbf{K}\cdot\mathbf{R}}}{\Omega^{1/2}_{2D} L^{1/2}_c} \sum_{\mathbf{G}} \phi^{(n)}_{\mathbf{G},\mathbf{K}}(z) \, \mathrm{e}^{\mathrm{i}\mathbf{G}\cdot\mathbf{r}} = \frac{1}{\Omega^{1/2}_{2D} L^{1/2}_c} \sum_{\mathbf{G}} \mathrm{e}^{\mathrm{i}(\mathbf{K}+\mathbf{G}_{\parallel})\cdot\mathbf{R}} \, \phi_{\mathbf{G},\mathbf{K}}(z) \mathrm{e}^{\mathrm{i}G_z z} \, . \tag{12.18}$$

We shall proceed retaining the nontrivial $z$-dependence of the wavefunctions. Of course, the case of scattering for electrons in a pure 2DEG will simplify considerably. We shall point out these simplifications at the end of the process.

Using Eq. (12.18), we can write the matrix element above as

$$\langle \mathbf{K}'n'|V_{\mathbf{q}}^{(\eta)}|\mathbf{K}n\rangle = \frac{1}{\Omega_{2D}L_c}\, V_{\mathbf{q}}^{(\eta)} \sum_{\mathbf{G}\mathbf{G}'} \int \mathrm{d}\mathbf{R}\; \mathrm{e}^{\mathrm{i}(\mathbf{K}-\mathbf{K}'+\mathbf{G}_{\|}-\mathbf{G}'_{\|}+\mathbf{Q})\cdot\mathbf{R}}$$
$$\times \int \mathrm{d}z\; \phi^{(n')*}_{\mathbf{G}',\mathbf{K}'}(z)\; \mathrm{e}^{\mathrm{i}(q_z+G_z-G'_z)z}\; \phi^{(n)}_{\mathbf{G},\mathbf{K}}(z)\,. \tag{12.19}$$

Now, proceeding as we did above in order to obtain Eq. (12.5), we write $\mathbf{R} = \mathbf{R}_l + \rho$, where $\mathbf{R}_l$ is a 2D lattice site and $\rho$ the 2D vector spanning the 2D cell. Thus:

$$\frac{1}{\Omega_{2D}} \sum_{\mathbf{G}\mathbf{G}'} \int \mathrm{d}\mathbf{R}\; \mathrm{e}^{\mathrm{i}(\mathbf{K}-\mathbf{K}'+\mathbf{G}_{\|}-\mathbf{G}'_{\|}+\mathbf{Q})\cdot\mathbf{R}}\, f_{\mathbf{G},\mathbf{G}',\mathbf{K},\mathbf{K}'}$$
$$= \frac{1}{\Omega_{2D}} \sum_{\mathbf{G}\mathbf{G}'} \sum_{l} \mathrm{e}^{\mathrm{i}(\mathbf{K}-\mathbf{K}'+\mathbf{G}_{\|}-\mathbf{G}'_{\|}+\mathbf{Q})\cdot\mathbf{R}_l} \int_{\Omega_{2D}} \mathrm{d}\rho\; \mathrm{e}^{\mathrm{i}(\mathbf{K}-\mathbf{K}'+\mathbf{G}_{\|}-\mathbf{G}'_{\|}+\mathbf{Q})\cdot\rho}\, f_{\mathbf{G},\mathbf{G}',\mathbf{K},\mathbf{K}'}\,, \tag{12.20}$$

where we have indicated with $f_{\mathbf{G},\mathbf{G}',\mathbf{K},\mathbf{K}'}$ the $z$-integral in Eq. (12.19). As before, the sum over 2D lattice sites yields a non-vanishing contribution only when $\mathbf{K}-\mathbf{K}'+\mathbf{Q}$ is equal to some vector of the 2D reciprocal lattice, $\mathbf{G}''_{\|}$. Then, relabeling the dummy variables $\mathbf{G}''$ as $\mathbf{G}$, $\mathbf{G}$ as $\mathbf{G}'$, and $\mathbf{G}'$ as $\mathbf{G}''$, Eq. (12.20) becomes

$$\sum_{\mathbf{G}_{\|}} \delta(\mathbf{K}-\mathbf{K}'+\mathbf{Q}+\mathbf{G}_{\|}) \sum_{\mathbf{G}'\mathbf{G}''} \int_{\Omega_{2D}} \frac{1}{\Omega_{2D}}\, \mathrm{d}\rho\; \mathrm{e}^{\mathrm{i}(\mathbf{G}_{\|}+\mathbf{G}'_{\|}-\mathbf{G}''_{\|})\cdot\rho}\, f_{\mathbf{G}',\mathbf{G}'',\mathbf{K},\mathbf{K}'}\,. \tag{12.21}$$

Now, using the result of Eq. (12.21) into Eq. (12.19), we obtain

$$\langle \mathbf{K}'n'|V_{\mathbf{q}}^{(\eta)}|\mathbf{K}n\rangle = \sum_{\mathbf{G}_{\|}} \delta(\mathbf{K}-\mathbf{K}'+\mathbf{Q}+\mathbf{G}_{\|})\; V^{(\eta)}_{\mathbf{Q},q_z} \sum_{\mathbf{G}'\mathbf{G}''} \frac{1}{\Omega_{2D}} \int_{\Omega_{2D}} \mathrm{d}\rho\; \mathrm{e}^{\mathrm{i}(\mathbf{G}_{\|}+\mathbf{G}'_{\|}-\mathbf{G}''_{\|})\cdot\rho}$$
$$\times \frac{1}{L_c} \int \mathrm{d}z\; \phi^{(n')*}_{\mathbf{G}'',\mathbf{K}'}(z)\; \mathrm{e}^{\mathrm{i}(q_z+G'_z-G''_z)z}\; \phi^{(n)}_{\mathbf{G}',\mathbf{K}}(z)\,. \tag{12.22}$$

Thus the scattering rate is given by

$$\frac{1}{\tau_\eta(\mathbf{K},n)} = \frac{2\pi}{\hbar} \sum_{\mathbf{K}'n'} \int \frac{\mathrm{d}q_z}{2\pi} \sum_{\mathbf{G}_{\|}} \Bigg| V^{(\eta)}_{\mathbf{K}-\mathbf{K}'+\mathbf{G}_{\|},q_z} \sum_{\mathbf{G}'\mathbf{G}''} \frac{1}{\Omega_{2D}} \int_{\Omega_{2D}} \mathrm{d}\rho\; \mathrm{e}^{\mathrm{i}(\mathbf{G}_{\|}+\mathbf{G}'_{\|}-\mathbf{G}''_{\|})\cdot\rho}$$
$$\times \frac{1}{L_c} \int \mathrm{d}z\; \phi^{(n')*}_{\mathbf{G}'',\mathbf{K}'}(z)\; \mathrm{e}^{\mathrm{i}(q_z+G'_z-G''_z)z}\; \phi^{(n)}_{\mathbf{G}',\mathbf{K}}(z) \Bigg|^2$$
$$\times\, \delta[E_n(\mathbf{K}) - E_{n'}(\mathbf{K}') \pm \hbar\omega^{(\eta)}_{\mathbf{K}-\mathbf{K}'+\mathbf{G}_{\|},q_z}]\,. \tag{12.23}$$

Finally, recalling that the equivalent of Eq. (12.8) holds also in two dimensions, we can rewrite this expression as

$$\frac{1}{\tau_\eta(\mathbf{K},n)} = \frac{2\pi}{\hbar} \sum_{\mathbf{K}'n'} \int \frac{dq_z}{2\pi} \sum_{\mathbf{G}_\parallel} \left| V^{(\eta)}_{\mathbf{K}-\mathbf{K}'+\mathbf{G}_\parallel,q_z} \right.$$

$$\times \left. \sum_{\mathbf{G}'_\parallel G'_z G''_z} \int \frac{dz}{L_c} \phi^{(n')*}_{\mathbf{G}_\parallel+\mathbf{G}'_\parallel,G''_z,\mathbf{K}'}(z)\, e^{i(q_z+G'_z-G''_z)z}\, \phi^{(n)}_{\mathbf{G}'_\parallel,G'_z,\mathbf{K}}(z) \right|^2$$

$$\times\, \delta[E_n(\mathbf{K}) - E_{n'}(\mathbf{K}') \pm \hbar\omega^{(\eta)}_{\mathbf{K}-\mathbf{K}'+\mathbf{G}_\parallel,q_z}] \,. \tag{12.24}$$

Considering as before only $N$ processes (i.e., collapsing the sum over $\mathbf{G}_\parallel$ to the lone term $\bar{\mathbf{G}}_\parallel$ needed to map $\mathbf{K}-\mathbf{K}'$ inside the first 2D BZ), we have

$$\frac{1}{\tau_\eta(\mathbf{K},n)} \approx \frac{2\pi}{\hbar} \sum_{\mathbf{K}'n'} \int \frac{dq_z}{2\pi} \left| V^{(\eta)}_{\mathbf{K}-\mathbf{K}',q_z} \sum_{\mathbf{G}_\parallel G_z G'_z} \int \frac{dz}{L_c} \phi^{(n')*}_{\mathbf{G}_\parallel,G'_z,\mathbf{K}'}(z) \right.$$

$$\times \left. e^{i(q_z+G_z-G'_z)z}\, \phi^{(n)}_{\mathbf{G}_\parallel,G_z,\mathbf{K}}(z) \right|^2 \delta[E_n(\mathbf{K}) - E_{n'}(\mathbf{K}') \pm \hbar\omega^{(\eta)}_{\mathbf{K}-\mathbf{K}',q_z}] \,. \tag{12.25}$$

To express this result in a more compact notation better suited to simplifications, it is convenient to define the functions:

$$\xi^{(n)}_{\mathbf{G}_\parallel\mathbf{K}}(z) = \frac{1}{L_c^{1/2}} \sum_{G_z} \phi^{(n)}_{\mathbf{G},\mathbf{K}}(z)\, e^{iG_z z} \,. \tag{12.26}$$

Then, from Eq. (12.25) we have

$$\mathscr{I}^{(2D)}_{\mathbf{K},\mathbf{K}',n,n',\mathbf{G}_\parallel}(q_z) = \sum_{\mathbf{G}'_\parallel} \sum_{G'_z G''_z} \int \frac{dz}{L_c} \phi^{(n')*}_{\mathbf{G}_\parallel+\mathbf{G}'_\parallel,G''_z,\mathbf{K}'}(z)\, e^{i(q_z+G'_z-G''_z)z}\, \phi^{(n)}_{\mathbf{G}'_\parallel,G'_z,\mathbf{K}}(z)$$

$$= \int dz \sum_{\mathbf{G}'_\parallel} \xi^{(n')*}_{\mathbf{G}_\parallel+\mathbf{G}'_\parallel,\mathbf{K}'}(z)\, e^{iq_z z}\, \xi^{(n)}_{\mathbf{G}'_\parallel,\mathbf{K}}(z) \,. \tag{12.27}$$

For $N$ processes this becomes simply:

$$\mathscr{I}^{(2D)}_{\mathbf{K},\mathbf{K}',n,n'}(q_z) = \int dz \sum_{\mathbf{G}_\parallel} \xi^{(n')*}_{\mathbf{G}_\parallel,\mathbf{K}'}(z)\, e^{iq_z z}\, \xi^{(n)}_{\mathbf{G}_\parallel,\mathbf{K}}(z) \,. \tag{12.28}$$

Using this expression for the overlap factor, Eq. (12.24) becomes

$$\frac{1}{\tau_\eta(\mathbf{K}),n} = \frac{2\pi}{\hbar} \sum_{\mathbf{K}'n'} \int \frac{dq_z}{2\pi} \sum_{\mathbf{G}_\parallel} \left| V^{(\eta)}_{\mathbf{K}-\mathbf{K}'+\mathbf{G}_\parallel,q_z}\, \mathscr{I}^{(2D)}_{\mathbf{K},\mathbf{K}',n,n';\mathbf{G}_\parallel}(q_z) \right|^2$$

$$\times\, \delta[E_n(\mathbf{K}) - E_{n'}(\mathbf{K}') \pm \hbar\omega^{(\eta)}_{\mathbf{K}-\mathbf{K}'+\mathbf{G}_\parallel,q_z}] \,. \tag{12.29}$$

With this notation, the scattering rate obtained by accounting for $N$ processes only reduces to

$$\frac{1}{\tau_\eta(\mathbf{K},n)} = \frac{2\pi}{\hbar} \sum_{\mathbf{K}'n'} \int \frac{dq_z}{2\pi} \left| V^{(\eta)}_{\mathbf{K}-\mathbf{K}',q_z} \mathscr{I}^{(2D)}_{\mathbf{K},\mathbf{K}',n,n'}(q_z) \right|^2$$
$$\times\ \delta[E_n(\mathbf{K}) - E_{n'}(\mathbf{K}') \pm \hbar\omega^{(\eta)}_{\mathbf{K}-\mathbf{K}',q_z}] \,. \tag{12.30}$$

Note that for *semiclassical* two-dimensional transport of a 2DEG (inversion layers, hetero-channels, etc.) with confinement handled using the supercell method, the scattering rate will be given exactly by Eqs. (12.24) or (12.30) using the wavefunctions $\xi^{(n)}_{\mathbf{G}_\parallel,\mathbf{K}}(z)$ given by Eq. (12.26) with constant (i.e., independent of $z$) Bloch coefficients $\phi^{(n)}_{\mathbf{G},\mathbf{K}}(z) = u^{(n)}_{\mathbf{G},\mathbf{K}}$ and overlap factors given by Eqs. (12.27) or (12.28).

Before seeing how these simplifications affect the expression for the scattering rate, we should remark that no additional simplifications are possible when the full wavefunctions $\xi^{(n)}_{\mathbf{G}_\parallel,\mathbf{K}}(z)$ [see Eqs. (12.26) and (12.27)] must be used. This case presents the obvious numerical difficulty caused by the size of the array $\xi^{(n)}_{\mathbf{G},\mathbf{K}}(z)$ we must store: Even in cases in which the $z$-dependence is absent (cases B1 and B2 above, for example), storing these (complex) wavefunctions for $\sim 10^3$ **G**-vectors at each of the $\sim 10^3$ **K**-points used to tabulate the band structure over the wedge of the BZ and for each of the $\sim 10$ bands $n$ requires $\sim 0.1$ GB of storage. However, when the "transverse" modes are bulk modes, as in cases B1 and B2, an additional simplification—employed throughout in the following—can be obtained by embracing the pure "envelope" approximation by ignoring the Bloch components $e^{i\mathbf{G}_\parallel\cdot\mathbf{R}}$ in Eq. (12.19) and considering only "envelope" wavefunctions averaged over a unit cell in the $(x,y)$ plane. This is fully equivalent to ignoring overlap-factor effects in 3D [i.e., setting $\mathscr{I}^{(3D)}_{\mathbf{k},\mathbf{k}',n,n';\mathbf{G}} = 1$ in Eqs. (12.3), (12.10), and (12.12)]. Then, from the full wavefunction $\psi_{\mathbf{K}n}(\mathbf{R},z)$ given by Eq. (12.18) we can extract $z$-only-dependent "envelopes":

$$\zeta^{(n)}_{\mathbf{K}}(z) = \frac{1}{L_c^{1/2}} \sum_{G_z} \phi^{(n)}_{\mathbf{G}_\parallel=\mathbf{0},G_z,\mathbf{K}}(z)\, e^{iG_z z} \,. \tag{12.31}$$

We have referred to these functions as "envelopes," since they represent the envelope approximation on the plane of the interface, varying—if at all—only over length-scales larger than the size of the in-plane unit cell. Thus, being cavalier about the fact that we neglect Bloch-overlap effects, this constitutes a satisfactory approximation whenever the perturbation $V^{(\eta)}_{\mathbf{Q},q_z}$ has negligible contributions at large spatial in-plane frequencies $\mathbf{Q}$. In practice this approximation amounts to selecting $\mathbf{G}'_\parallel = \mathbf{G}_\parallel = \mathbf{0}$ in Eq. (12.19) obtaining

$$\langle \mathbf{K}'n' | V^{(\eta)}_{\mathbf{q}} | \mathbf{K}n \rangle \approx \frac{1}{\Omega_{2D}} V^{(\eta)}_{\mathbf{q}} \sum_{G_z G'_z} \int d\mathbf{R}\, e^{i(\mathbf{K}-\mathbf{K}'+\mathbf{Q})\cdot\mathbf{R}} \tag{12.32}$$

$$\times \int \frac{dz}{L_c}\, \phi^{(n')*}_{\mathbf{0},G'_z\mathbf{K}'}(z)\, e^{i(q_z+G_z-G'_z)z}\, \phi^{(n)}_{\mathbf{0},G_z,\mathbf{K}}(z) \tag{12.33}$$

$$= V^{(\eta)}_{\mathbf{q}}\, \delta_{\mathbf{K}-\mathbf{K}+\mathbf{Q},\mathbf{0}} \int dz\, \zeta^{(n')*}_{\mathbf{K}'}(z)\, e^{iq_z z}\, \zeta^{(n)}_{\mathbf{K}}(z) \tag{12.34}$$

so that the 2D overlap factor $\mathscr{I}^{(2D)}_{\mathbf{K},\mathbf{K}',n,n';\mathbf{G}_{\|}}(q_z)$ becomes simply:

$$\tilde{\mathscr{I}}^{(2D)}_{\mathbf{K},\mathbf{K}',n,n'}(q_z) \approx \int \mathrm{d}z\ \zeta^{(n')*}_{\mathbf{K}'}(z)\ \mathrm{e}^{\mathrm{i}q_z z}\ \zeta^{(n)}_{\mathbf{K}}(z)\ , \tag{12.35}$$

which, except for the more complicated subband dispersion, allows us to formulate the scattering rate Eq. (12.30) in terms of its "usual" expression in 2D that can be derived, for example, using the effective-mass approximation. In this case, in Chap. 8, Sect. 8.2.1, we saw that the wavefunctions are given by Eq. (8.47) on page 174 and have the form $\Omega_{2D}^{-1/2}\ \mathrm{e}^{\mathrm{i}\mathbf{K}\cdot\mathbf{R}}\ \zeta(z)$. The $z$-dependence is absorbed into the function $\zeta(z)$ that takes the place of the appropriately named function $\zeta^{(n)}_{\mathbf{K}}(z)$ defined here. Scattering rates of the form Eq. (12.30) with "overlap factors" Eq. (12.35) have been extensively used in the literature to study electron transport in inversion layers and quantum wells. The original derivation dates back to work by Price [6] and Brian Ridley (results discussed in his more recent and excellent textbook [7]). These results are reviewed in Chap. 13, Sect. 13.4 for electron–phonon interactions, and in Chap. 14, Sect. 14.8, for electron-impurity scattering.

### 12.3.2 *Numerical Evaluation

In order to obtain a numerically computable expression, let's consider again the more generally valid expression, Eq. (12.29). In analogy with Eq. (12.12), having discretized the 2D BZ with squares centered at points $\mathbf{K}_j$, we can write

$$\frac{1}{\tau_\eta(\mathbf{K},n)} \approx \frac{2\pi}{\hbar} \sum_{jn'}{}' \int \frac{\mathrm{d}q_z}{2\pi} \sum_{\mathbf{G}_{\|}} \left| V^{(\eta)}_{\mathbf{K}-\mathbf{K}_j+\mathbf{G}_{\|},q_z}\ \mathscr{I}^{(2D)}_{\mathbf{K},\mathbf{K}_j,n,n';\mathbf{G}_{\|}}(q_z)\right|^2 \frac{1}{(2\pi)^2} \frac{L(w_{jn'})}{|\nabla_{2D}E_{jn'}|}\ . \tag{12.36}$$

Here $[1/(2\pi)^2]\ L(w_{jn})/\nabla_{2D}E_{jn}$ is the density of states on band $n'$ in the $j$th square of side $\Delta K$ with energy $E_{jn'}$ and gradient $\nabla_{2D}E_{jn'}$ at the center of the square, $\mathbf{K}_j$. A simple reduction of the Gilat–Raubenheimer scheme to two dimensions implies that

$$L(w) = \begin{cases} \dfrac{\Delta K}{\cos\alpha} (w \le w_0) \\ \\ \dfrac{w_1 - w}{\cos\alpha \sin\alpha} (w_0 \le w \le w_1), \end{cases} \tag{12.37}$$

where $w_0 = (\Delta K/2)(\cos\alpha - \sin\alpha)$, $w_1 = (\Delta K/2)(\cos\alpha + \sin\alpha)$, $\alpha$ is the angle between the $K_x$-axis and $\nabla_{2D}E_{jn}$, $w = (E_{\mathrm{final}} - E_{jn})/|\nabla_{2D}E_{jn}|$, and $E_{\mathrm{final}}$ is the final energy for a particular process and square $j$.

Note that, although not explicitly indicated here, the density of final states in the "energy-conserving" element $(j,n')$, $L(w_{jn'})/\nabla_{2D}E_{jn'}$, appears inside the integration over the perpendicular wavevector transfer $q_z$ since, in general, the energy transfer during the collision may depend also on $q_z$, as it is the case, for example, with scattering of electrons in an inversion layer with bulk phonons when retaining the full dispersion, $\hbar\omega_{\mathbf{Q},q_z}$, of the phonons. This leads to a very expensive numerical task.

## 12.4 Scattering in One Dimension

### *12.4.1 Scattering Rates*

As we have done in the previous section, let's consider how to handle scattering processes for electrons confined along two dimensions and free to move only along one direction, say along the $z$-axis. This amounts to considering semiclassical transport in 1D structures, such as nanowires or nanotubes with their axial direction on the $z$-axis. However, in complete analogy to the discussion with which we opened Sect. 12.3, we can take a more general approach and consider, for example, scattering for 3D electrons moving inside a structure that is inhomogeneous on the $(x,y)$ plane, but uniform along the $z$-axis.

The scattering rate must be evaluated considering the wavefunctions given by

$$\psi_{k_z n}(\mathbf{r}) = \frac{e^{ik_z z}}{\Omega_{2D}^{1/2} L_c^{1/2}} \sum_{\mathbf{G}} \phi_{\mathbf{G},k_z}^{(n)}(\mathbf{R})\, e^{i\mathbf{G}\cdot\mathbf{r}} = \frac{1}{\Omega_{2D}^{1/2} L_c^{1/2}} \sum_{\mathbf{G}} e^{i(k_z+G_z)z}\, \phi_{\mathbf{G},k_z}^{(n)}(\mathbf{R}) e^{i\mathbf{G}_{\parallel}\cdot\mathbf{R}}\,, \tag{12.38}$$

where $\Omega_{2D}$ now denotes the area of the supercell and $L$ is the normalization length. For a pure 1DEG (electrons in NWs or CNTs), $n$ labels the quantized states and the 1D wavevector $k_z$ represents the wavevector of the plane waves along the homogeneous direction $z$. In this case, the function $\phi_{\mathbf{G},k_z}^{(n)}(\mathbf{R})$ will *not* depend on the in-plane coordinate $\mathbf{R}$. As we have done in the previous section, we shall discuss, at the end of the laborious calculation, the simplification that this fact affords.

So, for an electron in subband (or continuum state) $n$ and longitudinal wavevector (in 1D it is actually a wavenumber) $k_z$:

$$\frac{1}{\tau_\eta(k_z,n)} = \frac{2\pi}{\hbar} \sum_{k_z'n'} \sum_{\mathbf{q}} \left| \langle k_z'n'|V_{\mathbf{q}}^{(\eta)}|k_z n\rangle \right|^2 \delta[E_n(k_z) - E_{n'}(k_z') \pm \hbar\omega_{\mathbf{q}}^{(\eta)}]\,. \tag{12.39}$$

The matrix element can be calculated in the usual way:

$$\langle k_z'n'|V_{\mathbf{q}}^{(\eta)}|k_z n\rangle = V_{\mathbf{q}}^{(\eta)} \sum_{G_z'G_z''} \int dz\, \frac{1}{L}\, e^{i(k_z-k_z'+G_z'-G_z''+q_z)z} \tag{12.40}$$

$$\times \int \frac{d\mathbf{R}}{\Omega_{2D}} \sum_{\mathbf{G}'\mathbf{G}''} \phi_{\mathbf{G}'',\mathbf{K}'}^{(n')*}(\mathbf{R})\, \phi_{\mathbf{G}',\mathbf{K}}^{(n)}(\mathbf{R})\, e^{i(\mathbf{Q}+\mathbf{G}'-\mathbf{G}'')\cdot\mathbf{R}} \tag{12.41}$$

$$= \sum_{G_z} V_{\mathbf{q}}^{(\eta)}\, \delta(k_z - k_z' + q_z + G_z)\, \mathscr{I}_{k_z,k_z',n,n';G_z}^{(1D)}(\mathbf{Q})\,. \tag{12.42}$$

In this expression $\Omega_{2D}$ can be either the normalization area (2D open boundary-conditions simulations) or the area of the supercell (1DEG transport), $L$ can be either the width of the supercell or the normalization length. Also, we have used the usual property of the sum of $e^{i(k_z-k_z'+G_z'-G_z''+q_z)z_l}$ over 1D-lattice sites $z_l$ to obtain the term $\sum_{G_z} \delta(k_z - k_z' + q_z + G_z)$ and the 1D overlap factor $\mathscr{I}_{k_z,k_z',n,n';G_z}^{(1D)}(\mathbf{Q})$ is given by

$$\mathscr{I}_{k_z,k_z',n,n'}^{(1D)}(\mathbf{Q}) = \sum_{G_z'G_z''} \frac{1}{L_c} \int_{L_c} d\rho\, e^{i(G_z'-G_z''+G_z)\rho}$$

$$\times \int \frac{d\mathbf{R}}{\Omega_{2D}} \sum_{\mathbf{G}'\mathbf{G}''} \phi_{\mathbf{G}'',k_z'}^{(n')*}(\mathbf{R})\, \phi_{\mathbf{G}',k_z}^{(n)}(\mathbf{R})\, e^{i(\mathbf{Q}+\mathbf{G}'-\mathbf{G}'')\cdot\mathbf{R}}\,. \tag{12.43}$$

We can express this overlap factor $\mathscr{I}^{(1D)}_{k_z,k_z',n,n';G_z}(\mathbf{Q})$ in terms of the following wavefunctions analogous to Eq. (12.26):

$$\xi^{(2D)(n)}_{G_z,k_z}(\mathbf{R}) = \frac{1}{\Omega_{2D}^{1/2}} \sum_{\mathbf{G}_\parallel} \phi^{(n)}_{\mathbf{G},k_z}(\mathbf{R}) e^{i\mathbf{G}_\parallel \cdot \mathbf{R}} . \tag{12.44}$$

Then:

$$\mathscr{I}^{(1D)}_{k_z,k_z',n,n';G_z}(\mathbf{Q}) = \int \mathrm{d}\mathbf{R} \sum_{G_z'} \xi^{(2D)(n')*}_{G_z+G_z',k_z'}(\mathbf{R})\, e^{i\mathbf{Q}\cdot\mathbf{R}}\, \xi^{(2D)(n)}_{G_z',k_z}(\mathbf{R}) , \tag{12.45}$$

which, for $N$ processes simplifies to

$$\mathscr{I}^{(1D)}_{k_z,k_z',n,n'}(\mathbf{Q}) = \int \mathrm{d}\mathbf{R} \sum_{G_z} \xi^{(2D)(n')*}_{G_z,k_z'}(\mathbf{R})\, e^{i\mathbf{Q}\cdot\mathbf{R}}\, \xi^{(2D)(n)}_{G_z,k_z}(\mathbf{R}) . \tag{12.46}$$

Using this expression for $\mathscr{I}^{(1D)}_{k_z,k_z',n,n';G_z}(\mathbf{Q})$, we have

$$\frac{1}{\tau_\eta(k_z,n)} = \frac{2\pi}{\hbar} \sum_{k_z'n'} \int \frac{\mathrm{d}\mathbf{Q}}{(2\pi)^2} \sum_{G_z} \left| V^{(\eta)}_{\mathbf{Q},k_z-k_z'+q_z+G_z}\, \mathscr{I}^{(1D)}_{k_z,k_z',n,n';G_z}(\mathbf{Q}) \right|^2$$
$$\times\ \delta[E_n(k_z) - E_{n'}(k_z') \pm \hbar\omega^{(\eta)}_{\mathbf{Q},k_z-k_z'+G_z}]. \tag{12.47}$$

Considering $N$ processes (i.e., retaining in the sum above only the term $G_z$ which brings $k_z' - k_z$ within the 1D BZ), we have

$$\frac{1}{\tau_\eta(k_z,n)} \approx \frac{2\pi}{\hbar} \sum_{k_z'n'} \int \frac{\mathrm{d}\mathbf{Q}}{(2\pi)^2} \left| V^{(\eta)}_{\mathbf{Q},k_z-k_z'+q_z}\, \mathscr{I}^{(1D)}_{k_z,k_z',n,n'}(\mathbf{Q}) \right|^2$$
$$\times\ \delta[E_n(k_z) - E_{n'}(k_z') \pm \hbar\omega^{(\eta)}_{\mathbf{Q},k_z-k_z'}]. \tag{12.48}$$

Note that in most cases this expression is deceptively simple: In the case of elastic scattering (namely, impurity scattering, edge- or surface-roughness scattering, or acoustic phonon scattering in the elastic approximation) or scattering with dispersionless excitations, usually only a small number of $k_z'$ are available as final states. In the following we shall leave the expressions for the scattering rates in their most general forms, although the algorithm described for their numerical evaluation will be implicitly simplified thanks to this consideration.

An important observation we should make is that Eq. (12.47) or (12.48) represents also the scattering rates in *semiclassical* 1D transport along $z$ with 2D confinement handled using the 2D supercell method. In this case, the Bloch coefficients in Eq. (12.44), $\phi^{(n)}_{\mathbf{G},k_z}(\mathbf{R})$, collapse to $u^{(n)}_{\mathbf{G},k_z}$, independent of $\mathbf{R}$.

Embracing the (in-plane) envelope approximation—which is a sensible approximation when we are dealing with semiclassical transport along $z$ and we can simplify Eqs. (12.45) or (12.46)—we can express the wavefunctions in analogy to Eq. (12.31) as

$$\zeta^{(2D)(n)}_{k_z}(\mathbf{R}) = \frac{1}{\Omega_{2D}^{1/2}} \sum_{\mathbf{G}_\parallel} \phi^{(n)}_{\mathbf{G}_\parallel,G_z=0,k_z}(\mathbf{R})\, e^{i\mathbf{G}_\parallel \cdot \mathbf{R}} . \tag{12.49}$$

so that the overlap factor $\mathscr{I}^{(1D)}_{k_z,k_z',n,n'}(\mathbf{Q})$ reduces to

$$\tilde{\mathscr{I}}^{(1D)}_{k_z,k_z',n,n'}(\mathbf{Q}) = \int \mathrm{d}\mathbf{R}\ \zeta^{(2D)(n')*}_{k_z'}(\mathbf{R})\ \mathrm{e}^{\mathrm{i}\mathbf{Q}\cdot\mathbf{R}}\ \zeta^{(2D)(n)}_{k_z}(\mathbf{R})\ . \tag{12.50}$$

### 12.4.2 *Numerical Evaluation

Regarding the numerical evaluation of the general expression—Eq. (12.47) or its *normal*-process approximation, Eq. (12.48), possibly with the overlap factor approximated as in Eq. (12.50)—let's assume that the $k_z$-axis has been discretized into bins of length $\Delta k_z$ indexed by an integer $j$ and centered around the point $k_{zj}$. The density of states into each bin will be $|\mathrm{d}E_n/\mathrm{d}k_{z,j}|$. Thus:

$$\frac{1}{\tau_\eta(k_z,n)} \approx \frac{2\pi}{\hbar}\sum_{jn'}^{*}\int\frac{\mathrm{d}\mathbf{Q}}{(2\pi)^2}\left|V^{(\eta)}_{\mathbf{Q},k_z-k_{zj}+q_z}\ \mathscr{I}^{(1D)}_{k_z,k_{zj},n,n'}(\mathbf{Q})\right|^2\frac{1}{2\pi}\left|\frac{\mathrm{d}E_{n'}}{\mathrm{d}k_{zj}}\right|^{-1}, \tag{12.51}$$

in analogy with the expressions given by Eq. (12.12) derived before [see also Eq. (13.63) below].

## Problems

**12.1. (Overlap Intergral)**

(a) Compute Eq. (12.9) for $\mathbf{k}$ and $\mathbf{k}'$ for $\mathbf{G}=0$ and $\mathbf{G}=\mathbf{b}_1$ for silicon using the code in Appendix D.
(b) Building on the result from problem the first part of the problem, compute $\sum_{\mathbf{G}}|V_{\mathbf{k}-\mathbf{k}'+\mathbf{G}}\mathscr{I}^{3D}|^2$ from Eq. (12.10) for $V(\mathbf{k}) = \mathrm{e}^{(-|\mathbf{k}|^2/|\mathbf{b}_1|^2)}$.

## References

1. M.V. Fischetti, S.E. Laux, Monte Carlo analysis of electron transport in small semiconductor devices including band-structure and space-charge effects. Phys. Rev. B **38**, 9721 (1988)
2. S.E. Laux, M.V. Fischetti, Numerical aspects and implementation of the DAMOCLES Monte Carlo device simulation program, in *Monte Carlo Device Simulation: Full Band and Beyond*, ed. by K. Hess (Kluwer, Norwell, MA, 1991), pp. 1–26
3. G. Gilat, L.J. Raubenheimer, Accurate numerical method for calculating frequency distribution functions in solids. Phys. Rev. **144**, 390 (1966)
4. G. Gilat, L.J. Raubenheimer, Accurate numerical method for calculating frequency distribution functions in solids: Erratum. Phys. Rev. **147**, 670 (1966)
5. M.V. Fischetti, B.Fu, S. Narayanan, J. Kim, Semiclassical and quantum electronic transport in nanometer-scale structures: empirical pseudopotential band structure, Monte Carlo simulations and Pauli master equation, in *Nano-Electronic Devices: Semiclassical and Quantum Transport Modeling*, ed. by D. Vasileska, S.M. Goodnick (Springer, New York, 2011), pp. 183–247
6. P.J. Price, Hot electrons in a GaAs heterolayer at low temperature. J. Appl. Phys. **53**, 6863 (1982)
7. B.K. Ridley, *Quantum Processes in Semiconductors* (Oxford University Press, Oxford, 2000)

# Chapter 13
# Electron–Phonon Interactions

In this chapter we consider the interaction between electrons and phonons. This subject, which is of utmost importance, since it deals with an intrinsic collision process that affects strongly electronic transport, is discussed very extensively in the literature. The early book by Ziman [1] remains one of the most complete treatment of this topic.

In general, two types of interactions can be distinguished: The first, called *deformation potential* interaction [2], is related to how the displacement of an atom changes the potential near the atom which is being displaced. The second, called *Fröhlich interaction*, describes the effect of the long-range electric fields caused by the displacement of the atoms. The former is active in all solids. The latter interaction, instead, depends on a long-range dipole field that can arise only when the displaced atoms have excess and opposite charges. Therefore, it is present only in ionic (polar) materials. We shall not discuss the additional polar coupling of electrons with acoustic phonons (*piezoelectric* coupling), since it affects electronic transport only weakly and at low temperatures.

In Sect. 13.1, we discuss how the strength of the deformation-potential interaction can be computed theoretically. The need to discuss it in greater depth is due to the fact that the "scattering potential" itself is a function of the ionic displacement and of the perturbation it causes to the electronic structure itself. This adds complexity, since its computation couples to the models used to describe both the lattice dynamics and the electronic spectrum. On the other hand, it presents the advantage of allowing *ab initio* calculations, or as close to "first-principles" as the models affords, such as DFT and the rigid-ion approximation based on empirical pseudopotentials. On the contrary, polar interactions are usually treated with the Fröhlich model that yields a scattering potential independent of the electronic structure itself. Therefore, in Sect. 13.2 we discuss in some depth the deformation-potential concept and how the deformation-potential interaction affects scattering. Fröhlich's approach to treat polar electron–phonon interactions is considered in Sect. 13.3. We then consider scattering with phonons for a 2DEG in Sect. 13.4 and for a 1DEG in Sect. 13.5. In both cases we consider nonpolar and polar interactions treated in a general way as well as, more specifically, with the effective-mass approximations, giving examples of each. We conclude in Sect. 13.6 with a brief discussion of scattering with coupled modes phonon-plasmon modes, important in polar semiconductors (or at interfaces with polar insulators) at the high densities at which many devices operate.



M. Fischetti, W.G. Vandenberghe, *Advanced Physics of Electron Transport in Semiconductors and Nanostructures*, Graduate Texts in Physics, DOI 10.1007/978-3-319-01101-1_13

## 13.1 General Formalism and the Rigid-Ion Approximation

The deformation-potential interaction is due to the effect which an ionic displacement, associated with a phonon, has on the ionic potential and consequently on the electron energy. It is called a *nonpolar* interaction, because the electrons do not feel any long-range electric (dipole) field, but feel simply the perturbation caused by the fact that the ions in the electronic Hamiltonian are not frozen in their equilibrium positions $\mathbf{R}_{l\gamma}$. Both acoustic and optical phonons cause this interaction, although there are subtle differences between the two.

### *13.1.1 General Formalism*

We have seen in Chap. 9, Eq. (9.24), that the electronic component of the Hamiltonian, Eq. (9.19), with a general potential energy $V(\mathbf{r})$ can be written as [see Eqs. (9.24) and (9.31)]:

$$\widehat{H}_{\mathrm{el}} = \int \mathrm{d}\mathbf{r} \left[ \left( -\frac{i\hbar}{2m} \right) \nabla \widehat{\pi}(\mathbf{r}) \cdot \nabla \widehat{\psi}(\mathbf{r}) - \frac{1}{\hbar} \widehat{\pi}(\mathbf{r}) V(\mathbf{r}) \widehat{\psi}(\mathbf{r}) \right] = \sum_{n\mathbf{k}} E_{n\mathbf{k}} \widehat{c}^{\dagger}_{n\mathbf{k}} \widehat{c}_{n\mathbf{k}} \, , \tag{13.1}$$

while the ionic component takes the form given by Eqs. (9.91) and (9.102):

$$\begin{aligned} \widehat{H}_{\mathrm{ion}} &= \frac{1}{2} \sum_{l\gamma} \left[ \frac{P_{l\gamma}^2}{M_\gamma} + \sum_{l' \neq l, \gamma'\gamma} \sum_{ij} \frac{\partial^2 U(\mathbf{R}_{l\gamma}^{(0)} - \mathbf{R}_{l'\gamma'}^{(0)})}{\partial R_{l\gamma i} \, \partial R_{l'\gamma' j}} \, \delta R_{l\gamma i} \, \delta R_{l'\gamma' j} \right] \\ &= \sum_{\eta\mathbf{q}} \hbar\omega_{\mathbf{q}}^{(\eta)} \, \widehat{b}^{\dagger}_{\mathbf{q}\eta} \widehat{b}_{\mathbf{q}\eta} \, . \end{aligned} \tag{13.2}$$

These two Hamiltonians have been decoupled thanks to the (adiabatic) Born–Oppenheimer approximation. In Chap. 4, when we were describing the approximations made to derive Eq. (13.1), we emphasized the fact that lattice potential energy $V^{(\mathrm{lat})}(\mathbf{r})$ entering the total potential energy $V(\mathbf{r})$ (that includes also the Hartree energy and exchange-correlation effects) was assumed to be due to the ions frozen in their equilibrium positions. This, indeed, is the assumption that allowed us to decouple the electron Hamiltonian from the Hamiltonian of the ions. In Sect. 4.2.1 we “promised” that we were going to consider in some perturbative way the effect that small deviation of the ions from their equilibrium positions. Here we keep our promise: Now it’s time to find the effect of the displacement of the ions away from their equilibrium position on the electron energy.

Proceeding as we have done in dealing with phonons, we consider the change of the ionic potential up to terms which depend linearly on the ionic displacement:

$$\delta U(\mathbf{r}) \approx \sum_{l\gamma} \frac{\partial U(\mathbf{r})}{\partial \mathbf{R}_{l\gamma}} \cdot \delta \mathbf{R}_{l\gamma} \, . \tag{13.3}$$

We will discuss a specific way to compute $\partial U(\mathbf{r})/\partial \mathbf{R}_{l\gamma}$ shortly, by invoking the rigid-ion approximation, but for now we assume that the change in the lattice potential with respect to a lattice displacement can be obtained in some way.

We can express the displacement of the ion $\gamma$ in cell $l$ in terms of the creation and annihilation operators of phonons of momentum $\mathbf{q}$ and branch $\eta$ via Eqs. (9.96) and (9.100):

$$\delta\widehat{U}(\mathbf{r}) \approx \sum_{l\gamma} \frac{1}{(N_{\text{cell}}M_\gamma)^{1/2}} \frac{\partial U(\mathbf{r})}{\partial \mathbf{R}_{l\gamma}} \cdot \sum_{\mathbf{q}\eta} \mathbf{e}_{\mathbf{q}\gamma}^{(\eta)} \, \widehat{Q}_{\mathbf{q}\gamma} \, e^{i\mathbf{q}\cdot\mathbf{R}_l}$$

$$= \sum_{\mathbf{q}\eta} \sum_{l\gamma} \frac{1}{(NM_\gamma)^{1/2}} \left(\frac{\hbar}{2\omega_{\mathbf{q}}^{(\eta)}}\right)^{1/2} (\widehat{b}_{\mathbf{q}\eta} + \widehat{b}^\dagger_{-\mathbf{q}\eta}) \, \mathbf{e}_{\mathbf{q}\gamma}^{(\eta)} \cdot \frac{\partial U(\mathbf{r})}{\partial \mathbf{R}_{l\gamma}} \, e^{i\mathbf{q}\cdot\mathbf{R}_l} \,. \tag{13.4}$$

The correction to the energy of the electrons due to the displacement Eq. (13.4) results simply from the electrostatic energy of the electron density $\widehat{\rho}(\mathbf{r}) = \widehat{\psi}^\dagger(\mathbf{r})\widehat{\psi}(\mathbf{r})$ interacting with the potential energy $\delta\widehat{U}(\mathbf{r})$. Thus:

$$\widehat{H}_{\text{ep}} = \int d\mathbf{r} \, \widehat{\psi}^\dagger(\mathbf{r}) \, \delta\widehat{U}(\mathbf{r}) \, \widehat{\psi}(\mathbf{r}) \,, \tag{13.5}$$

which should be regarded as the perturbation term due to the electron–phonon coupling. Using the expansions in Bloch functions

$$\widehat{\psi}(\mathbf{r}) = \frac{1}{\Omega^{1/2}} \sum_{n\mathbf{k}} \widehat{c}_{n\mathbf{k}} \, e^{i\mathbf{k}\cdot\mathbf{r}} \, u_{\mathbf{k}}(\mathbf{r}) \,, \tag{13.6}$$

and

$$\widehat{\psi}^\dagger(\mathbf{r}) = \frac{1}{\Omega^{1/2}} \sum_{n\mathbf{k}} \widehat{c}^\dagger_{n\mathbf{k}} \, e^{-i\mathbf{k}\cdot\mathbf{r}} \, u^*_{\mathbf{k}}(\mathbf{r}) \,, \tag{13.7}$$

we have (omitting the band index $n$ for simplicity)

$$\widehat{H}_{\text{ep}} = \frac{1}{\Omega} \sum_{\mathbf{k}\mathbf{k}'} \sum_{\mathbf{q}\eta} \sum_{l\gamma} \frac{1}{(NM_\gamma)^{1/2}} \left(\frac{\hbar}{2\omega_{\mathbf{q}}^{(\eta)}}\right)^{1/2}$$
$$\times e^{i\mathbf{q}\cdot\mathbf{R}_l} \, \widehat{c}^\dagger_{\mathbf{k}'} \, (\widehat{b}_{\mathbf{q}\eta} + \widehat{b}^\dagger_{-\mathbf{q}\eta}) \, \widehat{c}_{\mathbf{k}} \, \mathbf{e}_{\mathbf{q}\gamma}^{(\eta)} \cdot \mathbf{M}_{l\mathbf{k}\mathbf{k}'\gamma} \tag{13.8}$$

with

$$\mathbf{M}_{l\mathbf{k}\mathbf{k}'\gamma} = \int_\Omega d\mathbf{r} \, u^*_{\mathbf{k}'}(\mathbf{r}) \frac{\partial U(\mathbf{r})}{\partial \mathbf{R}_{l\gamma}} \, u_{\mathbf{k}}(\mathbf{r}) \, e^{i(\mathbf{k}-\mathbf{k}')\cdot\mathbf{r}} \,. \tag{13.9}$$

Because of the periodicity of $U(\mathbf{r})$, the matrix elements $\mathbf{M}_{l\mathbf{k}\mathbf{k}'}$ with different sub-indices $l$ are related. Specifically, we can rewrite

$$\mathbf{M}_{l\mathbf{k}\mathbf{k}'\gamma} = e^{i(\mathbf{k}-\mathbf{k}')\cdot\mathbf{R}_l} \int_\Omega d\mathbf{r} \, u^*_{\mathbf{k}'}(\mathbf{r}) \, \frac{\partial U(\mathbf{r})}{\partial \mathbf{R}_{l\gamma}} \, u_{\mathbf{k}}(\mathbf{r}) \, e^{i(\mathbf{k}-\mathbf{k}')\cdot(\mathbf{r}-\mathbf{R}_l)}$$
$$= e^{i(\mathbf{k}-\mathbf{k}')\cdot\mathbf{R}_l} \, \mathbf{M}_{\mathbf{k}\mathbf{k}'\gamma} \tag{13.10}$$

and the quantity

$$\mathbf{M}_{\mathbf{k}\mathbf{k}'\gamma} = \int_{\Omega} \mathrm{d}\mathbf{r}\, u^*_{\mathbf{k}'}(\mathbf{r}) \frac{\partial U(\mathbf{r})}{\partial \mathbf{R}_{l\gamma}}\, u_{\mathbf{k}}(\mathbf{r})\, \mathrm{e}^{\mathrm{i}(\mathbf{k}-\mathbf{k}')\cdot(\mathbf{r}-\mathbf{R}_l)} \tag{13.11}$$

is independent of $l$, since the integrand only depends on $\mathbf{r}-\mathbf{R}_{l\gamma}$ and the integration is taken over the entire volume.

In Eq. (13.8) the sum over $l$ only applies to the phase factor of the displacement vectors $\mathrm{e}^{\mathrm{i}\mathbf{q}\cdot\mathbf{R}_l}$ and the matrix elements $\mathbf{M}_{l\mathbf{k}\mathbf{k}'\gamma}$ and the sum can be evaluated

$$\sum_l \mathrm{e}^{\mathrm{i}\mathbf{q}\cdot\mathbf{R}_l}\, \mathrm{e}^{\mathrm{i}(\mathbf{k}-\mathbf{k}')\cdot\mathbf{R}_l}\, \mathbf{M}_{\mathbf{k}\mathbf{k}'\gamma} = \mathbf{M}_{\mathbf{k}\mathbf{k}'\gamma}\, N_{\mathrm{cell}} \sum_{\mathbf{G}} \delta(\mathbf{q}-(\mathbf{k}-\mathbf{k}')+\mathbf{G}) \,. \tag{13.12}$$

Processes where $\mathbf{G}$ is nonzero, while $\mathbf{q},\mathbf{k}$, and $\mathbf{k}'$ are inside the Brillouin Zone, are known as *Umklapp* processes. These occur when $\mathbf{k}-\mathbf{k}'$ falls outside the first Brillouin Zone.

The delta-function enables the evaluation of the sum over $\mathbf{k}',\mathbf{k}$, or $\mathbf{q}$. We choose to evaluate the sum over $\mathbf{k}'$, obtaining the expression for the electron–phonon interaction

$$\widehat{H}_{\mathrm{ep}} = \sum_{\mathbf{q}\mathbf{k}} \left(\frac{N_{\mathrm{cell}}}{M_{\mathrm{cell}}}\right)^{1/2} \sum_{\gamma\eta} \left(\frac{M_{\mathrm{cell}}}{M_\gamma}\right)^{1/2} \left(\frac{\hbar}{2\omega^{(\eta)}_{\mathbf{q}}}\right)^{1/2} \times \widehat{c}^{\dagger}_{\mathbf{k}+\mathbf{q}}(\widehat{b}_{\mathbf{q}\eta}+\widehat{b}^{\dagger}_{-\mathbf{q}\eta})\widehat{c}_{\mathbf{k}}\mathbf{e}^{(\eta)}_{\mathbf{q}\gamma}\cdot\mathbf{M}_{\mathbf{k}\mathbf{k}'\gamma} \,. \tag{13.13}$$

### 13.1.2 The Rigid-Ion Approximation

One possible way to treat the deformation-potential electron–phonon interaction is to assume that the total electron potential energy $U(\mathbf{r})$ is simply the lattice potential, $V^{(\mathrm{lat})}(\mathbf{r})$; that is, we assume that the total potential energy can be decomposed into a potential associated with each ion:

$$U(\mathbf{r}) \approx V^{(\mathrm{lat})}(\mathbf{r}) = \sum_{l\gamma} V^{(\gamma)}(\mathbf{r}-\mathbf{R}_{l\gamma}) \tag{13.14}$$

where $V^{(\gamma)}(\mathbf{r})$ is the potential associated with an atom of type $\gamma$ located at $\mathbf{R}=\mathbf{0}$. The gradient of the potential with respect to ion displacement can then be computed as

$$\frac{\partial V^{(\mathrm{lat})}(\mathbf{r})}{\partial \mathbf{R}_{l\gamma}} = -\nabla V^{(\gamma)}(\mathbf{r}-\mathbf{R}_{l\gamma}). \tag{13.15}$$

Note that here we are assuming that, when displaced, the ionic potential does not change but simply shifts rigidly. For this reason, this approximation is known as the *rigid-ion approximation*. It ignores the fact that valence electrons will redistribute themselves as the ions move away from their equilibrium positions (which is, in essence, dielectric screening). We may consider this as a higher-order effect ignored in our linear-response picture. But, clearly, this remains an approximation.

Expressing the ionic potential in terms of its Fourier components:

$$V^{(\gamma)}(\mathbf{r}-\mathbf{R}_{l\gamma}) = V^{(\gamma)}(\mathbf{r}-\mathbf{R}_l-\tau_\gamma) = \sum_{\kappa} V^{(\gamma)}_{\kappa}\, \mathrm{e}^{\mathrm{i}\kappa\cdot(\mathbf{r}+\mathbf{R}_l+\tau_\gamma)} \,, \tag{13.16}$$

we have

$$\nabla V^{(\gamma)}(\mathbf{r}-\mathbf{R}_l-\tau_\gamma)=\mathrm{i}\sum_{\kappa}\kappa\, V_{\kappa}^{(\gamma)}\,\mathrm{e}^{\mathrm{i}\kappa\cdot(\mathbf{r}+\mathbf{R}_l)}\,, \tag{13.17}$$

having absorbed the factor $\mathrm{e}^{\mathrm{i}\kappa\cdot\tau_\gamma}$ inside the Fourier component $V_{\kappa}^{(\gamma)}$.

We can repeat our analysis of the electron–phonon interaction starting from Eq. (13.8):

$$\begin{aligned}\widehat{H}_{\mathrm{ep}}=&\frac{1}{\Omega}\sum_{\mathbf{kk}'}\sum_{\mathbf{q}\eta}\sum_{l\gamma}\frac{1}{(NM_\gamma)^{1/2}}\left(\frac{\hbar}{2\omega_{\mathbf{q}}^{(\eta)}}\right)^{1/2}\\ &\times\,\mathrm{e}^{\mathrm{i}\mathbf{q}\cdot\mathbf{R}_l}\,\widehat{c}_{\mathbf{k}'}^{\dagger}\,(\widehat{b}_{\mathbf{q}\eta}+\widehat{b}_{-\mathbf{q}\eta}^{\dagger})\,\widehat{c}_{\mathbf{k}}\,\mathbf{e}_{\mathbf{q}\gamma}^{(\eta)}\cdot(\mathrm{i}\kappa)\,V_{\kappa}^{(\gamma)}\\ &\times\int d\mathbf{r}\,\mathrm{e}^{\mathrm{i}\kappa\cdot(\mathbf{r}-\mathbf{R}_l)}\,u_{\mathbf{k}'}^{*}(\mathbf{r})\,u_{\mathbf{k}}(\mathbf{r})\,\mathrm{e}^{\mathrm{i}(\mathbf{k}-\mathbf{k}')\cdot\mathbf{r}}\,.\end{aligned} \tag{13.18}$$

Again, the sum over $l$ can be replaced by a sum over Dirac deltas:

$$\sum_{l}\mathrm{e}^{\mathrm{i}(\mathbf{q}-\kappa)\cdot\mathbf{R}_l}=N_{\mathrm{cell}}\sum_{\mathbf{G}}\delta(\mathbf{q}-\kappa+\mathbf{G})\,, \tag{13.19}$$

while the integral involving Bloch functions vanishes unless $\mathbf{k}'-\mathbf{k}+\kappa=\mathbf{G}$ [see Eq. (8.11), page 165]. Thus:

$$\begin{aligned}\widehat{H}_{\mathrm{ep}}=&\sum_{\mathbf{kqG}}\left(\frac{N_{\mathrm{cell}}}{M_{\mathrm{cell}}}\right)^{1/2}\sum_{\gamma\eta}\left(\frac{M_{\mathrm{cell}}}{M_\gamma}\right)^{1/2}\left(\frac{\hbar}{2\omega_{\mathbf{q}}^{(\eta)}}\right)^{1/2}\mathrm{i}(\mathbf{q}+\mathbf{G})\cdot\mathbf{e}_{\mathbf{q}\gamma}^{(\eta)}\,V_{\mathbf{q}+\mathbf{G}}^{(\gamma)}\\ &\times\,\mathscr{I}_{\mathbf{k}+\mathbf{q}+\mathbf{G},\mathbf{k}}\,\widehat{c}_{\mathbf{k}+\mathbf{q}+\mathbf{G}}^{\dagger}\,(\widehat{b}_{\mathbf{q}\eta}+\widehat{b}_{-\mathbf{q}\eta}^{\dagger})\,\widehat{c}_{\mathbf{k}}\,.\end{aligned} \tag{13.20}$$

It is important to emphasize here that this equation implies that acoustic and optical phonons exhibit a vastly different behavior in how they couple with electrons at long wavelengths, that is, as $q\to 0$. Consider for simplicity Si, a crystal with two atoms per unit cell at positions $\tau_1=(0,0,0)$ and $\tau_2=(a_0/4)(1,1,1)$. Choosing the origin mid-way between the two atoms, we can write their positions as $\tau_1=(a_0/8)(1,1,1)=\tau$ and $\tau_2=-\tau$. At long wavelengths, the displacement of the two ions due to an acoustic phonons will be in phase and with the same magnitude, $\mathbf{e}_{\mathbf{0}1}^{(\mathrm{ac})}=\mathbf{e}e^{\mathrm{i}(\mathbf{q}+\mathbf{G})\cdot\tau}\approx\mathbf{e}e^{\mathrm{i}\mathbf{G}\cdot\tau}\approx\mathbf{e}_{\mathbf{0}2}^{(\mathrm{ac})}$. Considering in the sum over reciprocal-space wavevectors $\mathbf{G}$ in Eq. (13.20) a pair $\mathbf{G}$ and $-\mathbf{G}$, the sum over the ions yields a contribution $\sim\mathbf{e}\cdot\mathbf{G}\,\{\cos[(\mathbf{q}+\mathbf{G})\cdot\,\tau]-\cos[(\mathbf{q}-\mathbf{G})\cdot\,\tau]\}$. This vanishes as $q\to 0$. On the contrary, for long-wavelength optical phonons, the ionic displacements are out of phase, $\mathbf{e}_{\mathbf{0}1}^{(\mathrm{ac})}\approx\mathbf{e}e^{\mathrm{i}\mathbf{G}\cdot\tau}\approx-\mathbf{e}_{\mathbf{0}2}^{(\mathrm{ac})}$. Therefore, the sum over the same pair of opposite $\mathbf{G}$ vectors and over the ionic coordinates results in a term of the form $\sim\mathbf{e}\cdot\mathbf{G}\,\{\sin[(\mathbf{q}+\mathbf{G})\cdot\,\tau]-\sin[(\mathbf{q}-\mathbf{G})\cdot\,\tau]\}$. This now tends to a constant, as $q\to 0$. In more quantitative terms, the sum over each pair $\mathbf{G}$ and $-\mathbf{G}$ and over ionic coordinates yields a contribution

$$2\,\mathrm{i}\,V_{\mathbf{G}}\,\mathbf{e}_\eta\cdot[\mathbf{q}\,\cos(\mathbf{G}\cdot\tau)+\mathbf{G}\,(\mathbf{q}\cdot\tau)\,]\sum_{\mathbf{G}'}u_{\mathbf{K},\mathbf{G}'+\mathbf{G}}^{(n)*}\,u_{\mathbf{K},\mathbf{G}'}^{(n)}\,, \tag{13.21}$$

for acoustic phonons, and a contribution

$$2\,V_{\mathbf{G}}\sin(\mathbf{G}\cdot\tau)\,(\mathbf{q}+\mathbf{G})\cdot\mathbf{e}_\eta\sum_{\mathbf{G}'}u_{\mathbf{K},\mathbf{G}'+\mathbf{G}}^{(n)*}\,u_{\mathbf{K},\mathbf{G}'}^{(n)}\,, \tag{13.22}$$

for optical phonons. Here we have employed the symmetry property

$$\sum_{\mathbf{G}'} u^{(n)*}_{\mathbf{K},\mathbf{G}'+\mathbf{G}}\, u^{(n)}_{\mathbf{K},\mathbf{G}'} = \sum_{\mathbf{G}'} u^{(n)*}_{\mathbf{K},\mathbf{G}'-\mathbf{G}}\, u^{(n)}_{\mathbf{K},\mathbf{G}'} \,, \tag{13.23}$$

a consequence of the lattice symmetry. Equations (13.21) and (13.22) imply that at long wavelengths (small $q$) the coupling between electrons and acoustic phonons vanishes, whereas the coupling between electrons and optical phonons approaches a constant. Physically, this behavior originates from the fact that at long wavelengths the presence of acoustic phonons represents simply a uniform (adiabatic) translation of the entire lattice, so electrons remain unaffected. On the contrary, in the same limit $q \to 0$, optical phonons represent a uniform shift of one set of ions with respect to another, so that electrons are affected by the "strained" crystal structure. The quantities given by Eqs. (13.21) and (13.22) have dimensions of energy per unit length and, when summed over all pairs of $\mathbf{G}$ vectors, are usually called "deformation potentials" (for reasons that we shall see shortly) and are denoted by $\Delta_{\text{ac}}q$ and $(DK)_0$, respectively.

In order to simplify the notation, we shall adopt the convention that all phonon wavevectors must be mapped into the first BZ, whenever necessary. We now define the electron–phonon coupling constant as:

$$\mathscr{A}_{\mathbf{q}\eta} = \sum_{\mathbf{G}\gamma} \left( \frac{\hbar}{2\rho_{\text{x}}\omega^{(\eta)}_{\mathbf{q}}} \right)^{1/2} \mathrm{i}(\mathbf{q}+\mathbf{G}) \cdot \mathbf{e}^{(\eta)}_{\mathbf{q}\gamma}\, V^{(\gamma)}_{\mathbf{q}+\mathbf{G}} \,, \tag{13.24}$$

having absorbed the factor $(M_{\text{cell}}/M_{\gamma})^{1/2}$ as a different normalization of the Fourier transform of the ionic potential (so that it's normalized to the atomic volume rather than to the cell volume) and having converted the wavefunction (or field) normalization from a box-normalization to the infinite-volume normalization, so that the crystal mass density $\rho_{\text{x}}$ appears in this expression. Thus,

$$\widehat{H}_{\text{ep}} = \sum_{\mathbf{k}\mathbf{q}\eta} \mathscr{A}_{\mathbf{q}\eta}\, \mathscr{I}_{\mathbf{k}+\mathbf{q},\mathbf{k}}\, \widehat{c}^{\dagger}_{\mathbf{k}+\mathbf{q}} \left(\widehat{b}_{\mathbf{q}\eta} + \widehat{b}^{\dagger}_{-\mathbf{q}\eta}\right) \widehat{c}_{\mathbf{k}} \,. \tag{13.25}$$

Note the physical meaning of Eq. (13.25): It represents the annihilation of an electron of momentum $\mathbf{k}$, the creation (the term $\widehat{b}^{\dagger}_{-\mathbf{q}\eta}$) or annihilation (the term $\widehat{b}_{\mathbf{q}\eta}$) of a phonon of branch $\eta$ and momentum $\mathbf{q}$, and, finally, the creation of an electron with momentum $\mathbf{k}+\mathbf{q}$. Thus, it can be viewed as the transfer of momentum $\mathbf{q}$ from/to a phonons to/from an electron. Note that the only nonzero matrix elements of the Hamiltonian Eq. (13.25) are those which differ by one phonon and having an electron state $|\mathbf{k}\rangle$ swapped with a state $|\mathbf{k}+\mathbf{q}\rangle$.

* A technical but important discussion. *The last statement relies on the following considerations. The Hilbert space of the system, by axiom 9 of the Copenhagen interpretation (see Chap. 1, Sect. 1.4.4), is the tensor product of the Fock space associated with the electrons, and the Fock space associated with the phonons. If the electron–phonon coupling is weak, we may assume the general state of the system is just the tensor product of a state* $|\{n_{\mathbf{k}_i}\}\rangle$, *describing a configurations of electrons,* $n_{\mathbf{k}_i}$ *(=0 or 1, in absence of spin) of them in a single-electron state* $|\mathbf{k}_i\rangle$, *and of the state* $|\{N_{\mathbf{q}_j}\}\rangle$, *describing a configurations of phonons,* $N_{\mathbf{q}_j}$ *of them in a single-phonon state* $|\mathbf{q}_j\rangle$ *(ignoring for clarity the index* $\eta$ *labeling the phonon branches). Thus,*

$$\begin{aligned} |\{n_{\mathbf{k}_i}\},\{N_{\mathbf{q}_j}\}\rangle &= |\{n_{\mathbf{k}_i}\}\rangle|\{N_{\mathbf{q}_j}\}\rangle \\ &= \mathscr{N}\, (\widehat{c}^{\dagger}_{\mathbf{k}_1})^{n_{\mathbf{k}_1}} (\widehat{c}^{\dagger}_{\mathbf{k}_2})^{n_{\mathbf{k}_2}} (\widehat{c}^{\dagger}_{\mathbf{k}_3})^{n_{\mathbf{k}_3}} \ldots |0\rangle_{\text{el}} \\ &\quad \times (\widehat{b}^{\dagger}_{\mathbf{q}_1})^{N_{\mathbf{q}_1}} (\widehat{b}^{\dagger}_{\mathbf{q}_2})^{N_{\mathbf{q}_2}} (\widehat{b}^{\dagger}_{\mathbf{q}_2})^{N_{\mathbf{q}_3}} \ldots |0\rangle_{\text{ph}} \,, \end{aligned} \tag{13.26}$$

*where $\mathscr{N}$ accounts for the normalization constants and $|0\rangle_{\rm el}$ and $|0\rangle_{\rm ph}$ are the vacuum states for electrons and phonons, respectively. (Note such a "decoupling" is not rigorously correct. There are examples of systems, such as in semiconductors excited by sub-picosecond laser pulses and studied at the femto-second time scale, whose experimentally observed behavior can be understood only by describing them with fully coupled electron–phonon states. We shall not consider these peculiar cases.) From the form of the Hamiltonian given by Eq. (13.25) above, a general matrix element has the form*

$$\langle\{n_{\mathbf{k}_i}\}'\langle|\{N_{\mathbf{q}_j}\}]'|\widehat{c}^{\dagger}_{\mathbf{k}+\mathbf{q}}\,(\widehat{b}_{\mathbf{q}}+\widehat{b}^{\dagger}_{-\mathbf{q}})\,\widehat{c}_{\mathbf{k}}\,|\{n_{\mathbf{k}_i}\}\rangle|\{N_{\mathbf{q}_j}\}\rangle$$
$$=\langle\{n_{\mathbf{k}_i}\}'|\widehat{c}^{\dagger}_{\mathbf{k}+\mathbf{q}}\,\widehat{c}_{\mathbf{k}}|\{n_{\mathbf{k}_i}\}\rangle\,\langle\{N_{\mathbf{q}_j}\}'|(\widehat{b}_{\mathbf{q}}+\widehat{b}^{\dagger}_{-\mathbf{q}})|\{N_{\mathbf{q}_j}\}\rangle\,, \tag{13.27}$$

*having used the fact that electron and phonon operators commute.*

*Consider first the first term. It can only be nonzero when the state $|\{n_{\mathbf{k}_i}\}\rangle$ contains one electron in the single-electron state $|\mathbf{k}\rangle$, because the annihilation operator $\widehat{c}_{\mathbf{k}}$ will act on the electron vacuum and yield zero, unless it "finds" on its right an operator $\widehat{c}^{\dagger}_{\mathbf{k}}$ with which it can anticommute, yielding 1. Similarly, the state $\langle\{n_{\mathbf{k}_i}\}'|$ must contain an electron in state $|\mathbf{k}+\mathbf{q}\rangle$, otherwise the creation operator $\widehat{c}^{\dagger}_{\mathbf{k}+\mathbf{q}}$—that behaves as a annihilation operator when acting to the left—would annihilate the electron vacuum for that state. However, if the initial Fock state, $|\{n_{\mathbf{k}_i}\}\rangle$, contains already an electron in the state $|\mathbf{k}+\mathbf{q}\rangle$, the creator $\widehat{c}^{\dagger}_{\mathbf{k}+\mathbf{q}}$ would attempt to place another electron in the same state. The anticommutation rules prevent this, as we saw in Chap. 9, Eq. (9.51). Therefore, the matrix element vanishes in this case. This is a consequence of Pauli's exclusion principle and is called* Pauli blocking*: An electron can scatter only into an unoccupied state. Therefore, we see that the only nonzero matrix elements of the Hamiltonian Eq. (13.25) are those between an initial Fock state (a "ket" on the right) corresponding to an electron configuration containing an electron in the state* **k** *and no electrons in the state* $\mathbf{k}+\mathbf{q}$*, and a final Fock state (a "bra" on the left) with an electron in the state* $\mathbf{k}+\mathbf{q}$*, and no electrons in the state* **k***, all other electrons remaining unaffected. Accounting for the additional spin degree of freedom will permit a given state to be occupied by two electrons, provided they are in different spin states, but it will not change qualitatively these conclusions.*

*Dealing with the second factor in this equation is a bit more complicated. This is the sum of two terms: $\langle\{N_{\mathbf{q}_j}\}'|\,\widehat{b}_{\mathbf{q}}|\{N_{\mathbf{q}_j}\}\rangle$ and $\langle\{N_{\mathbf{q}_j}\}'|\,\widehat{b}^{\dagger}_{-\mathbf{q}}|\{N_{\mathbf{q}_j}\}\rangle$. Consider the first term. If the initial state $|\{N_{\mathbf{q}_j}\}\rangle$ contains no phonons of momentum* **q***, the annihilator $\widehat{b}_{\mathbf{q}}$ acts directly on the vacuum state for the single-phonon state $|\mathbf{q}\rangle$ and this term vanishes. If, instead, the state $|\{N_{\mathbf{q}_j}\}\rangle$ contains $N_{\mathbf{q}}$ phonons in the single-phonon state $|\mathbf{q}\rangle$, the annihilator $\widehat{b}_{\mathbf{q}}$ will convert this state into a state containing only $N_{\mathbf{q}}-1$ phonons. Therefore, the state $\langle\{N_{\mathbf{q}_j}\}'|$ must contain $N_{\mathbf{q}}-1$ phonons (for the matrix elements to be nonzero. If that were not the case, the two states would be orthogonal, since they would be eigenstates of the phonon Hamiltonian corresponding to different eigenvalues). For the same reason, the two states must contain the same number of phonons for all other states. Therefore, the only nonzero matrix elements (ignoring phonons with different wavevectors, they do not matter in this argument) will be of the form:*

$$\frac{1}{[(N_{\mathbf{q}}-1)!]^{1/2}\,[N_{\mathbf{q}}!]^{1/2}}\;{}_{\rm ph}\langle 0|(\widehat{b}_{\mathbf{q}})^{N_{\mathbf{q}}-1}\,\widehat{b}_{\mathbf{q}}\,(\widehat{b}^{\dagger}_{\mathbf{q}})^{N_{\mathbf{q}}}\,|0\rangle_{\rm ph}\,, \tag{13.28}$$

*where the pre-factor expresses the normalization of these states, as shown below. The "central" $\widehat{b}$ transforms the initial state of $N_{\mathbf{q}}$ phonons into a state with $N_{\mathbf{q}}-1$ phonons. This can be seen by shifting this central $\widehat{b}$ to the right. Each of the $N_{\mathbf{q}}$ steps requires a commutation that results in the inner product between the states that contain $N_{\mathbf{q}}-1$ phonons, whose norm is $\sqrt{(N_{\mathbf{q}}-1)!}$. By the* *time we shift it all the way to the right, thus annihilating the vacuum, we are left with a factor of* $N_{\mathbf{q}}\,(N_{\mathbf{q}}-1)!$*. Accounting for the initial normalization factors in the denominator of Eq. (13.28), the net result is that this matrix element is equal to* $N_{\mathbf{q}}(N_{\mathbf{q}}-1)!/\sqrt{(N_{\mathbf{q}}-1)!N_{\mathbf{q}}!}$*, or* $\sqrt{N_{\mathbf{q}}}$*.*

*For similar reasons, the second term is nonzero only when we have* $N_{\mathbf{q}}$ *phonons contained in the state* $|\{N_{\mathbf{q}_j}\}\rangle$ *and* $N_{\mathbf{q}}+1$ *in the state* $\langle\{N_{\mathbf{q}_j}\}'|$*. So, it has the similar form, except for a minus sign in front of* $\mathbf{q}$*. This reflects the fact that the emission processes subtract momentum from the electron. The symmetry of all crystals we consider makes this an irrelevant difference for the purpose of this discussion. Therefore, we shall ignore the "minus" sign and write the only non-vanishing matrix element as:*

$$\frac{1}{[(N_{\mathbf{q}}+1)!]^{1/2}\,[N_{\mathbf{q}}!]^{1/2}}\;{}_{\mathrm{ph}}\langle 0|(\widehat{b}_{\mathbf{q}})^{N_{\mathbf{q}}+1}\,\widehat{b}^{\dagger}_{\mathbf{q}}\,(\widehat{b}^{\dagger}_{\mathbf{q}})^{N_{\mathbf{q}}}\,|0\rangle_{\mathrm{ph}}\,. \tag{13.29}$$

*We can proceed as we just did for the first term, now shifting the "central"* $\widehat{b}^{\dagger}$ *all the way to the left, so that it will annihilate (since on the left annihilation and creation are swapped by the conjugation) the vacuum state. In so doing, we will be left with a factor of* $(N_{\mathbf{q}}+1)N_{\mathbf{q}}!$ *resulting from the steps required to shift the operator to the right and by the squared norm of the states containing* $N_{\mathbf{q}}$ *phonons. Therefore, this matrix element is equal to* $(N_{\mathbf{q}}+1)N_{\mathbf{q}}!/\sqrt{(N_{\mathbf{q}}+1)!N_{\mathbf{q}}!}$*, or* $\sqrt{N_{\mathbf{q}}+1}$*.*

*We are almost ready to conclude: If we assume that the "large" crystal is not perturbed significantly by the few free electrons moving around, we can also assume that its temperature will not increase too much. At thermal equilibrium, the number of phonons with wavevector* $\mathbf{q}$ *(or* $-\mathbf{q}$*, justifying the fact that we have ignored the minus sign) is given by the Bose–Einstein distribution function:*

$$\langle N_{\mathbf{q}}\rangle_{\mathrm{th}}=\frac{1}{\mathrm{e}^{(\hbar\omega_{\mathbf{q}})/(k_{\mathrm{B}}T)}-1}\,. \tag{13.30}$$

*Therefore, we conclude that the only nonzero matrix elements of the electron–phonon Hamiltonian, Eq. (13.25), are those between initial and final electron Fock states that are identical, except for the fact that they contain one electron in the initial state* $\mathbf{k}$ *and one in the final state* $\mathbf{k}+\mathbf{q}$*. Moreover, they represent the same phonon states, except for the fact that they contain* $N_{\mathbf{q}}$ *phonons in the initial state and* $N_{\mathbf{q}}\pm 1$ *phonons in the final states. These matrix elements represent absorption of a phonon, yielding a contribution* $\langle N_{\mathbf{q}}\rangle_{\mathrm{th}}^{1/2}$ *from the phonon states, or emission, yielding a contribution* $(1+\langle N_{\mathbf{q}}\rangle_{\mathrm{th}})^{1/2}$*. The squared matrix element is proportional to the scattering (transition) rate. Therefore, absorption processes are proportional to* $\langle N_{\mathbf{q}}\rangle_{\mathrm{th}}$*, the number of phonons presents at thermal equilibrium. This is a result that is intuitively clear: The probability per unit time of absorbing a phonon is proportional to the number of phonons available to be absorbed. On the contrary, the probability of emitting a phonon of wavevector* $\mathbf{q}$ *is proportional to* $N_{\mathbf{q}}+1$*. This is unexpected. The "1" represents* spontaneous *emission processes; the term* $\langle N_{\mathbf{q}}\rangle$ *represents* stimulated *emission processes.*

*These coefficients that depend on the occupation of boson states solve a very big problem with "conventional" Quantum Mechanics. Consider a H atom in an excited state 2p, say. This is a stationary state of the Schrödinger equation. As such, in absence of any perturbation, the electron should remain in that state forever. Yet, we know that the electron will eventually decay to the 1s ground state by emitting a photon. Why? We shall see that the electromagnetic field can be quantized. The rates at which electrons emit or absorb photons can be formulated exactly in the same way we have formulated phonon emission and absorption processes. The spontaneous emission term is what causes the electron decay. It is due to the fact that the absence of photons does not amount to absence of a perturbation. It simply means that the electromagnetic field is in its ground state (the vacuum state). But the electron field can "resonate" with this vacuum and spontaneous emission is the result.*

*A similar "resonance" explains the "stimulated" emission, a process that is the basic principle of which lasers (*Light Emission via Stimulated Emission Radiation*) are based.*

From this discussion it follows that for absorption processes only terms of the following form will be nonzero (ignoring the branch-index $\eta$ to simplify the notation):

$$\mathscr{A}_{\mathbf{q}} \langle N_{\mathbf{q}}-1|\widehat{b}_{\mathbf{q}}|N_{\mathbf{q}}\rangle\langle n_{\mathbf{k}+\mathbf{q}}+1;n_{\mathbf{k}}-1|\widehat{c}^{\dagger}_{\mathbf{k}+\mathbf{q}}\widehat{c}_{\mathbf{k}}|n_{\mathbf{k}+\mathbf{q}};n_{\mathbf{k}}\rangle$$
$$= \mathscr{A}_{\mathbf{q}} \sqrt{(1-n_{\mathbf{k}+\mathbf{q}})n_{\mathbf{k}}N_{\mathbf{q}}} \; . \tag{13.31}$$

We shall replace later the occupation numbers $N_{\mathbf{q}}$ with the thermal populations $\langle N_{\mathbf{q}}\rangle_{\rm th}$. For emission processes the only non-vanishing matrix elements will be of the form:

$$\mathscr{A}_{\mathbf{q}} \langle N_{-\mathbf{q}}+1|\widehat{b}^{\dagger}_{-\mathbf{q}}|N_{-\mathbf{q}}\rangle\langle n_{\mathbf{k}-\mathbf{q}}+1;n_{\mathbf{k}}-1|\widehat{c}^{\dagger}_{\mathbf{k}-\mathbf{q}}\widehat{c}_{\mathbf{k}}|n_{\mathbf{k}-\mathbf{q}};n_{\mathbf{k}}\rangle$$
$$= \mathscr{A}_{\mathbf{q}} \sqrt{(1-n_{\mathbf{k}-\mathbf{q}})n_{\mathbf{k}}(1+N_{-\mathbf{q}})} \; , \tag{13.32}$$

having indicated with $n_{\mathbf{k}}$ the electron number and with $N_{\mathbf{q}}$ the phonon number.

* *Regarding the normalization of Fock states for bosons mentioned above: Consider the state containing two phonons, $|2\rangle = \widehat{b}^{\dagger}\widehat{b}^{\dagger}|0\rangle$, having omitted the index $\mathbf{q}$. Let's compute its norm using the commutation rule $bb^{\dagger} = 1+\widehat{b}^{\dagger}\widehat{b}$:*

$$\langle 2|2\rangle = \langle 0|\widehat{b}\widehat{b\widehat{b}^{\dagger}}\widehat{b}^{\dagger}|0\rangle = \langle 0|\widehat{b}(1+\widehat{b}^{\dagger}\widehat{b})\widehat{b}^{\dagger}|0\rangle \; \langle 0|\widehat{b\widehat{b}^{\dagger}}|0\rangle \; + \; \langle 0|\widehat{b\widehat{b}^{\dagger}}\widehat{b\widehat{b}^{\dagger}}|0\rangle$$
$$= 1 + \langle 0|\widehat{b\widehat{b}^{\dagger}}(1+\widehat{b}^{\dagger}\widehat{b})|0\rangle = 1 + \langle 0|\widehat{b\widehat{b}^{\dagger}}|0\rangle = 2 \; . \tag{13.33}$$

*In general one can follow the same procedure to find (try it with $\langle 3|3\rangle$):*

$$\langle N|N\rangle = N! \; . \tag{13.34}$$

*Now consider, for example, the term $\langle N-1|\widehat{b}|N\rangle$ in Eq. (13.31) above:*

$$\langle N-1|\widehat{b}|N\rangle = \frac{1}{[(N-1)!N!]^{1/2}}\langle 0|\widehat{b}^{N-1}\widehat{b\widehat{b}^{\dagger N}}|0\rangle = \frac{1}{[(N-1)!N!]^{1/2}}\langle 0|\widehat{b}^{N}\widehat{b}^{\dagger N}|0\rangle$$
$$= \left(\frac{N!N!}{(N-1)!N!}\right)^{1/2}\langle N|N\rangle = N^{1/2} \; . \tag{13.35}$$

Note that the terms involving the electron number simply represent the effect of "Pauli blocking", we can create an electron with wavevector $\mathbf{k}+\mathbf{q}$ only if that state is available (that is: $n_{\mathbf{k}}+\mathbf{q}=0$). More interesting is the appearance of the phonon numbers that arise from the normalization mentioned at page 202. When squared inside Fermi's Golden Rule, they give rise to the "Einstein" coefficients $\langle N_{\mathbf{q}}\rangle_{\rm th}$ for absorption, $\langle N_{\mathbf{q}}\rangle_{\rm th}+1$ for emission: Stimulated emission is simply a result of the commutation rules!

The result of this long discussion is an expression for the nonpolar electron–phonon scattering rate. This is simply Eq. (12.10) (or its non-*Umklapp* approximation, Eq. (12.11)) of Chap. 12:

$$\frac{1}{\tau_{\eta}(\mathbf{k},n)} = \frac{2\pi}{\hbar}\sum_{\mathbf{k}'n'}\sum_{\mathbf{G}} |\mathscr{A}_{\mathbf{k}-\mathbf{k}'+\mathbf{G},\eta} \; \mathscr{I}^{(3D)}_{\mathbf{k},\mathbf{k}',n,n';\mathbf{G}}|^2$$

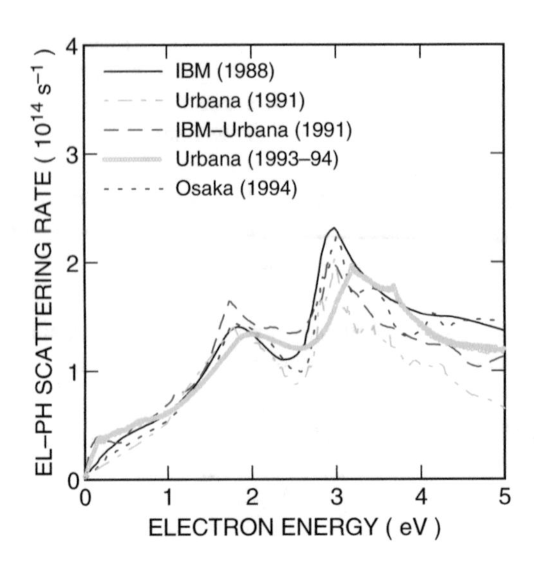


**Fig. 13.1** Total electron–phonon scattering rate at 300 K in Si calculated using the band structure obtained using nonlocal empirical pseudopotentials and the rigid-ion approximation (*dashed red line*, labeled "IBM-Urbana 1991," from [3]) compared to the scattering rates obtained by other groups. Data shown by the *thick yellow line* ("Urbana 1993–1994") are from [4], those shown by a *black solid line* ("IBM-1988") are from [5], those shown by a *dashed blue line* ("Osaka 1994") are from [6], and, finally, the data shown by *dashed cyan line* are from [7]

$$\times \delta[E_n(\mathbf{k}) - E_{n'}(\mathbf{k}') \pm \hbar\omega^{(\eta)}_{\mathbf{k}-\mathbf{k}'+\mathbf{G}}] . \tag{13.36}$$

A numerical evaluation can proceed as described in the previous chapter. Figure 13.1 shows the electron–phonon scattering rates in Si at 300 K calculated in [3] from Eq. (13.36) using empirical pseudopotentials and the rigid-ion approximation. Using the ensemble Monte Carlo method described in Chap. 19, Sects. 19.4.1 (19.4.1.2), they reproduce the experimentally measured mobility and drift-velocity field characteristics.

## 13.2 Deformation Potentials and the Harrison Interaction

Calculation of the nonpolar electron–phonon scattering rates using Eq. (13.36), just obtained with much pain, have become possible only recently, in the last couple of decades or so. Even more laborious are calculations based on the first-principles approaches outlined in Sect. 13.2.1 above. Historically, the problem of calculating the "coupling constant" $\mathscr{A}_\mathbf{q}$ was solved, at least for electrons at low energy, by employing the powerful connection between "deformation potentials" and the nonpolar interaction between electrons and acoustic phonons that was uncovered by Bardeen and Shockley [2] in 1950. Indeed, they have proposed an elegant method to estimate the electron/acoustic-phonon matrix element when electrons are close to the bottom of the conduction band (or holes close to the top of the valence band) from known parameters known as *deformation potentials*, hence the name of this type of interaction.

When we compress or dilate a crystal, say by changing the lattice constant $a \to a+u$, the band structure changes. In particular, the energy of the conduction band minimum moves by an amount

$$\Delta E_c = E_c(a+u) - E_c(a) \approx \frac{dE_c}{da}\, u .$$

In 3D, $\Delta E_c$ is proportional to the change of the volume $\Delta\Omega$ which, for a lattice displacement $\mathbf{u}$ is given by $\Delta\Omega = \Omega\nabla\cdot\mathbf{u}$, so that

$$\Delta E_c = \Omega\,\frac{dE_c}{d\Omega}\,\nabla\cdot\mathbf{u} \sim q\,\Delta_{ac}\,. \tag{13.37}$$

The constant $\Omega dE_c/d\Omega$ is called "deformation potential" for the conduction band. Let's denote it here by $\Delta_{ac}$, although, depending on the particular valley and valence/conduction band considered, several other symbols are used in the literature (such as $E_1$ or $\Xi$).

If we consider acoustic phonons of a sufficiently long wavelength (that is, spanning many cells), we can approximate the effect of the density-wave associated with acoustic phonons (consisting of alternating crests/troughs of compression/dilation, compression/dilation....) as having locally the same effect of a uniform, global compression or dilation of the crystal. Therefore, we can identify the uniform displacement $\mathbf{u}$, defined above, with the ionic displacement $\delta\mathbf{R}_{l\gamma}$ and the local shift of the band-edge with $\Delta_{ac}$. This "fluctuating" conduction-band edge is what causes the electrons to scatter.

With these simplifications, ignoring for now selection rules due to the symmetries, so replacing the dot-product $\mathbf{e}_{\mathbf{q}\gamma}^{(\eta)}\cdot(\mathbf{q}+\mathbf{G})$ simply with $q$, we can approximate the electron–phonon coupling constant $\mathscr{A}_{\mathbf{q}\eta}$ in Eq. (13.24) with

$$\mathscr{A}_{\mathbf{q}ac} \approx \Delta_{ac}\,q\left(\frac{\hbar}{2\rho_x\omega_{\mathbf{q}}^{(ac)}}\right)^{1/2}. \tag{13.38}$$

This is consistent with the discussion that had led us to Eq. (13.21) above. Thus, the squared matrix element for electron/acoustic-phonon scattering takes the form:

$$|\langle\mathbf{k}'|\widehat{H}_{ac}|\mathbf{k}\rangle|^2 = \frac{\hbar\Delta_{ac}^2 q^2}{2\rho_x\omega_{\mathbf{q}}^{(ac)}}\left(\langle N_{\mathbf{q}}\rangle_{th} + \frac{1}{2} \pm \frac{1}{2}\right),$$

where $\rho_x$ is the density of the crystal, the "plus" sign refers to emissions, the "minus" sign to absorption processes. Note that the frequency of acoustic phonons (in Si there are two transverse modes—TA—and a single longitudinal mode—LA) can be well approximated by a linear relation $\omega_{\mathbf{q}}^{(ac)} \approx c_s q$, where $c_s$ is the sound velocity, transverse or longitudinal, as appropriate. Therefore, the matrix element grows with momentum transfer and we see that scattering with acoustic phonons is a "large-angle scattering," very effective in "randomizing" the electron direction and momentum. This has a great effect on the mobility. A more correct expression for the deformation-potential interaction in the ellipsoidal valleys in Si and Ge has been given by Herring and Vogt in 1956 [8]. In terms of the uniaxial shear and dilation deformation potentials $\Xi_u$ and $\Xi_d$, respectively, the isotropic deformation potential $\Delta_{ac}$ appearing in Eq. (13.37) above is replaced by:

$$\Delta_{LA} = \Xi_d + \Xi_u\cos^2\theta_q\,, \tag{13.39}$$

for collisions assisted by LA phonons and

$$\Delta_{LA} = \Xi_u\cos\theta_q\sin\theta_q\,, \tag{13.40}$$

where $\theta_q$ is the angle between the longitudinal axis of the ellipsoid and the phonon momentum $\mathbf{q}$.

A similar expression, due to Harrison [9], holds for optical phonons simply replacing the frequency $\omega_{\mathbf{q}}$ and deformation potential $\Delta_{ac}$ with their "optical" counterparts $\omega$ (op) and $\Delta_{op}$. Note that the main difference lies in what we have discussed regarding Eqs. (13.20) and (13.22): Whereas for the acoustic strain the energy change depends on the gradient of the displacement (the strain $\propto \nabla \cdot \mathbf{u}$), for optical distortions of the lattice the energy change is affected mainly by the displacement itself, so the $q$-dependence in Eq. (13.37) above disappears. In the literature one finds many "optical deformation potentials." Usually the notation $(D_t K)_{\rm op}$ is used, so that:

$$|\langle \mathbf{k}'|\widehat{H}_{\rm op}|\mathbf{k}\rangle|^2 = \frac{(D_t K)^2_{\rm op}\hbar}{2\Omega\rho_{\rm x}\omega^{\rm (op)}}\left(\langle N_{\rm op}\rangle_{\rm th} + \frac{1}{2} \pm \frac{1}{2}\right) .$$

Within the effective-mass approximation, assuming a single isotropic valley with an effective mass $m^*$. The calculation of the scattering rates for emission and absorption of optical phonons is relatively simple, since the frequency of optical phonons is roughly independent of $q$, and so is the occupation number $\langle N_{\rm op}\rangle_{\rm th}$. Thus, with simple integrations:

$$\frac{1}{\tau_{\rm op}(E)} = \frac{(D_t K)^2_{\rm op} m^{*3/2}}{2^{1/2}\pi\hbar^3\rho_{\rm x}\omega^{\rm (op)}}\left[\langle N_{\rm op}\rangle_{\rm th}(E+\hbar\omega^{\rm (op)})^{1/2}\right.$$
$$\left. + (1+\langle N_{\rm op}\rangle_{\rm th})(E-\hbar\omega^{\rm (op)})^{1/2}\right] , \qquad (13.41)$$

where, of course, emission should be included only if $E > \hbar\omega^{\rm (op)}$.

For acoustic phonons things are more complicated, since the phonon energy depends on the phonon wavevector $q$, and so does the Bose factor $\langle N_{\mathbf{q}}\rangle_{\rm th}$. A common approximation embraced to simplify the problem is to assume that the phonon energy is smaller that the thermal energy $k_{\rm B}T$ (at sufficiently high $T$ and for sufficiently small $q$), so that $n_q \approx k_{\rm B}T/(\hbar\omega^{\rm (as)}) = k_{\rm B}T/(\hbar c_{\rm s} q)$ and to ignore also the energy lost or gained in a collision, if the phonon energy is much smaller than the electron energy. These two approximations are called the *equipartition* and the *elastic* approximations, respectively. Having made these approximations, the scattering rate becomes

$$\frac{1}{\tau_{\rm LA,TA}(E)} = \frac{2^{1/2}\Delta^2_{\rm LA,TA} m^{*3/2} k_{\rm B}T}{\pi\hbar^4\rho_{\rm x}c_{\rm s}^2} E^{1/2} . \qquad (13.42)$$

Nonpolar scattering with acoustic phonons controls the low-field electron mobility in nonpolar semiconductors, such as Si and Ge. Also important are processes in which short-wavelength phonons, both acoustic and optical, assist transition from one valley to another. Scattering among the regions close to the six equivalent minima in Si, scattering from the $\Gamma$ valley into the L or X valleys in GaAs are notable examples. Note also that some of the selection rules we have ignored before do play a role. For example, by symmetry considerations Harrison scattering [9] in the six Si X-valleys vanishes for electron states exactly on the $\Delta$ symmetry line from $\Gamma$ to X. Figure 13.1 shows the total electron–phonon and hole-phonon scattering rates at 300 K in Si calculated using the nonlocal empirical pseudopotential band structure and the numerical procedure described in Chap. 12, Sect. 12.2.2.

### *13.2.1 Deformation Potentials From First Principles in Bulk Si*

As an example of how to use "first principles" (i.e., DFT) calculations to calculate for the deformation potentials that may be used in simpler models, we consider here the calculation of the deformation potentials in bulk Si.

Starting from Eq. (13.13), DFT can provide all the required information, namely, the phonon displacement vectors, the gradient of the potential energy with respect to the ion displacement, and the wavefunctions:

- The phonons and their displacement vectors can be calculated as we have shown in Chap. 6.
- To obtain the gradient of the potential with respect to ion displacement, we can displace each atom from its equilibrium position ($\mathbf{R}_l^{(0)}$) in each direction $\alpha = x, y, z$ in the unit cell to a position $\mathbf{R}_l^{(0)} + \delta_\alpha$. The gradient of the potential energy with respect to atom displacement is then obtained using finite differences

$$\left[\frac{\partial U(\mathbf{r})}{\partial \mathbf{R}_{0l}}\right]_\alpha \approx \frac{U_{\mathrm{H}}(\mathbf{r})|_{\mathbf{R}_l=\mathbf{R}_l^{(0)}+\delta_\alpha} - U_{\mathrm{H}}(r)|_{\mathbf{R}_l=\mathbf{R}_l^{(0)}-\delta_\alpha}}{2|\delta_\alpha|}. \tag{13.43}$$

  We note that an advantage of using a DFT-based approach is that the screening of the ionic potential by the valence electrons is automatically accounted for.
- For the wavefunctions, we can take the Kohn–Sham wavefunctions for the unit cell where each atom is in its equilibrium position. We transform the wavefunctions from reciprocal to real space by performing an inverse Fourier transform onto the grid used for the potential energy.

We show the resulting deformation potentials obtained in [10] for an electron making a transition from the valence band to the conduction band in silicon in Fig. 13.2. The matrix elements have been translated into deformation potentials using

$$DK_{\mathbf{k}\mathbf{k}'}^{\nu} = \sum_l \mathbf{M}_{\mathbf{k}\mathbf{k}'l} \cdot \mathbf{e}_{\mathbf{q}l}^{\nu} \sqrt{\frac{M_{\mathrm{cell}}}{M_l}}. \tag{13.44}$$

The electron is taken to start at the top of the valence band ($\Gamma$) making a transition to somewhere in the conduction band along the $\mathrm{L} - \Gamma - \mathrm{X}$ path.

There is a first caveat in a numerical evaluation of the deformation potential related to the choice of the real-space grid for $\mathbf{r}$. In Eq. (13.11), the integration is taken over the volume of the entire lattice $\Omega$,

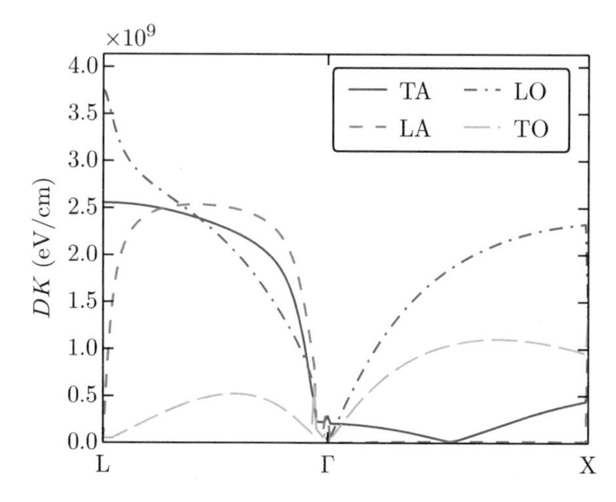


**Fig. 13.2** Silicon deformation potentials for tunneling from the *top* of the valence band towards the conduction band calculated from VASP using the PBE functional. For BTBT, the values are taken at 0.85 in the $\Gamma - X$ direction and measure $D_{\mathrm{TA}} = 3.5 \times 10^8\,\mathrm{eV/cm}$, $D_{\mathrm{TO}} = 1.0 \times 10^9\,\mathrm{eV/cm}$, and $D_{\mathrm{LO}} = 2.3 \times 10^9\,\mathrm{eV/cm}$. (Reprinted with permission from W.G. Vandenberghe and M.V. Fischetti, Appl. Phys. Lett. vol. 106, 013505 (2015). Copyright 2015, AIP Publishing LLP)

but numerically it is satisfactory to integrate over a small volume, since the gradient of the potential energy is screened by the valence electrons and quickly decays away from the atom that is being moved. Calculating the matrix element over a single cell and over multiple unit cells, one can see that an integration over a single unit cell is satisfactory, provided that the coordinates are chosen such that $\mathbf{r} = \mathbf{R}_l$ lies at the center of the unit cell.

A second caveat is related to the degeneracy of the initial electronic states ($m\mathbf{k}$), the final electronic states or ($n\mathbf{k}+\mathbf{q}$) or the phonon branches ($\nu\mathbf{q}$). In case of degeneracy, the band or branch index is not sufficient to determine the state uniquely and simply writing the deformation between conduction band and valence band as "$D^{\nu}_{\mathrm{cv}\mathbf{kq}}$" is ambiguous. For example the valence band at $\Gamma$ is threefold degenerate in absence of spin-orbit coupling and the different wavefunctions rotate like $p_x$, $p_y$ and $p_z$ orbitals. For the phonons, the TA and the TO phonons are degenerate: if we take the phonon wavevector along the $x$-direction, the transverse phonon displacements will lie in the $y-z$ plane. Therefore, depending on which linear combination of the valence band wavefunctions is taken and the direction of the phonon displacement vectors, different values for "$D^{\nu}_{\mathrm{cv}\mathbf{kq}}$" are obtained. This ambiguity can be removed without spending the effort to symmetrize the different degenerate wavefunctions and polarization vectors if the deformation potential between valence and conduction band $D^{\nu}_{\mathrm{cv}\mathbf{kq}}$ is defined as the largest deformation potential upon rotation of the wavefunctions and phonon displacement vectors in the subspace of the degeneracies. Numerically, $D^{\nu}_{\mathrm{cv}\mathbf{kq}}$ can be obtained by constructing a tensor out of the deformation potentials of the different degenerate bands and phonon branches and taking the largest multilinear singular value. In the case of silicon, looking at the valence band and phonon branch degeneracies is satisfactory and an ordinary singular value decomposition yields the deformation potentials.

As can be seen in Fig. 13.2, there is no interaction between the LA phonon along the (100) ($\Delta$) direction. This result can also be obtained directly from group theory: Along the (100) ($\Delta$) direction, the $O_h$ Si symmetry group reduces to $C_{4v}$ and the conduction band has $\Delta_1$ symmetry while the upper (heavy) valence band is twofold degenerate and has $\Delta_5$ symmetry and the lower (light) valence band has $\Delta_2'$ symmetry [11, 12]. For the phonons, the LA has $\Delta_1$ symmetry, the LO has $\Delta_2'$ symmetry and the TA and the TO have $\Delta_5$ symmetry. Transitions between the $\Delta_2'$ valence band and the conduction band cannot be mediated by an LA phonon since $\Delta_2' \otimes \Delta_1 \otimes \Delta_1 = \Delta_2'$ does not contain the trivial representation $\Delta_1$.

## 13.3 The Fröhlich Interaction

In polar semiconductors optical phonons do generate long-range dipole fields with which electrons interact. In order to evaluate the interaction Hamiltonian, first let's recall that only longitudinal modes give rise to a dipole field (see page 226). The dipole field associated with LO phonons will obviously be proportional to the displacement,

$$\mathbf{P} = \mathscr{B}_{\mathbf{q}} \left( \widehat{b}_{\mathbf{q}} + \widehat{b}^{\dagger}_{-\mathbf{q}} \right) , \tag{13.45}$$

having omitted the index $\eta$, since we deal only with the LO branch (there is only one LO branch in the semiconductors we are interested in, all having two ions in each unit cell).

In order to evaluate the coupling constant $\mathscr{B}_{\mathbf{q}}$, we can proceed as follows (see the analogous procedure described in Sect. 9.4.2 on page 205). Let's write the electrostatic potential due to the dipole field caused by the LO phonons as well as the (plasma) response of the carriers, if any:

$$\phi_{\mathbf{q}}(\mathbf{r},t) = \phi^{(0)}_{\mathbf{q}} \cos(\mathbf{q}\cdot\mathbf{r} - \omega^{(\mathrm{LO})}t) . \tag{13.46}$$

Let's now consider the energy associated with this potential. Since phonons and plasmons in the harmonic and linear-response approximations, respectively, are represented as harmonic oscillations, the time-averaged total energy associated with these excitations is simply twice the time-averaged potential energy, $\langle U_{\mathbf{q}}\rangle_{\text{time}}$. This, in turn, is the electrostatic (self)energy of the polarization charge density $\rho_{\mathbf{q}}(\mathbf{r},t)$ in the presence of the potential $\phi_{\mathbf{q}}(\mathbf{r},t)$ caused by the polarization charge itself. We may express this potential energy in two alternative equivalent ways: From the expression (13.46) for the potential, the density of the polarization charge associated with the LO mode can be obtained from the Poisson equation

$$\rho_{\mathbf{q}}(\mathbf{r},t) = \epsilon\nabla^2\phi_{\mathbf{q}}(\mathbf{r},t) \,, \tag{13.47}$$

so that over a volume $\Omega$:

$$\begin{aligned}\langle W_{\mathbf{q}}\rangle_{\text{time}} = 2\langle U_{\mathbf{q}}\rangle_{\text{time}} &= \frac{2}{\Omega}\int_{\Omega} d\mathbf{r}\ \phi_{\mathbf{q}}(\mathbf{r},t)\rho_{\mathbf{q}}(\mathbf{r},t) \\ &= \frac{2}{\Omega}\,\epsilon q^2\phi_{\mathbf{q}}^{(0)2}\int_{\Omega} d\mathbf{r}\ \cos^2(\mathbf{q}\cdot\mathbf{r}) = \epsilon q^2|\phi_{\mathbf{q}}^{(0)}|^2 \,.\end{aligned} \tag{13.48}$$

Alternatively, we can express $\langle W_{\mathbf{q}}\rangle_{\text{time}}$ in terms of the electrostatic energy of the field $\mathbf{E}_{\mathbf{q}} = -\nabla\phi_{\mathbf{q}}$:

$$\langle W_{\mathbf{q}}\rangle = \frac{2}{\Omega}\left\langle \int_{\Omega} d\mathbf{r}\ \epsilon\ |\mathbf{E}_{\mathbf{q}}(\mathbf{R},t)|^2 \right\rangle_{\text{time}} \,, \tag{13.49}$$

which yields the same result of Eq. (13.48) above. We now use a semiclassical argument as our final step: We set the quantity $\langle W_{\mathbf{q}}\rangle$ equal to the zero-point energy, $\hbar\omega^{(\text{LO})}/2$, of the quantized excitation. Thus:

$$\phi_{\mathbf{q}}^2 = \frac{\hbar\omega^{(\text{LO})}}{2q^2\epsilon} \,. \tag{13.50}$$

Before identifying this with the coupling constant $\mathscr{B}_{\mathbf{q}}$, we must recall that the field $\phi_{\mathbf{q}}$ is associated not only to the LO phonons, but it includes also the possible response of free carriers and of additional ionic modes. Whatever we include is determined by our choice of $\epsilon$. In order to isolate the contribution of the LO modes, note that if we use in Eq. (13.50) the static dielectric function $\epsilon_{\text{low}} = \epsilon(\omega_{\text{low}})$ evaluated at a frequency $\omega_{\text{low}} << \omega^{(\text{LO})}$, then we allow the LO phonons to respond fully. If, on the other hand, we set $\epsilon_{\text{hi}} = \epsilon(\omega_{\text{hi}})$ evaluated at a frequency $\omega_{\text{hi}} >> \omega^{(\text{LO})}$, then we prevent the LO phonons from responding. Therefore,

$$\phi_{\mathbf{q}}^2 = e^2\,\frac{\hbar\omega^{(\text{LO})}}{2q^2}\left(\frac{1}{\epsilon_{\text{hi}}} - \frac{1}{\epsilon_{\text{low}}}\right) \,, \tag{13.51}$$

accounts only for the fraction of the potential energy due to the LO phonons alone. Therefore, identifying the expression for $\phi_q$ above with the coupling constant $\mathscr{B}_q$ in Eq. (13.45), we arrive at the following expression for the interaction Hamiltonian [13]:

$$\widehat{H}_{\text{ep}}^{(\text{LO})} = e^2\sum_{\mathbf{kq}}\left[\frac{\hbar\omega^{(\text{LO})}}{2q^2}\left(\frac{1}{\epsilon_{\text{hi}}} - \frac{1}{\epsilon_{\text{low}}}\right)\right]^{1/2} c_{\mathbf{k}+\mathbf{q}}^{\dagger}(\widehat{b}_{\mathbf{q}} + \widehat{b}_{-\mathbf{q}}^{\dagger})c_{\mathbf{k}} \,. \tag{13.52}$$

When only a single LO mode is present and in the absence of free carriers able to screen the interaction, typically one sets $\epsilon_{\text{hi}} = \epsilon_{\infty}$ and $\epsilon_{\text{low}} = \epsilon_0$.

The numerical evaluation of the scattering rates for emission and absorption processes caused by the perturbing Hamiltonian given by Eq. (13.52) does not present the same level of difficulty we encountered when dealing with nonpolar electron–phonon interactions. As we have mentioned before, the Fröhlich potential does not depend on the lattice dynamics—except via the phonon frequency—or on the effect it has on the electronic states, as it was the case for the deformation-potential interaction. Therefore, it involves a straightforward use of Eq. (12.12) with matrix elements $V^{(\eta)}_{\mathbf{k}-\mathbf{k}'}$ replaced by $\delta U^{(\mathrm{pop})}_{\mathbf{k}-\mathbf{k}_j}$, with $\delta U^{(\mathrm{pop})}_{\mathbf{q}}$ given by Eq. (13.102).

Within the effective-mass approximation, the calculation of the scattering rate for this interaction proceeds in the usual way. Considering first emission processes, we have for the rate at which an electron with wavevector **k** emits any LO phonon:

$$\frac{1}{\tau^{(\mathrm{em})}_{\mathrm{LO}}(\mathbf{k})} = \frac{2\pi}{\hbar} \int \frac{\mathrm{d}\mathbf{q}}{(2\pi)^3} \, |\langle N_{\mathbf{q}}-1; n_{\mathbf{k}+\mathbf{q}}, 1-n_{\mathbf{k}}|\widehat{H}^{(\mathrm{LO})}_{\mathrm{ep}}|N_{\mathbf{q}}; 1-n_{\mathbf{k}+\mathbf{q}}, n_{\mathbf{k}}\rangle|^2$$
$$\times \delta[E(\mathbf{k}) - E(\mathbf{k}+\mathbf{q}) - \hbar\omega^{(\mathrm{LO})}] \,, \tag{13.53}$$

having considered the only matrix elements which will give a non-vanishing contribution and having assumed dispersionless LO phonons. Thus, inserting Eq. (13.52) into (13.53) we have

$$\frac{1}{\tau^{(\mathrm{em})}_{\mathrm{LO})}(\mathbf{k})} = \frac{2\pi}{\hbar} \frac{e^2\hbar\omega^{(\mathrm{LO})}}{2} \left(\frac{1}{\epsilon_{\mathrm{hi}}} - \frac{1}{\epsilon_{\mathrm{low}}}\right) (1 + \langle N_{\mathrm{LO}}\rangle_{\mathrm{th}})$$
$$\times \int \frac{\mathrm{d}\mathbf{q}}{(2\pi)^3} \frac{1}{q^2} \, \delta[E(\mathbf{k}) - E(\mathbf{k}+\mathbf{q}) - \hbar\omega^{(\mathrm{LO})}] \,. \tag{13.54}$$

It is convenient to express **q** in polar coordinates with the $z$-axis along **k**. Thus, in the simple case of parabolic, spherical bands with effective mass $m^*$, we can write the argument of the delta-function as:

$$E(\mathbf{k}) - E(\mathbf{k}+\mathbf{q}) = -\frac{\hbar^2 q^2}{2m^*} - \frac{\hbar^2 kq}{m^*} \cos\theta \,, \tag{13.55}$$

so that, after the trivial integration over the azimuthal angle $\phi$ we have

$$\frac{1}{\tau^{(\mathrm{em})}_{\mathrm{LO})}(\mathbf{k})} = \frac{e^2\omega^{(\mathrm{LO})}(1 + \langle N_{\mathrm{LO}}\rangle_{\mathrm{th}})}{4\pi} \left(\frac{1}{\epsilon_{\mathrm{hi}}} - \frac{1}{\epsilon_{\mathrm{low}}}\right)$$
$$\times \int_0^\infty \mathrm{d}q \int_0^\pi \mathrm{d}\theta \, \sin\theta \, \delta\left[\frac{\hbar^2 q^2}{2m^*} + \frac{\hbar^2 kq}{m^*}\cos\theta \, + \hbar\omega^{(\mathrm{LO})}\right] . \tag{13.56}$$

Now consider the integral over the polar angle: Changing the integration variable $\theta$ to $x = (\hbar^2 kq/m^*)\cos\theta$, this integral becomes

$$\int_0^\pi \mathrm{d}\theta \, \sin\theta \, \delta\left[\frac{\hbar^2 q^2}{2m^*} + \frac{\hbar^2 kq}{m^*}\cos\theta \, + \hbar\omega^{(\mathrm{LO})}\right]$$
$$= \frac{1}{a} \int_{-a}^{a} \mathrm{d}x \, \delta(x+b) \,, \tag{13.57}$$

with $a = \hbar^2 kq/m^*$ and $b = \hbar^2 q^2/(2m^*) + \hbar\omega^{(LO)}$. This integral yields $1/a = m^*/(\hbar^2 kq)$ if the argument of the delta-function vanishes for some $x$ in the interval $[-a, +a]$, that is for

$$|a| \geq |b| \quad \rightarrow \quad \frac{\hbar^2 q^2}{2m^*} + \hbar\omega^{(LO)} \leq \frac{\hbar^2 kq}{m^*} ,$$

or

$$q_- \leq q \leq q_+ , \tag{13.58}$$

with

$$q_\pm = k \pm \left( k^2 - \frac{2m^* \omega^{(LO)}}{\hbar} \right)^{1/2} = \left( \frac{2m^*}{\hbar^2} \right)^{1/2} [E^{1/2} \pm (E - \hbar\omega^{(LO)})^{1/2}] , \tag{13.59}$$

provided $E > \hbar\omega^{(LO)}$ (otherwise the integral vanishes, expressing the fact that the electron must have a kinetic energy larger than $\hbar\omega^{(LO)}$ to emit a phonon). Thus, finally, for $E > \hbar\omega^{(LO)}$:

$$\begin{aligned} \frac{1}{\tau_{LO}^{(em)}(\mathbf{k})} &= \theta(E - \hbar\omega^{(LO)}) \frac{e^2 m^* \omega^{(LO)} (1 + \langle N_{LO} \rangle_{th})}{4\pi\hbar^2 k} \left( \frac{1}{\epsilon_{hi}} - \frac{1}{\epsilon_{low}} \right) \int_{q_-}^{q_+} dq \, \frac{1}{q} \\ &= \theta(E - \hbar\omega^{(LO)}) \frac{e^2 m^* \omega^{(LO)} (1 + \langle N_{LO} \rangle_{th})}{4\pi\hbar^2 k} \left( \frac{1}{\epsilon_{hi}} - \frac{1}{\epsilon_{low}} \right) \\ &\quad \times \ln \left( \frac{\sqrt{E} + \sqrt{E - \hbar\omega^{(LO)}}}{\sqrt{E} - \sqrt{E - \hbar\omega^{(LO)}}} \right) , \end{aligned} \tag{13.60}$$

where the function $\theta(x)$ is the Heaviside step function, $\theta(x) = 1$ for $x > 0$, $\theta(x) = 0$ for $x \leq 0$. For absorption processes we find, following identical steps:

$$\frac{1}{\tau_{LO}^{(abs)}(\mathbf{k})} = \frac{e^2 m^* \omega^{(LO)} \langle N_{LO} \rangle_{th}}{4\pi\hbar^2 k} \left( \frac{1}{\epsilon_{hi}} - \frac{1}{\epsilon_{low}} \right) \ln \left( \frac{\sqrt{E + \hbar\omega^{(LO)}} + \sqrt{E}}{\sqrt{E + \hbar\omega^{(LO)}} - \sqrt{E}} \right) . \tag{13.61}$$

Note the typical behavior of Coulomb processes: The scattering rate increases at first with the density of states for small $E$, but at larger kinetic energy it decays as $1/k \approx 1/\sqrt{E}$. The reason is quite simple: At increasingly large kinetic energies, any Coulomb potential appears to be an increasingly smaller relative perturbation, thus affecting the electron to an increasingly smaller extent.

Fröhlich scattering, obviously absent in covalent materials like Si and Ge (for which the absence of ionic polarization implies $\epsilon_{low} = \epsilon_{hi}$), is the dominant scattering process affecting the mobility of III–V compound semiconductors. It also enters quite heavily in the dielectric breakdown of insulators (usually ionic materials—often amorphous) at very high electric fields: As long as the electric field is low enough, LO-scattering keeps the electrons at relatively low kinetic energies. In this low-energy range the scattering rates $1/\tau_{LO}(E)$ increases with increasing $E$. But if the electric field increases above some critical value, electrons will be accelerated to the range of higher kinetic energies in which $1/\tau_{LO}(E)$ decreases with increasing $E$: As they reach higher $E$, electrons will scatter less frequently, thus losing a smaller fraction of their kinetic energy to LO-phonons, thus gaining from the field even more kinetic energy,... etc. This diverging process, called *electron run-away*, could trigger dielectric breakdown in the absence of additional inelastic processes.

Finally, recall that we have so far ignored completely the overlap integral $\mathscr{I}_{\mathbf{k}+\mathbf{q}+\mathbf{G},\mathbf{k}}$, having taken it as equal to unity. Just to show that this approximation may be less than satisfactory in some important case, it is worth stating—without proof—a result due to Fawcett [14]: For polar materials characterized by a nonparabolic conduction band with nonparabolicity parameter $\alpha_{\rm np}$, the scattering rate given by Eq. (13.61) above should be replaced by the following expression:

$$\frac{1}{\tau_{\rm LO}^{\rm (em/abs)}(\mathbf{k})} = \theta(E')\,\frac{e^2\omega^{\rm (LO)}m^{*1/2}}{2^{5/2}\pi\hbar}\left(\frac{1}{\epsilon_{\rm hi}}-\frac{1}{\epsilon_{\rm low}}\right)\frac{1-2\alpha_{\rm np}E'}{\gamma(E)}\,c_{\rm pop}$$
$$\times\left\{a_{\rm pop}\,\ln\left[\frac{|\gamma(E)^{1/2}+\gamma(E')^{1/2}|}{|\gamma(E)^{1/2}-\gamma(E')^{1/2}|}\right]+b_{\rm pop}\right\}$$
$$\times(\langle N_{\rm LO}\rangle_{\rm th}+1/2\pm 1/2)\;\gamma(E')^{1/2}(1+2\alpha_{\rm np}E')\,, \tag{13.62}$$

where $E'=E(\mathbf{k})\mp\hbar\omega^{\rm (LO)}$ and $\gamma(E)$ is simply the parabolic dispersion $\hbar^2k^2/(2m^*)$. The effect of the overlap integral is seen in the terms:

$$a_{\rm pop} = 2(1-\alpha_{\rm np}E')(1-\alpha_{\rm np}E)\;-\;\alpha_{\rm np}[\gamma(E)+\gamma(E')]^2,$$
$$b_{\rm pop} = 2\alpha_{\rm np}[\gamma(E)\gamma(E')]^{1/2}[4(1-\alpha_{\rm np}E')(1-\alpha_{\rm np}E)$$
$$-\alpha_{\rm np}[\gamma(E)+\gamma(E')]\,,$$

and

$$c_{\rm pop} = [4(1-\alpha_{\rm np}E')(1-\alpha_{\rm np}E)\;(1-2\alpha_{\rm np}E')(1-2\alpha_{\rm np}E)]^{-1}\,.$$

The step function $\theta(E')$ is obviously always unity for absorption processes, but it accounts for the energy threshold for emission. This expression is rigorously valid for spherical valleys, while it can be derived for arbitrary ellipsoidal valleys by approximating

$$q\equiv|\mathbf{k}-\mathbf{k}'|\simeq\left(\frac{m_{\rm d}}{m_{\rm el}}\right)^{1/2}q^*,$$

in the denominator of the matrix element given in Eq. (13.52).

## 13.4 Electron–Phonon Interaction for a 2DEG

For a 2DEG confined by the geometry of the system—such as in quantum wells of two-dimensional structures like graphene—or by the geometry and a confining potential—such as in inversion layers—we can repeat exactly the discussion we have just presented for electrons in bulk crystals. The general expressions have been given in the previous chapter, Sect. 12.3.1. There we have considered a bulk perturbation of the form given by Eq. (12.2). We will give an important example of a two-dimensional perturbation of the form (12.13) in Sect. 13.4.2, namely, the interaction of a 2DEG with the two-dimensional phonons in graphene. Here we shall consider briefly the additional difficulties one must face when evaluating numerically the scattering rates using a full band structure. We shall then consider some specific cases that can be simplified when adopting the effective-mass approximation.

So, worrying first about calculations within the scheme of DFT or when employing the rigid-ion approximation, we should note that Eq. (12.36) still requires the numerical evaluation of a double integral: First over the $z$-component of the momentum transfer, $q_z$, then over the $z$-coordinate, hidden within the overlap factor $\mathscr{I}^{(2D)}_{\mathbf{K},\mathbf{K}_j,n,n';\mathbf{G}_{\parallel}}(q_z)$. The latter integration cannot be reduced to any closed-form expression since the wavefunctions are known only numerically. However, in some special cases the integration over $q_z$ can be performed analytically. We shall consider explicitly below, in Sect. 13.4.2, the complications we have to face in the case of graphene. Here we just observe that from a numerical perspective, considering only $N$-processes—and so ignoring *Umklapp* processes, is justified in many cases. We denote by $\delta U^{(\eta)}$ the change of the electron potential energy—nonpolar or polar—due to a phonon of branch $\eta$, and by $\delta U^{(\eta)}_{\mathbf{q}}$ its Fourier components. Note that these depend also on the frequency via the phonon dispersion $\omega^{\eta)}_{\mathbf{q}}$. Therefore, with this notation, Eq. (12.30) can be evaluated numerically as:

$$\frac{1}{\tau^{(\eta)}_n(\mathbf{K})} \approx \frac{2\pi}{\hbar} \sum_{jn'}{}' \int \frac{\mathrm{d}q_z}{2\pi} \left| \delta U^{(\eta)}_{\mathbf{K}-\mathbf{K}_j,q_z} \mathscr{I}^{(2D)}_{\mathbf{K},\mathbf{K}_j,n,n'}(q_z) \right|^2 \frac{1}{(2\pi)^2} \frac{L(w_{jn'})}{|\nabla_{2D} E_{jn'}|} . \tag{13.63}$$

We now consider two important examples: The calculation of the nonpolar electron–phonon scattering rates in Si inversion layers and in graphene. They both constitute, of course, cases of great practical interest, although they differ in the fact that in the first case we deal with bulk phonons, in the latter with two-dimensional modes. Since details about the numerical calculations are very similar, and since we have treated the interaction with bulk perturbations in the previous chapter, we provide them here only for the more subtle case of graphene, as we have already mentioned.

### *13.4.1 Electron Transport in Si Inversion Layers: Monte Carlo Results*

As an example of the implementation of this general scheme to a concrete case of technological interest, we consider what empirical pseudopotentials can tell us about the electron–phonon interaction and electron transport in Si inversion layers or Si thin films. These results have been presented in [15]. They have been obtained using the rigid-ion approximation described above in Sect. 13.1, assuming semiclassical transport using the Monte Carlo technique to solve the Boltzmann transport equation (BTE), described in Chap. 19, Sect. 19.4.1 below.

Consider the 12-cell-thick Si layer already considered in Chap. 7, Sect. 7.2.1. This may be viewed as a prototypical thin body of SOI FETs, of a FinFET or other double-gate device. The major difference is that the layer is bounded by vacuum, rather than by an insulator, so we expect a slightly stronger confinement due to the larger Si-vacuum barrier (that is, workfunction, ~4.5 eV).

Zunger's pseudopotential [16] without spin-orbit interaction is used to calculate the band structure. An external potential inducing a sheet electron charge of $10^{13}$ electrons/cm$^2$ is applied. Starting from a classical Poisson solution for the external potential and taking its Fourier transform as in Eq. (7.2), Eq. (7.4) is solved. From the wavefunctions, Eq. (12.31), averaged over a cell on the $(x,y)$-plane (i.e., on the plane of the surfaces), and their equilibrium Fermi occupation, the total charge can be obtained by summing over the (discretized) 2D BZ and Poisson's equation may then be solved in real space. The process is iterated until convergence is reached. The resulting band structure is shown in Fig. 13.3, left. Note once more the rather complicated structure near the $\overline{\mathrm{X}}$ symmetry point. In Chap. 7, Sect. 7.2.1.4, we have already emphasized this feature recalling the previous observation by Esseni and Palestri [17]. Here we will see the role this structure plays in determining the high-field transport properties.

The implementation of the rigid-ion approximation in two dimensions differs from the three-dimensional implementation we have discussed. We shall present it in detail for the case of graphene,

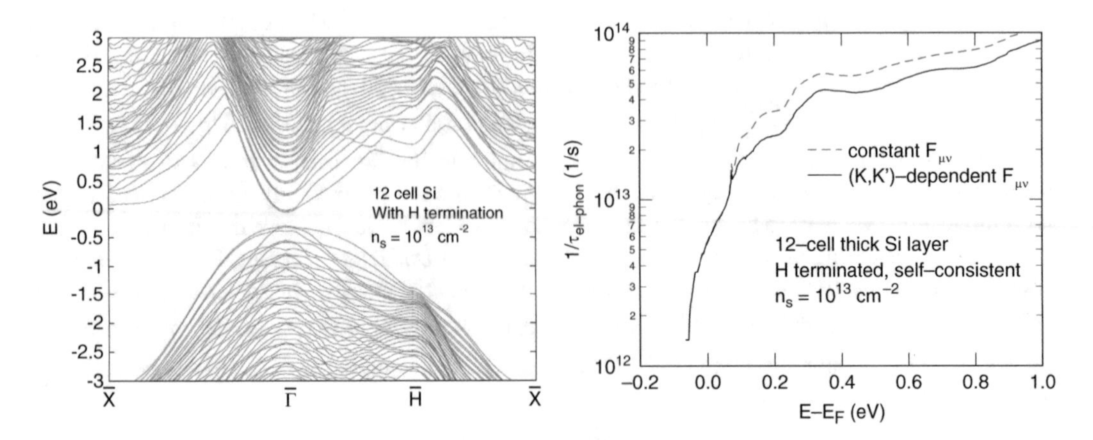


**Fig. 13.3** *Left*: Band structure of a 12-cell-thick H-terminated Si layer in the presence of a self-consistent potential inducing an electron sheet density of $10^{13}$ cm$^{-2}$. *Right*: Nonpolar electron–phonon scattering rates at 300 K in the layer at *left* calculated using either the full $(\mathbf{K},\mathbf{K}')$-dependent overlap factor, Eq. (13.104) (*black solid line and symbols*), or the approximated expression from Eq. (13.106) (*red dashed line, open symbols*). (Used from M.V. Fischetti, Bo Fu, S. Narayanan, and J. Kim, in *Nano-Electronic Devices: Semiclassical and Quantum Transport Modeling*, Dragica Vasileska and Stephen M. Goodnick Eds. (Springer, New York, 2011), pp. 183–247, with kind permission from Springer Science+Business Media)

in Sect. 13.4.2 below. Here we shall only mention that the nonpolar scattering rates with acoustic and optical phonons are computed at 300 K using either the full Eq. (13.104)—which accounts for the dependence of the overlap factor $\mathscr{F}^{(2D)}_{\mathbf{K}'\mathbf{K}nn'}$ on the initial and final wavevectors, $\mathbf{K}$ and $\mathbf{K}'$, respectively—or on the simplified expression Eq. (13.106)—where $\mathbf{K}_0$ is the wavevector at the valley minima, $\overline{\Gamma}$ for the "unprimed" ladder or $\overline{\mathrm{X}}$ for the primed ladder of states. Scattering with ionized impurities, fixed charges, and surface roughness are ignored, wishing to obtain the intrinsic band-structure-dependent transport properties. The scattering rates calculated with these two methods and averaged over electron energy are shown in Fig. 13.3, right. Note the smaller rates obtained when using the full $(\mathbf{K},\mathbf{K}')$-dependent overlap factor $\mathscr{F}^{(2D)}_{\mathbf{K}'\mathbf{K}nn'}$, since this accounts also for the overlap between the initial and final Bloch states.

An ensemble Monte Carlo method, discussed in Chap. 19, Sect. 19.4.1.2, and similar to the scheme described in [18] is employed to obtain the energy *vs.* field and drift-velocity *vs.* field characteristics shown in Fig. 13.4. Two features should be noted: First, when using either model to compute the scattering rates, the electron velocity saturates at a value ($\sim$ 5-to-6 $\times 10^6$ cm/s) much lower than the value of the saturated velocity of electrons in bulk Si at 300 K ($\approx 10^7$ cm/s). As discussed at length in [18], several experimental results have yielded such low values for the saturated velocity, but no study based on the effective-mass approximation, even with nonparabolic corrections, has been able to reproduce the experimental observations. Second, the use of the lower scattering rates obtained when using the full $(\mathbf{K},\mathbf{K}')$-dependent overlap factor, Eq. (13.104), results in a higher electron average energy (see right panel of Fig. 13.4), but in an even lower saturated electron velocity. As shown in Fig. 13.5, this is due to an enhanced population of the "primed" states near the higher-mass $\overline{\mathrm{X}}$ symmetry point. This constitutes a significant example of how the accurate evaluation of the band structure of small confined structures can produce results significantly different from those one may obtain using the effective-mass approximation.

We should note in passing that the Monte Carlo technique employed here can be easily extended to study 2D or 1D transport in other small structures. The main problems we must face consist in determining the correct physical models to be employed to handle electronic scattering, especially

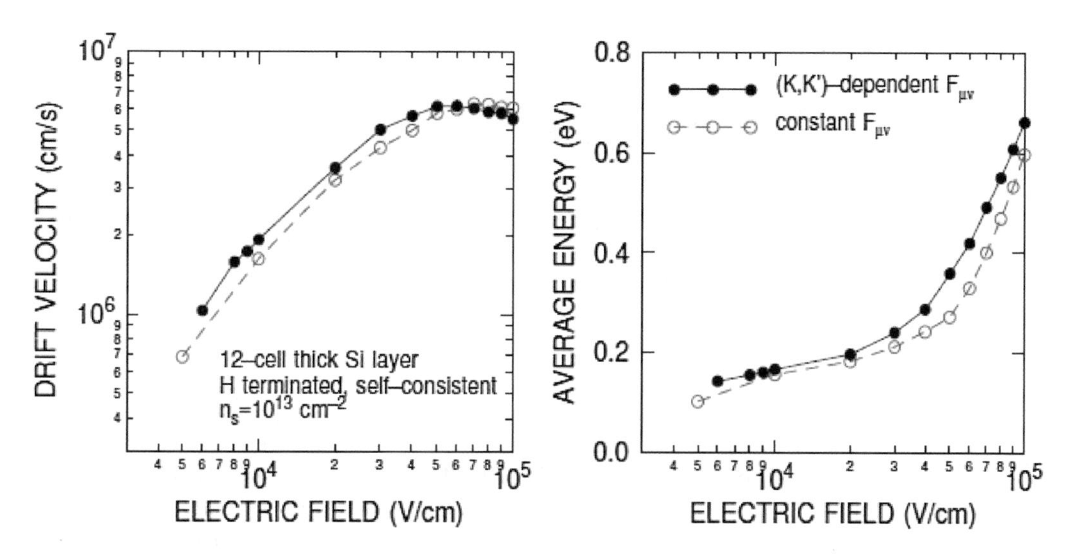


**Fig. 13.4** Electron average energy (*left*) and drift velocity (*right*) as a function of field on the plane of the layer using the two models for the scattering rates as in Fig. 13.3. In either case the electron velocity saturates at a value much smaller than in bulk Si, consistent with experimental results so far left unexplained by models based on the effective-mass subband structure. (Used from M.V. Fischetti, Bo Fu, S. Narayanan, and J. Kim, in *Nano-Electronic Devices: Semiclassical and Quantum Transport Modeling*, Dragica Vasileska and Stephen M. Goodnick Eds. (Springer, New York, 2011), pp. 183–247, with kind permission from Springer Science+Business Media)

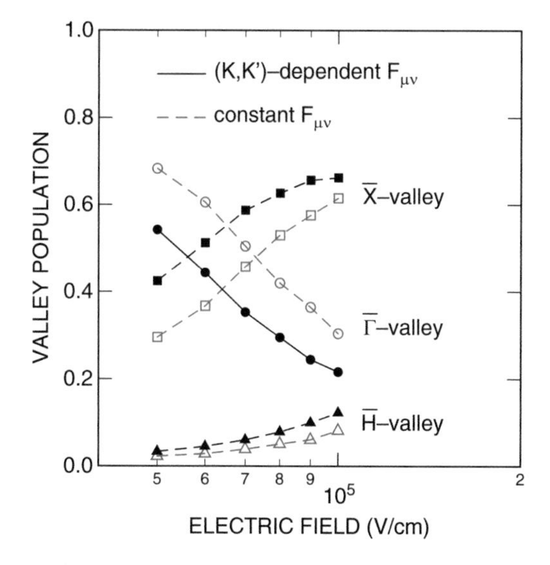


**Fig. 13.5** Population of—somewhat arbitrarily defined—regions of the 2D BZ around the $\overline{\Gamma}$ and $\overline{X}$ symmetry points, corresponding roughly to the unprimed and primed subband ladders. The larger occupation of the primed states—which exhibit a significantly larger conductivity mass—yields the lower saturated velocity and higher average energy observed at high fields in Fig. 13.4. (Used from M.V. Fischetti, Bo Fu, S. Narayanan, and J. Kim, in *Nano-Electronic Devices: Semiclassical and Quantum Transport Modeling*, Dragica Vasileska and Stephen M. Goodnick Eds. (Springer, New York, 2011), pp. 183–247, with kind permission from Springer Science+Business Media)

accounting for the presence of localized phonons in small structures, for scattering with interface or surface roughness, charges, and excitations and, of course, for long- and short-range carrier-carrier scattering. This last process will be discussed in Chap. 15.

### 13.4.2 *Electron–Phonon Interaction in Graphene*

As done in Sect. 13.4.1 for the case of Si films, we can use empirical pseudopotentials, apply the rigid-ion approximation (Sect. 13.1), assume semiclassical transport treated with the Monte Carlo technique (see Sect. 19.4.1), and study semiclassical electron transport in graphene (see Sect. 7.2.3). These results have been presented in [19]. A full list of references can be found there, in addition to many additional details.

#### 13.4.2.1 Electron–Phonon Hamiltonian

Here we present in its complicated details the entire derivation of the electron–phonon scattering rate using the rigid-ion approximation. The unavoidable repetition of the discussion of Sect. 13.1) is a necessary evil, since many important details and approximations concern explicitly the calculations in two dimensions, the interaction with two-dimensional phonons, particularly for the case of graphene.

As we saw in Eq. (13.5), the electron–phonon perturbation Hamiltonian due to a phonon of mode (or branch) $\eta$ can be expressed as:

$$\widehat{H}_{\rm ep}^{(\eta)} = \int \mathrm{d}\mathbf{r}\, \rho(\mathbf{r},t)\, \delta U^{(\eta)}(\mathbf{r},t) \,, \tag{13.64}$$

where $\rho$ is the electron charge and $\delta U^{(\eta)}$ is the change of the lattice (pseudo)potential due to the presence of phonons of branch $\eta$. Using the rigid-ion approximation, we set $\delta U^{(\eta)} = \delta V^{(\eta)}$, as before.

It is worth recalling here our general rule about the notation: We use lower-case bold characters for 3-vectors, upper-case bold characters for 2-vectors on the $(x,y)$ plane of the graphene sheet (or thin layer, the formalism is identical). The $z$ axis is along the direction perpendicular to the sheet. Exceptions to this rule are the vectors $\mathbf{G}$ of the reciprocal lattice of the supercell (since this is a universal notation), so $\mathbf{G}_{\parallel}$ will denote their projection on the plane of the sheet.

Using a supercell of volume $\Omega_{\rm c}$ (=$\Omega_{\rm 2D}\, L_{\rm c} = (3/2)\sqrt{3}\, a_C^2\, L_{\rm c}$, where $a_C \approx 0.142$ nm is the C-C bond length and $L_{\rm c}$ is the extension of the supercell along the direction $z$ perpendicular to the graphene sheet) to describe the electronic excitations, the change of the lattice potential can be written as:

$$\delta V^{(\eta)}(\mathbf{r},t) = \frac{1}{N_{\rm cell}\Omega_{\rm c}} \sum_{l,\gamma} \nabla V^{(\gamma)}(\mathbf{r}-\mathbf{r}_l-\tau_\gamma) \cdot \delta\mathbf{R}_{l,\gamma}^{(\eta)}(t) \,, \tag{13.65}$$

where the index $l$ labels the $N_{\rm cell}$ 3D (super)cells, $\gamma$ labels the ion in each cell, $\delta\mathbf{R}_{l,\gamma}^{(\eta)}$ is the displacement of ion $\gamma$ in cell $l$ due to the presence of a phonon of mode $\eta$, $V^{(\gamma)}$ is the pseudopotential of ion $\gamma$, $\mathbf{r}_l = (\mathbf{R}_l, z_l)$ is the coordinate of each (super)cell, and $\delta\mathbf{R}_{l,\gamma}^{(\eta)}(t)$ is the coordinate of ion $\gamma$ in each cell. Using the Fourier decomposition for the ionic (pseudo)potential,

$$V^{(\gamma)} = \sum_{\kappa} V_{\kappa}^{(\gamma)}\, \mathrm{e}^{\mathrm{i}\kappa\cdot(\mathbf{r}-\mathbf{r}_l-\tau_\gamma)} \,, \tag{13.66}$$

and the expression for the ionic displacement:

$$\delta\mathbf{R}_{l,\gamma}^{(\eta)}(t) = \sum_{\mathbf{Q}} \left( \frac{\hbar}{2\rho_{\rm x}\omega_{\mathbf{Q}}^{(\eta)}} \right)^{1/2} \mathbf{e}_{\mathbf{Q},\eta}^{(\gamma)}\, \mathrm{e}^{\mathrm{i}(\mathbf{Q}\cdot\mathbf{R}_l-\omega_{\mathbf{Q}}^{(\eta)}t)}\, (\widehat{b}_{-\mathbf{Q},\eta}^{\dagger} + \widehat{b}_{\mathbf{Q},\eta}), \tag{13.67}$$

(where $\rho_x$ is the mass density (per unit area) of the crystal, $\omega_{\mathbf{Q}}^{(\eta)}$ the frequency of a phonon of branch $\eta$ with 2D wavevector $\mathbf{Q}$, and $\mathbf{e}_{\mathbf{Q},\eta}^{(\gamma)}$ is its polarization and $\widehat{b}_{\mathbf{Q},\eta}$ and $\widehat{b}_{\mathbf{Q},\eta}^{\dagger}$ its annihilation and creation operators), Eq. (13.65) can be written as:

$$\delta V^{(\eta)}(\mathbf{r},t) = \frac{1}{N_{\text{cell}}\Omega_{\text{c}}} \sum_{l,\gamma} \sum_{\kappa} \mathrm{i}\kappa \, V_{\kappa}^{(\gamma)} \, \mathrm{e}^{\mathrm{i}\kappa\cdot(\mathbf{r}-\mathbf{r}_l-\tau_\gamma)}$$

$$\times \sum_{\mathbf{Q}} \mathbf{e}_{\mathbf{Q},\eta}^{(\gamma)} \left( \frac{\hbar}{2\rho_x \omega_{\mathbf{Q}}^{(\eta)}} \right)^{1/2} \mathrm{e}^{\mathrm{i}(\mathbf{Q}\cdot\mathbf{R}_l - \omega_{\mathbf{Q}}^{(\eta)} t)} \left( \widehat{b}_{-\mathbf{Q},\eta}^{\dagger} + \widehat{b}_{\mathbf{Q},\eta} \right). \tag{13.68}$$

Now let's recall that:

$$\frac{1}{N_{\text{cell}}} \sum_{\kappa} \sum_{l} \mathrm{e}^{\mathrm{i}(\mathbf{Q}-\kappa_{\parallel})\cdot\mathbf{R}_l} \, \mathrm{e}^{\mathrm{i}\kappa_z z_l} = \sum_{\kappa} \sum_{\mathbf{G}''} \delta_{\kappa,\mathbf{q}+\mathbf{G}''} \,, \tag{13.69}$$

having set $\mathbf{q} = (\mathbf{Q},0)$, having indicated with $\kappa_{\parallel}$ the projection of $\kappa$ on the $(x,y)$ plane of the graphene sheet and with $\kappa_z$ its out-of-plane component. As usual, $\mathbf{G}$ is the (3D) vector of the reciprocal lattice of the supercell. Thus, Eq. (13.64) can be written as:

$$\widehat{H}_{\text{ep}}^{(\eta)} = \sum_{n,n'} \sum_{\mathbf{Q},\mathbf{G}''} \mathrm{i}(\mathbf{q}+\mathbf{G}'') \cdot \sum_{\gamma} \mathbf{e}_{\mathbf{Q},\eta}^{(\gamma)} \frac{1}{\Omega_{\text{c}}} V_{\mathbf{q}+\mathbf{G}''}^{(\gamma)} \, \mathrm{e}^{\mathrm{i}\mathbf{G}''\cdot\tau_\gamma} \left( \frac{\hbar}{2\rho_x \omega_{\mathbf{Q}}^{(\eta)}} \right)^{1/2}$$

$$\times \mathrm{e}^{\mathrm{i}(\mathbf{q}+\mathbf{G}'')\cdot\mathbf{r}} \, \mathrm{e}^{\mathrm{i}\omega_{\mathbf{Q}}^{(\eta)} t} \, \widehat{c}_{\mathbf{K}+\mathbf{Q},n'}^{\dagger} \left( \widehat{b}_{-\mathbf{Q},\eta}^{\dagger} + \widehat{b}_{\mathbf{Q},\eta} \right) \widehat{c}_{\mathbf{K},n} \,, \tag{13.70}$$

where $\widehat{c}_{\mathbf{K},n}$ is the annihilator of a Bloch electron of 2D crystal momentum $\hbar\mathbf{K}$ in band $n$ and having absorbed a phase factor $\mathrm{e}^{\mathrm{i}\mathbf{q}\cdot\tau_\gamma}$ into the polarization vector $\mathbf{e}_{\mathbf{Q},\eta}^{(\gamma)}$. Since $\tau_\gamma$ is on the plane of the sheet ($z = 0$), for mono-atomic structures (so that $V^{(\gamma)}$ does not depend on $\gamma$) and absorbing the normalization to the cell volume and the form factor inside $V^{(\gamma)}$ itself, we can rewrite this expression as:

$$\widehat{H}_{\text{ep}}^{(\eta)} = \sum_{n,n'} \sum_{\mathbf{Q},\mathbf{G}''} \mathrm{i}(\mathbf{q}+\mathbf{G}'') \, V_{\mathbf{q}+\mathbf{G}''} \cdot \sum_{\gamma} \mathbf{e}_{\mathbf{Q},\eta}^{(\gamma)} \, \mathrm{e}^{\mathrm{i}\mathbf{G}_{\parallel}''\cdot\tau_\gamma} \left( \frac{\hbar}{2\rho_x \omega_{\mathbf{Q}}^{(\eta)}} \right)^{1/2}$$

$$\times \mathrm{e}^{\mathrm{i}(\mathbf{q}+\mathbf{G}'')\cdot\mathbf{r}} \, \mathrm{e}^{\mathrm{i}\omega_{\mathbf{Q}}^{(\eta)} t} \, \widehat{c}_{\mathbf{K}+\mathbf{Q},n'}^{\dagger} \left( \widehat{b}_{-\mathbf{Q},\eta}^{\dagger} + \widehat{b}_{\mathbf{Q},\eta} \right) \widehat{c}_{\mathbf{K},n} \,. \tag{13.71}$$

Note that, since $V_{\mathbf{q}}^{(\gamma)}$ does not depend on $\gamma$ (the pseudopotential is the same carbon pseudopotential for all atoms), we have taken it outside the sum over ions and dropped the superscript $\gamma$. Note also that coupling between electrons and the out-of-plane modes acoustic and optical modes (denoted by ZA and ZO) vanishes identically at first order: For these out-of-plane modes $\mathbf{e}_{\mathbf{Q},\eta}^{(\gamma)}$ has nonzero components along $z$ so $\mathbf{Q}\cdot\mathbf{e}_{\mathbf{Q},\eta}^{(\gamma)} = 0$ and the terms involving the $\mathbf{G}$ vectors will give a vanishing contribution when summed. The physical reason for this vanishing coupling is the inability of the electrons in the graphene sheet to modify their wavefunction along $z$, so the inability to support a change in physical momentum along the direction perpendicular to the layer. If perturbed wavefunctions were used, instead of the equilibrium unperturbed states, then a small coupling may arise. Higher-order processes—consisting of the simultaneous emission or absorption of two ZA phonons—are allowed.

However, their study requires accounting also for the higher-order, anharmonic coupling of out-of-plane (ZA) with in-plane (TA and LA) phonons. This renders the process relatively weak, therefore we shall not discuss this problem here.

#### 13.4.2.2 Electron–Phonon Matrix Element

If only a real-space expression for the wavefunction is available, the calculation of the matrix element of the form $\langle \mathbf{K}'n'|\widehat{H}_{\mathrm{ep}}^{(\eta)}|\mathbf{K}n\rangle$ (using an obvious simplified notation, as discussed in Chap. 12, page 257) must be performed employing the real-space form Eq. (13.70) of the Hamiltonian and performing a real-space integration. Thus the only non-vanishing matrix elements will be of the form:

$$\langle \mathbf{K}'n'|\widehat{H}_{\mathrm{ep}}^{(\eta)}|\mathbf{K}n\rangle = \sum_{\mathbf{G}} \mathscr{D}^{(\eta)}_{\mathbf{K}'-\mathbf{K},\mathbf{G}} \int \mathrm{d}\mathbf{r}\; \psi_{\mathbf{K}'}^{(n')*}(\mathbf{r})\; \mathrm{e}^{\mathrm{i}(\mathbf{q}+\mathbf{G})\cdot\mathbf{r}}\; \psi_{\mathbf{K}}^{(n)}(\mathbf{r})\,, \tag{13.72}$$

where now $\mathbf{q} = (\mathbf{K}' - \mathbf{K}, 0)$ and the "deformation potential" $\mathscr{D}^{(\eta)}_{\mathbf{QG}}$ is defined below.

If, on the contrary, a Fourier decomposition of the electronic Bloch states is available, the calculation of the matrix elements can be performed in $\mathbf{k}$-space. Then Eq. (13.72) takes the form:

$$\begin{aligned}\langle \mathbf{K}'n'|\widehat{H}_{\mathrm{ep}}^{(\eta)}|\mathbf{K}n\rangle = & \frac{1}{N_{\mathrm{cell}}\Omega_{\mathrm{c}}} \sum_{\mathbf{G}\mathbf{G}'\mathbf{G}''} \sum_{\mathbf{q}} \mathscr{D}^{(\eta)}_{\mathbf{Q},\mathbf{G}''} \int \mathrm{d}\mathbf{R}\; \mathrm{e}^{\mathrm{i}(\mathbf{K}-\mathbf{K}'+\mathbf{G}_{\parallel}-\mathbf{G}'_{\parallel}+\mathbf{Q}+\mathbf{G}''_{\parallel})\cdot\mathbf{R}} \\ & \times \int \mathrm{d}z\; \mathrm{e}^{\mathrm{i}(G_z-G'_z+G''_z)z}\; u^{(n')*}_{\mathbf{K}',\mathbf{G}'}\; u^{(n)}_{\mathbf{K},\mathbf{G}}\,,\end{aligned} \tag{13.73}$$

where the coefficients $u^{(n)}_{\mathbf{K},\mathbf{G}}$ are the Fourier coefficients of the periodic part of the Bloch functions $\psi_{\mathbf{K}}^{(n)}(\mathbf{r})$ and we have defined

$$\mathscr{D}^{(\eta)}_{\mathbf{Q},\mathbf{G}} = \mathrm{i}(\mathbf{q}+\mathbf{G}) \cdot \sum_{\gamma} \mathbf{e}^{(\gamma)}_{\mathbf{Q},\eta}\; V_{\mathbf{q}+\mathbf{G}}\; \mathrm{e}^{\mathrm{i}\mathbf{G}\cdot\tau_{\gamma}}\; \mathscr{A}_{\mathbf{Q},\eta}\,, \tag{13.74}$$

with

$$\mathscr{A}^2_{\mathbf{Q},\eta} = \left(\frac{\hbar}{2\rho_{\mathrm{x}}\omega_{\mathbf{Q}}^{(\eta)}}\right) \left\{ \begin{matrix} \langle N_{\mathbf{Q},\eta}\rangle_{\mathrm{th}} \\ 1+\langle N_{\mathbf{Q},\eta}\rangle_{\mathrm{th}} \end{matrix} \right\}, \tag{13.75}$$

where $N_{\mathbf{Q},\eta}$ is the occupation number of phonons of wavevector $\mathbf{Q}$ and branch $\eta$ and the upper (lower) symbol within curly bracket should be taken for absorption (emission) processes. Performing the spatial integrations this expression reduces to:

$$\langle \mathbf{K}'n'|\widehat{H}_{\mathrm{ep}}^{(\eta)}|\mathbf{K}n\rangle = \sum_{\mathbf{G},\mathbf{G}'} \mathscr{D}^{(\eta)}_{\mathbf{K}'-\mathbf{K},\mathbf{G}'-\mathbf{G}}\; u^{(n')*}_{\mathbf{K}',\mathbf{G}'}\; u^{(n)}_{\mathbf{K},\mathbf{G}}\,. \tag{13.76}$$

Rewriting this expression as:

$$\langle \mathbf{K}'n'|\widehat{H}_{\mathrm{ep}}^{(\eta)}|\mathbf{K}n\rangle = \sum_{\mathbf{G},\mathbf{G}'} \mathscr{D}^{(\eta)}_{\mathbf{K}'-\mathbf{K},\mathbf{G}'}\; u^{(n')*}_{\mathbf{K}',\mathbf{G}+\mathbf{G}'}\; u^{(n)}_{\mathbf{K},\mathbf{G}}\,, \tag{13.77}$$

the presence of $\mathbf{G}'$ in the last term reminds us of *Umklapp* processes. Thus, considering only *Normal* processes, for acoustic phonons one can approximate Eq. (13.77) as:

$$\langle \mathbf{K}'n'|\widehat{H}_{\mathrm{ep}}^{(\eta)}|\mathbf{K}n\rangle \approx \mathscr{D}_{\mathbf{K}'-\mathbf{K},\mathbf{0}}^{(\eta)} \sum_{\mathbf{G}} u_{\mathbf{K}',\mathbf{G}}^{(n')*} u_{\mathbf{K},\mathbf{G}}^{(n)} , \tag{13.78}$$

in which the first term plays the role of a deformation potential $\sim |\mathbf{K}'-\mathbf{K}|\Delta^{(\eta)}$ and the second term the role of a simple overlap integral between the "rigid ion wavefunctions" in each cell.

Note that in Eqs. (13.73)–(13.78) there appears to be a dependence on the size of the supercell via the normalization volume $\Omega_{\mathrm{c}} = \Omega_{2\mathrm{D}} L_{\mathrm{c}}$. In the limit of a supercell extending infinitely far along $z$ (that is, for $L_{\mathrm{c}} \to \infty$) this dependence is canceled by the sum over $G_z$ since

$$\frac{1}{L_{\mathrm{c}}} \sum_{G_z} \to \frac{1}{2\pi} \int \mathrm{d}G_z . \tag{13.79}$$

Thus, large cells will be required to guarantee this independence of the result on the size of the supercell. But in approximating Eq. (13.77) by ignoring *Umklapp* processes as in Eq. (13.78) this "theoretical" independence on supercell volume would be lost even in the infinite-volume limit. In order to retain this property the normalization volume must be modified. To see how to correctly normalize the pseudopotentials it is convenient to reach Eq. (13.78) by first setting $\mathbf{G}_{\parallel} = \mathbf{0}$ and then taking the infinite-volume limit, Eq. (13.79), obtaining for the "acoustic deformation potential"

$$\mathrm{i} \sum_{\gamma} \mathbf{Q} \cdot \mathbf{e}_{\mathbf{Q},\eta}^{(\gamma)} \mathscr{A}_{\mathbf{Q},\eta} \frac{1}{\Omega_{2\mathrm{D}}} \int_{-\infty}^{\infty} \frac{\mathrm{d}G_z}{2\pi} V(|\mathbf{Q}+G_z\widehat{\mathbf{z}}|) , \tag{13.80}$$

where we have put in evidence the normalization $1/\Omega_{\mathrm{c}}$ (so leaving $V$ un-normalized), we have used explicitly the fact that the atoms are on the plane of the sheet so that their coordinates $\tau_{\gamma}$ have vanishing $z$-components, and we have considered only the in-plane modes. One can now rewrite the last term in this expression as:

$$\frac{1}{\Omega_{2\mathrm{D}}} \int_{-\infty}^{\infty} \frac{\mathrm{d}G_z}{2\pi} V(|\mathbf{Q}+G_z\widehat{\mathbf{z}}|) \approx V(Q) \frac{1}{\Omega_{2\mathrm{D}} t_{\mathbf{Q}}} \tag{13.81}$$

where the quantity $t_{\mathbf{Q}}$ is defined as:

$$\frac{1}{t_{\mathbf{Q}}} = \frac{1}{V(Q)} \int_{-\infty}^{\infty} \frac{\mathrm{d}q}{2\pi} V(|\mathbf{Q}+q\widehat{\mathbf{z}}|) . \tag{13.82}$$

For small $Q$

$$\frac{1}{t_0} \approx \frac{2}{V(0)} \int_0^{\infty} \frac{\mathrm{d}q}{2\pi} V(q) . \tag{13.83}$$

is the reciprocal of the "range" of the pseudopotential or, equivalently, the reciprocal of the thickness of the graphene sheet (which for Kurokawa's C pseudopotentials is about 0.1713 nm). Using this last expression for any $Q$, in Eq. (13.83) the pseudopotential now should be normalized not to the supercell volume $\Omega_{2\mathrm{D}} L_{\mathrm{c}}$, but to the "effective" volume $\Omega_{2\mathrm{D}} t_0$ of a "3D graphene cell."

This discussion shows that an approximation probably better than Eq. (13.78) may be obtained by considering only *Normal* in-plane processes by setting $\mathbf{G}_{\parallel} = \mathbf{0}$ in Eq. (13.77), but retaining the sum over $G_z$ in the "deformation potential" $\mathscr{D}_{\mathbf{K}'-\mathbf{K},\mathbf{G}}^{(\eta)}$ appearing in Eq. (13.77):

$$\langle \mathbf{K}'n'|\widehat{H}_{\rm ep}^{(\eta)}|\mathbf{K}n\rangle \approx \sum_{G_z} \mathscr{D}^{(\eta)}_{\mathbf{K}'-\mathbf{K},\mathbf{G}_\parallel=\mathbf{0},G_z} \sum_{\mathbf{G}} u^{(n')*}_{\mathbf{K}',\mathbf{G}}\, u^{(n)}_{\mathbf{K},\mathbf{G}} \,, \tag{13.84}$$

an approximation which we shall call "in-plane no-*Umklapp*." See Fig. 13.7 for the scattering rates calculated using this approximation.

Finally, faster calculations may be used by considering acoustic and optical deformation potentials independent of the phonons wavevector **Q** fitted to more sophisticated approximations. In particular, in order to account correctly for the absence of the ZA modes, one may use Eq. (13.84) in place of the term $\sum_{\mathbf{G}} \mathscr{D}^{(\eta)}_{\mathbf{K}'-\mathbf{K},\mathbf{G}}$ the term:

$$\sum_{\mathbf{G}} \mathscr{D}^{(\eta)}_{\mathbf{K}'-\mathbf{K},\mathbf{G}} \approx \left| \sum_{\gamma} \mathbf{e}^{(\gamma)}_{\mathbf{Q},\eta} \cdot \mathbf{Q} \right| \Delta^{(\eta)}_{\rm ac} \tag{13.85}$$

for the acoustic phonons and a simple constant,

$$\sum_{\mathbf{G}} \mathscr{D}^{(\eta)}_{\mathbf{K}'-\mathbf{K},\mathbf{G}} \approx (DK)^{(\eta)}_{\rm op} \tag{13.86}$$

for the optical phonons. See Fig. 13.8 for the scattering rates calculated using these approximations with the same deformation potential $\Delta_{\rm ac}$ for all acoustic branches and the same value of $(DK)_{\rm op}$ for the optical branches.

It is interesting to express the matrix element using the wavefunctions

$$\xi^{(n)}_{\mathbf{K},\mathbf{G}_\parallel}(z) = \frac{1}{L_{\rm c}} \sum_{G_z} u^{(n)}_{\mathbf{K},\mathbf{G}}\, e^{{\rm i}G_z z} \,, \tag{13.87}$$

where $L_{\rm c}$ is the extension of the supercell along the out-of-plane direction. Then Eq. (13.76) can be rewritten as:

$$\langle \mathbf{K}'n'|\widehat{H}_{\rm ep}^{(\eta)}|\mathbf{K}n\rangle = \sum_{\mathbf{G}_\parallel,\mathbf{G}'_\parallel,G''_z} \mathscr{D}^{(\eta)}_{\mathbf{K}'-\mathbf{K},\mathbf{G}'_\parallel-\mathbf{G}_\parallel,G''_z} \int {\rm d}z\, \xi^{(n')*}_{\mathbf{K}',\mathbf{G}'_\parallel}(z)\, {\rm e}^{{\rm i}G''_z z}\, \xi^{(n)}_{\mathbf{K},\mathbf{G}_\parallel}(z) \,. \tag{13.88}$$

Ignoring Bloch-overlap effects we may replace the wavefunctions $\xi^{(n)}_{\mathbf{K},\mathbf{G}_\parallel}(z)$ with the cell-averaged functions $\zeta^{(n)}_{\mathbf{K}}(z) = \xi^{(n)}_{\mathbf{K},\mathbf{G}_\parallel=0}(z)$ so that

$$\langle \mathbf{K}'n'|\widehat{H}_{\rm ep}^{(\eta)}|\mathbf{K}n\rangle \approx \sum_{\mathbf{G}} \mathscr{D}^{(\eta)}_{\mathbf{K}'-\mathbf{K},\mathbf{G}} \int {\rm d}z\, \zeta^{(n')*}_{\mathbf{K}'}(z)\, {\rm e}^{{\rm i}G_z z}\, \zeta^{(n)}_{\mathbf{K}}(z) \,, \tag{13.89}$$

which is a simplified version of Eq. (13.78). (Note that we have assumed $\mathbf{G}_\parallel = 0$ for the wavefunctions but we cannot do this for the coupling constant $\mathscr{D}$, since the coupling strength depends strongly on the presence of the **G**-terms.) If the out-of-plane extension of the wavefunction is small compared to the in-plane lattice constant (as it is the case in graphene), then one can approximate $\zeta^{(n)}_{\mathbf{K}}(z) \sim \delta(z)$, as in a purely 2D sheet, so that, for the acoustic modes

$$\langle \mathbf{K}'n'|\widehat{H}_{\rm ep}^{(\eta)}|\mathbf{K}n\rangle \approx \sum_{\mathbf{G}} \mathscr{D}^{(\eta)}_{\mathbf{K}'-\mathbf{K},\mathbf{G}} \,, \tag{13.90}$$

which can be regarded once more as the "deformation potential" $\sim |\mathbf{K}'-\mathbf{K}|\Delta^{(\eta)}$.

#### 13.4.2.3 Dielectric Screening

We have already observed that the rigid-ion approximation assumes that the valence electrons shift rigidly as the ions are displaced. Therefore, their response, neglected by this approach, must be accounted for, somehow. The obvious path to follow is to screen the electron–phonon interaction accounting for the dielectric response of the electrons as we have done in Chap. 11. Graphene is a "pathological" case. Indeed, given the strong dielectric response of graphene (as evident, for example, from the reduction of the optical phonon energy, called *Kohn anomaly*, seen in the phonon spectrum and by the limited validity of the Born–Oppenheimer approximation), the theoretical situation is a bit "confused." The problem lies in the fact that graphene is neither a gapped dielectric (so that valence electrons do not respond at the phonon energies because these are usually much smaller than the gap), nor a metal (in which electrons respond instantaneously, the Born–Oppenheimer assumption): Graphene does not exhibit a gap, so electrons can respond at the frequency of the phonons. But the DOS at the Fermi level vanishes, so that this response is "sluggish." In other words: Detailed calculations are necessary. We may screen the electron–phonon matrix element using a model available in the literature, Wunsch's screening model [20], obtained in the random phase approximation (RPA):

$$\langle \mathbf{K}'n'|\widehat{H}_{\mathrm{ep}}^{(\eta)}|\mathbf{K}n\rangle \rightarrow \frac{\langle \mathbf{K}'n'|\widehat{H}_{\mathrm{ep}}^{(\eta)}|\mathbf{K}n\rangle}{1-[e^2/(2\kappa_0 Q)]\Pi_{\mathrm{RPA}}(Q,\omega_{Q\eta})}\,, \tag{13.91}$$

where $Q-|\mathbf{K}-\mathbf{K}'|$. The dynamic (RPA) polarizability of graphene is given by Wunsch as:

$$\Pi_{\mathrm{RPA}}(Q,\omega)=\frac{\Pi(Q,\omega)}{1-[e^2/(2\kappa_0 Q)]\Pi(Q,\omega)}\,, \tag{13.92}$$

with

$$\Pi(Q,\omega)=-\mathrm{i}\pi\frac{F(Q,\omega)}{\hbar^2 \upsilon_{\mathrm{F}}^2}+\Delta P^{(1)}(Q,\omega)\,, \tag{13.93}$$

where

$$\Delta P^{(1)}(Q,\omega)=-\frac{gE_{\mathrm{F}}}{2\pi\hbar^2\upsilon_{\mathrm{F}}^2}+\frac{F(Q,\omega)}{\hbar^2\upsilon_{\mathrm{F}}^2}\left\{G\left(\frac{\hbar\omega+2E_{\mathrm{F}}}{\hbar\upsilon_{\mathrm{F}}Q}\right)-\theta\left(\frac{2E_{\mathrm{F}}-\hbar\omega}{\hbar\upsilon_{\mathrm{F}}Q}-1\right)\right.$$
$$\left.\times\left[G\left(\frac{2E_{\mathrm{F}}-\hbar\omega}{\hbar\upsilon_{\mathrm{F}}Q}\right)-\mathrm{i}\pi\right]-\theta\left(\frac{\hbar\omega-2E_{\mathrm{F}}}{\hbar\upsilon_{\mathrm{F}}Q}+1\right)G\left(\frac{\hbar\omega-2E_{\mathrm{F}}}{\hbar\upsilon_{\mathrm{F}}Q}\right)\right\}. \tag{13.94}$$

The functions $F(Q,\omega)$ and $G(x)$ are defined by:

$$F(Q,\omega)=\frac{g}{16\pi}\frac{\hbar\upsilon_{\mathrm{F}}^2Q^2}{\sqrt{\omega^2-\upsilon_{\mathrm{F}}^2Q^2}}\,, \tag{13.95}$$

$$G(x)=x\sqrt{x^2-1}-\ln\left(x+\sqrt{x^2-1}\right)\,, \tag{13.96}$$

$E_{\mathrm{F}}$ is the Fermi level measured from the Dirac point, $\upsilon_{\mathrm{F}}$ the Fermi velocity, and $g=g_{\mathrm{v}}g_{\mathrm{s}}$ is the product of the spin and valley degeneracy. Note that for intrinsic (un-doped and un-gated) graphene, $\Delta P^{(1)}(Q,\omega)=0$.

#### 13.4.2.4 Electron–Phonon Scattering Rates

Now that we have obtained an expression for the matrix elements, we can now calculate the scattering rates. As we have already seen many times, the first-order rate at which a 2D electron of crystal momentum $\hbar\mathbf{K}$ in band $n$ will emit or absorb a phonon of branch $\eta$ can be written as:

$$\frac{1}{\tau_n^{(\eta)}(\mathbf{K})} = \frac{2\pi}{\hbar} \int \frac{d\mathbf{K}'}{(2\pi)^2} \left| \sum_{\mathbf{G},\mathbf{G}'} \mathscr{D}^{(\eta)}_{\mathbf{K}'-\mathbf{K},\mathbf{G}'-\mathbf{G}} \, u^{(n')*}_{\mathbf{K}',\mathbf{G}'} \, u^{(n)}_{\mathbf{K},\mathbf{G}} \right|^2 \times \delta(E_n(\mathbf{K}) - E_n(\mathbf{K}') \mp \hbar\omega_{\mathbf{K}-\mathbf{K}',\eta}) \,, \tag{13.97}$$

the upper (lower) sign in the energy-conserving delta describing absorption (emission) processes. To compute it numerically, we can follow the "usual" procedure: We first discretize the 2D BZ with squares centered at points $\mathbf{K}_j$, and write Eq. (13.97) as:

$$\frac{1}{\tau_n^{(\eta)}(\mathbf{K})} \approx \frac{2\pi}{\hbar} {\sum_{jn'}}' \left| \sum_{\mathbf{G},\mathbf{G}'} \mathscr{D}^{(\eta)}_{\mathbf{K}_j-\mathbf{K},\mathbf{G}'-\mathbf{G}} \, u^{(n')*}_{\mathbf{K}_j,\mathbf{G}'} \, u^{(n)}_{\mathbf{K},\mathbf{G}} \right|^2 \frac{1}{(2\pi)^2} \frac{L(w_{jn'})}{|\nabla_{2D} E_{jn'}|} \,. \tag{13.98}$$

Here $[1/(2\pi)^2]\, L(w_{jn})/\nabla_{2D}E_{jn}$ is the density of states on band $n'$ in the energy-conserving $j$th square of side $\Delta K$ with energy $E_{jn'}$ and gradient $\nabla_{2D}E_{jn'}$ at the center of the square, $\mathbf{K}_j$. Also, the "primed" sum over $j$ and $n'$ implies that only energy-conserving squares should be considered. This is simply the Gilat–Raubenheimer scheme in 2D, already discussed in Chap. 12, Sect. 12.3.2.

The last remaining piece of information we need has to do with the phonons. The calculations require knowledge of the phonon spectrum and of the polarization vectors. DFT or any of the empirical models discussed in Chap. 9. Section 9.4.1 can be used for this purpose. Figure 13.6 shows the phonon spectrum calculated employing the Born–von Kármán model.

To give an idea of the computational complexity involved, the results shown in Figs. 13.7, 13.8, and 13.9 have been obtained using three different meshes in **k**-space: A "coarse" mesh of $40 \times 40$ rectangular elements covering the rectangular section of the BZ including the (triangular) irreducible wedge of the 2D Brillouin Zone and a similar triangle completing the rectangle; a "fine" mesh of $40 \times 40$ rectangular elements covering a similar section of linear dimensions five times smaller around the K symmetry point; a "superfine" mesh covering the same section but with $80 \times 80$ elements. For the tabulation and interpolation of the phonon dispersion, a mesh of $300 \times 520$ rectangular elements covering the same section has been used.

The final results regarding electron transport in graphene are shown in Figs. 13.10 and 13.11. The calculated electron mobility is in rough agreement with the experimental data available. The experimental results are often affected by the poor quality of the graphene sheet available. It usually comes in the form of "flakes," several micrometer in size. This may be the unavoidable result of the Mermin–Wagner–Hohenberg theorem discussed in Sect. 7.2.3. Defects, impurities, charges in the supporting insulating substrates, and the problem of determining accurately the electron density (needed to extract the mobility from the measured conductivity) all contribute to a high level of uncertainty. Nevertheless, we should stress the fact that calculations that use as empirical parameters only the C pseudopotential (and these are fitted to the electronic structure of diamond, not even to graphene!) and the phonon spectra (that could be obtained theoretically from DFT) are able to predict electronic properties to a surprising level of accuracy.

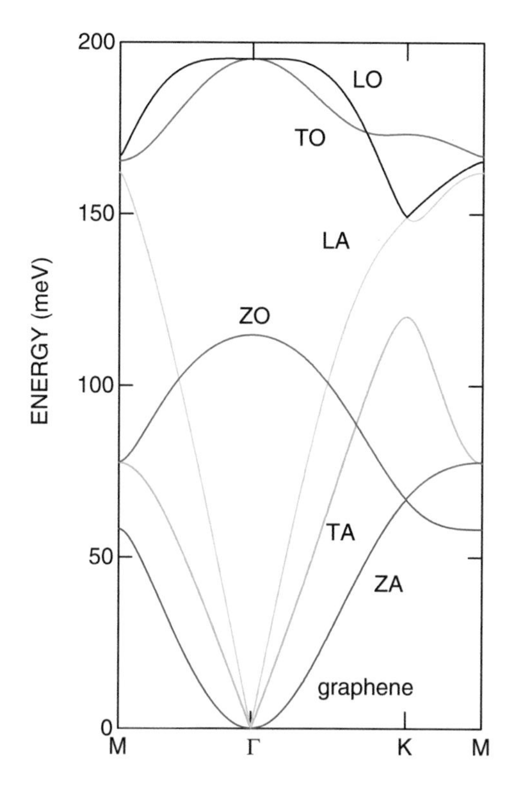


**Fig. 13.6** The phonon dispersion used here, calculated using a fourth-nearest-neighbors Born–von Kármán model. The modes are identified by their character, but zone-edge LO and LA modes are degenerate and a unique identification is impossible. Thus, in the following figures, labels such "LA" and "LO" should be interpreted somewhat loosely. (Reprinted with permission from M.V. Fischetti, J. Kim, S. Narayanan, Z.-Y. Ong, C. Sachs, D.K. Ferry, and S.J. Aboud, J. Phys.: Cond. Matter vol. 25, 473202 (2013). Copyright 2013, Institute of Physics)

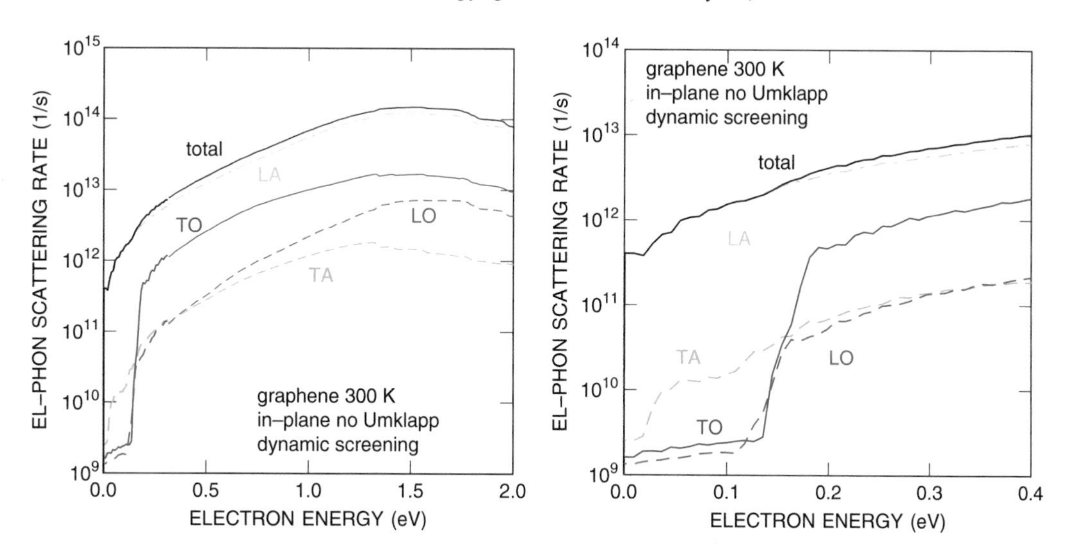


**Fig. 13.7** *Left*: Total electron–phonon scattering rates in graphene at 300 K using the "in-plane no-*Umklapp*" approximation, Eq. (13.84). *Right*: As in the *left frame*, but zooming on the low-energy range.

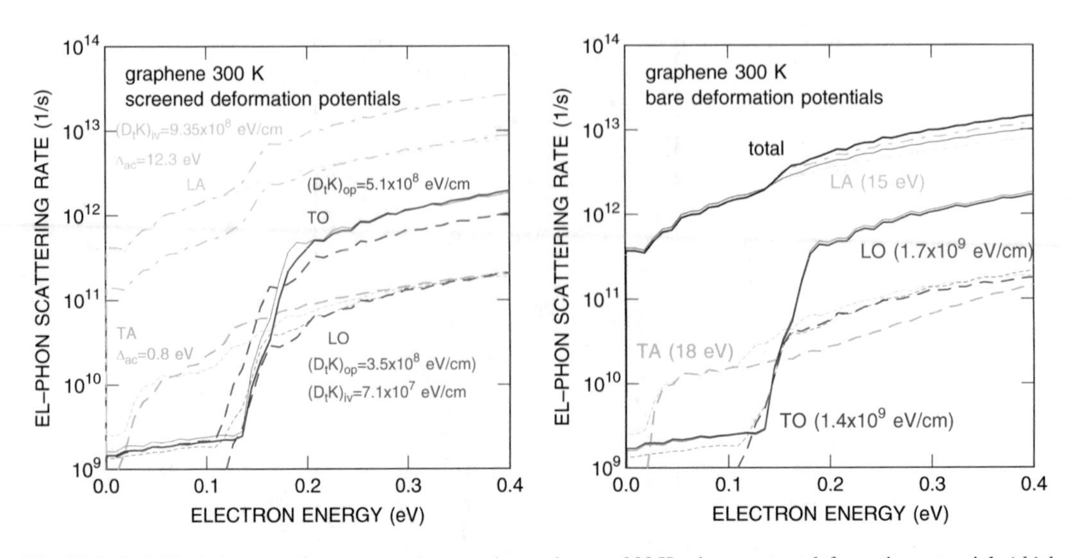


**Fig. 13.8** *Left*: Total electron–phonon scattering rates in graphene at 300 K using constant deformation potentials (*thick lines*) defined by Eqs. (13.85) and (13.86) without accounting for dielectric screening, for the phonon-polarization factor $\left|\sum_\gamma \mathbf{e}^{(\gamma)}_{\mathbf{Q},\eta} \cdot \mathbf{Q}\right|$ or for the overlap integral, $\sum_\mathbf{G} u^{(n')*}_{\mathbf{K}',\mathbf{G}}\, u^{(n)}_{\mathbf{K},\mathbf{G}}$, between initial and final electron wavefunctions, in order to reproduce the simple approximation often used in transport models. Note the "step" in the scattering rate observed as soon as intravalley $K - K$ and $K - K'$ emission processes become available. This step results from having ignored the $K - K'$ overlap factors (whose effect is seen at *right*). Fitting these rates to those calculated using the rigid-ion model with the in-plane no-*Umklapp* approximation (*thin lines*) requires different intra- and inter-valley deformation potentials, as indicated. *Right*: As in the *left frame*, but the "empirical" deformation potentials have now been screened and the phonon-polarization $\left|\sum_\gamma \mathbf{e}^{(\gamma)}_{\mathbf{Q},\eta} \cdot \mathbf{Q}\right|$ and overlap-integral factors have all been accounted for. Thus, these may be considered as much larger "bare" deformation potentials. Note the much reduced magnitude (or even absence) of the intervalley emission steps seen at *left*, since now the $K - K'$ overlap factor is included in the calculation

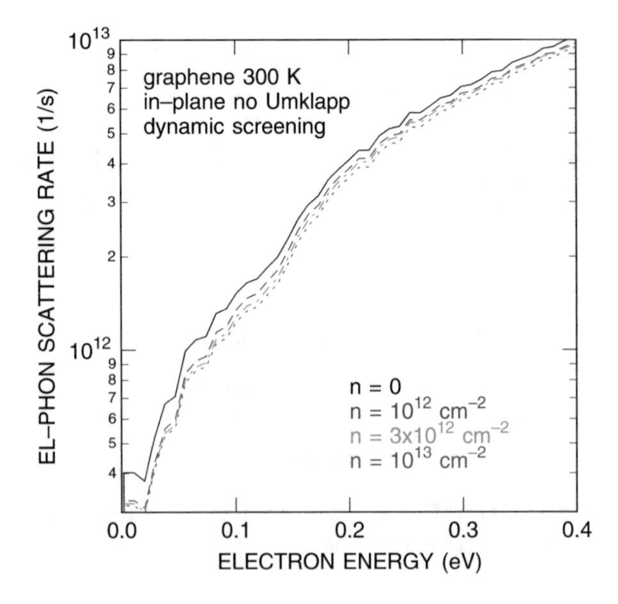


**Fig. 13.9** Comparison of the total electron–phonon scattering rates at the indicated electron densities. (Reprinted with permission from M.V. Fischetti, J. Kim, S. Narayanan, Z.-Y. Ong, C. Sachs, D.K. Ferry, and S.J. Aboud, J. Phys.: Cond. Matter vol. 25, 473202 (2013). Copyright 2013, Institute of Physics)

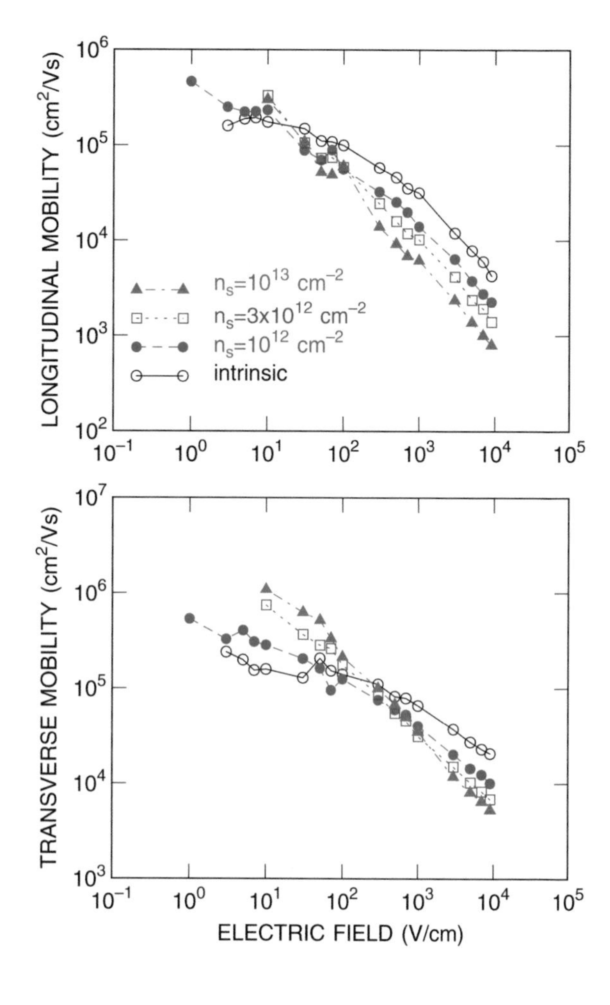


**Fig. 13.10** Longitudinal (*top*) and transverse (*bottom*) electron mobility in graphene at 300 K as a function of electric field along the [10] direction calculated using the Monte Carlo method. The various *curves* are parametrized by the electron density. (Reprinted with permission from M.V. Fischetti, J. Kim, S. Narayanan, Z.-Y. Ong, C. Sachs, D.K. Ferry, and S.J. Aboud, J. Phys.: Cond. Matter vol. 25, p. 473202 (2013). Copyright 2013, Institute of Physics)

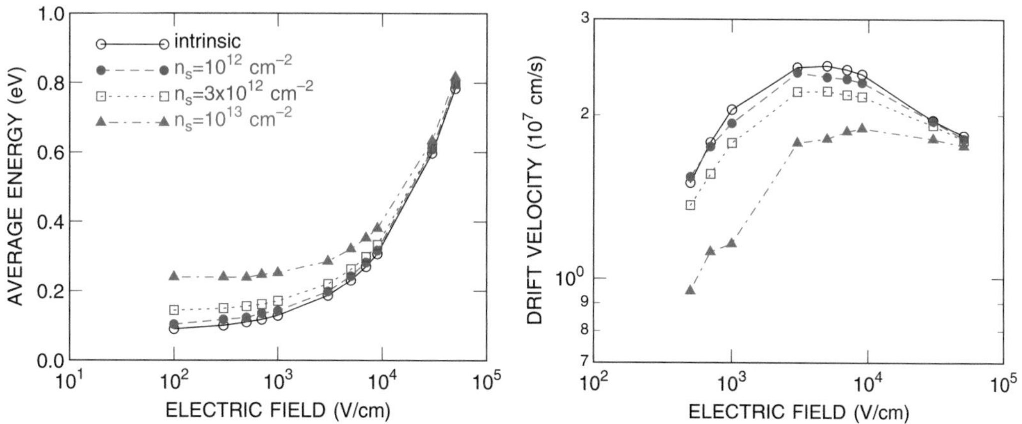


**Fig. 13.11** Average electron energy (*left*) and drift velocity (*right*) in graphene at 300 K as a function of electric field along the [10] direction calculated using the Monte Carlo method. The various *curves* are parametrized by the electron density. (Reprinted with permission from M.V. Fischetti, J. Kim, S. Narayanan, Z.-Y. Ong, C. Sachs, D.K. Ferry, and S.J. Aboud, J. Phys.: Cond. Matter vol. 25, 473202 (2013). Copyright 2013, Institute of Physics)

### 13.4.3 *Scattering Rates in the Effective-Mass Approximation*

When embracing the effective-mass approximation to describe the band structure, the task is greatly simplified, of course. The major difficulty lies in the evaluation of the overlap factors, $\mathscr{I}^{(2D)}_{\mathbf{K},\mathbf{K}_j,n,n'}(q_z)$, that enter the squared matrix element. More specifically the evaluation of the term

$$\int \frac{dq_z}{2\pi} \left| \delta U^{(\eta)}_{\mathbf{K}-\mathbf{K}_j,\,q_z}\, \mathscr{I}^{(2D)}_{\mathbf{K},\mathbf{K}_j,n,n'}(q_z) \right|^2 . \tag{13.99}$$

We now discuss specific examples in which this term can be simplified. We consider the following perturbation potentials $\delta U^{(\eta)}$:

$$\delta U^{(\mathrm{LA,TA})}_{\mathbf{q}} = \left[ \frac{\hbar \Delta^2_{\mathrm{LA,TA}}(\mathbf{q}) q^2}{2\rho_{\mathrm{x}}\, \omega^{(\mathrm{LA,TA})}_{\mathbf{q}}} \right]^{1/2} \left\{ \begin{array}{c} \langle N^{(\mathrm{LA,TA})}_{\mathbf{q}} \rangle^{1/2}_{\mathrm{th}} \\ (\langle N^{(\mathrm{LA,TA})}_{\mathbf{q}} \rangle_{\mathrm{th}} + 1)^{1/2} \end{array} \right\} , \tag{13.100}$$

for nonpolar scattering with acoustic phonons of branch $\eta$ in a crystal of mass density $\rho_{\mathrm{x}}$, with wavevector $\mathbf{q}$, frequency $\omega^{(\eta)}_{\mathbf{q}}$, Bose occupation number $\langle N^{(\eta)}_{\mathbf{q}} \rangle_{\mathrm{th}}$, deformation potential (possibly $\mathbf{q}$-dependent) $\Delta_\eta(\mathbf{q})$;

$$\delta U^{(\mathrm{op})}_{\mathbf{q}} = \left[ \frac{\hbar (DK)^2_{\mathrm{op}}}{2\rho_{\mathrm{x}}\, \omega^{(\mathrm{op})}} \right]^{1/2} \left\{ \begin{array}{c} \langle N^{(\mathrm{op})}_{\mathbf{q}} \rangle^{1/2}_{\mathrm{th}} \\ (\langle N^{(\mathrm{op})}_{\mathbf{q}} \rangle_{\mathrm{th}} + 1)^{1/2} \end{array} \right\} , \tag{13.101}$$

for nonpolar scattering with optical phonons with optical deformation potential $(DK)_{\mathrm{op}}$;

$$\delta U^{(\mathrm{pop})}_{\mathbf{q}} = \frac{e\mathscr{F}}{q\epsilon_r(\mathbf{q}, \omega^{(\mathrm{LO})})} \left\{ \begin{array}{c} \langle N^{(\mathrm{LO})}_{\mathbf{q}} \rangle^{1/2}_{\mathrm{th}} \\ (\langle N^{(\mathrm{LO})}_{\mathbf{q}} \rangle_{\mathrm{th}} + 1)^{1/2} \end{array} \right\} , \tag{13.102}$$

with $\mathscr{F}^2 = (\hbar\omega^{(\mathrm{LO})}_{\mathbf{q}}/2)(1/\epsilon_{\mathrm{hi}} - 1/\epsilon_{\mathrm{low}})$ for polar scattering with optical phonons ($\epsilon_r(\mathbf{q}, \omega)$ being the relative dielectric function, and $\epsilon_{\mathrm{hi}}$ and $\epsilon_{\mathrm{low}}$ the optical and static dielectric constants).

#### 13.4.3.1 Nonpolar Phonon Scattering, Momentum-Independent Matrix Element

Nonpolar scattering with acoustic and optical phonons can be simplified dramatically when the matrix element is assumed to be momentum-independent and the phonon energy is considered constant (zero for acoustic phonons, dispersionless for optical phonons). Thus neither $\hbar\omega^{(\eta)}_{\mathbf{q}}$ nor $\delta U^{\eta}_{\mathbf{q}}$ depend on $q_z$ and the only $q_z$ dependence occurs within the overlap factor. Thus the term (13.99) can be handled as follows:

$$\begin{aligned}
&|V^{(\eta)}|^2 \int \frac{dq_z}{2\pi} \left| \int dz\, \zeta^{(n')*}_{\mathbf{K}'}(z)\, e^{iq_z z}\, \zeta^{(n)}_{\mathbf{K}}(z) \right|^2 \\
&= |\delta U^{(\eta)}|^2 \int dz \int dz'\, \zeta^{(n')*}_{\mathbf{K}'}(z)\, \zeta^{(n)}_{\mathbf{K}}(z)\, \zeta^{(n')}_{\mathbf{K}'}(z')\, \zeta^{(n)*}_{\mathbf{K}}(z') \int \frac{dq_z}{2\pi}\, e^{iq_z(z-z')} \\
&= |\delta U^{(\eta)}|^2 \int dz\, \zeta^{(n')*}_{\mathbf{K}'}(z)\, \zeta^{(n)}_{\mathbf{K}}(z) \zeta^{(n')}_{\mathbf{K}'}(z)\, \zeta^{(n)*}_{\mathbf{K}}(z) \\
&= |\delta U^{(\eta)}|^2 \int dz \left| \zeta^{(n')*}_{\mathbf{K}'}(z) \right|^2 \left| \zeta^{(n)}_{\mathbf{K}}(z) \right|^2 = |\delta U^{(\eta)}|^2\, \mathscr{F}^{(2D)}_{\mathbf{K},\mathbf{K}',n,n'} .
\end{aligned} \tag{13.103}$$

In this case Eq. (12.30) simplifies to:

$$\frac{1}{\tau_n^{(\eta)}(\mathbf{K})} \approx \frac{2\pi}{\hbar} |\delta U^{(\eta)}|^2 \sum_{\mathbf{K}'n'} \mathscr{F}^{(2D)}_{\mathbf{K},\mathbf{K}',n,n'} \ \delta[E_n(\mathbf{K}) - E_{n'}(\mathbf{K}') \pm \hbar\omega^{(\eta)}] \,. \tag{13.104}$$

In the more complicated case in which the wavefunctions $\xi^{(n)}_{\mathbf{G}_{\parallel},\mathbf{K}}(z)$ given by Eq. (12.26) is required, then the factor $\mathscr{F}^{(2D)}_{\mathbf{K},\mathbf{K}',n,n'}$ in Eq. (13.103) becomes instead:

$$\mathscr{F}^{(2D)}_{\mathbf{K},\mathbf{K}',n,n'} == \int \mathrm{d}z \left| \sum_{\mathbf{G}_{\parallel}} \xi^{(n')*}_{\mathbf{G}_{\parallel}\mathbf{K}'}(z) \ \xi^{(n)}_{\mathbf{G}_{\parallel}\mathbf{K}}(z) \right|^2 \,. \tag{13.105}$$

If one is interested in transport not too far from equilibrium (as in mobility calculations), then one can assume that the carriers populate only regions of the first BZ not too far from a band extremum $\mathbf{K}_0$. Then one can ignore the $\mathbf{K}$-dependence of the wavefunctions and reduce the computational burden by having to calculate only a single overlap factor for each pair of (sub)bands $(n,n')$. So, the simplest possible expression for the scattering rate can be derived from Eq. (13.104):

$$\begin{aligned} \frac{1}{\tau_n^{(\eta)}(\mathbf{K})} &\approx \frac{2\pi}{\hbar} |V^{(\eta)}|^2 \mathscr{F}^{(2D)}_{\mathbf{K}_0,\mathbf{K}_0,n,n'} \sum_{\mathbf{K}'n'} \delta[E_n(\mathbf{K}) - E_{n'}(\mathbf{K}') \pm \hbar\omega^{(\eta)} \\ &= \frac{2\pi}{\hbar} |V^{(\eta)}|^2 \sum_{n'} \mathscr{F}^{(2D)}_{\mathbf{K}_0,\mathbf{K}_0,n,n'} \ \mathscr{D}^{(n')}[E_n(\mathbf{K}) \pm \hbar\omega^{(\eta)}] \,, \end{aligned} \tag{13.106}$$

where $\mathscr{D}^{(n)}(E)$ is the density of states in (sub)band $n$ at energy $E$. Clearly, Eq. (12.29) can be simplified by using any (or any combination) of the approximations we have considered here ($N$-process only, $\mathbf{q}$-independent scattering potential, $\mathbf{q}$-independent dispersion, $\mathbf{K}$-independent wavefunctions), depending on the particular physical system and conditions considered.

#### 13.4.3.2 Fröhlich Scattering

In the case of Fröhlich scattering, the perturbation $\delta U^{(\mathrm{pop})}$ can be approximated by a potential screened using a Debye–Hückel wavevector $\beta_{\mathrm{DH}}$, Eq. (11.52) in Chap. 11—approximated either by the 2D expression $(e^2/\epsilon_{\mathrm{s}})n_{\mathrm{s}}/(k_{\mathrm{B}}T)$ (we ignore for now multisubband screening, and the dynamic effects discussed in Sect. 11.3.2.4) or by the 3D expression, $[(e^2 n)/(k_{\mathrm{B}}T)]^{1/2}$—becomes

$$V_{\mathbf{q}}^{(\mathrm{pop,DH})} = \frac{e\mathscr{F}}{(q^2+\beta_{\mathrm{DH}}^2)^{1/2}} \left\{ \begin{array}{c} \langle N_{\mathrm{LO}} \rangle_{\mathrm{th}}^{1/2} \\ (\langle N_{\mathrm{LO}} \rangle_{\mathrm{th}} + 1)^{1/2} \end{array} \right\} = \frac{C_{\mathrm{pop}}}{(q^2+\beta_{\mathrm{DH}}^2)^{1/2}} \tag{13.107}$$

and therefore, Eq. (13.99) becomes

$$C_{\mathrm{pop}}^2 \int \mathrm{d}z \int \mathrm{d}z' \ \zeta_{\mathbf{K}'}^{(n')*}(z) \ \zeta_{\mathbf{K}'}^{(n')}(z') \ \zeta_{\mathbf{K}}^{(n)}(z) \ \zeta_{\mathbf{K}}^{(n)*}(z') \int \frac{\mathrm{d}q_z}{2\pi} \frac{\mathrm{e}^{\mathrm{i}q_z(z-z')}}{q_z^2+Q^2} \,, \tag{13.108}$$

where $Q = (|\mathbf{K}-\mathbf{K}'|^2 + \beta_{\mathrm{DH}}^2)^{1/2}$. The integral over $q_z$ above is easily evaluated:

$$\mathscr{I}_{\mathrm{pop}}^{(2\mathrm{D})} = \int \frac{\mathrm{d}q_z}{2\pi} \frac{\mathrm{e}^{\mathrm{i}q_z(z-z')}}{q_z^2+Q^2} = \pi\, \frac{\mathrm{e}^{-|z-z'|Q}}{Q}\,, \tag{13.109}$$

so that the scattering rate can be written as:

$$\frac{1}{\tau_n^{(\mathrm{pop})}(\mathbf{K})} \approx \frac{2\pi}{\hbar} \sum_{jn'}^{*} C_{\mathrm{pop}}^2 \frac{1}{(2\pi)^2} \frac{L(w_{jn'})}{|\nabla_{2\mathrm{D}} E_{jn'}|} \frac{1}{2Q_j}$$
$$\times \int \mathrm{d}z \int \mathrm{d}z'\, \zeta_{\mathbf{K}_j}^{(n')*}(z)\, \zeta_{\mathbf{K}_j}^{(n')}(z')\, \mathrm{e}^{-|z-z'|Q_j}\, \zeta_{\mathbf{K}}^{(n)}(z)\, \zeta_{\mathbf{K}}^{(n)*}(z')\,, \tag{13.110}$$

where $Q_j = (|\mathbf{K}-\mathbf{K}_j|^2 + \beta_{\mathrm{DH}}^2)^{1/2}$. Writing $g_{\mathbf{K},\mathbf{K}',n,n'}(z) = \zeta_{\mathbf{K}'}^{(n')*}(z)\zeta_{\mathbf{K}}^{(n)}(z)$, the overlap factor (the last factor in the equation above) can be written as:

$$\int \mathrm{d}z \int \mathrm{d}z'\, g_{\mathbf{K},\mathbf{K}',n,n'}(z')\, \mathrm{e}^{-|z-z'|Q_j}\, g_{\mathbf{K},\mathbf{K}',n,n'}^{*}(z)$$
$$= 2\,\mathrm{Re}\left\{ \int_0^{\infty} \mathrm{d}z\, g_{\mathbf{K},\mathbf{K}',n,n'}^{*}(z)\, \mathrm{e}^{-Qz} \int_0^{z} \mathrm{d}z'\, g_{\mathbf{K},\mathbf{K}',n,n'}(z')\, \mathrm{e}^{Qz'} \right\}, \tag{13.111}$$

an expression which shows explicitly that the overlap factor is real and provides a form more amenable to numerical integration. Indeed the numerical advantage provided by Eq. (13.111) consists in the fact that the "inner" integral over $z'$ depends on $z$ only via the upper integration limit as the integrand does not depend on $z$. Of course one can approximate $g_{\mathbf{K},\mathbf{K}',n,n'}(z)$ with $g_{\mathbf{K}_0,\mathbf{K}_0,n,n'}(z)$, where $\mathbf{K}_0$ is the location of the band extremum, in order to reduce the number of integrals to be evaluated and tabulated.

When the supercell method is employed and the wavefunctions $\xi_{\mathbf{G}_{\|},\mathbf{K}}^{(n)}(z)$ given by Eq. (12.26) should be used, bypassing the pure envelope approximation, in the Eq. (13.111) $g_{\mathbf{K},\mathbf{K}',n,n'}(z)$ should be replaced by

$$g_{\mathbf{K},\mathbf{K}',n,n'}(z) = \sum_{\mathbf{G}_{\|}} \xi_{\mathbf{G}_{\|},\mathbf{K}'}^{(n')*}(z)\, \xi_{\mathbf{G}_{\|},\mathbf{K}}^{(n)}(z)\,, \tag{13.112}$$

in order to obtain the correct matrix element.

## 13.5 Electron–Phonon Interaction for a 1DEG

Of course, having discussed electron–phonon scattering in bulk semiconductors and for a 2DEG, we must also at least mention its one-dimensional counterpart. As we have done in the previous section, we shall first see in general what a full-band calculation entails and then consider specific cases within the effective-mass approximation. Observations similar to those made before should be repeated here. We shall consider bulk perturbations, so we shall not consider the complications related to low-dimensional phonon modes. We should also emphasize again that the validity of any treatment of electronic transport based on the single-particle approximation is very doubtful in one dimension. In Chap. 15, Sect. 15.2.1, we shall mention that Coulomb interparticle interactions in one dimension usually result in the breakdown of the single-particle picture and in the appearance of a collective state called *Luttinger liquid*. We can only keep this in mind and regard results obtained in the single-particle picture as only "suggestive" at best.

In a full-band context, Eq. (12.51) requires the evaluation of a four-dimensional integral (2D in $\mathbf{Q}$ and 2D in $\mathbf{R}$ "hidden" within $\mathscr{I}^{(1D)}_{k_z,k_{zj},n,n'}(\mathbf{Q})$). In general cases—such as inelastic, anisotropic scattering with dispersive acoustic phonons—not much can be done to simplify the problem. One has to evaluate the overlap factor $\mathscr{I}^{(1D)}_{k_z,k_{zj},n,n'}(\mathbf{Q})$—possibly using some 2D fast Fourier transform (FFT) algorithm—and then perform the remaining 2D integral over the 2D momentum-transfer $\mathbf{Q}$.

We present two examples of these of DFT (and empirical pseudpotentials) to calculate the transport properties of a 1DEG: Silicane nanoribbons and graphene nanoribbons.

### *13.5.1 Deformation Potentials and Electron Mobility in Silicane NRs and AGNRs*

Silicane is almost the exact Si equivalent of graphene, but only "almost." Although not universally accepted, common terminology reserves the suffix "ene" to $sp^2$-coordinated atomic sheets. Thus, *silicene* would be the exact equivalent of graphene. The suffix "ane," instead, denotes a "buckled" $sp^3$-coordinated arrangement, as in silicane. This "buckling" has interesting consequences. Nanoribbons, as we saw for the case of GNRs, are desired because they exhibit a band gap. However, the presence of the Clar resonance causes the bang gap of AGNRs to exhibit "strange" claromatic oscillations. This would be very hard to control, should we desire to use them in electronics. Silicane nanoribbons, however, are not affected by Clar resonances because of their $sp^3$ bonding. For this reason, there is interest in Si NRs. Of course, we shall ignore complicated issues regarding their stability. As we saw for the case of graphene, two-dimensional crystals are always under the Damocles' sword of the Mermin–Wagner–Hohenberg theorem discussed at the beginning of Sect. 7.2.3. DFT can be used to study the atomic relaxation of Si NRs and, with additional techniques we shall not discuss here, the so-called ab initio thermodynamics, that permits the extension of zero-temperature DFT to finite temperatures. Results reported in [21] show that Si NRs would be the natural byproduct of the instability of an infinite silicane sheet: Under ambient conditions, two-dimensional silicane sheets will spontaneously break into stable one-dimensional ribbons.

Figures 13.12 and 13.13—also from [21]—show the atomic structure of Si NRs. We should note that this constitutes another example in which DFT plays a crucial role in determining how the structure "really looks like." Figure 13.14 shows the band structure calculated using the four possible combinations of unrelaxed/relaxed atomic coordinates and calculations using DFT or empirical pseudopotentials, as described in Chap. 7.

Having established the atomic structure, DFT can be used to calculate the electron–phonon matrix elements, scattering rates, and electron mobility (see Chap. 19). Figure 13.15 shows the resulting scattering rates, Fig. 13.16 the electron mobility. This quantity depends on the electron density and this is shown in the figure. Avoiding complicated details, it should be noted that the calculated low-field electron mobility shows a strong dependence on the crystallographic orientation of the edge of the ribbon, because of differences in the effective mass and momentum relaxation rates along the two transport directions (armchair and zigzag). Most important, that the electron mobility in zigzag-edge Si NRs is predicted to be approximately twenty times higher than in AGNRs. This case is illustrated in Figs. 13.17, 13.18, and 13.19.

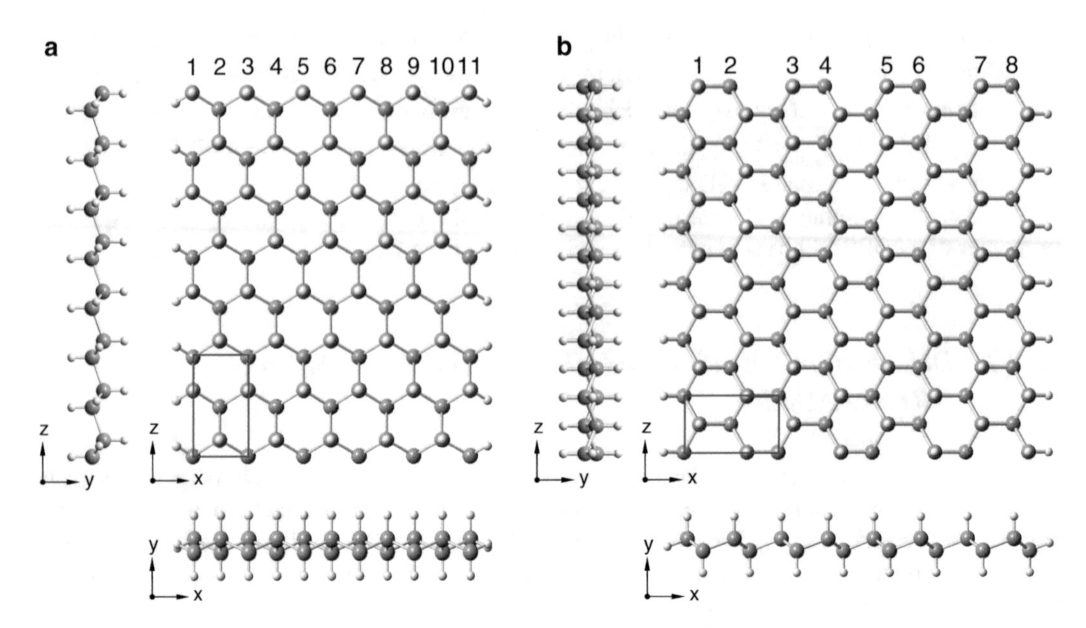


**Fig. 13.12** Atomic configurations of (**a**) 11-aSiNR and (**b**) 8-zSiNR. All the Si (*gold balls*) dangling bonds are passivated by H (*white balls*). Note the H atoms passivate the dangling bonds in alternating pairs at the *top and bottom sides* of the monolayer (*chair-like*). The charge carrier transport occurs along the $z$-direction which is equivalent to $[11\bar{2}]$ and $[10\bar{1}]$ of bulk Si for aSiNR and zSiNR, respectively. (Reprinted figure with permission from J. Kim, M.V. Fischetti, and S. Aboud, Phys. Rev. B vol. 86, 205323 (2012). Copyright 2012 by the American Physical Society)

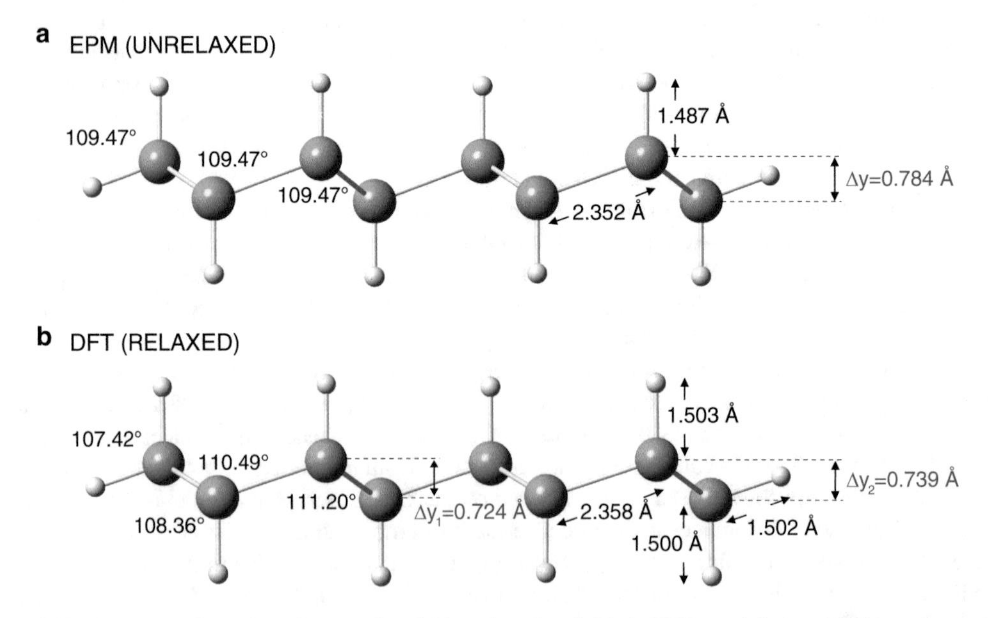


**Fig. 13.13** Atomic configuration of an unrelaxed (**a**) and a relaxed (**b**) 4-zSiNR used for empirical pseudopotential (EP) and DFT calculations, respectively. The buckling height is indicated by $\Delta y$ for the unrelaxed geometry, and by $\Delta y_1$ (inner Si-Si atoms) and $\Delta y_2$ (edge Si-Si atoms) for the relaxed geometry. (Reprinted figure with permission from J. Kim, M.V. Fischetti, and S. Aboud, Phys. Rev. B vol. 86, 205323 (2012). Copyright 2012 by the American Physical Society)

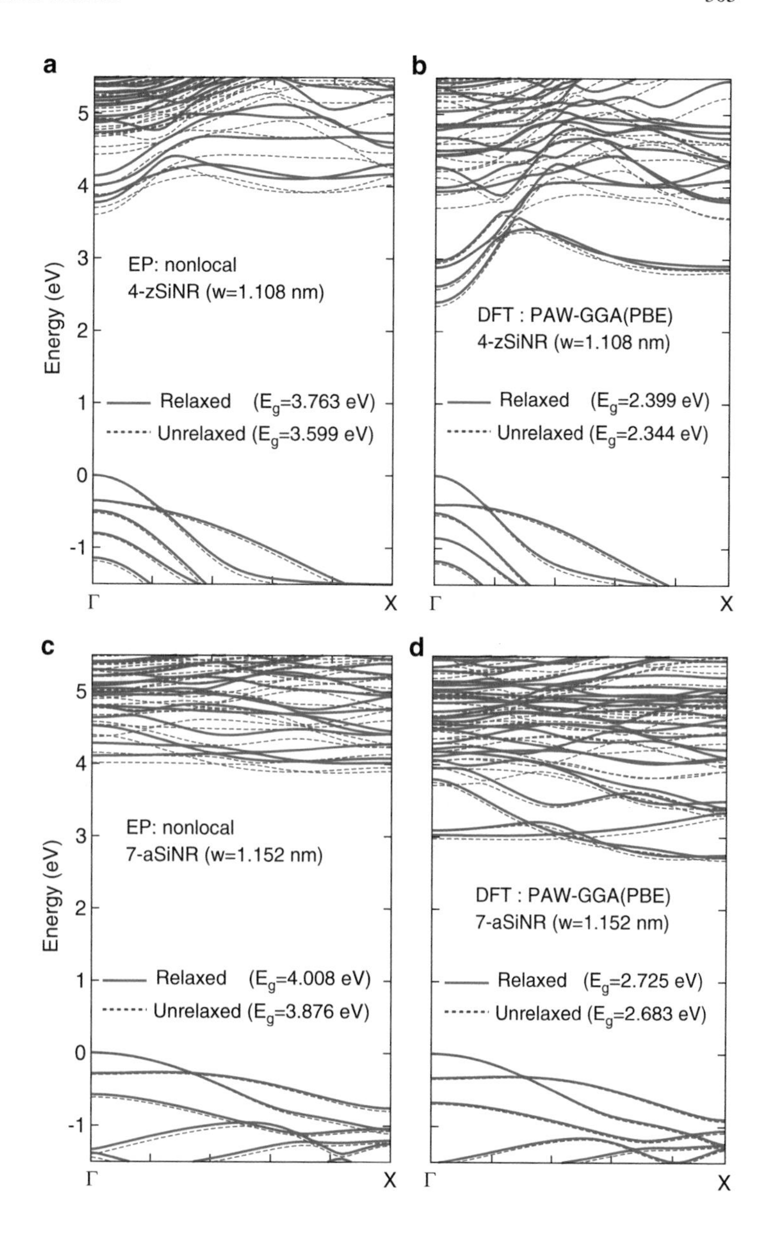


**Fig. 13.14** Electronic band structure of a 4-zSiNR and 7-aSiNR using EPs (**a**) and (**c**) and DFT (**b**) and (**d**) with unrelaxed (*dotted lines*) and fully relaxed (*solid lines*) geometry. The zero for the energy is set at the *top* of the valence band. The *insets* show in detail the structure of the conduction band the its *bottom*, set to zero in this case. (Reprinted figure with permission from J. Kim, M.V. Fischetti, and S. Aboud, Phys. Rev. B vol. 86, 205323 (2012). Copyright 2012 by the American Physical Society)

### 13.5.2 *Scattering Rates in the Effective-Mass Approximation*

We have seen in Sect. 13.4.3 that the use of the effective-mass approximation reduces enormously the complexity involved in the evaluation of the overlap factors entering the electron–phonon matrix elements for a 2DEG. The same is obviously true also in the case of 1DEG. The overlap factor $\mathscr{I}^{(1D)}_{k_z,k'_z,n,n'}(\mathbf{Q})$ appearing in Eq. (12.48) can be simplified following a similar approach. Considering specific examples for the $\mathbf{q}$-dependence of $\delta U^{(\eta)}_{\mathbf{q}}$, as done before for the 2D case, we can perform part of the task analytically in handling the term:

$$\int \frac{\mathrm{d}\mathbf{Q}}{(2\pi)^2} \left| \delta U^{(\eta)}_{\mathbf{Q},k_z-k_z+q_z} \, \mathscr{I}^{(1D)}_{k_z,k_z,n,n'}(\mathbf{Q}) \right|^2 . \tag{13.113}$$

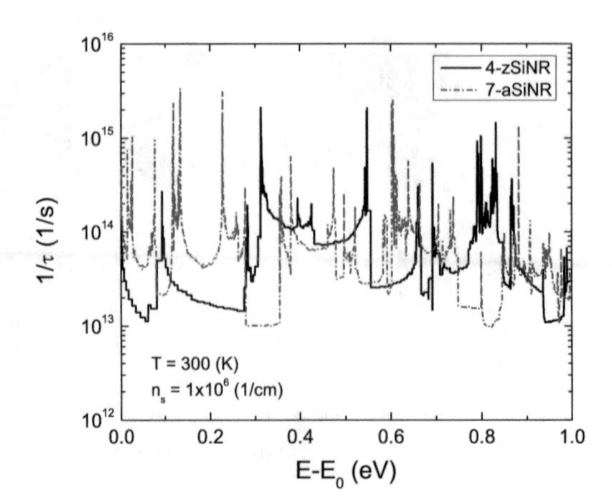


**Fig. 13.15** Total electron–phonon momentum relaxation rate accounting for emission and absorption of acoustic and optical phonons, the acoustic and optical emission and absorption processes. (Reprinted figure with permission from J. Kim, M.V. Fischetti, and S. Aboud, Phys. Rev. B vol. 86, 205323 (2012). Copyright 2012 by the American Physical Society)

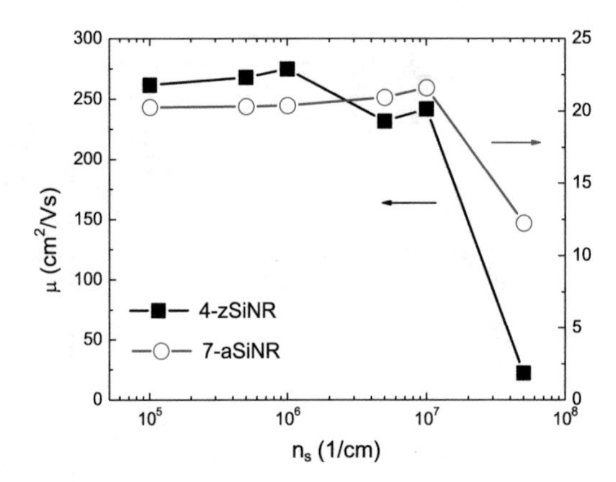


**Fig. 13.16** Phonon-limited electron mobility as a function of electron line-density $n_l$ for a 7-zSiNR (*squares*) and a 11-aSiNR (*circles*) at $T = 300$ K. (Reprinted figure with permission from J. Kim, M.V. Fischetti, and S. Aboud, Phys. Rev. B vol. 86, 205323 (2012). Copyright 2012 by the American Physical Society)

We consider the same cases we have considered above for scattering in a two-dimensional system

#### 13.5.2.1 Nonpolar Phonon Scattering, Momentum-Independent Matrix Element

Assuming (as we have done before in the 2D case) that the perturbation term $\delta U_{\mathbf{q}}^{(\eta)}$ given by Eq. (13.100) or by Eq. (13.101) does not depend on $\mathbf{q}$, and neither does $\hbar\omega_{\mathbf{q}}^{(\eta)}$, then the term (13.113) can be written as:

$$|\delta U^{(\eta)}|^2 \int \frac{\mathbf{Q}}{(2\pi)^2} \left| \int \mathrm{d}\mathbf{R}\ \zeta_{k_z'}^{(2D)(n')*}(\mathbf{R})\ \mathrm{e}^{\mathrm{i}\mathbf{Q}\cdot\mathbf{R}}\ \zeta_{k_z}^{(2D)(n)}(\mathbf{R}) \right|^2$$
$$= |\delta U^{(\eta)}|^2 \int \mathrm{d}\mathbf{R} \int \mathrm{d}\mathbf{R}'\ \zeta_{k_z'}^{(2D)(n')}(\mathbf{R}')\ \zeta_{k_z}^{(2D)(n)*}(\mathbf{R}')$$

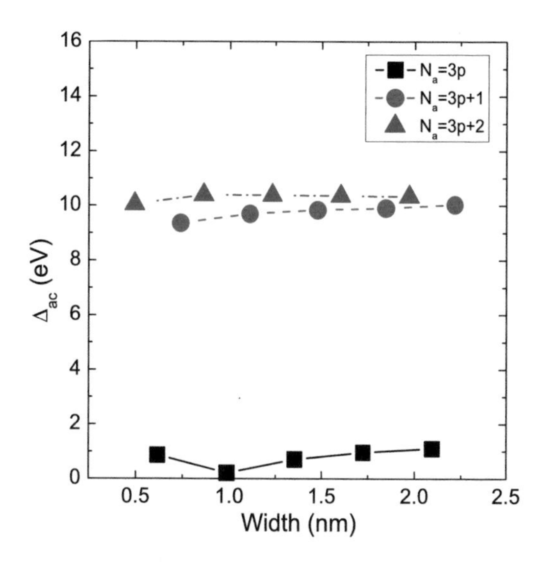


**Fig. 13.17** Acoustic deformation potential $\Delta_{\text{ac}}$ in AGNRs shown as a function of ribbon width calculated from DFT. (Reprinted with permission from M.V. Fischetti, J. Kim, S. Narayanan, Z.-Y. Ong, C. Sachs, D.K. Ferry, and S.J. Aboud, J. Phys.: Cond. Matter vol. 25, 473202 (2013). Copyright 2013, Institute of Physics)

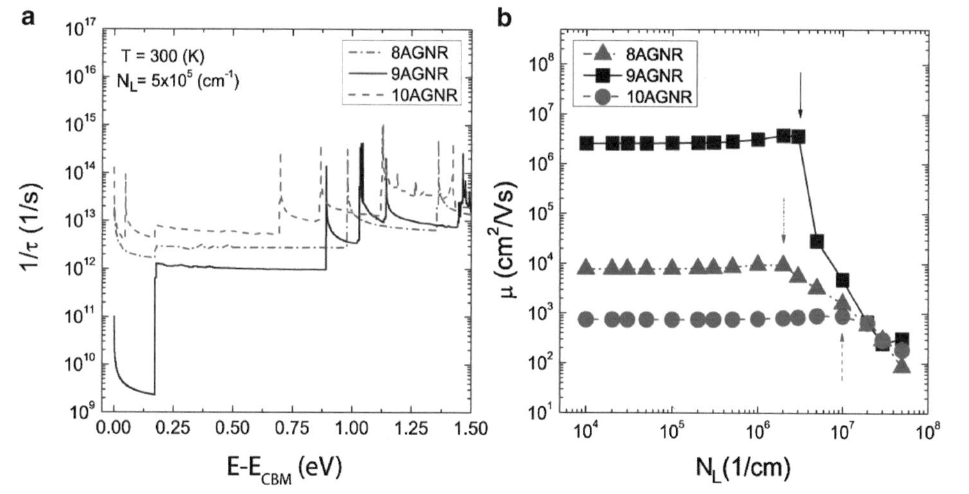


**Fig. 13.18** (**a**) Calculated total momentum relaxation rate in AGNRs shown as a function of electron kinetic energy measured from the bottom of the conduction band. (**b**) Low-field mobility vs. electron line density for an 8-, 9-, and 10-AGNR. (Reprinted with permission from M.V. Fischetti, J. Kim, S. Narayanan, Z.-Y. Ong, C. Sachs, D.K. Ferry, and S.J. Aboud, J. Phys.: Cond. Matter vol. 25, 473202 (2013). Copyright 2013, Institute of Physics)

$$\times\ \zeta_{k_z'}^{(2D)(n')*}(\mathbf{R})\ \zeta_{k_z}^{(2D)(n)}(\mathbf{R})\ \int \frac{\mathrm{d}\mathbf{Q}}{(2\pi)^2}\mathrm{e}^{\mathrm{i}\mathbf{Q}\cdot(\mathbf{R}-\mathbf{R}')}$$

$$= |\delta U^{(\eta)}|^2 \int \mathrm{d}\mathbf{R}\ \left|\zeta_{k_z'}^{(2D)(n')}(\mathbf{R})\right|^2\ \left|\zeta_{k_z}^{(2D)(n)}(\mathbf{R})\right|^2 = |\delta U^{(\eta)}|^2\, \mathscr{F}_{k_z,k_z',n,n'}^{(1D)}\,, \tag{13.114}$$

and the nonpolar electron–phonon scattering rate ($\eta$ = TA, LA, op) becomes simply:

$$\frac{1}{\tau_n^{(\eta)}(k_z)} \approx \frac{2\pi}{\hbar}\left|\delta U^{(\eta)}\right|^2 \sum_{k_z' n'} \mathscr{F}_{k_z,k_z',n,n'}^{(1D)}\ \delta[E_n(k_z) - E_{n'}(k_z') \pm \hbar\omega^{(\eta)}] \tag{13.115}$$

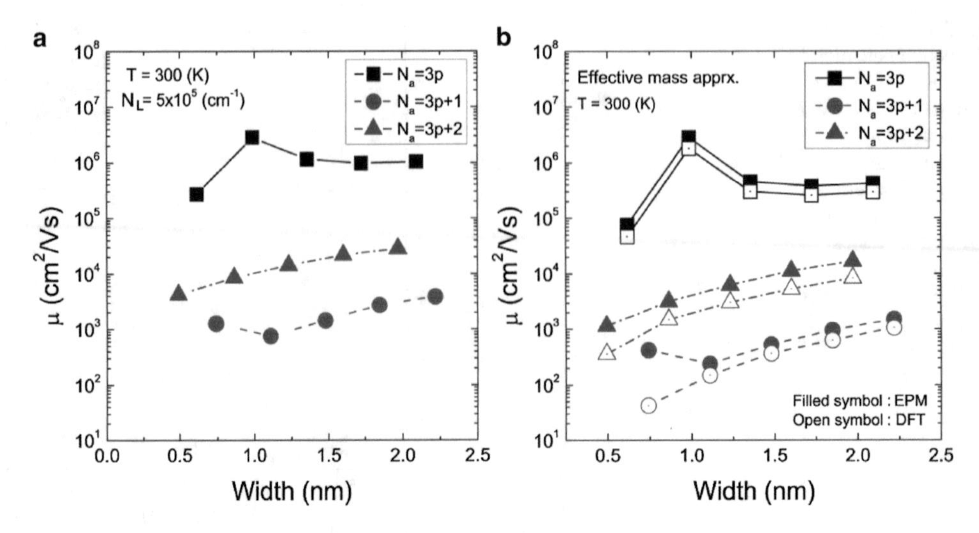


**Fig. 13.19** (**a**) Phonon-limited electron mobility in AGNRs using Kubo–Greenwood formula with full-band structure. (**b**) The same, but using the effective-mass approximation with effective masses calculated from EP (*solid symbols*) and DFT (*open symbols*). (Reprinted with permission from M.V. Fischetti, J. Kim, S. Narayanan, Z.-Y. Ong, C. Sachs, D.K. Ferry, and S.J. Aboud, J. Phys.: Cond. Matter vol. 25, 473202 (2013). Copyright 2013, Institute of Physics)

Ignoring the $k_z$-dependence of the envelope wavefunctions, this simplifies further to:

$$\frac{1}{\tau_n^{(\eta)}(k_z)} \approx \frac{2\pi}{\hbar} \left|\delta U^{(\eta)}\right|^2 \sum_{k_z' n'} \mathscr{F}^{(1D)}_{k_{z0},k_{z0},n,n'} \, \delta[E_n(k_z) - E_{n'}(k_z') \pm \hbar\omega^{(\eta)}]$$
$$= \frac{2\pi}{\hbar} \left|\delta U^{(\eta)}\right|^2 \sum_{n'} \mathscr{F}^{(1D)}_{k_{z0},k_{z0},n,n'} \, \mathscr{D}^{(n')}[E_n(k_z) \pm \hbar\omega^{(\eta)}] \,, \tag{13.116}$$

where $\mathscr{D}^{(n)}(E)$ is the density of states at energy $E$ in (sub)band $n$. However, if we wish to retain the full dependence of the wavefunctions on the **G** vectors avoiding any use of the envelope approximation, using Eqs. (12.44) and (12.46), all we have to do is to replace the form factor $\mathscr{F}^{(1D)}_{k_z,k_z',n,n'}$ with the more general expression [see Eq. (13.105)]:

$$\mathscr{F}^{(1D)}_{k_z,k_z',n,n'} = \int \mathrm{d}\mathbf{R} \left| \sum_{G_z} \xi^{(2D)(n')*}_{G_z,k_z'}(\mathbf{R}) \, \xi^{(2D)(n)}_{G_z,k_z}(\mathbf{R}) \right|^2 . \tag{13.117}$$

#### 13.5.2.2 Fröhlich Scattering

In the case of Fröhlich scattering, expressing the scattering potential as in Eq. (13.107) with $\beta_{\mathrm{DH}}$ approximated by the 3D bulk expression $[(e^2 n)/(k_{\mathrm{B}}T)]^{1/2}$, the term (13.113) can be rewritten as:

$$C_{\mathrm{pop}}^2 \int \mathrm{d}\mathbf{R} \int \mathrm{d}\mathbf{R}' \, g_{k_z,k_z',n,n'}(\mathbf{R}) \, g^*_{k_z,k_z',n,n'}(\mathbf{R}') \int \frac{\mathrm{d}\mathbf{Q}}{(2\pi)^2} \frac{\mathrm{e}^{i\mathbf{Q}\cdot(\mathbf{R}-\mathbf{R}')}}{Q^2 + \Delta Q^2} \,, \tag{13.118}$$

where $C_{\mathrm{pop}}$ is defined implicitly by Eq. (13.107), $\Delta Q = [(k_z - k_z')^2 + \beta_{\mathrm{DH}}^2]^{1/2}$, and

$$g_{k_z,k_z',n,n'}(\mathbf{R}) = \zeta^{(2D)(n')*}_{k_z'}(\mathbf{R}) \, \zeta^{(2D)(n)}_{k_z}(\mathbf{R}) \,.) \tag{13.119}$$

The integral over the 2D variable **Q** in Eq. (13.118) can be written as:

$$\mathscr{I}_{\rm pop}^{(1D)} = \frac{1}{(2\pi)^2} \int_0^\infty dQ \, \frac{Q}{Q^2 + \Delta Q^2} \int_0^{2\pi} d\phi \; e^{iQ\Delta R \cos\phi} \,, \tag{13.120}$$

where $\Delta R = |\mathbf{R} - \mathbf{R}'|$. Evaluating the angular integral gives us $2\pi J_0(Q\Delta R)$ so that:

$$\mathscr{I}_{\rm pop}^{(1D)} = \frac{1}{2\pi} \int_0^\infty dQ \, \frac{Q \, J_0(Q\Delta R)}{Q^2 + \Delta Q^2} = \frac{1}{2\pi} K_0(\Delta Q \Delta R) \,, \tag{13.121}$$

where $J_0$ and $K_0$ are Bessel functions. Thus, Eq. (12.51) can be expressed as:

$$\frac{1}{\tau_n^{\rm (pop)}(k_z)} \approx \frac{C_{\rm pop}^2}{\hbar} \sum_{jn'}^{*} \left| \frac{dE_{n'}}{dk_{zj}} \right|^{-1}$$
$$\times \int d\mathbf{R} \int d\mathbf{R}' \; g_{k_z,k_{zj},n,n'}(\mathbf{R}) \; g^*_{k_z,k_{zj},n,n'}(\mathbf{R}') \; K_0(\Delta Q_j |\mathbf{R} - \mathbf{R}'|) \,, \tag{13.122}$$

where $\Delta Q_j = [(k_z - k_{zj})^2 + \beta_{\rm DH}^2]^{1/2}$. The Bessel function $K_0(x)$ can be evaluated using its integral representation

$$K_0(x) = \int_1^\infty dt \, \frac{e^{-xt}}{\sqrt{t^2 - 1}} \,, \tag{13.123}$$

tabulated, and interpolated during the evaluation of Eq. (13.122). Notice that we have not reduced the number of integrations we must perform, but we have replaced the **Q**-integration involving a potentially fast-varying complex exponential with the integration over the "smooth" wavefunctions. In addition, the integrand is now a symmetric function of **R** and **R**′, so that it needs to be evaluated only over 1/2 of the 4D "hyperspace" $(\mathbf{R}, \mathbf{R}')$.

Whenever the use of the full wavefunctions given by Eq. (12.44) is required, we must replace the term $g_{k_z,k_z',n,n'}(\mathbf{R})$ given by Eq. (13.119) with the expression

$$g_{k_z,k_z',n,n'}(\mathbf{R}) = \sum_{G_z} \xi^{(2D)(n')*}_{G_z,k_z'}(\mathbf{R}) \; \xi^{(2D)(n)}_{G_z,k_z}(\mathbf{R}) \tag{13.124}$$

in the equations above.

## 13.6 Dielectric Screening and Coupled Plasmon/LO-Phonon Modes

We consider here another interesting example of electrons scattering with a *hybrid surface excitation*, that is, an elementary excitation that results from the simultaneous presence of two excitations, plasmons and optical phonons, whose electric fields and charges interact in a peculiar way to give rise to a new excitation. To make things even more interesting, these excitations exist in the 2DEG of a polar semiconductor present in a QW. These excitations are discussed by Ridley [22].

Specifically, consider a polar material (say, as usual, GaAs, to fix the ideas) in which electrons are affected by polar Fröhlich scattering. Suppose that the semiconductor is $n$-type doped, so that there is a uniform density of $n$ electrons per cubic centimeter. Clearly, the interaction between electrons and the LO phonons will be screened by these free electrons, since the electrons feel the dipole field associated with the LO phonons and respond to it by screening it. However, things are a little complicated.

For very small electrons densities, the interaction can be assumed to be essentially unscreened: The plasma frequency is very small (as it vanishes as $n^{1/2}$ as $n \to 0$), so that the free electrons cannot respond to the oscillating dipole field of the LO phonons. But if we consider increasing densities, we soon reach a situation where the plasma frequency, $\omega_P = [e^2 n/(\epsilon_\infty m^*)]^{1/2}$ and the frequency $\omega^{(LO)}$ of the phonons become comparable. What this means physically is that the field due to the LO phonons excites the plasma of the free electrons. This plasma responds near resonance causing, in turn, the phonons to oscillate differently under the action of the longitudinal field of the plasma oscillations. In other words: plasmons affect phonons and phonons affect plasmons. The two modes are not decoupled (or "independent") any longer. Rather, the system will exhibit oscillations at some new frequency and we cannot talk any more of LO phonons and plasmons as independent entities. The excitations of the systems are now "coupled phonon-plasmon modes."

In order to evaluate the frequency and the nature of these coupled modes we must go back to the basic definition of longitudinal excitations. We must recall that when we discussed the dielectric response of the ions, we saw that longitudinal excitations are those for which the dielectric function vanishes. In this case, indeed, we can have a nonzero electric field **E** with a vanishing macroscopic displacement field **D**. The full dielectric function of the system, accounting for both the ionic and the electronic response can be obtained from Eqs. (11.19) and (11.36) in the long-wavelength limit:

$$\epsilon(\omega) = \epsilon_\infty \left(1 - \frac{\omega_P^2}{\omega^2}\right) + (\epsilon_0 - \epsilon_\infty)\, \frac{\omega^{(TO)2}}{\omega^{(TO)2} - \omega^2} \,. \tag{13.125}$$

In order to calculate the frequency of the longitudinal modes we find that $\epsilon(\omega) = 0$ for $\omega = \omega_\pm$, where $\omega_\pm$ are the two solutions of the algebraic equation:

$$\omega^4 \;-\; \omega^2(\omega_P^2 + \omega^{(LO)2}) \;+\; \omega_P^2 \omega_{(TO)2} = 0, \tag{13.126}$$

or:

$$\omega_\pm^2 = \frac{1}{2}\left\{ (\omega_P^2 + \omega^{(LO)2}) \pm \left[ (\omega_P^2 + \omega^{(LO)2})^2 - 4\omega_P^2 \omega^{(TO)2} \right]^{1/2} \right\} . \tag{13.127}$$

Note that for $n \to 0$ we have

$$\omega_\pm^2 \;\to\; (1/2)(\omega^{(LO)2} \pm \omega^{(LO)2}) \,, \tag{13.128}$$

while in the opposite limit $n \to \infty$ we have

$$\omega_\pm^2 \;\to\; (1/2)[\omega_P^2 \pm (\omega_P^2 - 2\omega^{TO2}]) \,, \tag{13.129}$$

so that the lower-frequency mode has a dispersion $\omega_-$ vanishing in the limit of small density and asymptotically approaching the unscreened phonon frequency $\omega_{TO}$ in the limit of very large densities. In the same limits the dispersion of the high-frequency approaches $\omega^{(LO)}$ and $\omega_P$, respectively. The top-left frame of Fig. 13.20 shows the dispersion of the two modes as a function of electron density for the case of GaAs.

The calculation of the rate at which electrons scatter with these modes can proceed along the same path we have followed to reach Eq. (13.61), but a few comments and modifications are necessary. First, neither of the two modes is purely a phonon nor a plasmon. One must sort out the plasmon and phonon content of each coupled mode in order to evaluate its coupling with the electrons. The text by Ridley [22] (page 332) discusses this issue, as well as the scattering strength, from a vary nice "mechanical" perspective. Here we follow a more empirical approach. It may be shown (but we'll state the results without a rigorous proof) that the phonon content of each mode is given by:

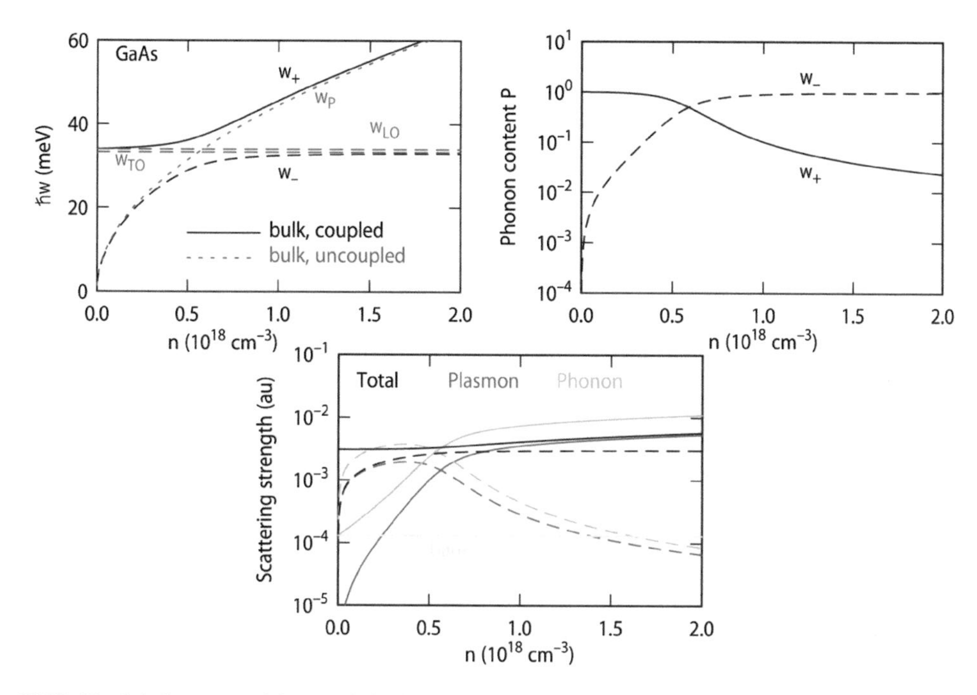


**Fig. 13.20** *Top left*: Spectrum of the coupled TO-phonon/plasmon modes as a function of electron density. *Top right*: The phonon and plasmon content of each mode. *Bottom*: Scattering rates for the electron/hybrid mode interaction

$$P(\omega_\pm) = \frac{\omega_\pm^2 - \omega_\mathrm{P}^2}{\omega_+^2 - \omega_-^2} . \tag{13.130}$$

The result should be at least "intuitively" correct: If the dispersion tracks exactly the plasma dispersion, the excitations are indeed a pure plasmon and its phonon content vanishes. Moreover, if we define similarly the plasmon content as

$$Q(\omega_\pm) = \frac{\omega_\pm^2 - \omega^{(\mathrm{LO})2}}{\omega_+^2 - \omega_-^2} , \tag{13.131}$$

we have $P(\omega_\pm) = 1 - Q(\omega_\pm)$ from Eq. (13.129), which shows that the plasmon and phonon content of each mode do indeed add up to unity, as it should be. The top-right frame of Fig. 13.20 shows that the high-frequency mode is mostly phonon-like at small densities, but it becomes mostly a plasmon at large densities. The opposite is true for the low-frequency mode.

The scattering strength can be estimated following the semiclassical approach on pages 282–285: We may calculate the total energy of the field associated with each mode and set it equal to the zero-point energy $\hbar\omega_\pm/2$ of the mode. We can then isolate the contribution of the phonon-component of each mode by considering the dielectric response when the phonon responds fully, the response when the phonon is assumed to be "frozen," and, finally, take the difference between these two fields and assign it exclusively to the phonon contribution to the field. Thus, the "bare" scattering strength

$$\hbar\omega^{(\mathrm{LO})} \left( \frac{1}{\epsilon_\mathrm{hi}} - \frac{1}{\epsilon_\mathrm{low}} \right) \tag{13.132}$$

of the "bare" (i.e., unscreened) LO-modes is replaced by:

$$S(\omega_\pm) = P(\omega_\pm)\, \hbar\omega_\pm \left( \frac{1}{\epsilon_{\rm hi}(\omega_\pm)} + \frac{1}{\epsilon_{\rm low}(\omega_\pm)} \right) , \tag{13.133}$$

where

$$\epsilon_{\rm hi}(\omega_\pm) = \epsilon_\infty \left( 1 - \frac{\omega_{\rm P}^2}{\omega_\pm^2} \right) , \tag{13.134}$$

(the limit $\omega \to \infty$ has been taken in the ionic part of Eq. (13.125) to account for the fact that the phonon does not respond) and

$$\epsilon_{\rm low}(\omega_\pm) = \epsilon_\infty \left( 1 - \frac{\omega_{\rm P}^2}{\omega_\pm^2} \right) + (\epsilon_0 - \epsilon_\infty) = \epsilon_0 \; - \; \epsilon_\infty \frac{\omega_{\rm P}^2}{\omega_\pm^2} , \tag{13.135}$$

having taken the limit $\omega \to 0$ in the ionic part of Eq. (13.125) to account for the fact that the phonon responds fully. The last (bottom) frame of the following figure shows the scattering strength for both modes. Of interest is the comparison with the bare scattering strength, Eq. (13.132) (yellow line): The "screened" scattering strength is larger than the unscreened one! To our knowledge, this phenomenon was first noticed by Rydley himself and it is called "anti-screening", for obvious reasons.

## Problems

**13.1.** Plot the nonpolar scattering rate with acoustic and optical phonons, given by Eq. (13.42) on page 280, as a function of electron energy in the range $0 < E \leq 1$ eV. Use the following values (roughly related to Si):

$(D_t K)_{\rm op} = 10^9$ eV/cm,
$\Delta_{\rm ac} = 10$ eV,
$m^* = 0.32\, m_0$ (where $m_0$ is the free-electron mass),
$\rho_{\rm x} = 2.33$ gr/cm$^3$,
$\hbar\omega^{(\rm op)} = 0.061$ eV,
$c_{\rm s} = 5 \times 10^5$ cm/sec.

**13.2.** **(a)** Do the same for the polar Fröhlich scattering, Eqs. (13.60) and (13.61). Use the following parameters (approximately correct for GaAs):

$m^* = 0.063\, m_0$,
$\hbar\omega^{(\rm LO)} = 0.036$ eV,
$\epsilon_{\rm low} = 10.92\, \epsilon_0$,
$\epsilon_{\rm low} = 12.9\, \epsilon_0$.

**(b)** Compare the result with the result of Problem 13.3.. Discuss how the different energy dependence will affect electron transport.

**13.3.** Consider the metal-vacuum interface located at $z = 0$, the metal filling the entire half-space $z \geq 0$, vacuum filling (!?) the half-space $z < 0$. The dielectric function in the metal in the long-wavelength limit is given by:

$$\epsilon_M(\omega) = \epsilon_0 \left(1 - \frac{\omega_P^2}{\omega^2}\right) , \tag{13.136}$$

where $\epsilon_0$ is the vacuum permittivity. In the metal a solution of Laplace's equation $\nabla^2\phi = 0$ is

$$\phi_M(x,z) = A\cos(kx)\, e^{-kz} , \tag{13.137}$$

so that $E_{Mz} = kA\cos(kx)\, e^{-kz}$ and $E_{Mx} = kA\sin(kx)\, e^{-kz}$ are the $z$ and $x$ components of the electric field.

**(a)** Show that in the vacuum (that is, for $z < 0$)

$$\phi_0(x,z) = A\cos(kx)\, e^{kz} , \tag{13.138}$$

satisfies the boundary condition that the tangential component of **E** be continuous at the metal-vacuum boundary; i.e., find $E_{0x}$.

**(b)** Recalling that $\mathbf{D}_M = \epsilon_M(\omega)\mathbf{E}_M$ and that $\mathbf{D}_0 = \epsilon_0\mathbf{E}_0$, show that the boundary condition that the normal component of **D** be continuous at the boundary requires $\epsilon_M(\omega) = -\epsilon_0$, so that, by Eq. (13.136), the frequency of the interface plasma oscillation, $\omega_s$, will be $\omega_P/\sqrt{2}$.

**13.4.** Consider now the interface between an insulator (for $z < 0$) and a semiconductor ($z \geq 0$) which at $z = 0$ behaves like a two-dimensional electron gas. The dielectric function of the insulator can be written as:

$$\epsilon_i(\omega) = \epsilon_{i\infty} + (\epsilon_{i0} - \epsilon_{i\infty}) \frac{\omega^{(TO)2}}{\omega^{(TO)2} - \omega^2} . \tag{13.139}$$

The dielectric function of the 2D electron gas can be written as:

$$\epsilon_{2D}(Q,\omega) = \epsilon_\infty \left(1 - \frac{\omega_P^2(Q)}{\omega^2}\right) , \tag{13.140}$$

where $Q$ is the 2D wavevector on the plane of the interface, and $\omega_P^2(Q) = e^2 n_{2D} Q/(\epsilon_\infty m^*)$ is the frequency of the two-dimensional plasma of density $n_{2D}$ (in units of carriers per unit area) which depends on $Q$ (unlike the bulk plasma). Following the same procedure of Problem 13.1. (that is: expressing the potential in the insulator and semiconductor with Eqs. (13.137) and (13.138) and imposing the continuity of $E_x$ and $D_z = \epsilon(\omega)E_z$ across the interface) find the dispersion of the coupled phonon-plasmon modes at the interface as a function of $Q$. Note that you should get something resembling the results shown in Fig. 13.20, page 311, but now $Q$ plays the role of the density $n$ in that figure.

## References


1. J. Ziman, *Electrons and Phonons: The Theory of Transport Phenomena in Solids* (Oxford University Press, Oxford, 1960)
2. J. Bardeen, W. Shockley, Deformation potentials and mobilities in non-polar crystals. Phys. Rev. **80**, 72 (1950)
3. M.V. Fischetti, J.M. Higman, Theory and calculation of the deformation potential electron-phonon scattering rates in semiconductors, in *Monte Carlo Device Simulation: Full Band and Beyond*, ed. by K. Hess (Kluwer Academic, Norwell, 1991), pp. 123–160

4. P.D. Yoder, J.M. Higman, J.D. Bude, K. Hess, Monte Carlo simulation of hot electron transport in Si using a unified pseudopotential description of the crystal. Semicond. Sci. Techn. **7**, B357–B359 (1992)
5. M.V. Fischetti, S.E. Laux, Monte Carlo analysis of electron transport in small semiconductor devices including band-structure and space-charge effects. Phys. Rev. B **38**, 9721–9745 (1988)
6. T. Kunikiyo, M. Takenaka, Y. Kamakura, M. Yamaji, H. Mizuno, M. Morifuji, K. Taniguchi, C. Hamaguchi, A Monte Carlo simulation of anisotropic electron transport in silicon including full band structure and anisotropic impact-ionization model. J. Appl. Phys. **75**, 297–312 (1994)
7. J.Y.-F. Tang, K. Hess, Impact ionization of electrons in silicon (steady state). J. Appl. Phys. **54**, 5139–5144 (1983)
8. C. Herring, E. Vogt, Transport and deformation-potential theory for many-valley semiconductors with anisotropic scattering. Phys. Rev. **101**, 944 (1956)
9. W. Harrison, *Electronic Structure and the Properties of Solids: The Physics of the Chemical Bond* (Dover, New York, 1989)
10. W.G. Vandenberghe, M.V. Fischetti, Deformation potentials for band-to-band tunneling in silicon and germanium from first principles. Appl. Phys. Lett. **106**, 013505 (2015)
11. M. Cardona, F.H. Pollak, Energy-band structure of Germanium and Silicon: the k·p method. Phys. Rev. **142**, 530 (1996)
12. M.S. Dresselhaus, G. Dresselhaus, A. Jorio, *Group Theory: Application to the Physics of Condensed Matter* (Springer, Berlin/Heidelberg, 2008)
13. H. Fröhlich, Electrons in lattice fields. Adv. Phys. **3**, 325 (1954)
14. W. Fawcett, A.D. Boardman, S. Swain, Monte Carlo determination of electron transport properties in gallium arsenide. J. Phys. Chem. Solids **31**, 1963–1990 (1970)
15. M.V. Fischetti, B. Fu, S. Narayanan, J. Kim, Semiclassical and quantum electronic transport in nanometer-scale structures: empirical pseudopotential band structure, Monte Carlo simulations and Pauli master equation, in *Nano-Electronic Devices: Semiclassical and Quantum Transport Modeling*, ed. by D. Vasileska, S.M. Goodnick (Springer, New York, 2011), pp. 183–247
16. S.B. Zhang, C.-Y. Yeh, A. Zunger, Electronic structure of semiconductor quantum films. Phys. Rev. B **48**, 11204 (1993)
17. D. Esseni, P. Palestri, Linear combination of bulk bands method for investigating the low-dimensional electron gas in nanostructured devices. Phys. Rev. B **72**, 165342 (2005)
18. M.V. Fischetti, S.E. Laux, Monte Carlo study of electron transport in silicon inversion layers. Phys. Rev. B **48**, 2244 (1993)
19. M.V. Fischetti, J. Kim, S. Narayanan, Z.-Y. Ong, C. Sachs, D.K. Ferry, S.J. Aboud, Pseudopotential-based studies of electron transport in graphene and graphene nanoribbons. J. Phys. Condens. Matter **25**, 473202 (2013)
20. B. Wunsch, T. Stauber, F. Sols, F. Guinea, Dynamical polarization of graphene at finite doping. New J. Phys. **8**, 318 (2006)
21. J. Kim, M.V. Fischetti, S. Aboud, Structural, electronic, and transport properties of silicane nanoribbons. Phys. Rev. B **86**, 205323 (2012)
22. B.K. Ridley, *Quantum Processes in Semiconductors* (Oxford University Press, Oxford, 2000)

# Chapter 14
# Scattering with Ionized Impurities

## 14.1 Overview

So far, we have considered almost exclusively ideal crystals, looking at their properties in their intrinsic form. Therefore, we have intentionally avoided the intricate task of dealing with defects and imperfections and how these may affect the electronic properties and transport. This is not to say that deviations from ideality are not important: quite the contrary. They have governed the progress of the microelectronics industry, the main concern behind this text, and they influence even more heavily other practical applications of physics and material science: Just think of metallurgy or polymer science and their applications to the steel or plastic industry. However, we have to limit the scope of our discussion. One exception to the "ideal crystal" philosophy we have followed so far is represented by dopants in semiconductors. They may be arguably considered one of their "intrinsic" features, given their importance. Therefore, in this chapter we shall look, albeit a bit too cursorily (this is one of the shortest chapters of the book), at the interaction between free carriers and ionized impurities.

In most applications, electrons and holes must carry current in the presence of a large concentration of dopants. By design, these impurities are ionized, so that the free carriers resulting from the ionization can contribute to the current-carrying process. Therefore, each impurity gives rise to a Coulomb field which, in the absence of screening, is long-range. The motion of the charge carriers, electrons and holes, will obviously be affected by this field. Here, we evaluate the scattering rate between electrons and the field due to these ionized impurities. A comprehensive, albeit a bit dated, discussion of the subject can be found in [1].

## 14.2 Electron-Impurity Hamiltonian

Assume that we have $N_{\rm D}$ ionized impurities in a volume $\Omega$ and that they are located at positions $\mathbf{r}_i$ for $i = 1, N_{\rm D}$. If $V^{(\rm imp)}(\mathbf{r}-\mathbf{r}_i)$ is the potential due to impurity $i$, then the Hamiltonian expressing the electron–impurity interactions will be, obviously,

$$\widehat{H}_{\rm imp} = \frac{e}{\Omega} \sum_{i=1}^{N_{\rm D}} \int \mathrm{d}\mathbf{r}\ \widehat{\psi}^{\dagger}(\mathbf{r})\ V^{(\rm imp)}(\mathbf{r}-\mathbf{r}_i)\ \widehat{\psi}(\mathbf{r})\ . \tag{14.1}$$

Before proceeding, we must pause and discuss a major difference between electron–phonon and electron-impurity scattering. Scattering with phonons—as with any other excitation with internal



M. Fischetti, W.G. Vandenberghe, *Advanced Physics of Electron Transport in Semiconductors and Nanostructures*, Graduate Texts in Physics, DOI 10.1007/978-3-319-01101-1_14

degrees of freedom—is inherently "phase-breaking." In other words, it is irreversible in time, as a single collision removes from the system information about its phase: We assume that the thermal bath to which phonons belong is so large (that is, it has so many degrees of freedom) that we cannot possibly keep track of the information carried by a single phonon. This information is lost—either because of our inability to record this information or because of the fast "decoherence" or loss of information intrinsic to any large system—so that the system behaves irreversibly, like any thermodynamic system. Impurities, on the other hand, do not carry any internal degree of freedom: The momentum transferred by an electron to the impurity is extremely small: Recall that each impurity is strongly bonded to the whole lattice, so that the momentum transferred to the lattice is of the order of $m_{\rm el}/M_{\rm lattice} \approx 0$. Thus, we can consider the impurity as a fixed (in space and time) potential. It follows that, in principle, we could solve the exact Schrödinger equation in the presence of the potential due to all of the impurities and we would have a perfectly reversible, non-dissipative system. In other words, we could lump the Hamiltonian (14.1) into the unperturbed Hamiltonian and stop talking about impurity scattering altogether. The effect of the presence of the impurities in transport would be that of modifying the wavefunctions, and so the transmission amplitude across the sample, via the coherent interaction between the electron wavefunction and the impurity potential.

However, some sort of loss of information is likely to arise from another consideration. In writing Eq. (14.1) we have assumed that the positions of all the impurities are precisely known. In experimental situations this is clearly not true: One must regard experiments on charge transport in a given sample as an ensemble average over many configurations $\alpha$, each configuration corresponding to a different set of random (but still yielding the same average impurity density) impurity positions, $\{\mathbf{r}_i\}_\alpha$. One may regard this either as an averaging over many different, equally "prepared" samples or as an average over different (microscopically large but macroscopically small) parts of the sample. In doing this average, all interference among partial waves scattered from different impurities is lost (or "averaged out"), as in the usual random phase approximation. Thus we are brought to an average expression for Eq. (14.1):

$$\langle \widehat{H}_{\rm imp} \rangle_{\rm config} = \frac{eN_{\rm D}}{\Omega} \int d\mathbf{r}\, \widehat{\psi}^\dagger(\mathbf{r})\, V^{(\rm imp)}(\mathbf{r})\, \widehat{\psi}(\mathbf{r}) = n_{\rm D} \int d\mathbf{r}\, \widehat{\psi}^\dagger(\mathbf{r})\, eV(\mathbf{r})\, \widehat{\psi}(\mathbf{r})\,, \tag{14.2}$$

where $n_{\rm D}$ is the average impurity concentration.

By now it should be clear how to calculate the scattering rate for the electron–impurity interaction. This seems to be a simple matter of inserting the Fourier components, $V_{\mathbf{q}}^{(\rm imp)}$, into Eqs. (12.10) or (12.11),

$$\frac{1}{\tau^{(\rm imp)}(\mathbf{k},n)} = \frac{2\pi}{\hbar} \sum_{\mathbf{k}'n'} \sum_{\mathbf{G}} |V^{(\rm imp)}_{\mathbf{k}-\mathbf{k}'+\mathbf{G}}\, \mathscr{I}^{(3D)}_{\mathbf{k},\mathbf{k}',n,n';\mathbf{G}}|^2\, \delta[E_n(\mathbf{k}) - E_{n'}(\mathbf{k}')] \tag{14.3}$$

and proceed as usual.

Things are more interesting than this. First, the scattering potential, $V^{(\rm imp)}(\mathbf{r})$, cannot be taken as the bare Coulomb potential of the ionized impurity: Both valence electrons and conduction electrons, when present in significant concentrations, will screen the potential. This actually helps, since the Coulomb potential has an infinite range and it is well known that the cross section for scattering with a bare Coulomb potential diverges. Moreover, when the impurities are sufficiently diluted, we can consider them as isolated and screen them one at a time with whatever model of static dielectric screening we choose from Chap. 11. However, as the concentration of the impurity grows (around the high $10^{18}$ impurities/cm$^3$ in Si, for example), this approach begins to fail, as discussed below in Sect. 14.5. At high impurity concentrations additional problems arise: Impurities may cluster as the solid-solubility limit is reached, which is the case in Si for impurity concentrations well above $10^{20}$/cm$^3$. The dopant atoms are also close enough for the bound states in the impurity potential, quite shallow (a few tens of meV), to delocalize, as electrons can tunnel among donor impurities, thus forming *impurity bands* that distort the band structure near the edge of the conduction band (or the valence band, for acceptors). Moreover, in Appendix A we see that the first Born approximation is valid only when the scattered wave is weak and does not affect significantly the incident wave. When impurities are close together, interference among the many waves scattered by the many impurities renders this assumption invalid. The Born approximation is also valid for electrons at an energy sufficiently large for their wavevector to satisfy the condition $k >> (2m^* U_0)/(\beta \hbar^2)$, where $m^*$ is the electron effective mass, $\beta$ is the screening parameter. The energy $U_0$ is some measure of the depth of the scattering "well" that, in our case, is the energy of the impurity states (only a few tens of meV, as just mentioned). It follows that the Born approximation fails for those low-energy electrons in which we are most interested when we wish to calculate their mobility, for example. Finally, we should recall once more the aforementioned issue about the "averaging" we need to perform over the spatial distribution of the impurities in order to reach the regime of dissipative semiclassical transport.

In this chapter we shall employ the first Born approximation: We are interested in the effect that ionized impurities have on electron transport when they affect it significantly. This happens at relatively high concentrations of impurities (donors/acceptors) and, usually by charge neutrality, of electrons/holes. The large density of free carriers will result in very localized impurity potentials, so that electrons can be considered as having a sufficiently "high-energy" even at thermal equilibrium. We shall consider a spatial distribution of impurities already averaged. As just mentioned, a discussion is given in Sect. 18.2.1. We shall discuss here mainly how to screen the impurity potentials and we will address briefly the other remaining concerns. We shall consider explicit cases within the effective-mass approximation. The main issue is the physics of the collision process itself and on the nature of the potential. Once this is understood, inserting the appropriate expression for $V_{\mathbf{q}}^{(\mathrm{imp})}$ into Eq. (14.3) is a straightforward matter of following the recipe given in Chap. 12.

## 14.3 Electron-Impurity Scattering Rates

To start, let's consider the impurity potential $V^{(\mathrm{imp})}(\mathbf{r})$ as the bare Coulomb potential

$$eV(r) = \frac{Ze^2}{4\pi\epsilon_s r} , \tag{14.4}$$

where $Z$ is the ionization of the impurity (typically $Z$=1) and $\epsilon_s$ is the static, valence band dielectric constant of the semiconductor. The Fourier components of $V^{(\mathrm{imp})}$ are

$$eV_q = \frac{Ze^2}{\epsilon_s q^2} . \tag{14.5}$$

Now we can follow the path we've followed in Eqs. (12.5) and (12.6) to obtain (dropping the symbol of ensemble average, we shall assume as implicitly taken in the following):

$$\widehat{H}_{\mathrm{imp}} = e\, n_{\mathrm{D}} \sum_{\mathbf{kqG}} \mathscr{I}_{\mathbf{k+q+G;k}}\, \widehat{c}^{\dagger}_{\mathbf{k+q+G}}\, V_q^{(\mathrm{imp})}\, \widehat{c}_{\mathbf{k}} , \tag{14.6}$$

having assumed that only intra-band processes are significant. This is indeed true, since the Coulomb potential is not sufficiently strong to trigger interband processes. Assuming *Normal* processes and ignoring the overlap factor, we may write more simply:

$$\widehat{H}_{\mathrm{imp}} \approx e\, n_{\mathrm{D}} \sum_{\mathbf{kq}} \widehat{c}^{\dagger}_{\mathbf{k}+\mathbf{q}}\, V_q^{(\mathrm{imp})}\, \widehat{c}_{\mathbf{k}} \,. \tag{14.7}$$

We are now ready to calculate the scattering rate:

$$\frac{1}{\tau^{(\mathrm{imp})}(\mathbf{k})} = n_{\mathrm{D}} \frac{2\pi}{\hbar} \sum_{\mathbf{q}} | < \mathbf{k}+\mathbf{q}|\widehat{H}_{\mathrm{imp}}|\mathbf{k} > |^2\, \delta[E(\mathbf{k}+\mathbf{q}) - E(\mathbf{k})] \tag{14.8}$$

$$= \frac{Z^2 e^4 n_{\mathrm{D}}}{4\pi^2 \epsilon_{\mathrm{s}}^2 \hbar} \int \mathrm{d}\mathbf{q}\, \frac{\delta[E(\mathbf{k}+\mathbf{q}) - E(\mathbf{k})]}{q^4} \tag{14.9}$$

$$= \frac{Z^2 e^4 n_{\mathrm{D}}}{2\pi \epsilon_{\mathrm{s}}^2 \hbar} \int_0^{\infty} \frac{\mathrm{d}q}{q^2} \int_0^{\pi} \mathrm{d}\theta \sin\theta\, \delta\left(\frac{\hbar^2 q^2}{2m^*} + \frac{\hbar^2 kq}{m^*}\cos\theta\right) . \tag{14.10}$$

Now, with the usual change of integration variable $x = (\hbar^2 kq/m^*)\cos\theta$ inside the angular integral we have

$$\int_0^{\infty} \frac{\mathrm{d}q}{q^2} \int_0^{\pi} \mathrm{d}\theta \sin\theta\, \delta\left(\frac{\hbar^2 q^2}{2m^*} + \frac{\hbar^2 kq}{m^*}\cos\theta\right) = \frac{m^*}{\hbar^2 k} \int_0^{2k} \frac{\mathrm{d}q}{q^3} \,. \tag{14.11}$$

The upper integration limit $2k$ results from the fact that we require the argument of the $\delta$-function to vanish in order to have a nonzero contribution. This happens only when $\cos\theta \leq 1$ or $q \leq 2k$, so that

$$\frac{1}{\tau^{(\mathrm{imp})}(\mathbf{k})} = \frac{Z^2 e^4 m^* n_{\mathrm{D}}}{2\pi \epsilon_{\mathrm{s}}^2 \hbar^3 k} \int_0^{2k} \frac{\mathrm{d}q}{q^3} = \frac{Z^2 e^4 m^{*1/2} n_{\mathrm{D}}}{2^{3/2} \pi \hbar^2 \epsilon_{\mathrm{s}}^2} \frac{1}{E^{1/2}} \int_0^{2k} \frac{\mathrm{d}q}{q^3} \,. \tag{14.12}$$

The obvious problem with this expression lies in the fact that the integral diverges. This is a well-known result: The quantum-mechanical cross section for scattering with a bare Coulomb potential diverges as a result of the infinite range of the potential. Clearly, we must consider the effect of dielectric screening of the bare potential of the impurities. Several models have been proposed in the past, each with its own merits and especially accurate in a limited range of impurity densities.

## 14.4 The Brooks–Herring Model

The first model we consider is based on the simple assumption that the impurity potential, Eq. (14.4), should be screened statically by the static free-carrier dielectric function with screening wavevector $\beta$ given by Eq. (11.50) or (11.52) [see also Eqs. (11.97) and (11.99)]. Thus [2, 3]:

$$eV_q = \frac{Ze^2}{\epsilon_{\mathrm{s}}(q^2 + \beta^2)} \,. \tag{14.13}$$

Equation (14.10) then becomes

$$\frac{1}{\tau_{\mathrm{BH}}(\mathbf{k})} = \frac{Z^2 e^4 n_{\mathrm{D}}}{2\pi \epsilon_{\mathrm{s}}^2 \hbar} \int_0^{\infty} \mathrm{d}q\, \frac{q^2}{[q^2 + \beta^2]^2} \int_0^{\pi} \mathrm{d}\theta \sin\theta\, \delta\left(\frac{\hbar^2 q^2}{2m^*} + \frac{\hbar^2 kq}{m^*}\cos\theta\right)$$

$$= \frac{2Z^2 e^4 m^* n_{\mathrm{D}}}{\pi \hbar^3 \epsilon_{\mathrm{s}}^2} \frac{k}{\beta^2(\beta^2 + 4k^2)} \,. \tag{14.14}$$

In our context, this form was first proposed by Brooks and Herring, hence the subscript BH in the formula above. Note that screening, via the term $\beta$, removes the divergence of the integral at $q \to 0$, that is, at long distances. Indeed screening makes the Coulomb potential a "short range" potential. The source of the divergence we have met before was due to the fact that electrons feel the potential of the impurity at all distances. Screening eliminates this problem.

## 14.5 The Conwell–Weisskopf Model

Mahan, however, has noticed an inconsistency in the way screening is employed in the Brooks–Herring approach [4]: On the one hand, in deriving the scattering rate (14.14), we assume that electrons are plane waves: Indeed, our initial and final states are pure plane waves and all electrons participate equally in the scattering process. On the other hand, we also assume that all electrons participate fully in the screening process. We are clearly double-counting the electrons. Having observed this inconsistency, Mahan proceeded with variational calculations trying to determine the actual electron wavefunctions in a homogeneous electron system in the presence of ionized impurities. The results showed that at small densities electrons do indeed "pile up" around the attractive potentials of donors (or away from the repulsive potential of acceptors), as implied by Debye–Hückel screening. But at larger densities, when the inter-impurity separations, the inter-electron separation, and the screening length $\beta^{-1}$ become comparable (about 1–3 nm in Si at densities of the order of $10^{18}$ to $10^{19}/\text{cm}^3$), the electron wavefunctions are spread almost uniformly, without "piling up" around the impurities: The average distance between the impurities is now smaller than the electron wavelength, so that these are not able to screen the potential. This effect is partially accounted for by the $q$-dependence of $\beta$ in the better model of Eqs. (11.63) and (11.64) on page 236. Therefore, at large densities it makes more sense to go back to the original model proposed by Conwell and Weisskopf [5]. They assumed that the Coulomb potential is unscreened. However, electrons can interact only with the nearest impurity, an assumption consistent with our earlier assumption of independent collisions with impurities implicit in Eq. (14.2). In order to see the implications of this model, it is convenient to define the average inter-impurity separation $r_0$ such that

$$n_{\rm D} = \frac{1}{4\pi r_0^3/3} \quad \to \quad r_0 = \left(\frac{3}{4\pi n_{\rm D}}\right)^{1/3} . \tag{14.15}$$

This will be the maximum "impact parameter" of the collision. The Fourier components of the Coulomb potential so "chopped" at large distances will be

$$V_q^{(\rm CW)} = \frac{Ze^2}{\epsilon_{\rm s} q} \int_0^{r_0} {\rm d}r \, \sin(qr) = \frac{Ze^2}{\epsilon_{\rm s} q^2} \left[1 - \cos(qr_0)\right] . \tag{14.16}$$

Thus:

$$\frac{1}{\tau_{\rm CW}(\mathbf{k})} = \frac{Z^2 e^4}{2\pi\hbar\epsilon_{\rm s}^2} \int_0^\infty {\rm d}q \, \frac{1}{q^2} \left[1 - \cos(qr_0)\right]^2 \int_0^\pi {\rm d}\theta \sin\theta \, \delta\left(\frac{\hbar^2 q^2}{2m^*} + \frac{\hbar^2 kq}{m^*}\cos\theta\right) . \tag{14.17}$$

Proceeding as usual with the azimuthal integral, we finally obtain

$$\frac{1}{\tau_{\rm CW}(\mathbf{k})} = \frac{Z^2 e^4 m^* n_{\rm D} r_0^3}{2\pi\hbar^3 \epsilon_{\rm s}^2} \mathscr{F}(2kr_0) , \tag{14.18}$$

where

$$\mathscr{F}(x) = \frac{1}{x}\int_0^x \mathrm{d}t\, \frac{(1-\cos t)^2}{t^3} .$$

Note that the scattering rate (14.18) does not depend on the impurity density $n_{\mathrm{D}}$, as can be seen from Eq. (14.15). There are several variations of these expressions, one such variation relying on the use of the bare Coulomb potential at infinity, but "chopping" the integration over $q$ above. However, the essence of the method remains unaltered: Rather than relying on dielectric screening to avoid the divergence at $q \to 0$ in Eq. (14.11), the cutoff at small $q$ is obtained from the maximum impact parameter allowed. If the impact parameter exceeds $r_0$ (however this might be defined), then another impurity will exhibit a smaller impact parameter and scattering will occur via that second impurity center.

## 14.6 Ridley's Statistical Screening Model

Ridley [6] (see also Ridley's text [7], pp. 148–152) has attempted to find a model reconciling the BH and CW approaches. He notices that in the CW approach the impurity with the smallest impact parameter is always chosen, even if it happens to be farther away from another impurity with a larger impact parameter. This is inconsistent with a true two-body, single-scattering-center collision. This leads us to consider the probability that another scattering center may be closer, even if it happens to be associated with a larger impact parameter. This method (which we shall not discuss further and is known as "Ridely's statistical screening" [6]) results in a correction to the BH scattering rate as follows

$$\frac{1}{\tau_{\mathrm{R}}(\mathbf{k})} = \frac{\upsilon(\mathbf{k})}{d}\left\{1-\exp\left[\frac{d}{\tau_{\mathrm{BH}}(\mathbf{k})\upsilon(\mathbf{k})}\right]\right\} , \qquad (14.19)$$

where $\upsilon(\mathbf{k})$ is the group velocity and $d = (2\pi n_{\mathrm{D}})^{-1/3}$ is the average inter-impurity distance. This model has been applied successfully to calculate the mobility in III-V hetero-structures. In the so-called high electron mobility transistors (HEMTs), electrons are induced in a narrow-gap channel (InGaAs, for example) by inserting a very narrow line of dopants in the larger-gap AlGaAs layer that acts as a gate insulator. The process of introducing this single-atom line of donor dopants, called "delta-doping," is easily performed during the MBE growth of the hetero-structure. Electrons resulting from these donors will transfer from the AlGaAs layer to the InGaAs channel and contribute to the conduction. Scattering now occurs between electrons in the channel and donors a few nanometers away in the adjacent AlGaAs layer. This process, called *remote impurity scattering*, is clearly strongly affected by Ridley's statistical screening.

## 14.7 *Additional Corrections to the Impurity Potential

There are several, even more sophisticated, models which attempt to find a better approximation to the potential of the substitutional impurity. An article by Chattopadhaya and Qeisser [1] discusses the state of the art. Since impurity scattering is important in determining the carrier mobility in heavily doped semiconductors, and since the average doping of semiconductors increases with the shrinking dimensions of the devices, it is easy to understand why there is such an interest. See [8].

A first correction is an attempt to account for wavelength-dependent screening by valence electrons. So far, screening by valence electrons has been treated simply via the static dielectric constant $\epsilon_s$, but some dependence on the wavelength may be included. Here we follow almost verbatim the discussion presented in [8], consider the case of donors in Si, and start by following Nara and Morita [9]. They consider the bare core potential, $V(Z_I, n; \mathbf{r})$, of a substitutional donor $n$-fold ionized, taken isotropic since only core electrons are considered. They express this as

$$V(Z_I, n; \mathbf{r}) = -\frac{e^2}{4\pi\epsilon_0 r}\left[n + (Z_I - n)\, e^{-\sigma_I r}\right], \tag{14.20}$$

where $\epsilon_0$ is the permittivity of vacuum and the atomic radii $1/\sigma_I$ are obtained from a fit to the Hartree–Fock potential. Their values are of the order of 0.01 nm for the species usually considered ($1/\sigma_I$ = 0.0124 nm for $Si^{4+}$, 0.0111 nm for $P^{5+}$, 0.0148 nm for $As^{5+}$, and 0.0156 nm for $Sb^{5+}$), consistent with the use of the vacuum permittivity $\epsilon_0$ in Eq. (14.20).

Valence electrons are now brought in first as frozen in their anisotropic, perfect-lattice distribution. Thus, the net potential of the substitutional impurity, $v_i(\mathbf{r})$, is obtained by replacing the $Si^{4+}$ ion with the $n$-fold ionized donor, that is

$$v_{imp}(\mathbf{r}) = V(Z_I, n; \mathbf{r}) - V(Z_{Si}, 4; \mathbf{r}) . \tag{14.21}$$

Finally, in the linear-response approximation, the valence electrons are allowed to readjust and screen $v_{imp}(\mathbf{r})$ via the static valence dielectric function of Si, $\epsilon_v(\mathbf{q})$. A possible model for $\epsilon_v(\mathbf{q})$—by no means unique, but approximated by a convenient analytic expression—has the following isotropic form:

$$\frac{\epsilon_0}{\epsilon_v(q)} = \frac{Aq^2}{q^2 + a^2} + \frac{Bq^2}{q^2 + b^2} + \frac{Cc^2}{q^2 + c^2}, \tag{14.22}$$

where $A = 1.175$, $B = -0.175$, $C = \epsilon_0/\epsilon_s \approx 0.08547$, $a = 0.7572$, $b = 0.3123$, and $c = 2.044$, $a$, $b$, and $c$ being in atomic units. Finally, the Fourier components of the screened isotropic impurity potential, $V^{(imp)}(q)$, become

$$V^{(imp)}(q) = \frac{e^2}{\epsilon_v(q)} \frac{H_I(q)}{q^2}, \tag{14.23}$$

where

$$H_I(q) = (n-4) + (Z_I - n)\frac{q^2}{q^2 + \sigma_I^2} - (Z_{Si} - 4)\frac{q^2}{q^2 + \sigma_{Si}^2}, \tag{14.24}$$

where $n = 3$ for acceptors, $n = 5$ for donors. Equation (14.23) should replace the Fourier transform of the impurity potential we have used before in Eq. (14.13) or, appropriately "chopped" at large distances, in Eq. (14.16).

In practice, these corrections do not matter too much quantitatively. Only at very large energies, when the electron approaches the impurities at very short distances (large $q$) one sees some differences between different donor or acceptor species, because of some differences in the core potentials of different ions. These are known as "central cell" corrections.

Friedel's sum rule [10] (discussed in Chap. 11, Sect. 11.3.2.3), as well as partial-waves approaches [11, 12], can tackle another interesting problem. The Born approximation we have used so far does not distinguish between repulsive and attractive potentials. Yet, we expect some differences: In the case of attractive potentials (electron-donors) the electronic wavefunction gets closer to the impurity core than in the case of repulsive potentials (electron-acceptors). Partial-wave analysis (which we saw above dealing with Friedel's sum rule, page 243) that goes beyond the first-order Born approximation, can account for this difference. The practical interest lies in the context of transport of electrons (holes) in the $p$-type ($n$-type) base of $n-p-n$ ($p-n-p$) bipolar transistors. These so-called minority carrier mobilities are indeed experimentally observed to be different (usually larger, as expected) than the majority carrier mobilities. Finally, note that connected to Friedel's sum rule, that essentially forces us to "count" electrons correctly, is the "double counting" problem noticed by Mahan [4] that we have already discussed.

## 14.8 *Electron-Impurity Scattering for a 2DEG

As we saw in Chap. 13, when dealing with scattering in low dimensions, Eq. (12.36) still requires the numerical evaluation of a double integral: First over the $z$-component of the momentum transfer, $q_z$, then over the $z$-coordinate, hidden within the overlap factor $\mathscr{I}^{(2D)}_{\mathbf{K},\mathbf{K}_j,n,n';\mathbf{G}_\parallel}(q_z)$. The latter integration cannot be reduced to any closed-form expression since the wavefunctions are known only numerically. As we saw in the case of the electron–phonon interaction, in some special cases the integration over $q_z$ can be performed analytically. So, for example, the term corresponding to Eq. (13.107) becomes

$$V_{\mathbf{q}}^{(\mathrm{imp,DH})} = \frac{e^2}{\epsilon_s(q^2+\beta_{\mathrm{DH}}^2)} = \frac{C_{\mathrm{imp}}}{q^2+\beta_{\mathrm{DH}}^2} , \tag{14.25}$$

where:

$$C_{\mathrm{imp}}^2 \int \mathrm{d}z \int \mathrm{d}z'\, \zeta_{\mathbf{K}'}^{(n')*}(z)\, \zeta_{\mathbf{K}'}^{(n')}(z')\, \zeta_{\mathbf{K}}^{(n)}(z)\, \zeta_{\mathbf{K}}^{(n)*}(z') \int \frac{\mathrm{d}q_z}{2\pi} \frac{\mathrm{e}^{\mathrm{i}q_z(z-z')}}{(q_z^2+Q^2)^2} . \tag{14.26}$$

The evaluation of the integral over $q_z$ is a bit more involved. Let's write this integral as

$$\mathscr{I}^{(2D)}_{\mathrm{imp}} = \frac{1}{2\pi} \int \mathrm{d}x \, \frac{\mathrm{e}^{ipx}}{(x^2+Q^2)^2} , \tag{14.27}$$

having set $p=z-z'$ and having renamed $x$ the dummy integration variable $q_z$. Let's consider the case $p>0$ and let's integrate by parts:

$$\mathscr{I}^{(2D)}_{\mathrm{imp}} = \frac{1}{2\pi} \int \mathrm{d}x \, \frac{1}{(x-\mathrm{i}Q)^2} \frac{\mathrm{e}^{ipx}}{(x-\mathrm{i}Q)^2} = \frac{1}{2\pi} \frac{-1}{x-\mathrm{i}Q} \frac{\mathrm{e}^{ipx}}{(x+\mathrm{i}Q)^2} \Bigg|_{-\infty}^{\infty}$$
$$+ \frac{1}{2\pi} \int \mathrm{d}x \, \frac{1}{x-iQ} \frac{\mathrm{d}}{\mathrm{d}x} \left[ \frac{\mathrm{e}^{ipx}}{(x+iQ)^2} \right] . \tag{14.28}$$

The first term on the right-hand side vanishes and we are left with an integrand which for $\mathrm{Im}(x)>0$ now has as singularity only a pole at $x=iQ$:

$$\mathscr{I}^{(2D)}_{\mathrm{imp}} = \frac{1}{2\pi} \int \mathrm{d}x \, \frac{\mathrm{e}^{ipx}}{x-\mathrm{i}Q} \frac{ip(x+\mathrm{i}Q)-2}{(x+\mathrm{i}Q)^3} . \tag{14.29}$$

Since we have assumed $p>0$, we can integrate over the upper half of the complex plane enclosing the single pole $x=iQ$, and obtain

$$\mathscr{I}_{\rm imp}^{(2D)} = \frac{1}{4} \frac{e^{-pQ}}{Q^3} (pQ+1) \, . \tag{14.30}$$

The case $p < 0$ can be treated similarly by integrating over the lower half of the complex plane. Thus, we can express the impurity scattering rate as

$$\frac{1}{\tau_n^{(\rm imp)}(\mathbf{K})} \approx N_{\rm imp} \frac{2\pi}{\hbar} \sum_{jn'}' C_{\rm imp}^2 \frac{1}{(2\pi)^2} \frac{L(w_{jn'})}{|\nabla_{2D} E_{jn'}|} \frac{1}{4Q_j^3}$$
$$\times \int {\rm d}z \int {\rm d}z' \, \zeta_{\mathbf{K}_j}^{(n')*}(z) \, \zeta_{\mathbf{K}_j}^{(n')}(z') \, e^{-|z-z'|Q_j} \left[|z-z'|Q_j + 1\right] \zeta_{\mathbf{K}}^{(n)}(z) \, \zeta_{\mathbf{K}}^{(n)*}(z') \, , \tag{14.31}$$

where $N_{\rm imp}$ is the concentration of impurities. The overlap factor entering the last equation can be split into a term identical to the factor appearing in Eq. (13.110) – which can be recast in the form of Eq. (13.111)—and another term,

$$\int {\rm d}z \int {\rm d}z' \, \zeta_{\mathbf{K}_j}^{(n')*}(z) \, \zeta_{\mathbf{K}_j}^{(n')}(z') \, e^{-|z-z'|Q_j} \, |z-z'|Q_j \, \zeta_{\mathbf{K}}^{(n)}(z) \, \zeta_{\mathbf{K}}^{(n)*}(z'), \tag{14.32}$$

which can be rewritten as

$$Q_j \int {\rm d}z \int {\rm d}z' \, g^*_{\mathbf{K},\mathbf{K}',n,n'}(z') \, e^{-|z-z'|Q_j} \, |z-z'| \, g_{\mathbf{K},\mathbf{K}',n,n'}(z)$$
$$= 2\, Q_j \, {\rm Re}\left\{ \int_0^{\infty} {\rm d}z \, g_{\mathbf{K},\mathbf{K}',n,n'}(z) \, e^{-Qz} \int_0^z {\rm d}z' \, g^*_{\mathbf{K},\mathbf{K}',n,n'}(z') \, e^{Qz'} (z-z') \right\}, \tag{14.33}$$

or:

$$= 2\, Q_j \, {\rm Re}\left\{ \int_0^{\infty} {\rm d}z \, g_{\mathbf{K},\mathbf{K}',n,n'}(z) \, z \, e^{-Qz} \int_0^z {\rm d}z' \, g^*_{\mathbf{K},\mathbf{K}',n,n'}(z') \, e^{Qz'} \right\}$$
$$- 2Q_j \, {\rm Re}\left\{ \int_0^{\infty} {\rm d}z \, g_{\mathbf{K},\mathbf{K}',n,n'}(z) \, e^{-Qz} \int_0^z {\rm d}z' \, g^*_{\mathbf{K},\mathbf{K}',n,n'}(z') \, z' \, e^{Qz'} \right\}. \tag{14.34}$$

As discussed in Chap. 13, the advantage of expressing the overlap factor in terms of Eq. (14.34) is that the inner integral over $z'$ depends on $z$ *only* via its upper integration limit. Thus, as we saw in Eq. (13.111), it can be evaluated as a discrete sum storing partial results and recalling these partial results when performing the "outer" integration over $z$. Thus, evaluating the double integral (14.34) actually requires the same number of operations required to perform two 1D integrals. In other words, the computational effort scales with $N_z$ (the number of $z$ points) rather than $N_z^2$, as it may at first appear from a look at Eq. (14.31), or $N_z \times N_{q_z}$, as in Eq. (13.99).

## 14.9 *Electron-Impurity Scattering for a 1DEG

We can obtain a similar simplification also when dealing with impurity scattering. Using Eq. (14.25) we can rewrite the term

$$\int \frac{{\rm d}\mathbf{Q}}{(2\pi)^2} \left| V^{(\rm imp)}_{\mathbf{Q},k_z-k_z'+q_z} \, \mathscr{I}^{(1D)}_{k_z,k_z',n,n'}(\mathbf{Q}) \right|^2 \tag{14.35}$$

as

$$C_{\rm imp}^2 \int d\mathbf{R} \int d\mathbf{R}'\ g_{k_z,k_z',n,n'}(\mathbf{R})\ g^*_{k_z,k_z',n,n'}(\mathbf{R}') \int \frac{d\mathbf{Q}}{(2\pi)^2} \frac{e^{i\mathbf{Q}\cdot(\mathbf{R}-\mathbf{R}')}}{(Q^2+\Delta Q^2)^2} , \tag{14.36}$$

where $\Delta Q = [(k_z - k_z')^2 + \beta_{\rm DH}^2]^{1/2}$ and

$$g_{k_z,k_z',n,n'}(\mathbf{R}) = \zeta_{k_z'}^{(2D)(n')*}(\mathbf{R})\ \zeta_{k_z}^{(2D)(n)}(\mathbf{R}) . \tag{14.37}$$

as defined in Chap. 13, after Eq. (13.118). The integral over $\mathbf{Q}$ can be rewritten as

$$\mathscr{I}_{\rm imp}^{(1D)} = \frac{1}{(2\pi)^2} \int_0^\infty dQ\ \frac{Q}{(Q^2+\Delta Q^2)^2} \int_0^{2\pi} d\phi\ e^{iQ\Delta R\cos\phi} . \tag{14.38}$$

Treating the angular integral as we did above in Eq. (13.120) we find

$$\mathscr{I}_{\rm imp}^{(1D)} = \frac{1}{2\pi} \int_0^\infty dQ\ \frac{QJ_0(Q\Delta R)}{(Q^2+\Delta Q^2)^2} . \tag{14.39}$$

Unlike what we found when dealing with Eq. (13.121), in this case there is no elegant closed-form expression for this integral. However, one can evaluate numerically, tabulate, and interpolate the function

$$K_{\rm imp}(x) = \int_0^\infty dt\ \frac{tJ_0(t)}{(t^2+x^2)^2} , \tag{14.40}$$

so that the overlap factor can be written in the compact form

$$\mathscr{I}_{\rm imp} = \frac{(\Delta R)^2}{2\pi}\ K_{\rm imp}(\Delta Q \Delta R) , \tag{14.41}$$

suitable for numerical evaluation and Eq. (12.51) can be expressed as

$$\frac{1}{\tau_n^{(\rm imp)}(k_z)} \approx \frac{N_{\rm imp} C_{\rm imp}^2}{\hbar} \sum_{jn'}{}' \frac{1}{2\pi} \left| \frac{dE_{n'}}{dk_{zj}} \right|^{-1}$$
$$\times \int d\mathbf{R} \int d\mathbf{R}'\ g_{k_z,k_{zj},n,n'}(\mathbf{R})\ g^*_{k_z,k_{zj},n,n'}(\mathbf{R}')\ K_{\rm imp}(\Delta Q_j |\mathbf{R}-\mathbf{R}'|)\ |\mathbf{R}-\mathbf{R}'|^2, \tag{14.42}$$

an expression which has the same properties (i.e., symmetry and smoothness) as Eq. (13.122). Once more, Eq. (13.124) instead of Eq. (13.119) should be used to evaluate the matrix elements above whenever the wavefunctions given by Eq. (12.44) are required.

## Problems

**14.1.** Assume a semiconductor described by a parabolic electron dispersion with an effective mass $m^*$. Plot the Brooks–Herring and the Conwell–Weisskopf electron-impurity scattering rates [Eqs. (14.14) and (14.18), respectively] as a function of electron kinetic energy in the range 0–1 eV.

Assume a density $n_d$ of $10^{16}$ donor impurities/cm$^3$. Redo the calculation assuming now a density of $10^{20}$/cm$^3$. Which model gives the largest scattering rate in each case? Can you explain your results in physical terms?

## References

1. D. Chattopadhyay, H.J. Queisser, Electron scattering by ionized impurities in semiconductors. Rev. Mod. Phys. **53**, 745 (1981)
2. H. Brooks, Scattering by ionized impurities in semiconductors. Phys. Rev. **83**, 879 (1951)
3. H. Brooks, Theory of the electrical properties of germanium and silicon. Adv. Electron. Electron Phys. **7**, 85–182 (1955)
4. G.D. Mahan, Energy gap in Si and Ge: impurity dependence. J. Appl. Phys. **51**, 2634 (1980)
5. E.M. Conwell, V.F. Weisskopf, Theory of impurity scattering in semiconductors. Phys. Rev. **77**, 388–390 (1950)
6. B.K. Ridley, Reconciliation of the Conwell-Weisskopf and Brooks-Herring formulae for charged-impurity scattering in semiconductors: third-body interference. J. Phys. C Solid State Phys. **10**, 1589 (1977)
7. B.K. Ridley, *Quantum Processes in Semiconductors* (Oxford University Press, Oxford, 2000)
8. M.V. Fischetti, S.E. Laux, Comment on "influence of the doping element on the electron mobility in n-type silicon" J. Appl. Phys. **83**, 3096 (1998); J. Appl. Phys. **85**, 7984 (1999)
9. H. Nara, A. Morita, J. Phys. Soc. Jpn. **21**, 1852 (1966)
10. J. Friedel, The distribution of electrons round impurities in monovalent metals. Philos. Mag. **43**, 153 (1952)
11. J.R. Meyer, F.J. Bartoli, Phase-shift calculation of ionized impurity scattering in semiconductors. Phys. Rev. B **23**, 5413 (1981)
12. H.S. Bennett, J.R. Lowney, Calculated majority- and minority-carrier mobilities in heavily doped silicon and comparisons with experiment. J. Appl. Phys. **71**, 2285 (1992)

# Chapter 15
# Coulomb Interactions Among Free Carriers

In Sect. 4.2.3 (pages 61 and following) we have considered a "mean field" approximation to treat the Coulomb interactions among valence electrons. The same approximation is usually embraced also when dealing with free carriers (say: electrons in the conduction band of an $n$-type doped semiconductor). This approximation is obtained by replacing the "fluctuating" interparticle Coulomb potential with a smoother Hartree potential obtained by averaging out (spatially and temporally) the instantaneous and local potential over all electron configurations. The Hartree–Fock approximation is similar, since it only adds exchange effects, but it employs the same averaging procedure. In order to re-capture the "fluctuating" nature of the interaction, we must consider the random deviations from the average ("fluctuations") of the electron positions and of the associated potential. A well-known theorem in condensed matter physics, the *fluctuation-dissipation theorem*, relates these fluctuations to the spectrum of the impedance of the system. We shall not discuss this theorem here, but it is obvious that these fluctuations will affect electronic transport.

We shall discuss first inter-carrier scattering processes with the conduction (electron–electron) or valence (hole–hole) bands, or even interband processes, in which electrons remain in the conductions bands and holes in the valence bands. We shall then consider higher-energy interband processes in which the energy exchange between two particles exceeds the energy of the band gap, so that an electron can be excited from the valence band across the gap to the conduction band (impact ionization).

Although not directly relevant to treat electronic transport, Coulomb interactions modify significantly some electronic properties of semiconductors when a very high density of free carriers is present. This is the case of very heavily doped semiconductors, in which the high density of dopants also causes interesting effects. Therefore, we shall conclude this chapter with a brief discussion of some of these effects. A correct treatment of the subject requires delving into the complicated subject of a quantum electron gas, a subject that goes beyond the scope of this text. Therefore, we shall mention a few basic concepts, thus keeping the discussion at a very basic level.

## 15.1 The Electron–Electron Hamiltonian

Consider a gas of interacting electrons in the absence of any other external perturbation. Ignoring spin, for now, the Hamiltonian has the "obvious" form:

$$\widehat{H}_{\mathrm{ee}} = \frac{1}{2} \int \mathrm{d}\mathbf{r} \int \mathrm{d}\mathbf{r}' \; \widehat{\psi}^{\dagger}(\mathbf{r}',t) \; \widehat{\psi}(\mathbf{r}',t) \; V(\mathbf{r}' - \mathbf{r}) \; \widehat{\psi}^{\dagger}(\mathbf{r},t) \; \widehat{\psi}(\mathbf{r},t) \; , \tag{15.1}$$



M. Fischetti, W.G. Vandenberghe, *Advanced Physics of Electron Transport in Semiconductors and Nanostructures*, Graduate Texts in Physics, DOI 10.1007/978-3-319-01101-1_15

where $V(\mathbf{r}) = e^2/(4\pi\epsilon_\infty r)$ is the Coulomb potential expressing the interaction among the electrons. The term $\widehat{\psi}^\dagger(\mathbf{r},t)\ \widehat{\psi}(\mathbf{r},t)$ is just the density operator which satisfies the expected property $\int d\mathbf{r}\ \widehat{\psi}^\dagger(\mathbf{r},t)\ \widehat{\psi}(\mathbf{r},t)\ = \sum_\mathbf{k} \widehat{c}_\mathbf{k}^\dagger(t)\ \widehat{c}_\mathbf{k}(t)$. Using Eqs. (9.29) and (9.30) (page 193), the Hamiltonian (15.1) can be written as

$$\widehat{H}_{ee} = \frac{1}{2}\sum_{\mathbf{k},\mathbf{p},\mathbf{q}} V_q\ \widehat{c}_{\mathbf{k}+\mathbf{q}}^\dagger\ \widehat{c}_{\mathbf{p}-\mathbf{q}}^\dagger\ \widehat{c}_\mathbf{k}\ \widehat{c}_\mathbf{p}\ , \tag{15.2}$$

where $V_q = e^2/(\epsilon_\infty q^2)$ is the Fourier transform of the Coulomb potential. We use the high-frequency dielectric response $\epsilon_\infty$ for now. We shall worry later about the response of the electrons to this potential (that is, dielectric screening). This expression has a simple interpretation: The Coulomb Hamiltonian destroys two electrons in the states **k** and **p** and creates two new electrons in states $\mathbf{k}+\mathbf{q}$ and $\mathbf{p}-\mathbf{q}$ via the exchange of momentum $\hbar\mathbf{q}$.

In the Hartree or Hartree–Fock approximation, we would look at a single electron of wavevector **k** and replace the effect of all "other" electrons by considering only the limit $q \to 0$ (that is, a spatial average over large distances), so that we would retain only the "diagonal" part of $\widehat{c}_{\mathbf{p}-\mathbf{q}}^\dagger \widehat{c}_\mathbf{p}$, that is, the number operator $\widehat{n}_\mathbf{p} = \widehat{c}_\mathbf{p}^\dagger \widehat{c}_\mathbf{p}$. We would then take a suitable ensemble average, $\langle \widehat{n}_\mathbf{p} \rangle = \mathrm{Tr}(\widehat{\rho}\widehat{n}_\mathbf{p})$, over the electron ensemble, obtaining just the average charge density $-e\langle \widehat{n}_\mathbf{p} \rangle$. This would result in the average "mean field" felt by each electron. On the contrary, now we want to isolate deviations from this mean field. Thus, we shall assume that the Hartree (or Hartree–Fock) potential is subtracted and isolated into another term $eV_{\mathrm{Hartree}}(\mathbf{r})$ in the Hamiltonian and assume that Eq. (15.2) represents only deviations from the mean field. Thus, we should consider in the Hamiltonian above the terms $\widehat{c}_{\mathbf{p}-\mathbf{q}}^\dagger \widehat{c}_\mathbf{p}$ as actually representing the "fluctuating" part $\widehat{c}_{\mathbf{p}-\mathbf{q}}^\dagger \widehat{c}_\mathbf{p} - \langle \widehat{c}_{\mathbf{p}-\mathbf{q}}^\dagger \widehat{c}_\mathbf{p} \rangle$. Similarly, the term $\widehat{c}_{\mathbf{k}+\mathbf{q}}^\dagger \widehat{c}_\mathbf{k}$ represents the fluctuating part $\widehat{c}_{\mathbf{k}+-\mathbf{q}}^\dagger \widehat{c}_\mathbf{k} - \langle \widehat{c}_{\mathbf{k}+\mathbf{q}}^\dagger \widehat{c}_\mathbf{k} \rangle$. In Chap. 18—that deals with the derivation of the Boltzmann transport equation from the quantum transport equation—we shall see how this is done "exactly" (see Sect. 18.2). For now, we shall simply consider the Hamiltonian Eq. (15.2) as expressing the effect of these fluctuations of the gas away from the mean field.

In a series of four beautiful articles published in the Physical Review in the early 1950s, David Bohm and his student David Pines studied the effects of Eq. (15.2) and reached the following general conclusions [1–4]:

When the magnitude $q$ of the momentum-transfer **q** is small, the interaction of an electron with all of the other electrons occurs via "collective excitations": The "initial" electron with wavevector **k** triggers (via emission or absorption) a plasma wave in the surrounding gas of free carriers. The interaction is an electron–plasmon collision. (Plasmons have been considered in Chap. 9, Sect. 9.3.) It may be characterized by the same procedure we have followed to study the Fröhlich interaction, simply replacing the coupling constant $(\hbar\omega_{LO}/2)[1/\epsilon_{hi} - 1/\epsilon_{low}]$ with $(\hbar\omega_P/2)(1/\epsilon_\infty)$ [see Eqs. (13.51), (13.53), (13.54) of Chap. 13.3]. Note, however, that in this process *momentum is not lost by the electron gas: It is simply re-distributed among electrons*.

On the contrary, when the magnitude $q$ of the momentum transfer grows beyond a certain limit, plasmons cease to be good excitations: They decay into single-particle excitations. Think of a surfer extracting energy from the wave as the surfer's velocity matches the (phase) velocity of the wave. Similarly, when the wavelength of the plasmon matches the Fermi wavelength, electrons at the Fermi surface will "surf" the plasmon, thus absorbing energy from it and damping it. This damping occurs via excitations of electrons from the "Fermi sea" which gain the plasma energy. This process is called *Landau damping*. Energy and momentum conservation allow this process to happen whenever the plasmon wavevector **q** satisfies the condition

$$E(\mathbf{k}_F + \mathbf{q})\ -\ E(\mathbf{k}_F) = \hbar\omega_P\ , \tag{15.3}$$

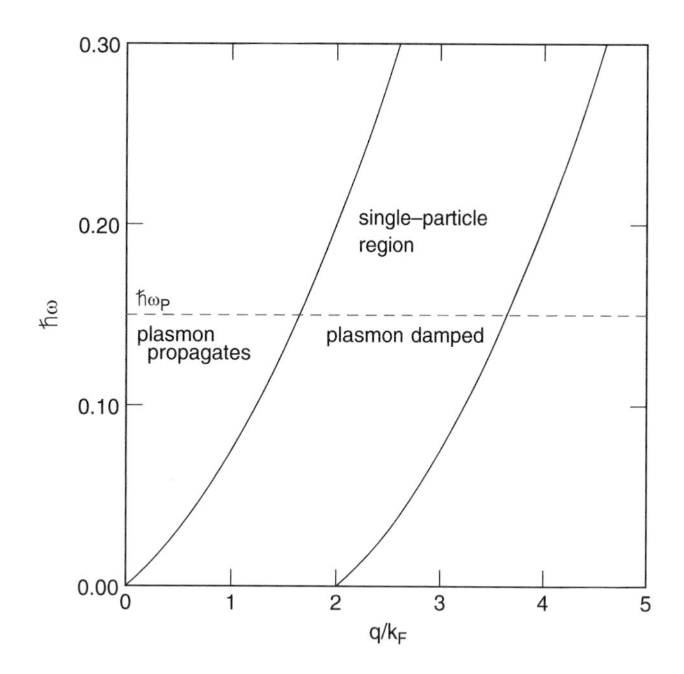


**Fig. 15.1** The two lines show the solutions of Eq. (15.4). When the gas is excited with a wavevector that lies outside the region between the two lines, the excitation is a plasmon. When wavevector lies in the region between the two lines, plasmons decay into single-particle excitations. This is the "single-particle region"

or, for parabolic and spherical bands,

$$\frac{\hbar^2}{2m^*}(q^2 - 2k_F q) \leq \hbar\omega_P \leq \frac{\hbar^2}{2m^*}(q^2 + 2k_F q) . \tag{15.4}$$

For $q$ in this range of values, the interaction between an electron and all other free electrons occurs via single-particle (two-body) collisions, as in the case of a classical collision between two charged particles.

These conclusions are illustrated in Fig. 15.1: The two lines show the solutions of Eq. (15.4). When the excitation of the electron gas occurs at a wavelength corresponding to a wavevector in the region to the upper-left (low-$q$, long-range interactions) or lower-right of the two lines, the excitation is a plasmon, i.e., a collective motion of the whole gas. But if the wavevector lies in the region between the two lines, then the velocity of the plasmon matches the Fermi velocity and plasmons decay into single-particle excitations. This is the "single-particle region."

We have already mentioned how to calculate the rate at which plasmons scatter electrons in the long-wavelength limit. Usually, this interaction has the effect of thermalizing the electrons, as all interparticle collisions do. This is a general result of detailed balance and it is discussed in Sect. 18.2.3. However, it is not a process that absorbs energy or momentum from the gas. So, in homogeneous situations, its effects are relatively minor (if we are willing to concede that thermalization is a "minor" effect). Not so in inhomogeneous situations: There, electrons that behave strongly off-equilibrium in a low-density region may scatter with plasmons that are presented elsewhere. For example, an electron in the channel of a transistor may excite plasmons in the source contact as it leaves the source itself. In this case, momentum is lost from the portion of the electron gas that carries the current and it is transferred to a region where inelastic collisions will ultimately damp the plasmon before it has the chance of "returning" the momentum to the current-carrying electrons in the channel. If we are interested exactly in studying the behavior of those low-density regions (think of the channel of a transistor), then electron–plasmon collisions can have a very strong effect on electron transport. These cases have been studied by Lugli and Ferry [5, 6], for example, and are also discussed in [7, 8]. So, we shall now consider electron–electron collisions in the single-particle region.

## 15.2 Single-Particle Short-Range Collisions

We must consider the Hamiltonian given by Eq. (15.2). Since we are considering single-electron states, we simplify the notation in the usual way, by writing the matrix elements between single-electron states, ignoring the complicated notation of states in Fock space. However, contrary to what we have done so far, we must consider initial states that refer to two-particle states. This will introduce some complications because we must consider the antisymmetrized form of these states, according to the anticommutation rules for Fermions.

Another complication we have to face is caused by the fact that collisions in which an amount $\Delta E$ of energy is exchanged are associated with a time $t \sim \hbar/\Delta E$, and so to a frequency $\Delta E/\hbar$. In our case, this can be understood when realizing that, as seen by one electron, the potential caused by the other electron appears time-dependent, as the other electron moves by. Therefore, if we want to account for the screening effects of the surrounding electron gas, we must use the full dynamic expression for the dielectric constant. This dynamic screening was discussed in Chap. 11, Sect. 11.3.2.2 [see Eq. (11.45) with Eq. (11.65), for example] and it reflects the fact that the electron gas must respond on the time scale of this varying interaction potential.

Note also that we consider semiclassical transport. Therefore, we consider a slow spatial dependence of all quantities and assume that, locally, we can approximate the electron gas with a gas locally in a homogeneous situation, although not necessarily at equilibrium. Therefore, a quantity such as the electron "distribution" in band $m$ and (crystal) momentum $\mathbf{p}$, $\langle \widehat{n_{\mathbf{p}}^{(m)}}(t)\rangle = \mathrm{Tr}[\widehat{\rho}\widehat{c_{\mathbf{p}}^{(m)\dagger}}(t)\widehat{c_{\mathbf{p}}^{(m)}}(t)]$ can be considered locally at each position $\mathbf{r}$ and we can define a distribution function in phase-space, $f(\mathbf{p},m,\mathbf{r},t)$, that coincides "locally" with each $\langle \widehat{n_{\mathbf{p}}^{(m)}}(t)\rangle$ at each position $\mathbf{r}$ at each time $t$.

Finally, for completeness, we must note that corrections to the first Born approximation are often required. The partial-wave formalism we have mentioned in the previous chapter can be used also in this case.

After this long introduction, we summarize our task by saying that we must evaluate the matrix element of the dynamically screened Coulomb potential between antisymmetrized wavefunctions of spin-1/2 particles, also accounting for the difference between repulsive and attractive interactions with phase-shift corrections.

Thus, for a "primary" particle of crystal momentum $\mathbf{k}$ in band $n$, the Coulomb scattering rate at location $\mathbf{r}$ at time $t$ is approximated by

$$\frac{1}{\tau^{(\mathrm{ee})}(\mathbf{k},\mathbf{r},n)} = \frac{2\pi}{\hbar}\sum_{\mathbf{G},m,n',m'}\int\frac{\mathrm{d}\mathbf{p}}{(2\pi)^3}f(\mathbf{p},m,\mathbf{r},t)\int\frac{\mathrm{d}\mathbf{k}'}{(2\pi)^3}|M(\mathbf{k}n,\mathbf{p}m;\mathbf{k}'n',\mathbf{p}'m';\mathbf{G})|^2$$
$$\times\delta[E_n(\mathbf{k})+E_m(\mathbf{p})-E_{n'}(\mathbf{k}')-E_{m'}(\mathbf{p}')][1-f(\mathbf{k}',n',\mathbf{r},t)]$$
$$[1-f(\mathbf{p}',m',\mathbf{r},t)]. \tag{15.5}$$

The explanation of the symbols and of the assumptions and approximations made to obtain this expression requires some attention: The final and initial states for two scattering particles "1" and "2" are

$$|\mathbf{k}n,\mathbf{p}m\rangle = \frac{1}{\sqrt{2}}\left[|\mathbf{k}_1 n_1\rangle|\mathbf{p}_2 m_2\rangle \pm |\mathbf{p}_1 m_1\rangle|\mathbf{k}_2 n_2\rangle\right] \tag{15.6}$$

and

$$|\mathbf{k}'n',\mathbf{p}'m'\rangle = \frac{1}{\sqrt{2}}\left[|\mathbf{k}_1' n1'\rangle|\mathbf{p}_2' m_2'\rangle \mp |\mathbf{p}_1' m1'\rangle|\mathbf{k}_2' n_2'\rangle\right]. \tag{15.7}$$

Momentum is of course conserved, so that the final state of the scattering "partner" has crystal momentum $\hbar\mathbf{p}' = \hbar\mathbf{k} + \hbar\mathbf{p} - \hbar\mathbf{k}'$. The $\pm/\mp$ sign entering these expressions should be clarified. In case of scattering between two *distinguishable* particles (such as in electron–hole collisions), the two-particle state is symmetric, so the "+" sign should be used. In the case of scattering between two indistinguishable spin-1/2 particles (electron–electron or hole–hole collisions), we have to consider the effect of spin. The total angular momentum, $S$, of the two-particle system is either zero, when the spins of the two particles are anti-parallel (up and down), or 1, when the spins are parallel (up–up or down–down). In the first case, the spin wavefunction is antisymmetric and there is only one state ($S$=0, $S_z$=0) of angular momentum. This is called a *singlet* state. The spatial dependence of the two-particle wavefunction must be symmetrized ["+" sign in Eqs. (15.6) and (15.7)], so that the total wavefunction (spatial dependence and spin) is antisymmetric. On the contrary, when the particles have parallel spins, the spin wavefunction now can only be symmetric and there are three possible angular momentum states: $S_z = 0$ and $S_z = \pm 1$. This is called a *triplet* state. Now the spatial component of the wavefunction must be antisymmetric ["minus" sign in Eqs. (15.6) and (15.7)], since the spin component is symmetric.

The quantity $f(\mathbf{k},m,\mathbf{r},t)$ is the "distribution function," i.e., the phase-space density at the phase-space location $(\mathbf{r},\mathbf{p})$ in band $m$, at time $t$, normalized so that the real-space density at $\mathbf{r}$ is

$$n(\mathbf{r},t) = 2\sum_m \int \frac{d\mathbf{k}}{(2\pi)^3} f(\mathbf{k},m,\mathbf{r},t) \ .$$

Note that in Eq. (15.5) the integral over the states of the "partner" electrons of crystal momentum $\hbar\mathbf{p}$ is done over 1/2 of them, since the sum over spin states is assumed to be already included in the matrix element (with the associated overlap factors), as we shall see shortly. Note that $f$ is, in principle, an unknown in our problem. Indeed, in general we want to study the effect of electron–electron interactions in off-equilibrium situations. Some self-consistent scheme must be employed to solve simultaneously the transport equations—in order to obtain information about $f$—and the collision rate that enters the transport equations themselves. We shall discuss this issue below.

The quantity $M(\mathbf{k}n,\mathbf{p}m;\mathbf{k}'n',\mathbf{p}'m';\mathbf{G})$ is the Coulomb matrix element. In the case of scattering between two indistinguishable particles (electron–electron or hole–hole), it is the antisymmetrized Coulomb matrix element obtained by summing over all spin states: If

$$M_{\mathrm{d}} = \frac{e^2}{\epsilon_{\mathrm{s}}} \frac{\mathscr{I}_{\mathbf{k}n,\mathbf{k}'n'} \mathscr{I}_{\mathbf{p}m,\mathbf{p}'m'}}{|\mathbf{k}-\mathbf{k}'+\mathbf{G}|^2 + \beta^2(q_{\mathrm{d}},\omega_{\mathrm{d}})} , \tag{15.8}$$

where $\beta(q,\omega)$ is defined below, is the matrix element for what we call the *direct* process $(\mathbf{k}m) \to (\mathbf{k}'m')$, $(\mathbf{p}n) \to (\mathbf{p}'n')$, and

$$M_{\mathrm{x}} = \frac{e^2}{\epsilon_{\mathrm{s}}} \frac{\mathscr{I}_{\mathbf{k}n,\mathbf{p}'m'} \mathscr{I}_{\mathbf{p}m,\mathbf{k}'n'}}{|\mathbf{k}-\mathbf{p}'+\mathbf{G}|^2 + \beta^2(q_{\mathrm{x}},\omega_{\mathrm{x}})} \tag{15.9}$$

is the matrix element for the *exchange* process $(\mathbf{k}m) \to (\mathbf{p}'n')$, $(\mathbf{p}n) \to (\mathbf{k}'m')$, the matrix element for scattering of spin-1/2 particles is given by the sum of the scattering probability in the singlet state and the scattering probability in the triplet state, i.e.

$$|M|^2 = \frac{1}{4}|M_{\mathrm{d}}+M_{\mathrm{x}}|^2 + \frac{3}{4}|M_{\mathrm{d}}-M_{\mathrm{x}}|^2 = |M_{\mathrm{d}}|^2 + |M_{\mathrm{x}}|^2 - \frac{1}{2}|M_{\mathrm{d}}^* M_{\mathrm{x}} + M_{\mathrm{d}} M_{\mathrm{x}}^*| . \tag{15.10}$$

The last "interference term," the "exchange term," is responsible for depressing the cross section at small relative energies, $\mathbf{k}-\mathbf{p} \ll \beta$.

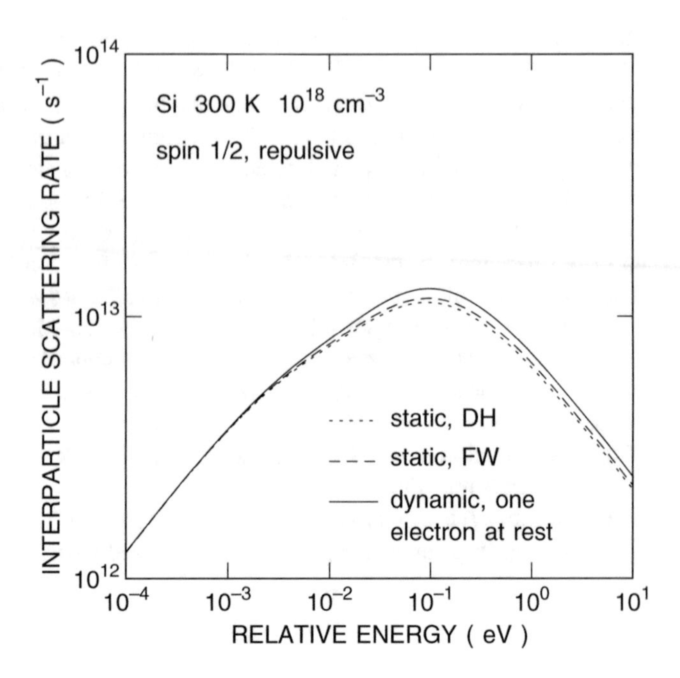


**Fig. 15.2** Effect of dynamic screening on the short-range interparticle scattering rate for a parabolic band model of Si at 300 K. Three cases are shown: the *dotted line* is for the case of static Debye–Hückel screening, the *dashed line* for a wavevector dependent static screening parameter, corresponding to a "head-on" collision. Finally, the *solid line* is relative to the case of a collision in which one electron is at rest in the frame of the electron gas

In the case of scattering between distinguishable particles (electron–hole scattering), these terms do not appear. Since, after integration, $|M_{\rm d}|^2$ and $|M_{\rm x}|^2$ yield the same result, one could obtain an equivalent result by ignoring spin altogether and replacing the integral over $\mathbf{p}$ above with twice that, i.e., by summing over all spin states. Figure 15.2 (right frame) shows the effect of (anti)symmetrization on the short-range scattering rate for a simple parabolic band of Si at 300 K.

The quantity $\beta(q,\omega)$ is the dynamic screening parameter, i.e. the quantity such that the dielectric function can be written as $\epsilon(q,\omega)/\epsilon_{\rm s} = 1+\beta(q,\omega)^2/q^2$. It must be evaluated at the crystal momentum transfer $q_{\rm d} = |\mathbf{k}-\mathbf{k}'+\mathbf{G}|$, and at the energy transfer $\hbar\omega_{\rm d} = E_n(\mathbf{k}) - E_{n'}(\mathbf{k}')$ for the "direct" process, at $q_{\rm x} = |\mathbf{k}-\mathbf{p}'+\mathbf{G}|$, and at $\hbar\omega_{\rm x} = E_n(\mathbf{k}) - E_{m'}(\mathbf{p}')$ for the exchange process. Figure 15.3 shows the ratio of the dynamic screening parameter to the conventional Debye–Hückel form (left frame) and the interparticle scattering rate as a function of electron energy obtained by adopting several approximations (right frame).

We have indicated with $\mathscr{I}_{\mathbf{k}n,\mathbf{k}'n'}$ the overlap integral between the Bloch states, accounting for the sum over spin states. Finally, phase-shift corrections—obtained via the partial-wave analysis we have seen before dealing with Friedel's sum rule—can be implemented and some of the following figures show the effect of these corrections (labeled as "phase shift"). They do indeed yield a different scattering rate for repulsive (weaker scattering) and attractive (stronger) interactions.

At low relative energies, Goodnick and Lugli [9], in dealing with scattering between indistinguishable particles, have assumed that only the direct term $|M_{\rm d}|^2$ matters (since the interaction between electron of parallel spins is reduced by exclusion) thus effectively obtaining a scattering rate equal to half the rate for indistinguishable particles.

## 15.2.1 *Numerical Evaluation and Scattering in Low-Dimensionality Systems

Before continuing our discussion, we should emphasize that, using a numerically computed band structure (e.g., from DFT or empirical pseudopotentials), the numerical evaluation of the short-range Coulomb interparticle scattering rate is quite laborious. The same Gilat–Raubenheimer scheme can

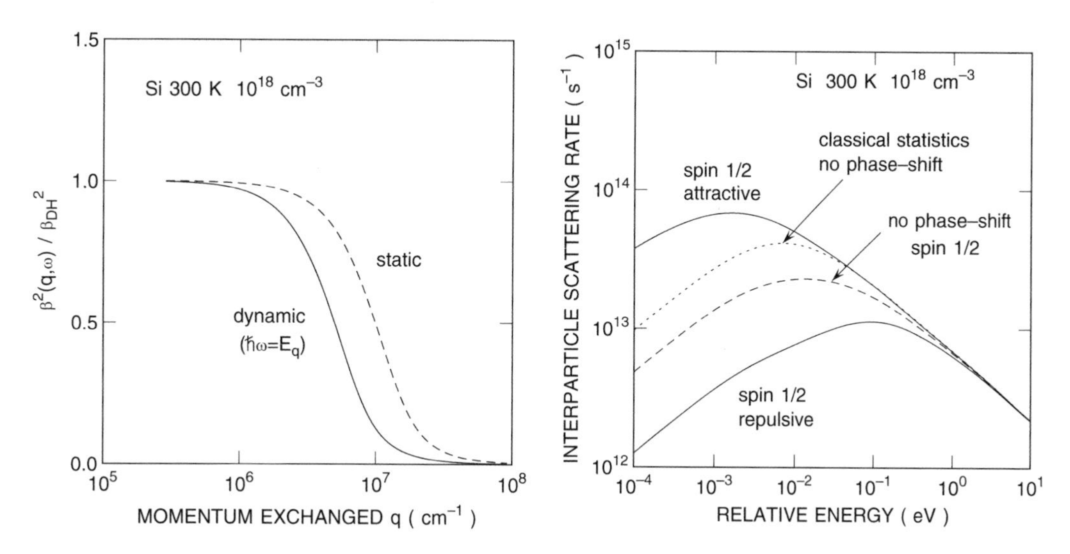


**Fig. 15.3** Ratio of the dynamic and static screening parameter to the Thomas–Fermi approximation $\beta$ as a function of momentum transfer (*left*) and the interparticle scattering rate as a function of electron energy in a variety of approximations (*right*): classical particles (no spin), spin-1/2 but no phase-shift corrections, and spin-1/2 with phase-shift corrections for attractive and repulsive potentials, the latter being of course appropriate to e–e scattering

be used, although now it will become more complicated because of a double search for two-particle states that conserve energy and crystal momentum. Details and examples concerning transport of energetic electrons in small Si devices are given in [10]. Given these complications, in the following we shall limit ourselves to discussing some commonly used approximations. As usual, interested readers should consult the references we have provided.

Problems arise also during the implementation of these scattering processes in numerical studies of electron transport. The first problem is due to the necessity of separating the short-range, single-particle Coulomb interactions from the long-range (plasmon) effects that may be already included semiclassically when the transport equations are solved self-consistently with Poisson's equation (a time-dependent Hartree term). Another problem is related to the fact that the distribution of scattering electrons, $f$, in Eq. (15.2), is an unknown. Conventional Monte Carlo methods can be extended to handle both difficulties. Suitable mesh sizes can be selected to avoid double-counting (or neglecting) long-range interactions. Suitable time steps can guarantee stability and proper inclusion of plasma effects. Finally, statistical sampling of neighboring electrons (in ensemble Monte Carlo simulations that deal with thousands of particles at a given time) permits a stochastic evaluation of the unknown as the simulation progresses. This is indeed one of the major advantages afforded by the Monte Carlo method. Algorithms based on expansions of the distribution function and partial differential equations (algorithms we shall review in the last chapter) cannot easily solve these problems.

Electron–electron scattering in two dimensions can also be treated on the same footing. While the formalism is exactly the same, a minor simplification comes from the reduced dimensionality of the phase-space for the two-particle states. The unscreened matrix elements can be calculated in a straightforward way, once the "confined" wavefunctions $\zeta_\nu(z)$ are known. Using the effective-mass approximation, discussed in Sect. 8.2.1, the task is eminently doable. The major numerical problem lies in computing the screened interaction potential. Since we must deal with electrons making also intersubband transitions, the matrix elements, that couple an initial state $|\mathbf{K}\nu,\mathbf{P}\mu\rangle$ to a final state $|\mathbf{K}'\nu',\mathbf{P}'\mu'\rangle$, are $N_{\rm sub}^2 \times N_{\rm sub}^2$ matrices for each set of $\mathbf{K},\mathbf{P},\mathbf{K}'$, and $\mathbf{P}'$. $N_{\rm sub}$ is the number of eigenstates (subbands) considered. This requires setting up and inverting an $N_{\rm sub}^2 \times N_{\rm sub}^2$ dielectric matrix given by Eq. (11.127) of Chap. 11, Sect. 11.3.2.4. It is easy to appreciate the complication since, even using as few as 10 subbands, these are $100 \times 100$ matrices and each of the $10^4$ elements of the bare potential and of the dielectric matrix has to be computed performing a numerical integration. This must be performed for each momentum-transfer $Q$ and energy-transfer $\hbar\omega$ needed. Doable, but not straightforward or painless!

Things are much more interesting in one dimension. An interacting 1DEG in most circumstances will behave as a collective system known as a *Luttinger liquid* [11, 12]. This causes a breakdown of the single-particle picture that constitutes the foundation of this discussion about short-range Coulomb interactions. All of these issues are extremely complicated, going beyond the scope of this textbook. We can only reiterate our invitation to consult the literature.

### 15.2.2 *Some Common Approximations

Even ignoring the complication due to the presence of the unknown distribution of partner particles, $f$, the calculation of the scattering rate given by Eq. (15.5) is extremely difficult. Therefore, several simplifications have been used. We review some of the most commonly used approximations.

The effect of degeneracy, represented by the factors $(1-f)$ in Eq. (15.5), is often included by replacing the distribution functions $f$ with their equilibrium values,

$$f(\mathbf{k},m,\mathbf{r},t) \simeq f_0[\mathbf{k},m,T_p,E_F], \tag{15.11}$$

where $f_0(\mathbf{k},m,T,E_F) = \{1+\exp[(E_m(\mathbf{k})-E_F)/(k_B T)]\}^{-1}$ is the Fermi–Dirac function at temperature $T$ for band $m$ and Fermi energy $E_F$, $T_p$ is the equivalent particle temperature. Note that both this degeneracy correction as well as the appearance of the distribution function in the screening corrections render the transport problem nonlinear: Typically, we use scattering rates to solve the transport problem in which $f$ is the unknown. But when the scattering rates depend on $f$ itself, either iterative or self-consistent schemes are devised, or the dependence of $f$ on the rates must be approximated. In typical cases, as we have seen here, a Fermi expression with an equivalent electron temperature is used.

Accordingly, *dynamic screening* may be treated in the high-temperature, nondegenerate limit valid for a single parabolic and spherical valley, as we have seen in Eqs. (11.63) and (11.64). We have already seen, in the left frame of Fig. 15.2, the ratio of the squared dynamic screening parameter to the squared Debye–Hückel parameter for the case of a "parabolic" relative energy transfer $\omega = \hbar q^2/(2m_d)$ in Si, where $m_d$ is the DOS effective mass. The same figure also shows the behavior of the static screening parameter $\beta^2(q,0)$.

Figure 15.4 shows—at the left—the short-range electron–electron scattering as a function of electron energy in Si at 300 K at a uniform field of 200 kV/cm, averaged self-consistently over the distribution function during Monte Carlo runs at the two densities indicated. At the right we see the short-range electron–electron scattering as a function of electron energy in Si at 300 K at a density of $10^{18}\,\text{cm}^{-3}$ at a field of 100 kV/cm. The effect of spin and dynamic screening is also shown. Figure 15.5 shows the same information, but at a field of 200 kV/cm and at a density of $10^{20}\,\text{cm}^{-3}$.

### 15.2.3 Electron–Electron Scattering in the Parabolic Band Approximation

In the simple case in which a parabolic approximation is used, the electron–electron scattering rate *between two electrons* can be evaluated analytically when ignoring degeneracy effects, *Umklapp* processes, and taking the overlap integrals as unity. Note the words in italic: Eq. (15.5) expresses the

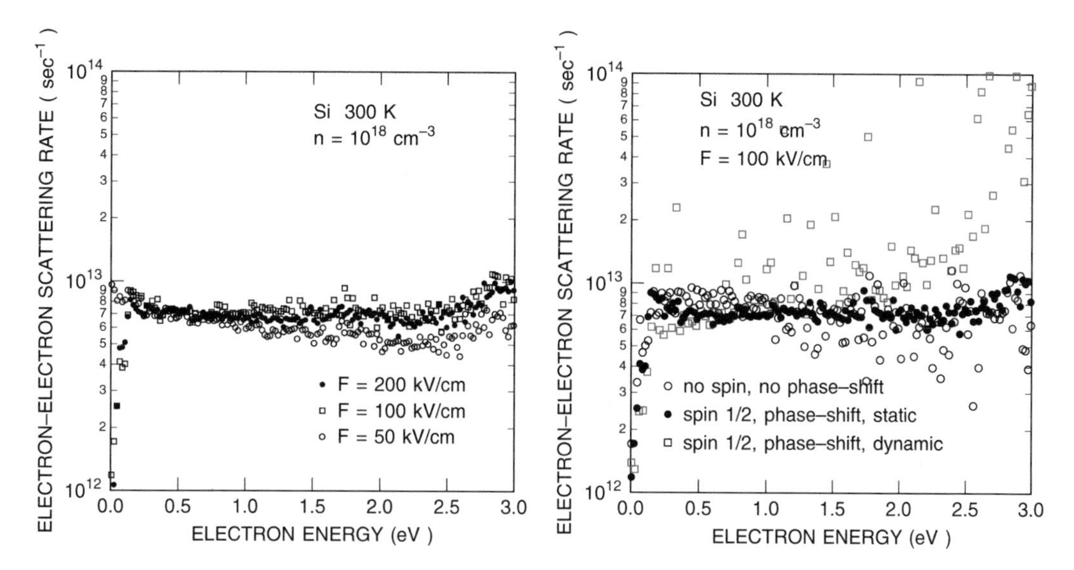


**Fig. 15.4** *Left*: short-range, single-particle electron–electron scattering rates in Si determined via Monte Carlo simulations in a homogeneous system with an applied electric field of 50 (*circles*), 100 (*open squares*), and 200 kV/cm (*dots*). The electron density is assumed to be $10^{18}$/cm$^3$. The rates are first calculated numerically via Eq. (15.2) using the Si band structure obtained using nonlocal empirical pseudopotentials. Dynamic RPA screening in the high-temperature approximation and phase-shift corrections are included. The scattering rates are calculated for each possible two-particle state. During the Monte Carlo simulation, for a given electron, a scattering partner is selected randomly within a sphere with radius equal to the screening length. This approximates stochastically the integration over the unknown distribution $f$ of "partner" electrons that appears in Eq. (15.2). Having reached steady state, the rates are cumulatively stored and are shown here as a scatter plot. The noise is due, of course, to the statistical error related to the stochastic sampling of the distribution function. *Right*: the same as in the left frame, but for a field of 100 kV/cm and employing the distinguishable-particles approximation (labeled "no spin") and ignoring phase-shift corrections (*circles*), or accounting for antisymmetrization and phase-shift corrections, but using static (*dots*) or dynamic screening (*pen red squares*). Note how assuming classical statistics of distinguishable particles decreases the scattering rate, while dynamic screening increases the collision frequency

collision rate for a given "primary" electron of wavevector $\mathbf{k}$ in band $n$ with any other electron in the system. Here we consider the collision rate between our primary electron and one particular "partner" electron with wavevector $\mathbf{p}$ in band $m$, lumping the effect of all other electrons into the electron density $n(\mathbf{r},t)$ appearing as a multiplicative factor. In other words, we approximate $f(\mathbf{p}',m',\mathbf{r},t)$ with $n(\mathbf{r},t)\delta_{mm'}\delta(\mathbf{p}-\mathbf{p}')$.

Therefore, Eq. (15.5) simplifies to

$$\frac{1}{\tau^{(\mathrm{ee})}(\mathbf{k}n,\mathbf{p}m,\mathbf{r},t)} = \frac{2\pi}{\hbar}\frac{e^4 n(\mathbf{r},t)}{4\epsilon_\infty^2}\sum_{\mathbf{k}'}\delta[E(\mathbf{k})+E(\mathbf{p})-E(\mathbf{k}')-E(\mathbf{p}')]$$

$$\times\left\{\frac{1}{(|\mathbf{k}'-\mathbf{k}|^2+\beta_\mathrm{d}^2)^2}+\frac{1}{(|\mathbf{k}'-\mathbf{p}|^2+\beta_\mathrm{x}^2)^2}-\frac{2}{(|\mathbf{k}'-\mathbf{k}|^2+\beta_\mathrm{d}^2)(|\mathbf{k}'-\mathbf{p}|^2+\beta_\mathrm{x}^2)}\right\}, \qquad (15.12)$$

where $\beta_\mathrm{d}$ is evaluated at $\mathbf{q}=\mathbf{k}'-\mathbf{k}$ and $\hbar\omega_\mathrm{d}=E(\mathbf{k}')-E(\mathbf{k})$ while $\beta_\mathrm{x}$ is evaluated at $\mathbf{q}=\mathbf{k}'-\mathbf{p}$ and $\hbar\omega_\mathrm{x}=E(\mathbf{k}')-E(\mathbf{p})$. Additional approximations are required in order to reach an analytic expression: First, dielectric screening must be considered only in the long-wavelength static limit. The justification lies in the fact that dynamic screening corrections are typically very small. Note, however, that this is not true in general: In cases where a "hot" electron interacts with a pool of "cold" electrons (such as an electron entering the drain region of a highly biased MOSFET), the large energy transfer renders

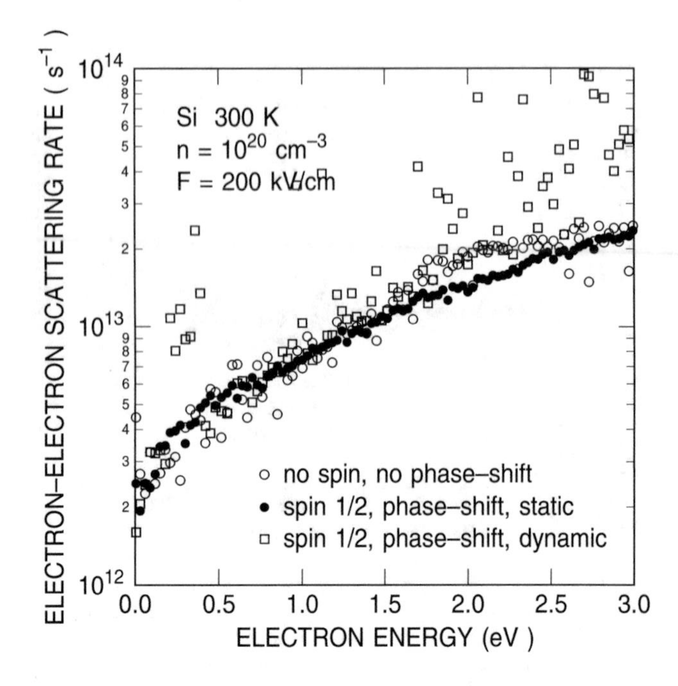


**Fig. 15.5** As in the *left frame* of Fig. 15.4, but at a higher field and density. Note how the scattering rate now increases with increasing electron energy, especially when using dynamic screening: higher-energy electrons on average exchange more energy, rendering dynamic screening effects weaker. In other words: they move too fast so that the surrounding electron gas becomes unable to respond and screen the interaction

either the direct or the exchange matrix element essentially unscreened, thus rendering the interaction stronger than what static screening would suggest. When static screening is considered, the screening parameter takes the form:

$$\beta^2 \simeq \beta_s^2 g_1(q\lambda),$$

where

$$\lambda = \left(\frac{2\pi\hbar^2}{m_{\mathrm{red}}k_B T}\right)^{1/2}$$

$$g_1(x) = \frac{2\pi^{1/2}}{x}\Phi\left(\frac{x}{4\pi^{1/2}}\right).$$

At the long wavelengths which dominate scattering when screening is not too strong, $\beta \simeq \beta_s$. Intervalley processes must also be ignored, since the valley separation would appear to the fourth power in the denominators of the Coulomb matrix elements, thus depressing their contribution.

Using these approximations, Eq. (15.12) can be integrated analytically. It is convenient to work in the center-of-mass frame by setting

$$\mathbf{g} = \mathbf{k} - \mathbf{p}\,, \qquad \mathbf{g}' = \mathbf{k}' - \mathbf{p}',$$

so that

$$|\mathbf{k} - \mathbf{k}'| = \frac{1}{2}|\mathbf{g} - \mathbf{g}'|,$$

keeping also in mind that $\mathbf{p}' = \mathbf{k} + \mathbf{p} - \mathbf{k}'$. Then:

$$E(\mathbf{k}) + E(\mathbf{p}) - E(\mathbf{k}') - E(\mathbf{p}') = \frac{\hbar^2 g^2}{2m^*} - \frac{\hbar^2 g'^2}{2m^*}\,.$$

Consider now just the first (direct) term inside the integral in Eq. (15.12):

$$\frac{\pi e^4 n(\mathbf{r},t)}{2\hbar \epsilon_\infty^2} \int \frac{\mathrm{d}\mathbf{g}'}{(2\pi)^3} \frac{\delta[E(\mathbf{g})-E(\mathbf{g}')]}{[|\mathbf{g}-\mathbf{g}'|^2/4+\beta^2]^2} . \tag{15.13}$$

This expression is identical to the integration we had to perform in order to calculate the electron-impurity scattering rate and we find

$$\frac{e^4 n(\mathbf{r},t) m^*}{2\pi\hbar^3\epsilon_\infty^3} \frac{g}{\beta^2(g^2+\beta^2)} . \tag{15.14}$$

Accounting for the other two terms, we finally obtain

$$\frac{1}{\tau_{\mathrm{self}}^{(\mathrm{ee})}(g)} \simeq \frac{e^4 m^* g n(\mathbf{r},t)}{2\pi\epsilon_{\mathrm{s}}^2\hbar^3} \left[ \frac{1}{\beta_{\mathrm{s}}^2(\beta_{\mathrm{s}}^2+4g^2)} - \frac{\ln(1+4g^2/\beta_{\mathrm{s}}^2)}{8g^2(\beta_{\mathrm{s}}^2+2g^2)} \right] . \tag{15.15}$$

Note that the cross section, $\sigma = 2m^*/(\tau n \hbar g)$, corresponding to this process equals the result reported by Ridley (see Eq. (4.136) of his textbook [13]). Replacing $2g$ with $k_{\mathrm{r}} = |\mathbf{k}-\mathbf{p}|$, Eq. (15.15) takes the form:

$$\frac{1}{\tau^{(\mathrm{ee})}(k_{\mathrm{r}})} \simeq \frac{e^4 m^* k_{\mathrm{r}} n(\mathbf{r},t)}{4\pi\epsilon_{\mathrm{s}}^2\hbar^3} \left[ \frac{1}{\beta_{\mathrm{s}}^2(\beta_{\mathrm{s}}^2+k_{\mathrm{r}}^2)} - \frac{\ln(1+k_{\mathrm{r}}^2/\beta_{\mathrm{s}}^2)}{2k_{\mathrm{r}}^2(\beta_{\mathrm{s}}^2+k_{\mathrm{r}}^2/2)} \right] . \tag{15.16}$$

For the case of spinless, distinguishable particles (i.e., classical statistics, unsymmetrized wavefunctions) and using the first Born approximation, the equation above reduces to the (familiar?) form:

$$\frac{1}{\tau^{(\mathrm{ee})}(k_{\mathrm{r}})} \simeq \frac{e^4 m^* k_{\mathrm{r}} n(\mathbf{r},t)}{4\pi\epsilon_{\mathrm{s}}^2\hbar^3} \frac{1}{\beta_{\mathrm{s}}^2(\beta_{\mathrm{s}}^2+k_{\mathrm{r}}^2)} , \tag{15.17}$$

since only the first term in Eq. (15.16) contributes. This is a result of "classical mechanics." This limit is reached in the limit of high relative energies: In this limit, $k_{\mathrm{r}}/\beta_{\mathrm{s}} \gg 1$, exchange does not play a dominant role, because particles can approach one another even if they have parallel spins, since the large relative energy pushes them sufficiently close. Not so in the opposite limit of small relative energies, $k_{\mathrm{r}}/\beta_{\mathrm{s}} \ll 1$. In this case Eq. (15.16) approaches half of the classical result,

$$\frac{1}{\tau_{\mathrm{self}}^{(\mathrm{ee})}(k_{\mathrm{r}})} \to \frac{e^4 m^* k_{\mathrm{r}} n(\mathbf{r},t)}{8\pi\epsilon_{\mathrm{s}}^2\hbar^3\beta_{\mathrm{s}}^4} \qquad \left(\frac{k_{\mathrm{r}}}{\beta_{\mathrm{s}}} \to 0\right) ,$$

since particles with parallel spins remain separated by the exclusion principle and their interaction vanishes. This is the limit considered by Goodnick and Lugli [9].

### 15.2.4 *Impact Ionization*

Another example of electron–electron collisions is "impact ionization." This is essentially a two-body collision, identical to what we have discussed above, but it involves *interband* transitions: Picture an electron traveling in the conduction band with a large kinetic energy. If this energy is sufficiently large, this "primary" electron can collide with an electron in the valence band, "kicking" this "partner" electron into the conduction band, leaving a "secondary" hole in the valence band. This process, which is the inverse of the Auger recombination mechanism, is responsible for avalanche breakdown phenomena in semiconductors and semiconductor devices.

**In 1922 Lise Meitner published a paper discussing a yet-not uncovered process by which an electron in an excited state of a many-electron atom can decay to a lower-energy state: Instead of following the most likely process of decaying by emitting a photon, it can drop into an empty core level and transfer its energy to another electron in the atom, kicking it out of its "orbit" [14]. This emitted electron is called an* Auger electron *and the process, for historical reason, takes its name from the French physicist Pierre Victor Auger, who discovered the effect the following year [15]. Unfortunately, it does not take its name from Lise Meitner, who proposed it first. This is not the first episode that has seen history treat Lise Meitner unfairly.*

The formal expression for the ionization rate can be obtained following a path very similar to the path we have followed above to calculate the collision rate for short-range electron–electron collisions. Within the Born approximation, the rate, $1/\tau^{(\mathrm{ii})}(\mathbf{k}n)$, at which an electron of quasi-momentum $\hbar\mathbf{k}$ in band $n$ creates an electron–hole pair is given by Sano et al. [16], Fischetti el al. [17]:

$$\frac{1}{\tau^{(\mathrm{ii})}(\mathbf{k}n)} = \frac{2\pi}{\hbar} \sum_{\mathbf{GG}'} \sum_{n_\mathrm{v} n'_\mathrm{c} n''_\mathrm{c}} \int \frac{\mathrm{d}\mathbf{k}'}{(2\pi)^3} \int \frac{\mathrm{d}\mathbf{p}'}{(2\pi)^3} |M_{\mathbf{GG}'}(\mathbf{k}n, \mathbf{p}n_\mathrm{v}; \mathbf{k}'n'_\mathrm{c}, \mathbf{p}'n''_\mathrm{c})|^2$$

$$\delta[E_n(\mathbf{k}) + E_{n_\mathrm{v}}(\mathbf{p}) - E_{n'_\mathrm{c}}(\mathbf{k}') - E_{n''_\mathrm{c}}(\mathbf{p}')] \ , \qquad (15.18)$$

where $E_n(\mathbf{k})$ is the energy of the state $\mathbf{k}$ in band $n$, and $M$ is the (anti)symmetrized screened Coulomb matrix element obtained by adding the scattering probabilities in the singlet and triplet states (as we saw before):

$$|M_{\mathbf{GG}'}|^2 = |M_\mathrm{d}|^2 + |M_\mathrm{x}|^2 - \frac{1}{2}|M_\mathrm{d}^* M_\mathrm{x} + M_\mathrm{x}^* M_\mathrm{d}| \qquad (15.19)$$

$M_\mathrm{d}$ and $M_\mathrm{x}$ being the "direct" and "exchange" matrix elements,

$$M_\mathrm{d} = e^2\, \epsilon^{-1}_{\mathbf{GG}'}(\mathbf{q}_\mathrm{d}, \omega_\mathrm{d}) \frac{I_\mathbf{G}(\mathbf{k}n; \mathbf{k}'n'_\mathrm{c}) I_{\mathbf{G}'}(\mathbf{p}n_\mathrm{v}; \mathbf{p}'n''_\mathrm{c})}{|\mathbf{q}_\mathrm{d} + \mathbf{G}|^2} \ , \qquad (15.20)$$

$$M_\mathrm{x} = e^2\, \epsilon^{-1}_{\mathbf{GG}'}(\mathbf{q}_\mathrm{d}, \omega_\mathrm{x}) \frac{I_\mathbf{G}(\mathbf{k}n; \mathbf{p}'n''_\mathrm{c}) I_{\mathbf{G}'}(\mathbf{p}n_\mathrm{v}; \mathbf{k}'n'_\mathrm{c})}{|\mathbf{q}_\mathrm{x} + \mathbf{G}|^2} \ . \qquad (15.21)$$

Here $\mathbf{q}_\mathrm{d}$, $\hbar\omega_\mathrm{d}$ and $\mathbf{q}_\mathrm{x}$, $\hbar\omega_\mathrm{x}$ are the direct and exchange quasi-momentum and energy transfers, respectively, defined as

$$\mathbf{q}_\mathrm{d} = \mathbf{k} - \mathbf{k}' \qquad \hbar\omega_\mathrm{d} = E_n(\mathbf{k}) - E_{n'_\mathrm{c}}(\mathbf{k}') \qquad (15.22)$$

$$\mathbf{q}_\mathrm{x} = \mathbf{k} - \mathbf{p}' \qquad \hbar\omega_\mathrm{x} = E_n(\mathbf{k}) - E_{n''_\mathrm{c}}(\mathbf{p}'). \qquad (15.23)$$

The vectors $\mathbf{k}'$, $\mathbf{p}$, and $\mathbf{p}'$ are the quasi-momenta of the final state of the primary particle and of the initial and final states of the "partner" particle, respectively. By momentum conservation they are related via $\mathbf{p}' = \mathbf{k} + \mathbf{p} - \mathbf{k}'$, up to a vector of the reciprocal lattice, already accounted for in the sums entering Eq. (15.18). The functions $I_\mathbf{G}(\mathbf{k}n; \mathbf{k}'n')$ are overlap integrals between Bloch states, summed over spin states $\sigma = \pm$:

$$|I_\mathbf{G}(\mathbf{k}n; \mathbf{k}'n')|^2 = \frac{1}{2} \sum_{\sigma\sigma'} |I_\mathbf{G}(\mathbf{k}n\sigma; \mathbf{k}'n'\sigma')|^2 \ , \qquad (15.24)$$

where:

$$I_{\mathbf{G}}(\mathbf{k}n\sigma;\mathbf{k}'n'\sigma') = \frac{1}{\Omega_c}\int_{cell} d\mathbf{r}\, u_{\mathbf{k}'n'}^{(\sigma')}(\mathbf{r})^*\, u_{\mathbf{k}n}^{(\sigma)}(\mathbf{r})\, e^{i\mathbf{G}\cdot\mathbf{r}} , \tag{15.25}$$

where $\Omega_c$ is the cell volume and $u_{\mathbf{k}n}^{(\sigma)}(\mathbf{r}) = \sum_{\mathbf{G}} u_{\mathbf{G}}^{(\sigma)}(\mathbf{k}n)e^{i\mathbf{G}\cdot\mathbf{r}}$ is the periodic part of the (pseudo)wavefunction. Finally, the sum over reciprocal lattice wavevectors $\mathbf{G}$ in Eq. (15.18) includes *Normal* ($\mathbf{G} = \mathbf{0}$) as well as *Umklapp* ($\mathbf{G} \neq \mathbf{0}$) processes (referred to as $N$- and $U$-processes in the following). A similar expression, with the obvious changes, describes the rate for hole-initiated processes.

The difficulties in the evaluation of Eq. (15.18) are of both a physical and numerical nature, since, given the large energies involved, any form of approximation for the electron dispersion is going to be less than satisfactory (except for very few cases, like small band-gap materials). The full-band structure must be used if we want to get realistic results and this requires intensive state-of-the-art numerical work.

Physically, the major difficulty lies in choosing an accurate yet manageable expression for the (dynamic) dielectric matrix $\epsilon_{\mathbf{GG}'}(\mathbf{q},\omega)$. Within the Random Phase Approximation (RPA), we saw that the dielectric function is given by the "usual" Lindhard expression:

$$\epsilon_{\mathbf{GG}'}(\mathbf{q},\omega) = \epsilon_0\, \delta_{\mathbf{GG}'} - \frac{e^2}{|\mathbf{q}+\mathbf{G}|^2}\sum_{nn'\mathbf{k}} \frac{f_{n'}(\mathbf{k}+\mathbf{q}) - f_n(\mathbf{k})}{E_{n'}(\mathbf{k}+\mathbf{q}) - E_n(\mathbf{k}) + \hbar\omega}$$
$$\times \langle \mathbf{k},n|e^{-i(\mathbf{q}+\mathbf{G})\cdot\mathbf{r}}|\mathbf{k}+\mathbf{q},n'\rangle\langle \mathbf{k}+\mathbf{q},n'|e^{i(\mathbf{q}+\mathbf{G}')\cdot\mathbf{r}}|\mathbf{k},n\rangle , \tag{15.26}$$

where $\epsilon_0$ is the permittivity of vacuum and $f_n(\mathbf{k})$ the occupation of the state $|\mathbf{k},n\rangle$ of quasi-momentum $\mathbf{k}$ in band $n$. It is customary to consider only the longitudinal response (i.e., the diagonal terms of the dielectric function) and it is convenient to separate Eq. (15.26) into the two terms representing the response of the valence electrons $\epsilon_{\mathbf{GG}}^{(v)}$ (simply written now as $\epsilon_{\mathbf{G}}^{(v)}$) and of the free carriers, by setting $f_n(\mathbf{k}) = f_n^0(\mathbf{k}) + p_n(\mathbf{k})$, $f_n^0(\mathbf{k})$ being unity for all states in the valence bands, zero otherwise, and $p_n(\mathbf{k})$ being the population of the free carriers. Thus, Eq. (15.26) can be recast as follows

$$\epsilon_{\mathbf{G}}(\mathbf{q},\omega) = \epsilon_{\mathbf{G}}^{(v)}(\mathbf{q},\omega)\left[1 + \epsilon_{\mathbf{G}}^{(v)-1}(\mathbf{q},\omega)\frac{\epsilon_0\beta^2(\mathbf{q}+\mathbf{G},\omega)}{|\mathbf{q}+\mathbf{G}|^2}\right], \tag{15.27}$$

having made use of a free-carrier screening parameter $\beta^2(\mathbf{q},\omega)$ defined as

$$\beta^2(\mathbf{q},\omega) = -\frac{e^2}{\epsilon_0}\sum_{nn'\mathbf{k}} \frac{p_{n'}(\mathbf{k}+\mathbf{q}) - p_n(\mathbf{k})}{E_{n'}(\mathbf{k}+\mathbf{q}) - E_n(\mathbf{k}) + \hbar\omega}\, |\langle \mathbf{k},n|e^{-i\mathbf{q}\cdot\mathbf{r}}|\mathbf{k}+\mathbf{q},n'\rangle|^2 . \tag{15.28}$$

An approximation usually tacitly made consists in considering only the $\mathbf{G}$-vector which minimizes the denominator of the Coulomb matrix elements, because of their small contribution (as $|\mathbf{q}+\mathbf{G}|^{-4}$) at short wavelengths in Eq. (15.18). Thus, one can drop the subscript $\mathbf{G}$ and write simply $\epsilon(\mathbf{q},\omega)$, understanding that all vectors in $\mathbf{k}$-space must be mapped to the first BZ. The calculation can proceed by ignoring the contribution of the free carriers, and employing the RPA and the pseudopotential band structure to evaluate $\epsilon^{(v)}(\mathbf{q},\omega)$ while retaining its frequency dependence. In general, though, in treating the response of the carriers in the valence bands, the static approximation [i.e., setting $\omega = 0$ in Eq. (15.26)] is a satisfactory simplification, since the energy exchanged in a pair-production process in realistic situations is relatively small (<3 eV or so), so that the static value is adequate. So, several approximations have been made, ranging from the use of the static, wavevector dependent expression for $\epsilon^{(\mathrm{v})}(\mathbf{q})$ to approximating $\epsilon_0\beta^2(\mathbf{q})/\epsilon^{(\mathrm{v})}(\mathbf{q})$ with $\beta_s^2$, the "usual" Thomas–Fermi screening parameter $(e^2/\epsilon_s)\partial n_f/\partial E_F$, where $E_F$ is the Fermi level , $\epsilon_s = \epsilon^{(\mathrm{v})}(\mathbf{q}=0)$ is the static, long-wavelength dielectric function of the semiconductor, and $n_f$ the density of free carriers, so that

$$\epsilon(\mathbf{q}) \simeq \epsilon^{(\mathrm{v})}(\mathbf{q}) \left[ 1 + \frac{\beta_s^2}{q^2} \right] . \tag{15.29}$$

This expression approaches correctly $\epsilon^{(\mathrm{v})}(\mathbf{q})$ at short wavelengths, and is generally valid at long wavelengths, independent of the form of $\epsilon^{(\mathrm{v})}(\mathbf{q})$. Note also that dynamic corrections—in principle important in treating the response of the free carriers whose plasma energy is much lower than the typical energy exchanged in ionization events—become negligible at long wavelengths, so that the use of a static $\beta(\mathbf{q},\omega)$ appears justified in this context. Finally, in an attempt to simplify the calculation as much as possible, one may consider the static, long-wavelength limit for the response of the valence carriers, by setting $\epsilon_v(\mathbf{q}) = \epsilon_s$ at all wavelengths, despite its questionable validity, and the Thomas–Fermi approximation for the free carriers. One possibility is the use of Eq. (15.29), using an expression proposed by Nara and Morita [18] [see page 321, Eq. (14.22)] for $\epsilon^{(\mathrm{v})}(\mathbf{q})$, and assuming a small density of free electrons (say, $n_f = 10^{16}\ \mathrm{cm}^{-3}$), so that the Thomas–Fermi expression for $\beta_s^2$ reduces to its nondegenerate Debye–Hückel form $e^2 n_f/(\epsilon_s k_B T)$.

From a numerical point of view, local [19] or nonlocal [20] empirical pseudopotentials have been used to evaluate the band structure and the pseudo-wavefunctions for 152 points in the irreducible wedge of the first Brillouin zone (BZ). The evaluation of Eq. (15.18) provides an anisotropic ionization rate as a function of the quasi-momentum $\mathbf{k}$ of the ionizing carrier and of the band index $n$.

In addition to the "correct" evaluation of the anisotropic ionization rate expressed by Eq. (15.18), other useful approximations can be employed. The first one, the "constant matrix element" (CME) approximation, emphasizes the role played (or, more likely, *not* played) by the matrix element in determining the energy dependence of the ionization rate. Accordingly, one assumes that the Coulomb matrix element $M$ in Eq. (15.18) is a constant, equal to its average value over all $\mathbf{k}$-points in the whole BZ, over all bands, and over both $N$- and $U$-processes. Thus:

$$|M_{\mathbf{GG}'}|^2 \simeq \langle M^2 \rangle = \frac{e^4 a_0^4}{(2\pi)^4\, \epsilon_s^2}\, m^2 , \tag{15.30}$$

where $m$ is a number of order 1 and:

$$\frac{1}{\tau_{\mathrm{CME}}^{(\mathrm{ii})}(\mathbf{k}n)} \simeq \frac{2\pi}{\hbar} \langle M^2 \rangle \sum_{\mathbf{G}} \sum_{n_v n_c' n_c''} \int \frac{d\mathbf{p}}{(2\pi)^3}\, \delta(\mathbf{k}+\mathbf{p}-\mathbf{k}'-\mathbf{p}'+\mathbf{G})$$

$$\times \int \frac{d\mathbf{k}'}{(2\pi)^3} \int \frac{d\mathbf{p}'}{(2\pi)^3}\, \delta[E_n(\mathbf{k}) + E_{n_v}(\mathbf{p}) - E_{n_c'}(\mathbf{k}') - E_{n_c''}(\mathbf{p}')] . \tag{15.31}$$

Note that we have explicitly re-introduced the delta-function expressing momentum conservation in the first (trivial) integral in the equation above, in order to make it clear in the following how the CME and the random-$k$ approximations can be related. The actual value of the average matrix element $\langle M^2 \rangle$ can be determined, whenever Eqs. (15.18) and (15.31) exhibit the same isotropic energy dependence, by adjusting its value in Eq. (15.31) in order to match the result of Eq. (15.18).

An even stronger approximation has been suggested by Kane in the case of electron-initiated ionizations in Si [21]. It consists in assuming that the kinematics of the ionization process is not controlled by momentum conservation, but mainly by the (joint) density of states available to the final (recoil and ionized) particles. This is expected to be approximately true when momentum randomization via $U$-processes is the most probable pair-production channel. In this case, many nonzero $\mathbf{G}$-vectors contribute in the sum in Eq. (15.31) above. Thus, it is sensible to make the approximation, known as the *random-k approximation* [21], of replacing the many "spikes" of the momentum-conserving delta-function with a uniform sampling of the entire Brillouin Zone, thus rewriting Eq. (15.31) as

$$\frac{1}{\tau^{(\mathrm{ii})}_{\mathrm{CME}\to\mathrm{relax})}(\mathbf{k}n)} = \frac{2\pi}{\hbar}\langle M^2\rangle \sum_{n_\mathrm{v}n'_\mathrm{c}n''_\mathrm{c}} \int_{\mathrm{BZ}} \frac{\mathrm{d}\mathbf{p}}{\Omega_{\mathrm{BZ}}} \times \int \frac{\mathrm{d}\mathbf{k}'}{(2\pi)^3} \int \frac{\mathrm{d}\mathbf{p}'}{(2\pi)^3}\, \delta[E_n(\mathbf{k}) + E_{n_\mathrm{v}}(\mathbf{p}) - E_{n'_\mathrm{c}}(\mathbf{k}') - E_{n''_\mathrm{c}}(\mathbf{p}')]\,, \quad (15.32)$$

where now, in place of the sum over $\mathbf{G}$-vectors, $\mathbf{p}$ runs over the Brillouin Zone, restricted only by energy conservation. Now recall that the volume of the Brillouin Zone, $\Omega_{\mathrm{BZ}}$, is given by $(4\pi/a)^3/2 = (2\pi)^3/\Omega_\mathrm{c}$, and employ the identity:

$$\int \mathrm{d}E\, \mathscr{D}_{\mathrm{c/v}}(E) = 2\sum_{n_{c/v}} \int \frac{\mathrm{d}\mathbf{k}}{(2\pi)^3}\,,$$

where $\mathscr{D}_\mathrm{c}$ and $\mathscr{D}_\mathrm{v}$ are the densities of states in the conduction and valence bands, respectively, accounting for spin degeneracy and always measuring energies from the band extrema with (kinetic hole) energies taken as positive in the valence bands. Thus, the ionization rate obtained within the CME approximation *and* having relaxed momentum conservation reduces to

$$\frac{1}{\tau^{(\mathrm{ii})}_{\mathrm{rk}}(E)} = \frac{2\pi}{\hbar}\frac{\Omega_\mathrm{c}}{8}\langle M^2\rangle \int_0^{E-E_\mathrm{g}} \mathrm{d}E_\mathrm{c} \int_0^{E-E_\mathrm{g}-E_\mathrm{c}} \mathrm{d}E_\mathrm{v}\, \mathscr{D}_\mathrm{c}(E_\mathrm{c})\mathscr{D}_\mathrm{v}(E_\mathrm{v})\mathscr{D}_\mathrm{c}(E - E_\mathrm{g} - E_\mathrm{c} - E_\mathrm{v}), \quad (15.33)$$

where $E_\mathrm{g}$ is the energy gap and noting that $(2\pi)^3/(8\,\Omega_{\mathrm{BZ}}) = \Omega_\mathrm{c}/8$ and having considered throughout electron-initiated processes, to fix the ideas. The quality of this approximation can be judged by whether the value of $\langle M^2\rangle$ in Eq. (15.33) required to fit quantitatively the CME results matches the value used in Eq. (15.31).

Finally, we should mention the popular Keldysh form of the ionization rate [22],

$$\frac{1}{\tau^{(\mathrm{ii})}(\mathbf{k}n)} = \frac{P}{\tau_{\mathrm{ep}}(E_{\mathrm{th}})}\left(\frac{E - E_{\mathrm{th}}}{E_{\mathrm{th}}}\right)^2, \quad (15.34)$$

where $E_{\mathrm{th}}$ is some ionization-threshold energy, $P$ a coupling parameter related to the Coulomb matrix element, and $\tau_{ep}(E_{\mathrm{th}})$ is either the carrier-phonon scattering rate at threshold, according to Keldysh's original formulation [22], or the energy relaxation rate at threshold, according to Ridley's lucky drift theory [23]. This expression has been used extensively in the past, treating the parameters $P$ and $E_{th}$ as fitting quantities.

In Figs. 15.6 and 15.7 we see the ionization rates for electrons in Si as well as the so-called ionization coefficient, $\alpha_{\mathrm{ii}}$, the number of pairs created per unit length in the presence of a uniform electric field $F$:

$$\alpha_{\mathrm{ii}}(F) = \frac{1}{v_\mathrm{d}(F)} \int \frac{\mathrm{d}\mathbf{k}}{(2\pi)^3} \frac{f_\mathrm{F}(\mathbf{k})}{\tau^{(\mathrm{ii})}(\mathbf{k})}\,, \quad (15.35)$$

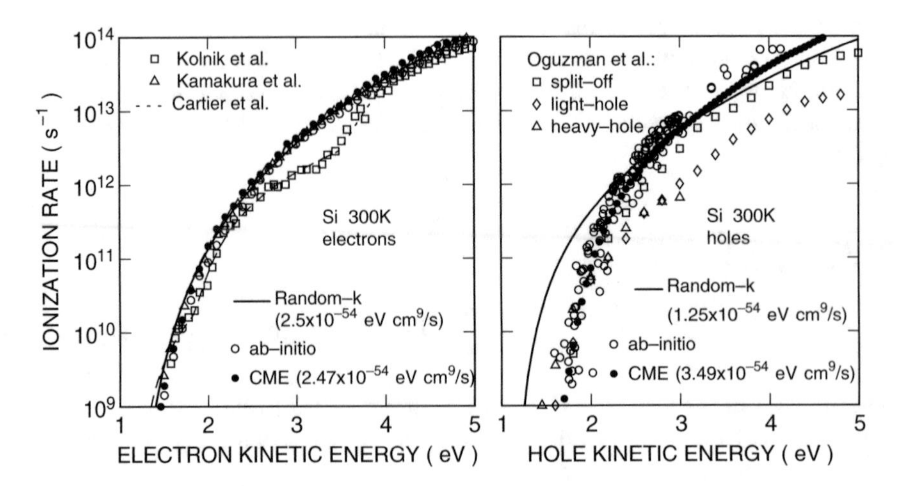


**Fig. 15.6** *Left*: electron-initiated impact ionization scattering rate in Si, as a function of electron kinetic energy, calculated using the several models described in the text and compared to some calculations performed using different models. References can be found in [17]. *Right*: the same for hole-initiated processes

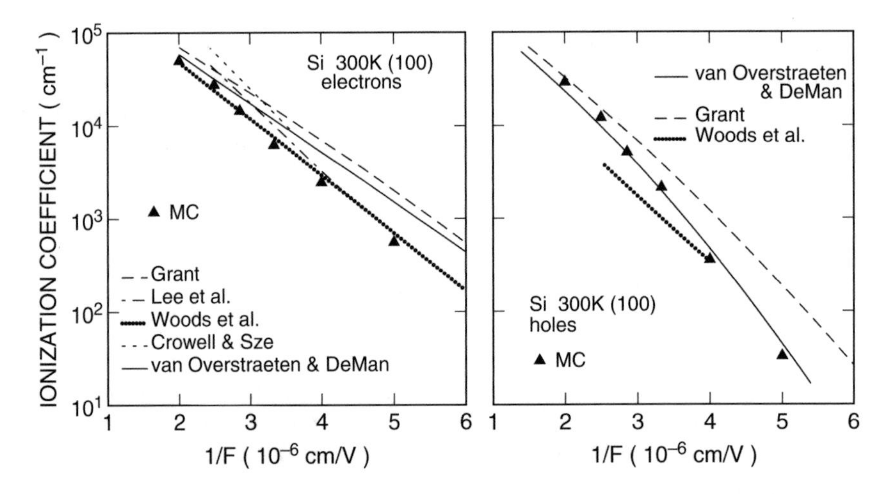


**Fig. 15.7** Ionization coefficients calculated using a Monte Carlo technique and the rates shown in the previous figure. Experimental data are also shown from the references cited in [17]

where $v_d(F)$ is the average electron drift velocity in the field $F$ and $f_F$ is the electron distribution. An empirical expression commonly used is

$$\alpha_{ii}(F) = \frac{eF_{th}}{E_{th}}\, e^{-F_{th}/F} ,$$

where $F_{th}$ is some threshold field and $E_{th}$ the energy threshold for pair generation.

## 15.3 *Long-Range Coulomb Interactions and Band-Gap Narrowing

We stated in the introduction of this chapter that when a high density of free carriers is present in a semiconductor, some electronic properties are strongly modified. A reduction of the band gap, a phenomenon called *band-gap narrowing*, clearly affects the operation of electron devices and the optical properties of the material. We conclude this chapter with a brief overview of the problem.

The subject of Coulomb interactions in an electron gas has been discussed for decades. Even for the "simple" case of a homogeneous electron gas the calculations are daunting, because of the complex nature of the many-body interactions with exchange and correlation effects, as well as of the interactions between the electrons and the "jellium" of the background positive charge required to keep the system at charge neutrality, without "exploding" as a result of Coulomb repulsion. Here we shall discuss the basic features of this system in a classical picture, mentioning later the main quantum-mechanical corrections [24–26]. We follow quite strictly the excellent account given by Mahan [27].

### 15.3.1 Classical Model

Consider a homogeneous electron gas of density $n = N/\Omega$, where $N$ is the number of electrons and $\Omega$ the volume occupied by the gas. Ignoring Coulomb interactions, the total energy $W$ per electron is purely the Fermi kinetic energy $K_{\rm F}$:

$$W = K_{\rm F} = \frac{g}{n} \int \frac{d\mathbf{k}}{(2\pi)^3} E_0(\mathbf{k}) f_{\rm FD}[E_0(\mathbf{k})] , \tag{15.36}$$

where $f_{\rm FD}(E) = \{1+\exp[(E_0 - E_{\rm F})/(k_{\rm B}T)]\}^{-1}$ is the equilibrium Fermi function at temperature $T$, $k_{\rm B}$ is the Boltzmann constant, $E_0(\mathbf{k})$ is the $g$-fold degenerate electron dispersion, and the Fermi energy $E_{\rm F}$ is determined so that

$$n = g \int \frac{d\mathbf{k}}{(2\pi)^3} f_{\rm FD}[E_0(\mathbf{k})] . \tag{15.37}$$

In the simplest possible picture, the Hartree model, the spatial distribution of both electrons and background positive charge (donors), required by charge neutrality, is assumed to be homogeneous and uniform. In this model the Coulomb interaction between electrons and donors vanishes. The first correction to this model consists in what Kittel calls the "modified Hartree" approximation. In this model one assumes that the electrons still constitute a uniform distribution of charge, but they are now embedded in a lattice of background, positively charged, point-like donors. Customarily, one thinks of the donor potential as statically screened by the electrons according to the Debye–Hückel theory, resulting in a Yukawa-like potential $e^{-\beta_s r}/(4\pi\epsilon_s^\infty r)$, where $\beta_s = [(e^2/\epsilon_s^\infty)(\partial n/\partial E_{\rm F})]^{1/2}$ is the screening parameter, $\epsilon_s^\infty$ being the static (valence) permittivity of Si, and $e$ the magnitude of the electron charge. However, as we already saw when dealing with the potential of an ionized impurity, Mahan, employing quantum-mechanical variational calculations, has shown that this picture is correct only at moderate densities ($\sim 10^{19}$ cm$^{-3}$ or below in Si), but at higher densities the assumption of a uniform electron distribution gives better agreement with the variational results. This is because the Yukawa-model over-counts the electrons: On the one hand, they are assumed to completely screen the donors, and so to be fully localized around each dopant ion. On the other hand, they are assumed to be uniformly distributed.

In order to compute the Coulomb energy of this model system, in principle one should follow a procedure similar to what is required to compute the Madelung energy in crystals and derive the positions of the donors by minimizing their total energy. Here we assume the impurities frozen in a cubic lattice, and consider spherical volumes of radius $r_0 = [3/(4\pi n)]^{1/3}$ around each donor. Thus, the total energy per dopant (i.e., in a volume $(4/3)\pi r_0^3$ around each impurity) of the electron gas and background point-like donors will be lowered by the electron-impurity Coulomb attraction,

$$\begin{aligned}\delta W_{\rm ed} &\approx -\frac{e^2}{(4/3)\pi r_0^3}\int_0^{r_0} dr \frac{r}{\epsilon_s^\infty} = -\frac{3}{2}\frac{e^2}{4\pi\epsilon_s^\infty r_0} \\ &= -\frac{3}{2}\left(\frac{4\pi}{3}\right)^{1/3}\frac{e^2 n^{1/3}}{4\pi\epsilon_s^\infty} \approx -2.418\ \kappa\, n^{1/3},\end{aligned} \tag{15.38}$$

where $\kappa = e^2/(4\pi\epsilon_s^\infty)$, and raised by the Coulomb repulsion of the uniform electron distribution with itself within an elemental spherical volume,

$$\begin{aligned}\delta W_{\rm ee} &\approx \frac{e^2}{[(4/3)\pi r_0^3]^2}\int_0^{r_0} dr \frac{4}{3}\pi r^3\, 4\pi r^2\, \frac{1}{4\pi\epsilon_s^\infty r} \\ &= \frac{3}{5}\frac{e^2}{4\pi\epsilon_s^\infty r_0} = \frac{3}{5}\left(\frac{4\pi}{3}\right)^{1/3}\frac{e^2 n^{1/3}}{4\pi\epsilon_s^\infty} \approx 0.967\ \kappa\, n^{1/3}.\end{aligned} \tag{15.39}$$

Note that the mutual donor–donor repulsion vanishes in this model, since the impurities in different volume elements are viewed as fully screened by the surrounding electrons when frozen in their minimum-energy configuration. Therefore, the total energy per impurity atom due to the Coulomb interaction will be

$$\delta W_{\rm C} = \delta W_{\rm ed} + \delta W_{\rm ee} \approx -1.451\ \kappa\, n^{1/3}. \tag{15.40}$$

Depending on the configuration of the impurities, one may actually obtain for the "Madelung coefficient" 1.451 above a slightly different value of 1.444 for close-packed fcc, hcp, or bcc lattices. Here, we shall not worry about these numerically small differences.

At densities low enough for the standard Debye–Hückel screening to be valid, an expression given by Lanyon and Tuft [28] may more appropriately replace Eq. (15.40),

$$\delta W_{\rm C} = -\frac{3}{4}\ \kappa\, \beta_{\rm s}. \tag{15.41}$$

As discussed by Mahan, the numerical values provided by Eqs. (15.40) and (15.41) are very similar in the range of densities of interest here, since the three characteristic lengths one can consider, the inverse wavevector at the Fermi surface $k_{\rm F}$, the screening length $\beta_{\rm s}^{-1}$, and the average donor separations $r_0$, are not too different in Si.

In the presence of random motion, induced by the Fermi kinetic energy and, at nonzero temperatures, by the electron–phonon interactions, the electrons will move around their "frozen" uniform configuration and set-up a fluctuating potential whose spatial Fourier components are identified with plasmons. In this case the total energy $\delta W_{\rm C}$ will re-distribute into a potential energy, $\delta U_{\rm C}$, and a kinetic energy component $\delta K_{\rm C}$. The virial theorem can be used to estimate the time-average of the kinetic energy $\delta K_{\rm C}$ once $\delta W_{\rm C}$ is known. Since for $N$ particles interacting via a central potential $V(r)$, a function only of the interparticle distance $r$,

$$\langle \delta K_{\rm C}\rangle = \frac{1}{2}\left\langle \sum_{i\neq j} \nabla_i V(\mathbf{r}_i - \mathbf{r}_j)\cdot \mathbf{r}_i \right\rangle_{\rm time}, \tag{15.42}$$

where the brackets $\langle \ldots \rangle_{\text{time}}$ denote time-average, for a purely Coulomb system

$$\langle \delta K_{\mathrm{C}} \rangle_{\text{time}} = -\frac{1}{2} \langle \delta U_{\mathrm{C}} \rangle_{\text{time}} \,, \tag{15.43}$$

so that (dropping the time-average brackets here and in the following, since we shall deal exclusively with time- or ensemble average, equivalent by the ergodic theorem)

$$\delta K_{\mathrm{C}} \approx 1.451 \; \kappa \, n^{1/3} \,. \tag{15.44}$$

The final quantity to consider is the shift of the chemical potential, $\delta\mu$, (in principle different from the Fermi energy $E_{\mathrm{F}}$ for interacting particles, as discussed below) induced by the Coulomb interactions among the $N$ electrons and the $N_{\mathrm{D}}$ donors. At zero temperature or, more generally, in degenerate situations, $\delta\mu$ gives the electron contribution to the narrowing of the energy gap. In general the chemical potential $\mu$ of a grand canonical ensemble is defined as (see any text on Statistical Mechanics, such as Kubo's 1965 book [29]):

$$\mu = \frac{1}{\Omega} \frac{\partial F}{\partial n} \,, \tag{15.45}$$

at constant temperature and volume $\Omega$, where $F = W_{\text{tot}} - TS$ is the (Helmholtz) free energy of the system, $W_{\text{tot}}$ being its total energy and $S$ its entropy. Therefore, at zero temperature the change of the chemical potential caused by the Coulomb interactions will be

$$\delta\mu = \frac{1}{\Omega} \frac{\partial \delta W_{\mathrm{C,tot}}}{\partial n} \,, \tag{15.46}$$

where $\delta W_{\mathrm{C,tot}} = N_{\mathrm{D}} \delta W_{\mathrm{C}}$ is the total Coulomb energy of the system. Keeping the volume $\Omega$ constant means that the identity $N_{\mathrm{D}} = N$ should be used only after having taken the derivative. Thus

$$\delta\mu = \frac{\partial}{\partial N} (N_{\mathrm{D}} \, \delta W_{\mathrm{C}}) = -0.481 \; \kappa \, n^{1/3} \,. \tag{15.47}$$

Ignoring band-tailing and renormalization of the electron dispersion, since the total number of particles remains constant when accounting for Coulomb interactions, this shift of the chemical potential must be accompanied by an equal downward shift of the reference energy, that is, the bottom of the conduction band, $\delta E_{\mathrm{CB}} = \delta\mu$.

### 15.3.2 Quantum Corrections

The modified Hartree model considered so far ignores the effect of the electron spin: Since Pauli's principle prevents electrons with parallel spins from getting too close to each other, their mutual Coulomb repulsion will be reduced by an amount known as the exchange energy, $\delta W_{\mathrm{x}}$. For Si the exchange energy per particle is given by Mahan [27]

$$\delta W_{\mathrm{x}} = -\frac{3}{4} \frac{e^2 \Lambda k_{\mathrm{F}}}{4\pi^2 \epsilon_{\mathrm{s}}^{\infty}} \approx -0.3861 \; \kappa \, n^{1/3} \,, \tag{15.48}$$

where $\Lambda = (m_L/m_T)^{1/3}$ $(\tan^{-1}\delta/\delta)$, $\delta = [(m_L/m_T) - 1]^{1/2}$, and $m_{t/l}$ denoting the transverse/longitudinal mass in the six ellipsoidal valleys of the Si conduction band. This second correction amounts to the Hartree–Fock results. The difference between this result and the energy of the ground state one would obtain performing an "exact" calculation is called the "correlation" energy [30]: The electrons are not uniformly distributed, but the energy of the system is lowered when accounting for a correlation of electron wavefunctions (or positions) which minimizes their mutual Coulomb repulsion. In the range of densities considered here, an expression for the correlation energy, $\delta W_{corr}$, has been given by Gell-Mann and Brueckner:

$$\delta W_{corr} = -0.096 + 0.00622 \ln r_s \tag{15.49}$$

while Nozières and Pines give [31]:

$$\delta W_{corr} = -0.115 + 0.0031 \ln r_s \ , \tag{15.50}$$

where the unit of energy is the Rydberg ($\approx$32.1 meV in Si) and $r_s$ is the average separation $r_0$ in units of the Bohr radius ($\approx$1.92 nm in Si), so that $r_s \approx 3.231\ (n/10^{18})^{-1/3}$, where the electron density $n$ is measured in cm$^{-3}$. Re-expressing Eq. (15.40) in atomic units, we have

$$\delta W_C = -\frac{1.798}{r_s}, \tag{15.51}$$

while the exchange term, Eq. (15.48), is

$$\delta W_x = -\frac{0.479}{r_s}, \tag{15.52}$$

so we see that the correlation energy can be neglected, while ignoring the exchange term, Eq. (15.48), amounts to underestimating the Coulomb energy by about 26 %. Correspondingly, the kinetic energy of the system will be modified only by a small term, usually called "correlation kinetic energy." We have discussed this energy before, in Chap. 4, Sect. 4.2.3, and commented on the extreme difficulty they pose to theoretical calculations. The texts by Mahan [24], Kittel [25], Fetter and Walecka [26], and two of the pioneering original papers on the subject, the articles by Gell-Mann and Brueckner [30] and by Nozières and Pines [31], give an idea of the complication. Theoretical work on the physics of a quantum electron gas continues to be actively performed to this day.

These corrections to the total energy of the electron gas at high densities are some of the causes of the well-known band-gap narrowing in semiconductors: We have already noticed above how the Coulomb interactions in a classical three-dimensional electron gas cause a shift $\delta E_{CB} = -0.481\kappa n^{1/3}$ of the bottom of the conduction band. Mahan has evaluated the shifts of the chemical potential $\delta\mu_x = [e^2/(4\pi^2\epsilon_s^\infty)]k_F\Lambda \approx -0.515\ \kappa\ n^{1/3}$ and $\delta\mu_{corr}$ associated with exchange and correlation effects, and so the corresponding shifts $\delta E_{CB}$. When accounting also for the contribution to the Coulomb energies due to the (minority) hole in the valence band (hole–donor and hole–electron interaction), Mahan has accounted for all of the contributions to the band-gap narrowing, with the exception of band-tailing, not discussed here. These are due to the close proximity of impurities at high density: As already mentioned in Sect. 14.2, the shallow electronic states in the impurity potential merge together to form a band of delocalized states, the so-called *impurity band*. Given the shallow energy of these states, in the gap but close either to the bottom of the conduction band (donors) or the top of the valence band (acceptors), they extend the density of states inside the gap with exponential "tails."

The major conclusion of this discussion is that the semiclassical model does indeed err in ignoring exchange and correlation, especially exchange, since its contribution can be dominant. Yet, the qualitative picture resulting from the classical model gives a clear physical picture insight and even correct trends.

Finally, regarding the possible implementation of these physical effects in "practical" simulations of electron transport in bulk solids devices, we mention that Monte Carlo simulations have been performed in "test cases" of periodic "chunks" of Si [7] and in devices [8, 32]. At high density, spatial fluctuations of the electron density oscillating at the plasma frequency cause an effective reduction of the total electron energy, equivalent to a narrowing of the band gap. This is shown in Fig. 15.8. Details about these calculations can be found in the references.

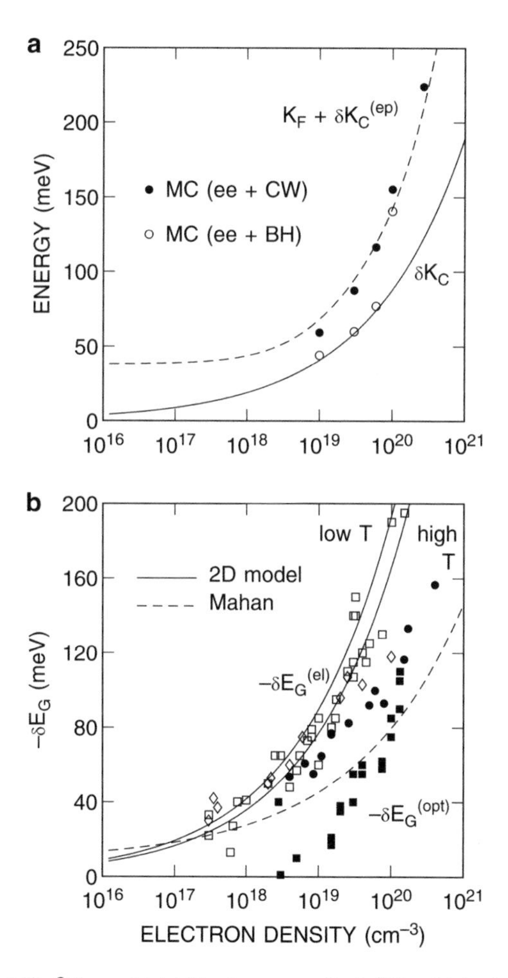


**Fig. 15.8** **(a)**: kinetic energy shift, $\delta K_C$, and total kinetic energy, $K_F + \delta K_C$, obtained from Monte Carlo simulations including the short-range electron-impurity scattering with the Conwell–Weisskopf (CW, *dots*) or Brooks–Herring (BH, *circles*) models. The result at the highest electron density has been extracted from the simulation of the source and drain regions of Si MOSFETs. **(b)**: Negative of the band-gap narrowing obtained from self-consistent Poisson/Monte Carlo simulations in the degenerate (*solid line* labeled "low-T") and nondegenerate (*solid line* labeled "high-T") cases, from Mahan's theory [27] (*dashed line*), and experimental data for the "apparent electrical gap" (*open symbols*) and optical gap (*solid symbols*) from the references cited in [7]. (Reprinted with permission from M.V. Fischetti, J. Appl. Phys. vol. 89, p. 1205 (2001). Copyright 2001, AIP Publishing LLC)

## Problems

### 15.1.

(a) Using a parabolic dispersion for the conduction and valence bands, assuming the same effective mass $m^*$ in both bands, and assuming a direct gap, derive Keldysh's expression, Eq. (15.34), for the impact ionization rate.
(b) Using, instead, the random-$k$ approximation and Eq. (15.33), what exponent $n$ would you obtain for a power-law $(E - E_{\text{th}})^n$ form of the impact ionization scattering rate?

## References


1. D. Bohm, D. Pines, A collective description of electron interactions. I. Magnetic interactions. Phys. Rev. **82**, 625 (1951)
2. D. Bohm, D. Pines, A collective description of electron interactions: II. Collective vs individual particle aspects of the interactions. Phys. Rev. **85**, 338 (1952)
3. D. Bohm, D. Pines, A collective description of electron interactions: III. Coulomb interactions in a degenerate electron gas. Phys. Rev. **92**, 609 (1953)
4. D. Pines, A collective description of electron interactions: IV. Electron interaction in metals. Phys. Rev. **92**, 626 (1953)
5. P. Lugli, D.K. Ferry, Effect of electron-electron and electron-plasmon interactions on hot carrier transport in semiconductors. Physica B+C **129**, 532 (1985)
6. P. Lugli, Electron-electron and electron-plasmon interaction in polar semiconductors. Physica B+C **134**, 364 (1985)
7. M.V. Fischetti, S.E. Laux, Long-range Coulomb interactions in small silicon devices. Part I: performance and reliability. J. Appl. Phys. **89**, 1205 (2001)
8. M.V. Fischetti, Long-range Coulomb interactions in small Si devices. Part II: effective electron mobility in thin-oxide structures. J. Appl. Phys. **89**, 1232 (2001)
9. S.M. Goodnick, P. Lugli, Effect of electron-electron scattering on nonequilibrium transport in quantum-well systems. Phys. Rev. B **37**, 2578 (1988)
10. M.V. Fischetti, S. E. Laux, E. Crabbé, Monte Carlo simulation of hot-electron transport in silicon devices: is there a short-cut? J. Appl. Phys. **78**, 1058 (1995)
11. F.D.M. Haldane, Luttinger liquid theory' of one-dimensional quantum fluids. J. Phys. C Solid State Phys. **14**, 2585 (1981)
12. G. Giuliani, G. Vignale, *Quantum Theory of the Electron Liquid* (Cambridge University Press, Cambridge, 2005)
13. B.K. Ridley, *Quantum Processes in Semiconductors* (Oxford University Press, Oxford, 2000)
14. L. Meitner, Über die Entstehung der $\beta$-Strahl-Spektren radioaktiver Substanzen, Z. Phys. **9**, 131 (1922)
15. P. Auger, Sur les rayons $\beta$ secondaires produits dans un gaz par des rayons X. C. R. Hebd. Seances Acad. Sci. **177**, 169 (1923)
16. N. Sano, T. Aoki, M. Tomizawa, A. Yoshii, Electron transport and impact ionization in Si. Phys. Rev. B **41**, 12122 (1990)
17. M.V. Fischetti, N. Sano, S.E. Laux, K. Natori, Full-band-structure theory of high-field transport and impact ionization of electrons and holes in Ge, Si, and GaAs. J. Technol. Comput. Aided Des. **1**, 1–50 (1997)
18. H. Nara, A. Morita, Shallow donor potential in silicon. J. Phys. Soc. Jpn. **21**, 1852 (1966)
19. M.L. Cohen, T.K. Bergstresser, Band structures and pseudopotential form factors for fourteen semiconductors of the diamond and zinc-blende structures. Phys. Rev. **141**, 789 (1966)
20. J.R. Chelikowsky, M.L. Cohen, Nonlocal pseudopotential calculations for the electronic structure of eleven diamond and zinc-blende semiconductors. Phys. Rev. **14**, 556–582 (1976)
21. E.O. Kane, Electron scattering by pair production in silicon. Phys. Rev. **159**, 624 (1967)
22. L.V. Keldysh, Concerning the theory of impact ionization in semiconductors. Sov. Phys. JETP **21**, 1135 (1965) [J. Exptl. Theoret. Phys. (U.S.S.R.) **48**, 1692 (June, 1965)].
23. B.K. Ridley, Lucky-drift mechanism for impact ionisation in semiconductors. J. Phys. C Solid State Phys. **16**, 3373 (1983)
24. G.D. Mahan, *Many Particle Physics* (Plenum, New York, 1991)
25. C. Kittel, *Quantum Theory of Solids*, 2nd edn. (Wiley, New York, 1987)
26. A.L. Fetter, J.D. Walecka, *Quantum Theory of Many-Particle Systems* (Dover, Mineola, New York, 2003)

27. G.D. Mahan, Energy gap in Si and Ge: impurity dependence. J. Appl. Phys. **51**, 2634 (1980)
28. H.P.D.Lanyon, R.A. Tuft, Bandgap narrowing in moderately to heavily doped silicon. IEEE Trans. Electron Dev. **ED-26**, 1014 (1979)
29. R. Kubo, H. Ichimura, T. Usui, N. Hashitsume, *Statistical Mechanics* (North-Holland, Amsterdam, 1965)
30. M. Gell-Mann, K. Brueckner, Correlation energy of an electron gas at high density. Phys. Rev. **106**, 364 (1957)
31. P. Nozières, D. Pines, Correlation energy of a free electron gas. Phys. Rev. **111**, 442 (1958)
32. T. Uechi, T. Fukui, N. Sano, 3D Monte Carlo simulation including full Coulomb interaction under high electron concentration regimes. Phys. Status Solidi C **5**, 102 (2009)

# Chapter 16
# Radiative Processes: The Dipole Approximation

The interaction of electromagnetic radiation with matter is another vast subject. We have already cited the early theoretical book by Heitler [1] that describes it in a second-quantized form. A more recent review of the state of the art can be found in the text edited by Singh [2]. A discussion more specific to semiconductors is given by Pankove's lectures [3], whereas a discussion of electron–photon interactions in nanostructures is given in the recent text by Weiner and Nunes [4]. Clearly, light-emitting diodes, semiconductor injection lasers, and detectors are all devices of great technological importance that rely on efficient electron–photon interactions.

Here we present only briefly the main process of photon emission and absorption by a crystal using perturbation theory, the full-band structure of the crystal, and the basic *dipole approximation*.

## 16.1 Electron–Photon Hamiltonian

So far we have considered processes in which electrons either did not gain or lose energy (elastic processes, as impurity scattering), or in which energy is exchanged within the lattice (with ions, as in electron–phonon interactions, or with other electrons). These are called "non-radiative" processes, since energy is not transferred to radiation (that is, electromagnetic waves, light). Here let's consider processes in which electrons interact with light. We shall focus on light absorption across the band gap and mention only the existence of other processes.

We have already used before [Eqs. (8.34) and (8.64) in Chap. 8] the fact that the Hamiltonian of an electron in the presence of an electromagnetic field can be obtained from the Hamiltonian in the absence of the field but replacing the electron momentum $\widehat{\mathbf{p}} = -\mathrm{i}\hbar\nabla$ with $\widehat{\mathbf{p}} + e\mathbf{A}$, where $\mathbf{A}$ is the vector potential. Therefore, the single-electron Schrödinger equation reads

$$\mathrm{i}\hbar\frac{\partial\psi}{\partial t} = \frac{1}{2m_{\mathrm{el}}}(\widehat{\mathbf{p}}+e\mathbf{A})^2\psi - e\varphi(\mathbf{r})\psi\,. \tag{16.1}$$

Assuming that the electrostatic potential $\varphi$ vanishes (so, assuming a purely electromagnetic field, such as in interactions with light) and retaining only terms first order in the "perturbation" $\mathbf{A}$, we have

$$\mathrm{i}\hbar\frac{\partial\psi}{\partial t} = -\frac{\hbar^2}{2m_{\mathrm{el}}}\nabla^2\psi + \mathrm{i}\hbar\frac{e}{m_{\mathrm{el}}}\mathbf{A}\cdot\nabla\psi\,. \tag{16.2}$$



M. Fischetti, W.G. Vandenberghe, *Advanced Physics of Electron Transport in Semiconductors and Nanostructures*, Graduate Texts in Physics, DOI 10.1007/978-3-319-01101-1_16

The Lagrangian density yielding Eq. (16.2) via the Euler–Lagrange equations is [see Eq. (9.8) on page 189]:

$$\mathscr{L}(\psi, \dot{\psi}, t) = \mathrm{i}\hbar\psi^* \frac{\partial \psi}{\partial t} - \frac{\hbar^2}{2m_{\mathrm{el}}} \nabla\psi^* \cdot \nabla\psi + \mathrm{i}\hbar \frac{e}{m_{\mathrm{el}}} \psi^* \mathbf{A} \cdot \nabla\psi , \tag{16.3}$$

so that we can identify the perturbation Lagrangian with the last term. Therefore the perturbation Hamiltonian density will be

$$\mathscr{H}_{\mathrm{em}}(\psi, \pi, t) = -\frac{e}{m_{\mathrm{el}}} \pi \, \mathbf{A} \cdot \nabla\psi = \mathrm{i}\hbar \frac{e}{m_{\mathrm{el}}} \psi^* \mathbf{A} \cdot \nabla\psi . \tag{16.4}$$

Using the second-quantized version of the vector potential [Eq. (9.137), page 212] in terms of the photon creation and annihilation operators, we can write the Hamiltonian as (using a finite-volume normalization):

$$\begin{aligned}
\widehat{H}_{\mathrm{em}} &= \int \mathrm{d}\mathbf{r} \, \widehat{\mathscr{H}}_{\mathrm{em}}(\widehat{\psi}, \widehat{\pi}, t) = \mathrm{i}\hbar \frac{e}{m_{\mathrm{el}}} \int \mathrm{d}\mathbf{r} \, \widehat{\psi}^{\dagger} \widehat{\mathbf{A}} \cdot \nabla \widehat{\psi} \\
&= \sum_{\mathbf{k}\mathbf{k}'\mathbf{q}\lambda nn'} \mathrm{i}\hbar \frac{e}{m_{\mathrm{el}}} \left( \frac{\hbar}{2\epsilon_\infty \omega_q} \right)^{1/2} \widehat{c}^{\dagger}_{\mathbf{k}'} (\widehat{b}_{\mathbf{q}\lambda} + \widehat{b}^{\dagger}_{-\mathbf{q}\lambda}) \widehat{c}_{\mathbf{k}} \, \mathbf{e}_{\mathbf{q}\lambda} \\
&\quad \cdot \frac{1}{\Omega} \int \mathrm{d}\mathbf{r} \, \mathrm{e}^{-\mathrm{i}\mathbf{k}' \cdot \mathbf{r}} \, u^*_{\mathbf{k}'n'}(\mathbf{r}) \, \mathrm{e}^{-\mathrm{i}\mathbf{q} \cdot \mathbf{r}} \, \nabla \, \mathrm{e}^{\mathrm{i}\mathbf{k} \cdot \mathbf{r}} \, u_{\mathbf{k}n}(\mathbf{r}) ,
\end{aligned} \tag{16.5}$$

where $n$ and $n'$ are band indices. This expression is identical to the expression we have derived before for the electron–phonon interaction. The only nonzero matrix elements are those in which the final state differs from the initial by the presence of one fewer or one additional photon of momentum $\mathbf{q}$ and polarization $\lambda$ (for absorption and emission processes, respectively, associated with the terms $\widehat{b}_{\mathbf{q}\lambda}$ and $\widehat{b}^{\dagger}_{-\mathbf{q}\lambda}$) and by the conversion of the wavevector of one electron from $\mathbf{k}$ to $\mathbf{k}'$. Such a matrix element will have the form:

$$\begin{aligned}
\langle \mathbf{k}', n' | \widehat{H}_{\mathrm{em}} | \mathbf{k}, n \rangle &= \frac{ie\hbar}{m_{\mathrm{el}}} \sum_{\mathbf{q}\lambda} \left\{ \begin{array}{c} N^{1/2}_{\mathbf{q}\lambda} \\ (1 + N_{\mathbf{q}\lambda})^{1/2} \end{array} \right\} \left( \frac{\hbar}{2\epsilon_\infty \omega_q} \right)^{1/2} \\
&\quad \times \mathbf{e}_{\mathbf{q}\lambda} \cdot \frac{1}{\Omega} \int \mathrm{d}\mathbf{r} \, \mathrm{e}^{-\mathrm{i}\mathbf{k}' \cdot \mathbf{r}} \, u^*_{\mathbf{k}'n'}(\mathbf{r}) \, \mathrm{e}^{-\mathrm{i}\mathbf{q} \cdot \mathbf{r}} \, \nabla \, \mathrm{e}^{\mathrm{i}\mathbf{k} \cdot \mathbf{r}} \, u_{\mathbf{k}n}(\mathbf{r}) ,
\end{aligned} \tag{16.6}$$

the factor $N_{\mathbf{q}\lambda}$ being valid for absorption, $1 + N_{\mathbf{q}\lambda}$ for emission.

### 16.1.1 The Dipole Approximation

If the wavelength of the photon is much longer than the electron wavelength (which is always the case up into the visible/UV portion of the spectrum), we can set $\mathrm{e}^{-\mathrm{i}\mathbf{q} \cdot \mathbf{r}} \approx 1$ and

$$\begin{aligned}
\langle \mathbf{k}', n' | \widehat{H}_{\mathrm{em}} | \mathbf{k}, n \rangle &\approx \sum_{\mathbf{q}\lambda} \left\{ \begin{array}{c} N_{\mathbf{q}\lambda} \\ 1 + N_{-\mathbf{q}\lambda} \end{array} \right\} \frac{ie\hbar}{m_{\mathrm{el}}} \left( \frac{\hbar}{2\epsilon_\infty \omega_q} \right)^{1/2} \\
&\quad \times \mathbf{e}_{\mathbf{q}\lambda} \cdot \frac{1}{\Omega} \int \mathrm{d}\mathbf{r} \, \mathrm{e}^{-\mathrm{i}\mathbf{k}' \cdot \mathbf{r}} \, u^*_{\mathbf{k}'n'}(\mathbf{r}) \, \nabla \, \mathrm{e}^{\mathrm{i}\mathbf{k} \cdot \mathbf{r}} \, u_{\mathbf{k}n}(\mathbf{r}) .
\end{aligned} \tag{16.7}$$

This is called the "dipole approximation." Now:

$$\int d\mathbf{r}\, e^{-i\mathbf{k}'\cdot\mathbf{r}}\, u^*_{\mathbf{k}'n'}(\mathbf{r})\, \nabla\, e^{i\mathbf{k}\cdot\mathbf{r}}\, u_{\mathbf{k}n}(\mathbf{r})$$
$$= \int d\mathbf{r}\, e^{i(\mathbf{k}-\mathbf{k}')\cdot\mathbf{r}}\, u^*_{\mathbf{k}'n'}(\mathbf{r})\, \nabla\, u_{\mathbf{k}n}(\mathbf{r}) + \int d\mathbf{r}\, e^{-i\mathbf{k}'\cdot\mathbf{r}}\, u^*_{\mathbf{k}'n'}(\mathbf{r})\, u_{\mathbf{k}n}(\mathbf{r}) \nabla e^{i\mathbf{k}\cdot\mathbf{r}}\,. \tag{16.8}$$

The second term contains an integral which is significantly different from zero only when $\mathbf{k}' \approx \mathbf{k}$. In this case the periodic parts of the Bloch functions $u_{\mathbf{k}'n'}(\mathbf{r})$ and $u_{\mathbf{k}n}(\mathbf{r})$ are orthogonal and the exponential is slowly varying. Therefore the second integral is usually neglected. The first term can be treated in the "usual" way, by splitting the integral over the entire volume into a sum of integrals over individual cells by setting $\mathbf{r} \to \mathbf{R}_l + \mathbf{r}$:

$$\langle \mathbf{k}'n'|\widehat{H}_{\mathrm{em}}|\mathbf{k}n\rangle \approx \frac{ie\hbar}{m_{\mathrm{el}}} \sum_{\mathbf{q}\lambda} \left\{ \frac{N_{\mathbf{q}\lambda}^{1/2}}{(1+N_{\mathbf{q}\lambda})^{1/2}} \right\} \left( \frac{\hbar}{2\epsilon_\infty \omega_q} \right)^{1/2}$$
$$\times\, \mathbf{e}_{\mathbf{q}\lambda} \cdot \sum_l e^{i(\mathbf{k}-\mathbf{k}')\cdot\mathbf{R}_l} \frac{1}{\Omega} \int_{\Omega_c} d\mathbf{r}\, u^*_{\mathbf{k}'n'}(\mathbf{r}) \nabla\, u_{\mathbf{k}n}(\mathbf{r})$$
$$= \frac{ie\hbar}{m_{\mathrm{el}}} \sum_{\mathbf{q}\lambda} \left\{ \frac{N_{\mathbf{q}\lambda}^{1/2}}{(1+N_{-\mathbf{q}\lambda})^{1/2}} \right\} \left( \frac{\hbar}{2\epsilon_\infty \omega_q} \right)^{1/2}$$
$$\times\, \mathbf{e}_{\mathbf{q}\lambda} \cdot \frac{1}{\Omega_c} \int_{\Omega_c} d\mathbf{r}\, u^*_{\mathbf{k}n'}(\mathbf{r}) \nabla\, u_{\mathbf{k}n}(\mathbf{r})\, \delta_{\mathbf{k}'\mathbf{k}}\,, \tag{16.9}$$

since $N_{\mathrm{cells}}/\Omega = 1/\Omega_c$. Thus, the optical matrix element vanishes unless $\mathbf{k}' = \mathbf{k}$. Only "vertical (or direct) transitions" can happen. Transitions which involve a change in (crystal) momentum can only happen at higher orders in perturbation theory, requiring extra momentum which must be provided by phonons, traps, and such.

Now let's define the "momentum matrix element" between the periodic part of two Bloch functions:

$$\langle n'|\widehat{\mathbf{p}}|n\rangle = -\frac{i\hbar}{\Omega_c} \int_{\Omega_c} d\mathbf{r}\, u^*_{\mathbf{k}n'}(\mathbf{r}) \nabla\, u_{\mathbf{k}n}(\mathbf{r})\,. \tag{16.10}$$

Then:

$$\langle \mathbf{k}'n'|\widehat{H}_{\mathrm{em}}|\mathbf{k}n\rangle \approx -\frac{e}{m_{\mathrm{el}}} \sum_{\mathbf{q}\lambda} \left\{ \frac{N_{\mathbf{q}\lambda}^{1/2}}{(1+N_{-\mathbf{q}\lambda})^{1/2}} \right\} \left( \frac{\hbar}{2\epsilon_\infty \omega_q} \right)^{1/2} \mathbf{e}_{\mathbf{q}\lambda} \cdot \langle n'|\widehat{\mathbf{p}}|n\rangle\,. \tag{16.11}$$

Note that this is the most notable example of a type of transition for which the form of the Bloch functions matters. In particular, well-known selection rules apply. For example, the momentum matrix between atomic orbitals is nonzero only when the azimuthal quantum number $m_J$ of the final and initial orbitals differs by zero or one. Similarly in a crystal, photon absorption can occur only between bands whose orbital character satisfies this selection rule. In addition, the knowledge of the Bloch functions is required to calculate the matrix element (16.11).

## 16.2 Transition Rates

The rate at which a photon of momentum $\mathbf{q}$ and polarization $\lambda$ is absorbed by an electron of wavevector $\mathbf{k}$ in band $n$ can be calculated from Fermi's Golden Rule:

$$\begin{aligned} W_{\text{abs},n} &= \frac{2\pi}{\hbar} \int \frac{d\mathbf{k}}{(2\pi)^3} \sum_{n'} \frac{\hbar e^2}{2\epsilon_\infty \omega_q m_{\text{el}}^2} N_{\mathbf{q}\lambda} \, |\mathbf{e}_{\mathbf{q}\lambda} \cdot \langle n'|\widehat{\mathbf{p}}|n\rangle|^2 \, \delta[E_n(\mathbf{k}) - E_{n'}(\mathbf{k}) - \hbar\omega_{\mathbf{q}\lambda}] \\ &\approx \frac{\pi e^2}{m_{\text{el}}^2 \omega \epsilon_\infty} N_\omega \sum_{n'} |\mathbf{e}_{\mathbf{q}\lambda} \cdot \langle n'|\widehat{\mathbf{p}}|n\rangle|^2 \, \mathscr{D}_{j,n,n'}(\hbar\omega) , \end{aligned} \tag{16.12}$$

where $\mathscr{D}_{j,n,n'}(E) = \int d\mathbf{k}/[(2\pi)^3]\delta[E_n(\mathbf{k}) - E_{n'}(\mathbf{k}) - E]$ is the joint density of states at energy $E$ in bands $n$ and $n'$. On the contrary, for the rate at which an electron of wavevector $\mathbf{k}$ in band $n$ emits a photon, we have (assuming spontaneous emission only):

$$\begin{aligned} W_{\text{emiss},n} &= \frac{2\pi}{\hbar} \int \frac{d\mathbf{q}}{(2\pi)^3} \sum_{n'} \frac{\hbar e^2}{2\epsilon_\infty \omega_q m_{\text{el}}^2} |\mathbf{e}_{\mathbf{q}\lambda} \cdot \langle n'|\widehat{\mathbf{p}}|n\rangle|^2 \, \delta[E_{n'}(\mathbf{k}+\mathbf{q}) - E_n(\mathbf{k}) - \hbar\omega_{\mathbf{q}\lambda}] \\ &= \frac{\pi e^2}{m_{\text{el}}^2 \epsilon_\infty} |\langle n'|\widehat{\mathbf{p}}|n\rangle|_{\text{ave}}^2 \int \frac{q^2 dq}{8\pi^3 \omega} = \frac{\pi e^2}{m_{\text{el}}^2 \epsilon_\infty} |\langle n'|\widehat{\mathbf{p}}|n\rangle|_{\text{ave}}^2 \int d(\hbar\omega)\mathscr{D}(\hbar\omega) , \end{aligned} \tag{16.13}$$

where $|\langle n'|\widehat{\mathbf{p}}|n\rangle|_{\text{ave}}$ is the dipole matrix element summed over all bands and averaged over the photon polarizations and $\mathscr{D}(\hbar\omega) = \omega^2/(2\pi^2 \upsilon_\omega^3)$ is the photon density of states expressed as a function of the phase velocity $\upsilon_\omega = \omega/q$.

### 16.2.1 Absorption Spectrum

In the particular case in which $n'$ labels the conduction band and $n$ the valence band, the matrix element for the process in which a photon of momentum $\mathbf{q}$ and polarization $\lambda$ is absorbed by the solid, triggering the excitation of an electron from the valence to the conduction band, will be

$$\langle \mathbf{k}c|\widehat{H}_{\text{em}}|\mathbf{k}v\rangle \approx -\frac{e}{m_{\text{el}}} N_{\mathbf{q}\lambda} \left(\frac{\hbar}{2\epsilon_\infty \omega_q}\right)^{1/2} \mathbf{e}_{\mathbf{q}\lambda} \cdot \langle c|\widehat{\mathbf{p}}|v\rangle . \tag{16.14}$$

The absorption rate will be, according to Eq. (16.12):

$$\begin{aligned} I(\hbar\omega) &= \frac{\pi e^2}{m_{\text{el}}^2 \epsilon_\infty \omega} N_\omega \int dE \, |\mathbf{e} \cdot \langle c|\widehat{\mathbf{p}}|v\rangle|^2 \mathscr{D}_j(E + \hbar\omega - E_{\text{gap}}) \\ &\quad \times [1 - f_c(E + \hbar\omega - E_{\text{gap}})] \, f_v(E) , \end{aligned} \tag{16.15}$$

where $\mathscr{D}_j$ is the joint DOS summed over all bands and $f_v$ and $f_c$ the Fermi occupation factors in the valence and conduction band, respectively. Assuming a completely full VB and empty CB,

$$I(\hbar\omega) \approx \frac{\pi e^2}{m_{\text{el}}^2 \epsilon_\infty \omega} N_\omega \int dE \, |\mathbf{e} \cdot \langle c|\widehat{\mathbf{p}}|v\rangle|^2 \mathscr{D}_{j,c}(E + \hbar\omega - E_{\text{gap}}) , \tag{16.16}$$

The absorption spectrum of a direct-gap semiconductor will show a sharp threshold at $\hbar\omega = E_{gap}$. For indirect-gap materials, instead, a weaker and softer threshold occurs at the gap, since the help of a phonon is necessary to conserve momentum and the matrix element in Eq. (16.15) is replaced by a second-order electron–phonon–photon matrix element.

One important feature of the absorption spectrum of several semiconductors is the appearance of "exciton peaks." When an electron–hole pair is formed by the absorption of a photon, the pair can remain weakly bound, as in a hydrogen atom. The binding energy of the pair will be exactly the binding energy of the H atom in its ground state, corrected for the different dielectric constant and for the reduced electron–hole mass, $1/m_{red} = 1/m_e + 1/m_h$, (where $m_e$ and $m_h$ are the masses of the conduction- and valence-band extrema, respectively):

$$\Delta E = \frac{e^2}{(4\pi\epsilon_\infty)^2} \frac{m_{red}}{2\hbar^2} \,. \tag{16.17}$$

Thus, the absorption spectrum shows a peak at a photon frequency $\hbar\omega = E_{gap} - \Delta E$ (see Fig. 5.4 in Ridley's book [5]).

Given the important application to photovoltaic solar cells, absorption spectra have been calculated for a variety of semiconductors (see the text by Haug and Koch [6], for example). In Fig. 16.1 we give a different example of what can be calculated using empirical pseudopotentials and the theory discussed in this chapter. This figure shows the energy distribution of electrons and holes generated (as pairs, obviously) in bulk GaAs at steady state under the illumination of the sun at sea level. These results have been obtained by calculating the absorption rates from the transition rates in the integrand of Eq. (16.12) for all bands $n$, by integrating over the solar spectrum, and integrating also over all electrons and holes generated at a particular energy. The solar spectrum is well known: The National Renewable Energy Laboratory provides solar spectra at all latitudes for length of path through the

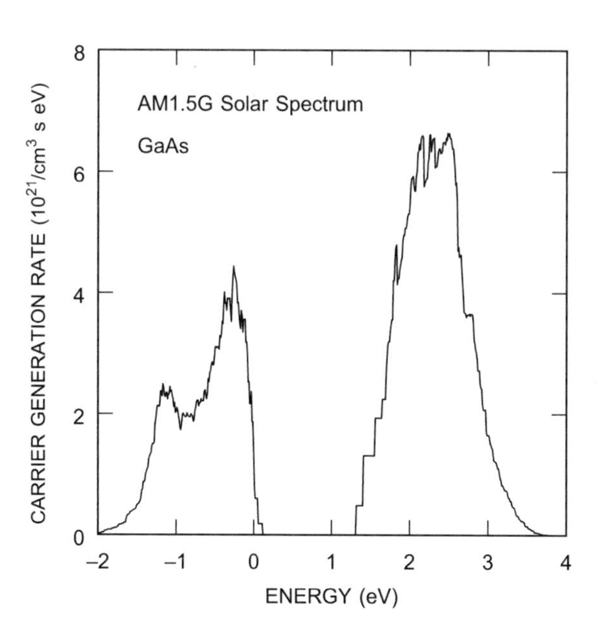


**Fig. 16.1** Energy distribution of electrons and holes generated (as pairs) in GaAs under illumination from the sun at sea level. These results have been obtained by calculating the absorption rates from the integrand of Eq. (16.12) for all bands $n$, integrating over the solar spectrum, and integrating over all electrons and holes generated at a particular energy. The solar spectrum is taken from well-known standards, in this case at sea level for a 1.5 air mass global, known as "AM1.5G"

atmosphere (that varies with time of the day and latitude, of course) [7]. This is known as the "air mass." The situation is highly idealized: Photons are assumed to be present everywhere in the bulk of the crystal whereas, in reality, they will be absorbed in the first few micrometers from the surface; generated carriers are assumed to be swept away as soon as they are generated, but, in reality, space-charge will build up while drift diffusion and thermalization will occur. Nevertheless, these effects can be accounted for in computer simulations in order to estimate the theoretical maximum quantum efficiency (that is, pairs generated per incident photon) and the actual efficiency of a device. Note how electrons and holes are generated at high energies. In conventional solar cells, most of this energy is lost to heat (phonons), as carriers decay to the extremum of their bands before being collected and contributing to the generation of electrical power. Harnessing this wasted energy is clearly one of the goals of current research and understanding carrier generation, recombination, and transport at this high energy is of paramount importance.

## 16.3 Radiative and Non-radiative Transitions

The theoretical scheme we have employed so far can be used in a variety of different situations. Electronic transitions can occur within the conduction band, from valence to conduction band, from valence or conduction band to localized defect or impurity states (traps). They can be radiative (when assisted by photons) or non-radiative. We have seen the cases of photon, phonon, and Coulomb matrix elements at first order. The general transition rate can be evaluated using first-order perturbation theory (Fermi's Golden Rule or Born approximation for direct transitions). The major cases we have not treated are transitions involving traps (which requires the knowledge of the localized wavefunction, often hard to estimate), transitions involving large energy exchange (such as a conduction electron being trapped into a localized state energetically located around the middle of the gap) which may require multi-phonon processes (such as emission of many phonons or processes relying on anharmonic corrections to the phonon Hamiltonian), or even more complicated processes in which the lattice rearranges itself. Ridley's text describes in Chap. 5 radiative transitions of types different from the valence-to-conduction direct absorption we have considered above: Free-carrier absorption (that is, absorption of a photon by a free electron in the conduction band, involving an analysis of the requirements posed by energy and momentum conservation), photo-ionization of localized states of exciton absorption (requiring some discussion of the localized or excitonic wavefunctions), indirect transitions (assisted by phonons and so requiring second-order perturbation theory), and light scattering by free electrons (which depends on the second-order term $e^2A^2/(2m_{\mathrm{el}})$, we have so far neglected in Eq. (16.1)). In Chap. 6 Ridley discusses non-radiative transitions. While the details (mainly dictated by energy/momentum conservation and by the particular form of the initial/final wavefunctions involved) change significantly from one case to the next, the general scheme employed is always the same. We shall not discuss these issues in detail, as the examples we have seen so far should be sufficient to give the gist of the matter.

## Problems

**16.1.** Given the AM1.5G solar spectrum, $\Phi(\lambda)$ (photons/(m$^2$s eV)), write the equations necessary to obtain the data shown in Fig. 16.1 starting from Eq. (16.12).

**16.2.** This is the inverse problem of the previous one: Given the data shown in Fig. 16.1, write the equations required to obtain the absorption spectrum of Si, knowing the AM1.5G solar spectrum, $\Phi(\lambda)$ (photons/(m$^2$s eV).

## References

1. W. Heitler, *The Quantum Theory of Radiation*, 3rd edn. (Dover, Mineola, NY, 2010)
2. J. Singh, *Optical Properties of Condensed Matter and Applications* (Wiley, London, 2006)
3. J. Pankove, *Optical Processes in Semiconductors* (Dover, Mineola, NY, 2010)
4. J. Weiner, F. Nunes, *Light-Matter Interaction: Physics and Engineering at the Nanoscale* (Oxford University Press, Oxford, 2012)
5. B.K. Ridley, *Quantum Processes in Semiconductors* (Oxford University Press, Oxford, 2000)
6. H. Haug, S.W. Koch, *Quantum Theory of the Optical and Electronic Properties of Semiconductors*, 5th edn. (World Scientific, Singapore, 2009)
7. Reference Solar Spectral Irradiance: Air Mass 1.5 National Renewable Energy Laboratory (2005)

# Part V
# Electronic Transport

# Chapter 17
# Overview of Quantum-Transport Formalisms

## 17.1 Overview

In Chap. 12, Sect. 12.2.1, we discussed the reason why it made sense to calculate scattering rates using Bloch functions. We explained that when an external bias results in a small change of the electron potential energy over the distance it takes the electron to lose memory of its phases, we can assume that "locally" the electron sees a homogeneous environment, with an electric field small enough to affect only minimally the band structure and the electron wavefunction. In other words, we restricted ourselves to situations in which "things change" over distances so long that quantum interference effects are small enough to be ignored. Collisions are so frequent that we cannot talk of a coherent "wavefunction" extending over distances so long. This led us to the regime of semiclassical transport. On the other hand, we have dedicated an entire long chapter, Chap. 7, to the study of the electronic structure in systems of nanometer dimensions. We have justified such a strong interest on structures with small "lateral" dimensions (thickness of a film or diameter of a NW, for example) by saying that such tiny dimensions are necessary when also the "length" decreases: Good electrostatic behavior of electronic devices requires scaling *all* dimensions at the same time. Therefore, in these tiny nanostructures, narrow/thin *and* short, the assumption that transport occurs semiclassically is likely to be unrealistic.

The term *Quantum Transport* is interpreted in many different ways. For example, clearly semiclassical transport in a narrow NW is affected by quantization effects. This may be viewed as "quantum," in some sense. However, if the NW is very long, then transport can still be treated semiclassically. Here by "quantum transport" we mean transport treated beyond the semiclassical approximation, retaining possible coherence of the electronic wavefunctions over distances comparable to the length of the region of interest (e.g., the length of a channel of a NW-FET).

In Quantum Field Theory (QFT), the full information of a system is contained into the expectation value of the quantum field on the vacuum state:

$$G(\mathbf{r},t;\mathbf{r}',t') = \langle 0|\hat{\psi}^{\dagger}(\mathbf{r}',t')\ \hat{\psi}(\mathbf{r},t)|0\rangle\ . \tag{17.1}$$

(We shall be cavalier about the time-ordering here, the problem will be considered below in Sect. 17.2.5.) This is called a *Green's function* and its Fourier transform, $\langle 0|\hat{c}^{\dagger}_{\mathbf{k}}\hat{c}_{\mathbf{k}}|0\rangle$, is called the *propagator*. Once this quantity is known, we can calculate every property of the system. Indeed the difficult part of the task is to calculate the field $\hat{\psi}(\mathbf{r},t)$. All interactions among particles and with



M. Fischetti, W.G. Vandenberghe, *Advanced Physics of Electron Transport in Semiconductors and Nanostructures*, Graduate Texts in Physics, DOI 10.1007/978-3-319-01101-1_17

external perturbations are included in the expression of the field itself. Starting from the known field of free particles, $\hat{\psi}_0(\mathbf{r},t)$, perturbation theory is usually employed to calculate some approximate expression for the interacting propagator.

In our case, a similar quantity plays the role of the propagator. Recall that our system is composed of many particles, usually interacting not only among themselves, but also with the surrounding "thermal bath" (i.e., the phonons of the crystal), with impurities and other elementary excitations. The number of degrees of freedom is so large, that we must seek help from Statistical Mechanics, as discussed in Chap. 10. Therefore, the equivalent of Eq. (17.1) will be the statistical average of the field:

$$G(\mathbf{r},t;\mathbf{r}',t') \;=\; \mathrm{Tr}\left[\hat{\rho}\ \hat{\psi}^\dagger(\mathbf{r}',t')\ \hat{\psi}(\mathbf{r},t)\right] = \langle \hat{\psi}^\dagger(\mathbf{r}',t')\ \hat{\psi}(\mathbf{r},t)\rangle\ . \tag{17.2}$$

For homogeneous systems at equilibrium, the grand canonical density matrix is often used. This allows for an elegant connection between Eqs. (17.1) and (17.2): The time evolution in the former is of the form $\exp[(\mathrm{i}/\hbar)\widehat{H}t]$, while the equilibrium density matrix has the form $\exp(-\beta\widehat{H})$, (where, to conform to the usual convention in the many-body/statistical-mechanics community, we temporarily use $\beta = 1/k_\mathrm{B}T$ to denote the inverse temperature). A rotation of the former to imaginary time transforms QFT to equilibrium quantum Statistical Mechanics, also know as Euclidean QFT, with the corresponding imaginary time Green's functions known as the "Matsubara" Green's functions. Results from QFT can be used to study equilibrium in many-body systems. The two-volume textbook by Zubarev, Morozov, and Ropke[1, 2] gives an ample and deep discussion of this topic. However, our case is more complicated. First, we are interested in systems that are driven off-equilibrium by the applied bias. Second, our systems are highly inhomogeneous, because, once more, of the applied bias and, in most cases, a compositional inhomogeneity due to different materials used for active layers—often heterogeneous as in III–V hetero-structures, contacts, or insulating layers. The former consideration requires going beyond Statistical Mechanics. Moreover, the inhomogeneity of the system means that, in most cases, we have to work hard just to calculate the unperturbed field (or wavefunctions in a simpler first quantization approximation). The inhomogeneity also prevents us from employing periodic boundary conditions, since our system is open. Therefore, we will have also to determine the proper boundary conditions to adopt.

In very general terms, approaches that deal with the strong off-equilibrium situation fall into two categories: The first category includes mainly the Non-Equilibrium Green's Function (NEGF) approach. This method was originally developed by Keldysh [3] and by Kadanoff and Baym [4] and today it is very popular. The second category consists in approaches based on the density matrix. This is the very same Green's function given by Eq. (17.2), but evaluated at *equal times*, $t' = t$. These methods lead to various Master-equation approaches—discussed in [5], including techniques to introduce irreversibility via the so called Lindblad equation. They also lead to approaches based on the Semiconductor Bloch Equations—formulation that is discussed in [6] and [7]. These are, more appropriately, formulations that deal more with the excited states of the system than with electronic transport *per se*; indeed, their main application has been in the study of ultra-fast optical excitations. Also based on the density matrix is the use of the transform of $\hat{\rho}$ called the *Wigner* or *Wigner–Weyl transform*. This leads to the so-called Wigner functions approach—discussed in [8, 9], and in [10], for example—whose major appeal relies on its intuitive similarity to the Boltzmann-equation formulation of semiclassical transport. Additional approaches such as the Bohm trajectory method [11] or on Feynman path integrals [12, 13] have been proposed, but NEGF- and density-matrix-based approaches remain the most popular choice. Jacoboni's [14] and Datta's [15] books brilliantly survey the state of the art, albeit with very different perspectives.

We shall proceed as follows: First, we shall worry about what is required to calculate the unperturbed, free fields (wavefunctions). As we have observed, this is a necessary and nontrivial task in our inhomogeneous systems. It also provides already some information about the current that a device will carry under ideal, *ballistic conditions*. The off-equilibrium conditions do not present any particular problem in this case. We shall present a general procedure, the *Quantum Transmitting Boundary Method* (QTBM) [16]—fully equivalent and even more efficient than the ballistic NEGF, to deal with open systems. The effective-mass approximation will be used for notational simplicity. Full-band treatments have been discussed by the NEMO group [17] using tight-binding, or by Choi and Ihm [18] or one of us [19] using empirical pseudopotentials. Since ballistic transport is an over-idealization of reality, we shall discuss briefly the NEGF and the Winger function approaches. In the next chapter we will deal with the density-matrix approach, motivated mainly by the (doomed) attempt to derive the semiclassical Boltzmann transport equation. However, in so doing, we shall derive the transport equation including dissipation in the restricted case of ultra-short devices and weak scattering.

## 17.2 Ballistic Transport

The system we have in mind is a general bounded region in real space. We may consider this as a region in empty space delimited by some boundary or, more appropriately for the subject of this book, a solid crystalline "chunk," of homogeneous or inhomogeneous composition, bounded by barriers that prevent electrons from "spilling" outside this region. To fix the idea, think of a Si slab or an InGaAs NW bounded by ideal insulators. In previous chapters we have discussed at length the electronic structure of such a "chunk." However, we have assumed that this region extended infinitely in three dimensions (bulk crystals), in two dimensions (thin films, graphene, 2DEG in general), or in one dimension (NWs, GNRs, CNTs). In the last two cases, these systems were "closed" in one or two dimensions, but open and periodic along the remaining dimension(s). Therefore, a NW is "closed" on the cross-sectional plane, but periodic along its axial direction.

The situation we wish to consider now is different. If our "chunk" were closed in all dimensions, we would treat it as a zero-dimensional electron gas, that is, a quantum dot, assuming that scattering does not break the coherence of the electrons within the chunk itself. However, now we assume that the system is "open" over a finite number of small regions on its boundary. These "openings" are thought to be in contact with conducting "wires" along which electrons can travel. We shall call these "openings" *contacts* and refer to our "chink" as the *device*. The system is now "open," in the sense that electrons can move in-and-out of the device. We should also assume that, in general, these contacts are kept at different chemical potentials (Fermi levels) by applying an external bias.

It is convenient to consider first electronic states in the contacts. We shall label them by $(\alpha,k)$. The label $k$ refers to a continuous quantum number; $\alpha$ is an additional quantum number labeling the different solutions with the same $k$. For example, if the contact is thought to be a straight nanowire, $k$ will label the electron wavevector along the axial direction of the NW, $\alpha$ will label the (sub)bands. If the system consists of a thin Si film on the $(x,y)$ plane, infinitely extended along $y$, but of finite length $L$ along $x$, with contacts at $x=0$ and $x=L$, the contacts themselves will be infinitely extended along $y$, In this case, $k$ will actually be a two-dimensional wavevector $\mathbf{K}$ (representing the wavevector of the electrons injected or extracted by the contacts) and $\alpha$ will label the subbands.

Since we are interested in calculating the current that flows in the device, we must first solve the Schrödinger equation:

$$\left[-\frac{\nabla^2}{2m} + V^{(\mathrm{tot})}(\mathbf{r})\right]\ \psi_\alpha(\mathbf{r};k) \ = \ E_\alpha(k)\ \psi_\alpha(\mathbf{r};k)\ , \tag{17.3}$$

where $V^{(\mathrm{tot})}$ is the total electron potential energy. In general, we may view this as the single-particle equation used to deal with a crystal, so that $V^{(\mathrm{tot})}$ will include the lattice potential, Hartree, and exchange-correlation contributions in addition to whatever external potential results from the applied bias. In this case, the mass $m$ will be the free-electron mass $m_{\mathrm{el}}$. Instead, when employing the simpler effective-mass approximation, the mass $m$ should be interpreted as some effective mass $m^*$ and $V^{(\mathrm{tot})}$ will represent only the Hartree potential (obtained by some self-consistent solution of Eq. (17.3) and Poisson equation, along the lines of the discussion of Sect. 8.2.1). Note that in considering Eq. (17.3) we are implicitly assuming that we are interested in the steady-state (stationary) behavior of the device. We shall worry later about the time dependence.

Equation (17.3) is, of course, a second-order, linear differential equation that requires boundary conditions in order to determine a unique solution. When dealing with bulk semiconductors or nanostructures in previous chapters, we have assumed periodic or closed boundary conditions. Now the situation is very different, exactly for the reasons mentioned before: The system is open. We shall consider below how to determine the proper open boundary conditions. Now we can see quite simply how, having solved Eq. (17.3), we can calculate the current flowing across the device. Recall that associated with each wavefunction $\psi_\alpha$ is a quantum-mechanical probability current:

$$\mathbf{J}_\alpha(\mathbf{r};k) = \frac{\hbar}{2mi}\left[\psi^*_\alpha(\mathbf{r};k)\nabla\psi_\alpha(\mathbf{r};k) - (\nabla\psi^*_\alpha(\mathbf{r};k))\,\psi_\alpha(\mathbf{r};k)\right] \tag{17.4}$$

$$= \frac{\hbar}{m}\mathrm{Im}\left[\psi^*_\alpha(\mathbf{r};k)\nabla\psi_\alpha(\mathbf{r};k)\right] . \tag{17.5}$$

Therefore, the probability current through a contact is obtained by integrating over a surface covering the contact:

$$J_\alpha(k) = \int_{\mathrm{contact}} \mathrm{d}\mathbf{S}\cdot\mathbf{J}_\alpha(\mathbf{r};k) . \tag{17.6}$$

The total electrical current flowing through the contact can be obtained by weighing each wavefunction with its probability of occupation $f_\alpha(k)$ and multiplying it by the elementary charge,

$$I = -e\sum_\alpha \int \frac{\mathrm{d}k}{2\pi} f_\alpha(k)\, J_\alpha(k) . \tag{17.7}$$

assuming that the wavefunctions are normalized:

$$\int \mathrm{d}\mathbf{r}\; \psi^*_\alpha(\mathbf{r};k)\; \psi_\alpha(\mathbf{r};k) = 2\pi\,\delta_{\alpha\alpha'}\,\delta(k-k') . \tag{17.8}$$

The current computed in this way is called the *ballistic current* since no additional scattering process, such as the electron–phonon interaction, is taken into account.

### *17.2.1 Quantum Conductance*

It is instructive to consider the very simple case of a one-dimensional system described by a single effective mass $m^*$ and identical contacts at the left and at the right and without a barrier in the device region. In this case, $k$ represents the wavevector of the injected waves and the index $\alpha$ is not necessary. Moreover, the wavefunctions in the contacts are plane waves $\psi(x) = \mathrm{e}^{ikx}$, each wavefunction carrying a probability current $J = \hbar k/m^*$. When a sufficiently small bias $V_{\mathrm{bias}}$ is applied between the left and right contacts, we can assume that the wavefunctions are still plane waves and carry a current $\hbar k/m$ across the device. While the effect of the bias on the wavefunctions is assumed to be negligible, the occupation of the states is still determined by the bias. Only states with $k$ in the range $k\in[k_{\mathrm{min}},k_{\mathrm{max}}]$ (with $E(k_{\mathrm{max}})-E(k_{\mathrm{min}}) = eV_{\mathrm{bias}}$) contribute to the current, since states above

$E(k_{\rm max})$ are unoccupied and the left-going and right-going fluxes of electrons in states below $E(k_{\rm min})$ cancel. As a consequence, the total current is

$$I^{\rm max} = -e\int_{k_{\rm min}}^{k_{\rm max}} \frac{{\rm d}k}{2\pi}\frac{\hbar k}{m} = -\frac{e}{2\pi}\frac{\hbar(k_{\rm max}^2 - k_{\rm min}^2)}{2m} = -e\frac{E(k_{max}) - E(k_{\rm min})}{h}. \tag{17.9}$$

Since $E(k_{\rm max}) - E(k_{\rm min}) = -eV_{\rm bias}$,

$$G_{\rm quantum} = 2\frac{{\rm d}I_{\rm max}}{{\rm d}V_{\rm bias}} = \frac{2e^2}{h} \approx 7.748\times 10^{-5}\ {\rm S} = 2/R_{\rm K} \tag{17.10}$$

where the factor 2 stems from spin degeneracy. Another more general and powerful way to formulate this expression is the make use of Landauer's formula [20]:

$$G_{\rm quantum} = \frac{e^2}{\pi\hbar}\sum_{\nu} T_{\nu}\,, \tag{17.11}$$

where $\nu$ now denotes the number of conduction "channels" and $T_{\beta}$ is the transmission probability of each channel. For ballistic transport in a one-dimensional system (a NW, a CNT, or a GNR, to fix the ideas) $\alpha$ will represent the number of quantized states (subbands) and $T_{\alpha} = 1$ is their transmission probability. We have made use of this expression in Chap. 7, when calculating the ballistic conductance at a given energy, Eq. (7.42). A related quantity is the von Klitzing constant $R_{\rm K} \approx 25813\,\Omega$ which gives the resistance of a channel without spin degeneracy.

### 17.2.2 *Open Boundary Conditions*

We now wish to return to the problem of establishing the correct open boundary conditions for Eq. (17.3) in the presence of an arbitrary potential. The method, commonly referred to as the *QTBM*, is described in general two-dimensional cases in [16] from a theoretical perspective, whereas [21] presents several interesting applications.

The first step is to investigate how a local change of the potential affects the ballistic current. Recall that we have divided the system into a *device region* and two or more *leads* (or *contacts*).

* *This separation of the system into a "device"—within which one assumes non-equilibrium—and "contacts"—kept at equilibrium—is artificial and ultimately not justified from a physics perspective: These external reservoirs are assumed to remain at thermal equilibrium, to absorb particles (waves) and inject them in an ideal fashion. However, they are assumed to remain at equilibrium and "field-free" even in the region adjacent to the device region. This violates current continuity, of course (equilibrium contacts carry no current, so that at the contact/device boundary the current density is discontinuous). Moreover, scattering is assumed to be infinitely strong within the reservoirs, so that we may justify why we keep them at equilibrium. However, some relaxation time and distance are required for waves entering the contacts to relax. Therefore, there will be a transition region between device and contact in which the device and the contacts interact. Several quantum-transport schemes account for these effects. For now, we shall consider this device/contact distinction as a mathematically necessary evil, since we need a boundary to study a finite region and we need boundary conditions to solve our transport equations, whatever form these may take.*

The potential is assumed to be uniform inside the leads but changing in the device region. Take, for example, a one-dimensional device with a left and a right lead, the left lead ending at $x = 0$ and the right lead starting at $x = L$:

$$V(x) = \begin{cases} V_{\rm L} & x \leq 0 \\ V_{\rm D}(x) & 0 < x < L \\ V_{\rm R} & x \geq L \end{cases} . \tag{17.12}$$

The solutions of the Schrödinger equation for a device with leads will obviously depend on the particular shape of $V_{\rm D}(x)$. However, in general two types of solutions can be distinguished: *extended* and *bound* states. Extended (or *propagating*) states are a linear combination of plane waves inside the leads and they may carry current, in a way similar to how plane waves carry current. Extended states have a continuous spectrum. Bound states, on the other hand, do not behave like plane waves inside the leads but, rather, decay exponentially into the leads, bound states can not carry current and they have a discrete spectrum.

In one dimension, a necessary and sufficient condition for the presence of bound states is $V_{\rm D} < \min(V_{\rm L}, V_{\rm R})$. For example, in the trivial case where $V_{\rm D}(x) = V_{\rm L} = V_{\rm R}$, the set of plane waves $e^{ikx}$ form a complete set of solutions labeled by a continuous quantum number $k$ and no bound states exist, since $V_{\rm D} \not< V_{\rm L,R}$. In higher dimensions, no simple criterion for the presence of bound states exists.

The solutions of interest for quantum transport are the extended states. The general procedure to determine the extended states is as follows:

- Choose a lead $s$, solve the Schrödinger equation in the lead, and select an *incoming wavefunction* $\psi_{s,\alpha}(\mathbf{r}; k_{\rm I})$ which has its current flowing towards the device. The wavefunction has an energy $E_{s\alpha}(k_{\rm I})$. In our one-dimensional example, the incoming wave is a plane wave $e^{ik_{\rm I}x}$ with $k_{\rm I} > 0$ if the wave is incoming in the left lead and $k_{\rm I} < 0$ otherwise.
- Determine an algebraic set of solutions with energy $E = E_{s,\alpha}(k_{\rm I})$ for the Schrödinger equation in all leads (including the chosen lead). The solutions need not be normalizable, so exponentially decaying or growing solutions are acceptable as well at this point. In practice, this is done by rewriting the Schödinger equation from an $E(k)$-eigenvalue problem to a $k(E)$-eigenvalue problem. Solving the latter is referred to as solving the complex band structure, since not only real $k$ but also complex $k$ are found. In our one-dimensional example, each lead has two solutions, $e^{\pm ik_{\rm L,R}x}$ with $\hbar^2 k_{\rm L,R}^2/(2m) + V_{\rm L,R} = E$ or $k_{\rm L,R} = \pm\sqrt{2m/\hbar^2 (E - V_{\rm L,R})}$.
- Divide the algebraic solutions obtained for each lead into four categories: (i) exponentially decaying into the lead (growing towards the device); (ii) current flowing into the lead (out of the device); (iii) current flowing out of the lead (into the device); (iv) exponentially growing into the lead (decaying towards the device). The incoming wave will belong to (iii). The number of independent extended states with energy $E$ is given by the total number of solutions in category (iii) for all leads.
- Solve the Schrödinger equation inside the device region with boundary conditions based on the algebraic solutions inside the leads. The boundary conditions should be chosen so that in all but the chosen lead, the solutions belong to categories (i–ii).This is referred to as imposing *transmitting boundary conditions*. In the chosen lead $l$, the solutions belong to categories (i–ii) in addition to the incoming wavefunction which belongs to category (iii). If a wavefunction is found which satisfies the device Schrödinger equation and only has solutions belonging to what we define as *category* (i–ii) in all leads, this wavefunction is a bound state.

#### 17.2.2.1 Rectangular Barrier

It is convenient at this stage to see how this method can be applied in practice to a simple one-dimensional system, determining the extended states for a potential that represents a rectangular barrier in one dimension.

Consider a potential that is the same in the left and right leads, $V_L = V_R = 0$, and a rectangular barrier obtained by setting the device potential $V_D(x) = V_D > 0$.

- We choose an incoming wave from the left $e^{ikx}$ with $k$, so that $V_D > E = \hbar^2 k^2/(2m) > 0$. This energy-range brings us to a situation of quantum-mechanical tunneling where transmission is classically prohibited.
- The solutions in the left and right lead are $\psi^{\pm}_{L,R}(x) = e^{\pm ikx}$ with $k_L = k_R = k = \sqrt{2m/\hbar^2 E}$.
- $\psi^-_L(x)$ and $\psi^+_R(x)$ belong to category (ii) and $\psi^+_L(x)$ and $\psi^-_R(x)$ belong to category (iii). In this case no elements belong to category (i) and (iv).
- The two independent solutions in the device region are $\psi^{\pm}_D(x) = e^{\pm\kappa x}$ with $\kappa = \sqrt{2m/\hbar^2(V_D - E)}$ which means $\psi(x) = d^+ e^{\kappa x} + d^- e^{-\kappa x}$. For an incoming component from the left, we must find a solution which is a linear combination of two wavefunctions $\psi^{\pm}_L(x)$ in the left lead and just the outgoing $\psi^+_R(x)$ in the right lead. Therefore, in the right lead, we have $\psi(x) = te^{ikx}$ which translates into a boundary condition $d\psi(x)/dx = ik\psi(x)$ at $x = L$:

$$\begin{aligned} \kappa(d^+ e^{\kappa L} - d^- e^{-\kappa L}) &= ik(d^+ e^{\kappa L} + d^- e^{-\kappa L}) \\ d^+(\kappa - ik)e^{\kappa L} &= d^-(\kappa + ik)e^{-\kappa L}. \end{aligned} \tag{17.13}$$

In the left lead $\psi(x) = e^{ikx} + re^{-ikx}$ and we require continuity of $\psi(x)$ and its first derivative:

$$\begin{bmatrix} 1 & 1 \\ ik & -ik \end{bmatrix} \begin{bmatrix} 1 \\ r \end{bmatrix} = \begin{bmatrix} 1 & 1 \\ \kappa & -\kappa \end{bmatrix} \begin{bmatrix} d^+ \\ d^- \end{bmatrix}. \tag{17.14}$$

Eliminating $r$ yields a second relation between $d^+$ and $d^-$:

$$2ik = (ik + \kappa)d^+ + (ik - \kappa)d^- \tag{17.15}$$

and using Eq. (17.13), we obtain

$$d^{\pm} = \frac{2ik(\kappa \pm ik)e^{\mp\kappa L}}{(ik + \kappa)^2 e^{-\kappa L} + (ik - \kappa)^2 e^{\kappa L}}. \tag{17.16}$$

The coefficient of the transmitted wave can now be determined

$$te^{ikL} = d^+ e^{\kappa L} + d^- e^{-\kappa L} = \frac{4ik\kappa}{(ik + \kappa)^2 e^{-\kappa L} + (ik - \kappa)^2 e^{\kappa L}}. \tag{17.17}$$

The reflected wave could similarly determined from $r = d^+ + d^- - 1$.

With the extended state fully determined, we can calculate its current most easily in the right lead:

$$J(k) = |t|^2 \hbar k/m = \frac{\hbar k}{m} \frac{1}{1 + V_D^2 \sinh(\kappa L)^2/(4E(V_D - E))}, \tag{17.18}$$

Current continuity ensures that an evaluation of $J = \hbar/m \,\mathrm{Im}(\psi(x) d\psi(x)/dx)$ in the left lead ($|t|^2 = 1 - |r|^2$) or inside the device yields the same result. For large barrier-widths, the tunneling current can be seen to exponentially decrease $\propto \exp(-2\kappa L)$.

### 17.2.3 Normalization

Setting the coefficient of the incoming wave to 1 means that the extended wavefunctions are normalized, that is

$$\int \mathrm{d}\mathbf{r}\ \psi^*_{s\alpha'}(\mathbf{r};k')\,\psi_{s\alpha}(\mathbf{r};k) \;=\; 2\pi\ \delta_{ll'}\ \delta_{\alpha\alpha'}\delta(k-k') \tag{17.19}$$

and the current is

$$I \;=\; -e\sum_{s\alpha}\int \frac{\mathrm{d}k}{2\pi}\ J_{s\alpha}(k)\, f_{s\alpha}(k) \tag{17.20}$$

where $f_{s\alpha}(k)$ gives the occupation of each state.

The wavevector of the incoming wave $k$ together with the band index $\alpha$ do not present a unique set of quantum numbers; an alternative set of quantum numbers is the energy $E$ together with an index $\beta$ distinguishing between the possible incoming wavefunctions with that energy. The index $\beta$ runs over all roots for $k$ satisfying $E-E_\alpha(k)=0$ corresponding to an incoming wave. To maintain normalization, the wavefunction has to be multiplied by a coefficient $\psi_\beta(\mathbf{r};E)=\sqrt{|\mathrm{d}k/\mathrm{d}E|}\,\psi_\alpha(\mathbf{r};k)$ so we now have

$$\int \mathrm{d}\mathbf{r}\ \psi^*_{\alpha'\beta'}(\mathbf{r};E')\ \psi_{\alpha\beta}(\mathbf{r};E) \;=\; 2\pi\ \delta_{\alpha\alpha'}\ \delta_{\beta\beta'}\delta(E-E') \tag{17.21}$$

and the current is

$$I=-e\sum_{s\alpha}\int \frac{\mathrm{d}E}{2\pi}\, f_{s\alpha}(E) J_{s\alpha}(E) \;=\; -e\sum_{s\alpha}\int \frac{\mathrm{d}E}{2\pi\hbar}\, f_{s\alpha}(E) T_{s\alpha}(E) \tag{17.22}$$

where $T_{s\alpha}(E)=\hbar J_{s\alpha}(E)$ is dimensionless, has a maximum value of 1 and is called the *transmission coefficient*. The transmission coefficient is the same as that in Landauer's formula [Eq. (17.11)] and can be interpreted as the probability that an electron will be able to pass through the potential barrier in the device. In the classical limit, $T=1$ if the incoming electron has an energy exceeding the barrier and $T=0$ otherwise.

### 17.2.4 Numerical Approach

With modern computing power, many quantum transport problems can be solved numerically in a very short time. In a numerical setting the QTBM presents the recipe for determining the extended states. In the following, we outline the general numerical procedure to find the extended states on the basis of a two-dimensional potential with a left and a right lead:

$$V(x,y)=\begin{cases} V_\mathrm{L}(y) & x\le 0 \\ V_\mathrm{D}(x,y) & 0<x<L \\ V_\mathrm{R}(y) & L\le x \end{cases}. \tag{17.23}$$

We determine the wavefunctions $\psi(x,y)$ assuming the device has a width $W$ so that $\psi(x,0)=\psi(x,W)=0$.

#### 17.2.4.1 Discretization

The first step in a numerical approach is to discretize the Hamiltonian and the wavefunction. We discretize on $n_x$ and $n_y$ points in the device region in the $x$ and $y$ direction, respectively. This gives the coordinates $x^{1D}_{j_X} = j_X\Delta x$ and $y_{j_Y} = (j_Y+1)\Delta y$ with $j_X = 0\ldots n_x-1, j_Y = 0\ldots n_y-1$ and $\Delta x = L/(n_x-1)$, $\Delta y = W/(n_y+1)$. But to write the Hamiltonian as a matrix acting on a vector, we need a single index $j$ rather than two indices $j_X$ and $j_Y$. This can be accomplished by taking $j = j_Y + n_y j_X$ and consequently $j_X(j) = \lfloor j/n_y \rfloor$ and $j_Y(j) = j \bmod n_x$.

In single-index notation, $x_j = x^{1D}_{j_X(j)}$ and $y_j = y^{1D}_{j_Y(j)}$, the wavefunction $\psi_j = \psi(x_j, y_j)$ and the potential $V_j = V(x_j, y_j) = V_D(x_j, y_j)$. The action of the Hamiltonian on the wavefunction can be written as a matrix-vector product and the Schrödinger equation becomes $\widehat{H}\psi = E\psi$. The Hamiltonian contribution can be separated into the kinetic energy in the $x$ direction ($\widehat{T}_x$), the kinetic energy in the $y$ direction ($\widehat{T}_y$), and the potential energy ($V$) so that $\widehat{H}\psi = (\widehat{T}_x + \widehat{T}_y + \widehat{V})\psi$.

The contribution of the potential energy $V\psi$ can be evaluated as the multiplication of the wavefunction $\psi$ with a diagonal matrix with $V_j$ on its diagonal. The kinetic energy in the $y$ direction can be approximated using finite differences and becomes

$$-\frac{\hbar^2}{2m}\frac{\mathrm{d}^2}{\mathrm{d}y^2}\psi(x,y)\bigg|_{x=x_j,y=y_j} \approx -E_{\Delta y}(\psi_{j-1} + \psi_{j+1} - 2\psi_j) \tag{17.24}$$

with $E_{\Delta y} = \hbar^2/(2m(\Delta y)^2)$. Equation (17.24) is not applicable for the points next to the $y = 0$ or $y = W$ boundary ($j_Y(j) = 0$ or $j_Y(j) = n_y - 1$). At these points the boundary condition $\psi(x,0) = \psi(x,W)$ applies and $\psi_{j-1}$ or $\psi_{j+1}$ needs to be removed from Eq. (17.24). Overall, the kinetic energy in the $y$-direction can be computed by multiplying $\psi_{j_X n_y \ldots (j_X+1)n_y-1}$ with a tridiagonal matrix of size $n_y \times n_y$:

$$\widehat{T}^{1D}_y = -E_{\Delta y}D_{2y} = -E_{\Delta y}\begin{bmatrix} -2 & 1 & 0 & \ldots & \cdot & \cdot & \cdot \\ 1 & -2 & 1 & \ldots & \cdot & \cdot & \cdot \\ \vdots & \vdots & \vdots & \ddots & \vdots & \vdots & \vdots \\ \cdot & \cdot & \cdot & \ldots & 1 & -2 & 1 \\ \cdot & \cdot & \cdot & \ldots & 0 & 1 & -2 \end{bmatrix} \tag{17.25}$$

where $\widehat{D}_{y2}$ is defined as a tridiagonal matrix with 1 on its off-diagonals and $-2$ on its diagonal. The matrix $\widehat{T}_y$ acting on $\psi$ is then a block-diagonal matrix replicating $\widehat{T}^{1D}_y$ $n_x$ times on its diagonal. Or employing the Krönecker product $\widehat{T}_y = \widehat{I}^{1D}_x \otimes \widehat{T}^{1D}_y$ where $\widehat{I}^{1D}_x$ is the identity matrix of dimension $n_x \times n_x$. The kinetic energy in the $x$ direction would similarly be $\widehat{T}_{x0} = \widehat{T}^{1D}_{x0} \otimes \widehat{I}_y$ with $\widehat{T}^{1D}_{x0} = -E_{\Delta x}\widehat{D}_{2x}$ if $\psi(-\Delta x, y)$ and $\psi(L+\Delta x, y)$ were to vanish.

The overall structure of the Hamiltonian $(\widehat{T}_x + \widehat{T}_y + \widehat{V})$ including the leads is

$$\widehat{H}^{\text{with leads}}\begin{bmatrix} \cdot \\ \psi_{-n_y \cdots -1} \\ \psi_{0 \ldots n_y-1} \\ \psi_{n_y \ldots 2n_y-1} \\ \cdot \end{bmatrix} = \begin{bmatrix} \ddots & \vdots & \vdots & \vdots & \ddots \\ \ldots & H_{-1,-1} & H_{-1,0} & 0 & \ldots \\ \ldots & H_{0,-1} & H_{0,0} & H_{0,1} & \ldots \\ \ldots & 0 & H_{1,0} & H_{1,1} & \ldots \\ \ddots & \vdots & \vdots & \vdots & \ddots \end{bmatrix}\begin{bmatrix} \vdots \\ \psi_{-n_y \cdots -1} \\ \psi_{0 \ldots n_y-1} \\ \psi_{n_y \ldots 2n_y-1} \\ \vdots \end{bmatrix} \tag{17.26}$$

where $H_{j_X, j'_X}$ are $n_y \times n_y$ matrices. If an expression for $\psi_{-n_y \cdots -1}$ is available, the Hamiltonian can be truncated to the device region

$$\widehat{H}\begin{bmatrix}\psi_{0\ldots n_y-1}\\ \psi_{n_y\ldots 2n_y-1}\\ \vdots\end{bmatrix}=\begin{bmatrix}H_{0,0} & H_{0,1} & \cdots\\ H_{1,0} & H_{1,1} & \cdots\\ \vdots & \vdots & \ddots\end{bmatrix}\begin{bmatrix}\psi_{0\ldots n_y-1}\\ \psi_{n_y\ldots 2n_y-1}\\ \vdots\end{bmatrix}+\begin{bmatrix}H_{1\mathrm{L}}\psi_{-n_y\cdots-1}\\ 0\\ \vdots\end{bmatrix}. \quad (17.27)$$

where we have indicated by $H_{1\mathrm{L}} = H_{-1,0}$ the part of the Hamiltonian lost through truncation.

#### 17.2.4.2 Complex Band Structure

To find all algebraic solutions in the leads, we make use of the periodicity of the potential inside the leads so all solutions $\{\ldots,\psi_{-2n_y\cdots-n_y-1},\psi_{-n_y\cdots-1},\psi_{0\ldots n_y-1},\ldots\}$ can be written as $\{\ldots,\lambda_{\mathrm{L}\beta}^2\phi,\lambda_{\mathrm{L}\beta}\phi_{\mathrm{L}\beta},\phi_{\mathrm{L}\beta},\ldots\}$ and $\phi_{\mathrm{L}\beta}$ must satisfy

$$(H_{1\mathrm{L}}\lambda_{\mathrm{L}\beta}^2+H_{0\mathrm{L}}\lambda_{\mathrm{L}\beta}+H_{1\mathrm{L}}^\dagger)\phi_{\mathrm{L}\beta}=E\lambda_{\mathrm{L}\beta}\phi_{\mathrm{L}\beta} \quad (17.28)$$

where $H_{1\mathrm{L}} = H_{-1,-2} = H_{0,-1}$, $H_{0\mathrm{L}} = H_{-1,-1} = H_{0,0}$, and $H_{1\mathrm{L}}^\dagger = H_{-1,0}$. Equation (17.28) is a second-order eigenvalue problem which can be rewritten as a generalized eigenvalue problem of twice the rank

$$\begin{bmatrix}0 & I\\ -H_{1\mathrm{L}}^\dagger & EI-H_{0\mathrm{L}}\end{bmatrix}\begin{bmatrix}\phi_{\mathrm{L}\beta}\\ \lambda_{\mathrm{L}\beta}\phi_{\mathrm{L}\beta}\end{bmatrix}=\lambda_{\mathrm{L}\beta}\begin{bmatrix}I & 0\\ 0 & H_{1\mathrm{L}}\end{bmatrix}\begin{bmatrix}\phi_{\mathrm{L}\beta}\\ \lambda_{\mathrm{L}\beta}\phi_{\mathrm{L}\beta}\end{bmatrix}. \quad (17.29)$$

#### 17.2.4.3 Quantum Transmitting Boundary Conditions

We first consider the case where we only apply transmitting boundary conditions on the left-hand side, e.g., if the incoming component comes from the right-hand side. This means that the wavefunction in the left lead has to be a linear combination of the $n_y$ outgoing/decaying waves (category i–ii) in the left lead:

$$\psi_{0\ldots n_y-1}=\sum_{\beta\in(\mathrm{i-ii})}c_{\mathrm{L}\beta}\phi_{\mathrm{L}\beta}=\Phi_{\mathrm{L}}c_{\mathrm{L}}. \quad (17.30)$$

where $\Phi_{\mathrm{L}}$ is the matrix composed of $\phi_{\mathrm{L}\beta}$ in its columns. $c_{\mathrm{L}\beta}$ are a set of coefficients and $c_{\mathrm{L}}$ is the vector composed of $c_{\mathrm{L}\beta}$ so that $\Phi_{\mathrm{L}}c_{\mathrm{L}}$ denotes a matrix-vector product.

Knowing that inside the leads, each mode $\beta$ is multiplied by $\lambda_{\mathrm{L}\beta}$ when moving to the next $x_j$, $\psi_{-n_y\cdots-1}=\sum_{\beta\in\mathrm{i-ii}}c_{\mathrm{L}\beta}\lambda_{\mathrm{L}\beta}\phi_{\mathrm{L}\beta}=\Phi_{\mathrm{L}}\Lambda_{\mathrm{L}}c_{\mathrm{L}}$ if $\Lambda_{\mathrm{L}}$ is a diagonal matrix with $\lambda_{\mathrm{L}\beta}$ as its elements. Together with Eq. (17.30), this establishes the relationship:

$$\psi_{-n_y\cdots-1}=\Phi_{\mathrm{L}}\Lambda_{\mathrm{L}}\Phi_{\mathrm{L}}^{-1}\psi_{0\ldots n_y-1} \quad (17.31)$$

resulting in the definition of the self-energy $\Sigma_{\mathrm{L}}$ imposing the transmitting boundary conditions:

$$H_{1\mathrm{L}}\psi_{-n_y\cdots-1}=H_{1\mathrm{L}}\Phi_{\mathrm{L}}\Lambda_{\mathrm{L}}\Phi_{\mathrm{L}}^{-1}\psi_{0\ldots n_y-1}\equiv\Sigma_{\mathrm{L}}\psi_{0\ldots n_y-1}. \quad (17.32)$$

If an extended state with an incoming component from the left-hand side is determined, the wavefunction is

$$\psi_{0\ldots n_y-1} = \phi_{\mathrm{I}} + \sum_{\beta\in \mathrm{i-ii}} c_{\mathrm{L}\beta}\phi_{\mathrm{L}\beta} \qquad = \phi_{\mathrm{I}} + \Phi_{\mathrm{L}} c_{\mathrm{L}} \tag{17.33}$$

$$\psi_{-n_y\cdots-1} = \lambda_{\mathrm{I}}\phi_{\mathrm{I}} + \sum_{\beta\in \mathrm{i-ii}} c_{\mathrm{L}\beta}\lambda_{\mathrm{L}\beta}\phi_{\mathrm{L}\beta} = \lambda_{\mathrm{I}}\phi_{\mathrm{I}} + \Phi_{\mathrm{L}}\Lambda_{\mathrm{L}} c_{\mathrm{L}} \tag{17.34}$$

with $\phi_{\mathrm{I}}$ the incoming component with a corresponding $\lambda_{\mathrm{I}} = \mathrm{e}^{\mathrm{i}k_{\mathrm{I}}\Delta x}$. The remainder after the truncation of the Hamiltonian is now

$$H_{1\mathrm{L}}\psi_{-n_y\cdots-1} = \Sigma_{\mathrm{L}}\psi_{0\ldots n_y-1} - (\Sigma_{\mathrm{L}} - H_{1\mathrm{L}}\lambda_{\mathrm{I}})\phi_{\mathrm{I}}. \tag{17.35}$$

#### 17.2.4.4 Determining the Extended States

We are now in a position to apply the general procedure to a numerical determination of the extended states:

- Choosing a wave with wavevector $k$ inside the leads and setting $\psi_{-n_y\cdots-1} = \phi$, $\psi_{-2n_y\cdots-n_y-1} = \mathrm{e}^{\mathrm{i}k\Delta x}\phi$, and $\psi_{0\ldots n_y-1} = \mathrm{e}^{-\mathrm{i}k\Delta x}\phi$. Substituting this wavefunction into the discretized Schrödinger equation results in an $E(k)$ eigenvalue problem:

$$\left(H_{1\mathrm{L}}\mathrm{e}^{\mathrm{i}k\Delta x} + H_{0\mathrm{L}} + H_{1\mathrm{L}}^{\dagger}\mathrm{e}^{-\mathrm{i}k\Delta x}\right)\phi = E\phi \tag{17.36}$$

  where $H_{1\mathrm{L}}, H_{0\mathrm{L}}$ are $H_{-1,0}, H_{0,0}$ for a wave incoming from the left and $H_{0\mathrm{R}}, H_{1\mathrm{R}}$ are $H_{n_x-1,n_x-1}, H_{n_x-1,n_x}$ for a wave incoming from the right.
- Determine all algebraic solutions $\phi_{\mathrm{L}\beta}$ for the left lead using Eq. (17.29) and similarly for the right lead substituting $H_{1\mathrm{L}}, H_{0\mathrm{L}}$ with $H_{1\mathrm{R}}, H_{0\mathrm{R}}$.
- The solutions with $|\lambda_\beta| > 1$ are exponentially growing into the lead (category iv), $|\lambda_\beta| < 1$ are decaying into the lead (category i) and $|\lambda_\beta| = 1$ are states which carry current (category ii–iii). The current can be calculated through the Hellmann-Feynman theorem (see Chap. 8, Sect. 8.1.3), by calculating the matrix element with

$$J_\beta = \frac{\mathrm{d}E}{\mathrm{d}\hbar k} = \frac{1}{\hbar}\phi_\beta^{\dagger}\frac{\mathrm{d}\widehat{H}}{\mathrm{d}k}\phi_\beta = \frac{\mathrm{i}\Delta x}{\hbar}\phi_\beta^{\dagger}(H_1\lambda_\beta - H_1^{\dagger}\lambda_\beta^{*})\phi_\beta. \tag{17.37}$$

  Assemble all $\phi_\beta$ corresponding to outgoing/decaying waves into a matrix $\Phi_{\mathrm{L,R}}$. In our example of a two-dimensional potential, $\Phi_{\mathrm{L,R}}$ will be orthonormal and $\Phi_{\mathrm{L,R}}^{-1} = \Phi_{\mathrm{L,R}}^{\dagger}$ but for a general Hamiltonian this is not the case and $\Phi_{\mathrm{L,R}}^{-1}$ has to be computed through matrix inversion.
- Take all outgoing/decaying $\phi_\beta$ in the left and the right lead and construct $\widehat{\Sigma}_{\mathrm{L}} = H_{1\mathrm{L}}\Phi_{\mathrm{L}}\Lambda_{\mathrm{L}\Phi_{\mathrm{L}}}^{-1}$ and similarly $\widehat{\Sigma}_{\mathrm{R}} = H_{1\mathrm{R}}\Phi_{\mathrm{R}}\Lambda_{\mathrm{R}}\Phi_{\mathrm{R}}^{-1}$. Solve the linear system

$$(E\widehat{I} - \widehat{H}_0 - \widehat{\Sigma})\psi = C \tag{17.38}$$

with

$$\widehat{\Sigma} = \begin{bmatrix} \Sigma_L & 0 & \dots & \cdot & \cdot \\ 0 & 0 & \dots & \cdot & \cdot \\ \vdots & \vdots & \ddots & \vdots & \vdots \\ \cdot & \cdot & \dots & 0 & 0 \\ \cdot & \cdot & \dots & 0 & \Sigma_R \end{bmatrix} \quad \text{and} \quad C = \begin{bmatrix} C_L \\ 0 \\ \vdots \\ 0 \\ C_R \end{bmatrix} \tag{17.39}$$

where for a wave incoming from the left $C_L = (H_{1L}e^{ik\Delta x}I_y - \Sigma_L)\phi_I$ and $C_R = 0$, or, for a wave incoming from the right $C_L = 0$ and $C_R = (H_R e^{ik\Delta x} - \Sigma_R)\phi_I$.

### 17.2.5 Green's Function Approach

An alternative approach to solve the previous problem is to employ the Green's functions of the Schrödinger equation. In this section, we will limit ourselves to the single-particle picture and the results we obtain will be equivalent to that of the previous section. For a complete treatment see Ref. [22].

* *Green's functions are not only used for solving the Schrödinger equation but also any inhomogeneous differential equation*

$$\widehat{L}f(x) = g(x) \tag{17.40}$$

*where L is a linear operator can be solved by first computing the Green's function*

$$LG(x,x') = \delta(x-x'). \tag{17.41}$$

*If we subsequently know $f_0(x)$, the homogeneous solution of $\widehat{L}f_0(x) = 0$, then a general solution of Eq. (17.40) is*

$$f(x) = f_0(x) + \int \mathrm{d}x' G(x,x')g(x') \tag{17.42}$$

*for any $g(x)$.*

*If the linear operator has translational invariance, the Green's function only depends on $x - x'$ and can be written using a Fourier transform*

$$G(x,x') = \int \frac{\mathrm{d}k}{2\pi} e^{ik(x-x')} G(k). \tag{17.43}$$

*Taking the case where the linear operator is $L = \mathrm{d}^2/\mathrm{d}x^2$, multiplying Eq. (17.41) with $e^{ikx}$ and integrating over x, the Green's function $G(k)$ must satisfy*

$$k^2 G(k) = 1 \tag{17.44}$$

*and the Green's function simply is: $G(k) = 1/k^2$. The differential equation in Eq. (17.41) has been turned into an algebraic equation, Eq. (17.44).*

#### 17.2.5.1 Green's Function of the Schrödinger Equation

In the single-particle picture, we can define the Green's function as an operator $\widehat{G}(E)$ satisfying

$$(E\widehat{I}-\widehat{H})\widehat{G}(E)=\widehat{I} \tag{17.45}$$

where $\widehat{I}$ is the identity operator $\widehat{I}\psi = \psi$ and $\widehat{H}$ the Hamiltonian. A true Green's *function* is obtained by writing the Green's function in a real-space basis

$$G(\mathbf{r},\mathbf{r}';E) = \langle \mathbf{r} \mid \widehat{G}(E) \mid \mathbf{r}' \rangle. \tag{17.46}$$

If the eigenvalues and eigenfunctions of the Hamiltonian are known

$$\widehat{H} = \sum_{\alpha} \mid \alpha \rangle E_{\alpha} \langle \alpha \mid, \tag{17.47}$$

computing the Green's function explicitly is straightforward

$$\widehat{G}_0(E) = \sum_{\alpha} \mid \psi_{\alpha} \rangle \frac{1}{E-E_{\alpha}} \langle \psi_{\alpha} \mid \tag{17.48}$$

provided the spectrum of $E_{\alpha}$ is discrete. For a closed system, this is indeed the case and Eq. (17.48) is an analytic function with poles for $E=E_{\alpha}$. If $\widehat{H}$ is represented by a matrix, the Green's function is simply the matrix inverse of the difference between the energy and the Hamiltonian

$$\widehat{G}_0(E) = (E\widehat{I}-\widehat{H})^{-1}. \tag{17.49}$$

If the system is open, the spectrum becomes continuous and the Green's function no longer has poles except in the presence of bound states. For plane waves in one dimension, we can diagonalize the Hamiltonian

$$\widehat{H} = \int_{-\infty}^{\infty} \frac{\mathrm{d}k}{2\pi} \mid k \rangle E(k) \langle k \mid \tag{17.50}$$

with $E(k) = \hbar^2 k^2/(2m)$ and we would straightforwardly write the Green's function as

$$\widehat{G}(E) = \int_{-\infty}^{\infty} \frac{\mathrm{d}k}{2\pi} \mid k \rangle \frac{1}{E-E(k)} \langle k \mid \tag{17.51}$$

but unfortunately the expression for the Green's function is a divergent integral for any $E > 0$. A way to circumvent this problem is to add or subtract an infinitesimal complex number so the integral over $k$ avoids the singularities

$$\lim_{\eta \to 0_+} \widehat{G}(E \pm \mathrm{i}\eta) = \widehat{G}^{\pm}(E). \tag{17.52}$$

For a continuous spectrum $\widehat{G}^{+}(E)$ and $\widehat{G}^{-}(E)$ are different and the former is called the retarded Green's function while the latter is called the advanced Green's function. The retarded and the advanced Green's functions are each other's Hermitian conjugate.

The retarded Green's function for the plane waves in one dimension is

$$G^{\mathrm{R,1D}}(x,x';E) = \int_{-\infty}^{\infty} \frac{\mathrm{d}k}{2\pi} \frac{\mathrm{e}^{\mathrm{i}k(x-x')}}{E+\mathrm{i}\eta - E(k)}. \tag{17.53}$$

The poles at $k = \pm k(E) = \pm\sqrt{E2m/\hbar^2}$ are shifted by an amount $\mathrm{i}\eta_k = \mathrm{i}\eta\mathrm{d}k/\mathrm{d}E(k)$, which means that the negative pole $k = -k(E)$ is shifted to below the real line since $\mathrm{d}E/\mathrm{d}k$ is negative and the positive pole is moved above the real line.

$$\left(E+\mathrm{i}\eta - \frac{\hbar^2}{2m}\frac{\mathrm{d}^2}{\mathrm{d}x^2}\right) G^+(x,x';E) = \delta(x-x'). \tag{17.54}$$

For $x > x'$, $\mathrm{e}^{\mathrm{i}(k(E)+\mathrm{i}\eta)(x-x')}$ decays for $x \to \infty$ while $\mathrm{e}^{\mathrm{i}(-k(E)-\mathrm{i}\eta)(x'-x)}$ grows for $x \to \infty$. As a consequence, only the positive root $k = k(E)$ is allowed for the retarded Green's function for $x > x'$: $G^+(x,x';E) = d^+\mathrm{e}^{\mathrm{i}k(E)(x-x')}$. For $x < x'$, only the solution decaying for $x \to -\infty$ is allowed and $G(x,x';E) = d^-\mathrm{e}^{-\mathrm{i}k(E)(x-x')}$. Calculating the second derivative at $x = x'$, the boundary conditions for $d^+$ and $d^-$ can be determined yielding the Green's function

$$G^+(x,x';E) = -\frac{im}{\hbar^2 k}\mathrm{e}^{\mathrm{i}k|x-x'|}. \tag{17.55}$$

#### 17.2.5.2 Spectral Function

The difference between the retarded and the advanced Green's function is called the spectral function

$$\widehat{A}(E) = \mathrm{i}(\widehat{G}^+(E) - \widehat{G}^-(E)) \tag{17.56}$$

and the spectral function is a Hermitian operator which is a solution of the Schrödinger equation

$$\begin{aligned} \widehat{H}\widehat{A}(E) &= \mathrm{i}\widehat{H}(\widehat{G}^+(E) - \widehat{G}^-(E)) \\ &= \mathrm{i}\left((E\widehat{G}^+(E) - \widehat{I} + \mathrm{i}\eta\widehat{H}) - (E\widehat{G}^-(E) - \widehat{I} - \mathrm{i}\eta\widehat{H})\right) \\ &\to E\widehat{A}. \end{aligned}$$

Going back to the expansion for the retarded and the advanced Green's function, the spectral function can be identified as an operator projecting into the subspace with a given energy

$$\begin{aligned} \widehat{A} &= \mathrm{i}\int \sum_\alpha \frac{\mathrm{d}k}{2\pi} \mid k\alpha\rangle \left(\frac{1}{E+\mathrm{i}\eta - E_\alpha(k)} - \frac{1}{E-\mathrm{i}\eta - E\alpha(k)}\right)\langle k\beta \mid \\ &= \sum_\alpha \int \mathrm{d}k \mid k\alpha\rangle\delta(E-E_\alpha(k))\langle k\alpha \mid \\ &= \sum_{k\alpha \text{ if } E_\alpha(k)=E} \mid k\alpha\rangle \left|\frac{\mathrm{d}k}{\mathrm{d}E_\alpha(k)}\right| \langle k\alpha \mid \end{aligned} \tag{17.57}$$

If a distribution function $f(E)$ is available, the density operator can be computed as

$$\widehat{\rho} = \int \frac{\mathrm{d}E}{2\pi} f(E)\,\widehat{A}(E). \tag{17.58}$$

#### 17.2.5.3 Numerical Approach

In a numerical approach, the Hamiltonian is discretized and becomes a matrix and similarly the Green's functions are discretized and become matrices. In a way similar to the transmitting boundary conditions in the Schrödinger approach, for those sites near the lead, the truncation of the matrix introduces contact terms.

For a system with leads, we define the retarded and the advanced Green's function to be the solution of

$$(E\widehat{I} - \widehat{H}_0 - \widehat{\Sigma}^{+,-}(E))G^{+,-}(E) = I. \tag{17.59}$$

The terms $\widehat{\Sigma}^{+,-}$(E) are called the retarded and the advanced self-energies and since $\widehat{G}^{-}(E) = \widehat{G}^{+}(E)^{\dagger}$, $\widehat{\Sigma}^{-}(E) = \widehat{\Sigma}^{+}(E)^{\dagger}$. As we explained in our derivation of the Green's function for plane waves in one dimension, the Green's functions have to decay inside the leads and adding $\mathrm{i}\eta$ to the energy selects the components with a velocity $\mathrm{d}E/\mathrm{d}(\hbar k)$ going into the leads. As a consequence, the expression for $\widehat{\Sigma}^{+}(E)$ is exactly the same as that obtained for $\Sigma$ in Eq. (17.39).

In general the spectral functions $\widehat{A}(E) = \mathrm{i}(\widehat{G}^{+}(E) - \widehat{G}^{-}(E))$ carry the same current to the left as to the right and do not carry total current. So to enable the description of a system which carries current a further subdivision is required. A natural way to subdivide the system is to separate the spectral function into a spectral function of extended states with an incoming component form the left $\widehat{A}_{\mathrm{L}}(E)$ and a separate function of extended states with an incoming component from the right $A_{\mathrm{R}}$(E). To enable this subdivision, we rewrite the spectral function using Eq. (17.59)

$$\widehat{A}(E) = \mathrm{i}(\widehat{G}^{+}(E) - \widehat{G}^{-}(E)) \tag{17.60}$$

$$= \mathrm{i}\left(\widehat{G}^{+}(E)(E\widehat{I} - \widehat{H}_0 - \widehat{\Sigma}^{-}(E))\widehat{G}^{-}(E) - \widehat{G}^{+}(E)(E\widehat{I} - \widehat{H}_0 - \widehat{\Sigma}^{+}(E))\widehat{G}^{-}(E)\right)$$

$$= \mathrm{i}\widehat{G}^{+}(E)(\widehat{\Sigma}^{+}(E) - \widehat{\Sigma}^{-}(E))\widehat{G}^{-} \tag{17.61}$$

$$\equiv \widehat{G}^{+}(E)\widehat{\Gamma}(E)\widehat{G}^{-}(E) \tag{17.62}$$

where $\widehat{\Gamma}(E)$ can be separated by inspecting $\widehat{\Sigma}$ in Eq. (17.39). Assigning $\widehat{\Gamma}_{\mathrm{L,R}}(E) = \mathrm{i}(\widehat{\Sigma}_{\mathrm{L,R}}(E) - \widehat{\Sigma}_{\mathrm{L,R}}(E)^{\dagger})$, we can write

$$\widehat{\Gamma}(E) = \widehat{\Gamma}_{\mathrm{L}}^{\mathrm{big}}(E) + \widehat{\Gamma}_{\mathrm{R}}^{\mathrm{big}}(E) = \begin{bmatrix} \widehat{\Gamma}_{\mathrm{L}}(E) & 0 & \dots & \cdot & \cdot \\ 0 & 0 & \dots & \cdot & \cdot \\ \vdots & \vdots & \ddots & \vdots & \vdots \\ \cdot & \cdot & \dots & 0 & 0 \\ \cdot & \cdot & \dots & 0 & 0 \end{bmatrix} + \begin{bmatrix} 0 & 0 & \dots & \cdot & \cdot \\ 0 & 0 & \dots & \cdot & \cdot \\ \vdots & \vdots & \ddots & \vdots & \vdots \\ \cdot & \cdot & \dots & 0 & 0 \\ \cdot & \cdot & \dots & 0 & \widehat{\Gamma}_{\mathrm{R}}(E) \end{bmatrix}. \tag{17.63}$$

where the superscript big denotes that the matrices $\widehat{\Gamma}_{\mathrm{R,L}}^{\mathrm{big}}$ have the dimension of the entire device compared to $\widehat{\Gamma}_{\mathrm{R,L}}$ which only have the size of the contact terms. The left and right spectral function are given by

$$\widehat{A}_{\mathrm{L,R}}(E) = \widehat{G}^{+}(E)\widehat{\Gamma}_{\mathrm{L,R}}^{\mathrm{big}}(E)\widehat{G}^{-}(E). \tag{17.64}$$

Assigning a separate distribution function to the left and the right lead, a non-equilibrium energy-resolved density matrix carrying current is obtained

$$\widehat{\rho}(E) = \int \frac{\mathrm{d}E}{2\pi} f_{\mathrm{L}}(E)\widehat{A}_{\mathrm{L}}(E) + f_{\mathrm{R}}(E)\widehat{A}_{\mathrm{R}}(E) \tag{17.65}$$

and its current is computed as

$$I = -e\mathrm{Tr}(\widehat{J\rho}) \tag{17.66}$$

with $J$ the current operator.

To compute the current at the left lead for an injected wave from the right, we have

$$J = \frac{\mathrm{i}}{\hbar}\left(\psi^{\dagger}_{0\dots n_y-1} H_{1\mathrm{L}} \psi_{-n_y\dots-1} - \mathrm{c.c.}\right) \tag{17.67}$$

$$= \frac{\mathrm{i}}{\hbar}\left(\psi^{\dagger}_{0\dots n_y-1} \Sigma_{\mathrm{L}} \psi_{0\dots n_y-1} - \mathrm{c.c.}\right) \tag{17.68}$$

$$= \frac{1}{\hbar}\psi^{\dagger}_{0\dots n_y-1} \Gamma_{\mathrm{L}} \psi_{0\dots n_y-1} \,. \tag{17.69}$$

Since the spectral function $A_{\mathrm{R}}(E)$ is only composed of extended states with an incoming component from the right, the current associated with $A_{\mathrm{R}}(E)$ can be calculated from $J(E) = \mathrm{Tr}(A_{\mathrm{R}}(E)\Gamma_L)/\hbar = T(E)/\hbar$ with the transmission coefficient

$$T^{\mathrm{R}\to\mathrm{L}} = \mathrm{Tr}(\widehat{A}_{\mathrm{R}}(E)\widehat{\Gamma}_{\mathrm{L}}(E)) = \mathrm{Tr}(\widehat{G}^{+}(E)\widehat{\Gamma}_{\mathrm{R}}(E)\widehat{G}^{-}(E)\widehat{\Gamma}_{\mathrm{L}}(E)) \;=\; T(E)\,, \tag{17.70}$$

which equals the current associated with $A_{\mathrm{L}}(E)$

$$T^{\mathrm{L}\to\mathrm{R}} = \mathrm{Tr}(\widehat{G}^{+}(E)\widehat{\Gamma}_{\mathrm{L}}(E)\widehat{G}^{-}(E)\widehat{\Gamma}_{\mathrm{R}}(E)) \;=\; \mathrm{Tr}\left((\widehat{G}^{+}(E)\widehat{\Gamma}_{\mathrm{L}}(E)\widehat{G}^{-}(E)\widehat{\Gamma}_{\mathrm{R}}(E))^{\dagger}\right) = T(E)\,. \tag{17.71}$$

The total current in the system can be computed from

$$I = -e\int \frac{\mathrm{d}E}{2\pi\hbar}\,(f_{\mathrm{L}}(E) - f_{\mathrm{R}}(E))T(E)\,. \tag{17.72}$$

Note that $T(E)$ defined in this way can exceed 1 if more than one injected wave from the left and right is available.

## 17.3 Non-equilibrium Green's Functions

In the previous section we have dealt with Green's functions but not accounted for any many-body effects making the Green's function approach in the previous section just a mathematical technique to solve the Schrödinger equation. In a many-body system, NEGFs can be defined yielding the NEGF formalism, also called the Keldysh–Kadanoff-Baym (KKB) formalism [3, 4].

Most commonly four Green's functions which are variations of Eq. (17.2) are used, paying particular attention to the time-ordering. They are called the "lesser," the "greater," the "retarded," and the "advanced" Green's function:

$$G^{<}(\mathbf{r},\mathbf{r}';t,t') \;=\; \frac{\mathrm{i}}{\hbar}\langle\widehat{\psi}^{\dagger}(\mathbf{r},t')\widehat{\psi}(\mathbf{r}',t)\rangle \tag{17.73}$$

$$G^{>}(\mathbf{r},\mathbf{r}';t,t') \;=\; -\frac{\mathrm{i}}{\hbar}\langle\widehat{\psi}(\mathbf{r},t)\widehat{\psi}^{\dagger}(\mathbf{r}',t')\rangle \tag{17.74}$$

$$G^{\mathrm{R}}(\mathbf{r},\mathbf{r}';t,t') = \theta(t-t')(G^{>}(\mathbf{r},\mathbf{r}';t,t') - G^{<}(\mathbf{r},\mathbf{r}';t,t')) \tag{17.75}$$

$$G^{\mathrm{A}}(\mathbf{r},\mathbf{r}';t,t') = \theta(t'-t)(G^{<}(\mathbf{r},\mathbf{r}';t,t') - G^{>}(\mathbf{r},\mathbf{r}';t,t')). \tag{17.76}$$

The field operators are in the Heisenberg picture and the expectation value ($\langle\rangle$) corresponds to the trace over the density matrix describing the system.

At steady state, the Green's functions depend only on the difference between the two time-arguments $t-t'$. Fourier transforming with respect to $t-t'$ results in the energy-dependent Green's functions $G^{<,>,R,A}(\mathbf{r},\mathbf{r}';E)$. The lesser Green's function is like an energy-resolved density matrix and determines where the electrons can be found: the charge density is computed from the diagonal elements of the lesser Green's function

$$n(\mathbf{r}) = -\int \frac{\mathrm{d}E}{2\pi}\, \mathrm{i}G^{<}(\mathbf{r},\mathbf{r};E)\,. \tag{17.77}$$

The greater Green's function determines where electrons cannot be found, or, where the holes can be found

$$p(\mathbf{r}) = \int \frac{\mathrm{d}E}{2\pi}\, \mathrm{i}G^{>}(\mathbf{r},\mathbf{r};E)\,. \tag{17.78}$$

The retarded Green's function is a measure for where electrons will flow if they are injected at $\mathbf{r}$ while the advanced Green's function is a measure of how electrons flow if current is extracted at $\mathbf{r}$.

In the absence of inelastic scattering, the relations between the Green's functions are

$$(E\widehat{I} - \widehat{H}_0)G^{\mathrm{R}}(\mathbf{r},\mathbf{r}';E) - \int \mathrm{d}\mathbf{r}_1\, \Sigma^{\mathrm{R}}(\mathbf{r},\mathbf{r}_1;E)G^{\mathrm{R}}(\mathbf{r}_1,\mathbf{r}';E) = \delta(\mathbf{r}-\mathbf{r}') \tag{17.79}$$

$$G^{<,>}(\mathbf{r},\mathbf{r}';E) = \int \mathrm{d}\mathbf{r}_1 \int \mathrm{d}\mathbf{r}_2\, G^{\mathrm{R}}(\mathbf{r},\mathbf{r}_1;E)\Sigma^{>,<}(\mathbf{r}_1,\mathbf{r}_2;E)G^{\mathrm{A}}(\mathbf{r}_2,\mathbf{r}';E)\,, \tag{17.80}$$

where the retarded, greater, and lesser self-energy $\Sigma^{\mathrm{R},<,>}(\mathbf{r},\mathbf{r}';E)$ account for the interactions in the system.

Ignoring all interactions, The Green's function formalisms become equivalent to the QTBM to solve the Schrödinger equation and $G^{\mathrm{R}}(\mathbf{r},\mathbf{r}';E) = G^{+}(\mathbf{r},\mathbf{r}';E)$, from Sect. 17.2.5. The only non-vanishing elements in the retarded self-energies are those associated with the truncation of the matrix at the left and the right lead $\widehat{\Sigma}^{\mathrm{R}}_{\mathrm{L,R}}(E)$. For $\widehat{\Sigma}^{\mathrm{R}}_{\mathrm{R}}(E)$, the subscript R refers to the retarded nature of the self-energy while the subscript refers to the right lead. Assigning a different Fermi level to the waves with an incoming component from the left or right corresponds to setting the lesser self-energy

$$\widehat{\Sigma}^{<}(E) = -f_{\mathrm{L}}(E)[\widehat{\Sigma}^{\mathrm{R}}_{\mathrm{L}}(E) - \widehat{\Sigma}^{\mathrm{R}}_{\mathrm{L}}(E)^{\dagger}] - f_{\mathrm{R}}(E)[\widehat{\Sigma}^{\mathrm{R}}_{\mathrm{R}}(E) - \widehat{\Sigma}^{\mathrm{R}}_{\mathrm{R}}(E)^{\dagger}] \tag{17.81}$$

yielding $\widehat{G}^{<}(E) = \mathrm{i}\widehat{\rho}(E)$ from Eq. (17.65).

In the NEGF formalism, the current can be calculated from the retarded Green's function:

$$\mathbf{J} = \frac{-e\hbar}{m}\int \frac{\mathrm{d}E}{2\pi}\, \lim_{\mathbf{r}\to\mathbf{r}'}(\nabla - \nabla')G^{<}(\mathbf{r},\mathbf{r}';E)\,. \tag{17.82}$$

To treat interactions such as the electron–phonon interaction, the self-consistent Born approximation can be invoked establishing a relationship between the self-energies and the Green's functions. The exact relationship is based on the interactions under consideration, e.g., the electron–phonon interaction but nearest neighbor, local or even more drastic approximations are often required to make a numerical solution possible. Indeed, from a numerical point of view, an exact treatment of collisions appears to be a daunting problem: In principle, the full inversion of a very large dense matrix is required. Exact solutions have been found in small molecular devices[23] or approximating the collisions as local, diagonal processes[24]. A general solution, though, remains unattainable. For this reason, the NEGF formalism is usually applied in its ballistic form. We have already observed that this is a fully equivalent—and numerically less efficient—way to implement the QTBM.

## 17.4 Wigner Distribution Functions

The Wigner function is a quantum-mechanical analogue of the Boltzmann distribution function and is defined as the Wigner–Weyl transformation, which is related to the Fourier transform, of the reduced density matrix $\rho(\mathbf{r},\mathbf{r}') = \langle \hat{\psi}^\dagger(\mathbf{r}t)\hat{\psi}(\mathbf{r}'t)\rangle$:

$$f(\mathbf{r},\mathbf{p},t) = \int_\Omega \mathrm{d}\mathbf{s}\, \mathrm{e}^{-\mathrm{i}\mathbf{s}\cdot\mathbf{p}/\hbar}\rho(\mathbf{r}+\mathbf{s}/2,\mathbf{r}-\mathbf{s}/2,t). \tag{17.83}$$

The meaning of the term "reduced" will be made clear in Sect. 18.1.2: Basically, only the degrees of freedom of a single electron are considered, having "traced out" all other degrees of freedom, such as those of the thermal bath. Note also that the density matrix is an equal-time Green's function. We shall not get into the spiny issue of what one loses by assuming $t = t'$ in Eq. (17.2).

Similar to the Boltzmann distribution function, the probability of finding a particle at a position $\mathbf{r}$ is given by

$$n(\mathbf{r},t) = \int \frac{\mathrm{d}\mathbf{p}}{(2\pi\hbar)^3} f(\mathbf{r},\mathbf{p},t) \tag{17.84}$$

while the electrical current can be computed from

$$\mathbf{J}(\mathbf{r},t) = -e \int \frac{\mathrm{d}\mathbf{p}}{(2\pi\hbar)^3}\, (\mathbf{p}/m) f(\mathbf{r},\mathbf{p},t). \tag{17.85}$$

The kinetic equation for the Wigner function is given by:

$$\frac{\partial}{\partial t} f(\mathbf{r},\mathbf{p},t) + \frac{\mathbf{p}}{m}\cdot\nabla_{\mathbf{r}} f(\mathbf{r},\mathbf{p},t) - \int \frac{\mathrm{d}\mathbf{k}'}{(2\pi\hbar)^3}\, V_{\mathrm{w}}(\mathbf{r},\mathbf{p}-\mathbf{p}') f(\mathbf{r},\mathbf{p}',t) = \left(\frac{\partial f}{\partial t}\right)_{\mathrm{coll}}, \tag{17.86}$$

where $V_{\mathrm{w}}(\mathbf{r},\mathbf{p})$ is a nonlocal kernel given by:

$$V_{\mathrm{w}}(\mathbf{r},\mathbf{p}) = \frac{1}{\mathrm{i}} \int \frac{\mathrm{d}\mathbf{s}}{(2\pi)^3} \left[V\left(\mathbf{r}+\frac{\mathbf{s}}{2}\right) - V\left(\mathbf{r}-\frac{\mathbf{s}}{2}\right)\right] \mathrm{e}^{-\mathrm{i}\mathbf{p}\cdot\mathbf{s}}. \tag{17.87}$$

The collision term $(\partial f_{\mathrm{w}}/\partial t)_{\mathrm{coll}}$ that has to be used to account for scattering events due to the electron–phonon or electron–electron interactions is not easily written. Usually, a Master-like collisional integral is used. This has the same form that appears in the Master equation and in the Boltzmann equation. It will be derived in the next chapter. However, finding its correct expression remains an elusive goal. If the NEGF approach finds in collisions a major computational stumbling block, the Wigner formalism has to yet find the correct way to even writing the effect of collisions.

One way to describe transport is by considering the steady-state solution under the application of boundary conditions corresponding to a set of leads. In one dimension with open boundaries at $x = 0$ and $x = L$, the boundary conditions of the Wigner distribution are given by

$$f(0,k)|_{k>0} = f_{\mathrm{L}}(k) \tag{17.88}$$

$$f(L,k)|_{k<0} = f_{\mathrm{R}}(k) \tag{17.89}$$

where $f_{\mathrm{L,R}}(k)$ give the distribution of the electrons inside the left and right contact.

The choice of the correct boundary conditions is another spiny issue that has raised concerns [25], since in their simplest formulation they may lead to non-uniqueness of the solution of the Wigner equation, possibly leading to non-physical solutions.

## References


1. D.N. Zubarev, V. Morozov, G. Ropke, *Statistical Mechanics of Nonequilibrium Processes: Basic Concepts, Kinetic Theory* (Wiley, New York, 1996)
2. D.N. Zubarev, V. Morozov, G. Ropke, *Statistical Mechanics of Nonequilibrium Processes: Relaxation and Hydrodynamic Processes* (Wiley, New York, 1997)
3. L.V. Keldysh, Diagram technique for nonequilibrium processes, Zh. Eksp. Teor. Fiz. **47**, 1515 (1964) [Sov. Phys. JEPT **20**, 1018 (1965)]
4. L.P. Kadanoff, G. Baym, *Quantum Statistical Mechanics* (Benjamin, New York, 1962)
5. B. Novakovic, I. Knesevic, Quantum master equations in electronic transport, in *Nano-Electronic Devices: Semiclassical and Quantum Transport Modeling*, ed. by D. Vasileska, S.M. Goodnick (Springer, New York, 2011), pp. 249–287
6. H. Haug, S.W. Koch, *Quantum Theory of the Optical and Electronic Properties of Semiconductors*, 5th edn. (World Scientific, Singapore, 2009)
7. K.M. Kira, S.W. Koch *Semiconductor Quantum Optics* (Cambridge University Press, Cambridge, 2011)
8. W.R. Frensley, Boundary conditions for open quantum systems driven far from equilibrium. Rev. Mod. Phys. **62**, 745 (1990)
9. C. Jacoboni, P. Bordone, The Wigner-function approach to non-equilibrium electron transport. Rep. Progr. Phys. **67**, 1033 (2004)
10. J.M. Sellier, S.M. Amoroso, M. Nedjalkov, S. Selberherr, A. Asenov, I. Dimov, Electron dynamics in nanoscale transistors by means of Wigner and Boltzmann approaches. Phys. A: Stat. Mech. Appl. **398**, 194 (2014)
11. X. Oriols, Quantum-trajectory approach to time-dependent transport in mesoscopic systems with electron–electron interactions. Phys. Rev. Lett. **98**, 066803 (2007)
12. K.K. Thornber, R.P. Feynman, Velocity acquired by an electron in a finite electric field in a polar crystal. Phys. Rev. B **4**, 674 (1971)
13. B.A. Mason, K. Hess, Quantum Monte Carlo calculations of electron dynamics in dissipative solid-state systems using real-time path integrals. Phys. Rev. B **39**, 5051 (1989)
14. C. Jacoboni, *Theory of Electron Transport in Semiconductors* (Springer, Berlin/Heidelberg, 2010)
15. S. Datta, *Electronic Transport in Mesoscopic Systems* (Cambridge University Press, Cambridge, 2014)
16. C.S. Lent, D.J. Kirkner, The quantum transmitting boundary method. J. Appl. Phys. **67**, 6353 (1990)
17. R. Lake, G. Klimeck, R.C. Bowen, D. Jovanovic, Single and multiband modeling of quantum electron transport through layered semiconductor devices. J. Appl. Phys. **81**, 7845 (1997)
18. H.J. Choi, J. Ihm, *Ab initio* pseudopotential method for the calculation of conductance in quantum wires. Phys. Rev. B **59**, 2267 (1999)
19. M.V. Fischetti, Bo Fu, S. Narayanan, J. Kim, Semiclassical and quantum electronic transport in nanometer-scale structures: empirical pseudopotential band structure, Monte Carlo simulations and Pauli master equation, in *Nano-Electronic Devices: Semiclassical and Quantum Transport Modeling*, ed. by D. Vasileska, S.M. Goodnick (Springer, New York, 2011), pp. 183–247
20. R. Landauer, Spatial variation of currents and fields due to localized scatterers in metallic conduction. IBM J. Res. Dev. **1**, 223 (1957)

21. S.E. Laux, A. Kumar, M.V. Fischetti, Analysis of quantum ballistic electron transport in ultra-small semiconductor devices including space-charge and geometric effects. J. Appl. Phys. **95** 5545 (2004)
22. E.N. Economou, *Green's Functions in Quantum Physics* (Springer, Berlin Heidelberg, 2006)
23. A. Pecchia, A. Di Carlo, Atomistic theory of transport in organic and inorganic nanostructures. Rep. Progr. Phys. **67**, 1497 (2004)
24. S. Jin, Y.J. Park, H.S. Min, A three-dimensional simulation of quantum transport in silicon nanowire transistor in the presence of electron–phonon interactions. J. Appl. Phys. **99**, 123719 (2004)
25. R. Rosati, F. Dolcini, R.C. Iotti, F. Rossi, Wigner-function formalism applied to semiconductor quantum devices: failure of the conventional boundary condition scheme. Phys. Rev. B **88**, 035401 (2013)

# Chapter 18
# *From Liouville–von Neumann to Boltzmann: The Semiclassical Limit

## 18.1 "Deriving" the Boltzmann Equation

Our goal is to derive the semiclassical Boltzmann transport equation (BTE) from the Liouville–von Neumann equation (10.33). We should immediately point out that this is an impossible task, at least at present. Even on purely classical terms, the "derivation" of the BTE is a goal which has eluded several attempts, including—and, obviously, starting with—Boltzmann himself. The crucial fact we should appreciate is the following: The physical content of the Liouville–von Neumann equation, as derived in Chap. 10, Eq. (10.33), is not different from the content of the Schrödinger equation. Indeed, we have "derived it" from it. As such, the Liouville–von Neumann equation represents purely ideal physical systems which behave in a fully reversible way: By switching $t$ to $-t$ in Eq. (10.32) we simply get the complex-conjugate equation that expresses the identical physical picture. Thus, the system can be "moved in time" in either direction. On the other hand, the BTE will express systems which behave "irreversibly." The introduction of the "arrow of time" (as it is often called in the literature) remains a matter of speculation even today. The book by Zurek summarizes the status of our understanding [1]. In the quantum-mechanical context, the issue is rendered even more complicated by considerations related to the point at which we can make a transition from the quantum mechanical, microscopic description of the world to the classical, macroscopic description. This is related to confusion regarding the interpretation of measurement in Quantum Mechanics (what differentiates the *measuring system* from the *measured system*? question known as the "von Neumann chain"), and so of the "collapse" of the wavefunction, the loss of coherence, ... In a few words: To the plethora of issues regarding the interpretation of Quantum Mechanics.

Therefore, it should be clear that we cannot "derive" the BTE. Our goal is to see how one may proceed along this path and see at which stage we must "throw our hands in the air" and invoke heuristic considerations. They will be plausible considerations, realistic, and sensible. But they will be "guesses" nevertheless.

Our plan of action is the following: We shall consider first a closed system of electrons (that is, no electron is allowed to leave or enter the volume) in thermal contact with a system of phonons (the "thermal bath"). The system will be assumed to occupy a finite volume, to be at finite temperature, and we shall consider the time evolution of that portion of the density matrix which is related only to the degrees of freedom of a single electron. This is the so-called *reduced density matrix* that we shall define below. It will be at this point—while considering this time evolution—that we will have to take the "huge plunge" in order to transform our description from reversible and non-dissipative to an irreversible and dissipative picture. Then we shall "open" the system, now allowing electrons



M. Fischetti, W.G. Vandenberghe, *Advanced Physics of Electron Transport in Semiconductors and Nanostructures*, Graduate Texts in Physics, DOI 10.1007/978-3-319-01101-1_18

to enter and leave the volume. Finally, under some approximations, we shall consider the effect of inhomogeneity (that is, consider what happens when a spatially varying potential is added to the picture). The role of other interactions—impurity and electron–electron—will be considered next.

Similar "derivations" have been given in 1955 in formidable pieces of work by Luttinger and Kohn [2] (accounting for electron-impurity interactions), by van Hove [3] (analyzing the form that interactions must have in order to yield irreversibility), and by Argyres [4] later on (accounting for electron–phonon interactions). On classical terms, a wonderful book by Carlo Cercignani on Ludwig Boltzmann's life (scientific and not) presents an account of Boltzmann's struggle to "prove" his equation (that is, irreversibility, statement technically known as the *H-theorem*) [5].

Before diving into this discussion, we should briefly mention the history of the Master equation, since our goal consists essentially in recovering a scattering operator of the form that appears not only in the BTE (the "collisional integral" at the right-hand side of the BTE), but also in the Master equation.

Wolfgang Pauli derived the Master equation in 1928 [6] looking at the equation of motion for the density matrix, Eq. (10.33), Chap. 10, and attempting to use it to describe how a system driven off-equilibrium approaches thermodynamic equilibrium. *Provided* we assume that at every time step the phases of states of the system are randomized, then the occupation $\rho_\alpha$ of any eigenstate $|\alpha\rangle$ of the Hamiltonian evolves in time according to the *Master equation*:

$$\frac{d\rho_\alpha}{dt} = \sum_\beta \left[W_{\alpha\beta}\rho_\alpha - W_{\beta\alpha}\rho_\beta\right] , \tag{18.1}$$

where $W_{\alpha\beta}$ is the transition rate from state $|\beta\rangle$ to state $|\alpha\rangle$. Note that $\rho_\alpha$ is just the diagonal element $\rho_{\alpha\alpha}$ of the density matrix of the system. The randomization of the phases required by Pauli to obtain Eq. (18.1) essentially ensures that all off-diagonal elements, $\rho_{\alpha\beta}$ with $\alpha \neq \beta$, disappear because of the cancellation of the random phases. Since the Master equation is irreversible, while it has been "derived" from the reversible Liouville–von Neumann equation, it is clear that this randomization of the phases is the key element that introduces irreversibility. This is not surprising: By ignoring the off-diagonal elements of the density matrix, that express interference between any two eigenstates $|\alpha\rangle$ and $|\beta\rangle$, we lose information and thus generate entropy.

In 1955, Léon Van Hove showed that Eq. (18.1) can be obtained without requiring a randomization of the phases at every time step [3]. If the interactions causing the transition rates $W_{\alpha\beta}$ are interactions with a system with a large number of degrees of freedom, it is sufficient to consider particular classes of initial states and Eq. (18.1) will follow. Interactions with phonons are of this type. Van Hove described two possible classes of initial states that satisfy his requirements: The class of states that initially are a superposition of energy eigenstates in a narrow energy range; or, on the opposite side, initial states that are linear superpositions of energy eigenstates in a very wide energy range. He went on to clarify that other such classes may exists, but he could not identify any other class. We shall revisit these ideas in the following, but it is convenient to keep them here, at the beginning of our discussion.

### 18.1.1 The Total Hamiltonian

The system we consider is a finite volume of semiconductor in thermal contact with a heat bath, so that—in the absence of external forces—electrons are in equilibrium with phonons within the volume. These phonons, in turn, are in equilibrium with the bath. The "bath" is a system so large that its energy is not appreciably altered when it attempts to keep the phonons in the system at equilibrium by exchanging energy with them. To simplify the notation, we shall ignore spin indices for now.

Thus, the total Hamiltonian of the system will be

$$\widehat{H} = \widehat{H}_{\mathrm{el}} + \widehat{H}_{\mathrm{ph}} + \widehat{H}_{\mathrm{ep}} . \tag{18.2}$$

Here $\widehat{H}_{\mathrm{el}}$ is the electron Hamiltonian, which is the second-quantized version of the single-electron Hamiltonian:

$$\widehat{H}_{\mathrm{el}}^{(1\ \mathrm{el})} = \widehat{\mathscr{T}} + V^{(\mathrm{H})}(\mathbf{r}) , \tag{18.3}$$

where $\widehat{\mathscr{T}}$ is the kinetic energy operator. It could be simply $-\hbar^2\nabla^2/(2m^*)$ in some effective-mass approximation, or it could be $E(-\mathrm{i}\nabla)$ in the Kohn–Luttinger equation, more generally. We shall leave it quite general for now. The term $V^{(\mathrm{H})}$ is the Hartree potential. To express the many-electron Hamiltonian, we denote by $|i\rangle$ and $E_i$ the eigenstates and eigenvalues of the free *single-electron* Hamiltonian. Thus:

$$\widehat{H}_{\mathrm{el}} = \sum_i E_i \, \hat{c}_i^\dagger \hat{c}_i . \tag{18.4}$$

To fix the ideas, the wavefunction $\psi_i(\mathbf{r}) = \langle \mathbf{r}|\hat{c}_i^\dagger|0\rangle$ is just the conventional single-particle wavefunction, i.e., an eigenstate of the non-second-quantized free Schrödinger equation in the presence of the potential energy $V^{(\mathrm{H})}(\mathbf{r})$. Note that the index $i$ actually denotes a set of 3 indices, such as the wavevector $\mathbf{k}$ when $V^{(\mathrm{H})}(\mathbf{r}) = 0$. Therefore, the index $i$ is a very general index that lumps all quantum numbers, discrete and continuous, in a compact way. It is equivalent to the pair of indices $(\alpha, k)$ used in the previous chapter.

The phonon Hamiltonian has the form

$$\widehat{H}_{\mathrm{ph}} = \sum_{\mathbf{q}\eta} \hbar\omega_{\mathbf{q}\eta} \, \hat{b}_{\mathbf{q}\eta}^\dagger \hat{b}_{\mathbf{q}\eta} , \tag{18.5}$$

and the electron–phonon Hamiltonian has the form [Eq. (9.102)]:

$$\widehat{H}_{\mathrm{ep}} = \mathrm{i}\alpha \sum_{ij\mathbf{q}\eta} \left[ \mathscr{F}_{ij}^{(\eta)}(\mathbf{q})\hat{c}_i^\dagger \, \hat{b}_{\mathbf{q}\eta}\hat{c}_j - \mathscr{F}_{ij}^{(\eta)*}(\mathbf{q})\hat{c}_j^\dagger \, \hat{b}_{-\mathbf{q}\eta}^\dagger \hat{c}_i \right] . \tag{18.6}$$

This expression can be derived easily from Eq. (13.25), Chap. 13, defining

$$\mathrm{i}\alpha\mathscr{F}_{ij}^{(\eta)}(\mathbf{q}) = \langle \mathrm{i}|\mathscr{A}_{\mathbf{q}\eta}\mathscr{I}_{ij} \, \mathrm{e}^{\mathrm{i}\mathbf{q}\cdot\mathbf{r}}|j\rangle . \tag{18.7}$$

This is just a change of representation going from the basis set of plane waves used to define the operators $\hat{c}_{\mathbf{k}}$ and $\hat{c}_{\mathbf{k}}^\dagger$ used, to the more general basis set of states $|i\rangle$ used here to define the creation and annihilation operators $\hat{c}_i^\dagger$ and $\hat{c}_i$. The dimensionless parameter $\alpha$ has been formally factored out in order to let us expand in powers of $\alpha$. It is just a mathematical device which allows us to use perturbation theory at various orders of $\alpha$.

### 18.1.2 The Reduced Density Matrix

Our system is constituted by electrons and ions (phonons). But what we are really interested in are electrons. Actually: Single electrons. We "admit our ignorance," being unable to get a full description of the large number of ionic and electronic degrees of freedom. Thus, we are really interested in what is known as the *reduced, single-electron density matrix*:

$$\hat{\rho}_{\rm el}(t) = \sum_{ij} \langle \hat{c}_i^\dagger \hat{c}_j \rangle_t \, |i\rangle\langle j| = \sum_{ij} {\rm Tr}_{\rm el}\, {\rm Tr}_{\rm ph}[\hat{\rho}(t)\hat{c}_i^\dagger \hat{c}_j] \, |i\rangle\langle j| \,. \tag{18.8}$$

Its matrix elements $(ij)$ are (as a function of time)

$$\rho_{{\rm el},ij}(t) = \langle \hat{c}_i^\dagger \hat{c}_j \rangle_t = {\rm Tr}_{\rm el}\, {\rm Tr}_{\rm ph}[\hat{\rho}(t)\hat{c}_i^\dagger \hat{c}_j] \,. \tag{18.9}$$

The reduced density matrix $\hat{\rho}_{\rm el}$ singles out the single-electron elements while we take the trace (that is, we average over) all other degrees of freedom due to ions and all other electrons represented by the full density matrix $\hat{\rho}(t)$.

### 18.1.3 Equations of Motion

Let's now work in the *interaction picture* (or *interaction representation*) in which observables evolve according to the "free" Hamiltonian $\widehat{H}_{\rm el} + \widehat{H}_{\rm ph}$ while states evolve according to the perturbation Hamiltonian $\widehat{H}_{\rm ep}$.

* *At this point it is appropriate to recall that the expectation value at time t of an observable $\widehat{A}$ on a state $|\psi\rangle$ is*

$$\langle \psi(t=0)| \, e^{-({\rm i}/\hbar)\widehat{H}_0 t} \, e^{-({\rm i}/\hbar)\widehat{H}' t} \, \widehat{A}(t=0) \, e^{({\rm i}/\hbar)\widehat{H}' t} \, e^{({\rm i}/\hbar)\widehat{H}_0 t} |\psi(t=0)\rangle \,,$$

*where we have denoted by $\widehat{H}_0$ the free Hamiltonian $\widehat{H}_{\rm el} + \widehat{H}_{\rm ph}$ and by $\widehat{H}'$ the perturbation $\widehat{H}_{\rm ep}$. In the Schrödinger picture, this expectation value is written as:*

$${}_{\rm S}\langle \psi(t)| \, \widehat{A}(t=0) \, |\psi(t)\rangle_{\rm S} \,,$$

*where the state evolves in time as $|\psi(t)\rangle_{\rm S} = e^{({\rm i}/\hbar)(\widehat{H}'+\widehat{H}_0)t} |\psi(t=0)\rangle$. In the Heisenberg picture, this is written as:*

$$\langle \psi(t=0)| \, \widehat{A}_{\rm H}(t) \, |\psi(t=0)\rangle \,,$$

*where the observable evolves in time as $\widehat{A}_{\rm H}(t) = e^{-({\rm i}/\hbar)(\widehat{H}_0+\widehat{H}')t} \, \widehat{A}(t=0) \, e^{({\rm i}/\hbar)(\widehat{H}'+\widehat{H}_0)t}$. Finally, in the interaction picture, the expectation value is written as:*

$${}_{\rm I}\langle \psi(t)| \, \widehat{A}_{\rm I}(t) \, |\psi(t)\rangle_{\rm I} \,,$$

*where $|\psi(t)\rangle_{\rm I} = e^{({\rm i}/\hbar)\widehat{H}' t} |\psi(t=0)\rangle$ and $\widehat{A}_{\rm I}(t)$ as $\widehat{A}_{\rm I}(t) = e^{-({\rm i}/\hbar)\widehat{H}_0 t} \, \widehat{A}(t=0) \, e^{({\rm i}/\hbar)\widehat{H}_0 t}$. We should also recall that the density matrix evolves in time with the dynamics that controls the time evolution of the states.*

Therefore, in this representation (a mix of the Schrödinger and Heisenberg representations) the time evolution of $\hat{\rho}$ is controlled only by $\widehat{H}_{\rm ep}$:

$$\frac{\partial \hat{\rho}^{(\rm I)}(t)}{\partial t} \;=\; -\,\frac{\rm i}{\hbar}\,[\widehat{H}^{(\rm I)}_{\rm ep}(t),\hat{\rho}^{(\rm I)}(t)]\,, \tag{18.10}$$

where $[.,.]$ denotes the conventional commutator and the superscripts $(I)$ remind us that we are indeed working in the interaction representation. Integrating once Eq. (18.10), we obtain

$$\hat{\rho}^{(\rm I)}(t) \;=\; \hat{\rho}^{(\rm I)}(0) \;-\; \frac{\rm i}{\hbar}\int_0^t {\rm d}t'\;[\widehat{H}^{(\rm I)}_{\rm ep}(t'),\hat{\rho}^{(\rm I)}(t')]\,. \tag{18.11}$$

Inserting this into the right-hand side of Eq. (18.10) to have

$$\frac{\partial \hat{\rho}^{(\rm I)}(t)}{\partial t} \;=\; -\,\frac{\rm i}{\hbar}\,[\widehat{H}^{(\rm I)}_{\rm ep}(t'),\hat{\rho}^{(\rm I)}(0)] \;-\; \frac{1}{\hbar^2}\int_0^t {\rm d}t'\;[\widehat{H}^{(\rm I)}_{\rm ep}(t),[\widehat{H}^{(\rm I)}_{\rm ep}(t'),\hat{\rho}^{(\rm I)}(t')]]\,. \tag{18.12}$$

The equivalent expression for the reduced density matrix can be obtained by tracing the expression above over the ionic and electronic degrees of freedom:

$$\begin{aligned}\frac{\partial\langle \hat{c}_i^\dagger\,\hat{c}_j\rangle_t}{\partial t} =& -\,\frac{\rm i}{\hbar}\;{\rm Tr}_{\rm el}\;{\rm Tr}_{\rm ph}\{\;[\widehat{H}^{(\rm I)}_{\rm ep}t),\hat{\rho}^{(\rm I)}(0)]\;\hat{c}_i^\dagger\;\hat{c}_j\}\\ &-\,\frac{1}{\hbar^2}\int_0^t {\rm d}t'\;{\rm Tr}_{\rm el}\;{\rm Tr}_{\rm ph}\;\{[\widehat{H}^{(\rm I)}_{\rm ep}(t),[\widehat{H}^{(\rm I)}_{\rm ep}(t'),\hat{\rho}^{(\rm I)}(t')]]\;\hat{c}_i^\dagger\;\hat{c}_j\}\,.\end{aligned} \tag{18.13}$$

We have to spend a few words about the initial density matrix $\hat{\rho}^{(\rm I)}(0)$. First, we shall assume that the electron–phonon coupling is weak enough to allow us to factorize the full density matrix into separate and diagonal electronic and ionic components:

$$\hat{\rho}^{(\rm I)}(t) \;=\; \hat{\rho}^{(\rm I)}_{\rm el}(t)\otimes\hat{\rho}^{(\rm I)}_{\rm ph}(t)\,, \tag{18.14}$$

where the phonon part of the density matrix is assumed to be diagonal, thanks to the phase-breaking occurring between the heat bath and the device. Thus:

$$\hat{\rho}^{(\rm I)}_{\rm ph}(t) \;=\; \sum_{\{n_{\mathbf q}\}}\;\hat{\rho}_{{\rm ph}\{n_{\mathbf q}\}}(t)\;|\{n_{\mathbf q}\}\rangle\langle\{n_{\mathbf q}\}|\,. \tag{18.15}$$

Here we have used the shorthand $|\{n_{\mathbf q}\}\rangle$ to indicate a general state in the phonon Fock space. The general state of the full electron+phonon Fock space will be denoted as $|\{N_n\}\{n_{\mathbf q}\}\rangle$.

Regarding the electronic component of $\hat{\rho}$, we should have in mind the situation in which our "chunk" of semiconductor is sealed off from the external world (that is, electrons cannot enter or exit the "chunk") for $t<0$. Let's call "contacts" the external world and "device" the chunk. At $t=0$ we let the contacts "flood" the device, close it again, and at later times we watch the device re-adjust to a new steady-state situation, in equilibrium with the reservoirs, the external bias, and its own internal dynamics. We always assume the contacts to be reservoirs in thermal equilibrium with the environment, and to be described by a purely diagonal density matrix, the occupation of each state being determined by the Fermi–Dirac factor. Actually, some off-diagonal terms cannot vanish, if we want to represent electrons in the reservoirs as characterized by a dephasing length $\lambda_\phi$. However, here we shall assume that the devices we are interested in have active regions $L<\lambda_\phi$, so that we can ignore the finite size of the injected electrons when analyzing transport over such a short length-scale and assume that they are described by fully delocalized wavefunctions. Therefore, we can assume that at $t = 0$ the electronic density matrix is diagonal of the form:

$$\hat{\rho}_{\rm el}^{(\rm I)}(0) = \sum_{\{N_n\}} \rho_{\{N_n\}} \, |\{N_n\}\rangle \, \langle\{N_n\}| \, , \tag{18.16}$$

where $|\{N_n\}\rangle$ is a state of Fock space in the notation given above and the occupation $\rho_{\{N_n\}}$ of each many-particle state $|\{N_n\}\rangle$ is just the Fermi function at the total energy $E_{\{N_n\}} = \sum_{N_n \in \{N_n\}} N_n \, E_n$.

Let's anticipate a bit what we are going to find out. First, note that we would like to be able to ignore the off-diagonal terms of the (reduced) density matrix. The first term at the rhs of Eq. (18.13), usually called a "memory" term (since it depends on the state of the system at $t = 0$), in general is quite troublesome. However, as we shall see, our assumption of a diagonal phonon density matrix will make it vanish.

The analysis of the second term at the rhs of Eq. (18.13) requires additional work and assumptions. In dealing with the electron–phonon [4] or electron-impurity interaction [2], the crucial argument has been given by van Hove [3]. It is essentially an argument based on the large number of degrees of freedom of either the thermal bath—for phonons—or the random distribution of very many dopants—as shown by Kohn and Luttinger—which shows that for off-diagonal elements of the density matrix the second-order terms of Eq. (18.13) give rise to "out-scattering" terms (at least, when we bypass the assumption of a diagonal density matrix at the initial time, as it will be clear below), but the "in-scattering" terms vanish. Therefore, off-diagonal elements can decay, but they cannot grow. If the initial configuration of the system (or, in our case, of the reservoirs) is diagonal, one can ignore the off-diagonal elements. Note that an additional source of off-diagonal elements does not conserve energy on a short time scale during a collision: In very fast transients as scattering occurs, the electrons can be in virtual intermediate states which are superpositions of eigenstates of the Hamiltonian. Their existence implies that there exist off-diagonal terms of the density matrix describing the "occupation" of the cross-terms arising from these superpositions. But here we shall take the "usual" approximation of completed collisions (Fermi's Golden Rule) and ignore these effects.

Now, the first term at the right-hand side of Eq. (18.13) can be explicitly written as:

$$\begin{aligned} {\rm Tr}_{\rm ph} \, {\rm Tr}_{\rm el} \, \{[\widehat{H}_{\rm ep}, \hat{\rho}(0)] \hat{c}_i^\dagger \hat{c}_j\} = {\rm Tr}_{\rm el} \Bigg\{ \sum_{\{n_{\mathbf{q}}\}} \hat{\rho}_{\rm el}(0) \, \Big[ &\langle\{n_{\mathbf{q}}\}| \widehat{H}_{\rm ep} \hat{\rho}_{\rm ph}(0) \, |\{n_{\mathbf{q}}\}\rangle \, \hat{c}_i^\dagger \, \hat{c}_j \\ &- \langle\{n_{\mathbf{q}}\}| \hat{\rho}_{\rm ph}(0) \, \widehat{H}_{\rm ep} \, |\{n_{\mathbf{q}}\}\rangle \, \hat{c}_i^\dagger \, \hat{c}_j \Big] \Bigg\} . \end{aligned} \tag{18.17}$$

Since $\widehat{H}_{\rm ep}$ changes one phonon occupation number by one unit, this term vanishes, as we had anticipated.

Let's now consider the more complicated second-order term in Eq. (18.13). We consider only second-order terms in $\alpha$, so we can replace $\hat{\rho}(t')$ with $\hat{\rho}(0)$. Indeed $\hat{\rho}$ changes in time only under the action of the perturbation Hamiltonian, so at any given time its deviation from the initial value is of the order $\alpha$ and we may ignore it. So, expanding the commutators, this term reads

$$\begin{aligned} -\frac{1}{\hbar^2} \int_0^t {\rm d}t' \sum_{\{M_m\}} \sum_{\{n_{\mathbf{q}}\}} \Big[ &\langle\{M_m\}\{n_{\mathbf{q}}\}| \, \widehat{H}_{\rm ep}^{(\rm I)}(t) \, \widehat{H}_{\rm ep}^{(\rm I)}(t') \, \hat{\rho}^{(\rm I)}(0) \, \hat{c}_i^\dagger \hat{c}_j \, |\{n_{\mathbf{q}}\}\{M_m\}\rangle \\ + &\langle\{M_m\}\{n_{\mathbf{q}}\}| \, \hat{\rho}^{(\rm I)}(0) \, \widehat{H}_{\rm ep}^{(\rm I)}(t') \, \widehat{H}_{\rm ep}^{(\rm I)}(t) \, \hat{c}_i^\dagger \hat{c}_j \, |\{n_{\mathbf{q}}\}\{M_m\}\rangle \\ - &\langle\{M_m\}\{n_{\mathbf{q}}\}| \, \widehat{H}_{\rm ep}^{(\rm I)}(t) \, \hat{\rho}^{(\rm I)}(0) \, \widehat{H}_{\rm ep}^{(\rm I)}(t') \, \hat{c}_i^\dagger \hat{c}_j \, |\{n_{\mathbf{q}}\}\{M_m\}\rangle \\ - &\langle\{M_m\}\{n_{\mathbf{q}}\}| \, \widehat{H}_{\rm ep}^{(\rm I)}(t') \, \hat{\rho}^{(\rm I)}(0) \, \widehat{H}_{\rm ep}^{(\rm I)}(t) \, \hat{c}_i^\dagger \hat{c}_j \, |\{n_{\mathbf{q}}\}\{M_m\}\rangle \Big] . \end{aligned} \tag{18.18}$$

Our assumption of a diagonal phonon density matrix shows that Hamiltonian terms of the form $\hat{b}^{\dagger}_{\mathbf{q}\eta}\hat{b}_{\mathbf{q}'\eta}$ contribute only if $\mathbf{q}=\mathbf{q}'$. Thus, in the following we will extract from the expectation values the phonon terms

$$\begin{aligned}\langle \hat{b}^{\dagger}_{\mathbf{q}\eta}\hat{b}_{\mathbf{q}'\eta}\rangle_t &\approx \sum_{\{n_{\mathbf{q}}\}} \langle\{n_{\mathbf{q}}\}|\, \hat{\rho}_{\mathrm{ph}}(0)\, \hat{b}^{\dagger}_{\mathbf{q}\eta}\hat{b}_{\mathbf{q}'\eta}|\{n_{\mathbf{q}}\}\rangle \\ &= \sum_{\{n_{\mathbf{q}}\}} \hat{\rho}_{ph\{n_{\mathbf{q}}\}}(0)\, \langle\{n_{\mathbf{q}}\}|\, \hat{b}^{\dagger}_{\mathbf{q}\eta}\hat{b}_{\mathbf{q}'\eta}|\{n_{\mathbf{q}}\}\rangle \;=\; n_{\mathbf{q}\eta}\,\delta_{\mathbf{q}\mathbf{q}'}\,, \end{aligned} \tag{18.19}$$

where $n_{\mathbf{q}\eta}$ is the Bose–Einstein occupation number for phonons of mode $\eta$. Similarly:

$$\langle \hat{b}_{\mathbf{q}\eta}\hat{b}^{\dagger}_{\mathbf{q}'\eta}\rangle_t \;=\; (n_{\mathbf{q}\eta}+1)\,\delta_{\mathbf{q}\mathbf{q}'}\,. \tag{18.20}$$

Note that the weak coupling approximation has allowed us to assume that the phonon bath remains always at thermal equilibrium, because we have used $\hat{\rho}_{\mathrm{ph}}(0)$ in place of $\hat{\rho}_{\mathrm{ph}}(t)$. Coherent phonon and electron–phonon effects are indeed higher-order corrections. A side benefit of this simplification is that the scattering kernel $W^{(\eta\pm)}_{im}$ below—see Eq. (18.27)—becomes a function of $t-t'$ rather than of $t$ and $t'$ separately, a fact which allows us to go to the van Hove limit.

The first term (an out-scattering term, since it will yield a negative contribution to $\partial\rho_{ii}/\partial t$, thus corresponding to scattering "out" of the state $|i\rangle$) at the right-hand side of Eq. (18.18) can be decomposed into a phonon emission term (containing the term $\hat{b}_{\mathbf{q}\eta}\hat{b}^{\dagger}_{\mathbf{q}\eta}$) and a phonon-absorption term (containing $\hat{b}^{\dagger}_{\mathbf{q}\eta}\hat{b}_{\mathbf{q}\eta}$). Mixed terms containing $\hat{b}^{\dagger}_{\mathbf{q}\eta}\hat{b}^{\dagger}_{\mathbf{q}\eta}$ or $\hat{b}_{\mathbf{q}\eta}\hat{b}_{\mathbf{q}\eta}$ do not contribute, as they vanish on diagonal terms. Using Eq. (18.20), the absorption term can be written as

$$\begin{aligned}&-\frac{\alpha^2}{\hbar^2}\int_0^t \mathrm{d}t' \sum_{\mathbf{q}\eta}\sum_{\{N_n\}} \rho_{\mathrm{el}\{N_n\}}\, n_{\mathbf{q}\eta}\, \langle\{N_n\}|\hat{c}^{\dagger}_i\,\hat{c}_j \sum_{lm}\hat{c}^{\dagger}_l\,\hat{c}_m \sum_{hr}\hat{c}^{\dagger}_h\,\hat{c}_r\,|\{N_n\}\rangle \\ &\times\, \mathscr{F}^{(\eta)}_{lm}(q)\,\mathscr{F}^{(\eta)*}_{hr}(q)\,\mathrm{e}^{-(\mathrm{i}/\hbar)(E_l-E_m+\hbar\omega_{\mathbf{q}\eta})t}\,\mathrm{e}^{-(\mathrm{i}/\hbar)(E_h-E_r-\hbar\omega_{\mathbf{q}\eta})t'}\,. \end{aligned} \tag{18.21}$$

The six-particle term $\langle\{N_n\}|\hat{c}^{\dagger}_i\,\hat{c}_j\,\hat{c}^{\dagger}_l\,\hat{c}_m\,\hat{c}^{\dagger}_h\,\hat{c}_r|\{N_n\}\rangle$ can now be handled in the following way intimately related to van Hove's original argument: Note that the expression above is just

$$\sum_{\{N_n\}} \rho_{\{N_n\}}\langle\{N_n\}|\widehat{H}^{(\mathrm{I})}_{\mathrm{ep}}(t)\,\widehat{H}^{(\mathrm{I})}_{\mathrm{ep}}(t')\,\hat{c}^{\dagger}_i\hat{c}_j|\{N_n\}\rangle\,. \tag{18.22}$$

Let us now insert a completeness into Eq. (18.22):

$$\sum_{\{M_m\}\{N_n\}} \rho_{\{M_m\}}\langle\{N_n\}|\widehat{H}^{(\mathrm{I})}_{\mathrm{ep}}(t)\,|\{M_m\}\rangle\langle\{M_m\}|\widehat{H}^{(\mathrm{I})}_{\mathrm{ep}}(t')\,\hat{c}^{\dagger}_i\hat{c}_j|\{N_n\}\rangle\,. \tag{18.23}$$

This term is now a genuine "van Hove term": It expresses the transition from an initial state $\widehat{H}^{(\mathrm{I})}_{\mathrm{ep}}(t')\hat{c}^{\dagger}_i\hat{c}_j|\{N_n\}\rangle$ to a final state $\widehat{H}^{(\mathrm{I})}_{\mathrm{ep}}(t)|\{N_n\}\rangle$ via a sum over a very large number of the many-electron intermediate states $|\{M_m\}\rangle$. When the phases of the initial and final states differ, the phases of the intermediate states will interfere destructively as we sum over very many (say $N$) electron states and the contribution to the double sum in Eq. (18.23) will be limited to "diagonal" processes, yielding a contribution $\sim O(N)$. If, instead, the initial and final states have identical phases, the phases of the intermediate states will interfere constructively and we will have a contribution $\sim O(N^2)$ from all terms to the sum. Therefore, in the limit of infinitely many electronic degrees of freedom, only states $\widehat{H}^{(\mathrm{I})}_{\mathrm{ep}}(t')\hat{c}^{\dagger}_i\hat{c}_j|\{N_n\}\rangle$ and $\widehat{H}^{(\mathrm{I})}_{\mathrm{ep}}(t)|\{N_n\}\rangle$, which have identical phases, will give a non-vanishing contribution to the sum in Eq. (18.21). The question now is to find out when this happens.

Since in Eq. (18.21) we consider transitions originating at time $t'$ ranging from the initial time 0 to the final time $t$, if we restrict our attention to final times $t$ large enough ($\sim O(\alpha^{-2})$) we see that in the multiple sums in Eq. (18.21) constructive phase interference will occur only for terms for which $E_l - E_m = -(E_h - E_r)$, and this constructive interference among the phases occurs at times $t' \sim t$. Only under these conditions do the phases of $\widehat{H}_{\text{ep}}^{(\text{I})}(t')|\{N_n\}\rangle$ and of $\widehat{H}_{\text{ep}}^{(I)\dagger}(t)|\{N_n\}\rangle$ match. Thus, in turn, in order for our phases of the initial and final states to match fully, we must have $i = j$, so that the phases of $|\{N_n\}\rangle$ and $\hat{c}_i^\dagger \hat{c}_j|\{N_n\}\rangle$ also match. Thus, we reach the conclusion that off-diagonal terms ($i \neq j$) drop out of the picture. Note that this depends on several assumptions: (1) we can truncate the perturbation expansion at second-order (relying on the weakness of the electron–phonon interaction), so that we may replace $\hat{\rho}^{(\text{I})}(t)$ with $\hat{\rho}^{(\text{I})}(0)$, (2) that $\hat{\rho}^{(\text{I})}(0)$ is diagonal, a condition we find in a small-enough region of the semiconductor, and (3) that we consider the dynamics only at large enough times. As we shall see below when taking the van Hove limit, this allows us to assume completed collisions. At short times off-diagonal elements matter (those corresponding to the off-shell virtual states present at short times during a collision process).

Now we look at each of the surviving "diagonal" terms and see that only terms $l = r$, $m = h$ should be retained, so that we are left with terms of the form $\langle\{N_n\}|\hat{c}_i^\dagger\, \hat{c}_i\, \hat{c}_l^\dagger\, \hat{c}_m\, \hat{c}_m^\dagger\, \hat{c}_l|\{N_n\}\rangle$. We must now have

$$\hat{c}_m\, \hat{c}_m^\dagger\, \hat{c}_l|\{N_n\}\rangle \;=\; \hat{c}_l\, \hat{c}_i^\dagger\, \hat{c}_i|\{N_n\}\rangle \,. \tag{18.24}$$

We should have either $i = m$ or $i = l$. The first set of terms will yield a vanishing contribution since the term $\langle\hat{c}_i^\dagger\, \hat{c}_i\, \hat{c}_i\hat{c}_i^\dagger\, \hat{c}_m\, \hat{c}_m^\dagger\rangle$ always vanishes (a state $|\{N_n\}\rangle$ either contains an electron in state $|i\rangle$, so that $\hat{c}_i\hat{c}_i^\dagger|\{N_n\}\rangle = 0$, or it does not, in which case $\hat{c}_i^\dagger\hat{c}_i|\{N_n\}\rangle = 0$). Thus, we are left with the second set of terms in the sum which yields terms of the form

$$\langle\{N_n\}|\hat{c}_i^\dagger\, \hat{c}_i\, \hat{c}_i^\dagger\, \hat{c}_m\, \hat{c}_m^\dagger\, \hat{c}_i|\{N_n\}\rangle \;=\; \langle\{N_n\}|\hat{c}_i^\dagger\hat{c}_i\, \hat{c}_m\, \hat{c}_m^\dagger|\{N_n\}\rangle \,. \tag{18.25}$$

Now we can use the "usual" Hartree–Fock factorization

$$\langle\hat{c}_i^\dagger\hat{c}_i\hat{c}_m\hat{c}_m^\dagger\rangle \;\approx\; \langle\hat{c}_i^\dagger\hat{c}_i\rangle[1 - \langle\hat{c}_m^\dagger\hat{c}_m\rangle] \;-\; \langle\hat{c}_i^\dagger\hat{c}_m\rangle[\delta_{im} - \langle\hat{c}_m^\dagger\hat{c}_i\rangle]. \tag{18.26}$$

Inserting this expression into the out-scattering term (18.21), the second term at the rhs of the expression above drops out and we can rewrite the out-scattering terms in the form:

$$-\int_0^t \mathrm{d}t' \,\frac{1}{2} \sum_{m\eta} K_{mi}^{(\eta+)}(t-t')\; \rho_i(t')\; [1 - \rho_m(t')] \,, \tag{18.27}$$

having defined the phonon-absorption rate:

$$K_{mi}^{(\eta+)}(t-t') \;=\; \frac{2}{\hbar^2}\,\alpha^2 \sum_{\mathbf{q}} |\mathscr{F}_{mi}^{(\eta)}(q)|^2\; n_{\mathbf{q}\eta}\; \mathrm{e}^{-(\mathrm{i}/\hbar)(E_i - E_m + \hbar\omega_{\mathbf{q}\eta})(t-t')} \,. \tag{18.28}$$

In these equations we have written simply $\rho_i(t)$ for the diagonal elements $\langle\hat{c}_i^\dagger\hat{c}_i\rangle_t$ of the reduced density matrix.

The second term in the integrand of Eq. (18.18) gives the same contribution, but with the "energy phase" of opposite sign, that is, it gives the complex conjugate of the term we have just derived.

Thus, adding these two terms we have

$$\left(\frac{\partial\rho_i}{\partial t}\right)_{\text{out}} \;=\; -\sum_{m\eta}\int_0^t \mathrm{d}t' \;\mathrm{Re}[K_{mi}^{(\eta+)}(t-t')]\; \rho_i(t')\; [1 - \rho_m(t')] \,, \tag{18.29}$$

The third and fourth terms in Eq. (18.18) are "in-scattering terms" which can be handled in an identical way. Thus, accounting for all of the terms second order in $\alpha$:

$$\left(\frac{\partial \rho_i}{\partial t}\right) = \sum_{m\eta} \int_0^t \mathrm{d}t' \left\{ \mathrm{Re}[K_{im}^{(\eta+)}(t-t')]\, \rho_m(t')\, [1-\rho_i(t')] \right.$$
$$\left. - \mathrm{Re}[K_{mi}^{(\eta+)}(t-t')]\, \rho_i(t')\, [1-\rho_m(t')] \right\} , \quad (18.30)$$

which has been called the "pre-Master equation" by Zwanzig [7].

Note that despite having forced the phonons to be at equilibrium, despite having taken the Hartree–Fock approximation, this equation is still symmetric under time-reversal and it shows how collisions at time $t$ depend on the history of the system at all earlier times $t'$. In order to proceed we need the equivalent of the assumption originally made by Boltzmann of "molecular chaos" (*Stosszahlansatz*: The assumption that collisions depend only on the instantaneous variables before the collisions, not on the previous history going back to the previous collision) is what is needed to avoid this "memory" in classical systems. Here we need another powerful and somewhat mysterious "trick": We now take the "van Hove limit": We set $\tau = \alpha^2(t-t')$ (where, recall, $\alpha$ is the dimensionless strength of the interaction), $G_{im}(\tau) = \mathrm{Re}[W_{im}^{(\eta+)}(t)]$, and $P_i(\tau) = \rho_i(t)$. Then, the out-scattering terms above can be rewritten as follows:

$$\left(\frac{\partial P_i}{\partial \tau}\right)_{\text{out}} = -\sum_{m\eta} \int_0^{\tau/\alpha^2} \mathrm{d}\tau'\, G_{im}(\tau')\, P_i(\tau-\alpha^2\tau')[1 - P_m(\tau-\alpha^2\tau')] . \quad (18.31)$$

Now, in the limit $t \to \infty$ and $\alpha^2 \to 0$, whilst $\tau$ remains constant, the factors containing the "populations" $P$ can be taken outside the time integral. The remaining factor $\int_0^\infty \mathrm{d}\tau' G(\tau')$ contains a time exponential [see the exponential in Eq. (18.28)] whose imaginary part (the principal part of the integral) is absorbed into the bare energy $E_i$ as a self-energy correction and whose real part gives the energy-conserving delta-function $\pi\hbar\delta(E_m - E_i)$, provided we regularize those integrals by adding an exponential damping term $\mathrm{e}^{-\eta\tau}$ and let $\eta \to 0$. *Note that this is the only step which breaks the time-reversal invariance of the problem!* Irreversibility does not arise from the Hartree–Fock factorization. It does not even arise from the random-phase assumption (infinite-volume limit, infinitely many degrees of freedom, etc.) used to invoke the van Hove "diagonal" argument. It only arises from the regularization of the complex time exponentials. Physically, the time scale $\eta^{-1}$ is a mysterious "decoherence" time scale which allows us to complete the collision and get the delta-function. Thus, Eq. (18.29) becomes

$$\left(\frac{\partial \rho_i}{\partial t}\right)_{\text{out}} = -\sum_m W_{mi}^{(\eta+)}\, \rho_i(t)\, [1 - \rho_m(t)] , \quad (18.32)$$

where the "golden-rule" transition rate is

$$W_{mi}^{(\eta\pm)} = \frac{2\pi}{\hbar}\, \alpha^2 \sum_{\mathbf{q}} |\mathscr{F}_{im}^{(\eta)}(q)|^2 \left(\frac{1}{2} \pm \frac{1}{2} + n_{\mathbf{q}\eta}\right) \delta(E_i - E_m \mp \hbar\omega_{\mathbf{q}\eta}) . \quad (18.33)$$

Accounting for the other terms (in-scattering) in Eq. (18.18), accounting for all phonon branches $\eta$ and both emission and absorption processes ($\pm$), we finally obtain the Master equation in the presence of electron–phonon interactions:

$$\frac{\partial \rho_i}{\partial t} = \sum_{m\eta\pm} \left\{ W_{im}^{(\eta\pm)} \rho_m(1-\rho_i) - W_{mi}^{(\eta\pm)} \rho_i(1-\rho_m) \right\} . \quad (18.34)$$

We should stress again that taking the van Hove limit by "completing" the collisions is a clear mathematical step required to regularize the integral. Yet, physically, its justification remains mysterious. Herein lies the mystery of the onset of the "arrow of time."

* *Although we have explicitly written the equation of motion for the diagonal elements of the density matrix, it is interesting to consider also the dynamics for the off-diagonal elements, $\rho_{ij}$ with $i \neq j$:*

$$\begin{aligned}\frac{\partial \rho_{ij}}{\partial t} &= \frac{1}{i\hbar}(E_i - E_j)\rho_{ij}(t) \\ &+ \sum_{m\eta\pm} \left\{ \delta_{ij}\, W_{im}^{(\eta\pm)} \rho_{mm}(1-\rho_{ij}) - \frac{1}{2}[W_{mi}^{(\eta\pm)} + W_{mj}^{(\eta\pm)}]\rho_{ij}(1-\rho_{mm}) + \right\}. \qquad (18.35)\end{aligned}$$

*Note that the off-diagonal elements can only decay, since only the first term on the rhs survives when $i \neq j$. The system "decoheres," unless some external driving force injects linear superpositions of eigenstates into the system.*

### 18.1.4 Boundary Effects: The "Contacts" of an Open System

So far we have considered a "closed" system: Our "device" has been populated at $t = 0$, but left closed thereafter, so that no electrons could leave or enter the device. Let's now consider an open system: The internal dynamics—so the "reshuffling" of the occupation $\rho_i$ of the states $|i\rangle$—is described by Eq. (18.34) above, but now we must somehow account for the fact that electrons can be exchanged between the device and some external reservoirs of electrons. Let's assume that these are ohmic "contacts": They are extremely large reservoirs of electrons kept at thermal equilibrium by the strong scattering occurring within them. Their overall occupation is essentially unchanged by subtraction or addition of a few electrons to or from any of their electronic states, thanks to their large size, and they are assumed to inject electrons whenever the occupation $\rho_i$ of the state $|i\rangle$ falls below the equilibrium occupation $f_i$ (the Fermi function in the contact) and absorb electrons without any reflection. For simplicity, let's start by considering a one-dimensional device of length $L$ (along the $z$-axis). We then have two contacts, one at the left (say, at $z = 0$) and one at the right (at $z = L$). Each state $|i\rangle$ will be doubly degenerate since at each energy we will have two states: One corresponding to an electron injected from the left, partially reflected back to the left and partially transmitted to the right, say $|i\rangle_+$ and one corresponding to an electron injected from the right contact, $|i\rangle_-$. The two contacts will be characterized by the Fermi levels $E_{\rm F,L}$ and $E_{\rm F,R}$, respectively. Thus, we can augment the Master equation (18.34) by the terms:

$$\left(\frac{\partial \rho_i}{\partial t}\right)_{\rm left} = \Phi_{i,\rm L}\,(\rho_i - f_{i,\rm L}) \qquad (18.36)$$

expressing the exchange of electrons with the left "contact," where $f_{i,\rm L} = \{1 + \exp[(E_i - E_{\rm F,L})/k_{\rm B}T)]\}^{-1}$ is the Fermi function of the left contact, and

$$\left(\frac{\partial \rho_i}{\partial t}\right)_{\rm right} = \Phi_{i,\rm R}\,(\rho_i - f_{i,\rm R}) \qquad (18.37)$$

for the right contact, where $f_{i,\rm R} = \{1 + \exp[(E_i - E_{\rm F,R})/k_{\rm B}T)]\}^{-1}$ is the Fermi function of the left contact. The terms $\Phi_{i,\rm L}$ and $\Phi_{i,\rm R}$ are flux terms. In our one-dimensional example they will have the form:

$$\Phi_{i,\mathrm{L}} = \frac{\upsilon_{i,z,\mathrm{L}}}{L_{i+}} , \tag{18.38}$$

where $\upsilon_{i,z}$ is the component of the velocity of an electron at energy $E_i$ perpendicular to the device-contact boundary and $L_i$ is the normalization required to normalize the wavefunction $\langle z|i\rangle_+$ so that

$$\frac{1}{L_{i+}} \int_0^L \mathrm{d}z \, |\langle z|i\rangle_+|^2 = 1 , \tag{18.39}$$

and similarly for the states $|i\rangle_-$. Thus, the total transport equation will be

$$\frac{\partial \rho_{is}}{\partial t} = \sum_{m\eta\pm ss'} \left\{ W^{(\eta\pm)}_{isms'} \rho_{ms'} (1-\rho_{is}) - W^{(\eta\pm)}_{ms'is} \rho_{is} (1-\rho_{ms'}) \right\} + \Phi_{i,s} \left(\rho_{is} - f_{i,s}\right) , \tag{18.40}$$

where $s, s' = \pm$ and the label $s$ of the contact term is L for $s = +$ and R for $s = -$.
Note that the treatment of "contacts" we have just given is phenomenological: It makes sense, but there is no formal justification available to prove the uniqueness (or even the correctness) of the expressions (18.36)–(18.38) above.

Figure 18.1 illustrates schematically ballistic transport in an open system. Figures 18.1 and 18.2 illustrate how the Master equation describes dissipative transport in the "small" open system.

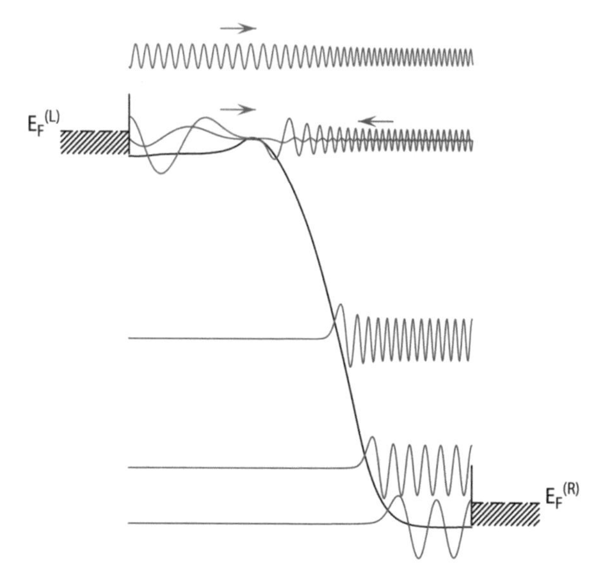


**Fig. 18.1** Schematic illustration of ballistic electron transport in an open system, in this case an $n^+ - i - n^+$ "diode": Waves are injected from the left contact, transmitted to the right and partially reflected (*red lines*). Similarly, waves are injected from the right contact (*blue lines*), partially transmitted to the left or reflected. The occupation of these states in the ballistic regime is controlled by the Fermi–Dirac distributions at the left and right contact. The left and right contacts (or reservoirs) are held at different chemical potentials (Fermi levels) by an externally applied bias. The electron potential energy (*solid black line*) is determined by solving Poisson equation with charge given by the dopants and the electron charge obtained from Schrödinger equation. Therefore, Poisson and Schrödinger equations must be solved self-consistently

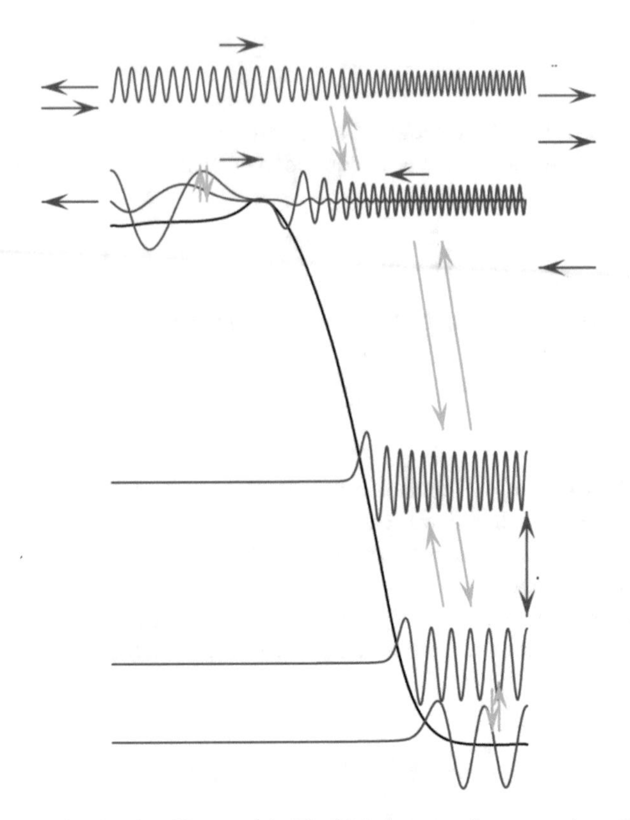

**Fig. 18.2** The same schematic situation illustrated in Fig. 18.1, but now the occupation of the electronic states is determined not only by Fermi–Dirac distribution of the contacts, but also by the Master scattering operator that accounts for electronic transitions among all electronic states (*green arrows*). This scattering operator accounts for collisions only to first order, so the procedure is correct only in small structures

## *18.1.5 Inhomogeneous Transport in Long Devices: The BTE*

Equation (18.40) is valid, as we have discussed, when the size L of the "device" is smaller than the dephasing length (the mean free path for inelastic collisions) $\lambda_\phi$. So, it may be our final transport equation if we were to deal exclusively with very small devices or "chunks" of semiconductor. If, instead, we have to deal with devices of much larger dimensions, we are in the opposite limit of contacts injecting wavepackets which are much smaller than the active region of the device, $L >> \lambda_\phi$. We can now proceed as follows.

Let's consider first Eq. (18.10), but now taking the matrix elements of $\hat{\rho}$ over eigenstates of the field-free (i.e., $V^{(\mathrm{H})}(\mathbf{r}) = 0$) free-electron Hamiltonian, i.e., plane waves such that $\widehat{\mathscr{T}}|\mathbf{k}\rangle = E_{\mathbf{k}}|\mathbf{k}\rangle$, where $\mathbf{k} = (\mathbf{K}, k_z)$. Then, in place of Eq. (18.40), in the nondegenerate limit (that is, $1 - \rho_{\mathbf{k}'\mathbf{k}'} \approx 1$) one obtains

$$\frac{\partial \rho_{\mathbf{kk}}}{\partial t} = \sum_{\mathbf{k}' \neq \mathbf{k}} [W_{\mathbf{k};\mathbf{k}'}\, \rho_{\mathbf{k}'\mathbf{k}'} - W_{\mathbf{k}';\mathbf{k}}\, \rho_{\mathbf{kk}}] + \frac{\mathrm{i}}{\hbar}\langle \mathbf{k}|[\hat{\rho}^{(\mathrm{I})}(t), -e\varphi]|\mathbf{k}\rangle + \left(\frac{\partial \rho_{\mathbf{kk}}}{\partial t}\right)_{\mathrm{res}}, \tag{18.41}$$

where $-e\varphi(\mathbf{r})$ is the electrostatic potential associated with the Hartree potential $V^{(\mathrm{H})}(\mathbf{r})$ introduced in Eq. (18.3) and the last term is just the reservoir (or "contact") term discuss above. Let's now divide the region $(0, L)$ into intervals of length $\delta z << \lambda_\phi$, so that in each interval we are brought back to the limit of delocalized packets and negligible off-diagonal elements of the density matrix. Under the assumption that the potential $\varphi$ varies slowly over the average electron wavelength, in each interval we can consider the field as constant, i.e., $-e\varphi_j(z) \approx -e\varphi(z_j) + eF_j(z - z_j)$ in the $j$th interval. Note that now the index $j$ labels spatial positions, not electronic states, since these are now labeled by the wavevector $\mathbf{k}$. Therefore, to first order in $\alpha$ *and* in the electric field $F_j$ (denoted by $\mathbf{F}$ in this and in the following chapter, to avoid confusion with the energy $E$):

$$\frac{\mathrm{i}}{\hbar}\langle\mathbf{k}|[\rho, -e\varphi_j]|\mathbf{k}\rangle = -\frac{eF_j}{\hbar}\frac{\partial \rho_{\mathbf{kk}}}{\partial k_z} \,. \tag{18.42}$$

This expression follows from the fact that as the momentum operator is just the derivative with respect to spatial coordinates (i.e., $p_z = -\mathrm{i}\hbar\partial/\partial z$), so the operator position is just the derivative with respect to momentum coordinates, that is $z = i\partial/\partial k_z$, as it follows either from simple considerations on Fourier transforms or from the commutation rule $[z, p_z] = \mathrm{i}\hbar$. In addition, embracing the semiclassical limits, one can assume that the fluxes in and out of each interval consist of particles with completely randomized phases, so that, in analogy with Eq. (18.38):

$$\left(\frac{\partial \rho_{\mathbf{kk}}}{\partial t}\right)_{\mathrm{res},\, j} = [\,\rho_{\mathbf{kk}}^{j+1} - \rho_{\mathbf{kk}}^{j}]\,\frac{\upsilon_{z,j}}{\delta z} \,. \tag{18.43}$$

(where, in the case of simple parabolic bands $\upsilon_{z,j} = \hbar k_{z,j}/m^*$ and $\hbar k_{z,j} = [2m^*(E_{\mathbf{k}} + e\varphi_j)]^{1/2}$) for all intervals not "touching" the contacts. The factor $\delta z$ in the denominator originates from the normalization of the plane waves to unity in each interval. In the limit of vanishing $\lambda_\phi/L$, $\delta z \to 0$, the diagonal elements $\rho_{\mathbf{kk}}^{j}$ become the "distribution function" $f(\mathbf{k}, z_j, t)$, the discrete index $j$ can be replaced by continuous variables, and Eqs. (18.41)–(18.43) yield

$$\frac{\partial \rho_{\mathbf{kk}}}{\partial t} = \sum_{\mathbf{k}'\neq\mathbf{k}} [W_{\mathbf{k};\mathbf{k}'}\,\rho_{\mathbf{k}'\mathbf{k}'} - W_{\mathbf{k}';\mathbf{k}}\,\rho_{\mathbf{kk}}] - \frac{eF(z)}{\hbar}\frac{\partial \rho_{\mathbf{kk}}}{\partial k_z} + \upsilon_z\frac{\partial \rho_{\mathbf{kk}}}{\partial z}, \tag{18.44}$$

which is the Boltzmann equation for an inhomogeneous (in the $z$ direction) system, subject to the boundary conditions:

$$\left(\frac{\partial \rho_{\mathbf{kk}}}{\partial t}\right)_{\mathrm{res,L}} = \upsilon_{zj+}\left(\frac{\partial \rho_{\mathbf{kk}}}{\partial z}\right)_{z=0} \,. \tag{18.45}$$

and

$$\left(\frac{\partial \rho_{\mathbf{kk}}}{\partial t}\right)_{\mathrm{res,R}} = -\,\upsilon_{zj-}\left(\frac{\partial \rho_{\mathbf{kk}}}{\partial z}\right)_{z=L} \,. \tag{18.46}$$

We can generalize Eq. (18.44) to three dimensions:

$$\begin{aligned}\frac{\partial f(\mathbf{r},\mathbf{k},t)}{\partial t} &= -\frac{\mathrm{d}\mathbf{r}}{\mathrm{d}t}\cdot\nabla_{\mathbf{r}} f(\mathbf{r},\mathbf{k},t) - \frac{\mathrm{d}\mathbf{k}}{\mathrm{d}t}\cdot\nabla_{\mathbf{k}} f(\mathbf{r},\mathbf{k},t) \\ &\quad + \sum_{\mathbf{k}'}[\,W(\mathbf{k},\mathbf{k}')f(\mathbf{r},\mathbf{k}',t) - W(\mathbf{k}',\mathbf{k})f(\mathbf{r},\mathbf{k},t)\,]\,,\end{aligned} \tag{18.47}$$

which, using the equations of motion and introducing the electrostatic potential $\phi(\mathbf{r})$ such that $-\nabla_{\mathbf{r}}\phi(\mathbf{r}) = \mathbf{F}$, becomes

$$\begin{aligned}\frac{\partial f(\mathbf{r},\mathbf{k},t)}{\partial t} &= -\frac{1}{\hbar}\nabla_{\mathbf{k}}E(\mathbf{k})\cdot\nabla_{\mathbf{r}} f(\mathbf{r},\mathbf{k},t) + e\nabla_{\mathbf{r}}\varphi(\mathbf{r},t)\cdot\nabla_{\mathbf{k}} f(\mathbf{r},\mathbf{k},t) \\ &\quad + \sum_{\mathbf{k}'}[\,W(\mathbf{k},\mathbf{k}')f(\mathbf{r},\mathbf{k}',t) - W(\mathbf{k}',\mathbf{k})f(\mathbf{r},\mathbf{k},t)\,]\,.\end{aligned} \tag{18.48}$$

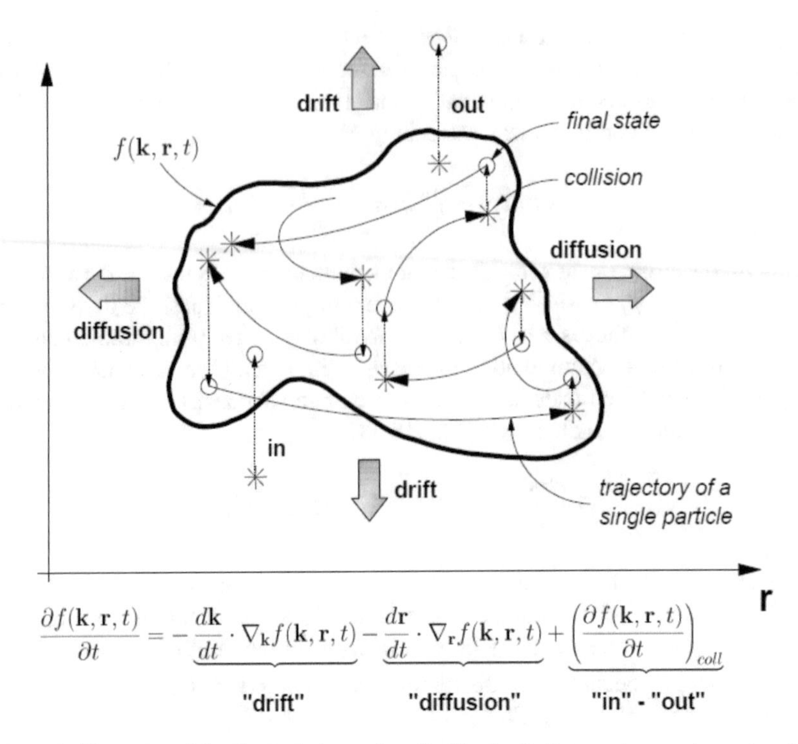


**Fig. 18.3** Schematic illustration of the physical picture described by the Boltzmann transport equation. In phase-space, the horizontal axis represents the spatial coordinates of a particle, the vertical axis their momenta. Electrons in a system are distributed in phase-space. Drift due to an external spatially dependent driving force (an electric field) has the main effect of pushing the distribution along the vertical axis, by accelerating them, as well as along the horizontal axis as particles move. Diffusion, instead, spreads the distribution spatially. Collisions "reshuffle" electrons within the distribution

Figure 18.3 illustrates the phenomenological interpretation of the BTE: The total variation with respect to time of the distribution function is due to drift (electrons exit the infinitesimal phase-space volume $\mathbf{drdk}$ or enter it from the outside because of the electric field which moves electrons along the $\mathbf{k}$-axis), diffusion (in or out the infinitesimal phase-space volume $\mathbf{drdk}$ as electron diffuse in real space, so along the $\mathbf{r}$-axis), or scattering.

### *18.1.6 Review of the Assumptions Behind the BTE*

It is useful to recall the several approximations we have been forced to make in order to reach the BTE:

1. Plane waves of the form $e^{i\mathbf{k}\cdot\mathbf{r}}$ (or the corresponding plane-wave "envelopes" times a Bloch factor in a semiconductor) describe an electron with well-defined momentum but uniformly distributed in space. Instead, an electron localized in space must be described by a "wave packet" which consists of a linear superposition of plane waves. Thus, as a consequence of Heisenberg's principle, we must have some uncertainty about the electron position as well as about its momentum. However, we have assumed that the size of the device is much larger than the size of the electron wave packet (given heuristically by the dephasing length $\lambda_\phi$) and that the energy scales of interest (mainly, the applied bias) are much larger than the uncertainty implied by the "spread" of the electron momentum. Therefore, we have assumed that the electron can be described by its well-defined position $\mathbf{r}$ and its well-defined (crystal) momentum $\hbar\mathbf{k}$, as it if were a classical particle.

2. The electric field and all other perturbations are weak enough so that we can use the envelope approximation to describe the motion of electrons (or holes) in a semiconductor.
3. All collision processes are independent, instantaneous, local in space, and are perturbations weak enough to justify the use of first-order perturbation theory, Fermi's Golden Rule or the first Born approximation.
4. As a consequence, all collision processes depend only on the local, instantaneous electron variables, without retaining memory of where and when the previous collision happened. This is the equivalent of Boltzmann's assumption of "molecular chaos" which is essential to arrive at irreversible dynamics.

## 18.2 Coulomb Interactions

In the previous discussion we have considered a system consisting of an electron gas and phonons. We have seen that the interactions between the two systems are of the van Hove type, a property that is necessary (but not sufficient!) to obtain a Master-type collisional operator and irreversibility. Here we wish to consider Coulomb interactions and see whether or not these are also of the van Hove type.

We have already emphasized several times that scattering with impurities is, by itself, *not* of the van Hove type: Ignoring phonons, in a small structure with a few dopants, we can describe electronic transport exactly solving self-consistently the Schrödinger equation with open boundary conditions and Poisson equation. This is routinely done using the ballistic version of the non-equilibrium Green's functions (NEGFs) [8, 9] or the quantum transmitting boundary method (QTBM) [10]. This is a fully reversible problem. Indeed, the conductance observed in systems at the nanoscale, such as in point-contacts, exhibits experimentally a fully reproducible "noise," which is just a reflection of the electrons being transmitted across the potential due to a particular spatial configuration of impurities. What is necessary to render this process irreversible is an average over various spatial configurations of the impurities. This is indeed the main result obtained by Kohn and Luttinger [2]. We shall see here how this is done.

Electron–electron interactions, on the other hand, are of the van Hove type, as long as the system contains a "large" number of electrons. This resembles the requirement we made before regarding phonons: In this case, it is the large number of degrees of freedom of the electron gas that results in the randomization of the phases required to reach a collision operator of the Master type. As important byproducts of our discussion on electron–electron interactions, we will see how the mean-field approximations (Hartree and Hartree–Fock) emerge and how these interactions drive the system towards a Fermi–Dirac distribution.

### 18.2.1 Scattering with Ionized Impurities

In order to account for electrons scattering with ionized impurities in our transport model, we must consider again the Hamiltonian given by Eq. (14.6) in Chap. 15. The various forms for the Fourier components of the potential that we may consider are given by Eq. (14.5) (unscreened), by Eq. (14.13) (screened, with a suitable expression for the screening parameter $\beta$), or, more generally, by Eq. (14.23). In the representation provided by the basis states $|i\rangle$ the Hamiltonian takes the form:

$$\widehat{H}_{\mathrm{ei}} = \sum_{\alpha=1}^{N_{\mathrm{imp}}} V_{ji}^{(\mathrm{imp},\alpha)} \, \hat{c}_j^{\dagger} \, \hat{c}_i \, , \tag{18.49}$$

where

$$V_{ji}^{(\mathrm{imp},\alpha)} = \sum_{\mathbf{q}} \langle j|V^{(\mathrm{imp},\alpha)}(\mathbf{q})\mathrm{e}^{\mathrm{i}\mathbf{q}\cdot(\mathbf{r}-\mathbf{r}_\alpha)}|i\rangle \ . \tag{18.50}$$

The superscript $\alpha$ appearing in these terms is related to the fact that we are assuming that the impurities are distributed in the system in a particular configuration characterized by this index. As already mentioned before, we will have to average over all possible configurations $\alpha$ in order to obtain an irreversible physical picture.

The equation of motion for the reduced density matrix, Eq. (18.8) is similar to Eq. (18.13):

$$\begin{aligned}\frac{\partial\langle\hat{c}_i^\dagger\,\hat{c}_j\rangle_t}{\partial t} =& -\frac{\mathrm{i}}{\hbar}(E_\mathrm{i}-E_j)\langle\hat{c}_i^\dagger\,\hat{c}_j\rangle_t - \frac{\mathrm{i}}{\hbar}\,\mathrm{Tr}_\mathrm{el}\,\{\,[\widehat{H}_\mathrm{ei}^{(\mathrm{I})}(t),\hat{\rho}^{(\mathrm{I})}(0)]\,\hat{c}_i^\dagger\,\hat{c}_j\} \\ &-\frac{1}{\hbar^2}\int_0^t \mathrm{d}t'\,\mathrm{Tr}_\mathrm{el}\{\,[\widehat{H}_\mathrm{ei}^{(\mathrm{I})}(t),[\widehat{H}_\mathrm{ei}^{(\mathrm{I})}(t'),\hat{\rho}^{(\mathrm{I})}(t')]]\,\hat{c}_i^\dagger\,\hat{c}_j\}\ .\end{aligned} \tag{18.51}$$

With the usual assumptions, the first-order (memory) term can be written as:

$$-\frac{\mathrm{i}}{\hbar}\sum_{\{N_n\}}\sum_{lm}\sum_{\alpha}\rho_{\{N_n\}}\,V_{ml}^{(\mathrm{imp},\alpha)}\,\langle\{N_n\}|[\hat{c}_m^\dagger\hat{c}_l\hat{c}_i^\dagger\hat{c}_j-\hat{c}_i^\dagger\hat{c}_j\hat{c}_m^\dagger\hat{c}_l]|\{N_n\}\rangle\ . \tag{18.52}$$

For diagonal elements ($i=j$) only terms for which $m=l$ give nonzero contributions:

$$-\frac{\mathrm{i}}{\hbar}\sum_{\{N_n\}}\sum_{l}\sum_{\alpha}\rho_{\{N_n\}}\,V_{ll}^{(\mathrm{imp},\alpha)}\,\langle\{N_n\}|\hat{c}_l^\dagger\hat{c}_l\hat{c}_i^\dagger\hat{c}_i|\{N_n\}\rangle\ . \tag{18.53}$$

But these terms vanish. For off-diagonal terms we must have $m=j$ and $l=i$, so that the sum (18.52) becomes

$$-\frac{\mathrm{i}}{\hbar}\sum_{\{N_n\}}\sum_{\alpha}\rho_{\{N_n\}}\,V_{ji}^{(\mathrm{imp},\alpha)}\,\langle\{N_n\}|[\hat{c}_j^\dagger\hat{c}_i\hat{c}_i^\dagger\hat{c}_j-\hat{c}_i^\dagger\hat{c}_j\hat{c}_j^\dagger\hat{c}_i]|\{N_n\}\rangle\ , \tag{18.54}$$

that is

$$-\frac{\mathrm{i}}{\hbar}\sum_{\alpha}V_{ji}^{(\mathrm{imp},\alpha)}\,[\langle\hat{c}_j^\dagger\hat{c}_j\rangle-\langle\hat{c}_i^\dagger\hat{c}_i\rangle]\ . \tag{18.55}$$

Considerations about averaging over an ensemble of impurity configurations, $\{\mathbf{r}_\alpha\}$, will be discussed at length below. Here note that summing incoherently over many impurity positions will result in random cancellations of the phase of Eq. (18.50) in Eq. (18.55), unless $i=j$. Thus, once more, we can ignore the memory term in Eq. (18.51).

The second-order terms can be handled in the usual way we have already followed twice, but there are two noteworthy differences: In first place, the Coulomb potential is not a weak perturbation and halting the perturbative expansion at the second order must be justified on the grounds of "diluted impurities." Second, when expressing van Hove's argument as in Eq. (18.23), we can recover the usual "random phase" argument provided we average these second-order terms over a large ensemble of configurations $\{\mathbf{r}_\alpha\}$ of the impurities. That is, Eq. (18.23) is replaced by

$$\sum_{\{\mathbf{r}_\alpha\}}\sum_{\{M_m\}\{N_n\}}\rho_{\{M_m\}}\sum_{\alpha\beta}\langle\{M_m\}|\hat{c}_j\,\widehat{H}_\mathrm{ei}^{(I)(\alpha)}(t)\,|\{N_n\}\rangle\langle\{N_n\}|\widehat{H}_\mathrm{ei}^{(I)(\beta)}(t')\,\hat{c}_i^\dagger|\{M_m\}\rangle\ . \tag{18.56}$$

For a given configuration $\{\mathbf{r}_\alpha\}$ of impurities in a given device, van Hove's argument cannot apply. Rather, we must think of averaging Eq. (18.56) over $N$ configurations $\{\mathbf{r}_\alpha\}$. In so doing, "diagonal" terms $\alpha = \beta$ yield a contribution $\sim O(N^2)$ to the sum, in which case we must have $i = j$ by the usual van Hove argument. If, instead, $\alpha \neq \beta$, only $\sim O(N)$ terms contribute to the sum, and they will have a non-vanishing contribution also if $i \neq j$. Thus, in the limit of an infinitely large ensemble of configurations ($N \to \infty$), we can ignore off-diagonal terms. In other words: Now it is not the large number of degrees of freedom (electronic or of the phonon bath) which determines the dominant role played by diagonal elements. Rather, it is the sampling over a larger number of different configurations of the impurities. Each device—in other words—remains dissipation free and the dynamic remains time-reversal invariant. But we recover dissipation when averaging over many macroscopically identical but microscopically different devices.

Assuming a diagonal density matrix at $t = 0$, employing the arguments above, and going to the usual van Hove limit, we reach the Master equation:

$$\frac{\partial \rho_i}{\partial t} = \sum_j \left[ W_{ij}^{(\mathrm{imp})} \rho_j(1-\rho_i) - W_{ji}^{(\mathrm{imp})} \rho_i(1-\rho_j) \right], \tag{18.57}$$

(up to injection/extraction terms due to the contacts), where the scattering rate $W_{im}^{(\mathrm{imp})}$ is given by the first Born approximation:

$$W_{ji}^{(\mathrm{imp})} = \frac{2\pi}{\hbar} \delta(E_i - E_j) \left| \sum_{\alpha}^{N_{\mathrm{imp}}} V_{ij}^{(\mathrm{imp},\alpha)} \right|^2 \to \frac{2\pi}{\hbar} \delta(E_i - E_j) \sum_{\alpha=1}^{N_{\mathrm{imp}}} \left| V_{ij}^{(\mathrm{imp},\alpha)} \right|^2, \tag{18.58}$$

where the last step relies on the configuration average. In the special case of uniform doping and employing plane waves (that is, $|i\rangle = |\mathbf{k}\rangle$),

$$V_{\mathbf{k},\mathbf{k}'}^{(\mathrm{imp},\alpha)} = \mathrm{e}^{\mathrm{i}(\mathbf{k}-bfk')\cdot\mathbf{r}_\alpha} v_{\mathbf{k}-\mathbf{k}'}^{(\mathrm{imp})}, \tag{18.59}$$

where we have indicated with $v_{\mathbf{q}}^{(\mathrm{imp})}$ the Fourier coefficients of the potential of a single impurity. Therefore,

$$W_{\mathbf{k},\mathbf{k}'}^{(\mathrm{imp})} = \frac{2\pi}{\hbar} \delta(E_{\mathbf{k}} - E_{\mathbf{k}'}) \frac{N_{\mathrm{imp}}}{\Omega} |v_{\mathrm{imp}}(\mathbf{k}-\mathbf{k}')|^2, \tag{18.60}$$

as we had found before [see Eq. (14.10), page 318] noticing that $n_D = N_{\mathrm{imp}}/\Omega$. Note now that Eq. (18.57) is valid for devices short enough so that $L << \lambda_\phi$. For longer devices we can follow the same procedure we have followed above to reach the BTE. The rhs of Eq. (18.57) will then become simply the rhs of the BTE.

### 18.2.2 Electron–Electron Scattering

The treatment of the electron–electron interaction is more cumbersome, but we can follow a logical path completely analogous to what we have done for the electron–phonon interaction. We should emphasize that we are mainly interested in establishing whether the effect of interparticle interactions on the time evolution of the density matrix can also be described by a scattering operator of the Master type. Therefore, we restrict our attention to single-particle processes and ignore the collective plasma excitations. These cannot be captured by our perturbation-theory approach that does not account explicitly for the feedback of the electron-density fluctuations on the electric field.

In our basis of states $|i\rangle$ the electron–electron Coulomb Hamiltonian takes the form

$$\widehat{H}_{ee} = \frac{1}{2} \sum_{i',j',i,j} V_{i'j'ij}\, \hat{c}^{\dagger}_{i'}\, \hat{c}^{\dagger}_{j'}\, \hat{c}_i\, \hat{c}_j \,, \tag{18.61}$$

where the matrix element $V_{i'j'ij}$ is given by:

$$V_{i'j'ij} = \sum_{\mathbf{q}} \langle i'j'|V_{\mathbf{q}}|ij\rangle \,, \tag{18.62}$$

in analogy with Eq. (18.7). Of course, $V_{\mathbf{q}} = e^2/(\epsilon q^2)$ are the Fourier coefficients of the Coulomb potential. The dielectric constant (or function) in this expression represents the dielectric response of the electrons in the valence band, $\epsilon_{\rm v}(q) \approx \epsilon_{\rm s}$, since we are considering transport of free carriers.

The equation of motion for the density matrix will be

$$\frac{\partial \hat{\rho}^{(\rm I)}(t)}{\partial t} = -\frac{\rm i}{\hbar} [\widehat{H}^{(\rm I)}_{\rm ee}(t), \hat{\rho}^{(\rm I)}(t)] \,, \tag{18.63}$$

which, to second order in the perturbation term can be rewritten as:

$$\frac{\partial \hat{\rho}^{(\rm I)}(t)}{\partial t} = -\frac{\rm i}{\hbar} [\widehat{H}^{(\rm I)}_{\rm ee}(t), \hat{\rho}^{(\rm I)}(0)] - \frac{1}{\hbar^2} \int_0^t {\rm d}t' \, [\widehat{H}^{(\rm I)}_{\rm ee}(t), [\widehat{H}^{(\rm I)}_{\rm ee}(t'), \hat{\rho}^{(\rm I)}(t')]] \,. \tag{18.64}$$

The equation of motion for the reduced density matrix will be

$$\begin{aligned} \frac{\partial \langle \hat{c}^{\dagger}_i\, \hat{c}_j \rangle_t}{\partial t} = & -\frac{\rm i}{\hbar} {\rm Tr}_{\rm el} \left\{ [\widehat{H}^{(\rm I)}_{\rm ee}(t), \hat{\rho}^{(\rm I)}(0)]\, \hat{c}^{\dagger}_i\, \hat{c}_j \right\} \\ & - \frac{1}{\hbar^2} \int_0^t {\rm d}t' \, {\rm Tr}_{\rm el} \left\{ [\widehat{H}^{(\rm I)}_{\rm ee}(t), [\widehat{H}^{(\rm I)}_{\rm ee}(t'), \hat{\rho}^{(\rm I)}(t')]]\, \hat{c}^{\dagger}_i\, \hat{c}_j \right\} . \end{aligned} \tag{18.65}$$

As we have done previously, it is useful to anticipate the physical picture that we are going to uncover. The first term at the rhs of Eq. (18.65) (the "memory" term), is in general quite troublesome. However, the choice of basis functions we make (eigenstates of the self-consistent Poisson–Schrödinger problem), makes this term vanish in the Hartree–Fock (or "mean field," or also "random-phase") approximation. Had we used a different set of basis functions (plane waves, as commonly done), the first term would be responsible for a strong growth of off-diagonal elements $\rho_{\mathbf{kk}'}$ since, obviously, any inhomogeneous potential would mix different plane waves via acceleration terms. This term is, indeed, the field-term appearing in the Boltzmann equation. So, we pay the (CPU-time) price of solving the self-consistent Poisson–Schrödinger problem, but we get rewarded by the lack of the first-order term (in $\widehat{H}^{(\rm I)}_{\rm ee}(t)$) of Eq. (18.65).

The analysis of the last term at the rhs of Eq. (18.65) requires additional thought. When dealing with the electron–phonon or electron-impurity interaction, we had used van Hove's argument to show that, for off-diagonal elements of the density matrix, the second-order terms of Eq. (18.65) are negligible when compared to the diagonal elements, thanks to the many degrees of freedom of the heat bath or to the incoherent averaging over many impurity configurations. In the present context, we must make a similar approximation of "infinitely many degrees of freedom" on the number of electrons in the system. Here the discussion becomes conceptually very difficult: We must face the problem of determining how fast a system of how many degrees of freedom will decohere and exhibit irreversibility. We shall assume our devices are "big" enough (have sufficiently many electrons) to benefit from the randomization of phases when we look at one electron interacting with many others.

This will lead us to a situation similar to what was found in the case of systems interacting only with the phonon bath. Now the major difference consists in the fact that we must make throughout the assumption of a diagonal density matrix at the initial time. Indeed, off-diagonal terms of the density matrix *do not* decay when only interparticle Coulomb interactions are accounted for, but they do not grow either. So, for a system containing "many" electrons, initially assumed "diagonal," and simultaneously in contact with the phonon bath, we can consider only the diagonal elements. On the other side, there are cases in which this assumption is violated: Single-electron transistors, for example, would have to be treated with the inclusion of off-diagonal terms as well.

We can now write explicitly the second term at the rhs of Eq. (18.65). Omitting for simplicity a phase $e^{-(i/\hbar)(E_{l'}+E_{m'}-E_l-E_m)t}$ which does not modify our arguments, we have

$$
\begin{aligned}
&\mathrm{Tr}_{\mathrm{el}}\,\{[\widehat{H}_{\mathrm{ee}}^{(\mathrm{I})}(t),\hat{\rho}^{(\mathrm{I})}(0)]\,\hat{c}_i^\dagger\,\hat{c}_j\} \\
&= \frac{1}{2}\sum_{\{N_n\}}\sum_{\{M_m\}}\rho_{\{N_n\}}\sum_{l'm'lm}V_{l'm'lm}\,[\langle\{M_m\}|\hat{c}_{l'}^\dagger\hat{c}_{m'}^\dagger\hat{c}_l\hat{c}_m|\{N_n\}\rangle\,\rangle\{N_n\}|\hat{c}_i^\dagger\hat{c}_j|\{M_m\}\rangle \\
&\quad -\langle\{M_m\}|\hat{c}_i^\dagger\hat{c}_j|\{N_n\}\rangle\,\langle\{N_n\}|\hat{c}_{l'}^\dagger\hat{c}_{m'}^\dagger\hat{c}_l\hat{c}_m|\{M_m\}\rangle] \\
&= \frac{1}{2}\sum_{\{N_n\}}\rho_{\{N_n\}}\sum_{l'm'lm}V_{l'm'lm}\,\langle\{N_n\}|[\hat{c}_i^\dagger\hat{c}_j\hat{c}_{l'}^\dagger\hat{c}_{m'}^\dagger\hat{c}_l\hat{c}_m\;-\;\hat{c}_{l'}^\dagger\hat{c}_{m'}^\dagger\hat{c}_l\hat{c}_m\hat{c}_i^\dagger\hat{c}_j]|\{N_n\}\rangle\,,
\end{aligned}
\tag{18.66}
$$

where in the last step we have switched the order of the factors inside the sums and have removed the completeness $\sum_{\{M_m\}}|\{M_m\}\rangle\langle\{M_m\}|$.

Before proceeding, note one important (and troublesome) complication: In the following we will have to assume that the Fock states $(\hat{c}_1^\dagger)^{n_1}(\hat{c}_2^\dagger)^{n_2}\ldots(\hat{c}_N^\dagger)^{n_N}|0\rangle$ are all orthogonal (we'll talk about normalization later on). This is not generally true for scattering states. We shall ignore this complication, as done so many times before in the literature: Since these mixed (off-diagonal) terms are proportional to the product of the reflection and transmission coefficients of the states, the cross-terms are non-negligible only in a small range of energies, when scattering states are half transmitted and half reflected. For large currents (small barriers, almost all states transmitted) or small currents (large barriers, or off-state for FETs), this approximation is satisfactory.

With this assumption, now notice that each term $\langle\{N_n\}|\hat{c}_i^\dagger\hat{c}_j\hat{c}_{l'}^\dagger\hat{c}_{m'}^\dagger\hat{c}_l\hat{c}_m|\{N_n\}\rangle$ in the sum above is an inner product between the two states $\hat{c}_{m'}^\dagger\hat{c}_l\hat{c}_m|\{N_n\}\rangle$ and $\hat{c}_{l'}\hat{c}_j^\dagger\hat{c}_i|\{N_n\}\rangle$. The only non-vanishing terms in the sums will be those for which $m'=j$ (the only creation terms) and either $m=i$, $l=l'$ or $m=l'$, $l=i$. Thus we have terms of the form

$$
\langle\{N_n\}|\hat{c}_i^\dagger\hat{c}_j\hat{c}_l^\dagger\hat{c}_j^\dagger\hat{c}_l\hat{c}_i|\{N_n\}\rangle \quad \text{and} \quad \langle\{N_n\}|\hat{c}_i^\dagger\hat{c}_j\hat{c}_m^\dagger\hat{c}_j^\dagger\hat{c}_i\hat{c}_m|\{N_n\}\rangle\,. \tag{18.67}
$$

For diagonal terms of the density matrix (i.e., $i=j$) these terms can be recast in the form $-\langle\{N_n\}|\hat{c}_i^\dagger\hat{c}_i\hat{c}_l^\dagger\hat{c}_l|\{N_n\}\rangle$ and $\langle\{N_n\}|\hat{c}_i^\dagger\hat{c}_i\hat{c}_m^\dagger\hat{c}_m|\{N_n\}\rangle$ (having used the fact that $n_in_i=n_i$ for Fermions) whenever $i\neq l$ or $i\neq m$, respectively. (The terms $i=l$ or $i=m$ are just self-energy terms which we assume lumped into the dispersion of the free electrons.) In the Hartree–Fock approximation (required to close the hierarchy of infinite equations involving Green's function of an increasing number of particles, as it is already clear), we shall factor the expectation values $\langle\hat{c}_m^\dagger\hat{c}_m\hat{c}_l^\dagger\hat{c}_l\rangle_t$ as $\langle\hat{c}_m^\dagger\hat{c}_m\rangle_t\,\langle\hat{c}_l^\dagger\hat{c}_l\rangle_t$. Then, accounting for both types of terms of the form (18.67) and for both terms in the commutator, we get an overall contribution of the form $\sum_l(V_{lili}-V_{liil})\rho_i(t)\rho_l(t)$. This is just the Hartree and exchange component of the potential felt by electrons in the single-particle state $|i\rangle$ under the electrostatic influence of a mean field caused by all other electrons in states $|l\rangle$. In other words, this is the Hartree (with exchange corrections, so, actually, Hartree–Fock) energy included in the self-consistent potential entering the electron Hamiltonian [Eq. (18.3)]. Therefore we must ignore these terms in order to avoid double-counting them. As a parenthetical remark, at this point we would recover plasma excitations by retaining the small-$q$ fluctuations $\hat{c}_i^\dagger \hat{c}_i - \langle \hat{c}_i^\dagger \hat{c}_i \rangle$, accounting for their effect on the Hartree potential and for the feedback of the potential on the fluctuations themselves. This was indeed the main result of Bohm's and Pines' work [11].

Off-diagonal terms of the density matrix, terms of the form (18.67), after factorization into mean-field (Hartree–Fock) single-particle expectation values, have the same physical meaning, and shall be similarly ignored. In conclusion, if we are willing to ignore the exchange interaction—or if we assume that we have already accounted for it in the Hartree–Fock term of the unperturbed electron Hamiltonian $\widehat{H}_{\rm el}$—, the first-order term in Eq. (18.65) vanishes for our choice of basis functions.

Consider now the third term at the right-hand side of Eq. (18.65). First of all, note that we shall consider only the leading-order (first Born) approximation of the Coulomb interactions. Therefore, to second order in $\widehat{H}_{\rm ee}^{(\rm I)}$, we can replace $\hat{\rho}^{(\rm I)}(t')$ with $\hat{\rho}^{(\rm I)}(0)$. This is the "weak coupling limit" also employed by van Hove to derive the Master equation. Expanding the commutator, this takes the form

$$
\begin{aligned}
- \frac{1}{\hbar^2} \int_0^t \mathrm{d}t' \sum_{\{M_m\}} \Big[ & \langle \{M_m\}| \, \widehat{H}_{\rm ee}^{(\rm I)}(t) \, \widehat{H}_{\rm ee}^{(\rm I)}(t') \, \hat{\rho}^{(\rm I)}(0) \, \hat{c}_i^\dagger \hat{c}_j \, |\{M_m\}\rangle \\
& + \langle \{M_m\}| \, \hat{\rho}^{(\rm I)}(0) \, \widehat{H}_{\rm ee}^{(\rm I)}(t') \, \widehat{H}_{\rm ee}^{(\rm I)}(t) \, \hat{c}_i^\dagger \hat{c}_j \, |\{M_m\}\rangle \\
& - \langle \{M_m\}| \, \widehat{H}_{\rm ee}^{(\rm I)}(t) \, \hat{\rho}^{(\rm I)}(0) \, \widehat{H}_{\rm ee}^{(\rm I)}(t') \, \hat{c}_i^\dagger \hat{c}_j \, |\{M_m\}\rangle \\
& - \langle \{M_m\}| \, \widehat{H}_{\rm ee}^{(\rm I)}(t') \, \hat{\rho}^{(\rm I)}(0) \, \widehat{H}_{\rm ee}^{(\rm I)}(t) \, \hat{c}_i^\dagger \hat{c}_j \, |\{M_m\}\rangle \Big] \, . 
\end{aligned} \tag{18.68}
$$

Let us consider the first term (an out-scattering term) within the square brackets at the right-hand side. Ignoring a phase factor $\exp[\mathrm{i}(E_i - E_j)t]$ which does not affect our arguments and which collapses to unity for diagonal terms, the integrand of this term has the form:

$$
\begin{aligned}
& \frac{1}{4} \sum_{\{M_m\}\{N_n\}} \sum_{l'm'lm} \sum_{h'r'hr} \rho_{\{M_m\}} V_{l'm'lm} V_{h'r'hr} \, \mathrm{e}^{-(\mathrm{i}/\hbar)[E_{l'}+E_{m'}-E_l-E_m] \, t} \\
& \times \mathrm{e}^{-(\mathrm{i}/\hbar)[E_{h'}+E_{r'}-E_h-E_r] \, t'} \, \langle \{N_n\}|\hat{c}_{l'}^\dagger \hat{c}_{m'}^\dagger \hat{c}_l \hat{c}_m \hat{c}_{h'}^\dagger \hat{c}_{r'}^\dagger \hat{c}_h \hat{c}_r|\{M_m\}\rangle \langle \{M_m\}|\hat{c}_i^\dagger \hat{c}_j|\{N_n\}\rangle \, . 
\end{aligned} \tag{18.69}
$$

Now we switch the order of the factors and eliminate the completeness to obtain:

$$
\begin{aligned}
& \frac{1}{4} \sum_{\{M_m\}} \sum_{l'm'lm} \sum_{h'r'hr} \rho_{\{M_m\}} V_{l'm'lm} V_{h'r'hr} \, \mathrm{e}^{-(\mathrm{i}/\hbar) \, [E_{l'}+E_{m'}-E_l-E_m] \, t} \\
& \times \mathrm{e}^{-(\mathrm{i}/\hbar) \, [E_{h'}+E_{r'}-E_h-E_r] \, t'} \, \langle \{M_m\}|\hat{c}_i^\dagger \hat{c}_j \hat{c}_{l'}^\dagger \hat{c}_{m'}^\dagger \hat{c}_l \hat{c}_m \hat{c}_{h'}^\dagger \hat{c}_{r'}^\dagger \hat{c}_h \hat{c}_r|\{M_m\}\rangle \, . 
\end{aligned} \tag{18.70}
$$

We can now proceed in a way which is intimately related to van Hove's original argument: Note that the expression above is just

$$
\sum_{\{M_m\}} \rho_{\{M_m\}} \langle \{M_m\}|\widehat{H}_{\rm ee}^{(\rm I)}(t) \, \widehat{H}_{\rm ee}^{(\rm I)}(t') \, \hat{c}_i^\dagger \hat{c}_j|\{M_m\}\rangle \, . \tag{18.71}
$$

Let us now insert a completeness into Eq. (18.71):

$$
\sum_{\{M_m\}\{N_n\}} \rho_{\{M_m\}} \langle \{M_m\}|\widehat{H}_{\rm ee}^{(\rm I)}(t) \, |\{N_n\}\rangle \langle \{N_n\}|\widehat{H}_{\rm ee}^{(\rm I)}(t') \, \hat{c}_i^\dagger \hat{c}_j|\{M_m\}\rangle \, . \tag{18.72}
$$

This term is now of the "genuine van Hove form" we have encountered before: Stating the same considerations we stated before when dealing with electron–phonon scattering, we note that this term expresses the transition from an initial state $\widehat{H}_{\text{ee}}^{(\text{I})}(t')\hat{c}_i^\dagger\hat{c}_j|\{M_m\}\rangle$ to a final state $\widehat{H}_{\text{ee}}^{(I)\dagger}(t)|\{M_m\}\rangle$ via a sum over a very large number of the many-electron intermediate states $|\{N_n\}\rangle$. When the phases of the initial and final states differ, the phases of the intermediate states will interfere destructively as we sum over very many (say $N$) electron states and the contribution to the double sum in Eq. (18.72) will be limited to "diagonal' processes, yielding a contribution $\sim O(N)$. If, instead, the initial and final states have identical phases, the phases of the intermediate states will interfere constructively and we will have a contribution $\sim O(N^2)$ from all terms to the sum. Therefore, in the limit of infinitely many electronic degrees of freedom, only states $\widehat{H}_{\text{ee}}^{(\text{I})}(t')\hat{c}_i^\dagger\hat{c}_j|\{M_m\}\rangle$ and $\widehat{H}_{\text{ee}}^{(I)\dagger}(t)|\{M_m\}\rangle$ which have identical phases will give a non-vanishing contribution to the sum in Eq. (18.70).

This "phase matching" considerations have to account for the time-dependent phases in Eq. (18.70). Again, let's proceed as we did before. Since in Eq. (18.69) we consider transitions originating at time $t'$ ranging from the initial time 0 to the final time $t$, if we restrict our attention to final times $t$ large enough ($\sim O(\alpha_{\text{ee}}^{-2})$, where $\alpha_{\text{ee}}$ is some measure of the strength of the Coulomb interaction among electrons), we see that in the multiple sums in Eq. (18.70) constructive phase interference will occur only for terms for which $E_{l'}+E_{m'}-E_l-E_m=-(E_{h'}+E_{r'}-E_h-E_r)$, and this constructive interference among the phases occurs at times $t'\sim t$. Only under these conditions do the phases of $\widehat{H}_{\text{ee}}^{(\text{I})}(t')|\{M_m\}\rangle$ and of $\widehat{H}_{\text{ee}}^{(I)\dagger}(t)|\{M_m\}\rangle$ match. Thus, in turn, in order for our phases of the initial and final states to match fully, we must have $i=j$, so that the phases of $|\{M_m\}\rangle$ and $\hat{c}_i^\dagger\hat{c}_j|\{M_m\}\rangle$ also match. Thus, we reach the conclusion that also in the case of interparticle Coulomb interactions the off-diagonal terms ($i\neq j$) drop out of the picture. Note that this depends on the same assumptions (1)–(3) stated on page 388. In addition, we have had to assume that our system consists of a number of electrons large enough to make the "infinite volume limit" valid, so that the sum in Eq. (18.72) is over so many incoherent intermediate states as to cancel destructively all their random phases.

Having established that only terms such that $i=j$ survive in the sum in Eq. (18.70), let's consider the many-particle Green's functions in the sum (18.70) to derive an expression for the electron–electron transition rate. We already know that we must have $E_{l'}+E_{m'}-E_l-E_m=-(E_{h'}+E_{r'}-E_h-E_r)$ in Eq. (18.70). This is not sufficient to "collapse" any indices in the sum when the energy levels $E_i$ are degenerate. Thus, let's consider again each term of the sum in Eq. (18.70). These have the form

$$\langle\{M_m\}|\hat{c}_i^\dagger\hat{c}_i\hat{c}_{l'}^\dagger\hat{c}_{m'}^\dagger\hat{c}_l\hat{c}_m\hat{c}_{h'}^\dagger\hat{c}_{r'}^\dagger\hat{c}_h\hat{c}_r|\{M_m\}\rangle\ . \tag{18.73}$$

First note that terms such that $l=m$ and $l'=m'$ should be ignored because they are Hartree contributions already included in the self-consistent potential $V^{(\text{H})}(\mathbf{R})$ in Eq. (18.3). Similarly, terms $l=l'$ and $m=m'$ are self-energy (or "correlation") terms and should also be ignored. Note that this is completely equivalent to the statements made by Haas, Rossi, and Kuhn [12]: They state that scattering terms arise only at second order when one looks at the deviation of the two-particle Green's functions from their factorized Hartree–Fock form, that is, terms of the form $\langle\hat{c}^\dagger cc^\dagger\hat{c}\rangle$ - $\langle\hat{c}^\dagger\hat{c}\rangle\langle\hat{c}^\dagger\hat{c}\rangle$. This is indeed equivalent to ignoring the Hartree terms.

Now, using the fact that $\langle\{M_m\}|\hat{c}_i^\dagger\hat{c}_i=n_i\langle\{M_m\}|$, we see that Eq. (18.73) expresses the inner product of the two states $\hat{c}_{h'}^\dagger\hat{c}_{r'}^\dagger\hat{c}_h\hat{c}_r|\{M_m\}\rangle$ and $\hat{c}_m^\dagger\hat{c}_l^\dagger\hat{c}_{m'}\hat{c}_{l'}|\{M_m\}\rangle$. Non-vanishing terms will belong to four possible classes: Those for which $l'=h$, $m'=r$, $l=h'$, $m=r'$ correspond to "direct" processes; those for which $l'=r$, $m'=h$, $l=r'$, and $m=h'$ correspond to "exchange" processes; finally, two mixed terms, corresponding to $l'=h$, $m'=r$, $l=r'$, $m=h'$, and $l'=r$, $m'=h$, $l=h'$, $m=r'$, appear with a negative sign because they can be reduced to the same form of the previous terms via an additional anticommutation. Let's consider only the direct terms for which $l'=h$, $m'=r$, $l=h'$, and $m=r'$, adding the contribution of the other terms later. Then, from Eq. (18.70) we have

$$\frac{1}{4} \sum_{\{M_m\}} \widehat{\sum}_{l'm'lm} \rho_{\{M_m\}} \, |V_{l'm'lm}|^2 \, e^{-(i/\hbar)\,[E_i+E_m-E_{l'}-E_{m'}]\,(t-t')}$$
$$\times \langle\{M_m\}|\hat{c}_i^\dagger \hat{c}_i \hat{c}_{l'}^\dagger \hat{c}_{m'}^\dagger \hat{c}_l \hat{c}_m \hat{c}_l^\dagger \hat{c}_m^\dagger \hat{c}_{l'} \hat{c}_{m'}|\{M_m\}\rangle, \qquad (18.74)$$

where the symbol $\widehat{\sum}$ reminds us that the sum extends over $l \neq l', m$, $m' \neq l, m$. We shall write simply $\sum$ in place of $\hat{\sum}$ in the following, but we should keep in mind that Hartree and self-energy terms must be excluded from the sums. Performing an even number of anticommutations,

$$\langle\{M_m\}|\hat{c}_i^\dagger \hat{c}_i \hat{c}_{l'}^\dagger \hat{c}_{m'}^\dagger \hat{c}_l \hat{c}_m \hat{c}_l^\dagger \hat{c}_m^\dagger \hat{c}_{l'} \hat{c}_{m'}|\{M_m\}\rangle \rightarrow \langle\{M_m\}|\hat{c}_i^\dagger \hat{c}_i \hat{c}_{l'} \hat{c}_{l'}^\dagger \hat{c}_{m'} \hat{c}_{m'}^\dagger \hat{c}_m^\dagger \hat{c}_m \hat{c}_l^\dagger \hat{c}_l|\{M_m\}\rangle,$$

having switched the primed and unprimed indices for convenience (which we are allowed to do under the multiple sums). Proceeding as before, non-vanishing contributions will be given only by states $|\{M_m\}\rangle$ containing single-electron states $|m\rangle$ but not $|l'\rangle$ such that:

$$\hat{c}_{m'}^\dagger \hat{c}_m^\dagger \hat{c}_m \hat{c}_l^\dagger \hat{c}_l|\{M_m\}\rangle = \hat{c}_{m'}^\dagger \hat{c}_{l'} \hat{c}_{l'}^\dagger \hat{c}_i^\dagger \hat{c}_i|\{M_m\}\rangle, \qquad (18.75)$$

so that only terms for which $i = l$ or $i = m$ will contribute equally to the sum, i.e., we can consider only one class of these terms and multiply the result by a factor 2 at the end. The remaining three pairings of the indices $\{m'l'ml\}$ with $\{r'h'rh\}$ discussed above give terms essentially identical, but with different permutations of the indices of the matrix elements $V_{m'l'ml}$. Collecting all four terms and recalling that $\hat{c}_i^\dagger \hat{c}_i \hat{c}_i^\dagger \hat{c}_i = \hat{c}_i^\dagger \hat{c}_i$, finally we have for the first term of Eq. (18.68):

$$-\frac{1}{\hbar^2} \int_0^t dt' \sum_{l'm'm} \frac{1}{2} \, |V_{l'm'im} - V_{m'l'im}|^2$$
$$\times \, e^{-(i/\hbar)[E_i+E_m-E_{l'}-E_{m'}](t-t')} \, \langle \hat{c}_{m'} \hat{c}_{m'}^\dagger \hat{c}_{l'} \hat{c}_{l'}^\dagger \hat{c}_m^\dagger \hat{c}_m \hat{c}_i^\dagger \hat{c}_i \rangle_{t'} \, \delta_{ij}$$
$$= -\frac{1}{2} \int_0^t dt' \sum_{l'm'im} K_{l'm'il}(t-t') \, \langle \hat{c}_{m'} \hat{c}_{m'}^\dagger \hat{c}_{l'} \hat{c}_{l'}^\dagger \hat{c}_m^\dagger \hat{c}_m \hat{c}_i^\dagger \hat{c}_i \rangle_{t'} \, \delta_{ij} \, . \qquad (18.76)$$

Note that the term $|V_{l'm'im} - V_{m'l'im}|^2$ takes this form thanks to the antisymmetrization of the two-particle states. Had we included spin in our formulation, we would have found identical terms for spin-triplet states, but terms containing instead $|V_{l'm'im} + V_{m'l'im}|^2$ for spin-singlet states, since the spatial component of the two-electron wavefunction must be symmetrized for antisymmetric spin (singlet) states. Let's now invoke the usual Hartree–Fock factorization, the usual Markov approximation, and write simply $\rho_i(t)$ for the diagonal elements $\langle \hat{c}_i^\dagger \hat{c}_i \rangle_t$ of the reduced density matrix. Then Eq. (18.76) can be recast in the "intuitive" form:

$$-\frac{1}{2} \int_0^t dt' \sum_{l'm'm} K_{l'm'il}(t-t') \, \rho_i(t') \, \rho_l(t') \, [1 - \rho_{l'}(t')] \, [1 - \rho_{m'}(t')] \, \delta_{ij} \, . \qquad (18.77)$$

The second term in the integrand at the right-hand side of Eq. (18.68) gives a similar contribution, but with the "energy phase" of opposite sign, that is, it gives the complex conjugate of the term we just derived. Thus, the "out-scattering" component [the first two terms at the right-hand side of Eq. (18.68)] of the equation describing the time evolution of the diagonal elements of the reduced density matrix will have the form:

$$\left(\frac{\partial \rho_i}{\partial t}\right)_{\text{out}} = -\sum_{l'm'm} \int_0^t dt' \, \text{Re} \, [\, K_{l'm'il}(t-t') \,] \, \rho_i(t') \, \rho_l(t') \, [1 - \rho_{l'}(t')] \, [1 - \rho_{m'}(t')] \, . \qquad (18.78)$$

As we have done before, we now take the van Hove limit: We set $\tau = \alpha^2(t-t')$ (where, recall, $\alpha$ is the dimensionless strength of the interaction), $G_{lmij}(\tau) = \mathrm{Re}[K_{lmij}(t)]$, and $P_i(\tau) = \rho_i(t)$. Then:

$$\left(\frac{\partial P_i}{\partial \tau}\right)_{\mathrm{out}} = -\sum_{l'm'm} \int_0^{\tau/\alpha^2} \mathrm{d}\tau' \; G_{l'm'il}(\tau') \\ \times P_i(\tau-\alpha^2\tau')\, P_l(\tau-\alpha^2\tau')\, [1 - P_{l'}(\tau-\alpha^2\tau')]\, [1 - P_{m'}(\tau-\alpha^2\tau')] \,, \tag{18.79}$$

we take the limit $t \to \infty$ and $\alpha^2 \to 0$, whilst $\tau$ remains constant, and Eq. (18.78) becomes

$$\left(\frac{\partial \rho_i}{\partial t}\right)_{\mathrm{out}} = -\sum_{l'm'm} W^{(\mathrm{ee})}_{l'm'il}\, \rho_i(t)\, \rho_l(t)\, [1 - \rho_{l'}(t)]\, [1 - \rho_{m'}(t)] \,, \tag{18.80}$$

where, assuming that the total spin of 2-electron states is randomly distributed among singlet and triplet states, the transition rate is

$$W^{(\mathrm{ee})}_{lmij} = \frac{2\pi}{\hbar} \left[\frac{1}{4}\left|V^{(\mathrm{d})}_{lmij} + V^{(x)}_{lmij}\right|^2 + \frac{3}{4}\left|V^{(\mathrm{d})}_{lmij} - V^{(x)}_{lmij}\right|^2\right] \delta(E_m + E_l - E_i - E_j) \,, \tag{18.81}$$

that is, the second-order Coulomb matrix element (first Born approximation). In Eq. (18.81) we have defined the "direct" matrix element $V^{(\mathrm{d})}_{lmij}$ as in Eq. (18.62), while the "exchange" matrix element is similarly defined as:

$$V^{(x)}_{i'j'ij} = \sum_{\mathbf{q}} \langle i'j|V_{\mathbf{q}}|ij'\rangle \,. \tag{18.82}$$

Now let's consider the first of the in-scattering terms, that is, the third term inside the integral at the right-hand side of Eq. (18.68). This integrand has the form:

$$\frac{1}{4} \sum_{\{M_m\}\{N_n\}} \sum_{l'm'lm} \sum_{h'r'hr} \rho_{\{M_m\}}\, V_{l'm'lm}\, V_{h'r'hr} \\ \times \mathrm{e}^{-(\mathrm{i}/\hbar)[E_{l'}+E_{m'}-E_l-E_m]\,t}\, \mathrm{e}^{-(\mathrm{i}/\hbar)[E_{h'}+E_{r'}-E_h-E_r]\,t'} \\ \times \langle\{N_n\}|\hat{c}^\dagger_{l'}\hat{c}^\dagger_{m'}\hat{c}_l\hat{c}_m|\{M_m\}\rangle\langle\{M_m\}|\hat{c}^\dagger_{h'}\hat{c}^\dagger_{r'}\hat{c}_h\hat{c}_r\hat{c}^\dagger_i\hat{c}_j|\{N_n\}\rangle \,. \tag{18.83}$$

Proceeding in the usual way, van Hove's argument tells us that only diagonal terms survive (i.e., $i=j$) and we will have a non-vanishing contribution to the sum only from four classes of terms. Let's consider now, as an example, only one class of these four (direct, exchange, and two mixed processes) classes of terms, the direct terms for which $h=l'$, $r=m'$, $h'=l$, and $r'=m$, as done above. We can switch the order of the sums, get rid of the completeness and obtain, in full analogy with Eq. (18.74):

$$\frac{1}{4} \sum_{\{M_m\}} \sum_{l'm'lm} \rho_{\{M_m\}}\, |V_{l'm'lm}|^2\, \mathrm{e}^{-(\mathrm{i}/\hbar)\,[E_{l'}+E_{m'}-E_l-E_m]\,(t\,-\,t')} \\ \times \langle\{M_m\}|\hat{c}^\dagger_l\hat{c}^\dagger_m\hat{c}_{l'}\hat{c}_{m'}\hat{c}^\dagger_i\hat{c}_i\hat{c}^\dagger_{l'}\hat{c}^\dagger_{m'}\hat{c}_l\hat{c}_m|\{M_m\}\rangle \,. \tag{18.84}$$

We can see that only terms for which $i=l'$ or $i=m'$ will give non-vanishing contributions (because if it were $i=l$ or $i=m$ we would have two identical annihilation operators acting on the state $|\{M_m\}\rangle$).

Both cases can be lumped into terms of identical forms. For example, for $i = l'$ we have terms like

$$\langle \{M_m\} | \hat{c}_l^\dagger \hat{c}_m^\dagger \hat{c}_i \hat{c}_{m'} \hat{c}_i^\dagger \hat{c}_i \hat{c}_i^\dagger \hat{c}_{m'}^\dagger \hat{c}_l \hat{c}_m | \{M_m\} \rangle \ . \tag{18.85}$$

With a few anticommutations this term can be rewritten as:

$$\langle \{M_m\} | [1 - \hat{c}_i^\dagger \hat{c}_i] \hat{c}_i \hat{c}_i^\dagger \hat{c}_{m'} \hat{c}_{m'}^\dagger \hat{c}_m^\dagger \hat{c}_m \hat{c}_l^\dagger \hat{c}_l | \{M_m\} \rangle \ . \tag{18.86}$$

But a term containing $\hat{c}_i \hat{c}_i$ is identically zero, so only the term

$$\langle \{M_m\} | \hat{c}_i \hat{c}_i^\dagger \hat{c}_{m'} \hat{c}_{m'}^\dagger \hat{c}_m^\dagger \hat{c}_m \hat{c}_l^\dagger \hat{c}_l | \{M_m\} \rangle \tag{18.87}$$

survives. Invoking once more the Hartree–Fock and Markov approximations, going to the large-time van Hove limit, and noticing that the contribution of the remaining term in Eq. (18.68) is the complex conjugate of Eq. (18.84), we finally obtain the Master equation:

$$\begin{aligned}\frac{\partial \rho_i}{\partial t} = \sum_{lmm'} \Big\{ & W^{(\mathrm{ee})}_{im'lm} \, \rho_l(t) \, \rho_m(t) \, [1 - \rho_i(t)] \, [1 - \rho_{m'}(t)] \\ & - W^{(\mathrm{ee})}_{lm'im} \, \rho_i(t) \, \rho_m(t) \, [1 - \rho_l(t)] \, [1 - \rho_{m'}(t)] \Big\} \ . \end{aligned} \tag{18.88}$$

Similarly to the previous discussion regarding impurity scattering, Eq. (18.88) represents the collisional integral of the BTE when considering large "devices" requiring the use of the BTE instead of the Master equation.

### 18.2.3 Equilibrium and Detailed Balance

We now want to verify that the BTE satisfies the basic requirement of yielding the equilibrium distribution in the absence of an external field or for a closed system. Let's ignore here scattering with ionized impurities and other elastic processes (namely, acoustic phonons in our approximation) since, clearly, they have no effect on the energy distribution of the carriers. Let's also assume a small device, so that we can use the Master equation (but the analysis carries identically to the case of the BTE in the absence of a field). First of all, let's write the Master Equation (ME) in its full glory in terms of the occupation $\rho_i$ per single state (and spin):

$$\begin{aligned}\frac{\partial \rho_i}{\partial t} = & \ \Phi_i \, [f_i - \rho_i] + \sum_j \Big\{ W^{(\mathrm{ph})}_{ij} \rho_j [1-\rho_i] - W^{(\mathrm{ph})}_{ji} \rho_i [1-\rho_j] \Big\} \\ & + \sum_{jlm} \Big\{ W^{(\mathrm{ee})}_{ijlm} \, \rho_l \rho_m [1-\rho_i][1-\rho_j] - W^{(\mathrm{ee})}_{lmij} \, \rho_i \rho_j [1-\rho_l][1-\rho_m] \Big\} \ , \end{aligned} \tag{18.89}$$

where $f_i = 1/\{1 - \exp[(E_i - E_F^{(s)}]/(k_\mathrm{B}T)\}$ is the Fermi function of the $s$th contact at energy $E_i(k)$, and $\Phi_i$ is the flux of the state $i$ into the device from contact $s$.

At equilibrium the injection term $\Phi_i \, [f_i - \rho_i]$ clearly vanishes for $\rho_i = f_i$. This is the situation found in ballistic transport.

Let's consider now the inelastic (optical) phonon-scattering term. Detailed balance requires

$$W^{(\mathrm{ph})}_{ij} \rho_j [1-\rho_i] = W^{(\mathrm{ph})}_{ji} \rho_i [1-\rho_j] \ . \tag{18.90}$$

Let's assume $E_i < E_j$. Then the transition $j \to i$ must be assisted by phonon emission, so that $W_{ij} = D(1+N_{\rm op})$, where $D$ is a constant, while, obviously, we must have $W_{ji} = DN_{\rm op}$. Thus:

$$\frac{\rho_j[1-\rho_i]}{\rho_i[1-\rho_j]} = \frac{W_{ji}^{(\rm ph)}}{W_{ij}^{(\rm ph)}} = \frac{N_{\rm op}}{1+N_{\rm op}} = {\rm e}^{(E_i-E_j)/(k_{\rm B}T)} \,. \tag{18.91}$$

So, we must have

$$\frac{\rho_i {\rm e}^{E_i/(k_{\rm B}T)}}{1-\rho_i} = \frac{\rho_j {\rm e}^{E_j/(k_{\rm B}T)}}{1-\rho_j} \,. \tag{18.92}$$

Since this must be true for every $i$ and $j$, each of these terms must be equal to a constant independent of $i$ or $j$. Let's call it $K^{-1}$. This implies that

$$\rho_i = \frac{1}{1+K{\rm e}^{E_i/(k_{\rm B}T)}} \,. \tag{18.93}$$

The constant $K$ must be determined by the normalization of $\rho_i$ so that $\sum_i \rho_i = N$, the total number of electrons in the system. It follows that $K = {\rm e}^{-E_F/(k_{\rm B}T)}$. This shows that the inelastic phonon scattering term of the ME vanishes only if $\rho_i = \alpha_i f_i$ at the temperature set by the phonon bath.

Let's consider now the electron–electron scattering term. Detailed balance requires for any two states $i$ and $l$:

$$T_{il}\, \rho_l\, [1-\rho_i] = T_{li}\, \rho_i\, [1-\rho_l] \,, \tag{18.94}$$

where $T_{il} = \sum_{jm} W_{ijlm}^{(\rm ee)} \rho_m [1-\rho_j]$. Therefore:

$$\frac{\rho_l[1-\rho_i]}{\rho_i[1-\rho_l]} = \frac{\sum_{jm}\, W_{ijlm}^{(\rm ee)}\, \rho_m\, [1-\rho_j]}{\sum_{jm}\, W_{lmij}^{(\rm ee)}\, \rho_j\, [1-\rho_m]} \,. \tag{18.95}$$

Unlike what we found for the electron–phonon interaction, we cannot "derive" the equilibrium function from the properties of the scattering kernels, since there is no thermal bath setting the temperature of the system. Rather, we now assume that $\rho_i$ has the desired form, $f_i$ and see whether this is consistent with the Coulomb term of the ME and detailed balance. If $\rho_i = f_i$, then $\rho_m[1-\rho_j] = \rho_j[1-\rho_m]{\rm e}^{[E_m-E_j]/(k_{\rm B}T)}$. Thus, the left-hand side of Eq. (18.95) above will be ${\rm e}^{(E_i-E_l)/(k_{\rm B}T)}$. The right-hand side will be

$$\frac{\sum_{jm}\, W_{ijlm}^{(\rm ee)}\, \rho_j\, [1-\rho_m]\, {\rm e}^{[E_m-E_j]/(k_{\rm B}T)}}{\sum_{jm}\, W_{lmij}^{(\rm ee)}\, \rho_j\, [1-\rho_m]} \,. \tag{18.96}$$

Since $W_{ijlm}^{(\rm ee)} = W_{lmij}^{(\rm ee)}$, and using the fact that $E_m - E_j = E_i - E_l$ since Coulomb interactions conserve the total energy of the pair, Eq. (18.96) is also equal to ${\rm e}^{(E_i-E_l)/(k_{\rm B}T)}$. Thus, the electron–electron scattering term of the ME also vanishes at equilibrium. Note that this interparticle scattering term vanishes when $\rho_i$ is a (Fermi–Dirac) equilibrium solution at *any* temperature. Thus, it is easy to understand that electron–electron collisions will drive off-equilibrium distribution functions towards a Fermi–Dirac distribution. Since these processes conserve total energy, the temperature defining this distribution will be the "thermal" temperature at the average energy of the off-equilibrium state.

## References

1. W.H. Zurek, Information transfer in quantum measurements: irreversibility and amplification, in *Quantum Optics, Experimental Gravity, and Measurement Theory*. NATO Advanced Science Institutes Series, vol. 94 (Plenum, New York, 1983), pp. 87–116
2. W. Kohn, J.M. Luttinger, Quantum theory of electrical transport phenomena. Phys. Rev. **108**, 590 (1957)
3. L. Van Hove, Quantum-mechanical perturbations giving raise to a statistical transport equation. Physica **XXI**, 517 (1955)
4. P. N. Argyres, Quantum theory of kinetic equations for electrons in phonon fields. Phys. Rev. **132**, 1527 (1963)
5. C. Cercignani, *Ludwig Boltzmann: The Man Who Trusted Atoms* (Oxford University Press, New York, 1998)
6. W. Pauli, in *Festschrift zum 60. Geburtstage A. Sommerfeld* (Hirzel, Leipzig, 1928), p. 30
7. R.W. Zwanzig, Statistical mechanics of irreversibility, in *Quantum Statistical Mechanics*, ed. by P.H.E. Meijer (Gordon and Breach, New York, 1966), p. 139
8. L.V. Keldysh, Diagram technique for nonequilibrium processes. Zh. Eksp. Teor. Fiz. **47**, 1515 (1964) [Sov. Phys. JETP **20**, 1018 (1965)]
9. L.P. Kadanoff, G. Baym, *Quantum Statistical Mechanics* (Benjamin, New York, 1962)
10. C.S. Lent, D.J. Kirkner, The quantum transmitting boundary method. J. Appl. Phys. **67**, 6353–6359 (1990)
11. D. Bohm, D. Pines, A collective description of electron interactions: II. Collective vs individual particle aspects of the interactions. Phys. Rev. **85**, 338 (1952)
12. S. Haas, F. Rossi, T. Kuhn, Generalized Monte Carlo approach for the study of the coherent ultrafast carrier dynamics in photoexcited semiconductors. Phys. Rev. B **53**, 12855 (1996)

# Chapter 19
# Solution Methods for Semiclassical Transport

In the previous chapter, Chap. 18, we have "derived" the Boltzmann transport equation (BTE) for the particular system of electrons in a crystal. The quotes bracketing the word "derive" are required for reasons that should now be obvious. However, this equation has been originally written by Ludwig Boltzmann in 1872 in a much more general context: After all, in 1872 physicists were still arguing whether or not atoms exist and the electron had not been yet discovered (it was discovered by J.J. Thomson only 25 years later, in 1897). The captivating book by Carlo Cercignani [1] reads as a mystery novel while telling Boltzmann's life and his search for the elusive proof of the so-called *H-theorem*; that is, for the identification of the cause of irreversibility, of what makes time flow only in one direction.

The BTE has been applied to a variety of kinetic problems, from gases to neutron transport. Therefore, there exists a vast literature, both in the mathematics as well as in the physics community. Here we discuss how one may solve it only in the limited context of our perspective, electron transport in crystals.

Of course, interest in searching for a solution in this context has been largely triggered by the microelectronics industry. Therefore, analytical and numerical methods devised to solve the BTE are often intrinsically and tightly intertwined with methods to simulate the electrical behavior of semiconductor devices. In this context, general overviews can be found in the texts by Jacoboni [2] that goes beyond semiclassical transport; by Hess, that emphasizes the physics [3]; and by Taur and Ning [4], more specific to electron devices.

We should add that the "pioneering" period, spanning the years from the 1950s to the mid-1990s, has seen tremendous activity, obviously related to the exponential growth of the VLSI industry and the associated miniaturization of devices. The "hot electron" crisis of the 1980s–1990s has also stimulated the search for more sophisticated, computationally demanding physical models that stimulated even more activity. This "golden period" is now over: Devices are so small, the applied bias reduced to such a level, that interest is shifting to quantum transport. Nevertheless, semiclassical transport, described by the BTE and treated numerically with the methods described below, remains a work-horse in the VLSI industry, even though active research has shifted away from it as the many device-simulation software packages available have rendered this field a "commodity."

As an important matter of notation, note that in this chapter we use the symbol $E$ for the kinetic energy of the charge carriers, the symbol $\mathbf{F}$ for the electric field. In doing this, we follow the usual convention in the scientific and technical literature that deals with approximate methods to treat semiclassical transport in semiconductor devices.



M. Fischetti, W.G. Vandenberghe, *Advanced Physics of Electron Transport in Semiconductors and Nanostructures*, Graduate Texts in Physics, DOI 10.1007/978-3-319-01101-1_19

## 19.1 Overview

The BTE, Eq. (18.47), is a partial integro-differential equation, possibly also nonlinear when the collision rates $W$ depend on the distribution function $f$ itself. This is indeed the case when dealing with Coulomb interactions that must be screened via the Lindhard formula in which $f(\mathbf{k})$ enters. It is no surprise that a general solution in closed form does not exist, except in some special cases, like at equilibrium. Mathematically speaking, existence and uniqueness proofs have not been provided, except in particular oversimplified cases. Therefore solutions of the BTE must be obtained either under simplifying approximations (for example, under homogeneous conditions and small electric fields, which allow us to linearize the BTE with respect to the field), or numerically (such as when using the Monte Carlo method) or, often and sadly, numerically *and* under simplifying approximations.

Below we shall consider a few methods of solution. The first one, the *method of moments*, consists in giving up on the idea of obtaining the full distribution function $f(\mathbf{r}, \mathbf{k}, t)$ being satisfied, instead, with the knowledge of averages of physical quantities (such as carrier density, velocity, energy, etc.) over the distribution function. These averages—that we shall call *ensemble averages* and denote by the symbol $\langle \ldots \rangle_{\text{ens}}$—are essentially averages of powers of $\mathbf{k}$, such as $\langle k^m(\mathbf{r}, t) \rangle_{\text{ens}} = [1/n(\mathbf{r}, t)] \int d\mathbf{k} f(\mathbf{r}, \mathbf{k}, t) k^m$ (where $n$ is the carrier density), which is called the $m$-th moment of $f$. This results in the reduction of the partial integro-differential equation (18.47) to a system of partial differential equations (PDEs). The limiting cases in which equations are sought only for density and velocity (*drift-diffusion* approximation, DD) or density, velocity, and average energy (*hydrodynamic* or *energy transport* model) are of significant importance because of their widespread use in the simulation of electronic devices. While the advantage of this method lies obviously in its reduced computational cost, the disadvantage lies in the fact that we must consider only a few moments in order to deal with a numerically tractable system of equations. The neglected terms (higher-order moments) may become significant in some situations, but we have no way of accounting for them. The appearance of hard-to-define "relaxation times" in the model and the issue of the "closure" (as we shall see below) are also causes of concern.

The second method (or set of methods, which we shall label as *expansion methods*) is of a similar nature: The unknown $f(\mathbf{r}, \mathbf{k}, t)$ is expanded into suitable basis functions (spherical harmonics or other orthogonal polynomials) and, having inserted these expansions into the BTE, equations are found for the coefficients of the expansion, coefficients that are functions of position and time. This also results in PDEs, so that the problem is simplified. Clearly, the expansion must be truncated at some level. As it was the case for the method of moments, any truncation will result in errors in some cases.

Solutions of the full BTE can be also obtained by *iterative techniques*. These techniques consist in starting from a "guess" $f_0$ for the solution, in inserting this guess into the collision integral, and in solving the resulting PDE, thus obtaining a new solution $f_1$ and iterating the process. One can also seek a *direct solution* by discretizing the BTE, thus converting it into a large matrix-inversion problem. Finally, solutions can be obtained by *particle methods*. We will discuss the *weighted particle method*, the *scattering matrix method*, and the *cellular automata method*. Of course, it is the *Monte Carlo method* that provides the physically most accurate solution, albeit a stochastic one.

## 19.2 The Moments Method

The *moments method* is arguably the most popular method used to solve the BTE. Historically, the first theoretical results about the properties of charge transport in semiconductors, how these properties could be used in diodes and transistors, and the first quantitative simulations of electron devices have all been based on this method. Shockley's early paper [5] gives such an example. The literature is so vast that we can only refer to texts that review some aspects of the problems, such as Selberherr's [6] and Hänsch's [7] books on the drift-diffusion approximation. A text edited by Hess, Leburton, and Ravaioli in 1991 [8] gives a picture of the state of the art at that time during the "golden period."

The moments method consists in deriving from Eq. (18.47) PDEs for "ensemble averages" of various quantities $Q(\mathbf{k})$,

$$\langle Q(\mathbf{r},t)\rangle_{\text{ens}} = \frac{2}{n(\mathbf{r},t)} \int \frac{d\mathbf{k}}{(2\pi)^3} Q(\mathbf{k}) f(\mathbf{k},\mathbf{r},t) \, , \tag{19.1}$$

where $n(\mathbf{r},t) = 2\int d\mathbf{k}\, f(\mathbf{k},\mathbf{r},t)/(2\pi)^3$ is the carrier density (note the appearance of a factor of 2 for spin), and where the functions $Q$ are powers of the group velocity $\mathbf{v}(\mathbf{k}) = \hbar\mathbf{k}/m^*$, which, for parabolic and spherical bands, are equivalent to powers of the crystal momentum $\mathbf{k}$.

Taking the first moment amounts to integrating the BTE over $\mathbf{k}$-space, i.e., $Q(\mathbf{k}) = 1$. Thus, we obtain simply the *continuity equation*:

$$\frac{\partial n}{\partial t} = \frac{1}{e} \nabla \cdot \mathbf{j} \, , \tag{19.2}$$

where the current density $\mathbf{j}$ is defined as

$$\mathbf{j} = -2e \int \frac{d\mathbf{k}}{(2\pi)^3} \mathbf{v}(\mathbf{k}) f(\mathbf{k},\mathbf{r},t) = -en\langle \mathbf{v}\rangle_{\text{ens}} \, . \tag{19.3}$$

Two alternative approaches can be followed in order to derive a continuity equation for the second moment.

The first approach is a straightforward application of the "moment method": we multiply the BTE by the group velocity, $\hbar\mathbf{k}/m^*$, and integrate, obtaining the *current continuity* equation:

$$\begin{aligned} j_i &= e\, \tilde{\tau}_{\text{p},i} \sum_j \partial_j (n\langle \upsilon_i \upsilon_j \rangle_{\text{ens}}) - e^2 n \frac{\tilde{\tau}_{\text{p},i}}{m^*} \sum_j F_j \left\langle \upsilon_i \frac{\partial f}{\partial \upsilon_j} \frac{1}{f} \right\rangle_{\text{ens}} + \tilde{\tau}_{\text{p},i} \frac{\partial j_i}{\partial t} \\ &= e\tilde{\tau}_{\text{p},i} \sum_j \partial_j (n\langle \upsilon_i \upsilon_j \rangle_{\text{ens}}) + e^2 n \frac{\tilde{\tau}_{\text{p},i}}{m^*} F_i + \tilde{\tau}_{\text{p},i} \frac{\partial j_i}{\partial t} \, , \end{aligned} \tag{19.4}$$

having assumed a reasonable boundary condition for $f$ as $|\mathbf{v}| \to \infty$, so that $\langle \upsilon_i (\partial f/\partial \upsilon_j)/f\rangle_{\text{ens}} = -\delta_{ij}$. The particular form of momentum (or velocity) relaxation time, $\tilde{\tau}_{\text{p},i}$, which we must employ in order to be consistent with the BTE is given by

$$\frac{1}{\tilde{\tau}_{\text{p},i}} \equiv \frac{\langle \upsilon_i/\tau_{\text{p},i}\rangle_{\text{ens}}}{\langle \upsilon_i\rangle_{\text{ens}}} = \frac{2}{n\langle \upsilon_i\rangle_{\text{ens}}} \int \frac{d\mathbf{k}'}{(2\pi)^3} \int \frac{d\mathbf{k}}{(2\pi)^3} [\upsilon_i(\mathbf{k}) - \upsilon_i(\mathbf{k}')] W(\mathbf{k}',\mathbf{k}) f(\mathbf{k},\mathbf{r},t) \, , \tag{19.5}$$

where the "microscopic" momentum relaxation time $\tau_{\text{p},i}$ is defined by

$$\frac{1}{\tau_{\text{p},i}(\mathbf{k})} = \int \frac{d\mathbf{k}'}{(2\pi)^3} \left[1 - \frac{\upsilon_i(\mathbf{k}')}{\upsilon_i(\mathbf{k})}\right] W(\mathbf{k}',\mathbf{k}) \, . \tag{19.6}$$

It is useful to recall here that the quantities $W(\mathbf{k}',\mathbf{k})$ are the probabilities per unit time that a carrier will make a transition from a state $|\mathbf{k}\rangle$ to a state $|\mathbf{k}'\rangle$ via any of the collision processes we have considered in previous chapters. Here we shall use the term *microscopic* to indicate relaxation times that are functions of $\mathbf{k}$. They can be directly obtained from the knowledge of the scattering kernels $W(\mathbf{k}',\mathbf{k})$, without taking any average over the distribution function. With the term *macroscopic*, instead, we refer to relaxation times obtained by averaging the microscopic counterpart over the ensemble, objects which are more directly related to measurable quantities, such as mobility, diffusion constant, or thermal diffusivity, among others.

The second approach has a more rigorous appearance, although it does not always lead to a simple concept of "momentum relaxation time." Instead of multiplying the BTE by the group velocity, we first define a "formal" relaxation time as the solution of the integral equation:

$$\frac{1}{\tau_{\mathrm{p},i}(\mathbf{k})} = \int \frac{\mathrm{d}\mathbf{k}'}{(2\pi)^3} \left[ 1 - \frac{\upsilon_i(\mathbf{k}')\tau_{\mathrm{p},i}(\mathbf{k}')}{\upsilon_i(\mathbf{k})\tau_{\mathrm{p},i}(\mathbf{k})} \right] W(\mathbf{k}',\mathbf{k}) \,. \tag{19.7}$$

This cumbersome definition is the rigorous expression one can obtain from a linearization of the BTE in the applied field [3, 9–13], as illustrated by Eq. (19.42) below [10]. Of course, the concept of relaxation time itself is useful only when it can be evaluated easily from the microscopic models employed to obtain the "scattering rate"

$$\frac{1}{\tau(\mathbf{k})} = \int \frac{\mathrm{d}\mathbf{k}'}{(2\pi)^3} W(\mathbf{k}',\mathbf{k}) \,. \tag{19.8}$$

For elastic and isotropic scattering, Eq. (19.7) can be evaluated easily, since it reduces to the familiar form

$$\frac{1}{\tau_{\mathrm{p},i}(\mathbf{k})} = \int \frac{\mathrm{d}\mathbf{k}'}{(2\pi)^3} (1-\cos\theta) W(\mathbf{k}',\mathbf{k}) \,, \tag{19.8}$$

where $\theta$ is the angle between $\mathbf{k}$ and the $i$-axis. Even for isotropic but inelastic scattering (e.g., intervalley scattering with optical and acoustic phonons) a proper relaxation time can be defined and Eq. (19.8) coincides with Eq. (19.6). In general, though, for inelastic and anisotropic collisions (such as polar scattering with optical phonons in polar semiconductors), the relaxation time is defined only "formally" by Eq. (19.7). More sophisticated approaches are needed to derive even a simple expression for the mobility [10, 11, 14], as we shall mention below.

We now multiply the BTE by $\upsilon_i(\mathbf{k})\tau_{\mathrm{p},i}(\mathbf{k})$ and integrate. Thus, assuming that $W(\mathbf{k}',\mathbf{k})$ does not depend on position, we get the differential equation for the second moment, expressing again current continuity:

$$j_i = e\sum_j \langle \tau_{\mathrm{p},i}\partial_j(n\upsilon_i\upsilon_j)\rangle_{\mathrm{ens}} - \frac{e^2 n}{\hbar}\sum_j F_j \left\langle \upsilon_i\tau_{\mathrm{p},i}\frac{\partial f}{\partial k_j}\frac{1}{f}\right\rangle_{\mathrm{ens}} + \frac{\partial}{\partial t}\langle j_j\tau_{p,i}\rangle_{\mathrm{ens}} \,. \tag{19.9}$$

Note that Eqs. (19.4) and (19.9) are identical. Yet, macroscopic quantities are represented as averages of different microscopic variables and each form is better suited to a different application, as we shall see below.

Coming to the third moment, multiplying the BTE by the total "kinetic energy tensor," $E_{ij}(\mathbf{k}) = \hbar^2 k_i k_j/(2m^*)$, and integrating, one obtains

$$\frac{\partial}{\partial t}(n\langle E_{ij}\rangle_{\mathrm{ens}}) = -\sum_l \partial_l(n\langle E_{ij}\upsilon_l\rangle_{\mathrm{ens}}) + \frac{en}{\hbar}\sum_l F_l \left\langle E_{ij}\frac{\partial f}{\partial k_l}\frac{1}{f}\right\rangle_{\mathrm{ens}} - \frac{n\langle E_{ij}\rangle_{\mathrm{ens}}}{\tilde{\tau}_{w,ij}} \,, \tag{19.10}$$

where the relaxation times $\tilde{\tau}_{\mathrm{w},ij}$ are defined as

$$\frac{1}{\tilde{\tau}_{\mathrm{w},ij}} = \frac{2}{n\langle E_{ij}\rangle_{\mathrm{ens}}}\frac{\hbar^2}{2m^*} \int \frac{\mathrm{d}\mathbf{k}'}{(2\pi)^3} \int \frac{\mathrm{d}\mathbf{k}}{(2\pi)^3} (k_i k_j - k_i' k_j') W(\mathbf{k}',\mathbf{k}) f(\mathbf{k},\mathbf{r},t) \,.$$

For simplicity, only the trace of the equations above is considered, thus obtaining the much simpler-looking expression for conservation of the kinetic energy $E(\mathbf{k}) = \hbar^2 k^2/(2m^*)$:

$$\frac{\partial}{\partial t}(n\langle E\rangle_{\mathrm{ens}}) = -\nabla\cdot(n\langle E\mathbf{v}\rangle_{\mathrm{ens}}) + en\mathbf{F}\cdot\langle\mathbf{v}\rangle_{\mathrm{ens}} - \frac{n\langle E\rangle}{\tilde{\tau}_{\mathrm{w}}}, \tag{19.11}$$

where the trace of the average "energy relaxation time" $\tilde{\tau}_{\mathrm{w}}$ is given by

$$\frac{1}{\tilde{\tau}_{\mathrm{w}}} = \frac{2}{n\langle E\rangle_{\mathrm{ens}}}\int\frac{d\mathbf{k}'}{(2\pi)^3}\int\frac{d\mathbf{k}}{(2\pi)^3}\,[E(\mathbf{k}) - E(\mathbf{k}')]\;W(\mathbf{k}',\mathbf{k})\,f(\mathbf{k},\mathbf{r},t)\,. \tag{19.12}$$

Even higher-order moments can be considered, but we shall stop here, since the system of Eqs. (19.2), (19.9), and (19.12), coupled also to the Poisson equation

$$\nabla\cdot(\epsilon\mathbf{F}) = -en + \rho_{\mathrm{d}}\,, \tag{19.13}$$

contains the essential ingredients needed to discuss the main approximations made by the moments method to handle electronic transport. Also, it illustrates nicely the main approximations which are required to make the problem tractable and which are widely employed in device simulations.

It should be noticed that, so far, we have accomplished nothing: The moment equations have the exterior appearance of simpler equations than the original BTE, since we have converted an integro-differential equation in seven variables into PDEs in four variables. Yet, the averages $\langle\ldots\rangle$ appearing in the moment equations require the knowledge of the unknown $f$. This is the first hurdle that the moments methods have to overcome in order to render the problem tractable: one must be able to downgrade these *functionals* of $f$ to simple functions of one (or more) of the variables entering the problem. This transition is the step where most of the physical information is lost. A second hurdle is also quite evident: Were we wishing to consider only the first moment, Eq. (19.2), we would have no way of coupling transport (i.e., $\mathbf{j}$) into the Poisson equation. Thus, one must also consider (at least) Eq. (19.9). As it stands, the first term at the right-hand side involves the moment of higher order, whose behavior is controlled by Eq. (19.10) or (19.11). Similarly, if we wanted to stop at the third moment, Eq. (19.10) or (19.11), we would have to find some way of handling the term $\langle E\mathbf{v}\rangle_{\mathrm{ens}}$, which is the fourth moment obeying a differential equation we have not even derived. In general, the hierarchy of the moment equations leads unavoidably to equations for the $n$-th moment involving the $(n+1)$-st moment, because of the "drift" term $\mathbf{v}\cdot\nabla f$ in the BTE, which leads to a higher power of $\mathbf{v}$ when taking moments. The chain must be somehow closed. The search for these *closure relations* is the second major hurdle the moments methods must face. Two "popular" choices are the so-called *drift-diffusion* and *hydrodynamic* approximations.

### 19.2.1 The Drift-Diffusion Approximation

In the drift-diffusion (DD) approximation [5, 7, 15], only the first two moments are considered. The continuity equation, Eq. (19.2), and the Poisson equation, Eq. (19.13), pose no concerns. Care must be exercised when dealing with the current conservation equation, Eq. (19.4) or (19.9).

When Eq. (19.9) is chosen as second moment, the system can be "closed" (i.e., rendered independent from higher moments) by noticing that, if we assume that $f$ is a Boltzmann distribution at temperature $T$ (i.e., $f \simeq \exp\{-[E(\mathbf{k}) - E_{\mathrm{F}}]/(k_B T)\}$, where $E_F$ is the local Fermi level), then the tensor $D_{ij} = \langle\tau_{\mathrm{p},i}\upsilon_i\upsilon_j\rangle_{\mathrm{ens}}$ can be related to the other tensor entering the same equation, $\mu_{ij} = -(e/\hbar)\langle\tau_{\mathrm{p},i}\upsilon_i(\partial f/\partial k_j)/f\rangle_{\mathrm{ens}}$, by the well-known Einstein relation,

$$D_{ij} = \frac{k_B T}{e} \mu_{ij} \,, \tag{19.14}$$

where $k_B$ is the Boltzmann constant and $T$ is the "electron temperature," which in the DD approximation is always taken to be the lattice temperature $T_0$. This relation also suggests the way the "mobility" functional $\mu_{ij}$ could be reduced to a simple constant (or, at most, a function of the variables $\mathbf{r}$, $n$, $\mathbf{j}$, or $\mathbf{F}$): If we restrict ourselves to the analysis of devices in which the carriers are always not too far from thermal equilibrium, we can adopt Eq. (19.14) as our closure relation, assume that both $D$ and $\mu$ (the diffusion constant and mobility of the carriers, respectively) are constant quantities (at most functions of position via the spatially varying doping density), whose values are given by the ohmic (i.e., zero-field limit) values, and we can write Eq. (19.2) in the solvable form:

$$\mathbf{j} = e\nabla(Dn) + en\mu\mathbf{F} \,, \tag{19.15}$$

having also assumed steady state (no explicit time dependence) to simplify the picture. We have also assumed that $\tau_p$ does not depend explicitly on position, so that we could bring the diffusion constant inside the gradient. This is a tricky issue, since $D$ (and $\tilde{D}$ as well) may depend on position either via the average $\langle \ldots v_i v_j \rangle_{\mathrm{ens}}$, which should remain inside the spatial gradient, or via the relaxation time, which should remain outside the gradient. Note also that we have replaced tensors with scalars (their diagonal elements, assumed to be equal) thanks to the spherical symmetry of the distribution function at equilibrium. This is not necessarily true at high fields. It should be noted that under the same approximations of thermal equilibrium at the lattice temperature, the mobility $\mu$ is given by the well-known Kubo–Greenwood formula [12, 13], after integrating over the polar angle and converting the integration over $\mathbf{k}$ to an integration over the energy $E$, via the DOS, $\mathscr{D}(E)$:

$$\mu = -\frac{2e}{3nm^*} \int_0^\infty \mathrm{d}E\, \mathscr{D}(E)\, E\, \tau_\mathrm{p}(E) \frac{\partial f_0}{\partial E} = \frac{2e}{3nm^* k_B T_0} \int_0^\infty \mathrm{d}E\, \mathscr{D}(E)\, E\, \tau_\mathrm{p}(E)\, f_0(E) \,, \tag{19.16}$$

where $f_0$ is the Boltzmann equilibrium distribution function at the lattice temperature $T_0$, and $\tau_p$ is given by Eq. (19.7) or (19.8).

When Eq. (19.4) is chosen as second moment, closure is obtained in an identical way, by noticing that the Einstein relation still relates the diffusion tensor, now defined as $\tilde{D}_{ij} = \tilde{\tau}_{\mathrm{p},i} \langle v_i v_j \rangle_{\mathrm{ens}}$, to the mobility tensor, $\tilde{\mu}_{ij} = e\tilde{\tau}_{\mathrm{p},i}/m^*$, at equilibrium. Taking the isotropic approximation (i.e., considering only the diagonal elements of these tensors and assuming they are all equal), we have

$$\tilde{D} \equiv \frac{\tilde{\tau}_\mathrm{p}}{3} \langle v^2 \rangle_{\mathrm{ens}} = \frac{2}{3} \frac{\tilde{\tau}_\mathrm{p}}{m^*} \langle E \rangle_{\mathrm{ens}} \simeq \frac{\tilde{\tau}_\mathrm{p}}{m^*} k_B T = \frac{k_B T}{e} \tilde{\mu} \,. \tag{19.17}$$

The "third moment," given by the average kinetic energy, $\langle E \rangle_{\mathrm{ens}}$, or carrier temperature $T$, has been simply replaced by its value, $(3/2)k_B T$, for a Maxwell–Boltzmann distribution, in order to obtain the last expression at the right-hand side of Eq. (19.17). From this expression we obtain $\tilde{D} = \tilde{\tau}_\mathrm{p} k_B T/m^*$. Substituting these expressions into Eq. (19.4), we find

$$\mathbf{j} = e\, \nabla(\tilde{D}n) + e\, n\, \tilde{\mu}\mathbf{F} \,, \tag{19.18}$$

where $\tilde{D}$ is actually neither inside nor outside the gradient. Usually, it is placed either inside or outside the gradient, depending on the assumptions made (i.e., no position dependence of $\tilde{\tau}_\mathrm{p}$ and no position dependence of the average energy $\langle v^2 \rangle_{\mathrm{ens}}$, respectively), as discussed before.

Note that the definition Eq. (19.17) for the mobility $\tilde{\mu}$ does not lead immediately to the Greenwood–Kubo definition of mobility [12, 13] in the linear regime. To see that the two definitions given by Eqs. (19.16) and (19.17) coincide, we restrict ourselves to the linear regime, and employ the linearized solution of the BTE given by Eq. (19.42) below. Then:

$$\mu_{ij} \equiv -\frac{e}{m^*}\left\langle \tau_{p,i} v_i \frac{\partial f}{\partial v_j}\frac{1}{f}\right\rangle_{ens} \simeq -\frac{e}{k_B T}\langle \tau_{p,i} v_i v_j\rangle_{ens} \to \frac{e}{m^*}\frac{\langle \tau_p E\rangle_{ens}}{\langle E\rangle}, \tag{19.19}$$

where the last step involves taking the diagonal elements $\mu_{ii}$ to be equal. On the other side, we have, by our definition:

$$\tilde{\mu}_{ij} \equiv \frac{e}{m^*}\tilde{\tau}_{p,i} \equiv \frac{e}{m^*}\frac{\langle v_i\rangle_{ens}}{\langle v_i/\tau_{p,i}\rangle_{ens}}. \tag{19.20}$$

Now, at equilibrium and using the linearized solution of the BTE, Eq. (19.42) below,

$$\frac{\langle v_i\rangle_{ens}}{\langle v_i/\tau_{p,i}\rangle_{ens}} \to \frac{\langle \tau_p E\rangle_{ens}}{\langle E\rangle_{ens}},$$

in agreement with expression (19.19). The choice of either formulation of the current continuity equation is simply a matter of preference. The formulation given by Eqs. (19.9) and (19.15) emphasizes the possibility of calculating directly the macroscopic parameters $D$ and $\mu$ from the knowledge of the microscopic scattering kernels, $W(\mathbf{k}',\mathbf{k})$, although this is only possible in the linear regime. Also, this formulation shows very clearly the conditions under which a momentum relaxation time can be defined in a useful way, i.e., calculable from Eq. (19.7). For these reasons, this is the preferred approach when placing emphasis on the physics of the approximation [3, 9–13]. The alternative formulation, given by Eqs. (19.4) and (19.18), involves a definition of the momentum relaxation time, $\tilde{\tau}_p$, Eq. (19.5), which is more cumbersome to evaluate from "first principles." However, this is the single ensemble average entering the model, since both $\tilde{D}$ and $\tilde{\mu}$ are defined in terms of $\tilde{\tau}_p$. Once a model for the relaxation time is provided, the entire problem is specified. This is the reason why this formulation is (often implicitly) chosen in the engineering community.

Despite the dramatic approximations we have adopted to arrive at the simple form Eq. (19.15), the time-dependent term we have ignored involves the time derivative of the average $\langle \mathbf{j}\tau_p\rangle_{ens}$. We do not know how to evaluate this term without invoking, once more, the approximation of thermal equilibrium for $f$ and some specific form for the (*not* ensemble averaged) relaxation time $\tau_p(\mathbf{k})$. Notice that the approximation made so far assumes that the Einstein relation holds "locally," i.e., at every position. This will be true only if the device is so large that in the neighborhood of every point the carriers are always so strongly thermalized as to be always at local equilibrium with the field. Thus, mobility and diffusion are always related by the Einstein equation. In addition, in a homogeneous situation, Eq. (19.15) implies

$$\mathbf{j} = en\mu\mathbf{F},$$

so that the carrier velocity is given by $\mu\mathbf{F}$, as it follows from the definition of current density, Eq. (19.3). Again, this implies that the carriers must scatter so often in the neighborhood of any given point, as to achieve local equilibrium with the field. In summary, the DD approximation outlined so far can be expected to represent correctly the dynamics of the carriers whenever one can neglect the variation of the field over some multiple of the mean distance a carrier covers between successive collisions (the *mean free path*, $\lambda$), and whenever fields are small enough to maintain the carriers at near equilibrium with the lattice, i.e.,

$$eF\lambda \le \hbar\omega \sim k_B T, \tag{19.21}$$

where $\hbar\omega$ the average energy lost per collision, which is usually of the order of the thermal energy, $k_B T$, or of the phonon energy.

"Hot-carrier effects," such as velocity saturation, and a decrease of the diffusion constant and mobility at fields larger than the ohmic limit (19.21), have been empirically introduced by rendering $D$ and $\mu$ functions of the local field [16]. For example, a very "popular" model is the Caughey–Thomas model [17],

$$\mu(F) = \mu_0[1 + (\mu_0 F/\upsilon_{\mathrm{sat}})^2]^{-1/2} ,$$

where $\mu_0$ is the ohmic mobility, and $\upsilon_{\mathrm{sat}}$ is the saturated velocity. Similarly, the modeling of the mobility in Si inversion layers, of utmost practical importance in the simulation of metal-oxide-semiconductor field-effect transistors (MOSFETs), requires some modification, because of the presence of a large "transverse" field which pushes the carriers towards the interface with the gate insulator. The choice of the best "mobility models" has become an art and correct choices have led to remarkable successes in explaining the electrical behavior of devices small enough and under electric fields large enough to violate most of the approximations under which DD applies [18]. Attempts have also been made to extend the range of applicability of DD models to situations in which relatively large field gradients cause velocity overshoot. A one-dimensional model proposed by Thornber [19] consists in adding to the expression $en\mu\mathbf{F}$ for the current density, Eq. (19.15), a term proportional to the gradient of the field, $n\mu L\partial F/\partial x$, where $L$ is a parameter having dimension of length expressing some sort of momentum relaxation length. Price [20] and Artaki [21] have proposed analytic and Monte Carlo methods to evaluate the coefficient $L$.

It is difficult to establish the history of the DD approximation, since it was born with semiconductor physics itself. Shockley's book [5] is probably one of the preferred sources for the *semiconductor equations* (19.2) and (19.15). An equally notably "ancient" formulation is given in [22]. Application to device simulations has flourished in the late 1960s, leading to a well established and numerically sophisticated sets of algorithms. The DD approach remains the foundation of device simulation even at present. The ultimate reason why the simple DD approximation is still so useful lies in the fact that transport parameters are easily fit to experimental data or extrapolated "from the past." Its predictive power is limited to the electrostatic behavior of the device in the "off" state. This is of paramount importance from a power-dissipation point of view and Poisson equation remains always valid, regardless of how poorly electron transport is treated.

### *19.2.2 Hydrodynamic Approximation*

As we mentioned in the introduction of this chapter, hot-carrier effects have been a major source of concern in the 1980s–1990s, as devices were miniaturized but the applied voltage was kept constant at 5 V, thus causing increasingly higher electric fields that drive electrons to strong off-equilibrium conditions. In the much shorter, nanometer-scale devices that are of interest today, these effects are also important, since transport approaches the ballistic regime. The electron kinetic energy is now reduced, since the applied bias has been scaled to below 1 V. As a consequence, the violation of the "locality rule" discussed above can be included in the model by going one step further and considering the third moment, Eq. (19.10) or (19.11). Assuming again that the distribution function is mainly isotropic, so using Eq. (19.10) or (19.11), this equation can be recast in the form

$$\frac{\partial}{\partial t}(n\langle E\rangle_{\mathrm{ens}}) = -\nabla \cdot \mathbf{S} + \mathbf{j} \cdot \mathbf{F} - \frac{n\langle E\rangle_{\mathrm{ens}}}{\tilde{\tau}_{\mathrm{w}}} , \tag{19.22}$$

where the energy flux $\mathbf{S} = n\langle E\mathbf{v}\rangle_{\mathrm{ens}}$ is the troublesome term involving a higher moment. In this form, this equation has a very transparent meaning: the variation per unit time of the energy contained in an infinitesimal volume of real space, $n\langle E\rangle_{\text{ens}}$, is given by the "Joule heating" caused by the field which accelerates the carriers within the volume, $\mathbf{j}\cdot\mathbf{F}$, less the energy lost to the lattice, corresponding to the collision term $n\langle E\rangle/\tilde{\tau}_{\text{w}}$, expressed in the relaxation time approximation, less again the energy flux, $\nabla\cdot\mathbf{S}$, that represents the energy density which leaves the volume in a unit time. In order to overcome the "impasse" caused by the energy flux, a fourth moment, one can decompose the average total kinetic energy $\langle E\rangle$ into a "drift" and a "thermal" component:

$$\langle E\rangle_{\text{ens}} = \frac{1}{2}m^*\langle \upsilon\rangle_{\text{ens}}^2 + \frac{1}{2}m^*\langle \upsilon^2 - \langle \upsilon\rangle_{\text{ens}}^2\rangle_{\text{ens}} \equiv \frac{1}{2}m^*\langle \upsilon\rangle_{\text{ens}}^2 + \frac{3}{2}k_{\text{B}}T\,, \tag{19.23}$$

where what should have been a temperature tensor, $T_{ij} = m^*\langle(\upsilon_i - \langle\upsilon_i\rangle_{\text{ens}})(\upsilon_j - \langle\upsilon_j\rangle_{\text{ens}})\rangle_{\text{ens}}/(3k_{\text{B}})$, has been reduced to a scalar by taking its trace. Thus, the energy flux vector $\mathbf{S}$ can be decomposed into a "drift" term, which is only related to the second moment, and a "heat" term, $\mathbf{Q}$, still containing troublesome fourth moments:

$$\mathbf{S} = \frac{1}{2}nm^*\langle\upsilon\rangle_{\text{ens}}^2\langle\mathbf{v}\rangle_{\text{ens}} + \frac{3}{2}nk_{\text{B}}T\langle\mathbf{v}\rangle_{\text{ens}} + \mathbf{Q}\,, \tag{19.24}$$

where

$$\mathbf{Q} = \frac{1}{2}nm^*\langle(\mathbf{v} - \langle\mathbf{v}\rangle_{\text{ens}})(\upsilon^2 - \langle\upsilon\rangle_{\text{ens}}^2)\rangle_{\text{ens}}\,. \tag{19.25}$$

Finally, the heat flux, representing the flux of the "random" component of the total kinetic energy, can be related to a moment of lower order, the electron temperature, $T$, via a phenomenological Fourier heat diffusion law, which constitutes the desired closure relation:

$$\mathbf{Q} = -\kappa\nabla T\,. \tag{19.26}$$

This relation plays the role of the Einstein relation, Eq. (19.14), employed to express the third moment $D$ in terms of the second moment $\mu$. The thermal conductivity $\kappa$ is given by the Wiedemann–Franz law:

$$\kappa = \left(\frac{5}{2} + c_{\text{hd}}\right)\left(\frac{k_{\text{B}}}{e}\right)^2 ne\mu T\,, \tag{19.27}$$

where the value of the parameter $c_{\text{hd}}$ is still subject of investigation, ranging from 0 to $-2.1$ (or even $-2.5$, which amounts to suppressing altogether the role of $\mathbf{Q}$).

We have already seen that we have different choices to derive the current continuity equation, Eq. (19.15) or (19.18), depending on whether we take strictly the second moment, or the slightly different averages considered in Eq. (19.9). When dealing with the energy continuity equation, there is an even wider range of choices. The algebraic manipulations, approximations, and even the choice of closure relation leading to the set of equations obtained from the first three moments of the BTE differ significantly among published works. In view of the absence of a "standard" model, we shall present here one of the most general set of equations, known as the *hydrodynamic* approximation (see, for example, [23] for some additional details). Some discussion about alternative forms will follow.

The expression we are going to reproduce below has been obtained rigorously by Rudan and Odeh [23]. We start by revisiting the current continuity equation in the form of Eq. (19.4) and express the diffusion constant, $\tilde{\tau}_{\text{p},i}\langle\upsilon_i\upsilon_j\rangle_{\text{ens}}$, in terms of the second-order moment, $T$, by using the decomposition into "thermal" and "drift" components of the velocity. Thus,

$$\begin{aligned}\tilde{\tau}_{p,i}\langle \upsilon_i \upsilon_j \rangle_{ens} &= \tilde{\tau}_{p,i}\langle (\upsilon_i - \langle \upsilon_i \rangle_{ens})(\upsilon_j - \langle \upsilon_j \rangle_{ens}) \rangle_{ens} + \tilde{\tau}_{p,i}\langle \upsilon_i \rangle_{ens}\langle \upsilon_j \rangle_{ens} \\ &\equiv \tilde{D}_{ij} + \tilde{\tau}_{p,i}\frac{j_i j_j}{e^2 n} \to \tilde{D}_0\left(\frac{T}{T_0}\right) + \tilde{\tau}_p \frac{\mathbf{j}\cdot\mathbf{j}}{e^2 n} ,\end{aligned}$$

the last expression being obtained within the isotropic approximation by taking the trace. The "diffusion constant" $D_{ij}$ has been defined in analogy with the equilibrium case, and it has been decomposed into its linear dependence on carrier temperature and a function $\tilde{D}_0$, which we can write as $\tilde{D}_0 = k_B T_0 \tilde{\tau}_p / m^*$, so that for the mobility we have $\tilde{\mu} = e\tilde{\tau}_p/m^* = \tilde{D}_0/(k_B T_0)$, where the subscript 0 refers to quantities evaluated at equilibrium at the lattice temperature. Then, employing the closure Eq. (19.26), the set of equations (19.2), (19.4), and (19.10), coupled to the Poisson equation (19.13), in the by now familiar isotropic and time-independent form, can be written as

$$\nabla \cdot (\epsilon \mathbf{F}) = -en + \rho_d , \tag{19.28}$$

$$\nabla \cdot \mathbf{j} = 0 , \tag{19.29}$$

$$\mathbf{j} - \frac{\tilde{\tau}_p}{e}(\mathbf{j}\cdot\nabla)\frac{\mathbf{j}}{n} = e\tilde{D}_0(\frac{T}{T_0})\nabla(n) + \frac{en\tilde{D}_0}{k_B T_0}\nabla(k_B T - e\varphi) , \tag{19.30}$$

$$\nabla \cdot \mathbf{Q} \equiv -\nabla \cdot (\kappa \nabla T) = \frac{k_B}{e}\mathbf{j}\cdot\left[\frac{3}{2}\nabla T - \frac{T}{n}\nabla n\right] + \frac{n k_B T_0}{\tilde{\tau}_{p,w}} , \tag{19.31}$$

in the unknowns $n$, $\varphi$, and $\mathbf{j}$ (or, equivalently, $\mathbf{v} = -\mathbf{j}/(en)$), and the electron temperature $T$. Note how the spatial variation of $\tilde{D}$ has been handled correctly: the position dependence of $\tilde{\tau}_p$ is lumped into $\tilde{D}_0$, which is outside the gradient, while the variation of the "temperature" $\langle (\upsilon - \langle \upsilon \rangle_{ens})^2 \rangle_{ens}$ is retained inside the gradient. The momentum relaxation time $\tilde{\tau}_p$ is given by Eq. (19.5), for any component $i$, and the combined relaxation rate, $1/\tilde{\tau}_{p,w}$, above is given by

$$\frac{1}{\tilde{\tau}_{p,w}} = \frac{m^* \langle \upsilon \rangle_{ens}^2}{k_B T_0}\frac{1}{\tilde{\tau}_p} - \frac{1}{2}\left[\frac{m^* \langle \upsilon \rangle_{ens}^2}{k_B T_0} + 3\left(\frac{T}{T_0} - 1\right)\right]\frac{1}{\tilde{\tau}_w} , \tag{19.32}$$

where the relaxation rate, $1/\tilde{\tau}_w$, is given by Eq. (19.12). At this point, one must employ some independent models in order to express the relaxation times as functions of the set of variables at our disposal, now richer than the set available in drift diffusion: $\mathbf{r}$, $n$, $\mathbf{j}$, $\mathbf{F}$, and $T$. Since "nonlocality" is one of the major effects that the hydrodynamic model attempts to include, typically one chooses a dependent variable, such as $T$, to model the collision dynamics via the relaxation times. This contrasts the dependence of $\mu$ and $D$ on the field $\mathbf{F}$ (an independent variable, as far as transport is concerned), which rendered them purely local functions. Bypassing this "local" approximation is what renders the hydrodynamic model capable of tackling situations with strong fields, so that the carriers are not necessarily assumed to be at local equilibrium with the field.

The hydrodynamic approximations represent a significant improvement over the DD approximation, since, as we have already noted, carriers are allowed to acquire energy in the field and their dynamics is not necessarily described by local quantities. Yet, in going from the BTE to the hydrodynamic formulation (or to any other higher-order moments approximation) we have clearly paid a price in three ways.

First, we have lost a large amount of information: We must content ourselves with the knowledge of some averages (the moments) over the distribution function $f$. Detailed information about, say, the high-energy part of the distribution is lost, since most averages are insensitive to this component. Yet, some applications (such as predictions of breakdown voltage) may depend crucially on this lost information.

Second, the closure relation (19.26) has been criticized repeatedly. Indeed, it represents the major issue in the moments method: Each moment is always coupled to the next higher moment and, as higher moments are considered, the physical meaning of a "closure" becomes less obvious. As in the DD equation (19.15), very large density gradients would result in unreasonably large carrier "diffusion" velocities. This has been clearly identified as a big problem in simple non-self-consistent simulations of short bases of bipolar transistors. Therefore, in Eq. (19.31) large temperature gradients are consistent with unreasonably large heat currents. This feeds back into the carrier velocity via Eq. (19.31) and "anomalous velocity overshoot" is observed in very small $n^+$-$n$-$n^+$ diodes, called "ballistic" diodes only for historical reasons. More rigorous corrections to the Wiedemann–Franz law have been obtained by utilizing the principles of extended thermodynamics, accounting for viscosity effects [24]. Good results have also been obtained by either ignoring the convective term $(\tilde{\tau}_p/e)(\mathbf{j}\cdot\nabla)(\mathbf{j}/n)$ in Eq. (19.30), leading to the class of approximations usually known as *energy transport* (ET) models (discussed below), or by ignoring the "drift" energy $m^*\upsilon^2/2$ with respect to the thermal energy, $(3/2)k_BT$. This is indeed an excellent approximation in Si at room temperature, down to very small distance scales and large fields. Only in the presence of very large field gradients (i.e., large variations of the electric field over a mean free path) and/or of strongly anisotropic scattering processes, such as Fröhlich scattering, does the drift component of the kinetic energy becomes significant, which happens in materials with a long mean free path (like most III–V compound semiconductors) and at sufficiently low temperatures. More recent developments have resulted in a rigorous formulation of the problem using the entropy-maximization principle [25].

Third, we are forced to employ approximate models for the relaxation times. These models clearly imply, unfortunately in a very indirect way, some particular property of the distribution function, but it is nearly impossible to identify uniquely their "shape," not to mention the functional dependence on $\mathbf{k}$ or $E$ and $T$, of $f$, as various models are proposed. Situations in which the distribution deviates significantly from whatever "shape" is tacitly assumed by the models, will be mishandled. This is particularly true for the momentum relaxation rate, whose energy dependence is much stronger than the dependence exhibited by $\tau_w(E)$. Indeed, Monte Carlo (MC) simulations show it to be practically "flat" above the threshold of optical phonon emission, at least in covalent materials. Popular models for the relaxation times employed in various hydrodynamic simulations include:

- The simple extraction of this quantity from the MC simulation of transport in homogeneous situations [26],

$$\tilde{\tau}_w = \frac{w - w_0}{eF(T)\upsilon_d(T)}, \tag{19.33}$$

$$\tilde{\tau}_p = \frac{m^*\upsilon_d(T)}{eF(T)}, \tag{19.34}$$

  where $T$ is the average temperature of the carriers at the field $F$. This is taken to be $2\langle E\rangle/(3k_B)$ to a good degree of approximation, at least in the absence of strong overshoot or streaming motion. The quantities $\upsilon_d(T)$ and $F(T)$ are the drift velocity and the field itself, expressed as a function of $T$ by inverting the $\upsilon_d$-vs.-$F$ and $T$-vs.-$F$ relations obtained from the MC simulations. Clearly, situations in which the distribution function deviates significantly from the homogeneous case will pose some problems, as discussed in [27].
- The model by Baccarani and Wordeman [28],

$$\tilde{\tau}_w = \frac{m^*\mu_0}{2e}\frac{T_0}{T} + \frac{3}{2}\frac{k_B\mu_0}{e\upsilon_{sat}^2}\frac{TT_0}{T+T_0}, \tag{19.35}$$

$$\tilde{\tau}_p = \frac{m^*\mu_0}{e}\frac{T_0}{T}, \tag{19.36}$$

  where $\mu_0$ is the ohmic mobility at the lattice temperature $T_0$,

- The model by Hänsch and Miura-Mattausch [7, 29], for the momentum relaxation rate, expressed as a function of the current density, **j**, and of the energy flux density, **S**:

$$\tilde{\tau}_{\rm p} = \left( A + B\,\frac{\mathbf{j}\cdot\mathbf{S}}{j^2} \right)^{-1} , \tag{19.37}$$

where $A$ and $B$ are adjustable constants, in principle functions of the lattice temperature.
- The model proposed by Lee and Tang [30], which fixes the constants $A$ and $B$ in the previous equation in terms of the energy relaxation time, $\tilde{\tau}_{\rm w}$, the saturated velocity, $\upsilon_{\rm sat}$, and the low-field mobility, $\mu_0$:

$$\tilde{\tau}_{\rm p} = \frac{m^* \mu_0}{e} \left[ 1 - \frac{3\mu_0}{2\tilde{\tau}_{\rm w}\upsilon_{\rm sat}^2} \left( \frac{k_{\rm B}T_0}{e} + \frac{2}{5}\frac{S}{j} \right) \right]^{-1} . \tag{19.38}$$

The wealth of models available is a clear indication of the seriousness of the problem and of the importance of a correct selection of the relaxation times. Let's consider again the "anomalous overshoot problem" in $n^+$-$n$-$n^+$ diodes. While, as we said above, the Wiedemann–Franz law has been almost universally blamed, one should also note that the large temperature gradient, $\nabla T$, observed around the $n$-$n^+$ junctions, is not due to a fast energy relaxation, but simply to the fact that the concentration of "cold" carriers in the contact (the $n^+$ region) increases sharply, while the relaxation of the carriers coming from the $n$-channel is washed out in the sea of cold carriers. In other words, the distribution at the junction and for some distance inside the "drain" is strongly "bimodal": a cold component due to the carriers at equilibrium in the drain is added to the hot component of carriers arriving from the channel. No moment method (and no other methods based on expansion, as those described in the next section) can handle such a situation and the "bimodality" of the distribution is lost. This loss of information is reflected in an artificially high $\nabla T$. The real solution of the problem appears to consist not so much in altering Eq. (19.27) at large $\nabla T$, but in including the "correct" $\nabla T$ (i.e., only the component of the hot carriers) in the equation and the correct $\tilde{\tau}_{\rm p}$ appropriate to the bimodal distribution. This has been shown by Anile and Mascali [31], who have accounted for the presence of these two (hot and cold) fluids in the system.

The model given by Eq. (19.37) above is perhaps the most successful attempt to obviate the problem of the anomalous overshoot with a redefinition of the momentum relaxation time, since it relates $\tilde{\tau}_{\rm p}$ only to the energy-carrying component of the distribution, but it really does not address the real physical problem. In general, transport situations in which $f$ exhibits a strong bimodal (or quasi-ballistic) behavior cannot be handled by any moment scheme. Perturbing away from ballistic transport (rather than away from equilibrium) would be the correct way to go and has been attempted in particular cases [32]. The major question remains of establishing the length-scale and the energy scale at which the moments methods would begin to show their limits. In a very "hand-waving" fashion, we can express the range of validity of the hydrodynamic approximation (or of ET models) as a requirement on the gradient of the electric field, rather than on the (stronger) conditions on the strength of the field itself which applies to DD models, Eq. (19.21)

$$\frac{1}{|F|} \left| \frac{\partial F}{\partial x} \right| \lambda \leq 1 , \tag{19.39}$$

i.e., the electric field should vary significantly over a length-scale not too much smaller than a few mean free paths.

### *19.2.3 Energy Transport*

For completeness, we should mention a method which, historically, led to the hydrodynamic or energy transport models, the *drifted Maxwellian* approximation [9]. If we assume *a priori* that the unknown $f$ has the form:

$$f \sim \exp\left[-\frac{\hbar^2(\mathbf{k}-\mathbf{k}_\mathrm{d})^2}{2m^* k_\mathrm{B} T_e}\right], \tag{19.40}$$

and substitute this expression into the BTE, we derive equations for the "drift wavevector" $\mathbf{k}_\mathrm{d}$ and the electron temperature $T_e$, which are very similar to the hydrodynamic equations for the first- and second-order moments, $\langle \mathbf{k} \rangle_\mathrm{ens}$ and $T$, respectively. This method has a very transparent physical interpretation and exhibits very clearly the form of the distribution function which is assumed, with all the limitations it carries. This approximation is probably a reasonable one at large carrier densities, when strong interparticle scattering is very effective in thermalizing the distribution function, as we have shown in Sect. 18.2.3. The power of this method consists in providing qualitative information about hot-carrier transport, but it depends on such a strong approximation on the final solutions that quantitative information should not be sought. It is in particular the concept of "electron temperature" as a specifier for the shape of the distribution function which should be analyzed carefully.

The historical development of the moments method originated in 1962 with a paper by Stratton [9], applicable to the case of uniformly doped semiconductors, who considered the energy balance beyond the ohmic regime derived from the *Ansatz* (19.40), thus effectively considering the third moment of the BTE. Bløtekjær [33] extended the method to nonuniform situations, in 1970. Over the years the method was subject to modifications and applied to a variety of situations.

## 19.3 *Expansion Methods

Another possible strategy one may employ to solve the BTE is to expand the unknown distribution function into a complete basis of orthogonal functions, insert this expansion into the BTE, and derive PDEs for the coefficients of each term of the expansion. Therefore, the main task consists in finding some sound physical argument that permits to halt the expansion to a finite, preferably small, number of basis functions, showing that only the first few terms of the expansion are sufficient to capture qualitatively the physics of the problem. For example, the distribution function is spherically symmetric at equilibrium if one assumes isotropic bands. The presence of a weak electric field should distort the distribution only slightly away from spherical symmetry. In the limit of a vanishing field, this is simply the linear-response theory on which the Kubo–Greenwood formula relies. Therefore, expanding the unknown distribution function over spherical harmonics, one expects that only a few terms will be sufficient to capture the deviations from spherical symmetry.

### *19.3.1 The Spherical Harmonics Expansion*

Indeed, historically, the *Spherical Harmonics Expansion* (SHE) was the first expansion proposed in the literature. The first applications of the BTE to electronic transport in semiconductors dealt with a study of the carrier mobility in the ohmic regime (i.e., the $\mathbf{F} \to 0$ limit) [10, 11]. In this limit, one assumes that the distribution function is essentially the spherically symmetric Boltzmann equilibrium function, $f_0$, with small non-spherical corrections proportional to $\cos\theta$, where $\theta$ is the angle between the field $\mathbf{F}$ and the crystal momentum $\mathbf{k}$. In a steady-state, homogeneous situation, one can write

$$f(\mathbf{k}) \simeq f_0(E) + f_1(E)\cos\theta \ . \tag{19.41}$$

Substituting this expression into the BTE and retaining only terms linear in $F$, one finds [10]:

$$f_1(E) = -\frac{eF\hbar k}{m^*}\tau_{\rm p}(E)\frac{\partial f_0}{\partial E} \ , \tag{19.42}$$

where $\tau_{\rm p}(E)$ is the total momentum relaxation time, obtained from Eq. (19.7), after having integrated over all angles in the isotropic approximation and having expressed the remaining integral over the magnitude of $k$ as an integral over the energy $E$. Note that the Kubo–Greenwood formula, Eq. (19.16), is immediately derived from Eq. (19.42) above. As we have already remarked, for isotropic and/or elastic scattering processes $\tau_{\rm p}(E)$ can be rigorously defined, as it does not depend on $f$ itself. In other cases (most notably, polar optical scattering, which is neither isotropic nor elastic), such a simple expansion, while formally possible, does not accomplish anything, since $\tau_{\rm p}(E)$ remains a full functional of $f$, and alternative, more sophisticated strategies are required to obtain the ohmic mobility [14, 34, 35].

The basic idea leading to Eq. (19.41) can be extended naturally to a more general situation. The $\mathbf{k}$-space dependence of the distribution function can be expanded over a basis of orthogonal functions, the coefficients of each term being functions of position in real space (and, possibly, time). The scattering kernels $W(\mathbf{k}',\mathbf{k})$ in the collision integrals can also be expanded over the same basis. Finally, the expansions can be introduced into the BTE, and the coefficients of each of the basis functions set equal in order to derive a hierarchy of PDEs in the variables $\mathbf{r}$ and $t$. By truncating the expansion at a suitable level, the solutions of these coupled differential equations provide a solution of the BTE.

Many different implementations of the SHE method have been proposed, even accounting for full-band effects [36–39]. In the case of a stationary but space-dependent problem, a variable reference frame of unit vectors, $\{\hat{\mathbf{J}}(\mathbf{r}), \hat{\mathbf{e}}_1(\mathbf{r}), \hat{\mathbf{e}}_2(\mathbf{r})\}$, is defined at every point in space. The vector $\hat{\mathbf{J}}$ lies along the direction of the current density $\mathbf{j}$, $\hat{\mathbf{e}}_1$ lies in the plane defined by $\hat{\mathbf{J}}$ and the electric field and is normal to $\hat{\mathbf{J}}$, while $\hat{\mathbf{e}}_2$ is normal to the other two vectors. Spherical bands are assumed for the semiconductor, but an arbitrary "band shape" $E(k)$ can be easily considered. Then, the distribution function can be expanded as

$$f(\mathbf{k},\mathbf{r}) = f(k,\theta,\phi,\mathbf{r}) = \sum_l \sum_{m=-l}^{l} f_l^m(\mathbf{r},k)Y_l^m(\theta,\phi) \ , \tag{19.43}$$

where the functions $Y_l^m$ are the spherical harmonics, and $\theta$ and $\phi$ are the polar and azimuthal angles, respectively, in the frame whose polar axis is along $\hat{\mathbf{J}}$. It is also assumed that the scattering kernels $W(\mathbf{k}',\mathbf{k})$ depend *only* on the magnitudes $k$ and $k'$ of $\mathbf{k}$ and $\mathbf{k}'$, and on the angle $\xi$ between them. Thus, one can expand the total kernel as

$$W(\mathbf{k}',\mathbf{k}) = W(k,\xi) = \sum_l W_l(k)P_l(\cos\xi) \ , \tag{19.44}$$

where $P_l$ is the Legendre polynomial of order $l$. Many scattering processes can easily be considered, such as the isotropic and elastic scattering with acoustic phonons, the inelastic isotropic scattering with optical phonons, the elastic anisotropic scattering with ionized impurities, and impact ionization. Considering only nonpolar materials, in order to avoid the "troublesome" inelastic and anisotropic Fröhlich scattering, and assuming that the distribution is azimuthally symmetric, so that only the $m = 0$

harmonics are retained, truncating the expansion to the $f_2$ term, substituting Eqs. (19.43) and (19.44) into the BTE, and equating coefficients of polynomials of the same order $l$, one obtains the equations:

$$\begin{aligned}
&\hat{\mathbf{J}}\cdot\nabla f_1 + f_1\nabla\cdot\hat{\mathbf{J}} + e\mathbf{F}\cdot\hat{\mathbf{J}}\left(\nabla_E f_1 + \frac{\gamma'}{\gamma}f_1\right) \\
&= \frac{3c_{\mathrm{op}}}{\upsilon}\{\mathscr{D}(E+\hbar\omega^{(\mathrm{op})})[(\langle N_{\mathrm{op}}\rangle_{\mathrm{th}}+1)f_0(E+\hbar\omega^{(\mathrm{op})}) - \langle N_{\mathrm{op}}\rangle_{\mathrm{th}} f_0(E)] \\
&\quad -\mathscr{D}(E-\hbar\omega^{(\mathrm{op})})[(\langle N_{\mathrm{op}}\rangle_{\mathrm{th}}+1)f_0(E) - \langle N_{\mathrm{op}}\rangle_{\mathrm{th}} f_0(E-\hbar\omega^{(\mathrm{op})})]\}\,,
\end{aligned} \tag{19.45}$$

$$\begin{aligned}
&\hat{\mathbf{J}}\cdot\nabla f_0 + \frac{2}{5}\hat{\mathbf{J}}\cdot\nabla f_2 + \frac{3}{5}f_2\nabla\cdot\hat{\mathbf{J}} \\
&\quad = -e\mathbf{F}\cdot\hat{\mathbf{J}}\left(\nabla_E f_0 + \frac{2}{5}\nabla_E f_2 + \frac{3\gamma'}{5\gamma}f_2\right) - \frac{f_1}{\lambda}\,,
\end{aligned} \tag{19.46}$$

$$2\hat{\mathbf{J}}\cdot\nabla f_1 + f_1\nabla\cdot\hat{\mathbf{J}} + e\mathbf{F}\cdot\hat{\mathbf{J}}\left(2\nabla_E f_1 - \frac{\gamma'}{\gamma}f_1\right) + \frac{3f_1}{\lambda} = 0\,, \tag{19.47}$$

$$\hat{\mathbf{e}}_1\cdot\nabla f_1 - f_1\mathbf{a}\cdot\hat{\mathbf{J}} = e\mathbf{F}\cdot\hat{\mathbf{e}}_1\left(\nabla_E f_1 - \frac{\gamma'}{2\gamma}f_1\right)\,, \tag{19.48}$$

$$\hat{\mathbf{e}}_2\cdot\nabla f_1 + f_1\mathbf{b}\cdot\hat{\mathbf{J}} = 0\,. \tag{19.49}$$

In these equations, $\gamma(E) = \hbar^2k^2/(2m^*)$, the total mean free path, $\lambda(E) = \upsilon(E)\tau(E)$, is defined in terms of the group velocity $\upsilon$ and of the total scattering rate, $1/\tau(E)$, as a function of electron energy, and the symbol $'$ means differentiation with respect to $E$. The quantities $c_{\mathrm{op}}$, $\langle N_{\mathrm{op}}\rangle_{\mathrm{th}}$, and $\hbar\omega^{(\mathrm{op})}$ are, respectively, the coupling constant, the Bose–Einstein phonon occupation number, and the energy of the optical phonon. They have been shown explicitly to emphasize the ability of the model to account for inelastic processes. The vectors **a** and **b** above are defined by

$$\frac{\partial \hat{J}_j}{\partial x_i} = a_i\hat{e}_{1,j} + b_i\hat{e}_{2,j}\,. \tag{19.50}$$

The five equations of system (19.43)–(19.44) and Eq. (19.50) are sufficient to determine fully the three unknowns $f_i(\mathbf{r})$ and the two degrees of freedom for the unit vector $\hat{\mathbf{J}}(\mathbf{r})$. Note that only the function $f_0$ is needed to obtain the averages $n$ and $\langle E\rangle_{\mathrm{ens}}$ of the carrier density and kinetic energy, while only $f_1$ is needed to obtain the average velocity $\langle\mathbf{v}\rangle_{\mathrm{ens}}$, or, equivalently, the current density $\mathbf{j}$. The effort to obtain these quantities is thus comparable to the effort required when employing the moments method.

### 19.3.2 Other Expansions and Their Validity

Alternative expansions may be devised, such as expansions over the Hermite polynomials, and have been attempted [7, 40, 41].

The question which we must ask is what advantage these expansion methods give us in solving device-related problems. It is useful to compare *vis-à-vis* the moments method with the expansion method. On first thought, the expansion method presents a major improvement over the moments method, since it solves the two major difficulties of the former approach: We do not have to look for a closure relation, and we do not have to resort to phenomenological models in order to express the relaxation times as functions of the available variables. This is accomplished essentially by accounting for the (hopefully) dominant traces (i.e., the sum of the diagonal or "near diagonal" entries) of all even order moments via $f_0$, the traces of all odd order moments via $f_1$, thus neglecting only the entries far-off the diagonal of higher-order moments. These should vanish for moderately symmetric distribution functions. On a deeper level, however, one realizes that truncating the expansion at some arbitrary order amounts to some form of "closure." It also appears that both methods rely on essentially the same approximation of the distribution function. Indeed, what the expansion method requires is a fast convergence of the expansion, so that ignoring higher-order terms, such as $f_3$ in the system (19.43)–(19.44), does not cause major errors in the evaluation of the lower-order coefficients $f_0$, $f_1$, and $f_2$. We are lacking, at present, a rigorous mathematical criterion defining the range of convergence of the expansion (19.43) (or similar expansions over different basis functions). It appears that while some "functional" convergence may be obtained in most interesting cases, point-wise convergence may be questionable. For example, even in a time-independent and homogeneous situation, the high-energy components of the coefficients $f_i$ seem to be roughly constant, thus implying a lack of convergence at high energies. If we consider again the case of a bimodal distribution, such as what is observed in very short $n^+$-$n$-$n^+$ diodes, it is clear that many terms in any expansion are required to represent correctly the distribution and even the more relaxed "functional" convergence (such as the convergence of mean values, such as $\langle E \rangle_{ens}$) may require too many terms in the expansion to make the method practical. Thus, one can expect that the expansion method will exhibit the same problems also exhibited by the moments method. Indeed, while within the context of the moments methods we implicitly ask that the distribution function be "well behaved" (i.e., not too far from a Maxwellian of some sort, so that the first moments may characterize it well enough), a roughly equivalent form of requirement is also needed by the expansion method. It appears that, after all, the expansion method may simply allow to establish more rigorously which functional form of the distribution function is required in order to be able to describe transport in a simpler way. The moments method hides these requirements under the thick cover of the relaxation rates, thus obscuring this important issue. But, from a practical point of view, the two methods appear to be similar.

## 19.4 Particle-Based Methods

Both of the methods we have seen so far, the moments and the expansion methods, essentially make use of the same basic approximation: They both start from the premise that the distribution function does not deviate "too much" from its equilibrium form. Here "too much" means that, whereas *in principle* these methods are "exact" (just retain an infinite number of moments or terms in the expansion), *in practice* we can obtain numerically a solution in an efficient way only if we retain just a few terms. Moreover, these models handle complicated band structures and device geometries in a rather cumbersome fashion, often implementing the full-band electron dispersion only via the density of states and group velocity expressed as functions of the electron energy, thus in some angle-averaged form. This is a major drawback: As interest shifts to the study of electronic transport in small nanostructures, band-structure effects become dominant (see Chap. 7, Sect. 7.3.1, to see how much the electron dispersion in a Si nanowire differs from that of bulk Si, for example) and transport begins to approach the ballistic regime, dramatically different from equilibrium. Historically, we have already mentioned that hot-electron effects, so important in the VLSI technology of the 1980s and 1990s, also demanded the use of better models to represent the band structure and electronic transport at energies much higher than a few $k_B T$s. Now, these considerations, origination from our interest in smaller and smaller structures, make it necessary to consider methods of solutions that do not rely on any assumption of being near equilibrium and that can fully account for band-structure effects. Additional complications related to our concerns about the validity of semiclassical transport have been discussed in the previous chapter.

Therefore, we discuss here methods that attempt to solve the BTE even in strong off-equilibrium situations. Most of these methods are based on the concept of "particles." In this context, these should be considered numerical entities that "probe" the distribution function in phase-space. The Monte Carlo method is the prototypical embodiment of these methods. But we shall also discuss methods like the *weighted particle* scheme, *cellular automata*, and even the *transmission (or scattering) matrices* methods, that we shall consider "particle-based," since the matrices themselves have to be calculated and stored using some particle-based method.

For the reader who wishes to go deeper into the physical foundation and formulation of these methods, the text by Hockney and Eastwood constitutes a basic reference [42]. The Monte Carlo method applied to electronic transport in semiconductors is clearly presented in the text by Jacoboni and Lugli [43] and in the review articles by Price [44] and Jacoboni and Reggiani [45].

## *19.4.1 The Monte Carlo Method*

### 19.4.1.1 Overview

The MC method is based on a stochastic solution of the BTE. It was first devised in the early days of digital computers to study neutron transport in nuclear fission [46]. In the context of electron transport in crystals, it was first introduced by Kurosawa in 1966 [47]. It was shown to be fully equivalent to a solution of the BTE by Fawcett and co-workers [48]. The Malvern group (for example, see [49]) soon afterwards applied it to the study of high-field electron transport in III–V compound semiconductors. The Modena group (see [45] and references therein) applied it to the study of electron and hole transport in Si and Ge.

In order to get a quick idea of how the method works, it is convenient to look at the way the BTE can be reformulated as a path integral, known as the *Chamber's formulation* [50, 51]. For simplicity, we express it here in the simpler form valid in homogeneous situations:

$$f(\mathbf{k},t) = \frac{1}{(2\pi)^3} \int_0^{\infty} \mathrm{d}s \, \exp\left[-\int_0^s \mathrm{d}s' \, \tau^{-1}\left(\mathbf{k} - \frac{e\mathbf{F}}{\hbar}s'\right)\right] \times \int \mathrm{d}\mathbf{k}' \, f(\mathbf{k}',t-s) \, W\left(\mathbf{k} - \frac{e\mathbf{F}}{\hbar}s, \mathbf{k}'\right) , \tag{19.51}$$

where $\tau^{-1}(\mathbf{k})$ is the total scattering rate. This equation can be interpreted as follows: the distribution function at $\mathbf{k}$ and time $t$ can be traced back to the sum over all "past" distributions at any past time $t-s$, by considering all particles which were at $\mathbf{k}'$ at that time, scattered into the state $\mathbf{k} - e\mathbf{F}s/\hbar$, and survived without scattering from time $s$ to time $t$, which happens with probability

$$P(t-s,t) = \exp\left[-\int_0^s \mathrm{d}s' \, \tau^{-1}\left(\mathbf{k} - \frac{e\mathbf{F}}{\hbar}s'\right)\right] . \tag{19.52}$$

We can reproduce this sequence of events by proceeding as follows:

1. We start with a particle in some initial **k**-state and position in real space. These initial conditions reflect the physical initial condition of the system.
2. We calculate, for all scattering processes of type $r$, the rate

$$\frac{1}{\tau_r(\mathbf{k})} = \int \frac{d\mathbf{k}'}{(2\pi)^3} W_r(\mathbf{k},\mathbf{k}') \, , \tag{19.53}$$

and the "total" scattering rate:

$$\frac{1}{\tau(\mathbf{k})} = \sum_r \frac{1}{\tau_r(\mathbf{k})} \, . \tag{19.54}$$

3. We calculate the duration $s$ of the ballistic flight of the particle, according to the probability (19.52) above. We do this "stochastically," by selecting a random number $\xi$ in the interval $(0,1)$, and setting $\xi = P(s)$. Since the integral in the exponent can be quite troublesome to evaluate numerically, the "self-scattering" algorithm due to Rees can be employed [51, 52]: One adds to the "real" scattering rate a fictitious scattering process with renders $1/\tau(\mathbf{k})$ constant, i.e., independent of **k**. Calling $\Gamma$ this constant, one simply has $P(s) = \exp(-\Gamma s)$, so that $s = -\ln(\xi)/\Gamma$. Alternatively, one only considers very small time steps, so that $P(s) \simeq 1 - s/\tau(\mathbf{k})$, and **k** can be considered constant.
4. Once $s$ is known, the equations of motion for a free particle under the action of the field are integrated from the initial time to time $s$.
5. At the end of the flight, we select the scattering process with a probability given by their relative weight $\tau(\mathbf{k})/\tau_r(\mathbf{k})$. One of these scattering processes may be a self-scattering process, if this is the chosen algorithm, or the particle may not scatter at all, if a fixed time step $s$ is chosen and $\xi > s/\tau(\mathbf{k})$. In either case, the particle remains in its original **k**-state. If a "real" scattering process is chosen, we let the particle scatter to a new state, according to the probability $W(\mathbf{k},\mathbf{k}')$.
6. We store the desired "ensemble averages," by updating the appropriate statistical estimators, update the time, and proceed to step 1 above.

#### 19.4.1.2 Single Particle and Ensembles

Two basic schemes may be used: a single-particle simulation, in which only one particle at a time is simulated, or an "ensemble" MC simulation, in which the time evolution of an ensemble of particles is simulated. Typically, in device simulations the latter scheme is preferred. If a self-consistent solution of the Poisson equation is desired, before closing the loop above, a few additional steps are necessary to feed the new particle positions into the updated field:

1. We assign the charge of each particle to the mesh used to solve the Poisson equation. Several schemes are possible. Hockney's and Estwood's book discusses them in detail [42].
2. We solve the Poisson equation in the mesh, by solving the associated finite element or finite difference discretized matrix problem.
3. We compute the components of the field on the nodes of the mesh.
4. We interpolate the field back to the particle positions. Here the subtle issue of *self-forces* enters the picture: we have to make sure that the field due to a particle, when interpolated back to the position of the same particle, does not give rise to any unphysical force.

As stated in this simple test case we have employed to outline all methods of solution, the MC method appears quite simple and intuitive. Indeed, while a rigorous proof can be provided that such a "direct" simulation is actually equivalent to the solution of the BTE [48], the method amounts to a conceptually simple simulation of the motion of semiclassical particles and their collisions. This intellectual simplicity is the main advantage of the method: However, others may find the method unsatisfactory, because of the stochastic (i.e., not *exact* in this sense) nature of the solution. Many engineers will not tolerate the long computation times caused by this stochastic nature in order to reduce the "noise" (or, more properly, the "variance") of the desired information. On the other side, the entire emphasis is on the physical models. The scattering kernels $W(\mathbf{k},\mathbf{k}')$ can be rendered explicit functions of position, of time, functionals of the distribution function, and still the resulting BTE is solved *exactly* (from the viewpoint of a physicist), without invoking approximations on the shape of the distribution function, on the symmetry of the band structure, on the existence of relaxation times, etc. In this sense, the MC method provides a "physically" exact solution of the BTE. We are only limited by the amount of physics we are able to put into the BTE itself (and by our patience in waiting for the evaluation of the model!).

#### 19.4.1.3 Comparison with Other Techniques

As just stated, the main advantage of the MC method in device simulation is its simple implementation and the possibility of keeping the number of approximations to a minimum. The high computational cost is its major drawback. It is hard for the authors of this book to maintain an unbiased view of the issue and compare in an absolutely objective way various approaches. Yet, when comparing the MC method to other methods we have reviewed in this chapter, a few considerations should be kept in mind.

Very frequently comparisons based on performance, measured by speed of execution per "bias point," are reported. It is important to make these comparisons on an "equi-physics" basis, by comparing different methods using the same physical models. The MC method excels in its ability to incorporate physical models beyond what other techniques can handle. Thus, a "fair" comparison is sometimes impossible. An example is provided by the fact that most MC programs handle with relative ease realistic band-structure models, such as ellipsoidal valleys (as for Si and Ge) (see the original work cited in [45]), including nonparabolic corrections, multi-valley materials, such as GaAs [49], ternary alloys of III–V compound semiconductors, strained alloys, or even band structures derived from pseudopotential calculations [53–60]. We have indeed seen in Chap. 13 examples of ensemble Monte Carlo simulations used to calculate the mobility and velocity field characteristics in thin Si layers (Sect. 13.4.1, Figs.13.4 and 13.5) and in graphene (Sect. 13.4.2, Figs. 13.10 and 13.11). Equally important, we should also emphasize that ensemble MC simulations can handle Coulomb carrier–carrier interactions, processes that play an increasingly important role in nanometer-scale devices. Alternative methods struggle dealing with these processes, that are therefore regarded as "negligible effects."

Methods based on the moments of the BTE, or on expansions over orthogonal functions, make many simplifying assumptions based on spherical and/or parabolic and single valley band structures. For example, the hydrodynamic model, when applied to multi-valley semiconductors, requires a separate set of equations for each valley, thus increasing the time required to obtain a solution and creating nontrivial convergence problems. Expansion methods rely heavily on quasi-spherical valleys (i.e., valleys for which a simple transformation of **k**-space coordinates, due to Herring and Vogt [61], can be applied to reduce the problem to spherical symmetry), with more general (and realistic) band structures representing rather formidable implementation tasks. Indeed, too many terms would be required by the SHE to account for the high anisotropy of the full-band structure. This is the reason why full-band effects are usually included only in some angle-averaged fashion, by expressing energy, density of states, and group velocity as functions of the carrier energy [36, 37].

MC methods are intrinsically amenable to geometries of higher dimensionality, which represent almost exclusively a memory management problem rather than a problem of execution speed. An example may be a comparison between the MC method and any other method based on PDEs in a one-dimensional, homogeneous, steady-state situation. The MC technique is intrinsically three-dimensional both in real space and in **k**-space, so that such a simple problem is a poor test of the MC method. A homogeneous situation favors alternative methods, while the MC algorithm gains from the space-independence of the problem only marginally (averages can be taken over the entire simulated region and noise can be brought down to acceptable levels in a shorter time), yet it solves the BTE in its full seven-dimensional form anyway. But if we now consider space-dependent problems, and move from a one-dimensional to a three-dimensional geometry, only the Poisson solver—a relatively minor item in the overall computing budget—represents an increased computational burden in self-consistent Monte Carlo/Poisson simulations. This is not so for deterministic methods based on PDEs, whose computational demands increase in a very significant way. Therefore, as the use of simulation programs handling 2 or 3 real-space dimensions becomes mandatory in the engineering community, due to the reduction of linear and vertical dimensions, the computational cost of MC simulations becomes comparable to that of alternative methods.

The real limit of MC methods can be inferred by recalling the "intuitive" correspondence between simulation and "reality" that these methods embrace. Slow devices will be "harder" to simulate using MC techniques than fast devices. Thus, paradoxically, a 10 μm MOSFET is prohibitive: since at best we can expect a 0.1 ns source-to-drain transit time, we are about 2 orders of magnitude beyond what typical ensemble MC schemes can afford to do (typically simulation times of the order of 1–10 ps). Similarly, the full switch-off of a non-field-aided *n-p-n* bipolar device requires that low mobility holes leave the base under a very low field. The resulting time constant is, again, of the order of several nanoseconds. It should be noted, however, that "slow" devices are usually "large" devices, for which the DD approximation should be fully satisfactory. Thus, the MC method becomes impractical exactly in those situations for which the method is not needed anyway. Low current situations, usually controlled by electrostatics (e.g., the subthreshold characteristics of the device), are also hard to tackle with MC techniques, because of the inherently large noise in this situation. Again, when electrostatics dominate, MC methods are not necessary and should be avoided.

Related to this is the issue of "rare events." Because of the finite number of particles that can be used in the simulation, events that occur with a low probability (e.g., that depend on the high-energy tail of the electron energy distribution) cannot be captured, or are affected by an unacceptably high statistical noise. For example, if we are interested in studying the injection of hot electrons from Si into $SiO_2$, we must simulate electrons at energies approaching or exceeding the conduction-band discontinuity between Si and $SiO_2$, approximately 3 eV. Even if very high electric fields are present in the semiconductor, only a fraction $\sim 10^{-9}$-to-$10^{-6}$ of the electrons will be present at these energies. Clearly, one cannot employ $10^9$ particles in the simulation to have 1 of them at these energies. *Variance reduction* techniques exist that can bypass this problem, by assigning different statistical weights to particles occupying different regions of phase-space (or even real space, as in low-density regions). These methods, already known in the "early days" of MC simulations [62, 63], have been successfully implemented in our context [64]. The problem remains that these techniques require *a priori* knowledge of the rare event that requires statistical amplification. Unknown, unexpected rare events are likely to remain hidden.

These criteria notwithstanding, MC simulation should not be used to mindlessly generate sets of current–voltage characteristics. As we noted above, and as it has been stressed in the literature, the DD approximation is often sufficient to obtain terminal currents. The main strength of MC simulators lies in their physical basis. Given the sometimes blind approximations made by faster simulation approaches, MC programs should be used to calibrate both the parameters entering those simpler approaches as well as our mind-set about the internal physics of electronic transport inside the device.

### 19.4.2 *Cellular Automata

Despite the high degree of nonlocality exhibited by the BTE, it has been shown by the Munich group [65, 66] that it can be translated into a lattice gas automaton with only local interactions in space and probabilistic rules. The major step is the replacement of the physical ballistic trajectories of the carriers between collisions with probabilistic rules, so that the drift terms in the BTE become local in space. This method has been termed *Cellular Automata* (CA).

Following Kometer and co-workers [65, 66], the starting point is the discretization of the BTE in phase-space and time. Real space is divided into cells, centered around sites $\{\mathbf{R}_i\}$. Assuming only two-dimensional geometries, the cells are taken to be hexagons with a lattice spacing $a$ of a few nm, so that the index $i$ runs from 1 to $10^5$–$10^6$. Similarly, **k**-space is divided into shells of thickness $\Delta K$ centered around the values $\{K_s\}$, with $s$ running from 1 to 20 or more, and divided into 60° segments, each labeled by an index $q$, corresponding to group velocities pointing to nearest-neighbor real-space cells. The spacing $\Delta K$ is chosen so that the lowest energy spacing is of the order of the optical phonon energy. Each cell in discretized phase-space can be labeled formally by $(\mathbf{R},\mathbf{K})$, where $\mathbf{R}$ stands for the index $i$ of the real-space cell, and $\mathbf{K}$ stands for the pair of indices $s$ and $q$.

The BTE, discretized in this fashion, can be written as

$$\begin{aligned} n(\mathbf{R},\mathbf{K},t+\Delta t) &= n(\mathbf{R},\mathbf{K},t) \\ &+ \sum_{\mathbf{R}'=\mathbf{R}_{\mathrm{NN}}} \left[p_{\mathrm{T}}(\mathbf{R}',\mathbf{R},\mathbf{K})n(\mathbf{R}',\mathbf{K},t) - p_{\mathrm{T}}(\mathbf{R},\mathbf{R}',\mathbf{K})n(\mathbf{R},\mathbf{K},t)\right] \\ &+ \sum_{\mathbf{R}'=\mathbf{R}_{\mathrm{NN}}} \left\{\left[p_{\mathrm{F}}(\mathbf{R},\mathbf{K}',\mathbf{K}) + p_{\mathrm{coll}}(\mathbf{R},\mathbf{K}',\mathbf{K})\right] n(\mathbf{R},\mathbf{K}',t)\right. \\ &\qquad \left. - \left[p_{\mathrm{F}}(\mathbf{R},\mathbf{K},\mathbf{K}') + p_{\mathrm{coll}}(\mathbf{R},\mathbf{K},\mathbf{K}')\right] n(\mathbf{R},\mathbf{K},t) \right\}, \end{aligned} \tag{19.55}$$

where the first sum is performed over the cells $\mathbf{R}_{\mathrm{NN}}$, the nearest-neighbor cells, $n(\mathbf{R},\mathbf{K},t)$ is an integer expressing the occupancy of the cell $(\mathbf{R},\mathbf{K})$, defined in such a way that the average over an ensemble of lattice gas automata—indicated by $\langle\ldots\rangle_{\mathrm{CA}}$ and realized by averaging over 100 cells or so—yields the usual ensemble average:

$$\langle n(\mathbf{R},\mathbf{K},t)\rangle_{\mathrm{CA}} = 2\int_{\Omega(\mathbf{R})} \mathrm{d}\mathbf{r} \int_{\Omega(\mathbf{K})} \frac{\mathrm{d}\mathbf{k}}{(2\pi)^3} f(\mathbf{k},\mathbf{r},t)\,. \tag{19.56}$$

The volumes $\Omega(\mathbf{R})$ and $\Omega(\mathbf{K})$ are the volumes of a cell in real and **k**-space, respectively. The scattering probabilities $p_{\mathrm{T}}$, $p_{\mathrm{F}}$, and $p_{\mathrm{coll}}$ entering Eq. (19.55) are defined as follows: The ensemble average of the collision probability $p_{\mathrm{coll}}$ is the integral of the scattering kernel $W_{\mathbf{r}}(\mathbf{k},\mathbf{k}')$ (note that we have explicitly expressed its possible dependence on position) between the cells $\mathbf{K}$ and $\mathbf{K}'$:

$$\langle p_{\mathrm{coll}}(\mathbf{R},\mathbf{K},\mathbf{K}')\rangle_{\mathrm{CA}} = \frac{\Omega}{\Omega(\mathbf{K})} \int_{\Omega(\mathbf{K}')} \mathrm{d}\mathbf{k}' \int_{\Omega(\mathbf{K})} \frac{\mathrm{d}\mathbf{k}}{(2\pi)^3} W_{\mathbf{r}}(\mathbf{R},\mathbf{k},\mathbf{k}')\Delta t\,. \tag{19.57}$$

$\Omega$ is the normalization volume. The diffusion term $(\mathrm{d}\mathbf{r}/dt)\cdot\nabla_{\mathbf{r}}f$ in the BTE yields the "transfer" probability from one cell in real space to the one of the six nearest neighbors:

$$\langle p_{\mathrm{T}}(\mathbf{R},\mathbf{R}',\mathbf{K})\rangle_{\mathrm{CA}} = \frac{\upsilon(\mathbf{K})\Delta t}{a}\,\delta\!\left(\frac{(\mathbf{R}'-\mathbf{R})}{a} - \frac{\mathbf{K}}{|\mathbf{K}|}\right), \tag{19.58}$$

where $\upsilon(\mathbf{K})$ is the group velocity. This is actually a Boolean variable, taking values 0 or 1, implying that the particle will remain in the same cell in real space after time $\Delta t$ with probability $1-\upsilon\Delta t/a$, or will move to the nearest neighbor in the direction $\mathbf{K}/|\mathbf{K}|$ with probability $\upsilon\Delta t/a$. On average, the particle will travel at its group velocity $\upsilon$. Finally, the drift term, $-e\mathbf{F}\cdot\nabla_{\mathbf{k}}f$, in the BTE gives rise to an effective scattering probability

$$\langle p_{\mathrm{F}}(\mathbf{R},\mathbf{K},\mathbf{K}')\rangle_{\mathrm{CA}} = \frac{\Delta t}{\tau_{\mathrm{free}}(\mathbf{R},\mathbf{K})}\frac{O_{\Delta\mathbf{k}}(\mathbf{K},\mathbf{K}')}{\Omega(\mathbf{K})}\,, \tag{19.59}$$

where $O_{\Delta\mathbf{k}}(\mathbf{K},\mathbf{K}')$ is the volume of the cell $\mathbf{K}$ which overlaps with the cell $\mathbf{K}'$ after the shift $\mathbf{K}\to\mathbf{K}+e\mathbf{F}\tau_{\mathrm{free}}/\hbar$ under the action of the field for a time $\tau_{\mathrm{free}}$ given by the mean free time of the electron in cell:

$$\frac{1}{\tau_{\mathrm{free}}(\mathbf{R},\mathbf{K})} = \frac{1}{\Delta t}\sum_{\mathbf{K}'}\langle p_{\mathrm{coll}}(\mathbf{R},\mathbf{K},\mathbf{K}')\rangle_{\mathrm{CA}}\,, \tag{19.60}$$

provided $\tau$ does not change too much in the shift above.

The basic algorithm for evolution through a single time step of duration $\Delta t$ is:

1. Initialize the particles in the device, label them in phase- space.
2. For each particle, evaluate **k**-space scattering at each site **R**, by selecting a final state $\mathbf{K}'$ with probability $p_{\mathrm{F}}(\mathbf{R},\mathbf{K},\mathbf{K}')+p_{\mathrm{coll}}(\mathbf{R},\mathbf{K},\mathbf{K}')$, and transfer the particle from $(\mathbf{R},\mathbf{K})$ to $(\mathbf{R},\mathbf{K}')$.
3. Evaluate real-space "scattering" by selecting a final cell with probability $p_{\mathrm{T}}(\mathbf{R},\mathbf{R}',\mathbf{K}')$, remaining in the same site with probability $1-p_{\mathrm{T}}(\mathbf{R},\mathbf{R}',\mathbf{K}')$, moving to that site with probability $p_{\mathrm{T}}(\mathbf{R},\mathbf{R}',\mathbf{K}')$. In the latter case, transfer to the state $(\mathbf{R}',\mathbf{K}')$.
4. Evaluate the cell occupation $n(\mathbf{R},\mathbf{K},t)$.
5. Solve the Poisson equation.

The procedure continues through a loop over time steps, as in any other conventional simulation. The basic criteria for an accurate simulation are related to the choice of the quantities $\Delta t$ and $a$: The "velocity" $a/\Delta t$ must be larger than any physical velocity in the system, so that the kinematics are treated correctly. The time step $\Delta t$ implies, via the probabilistic treatment of the kinematics, a fictitious diffusion constant which must be much smaller than any physical diffusion in the system. Typically, for GaAs at room temperature, $\Delta t = 1$ fs and $a$= 1 nm are satisfactory choices.

This method has been applied to the study of complex devices, especially scaled MOSFETs [67] and two-dimensional GaAs MESFETs [68, 69]. Yet, from a "philosophical" point of view, it appears to be very similar to the MC method, in that it shifts the emphasis away from a complex algorithmic/mathematical formulation of the BTE (even oversimplifying the kinematics, with simplistic band-structure models, and the dynamics, with simplistic collision terms) to a perhaps not-so-elegant technique which, *on the other side*, gives more intellectual freedom to enhance the physical models.

### *19.4.3 *Weighted Particle Method*

A deterministic particle simulation approach has been proposed by Degond and co-workers [70, 71]. It can be considered a variant of the Monte Carlo technique, in the sense that the ballistic transport of (super)particles is explicitly simulated between collisions. However, while the MC method evaluates the collision integral stochastically, this method employs the particles as quadrature points to evaluate the collision integral itself. In order to do so, the effect of the collisions is not accounted for by changing the particle trajectories, but rather by changing their "weights," which thus carry the whole information about the distribution function.

The starting point of the method is the approximation of the distribution function by its particle representation: Each particle, labeled by the index $i$, is assigned a **k**-space volume $\omega_i$ and a weight $w_i$. The distribution function is then approximated as a sum over delta functions at the particle locations. Assuming, for simplicity, a homogeneous situation,

$$f(\mathbf{k},t) \simeq \sum_i \omega_i \, w_i(t) \, \delta[\mathbf{k}-\mathbf{k}_i(t)] \,, \tag{19.61}$$

where $\mathbf{k}_i(t)$ evolves according to the free equations of motion (i.e., $\mathbf{k}_i(t) = \mathbf{k}_i(0) - e\mathbf{F}t/\hbar$). Using this approximation, the collision integral becomes

$$Q_i(t) \simeq \sum_j \omega_j \{W[\mathbf{k}_j(t),\mathbf{k}_i(t)]w_j(t) - W[\mathbf{k}_i(t),\mathbf{k}_j(t)]w_i(t)\} \,, \tag{19.62}$$

and the BTE reduces to a system of ordinary differential equations for the weights:

$$\frac{\mathrm{d}w_i(t)}{dt} = Q_i(t) \,. \tag{19.63}$$

One can now start from an "educated guess" on the initial weights (such as those associated with a Maxwellian), assume the appropriate boundary conditions on a closed domain for the **k**-vectors, and solve Eq. (19.63) above. All "ensemble averages" can be easily obtained from the known weights. In order to make the evaluation of the collision integrals $Q_i(t)$ as fast as possible, approximations to the scattering kernels using velocity-randomization (i.e., isotropic) are preferred. A very careful analysis of the stability, accuracy, and error estimates of the method has been provided [70, 71].

### 19.4.4 *The Scattering Matrix Approach

The "Scattering Matrix Approach" (SMA) originates from the "flux method," proposed by McKelvey and co-workers in 1961 [72]. It is not a "particle-based" method, but we discuss it here because the basic building blocks of the methods, the scattering matrices themselves, are usually calculated and stored using the Monte Carlo method. In some very broad sense, it can be viewed as a variation of the Monte Carlo method and of the cellular automata method.

Consider a one-dimensional geometry and divide the semiconductor into thin slabs with a given field, doping density, etc. defined in each slab. Let's denote by $a^+$ ($b^-$) the flux of carriers entering the slab from the left (right), and $a^-$ ($b^+$) the flux of carriers leaving the slab on the left (right), so that $a$ refers to fluxes at the left, $b$ to fluxes at the right, the superscript "+" to right-traveling fluxes, and the superscript "−" to left-going fluxes. In each slab, define a matrix,

$$\mathbf{S} = \begin{bmatrix} t & r' \\ r & t' \end{bmatrix}, \tag{19.64}$$

where $t$, $t'$, $r$, and $r'$ are the transmission and reflection probabilities associated with the slab for the fluxes incident from the left and from the right, respectively. Then, an intuitive formulation of current continuity requires

$$\begin{bmatrix} b^+ \\ a^- \end{bmatrix} = \mathbf{S} \begin{bmatrix} a^+ \\ b^- \end{bmatrix}, \tag{19.65}$$

together with the conditions $t+r=1$ and $t'+r'=1$. One can now analyze the whole device: the $i$-th slab is centered at the coordinate $z_i$, and it is characterized by a doping $N_i$ and an electric field $F_i$. Two scattering matrices associated with adjacent slabs $i$ and $i+1$ are combined in an algebraic rule called "cascading": if $\mathbf{S}_c$ is the "cascaded" matrix, its elements are given by

$$t_c = \frac{t_i t_{i+1}}{1-r'_i r_{i+1}}\,, \tag{19.66}$$

$$r'_c = r'_{i+1} + \frac{r'_i t_{i+1} t'_{i+1}}{1-r'_i r_{i+1}}\,, \tag{19.67}$$

$$r_c = r_i + \frac{t_i t'_i r_{i+1}}{1-r'_i r_{i+1}}\,, \tag{19.68}$$

$$t'_c = \frac{t'_i t'_{i+1}}{1-r'_i r_{i+1}}\,. \tag{19.69}$$

By cascading all the matrices associated with the slabs, one obtains

$$\begin{bmatrix} b_R^+ \\ a_L^- \end{bmatrix} = \mathbf{S}_c \begin{bmatrix} a_L^+ \\ b_R^- \end{bmatrix}\,, \tag{19.70}$$

where the subscript L represents the left contact, the subscript R the right contact. Assuming local thermodynamic equilibrium at the contacts, one can easily evaluate the incoming fluxes:

$$a_L^+ = 2\int_{\upsilon_z^{(g)}>0} \frac{d\mathbf{k}}{(2\pi)^3}\, \upsilon_z^{(g)}(\mathbf{k})\, f_{0,L}(\mathbf{k})\,, \tag{19.71}$$

$$b_R^- = 2\int_{\upsilon_z^{(g)}<0} \frac{d\mathbf{k}}{(2\pi)^3}\, \upsilon_z^{(g)}(\mathbf{k})\, f_{0,R}(\mathbf{k})\,, \tag{19.72}$$

where $f_{0,L}$ and $f_{0,R}$ are Fermi–Dirac distributions evaluated at the Fermi level at the left and right contacts, as specified by the boundary conditions. The problem (19.70) can be solved and various quantities of interest can be derived. For example, the current at the left contact can be obtained, assuming the case of electrons, as

$$J_L = -e(a_L^+ - a_L^-)\,, \tag{19.73}$$

and similar for the right contact. The electron density follows trivially:

$$n_L = \frac{a_L^+ + a_L^-}{\upsilon_z^{(g)}}\,, \tag{19.74}$$

where $\upsilon_z^{(g)}$ is the velocity with which carriers are injected from the left at the contact, as averaged over the distribution $f_{0,L}$ considered in Eq. (19.71) above. In order to evaluate carrier and current densities within the device, one defines "transfer matrices" $\mathbf{T}_i$, which relate the left and right fluxes in each slab:

$$\begin{bmatrix} b_i^+ \\ b_i^- \end{bmatrix} = \mathbf{T}_i \begin{bmatrix} a_i^+ \\ a_i^- \end{bmatrix} = \begin{bmatrix} t_i - r'_i r_i/t'_i & r'/t'_i \\ -r/t'_i & 1/t'_i \end{bmatrix} \begin{bmatrix} a_i^+ \\ a_i^- \end{bmatrix}\,. \tag{19.75}$$

Since the contact fluxes $a_{\mathrm{L}}^{+}$ and $a_{\mathrm{L}}^{-}$ are known from the procedure described above, one can obtain the fluxes in and out of an arbitrary slab by working from the left contact to that slab via repeated applications of the transfer matrices. Thus:

$$\begin{bmatrix} b_i^{+} \\ b_i^{-} \end{bmatrix} = \mathbf{T}_1 \mathbf{T}_2 \ldots \mathbf{T}_i \begin{bmatrix} a_{\mathrm{L}}^{+} \\ a_{\mathrm{L}}^{-} \end{bmatrix} . \tag{19.76}$$

The algorithm just described has been extended by the group at Purdue University [73, 74], in order to evaluate not only ensemble averages but also a discretized form of the distribution function. The idea is to decompose each flux into its **k**-modes. If the BZ is divided into $M$ elemental volumes, the transmission and reflection coefficients become $M \times M$ matrices, and each scattering matrix becomes a $2M \times 2M$ matrix. The entire algorithm proceeds in exactly the same way, but now one is able to obtain at each slab the carrier density in each of the $M$ modes, i.e., the distribution function in the elemental volume in the BZ corresponding to the given mode. It can be shown that the SMA is fully equivalent to a solution of the BTE [75].

The amount of memory required to store the scattering matrices can be minimized by discretizing the longitudinal modes (i.e., the **k**-vector along the direction of the field) and only one of the transverse modes, averaging it over the azimuthal angle, thanks to the azimuthal symmetry of the problem. Iterative schemes can be employed to solve the problem. In the simplest case, one imposes periodic boundary conditions at the left and the right contacts. Thus, Eq. (19.70) becomes an eigenvalue problem. This is solved by "guessing" an incident flux from the left computing the outgoing flux to the right, using the result as a new incoming flux from the left, and repeating the procedure until the desired convergence is reached.

It is clear that both the kinematics and the dynamics are hidden within the scattering matrices. Nothing has been said about the physics of the problem (band structure, collision kernels, etc.) until the scattering matrices are specified. Typically, Monte Carlo simulations are employed to evaluate the scattering matrices and build a "library" of them, each for a given field, doping concentration, etc. It is clear that memory requirements [76] can easily exceed any practical limit. Schemes have been devised to minimize the storage problem, by assuming field-free slabs coupled to voltage steps between adjacent slabs. Any field can be mimicked by an appropriate slab thickness and step height.

The method has the advantage of being intuitive, and of requiring no approximations of the physics, being gated in this respect by the physics employed in the MC code used to generate the library of scattering matrices. It is also amenable to the study of some quantum effects, such as tunneling and, more generally, "vertical transport," by promoting the transmission and reflection coefficients to complex numbers obtained by computing the quantum-mechanical transmission and reflection amplitudes for step potentials or across potential barriers. There are, however, severe problems with the method in general. First, it is easy to understand that the simulation of two-dimensional geometries would require an impractical effort in terms of memory management. Attempts have been made to couple this method to the expansion method, described above, in order to minimize the memory requirements. Secondly, numerical problems of the form of accuracy and propagation of round-off errors may plague the cascading of many scattering matrices and the multiplication of many transfer matrices. Related to this problem is the possible propagation of the intrinsic "noise" (or "variance") of the MC method, when used to generate the library of scattering matrices, which may be amplified by the cascading process. Finally, the convergence of the iterative scheme may become a source of concerns.

## 19.5 *Direct Solutions

Several techniques have been proposed to solve the BTE without the help of moments, relaxation times, or truncated expansions. This is partially due to the justifiable desire to assess how well those methods approximate the "exact" solution. Among the various methods employed in the relative long history of the BTE applied to electronic transport in solids, the most interesting are *iterative techniques*, the *direct solutions* of the BTE attempting to express it in a finite difference form.

### *19.5.1 The Iterative Technique*

For uniform systems, we have seen before that the BTE can be recast into the Chamber's path integral form, Eq. (19.51) [50, 51]. One can now attempt to solve this equation by iterations, as first suggested by Budd [77] (curiously, at the same conference at which the MC method was first proposed). As it is usually done in iterative approaches, one can substitute to the unknown $f$ by an arbitrary function, $f_0$, at the right-hand side, calculate a new $f_1$ at the left-hand side by performing the integrations indicated over the paths $\mathbf{k} - e\mathbf{F}t/\hbar$, substitute the newly found $f_1$ in the right-hand side, and proceed with the iteration until some convergence criterion is met. A particular application of the iterative technique is the study of time transients, starting from a known initial distribution $f_0$. The self-scattering algorithm can be employed also in this case: A fictitious scattering kernel, $W_{\text{self}}(\mathbf{k}, \mathbf{k}')$, is introduced, so that the total scattering rate, $1/\tau(\mathbf{k})$, is independent of $\mathbf{k}$ and equal to a constant, $\Gamma$ [51, 52]. In this fictitious self-scattering process, the state of the particle does not change, i.e., nothing happens to the particle. If $\Gamma \gg 1/\tau(\mathbf{k})$, successive iterations exhibit now the property [78] that, if $f_n$ and $f_{n+1}$ are the results of two successive iterations,

$$f_{n+1}(\mathbf{k}) \simeq f_n(\mathbf{k}) + \frac{1}{\Gamma} \frac{\partial f_n(\mathbf{k};t)}{\partial t} , \tag{19.77}$$

and the iteration acquires the physical meaning of time evolution. Among the advantages of the iterative procedure is the possibility of employing arbitrarily complicated scattering kernels and band structures. The major difficulty lies in the solution of spatially inhomogeneous problems. Caution should also be exercised in the analysis of error propagation through the iteration, since round-off errors and algorithmic errors in the numerical evaluation of the many integrals may be hard to estimate and to keep under control. A slow convergence is also a potential problem. These issues have been analyzed in [51].

The iterative method has been applied only to a limited set of problems [79], because of the aforementioned limitations. Most interestingly, one of the most well-known applications has been on the study of the time dependence of polar optical runaway of electrons in $SiO_2$ [80]. In the mid-1990s, the method was revived by Leung and Childs [81] to study spatially transient distributions in silicon. A variation of the iterative technique, the "transaction method," has been proposed by Iizuka and Fukuma [82]. In some sense, this is a method bridging the iterative technique to the direct solution methods mentioned below: The BTE is recast into the integral Chambers form, **k**-space is discretized, and the scattering term [the last factor in the integrand in Eq. (19.51) above] is treated as a set of discrete transitions (or "transactions") between states, now represented as "cells." Starting from a "zero-time" (or "last-minute," in the language of Iizuka and Fukuma) distribution, the distribution is evolved by one time step (of the order of an appropriate inverse self-scattering time) by first accounting for the "transactions" in **k**-space, and by translating each cell under the action of the field, until the required final time is reached. Once more, only homogeneous cases with azimuthal symmetry have been considered so far.

We should note that finding directly a solution of the BTE in inhomogeneous situations in three spatial dimensions and bulk transport remains computationally prohibitive. However, structures in which quantum confinement forces transport to occur in one dimension are within reach. Nanowires, nanoribbons, and ultra-narrow fins in FinFETs belong to this class. In these cases, the transport variables are fewer, just one spatial coordinate, $z$, the wavenumber $k_z$, and time. Scattering is also simplified, since a carrier can only scatter either forward or backward. Such a major reduction in the phase-space allows us to solve directly the BTE. The work by Jin et al. [83] is just an example in which transport in Si NWs has been studied in this way.

### 19.5.2 Direct Matrix Inversion

The starting BTE, Eq. (18.47), may be discretized in phase-space and time and, similarly to the Chambers formulation above, may be reduced to a large matrix-inversion problem. For a detailed description, see the text by Nag [84]. It is quite evident that, depending on the level of coarseness adopted to discretize **k**-space and real space, the matrix may easily reach impractical sizes. So far, only spatially homogeneous problems have been handled. Even in these cases, practical storage and execution speed requirements force a poor discretization of **k**-space. Perhaps, the speed and memory capabilities of future computers will render this method practical also for interesting problems.

## Problems

**19.3. (Electron Mobility in the Effective-Mass Approximation and Matthiessen's Rule)**

(a) Use the Kubo–Greenwood formula to calculate the electron mobility at 300 K, $\mu_{ac}$, limited by scattering with acoustic phonons in Si. Use Eq. (13.42), obtained in the elastic, equipartition approximation. For simplicity, use an isotropic effective mass $m^* = 0.32 m_{el}$, consider scattering only with LA phonons with $\Delta_{LA} = 10$ eV, longitudinal sound velocity $c_L = 9 \times 10^3$ m/s, and a crystal mass density $\rho = 5.33$ g/cm$^3$. Compare your result with the experimental value of about 1,500 cm$^2$/V·s. Why do you think your result is different?

(b) Repeat the same calculation to calculate the electron mobility, $\mu_{op}$, limited by nonpolar scattering with optical phonons. Use Eq. (13.41) with $(DK)_{op} = 5 \times 10^8$ eV/cm, and a phonon energy $\hbar\omega_{op}$ = 61 meV. Use *Matthiessen's rule* to calculate the total phonon-limited mobility, $\mu_{ph}$:

$$\frac{1}{\mu_{ph}} = \frac{1}{\mu_{ac}} + \frac{1}{\mu_{op}} . \tag{19.78}$$

Has the agreement with the experimental value improved?

(c) Finally, calculate the electron mobility, $\mu_{imp}$, limited by scattering with ionized impurities. Use Eq. (14.14) with Debye–Hückel screening, Eq. (11.52). Calculate also how this mobility depends on the concentration of impurities $n_D$. Above which impurity density does impurity scattering become more important than phonon scattering?

## References


1. C. Cercignani, *Ludwig Boltzmann: The Man Who Trusted Atoms* (Oxford University Press, New York, 1998)
2. C. Jacoboni, *Theory of Electron Transport in Semiconductors* (Springer, Berlin/Heidelberg, 2010)
3. K. Hess, *Advanced Theory of Semiconductor Devices* (Wiley-IEEE Press, Piscataway, NJ, 1999)
4. Y. Taur, T.H. Ning, *Fundamentals of Modern VLSI Devices* (Cambridge University Press, Cambridge, 2009)
5. W. Shockley, *Electrons and Holes in Semiconductors, with Applications to Transistor Electronics*, Bell Telephone Laboratories Series (Van Nostrand, New York, 1950)
6. S. Selberherr, *Analysis and Simulation of Semiconductor Devices* (Springer, Vienna, 1984)
7. W. Hänsch, *The Drift Diffusion Equation and Its Application to MOSFET Modeling*. Series in Computational Microelectronics. (Springer, Vienna, 1991)
8. K. Hess, J.P. Leburton, *Computational Electronics. Semiconductor Transport and Device Simulation* (Kluwer Academic, Boston, MA, 1991), p. 47
9. R. Stratton, Diffusion of hot and cold electrons in semiconductor barriers. Phys. Rev. **126**, 2002–2014 (1962)
10. W. Jones, N.H. March, *Theoretical Solid State Physics*. Interscience Monographs and Texts in Physics and Astronomy, vol. 27 (Wiley-Interscience, New York, 1973)
11. F.J. Blatt, Theory of mobility of electrons in solids, in *Solid State Physics*, ed. by F. Seitz, D. Turnbull, vol. 4 (Academic, New York, 1957), pp. 199–366
12. R. Kubo, Statistical–mechanical theory of irreversible processes. I. General theory and simple applications to magnetic and conduction problems. J. Phys. Soc. Jpn. **12**, 570 (1957)
13. D.A. Greenwood, The Boltzmann equation in the theory of electrical conduction in metals. Proc. Phys. Soc. Lond. **71**, 585 (1958)
14. D.L. Rode, Electron mobility in direct-gap polar semiconductors. Phys. Rev. B **2**, 1012 (1970)
15. S. Sze, M.K. Lee, *Semiconductor Devices: Physics and Technology* (Wiley, New York, 2012)
16. D.L. Scharfetter, H.K. Gummel, Large-signal analysis of a silicon Read diode oscillator. IEEE T. Electron Dev. **ED-16**, 64 (1969)
17. D.M. Caughey, R.E. Thomas, Carrier mobilities in silicon empirically related to doping and field. IEEE Proc. **55**, 2192 (1967)
18. S. Selberherr, W. Hänsch, M. Seavey, J.W. Slotboom, The evolution of the MINIMO mobility model. Solid State Electron. **33**, 1425 (1990)
19. K.K. Thornber, Current equations for velocity overshoot. Electron Device Lett. **EDL-3**, 69 (1982)
20. P.J. Price, On the flow equation in device simulation. J. Appl. Phys. **63**, 4718 (1988)
21. M. Artaki, Hot-electron flow in an inhomogeneous field. Appl. Phys. Lett. **52** 141 (1988)
22. W. van Roosbroeck, Theory of the flow of electrons and holes in germanium and other semiconductors. Bell Syst. Tech. J. **29**, 560 (1950)
23. M. Rudan, F. Odeh, Multi-dimensional discretization scheme for the hydrodynamic model of semiconductor devices. COMPEL **5**, 149 (1986)
24. A.M. Anile, S. Pennisi, Thermodynamic derivation of the hydrodynamic model for charge transport in semiconductors, Phys. Rev. B **46**, 13186 (1992)
25. M. Trovato, Maximum entropy principle within a total energy scheme: application to hot-carrier transport in semiconductors. Phys. Rev. B **61**, 16667 (2000)
26. M. Shur, Influence of nonuniform field distribution on frequency limits of GaAs field-effect transistors. Electron. Lett. **12**, 615 (1976)
27. S.-C. Lee, T.-W. Tang, A study of the relaxation-time model based on the Monte Carlo simulation, in *Computational Electronics. Semiconductor Transport and Device Simulation*, ed. by K. Hess, J.P. Leburton, U. Ravaioli (Kluwer Academic, Boston, MA, 1991), p. 127
28. G. Baccarani, M.R. Wordeman, An investigation of steady-state velocity overshoot in silicon. Solid State Electron. **28**, 407 (1985)
29. W. Hänsch, M. Miura-Mattausch, The hot-electron problem in small semiconductor devices, J. Appl. Phys. **60**, 650 (1986)
30. S.-C. Lee, T.-W. Tang, Transport coefficients for a silicon hydrodynamic model extracted from inhomogeneous Monte-Carlo calculations. Solid State Electron. **35**, 561 (1992)
31. A.M. Anile, G. Mascali, A two-population model for electron transport in silicon, in *Proceedings "WASCOM 2003": 12th Conference on Waves and Stability in Continuous Media* (World Scientific, Singapore, 2003)
32. H.U. Baranger, J.W. Wilkins, Ballistic structure in the electron distribution function of small semiconducting structures: general features and specific trends. Phys. Rev. B **36**, 1487 (1987)
33. K. Bløtekjær, Transport equations for electrons in two-valley semiconductors. IEEE Trans. Electron Devices **17**, 38 (1970)

34. D.J. Howarth, E.H. Sondheimer, The theory of electronic conduction in polar semi-conductors. Proc. R. Soc. Lond. A Mat. **219**, 5374 (1953)
35. J. Appel, "Polarons", in *Solid State Physics*, ed. by F. Seitz, D. Turnbull, H. Ehrenreich, vol. 21 (Academic, New York, 1968), pp. 193–391
36. S. Reggaini, M.C. Vecchi, M. Rudan, Investigation on electron and hole transport properties using the full-band spherical-harmonics expansion method. IEEE Trans. Elec. Dev. **45**, 2010 (1998)
37. S. Jin, A. Wettstein, W. Choi, F.M. Bufler, E. Lyumkis, Gate current calculations using spherical harmonic expansion of Boltzmann equation, in *Proceedings of SISPAD 2009* (IEEE, Piscataway, NJ, 2009), pp. 202–205
38. G. Matz, S.-M. Hong, C. Jungemann, Spherical harmonics expansion of the conduction band for deterministic simulation of SiGe HBTs with full band effects, in *Proceedings of the SISPAD 2010* (IEEE, Piscataway, NJ, 2010), pp. 167–170
39. H. Sung-Min, G. Matz, C. Jungemann, A deterministic Boltzmann equation solver based on a higher order spherical harmonics expansion with full-band effects. IEEE T. Electron Dev. **57**, 2390 (2010)
40. B.H. Floyd, Y.L.L. Coz, Iterative spectral solution of Boltzmann's equation for semiconductor devices, in *Computational Electronics. Semiconductor Transport and Device Simulation*, ed. by K. Hess, J.P. Leburton, U. Ravaioli (Kluwer Academic, Boston, MA, 1991), p. 93
41. S. Krishnamurthy, M. van Schilfgaarde, Eigenvalue solution to steady-state Boltzmann equation, in *Computational Electronics. Semiconductor Transport and Device Simulation*, K. Hess, J.P. Leburton, U. Ravaioli (Kluwer Academic, Boston, MA, 1991), p. 119
42. R.W. Hockney, J.W. Eastwood, *Computer Simulation Using Particles* (Taylor and Francis, New York, 1988)
43. C. Jacoboni, P. Lugli, *The Monte Carlo Method for Semiconductor Device Simulation*. Series in Computational Microelectronics (Springer, Vienna, 1989)
44. P.J. Price, Monte Carlo calculation of electron transport in solids, in *Semiconductors and Semimetals*, ed. by R.K. Willardson, A.C. Beer. Lasers, Junctions, Transported, vol. 14 (Academic, New York, 1979), pp. 249–308
45. C. Jacoboni, L. Reggiani, The Monte Carlo method for the solution of charge transport in semiconductors with applications to covalent materials. Rev. Mod. Phys. **55**, 645 (1983)
46. N. Metropolis, A.W. Rosenbluth, M.N. Rosenbluth, A.H. Teller, E. Teller, Equations of state calculations by fast computing machines. J. Chem. Phys. **21**, 1087 (1953)
47. T. Kurosawa, Monte Carlo calculation of hot electron problems, *International Conference on the Physics of Semiconductors*, Kyoto – J. Phys. Soc. Jpn **21**, Supplement (1966), pp. 424–426
48. A.D. Boardman, W. Fawcett, H.D. Rees, Monte Carlo calculation of the velocity-field relationship for gallium arsenide. Solid State Commun. **6**, 305 (1968)
49. W. Fawcett, A.D. Boardman, S. Swain, Monte Carlo determination of electron transport properties in gallium arsenide. J. Phys. Conf. Ser. **31** (9), 1963–1990 (1970)
50. R.G. Chambers, The kinetic formulation of conduction problems. Proc. Phys. Soc. Lond. **65**, 458 (1952)
51. H.D. Rees, Calculation of distribution functions by exploiting the stability of the steady state. J. Phys. Conf. Ser. **30**, 643 (1969)
52. H.D. Rees, Calculation of steady state distribution functions by exploiting stability. Phys. Lett. **26A**, 416 (1968)
53. H. Shichijo, K. Hess, Band-structure-dependent transport and impact ionization in GaAs. Phys. Rev. B **23**, 4197 (1981)
54. J.Y.-F. Tang, K. Hess, Impact ionization of electrons in silicon (steady state). J. Appl. Phys. **54**, 5139 (1983)
55. J.Y.-F. Tang, K. Hess, Theory of hot electron emission from silicon into silicon dioxide. J. Appl. Phys. **54**, 5145 (1983)
56. M.V. Fischetti, S.E. Laux, Monte Carlo analysis of electron transport in small semiconductor devices including band-structure and space-charge effects. Phys. Rev. B **38**, 9721 (1988)
57. M.V. Fischetti, Monte Carlo simulation of transport in technologically significant semiconductors of the diamond and zinc-blende structures – Part I: Homogeneous transport. IEEE T. Electron Dev. **38**, 634 (1991)
58. P.D. Yoder, J.M. Higman, J.D. Bude, K. Hess, Monte Carlo simulation of hot electron transport in Si using a unified pseudopotential description of the crystal. Semicond. Sci. Technol. **7**, B357 (1992)
59. T. Kunikiyo, M. Takenaka, Y. Kamakura, M. Yamaji, H. Mizuno, M. Morifuji, K. Taniguchi, C. Hamaguchi, A Monte Carlo simulation of anisotropic electron transport in silicon including full band structure and anisotropic impact? Ionization model. J. Appl. Phys. **75**, 297 (1994)
60. Y. Kamakura, H. Mizuno, M. Yamaji, M. Morifuji, K. Taniguchi, C. Hamaguchi, T. Kunikiyo, M. Takenaka, Impact ionization model for full band Monte Carlo simulation J. Appl. Phys. **75**, 3500 (1994)
61. C. Herring, E. Vogt, Transport and deformation-potential theory for many-valley semiconductors with anisotropic scattering. Phys. Rev. **101**, 944 (1956)
62. H. Kahn, A.W. Marshall, Methods of reducing sample size in Monte Carlo computations. J. Oper. Res. Soc. **1**, 263 (1953)
63. J.M. Hammersley, D.C. Handscomb, *Monte Carlo Methods* (Methuen, London, 1964)

64. U. Ravaioli, A. Duncan, A. Pacelli, C. Wordelman, K. Hess, Hierarchy of full band structure models for Monte Carlo simulation. VLSI Des. **6**, 1–4 (1998)
65. K. Kometer, G. Zandler, P. Vogl, Cellular automata simulation of stationary and transient high-field transport in submicron Si and GaAs devices. Semicond. Sci. Technol. **7**, B559 (1992)
66. K. Kometer, G. Zandler, P. Vogl, Lattice-gas cellular-automaton method for semiclassical transport in semiconductors. Phys. Rev. B **46**, 1382 (1992)
67. M. Saraniti, G. Zandler, G. Formicone, S. Wigger, S. Goodnick, Cellular automata simulation of nanometre-scale MOSFETs. Semicond. Sci. Technol. **13**, A177 (1998)
68. G. Zandler, A.D. Di Carlo, K. Kometer, P. Lugli, P. Vogl, E. Gornik, A comparison of Monte Carlo and cellular automata approaches for semiconductor device simulation. Electron Device Lett. **14**, 77 (1993)
69. M. Saraniti, S.M. Goodnick, Hybrid fullband cellular automaton/Monte Carlo approach for fast simulation of charge transport in semiconductors. IEEE Trans. Electron Dev. **47**, 1909 (2000)
70. B. Niclot, P. Degond, F. Poupaud, Deterministic particle simulations of the Boltzmann transport equation of semiconductors. J. Comput. Phys. **78**, 313 (1988)
71. P. Degond, F. Guyot-Delaurens, Particle simulations of the semiconductor Boltzmann equation for one-dimensional inhomogeneous structures. J. Comput. Phys. **90**, 65 (1990)
72. J.P. McKelvey, R.L. Longini, T.P. Brody, Alternative approach to the solution of added carrier transport problems in semiconductors. Phys. Rev. **123**, 51 (1961)
73. A. Das, M.S. Lundstrom, A scattering matrix approach to device simulation. Solid State Electron. **33**(10), 1299 (1990)
74. M.A. Stettler, M.S. Lundstrom, Self-consistent scattering matrix calculation of the distribution function in semiconductor devices. Appl. Phys. Lett. **60**, 2908 (1992)
75. M.A. Alam, M.A. Stettler, M.S. Lundstrom, Formulation of the Boltzmann equation in terms of scattering matrices. Solid State Electron.**36**, 263 (1993)
76. M.A. Alam, M.A. Stettler, M.S. Lundstrom, A spectral flux method for solving the Boltzmann equation. J. Appl. Phys. **73**, 4998 (1993)
77. H. Budd, Hot carriers and the path variable method, in *International Conference on the Physics of Semiconductors*, Kyoto – J. Phys. Soc. Jpn **21**, Supplement, 420 (1966)
78. W. Fawcett, Non-ohmic transport in semiconductors, in *Electrons in Crystalline Solids*, ed. by A. Salam (International Atomic Energy Agency, Vienna, 1973), p. 531
79. P.J.Price, The theory of hot electrons. IBM J. Res. Dev. **14**, 12 (1970)
80. D.K. Ferry, Electron transport and breakdown in $SiO_2$. J. Appl. Phys. **50**, 1422 (1979)
81. C.C.C. Leung, P.A. Childs, Spatially transient hot electron distributions in silicon determined from the Chambers path integral solution of the Boltzmann transport equation. Solid State Electron. **36**, 1001 (1993)
82. T. Iizuka, M. Fukuma, Carrier transport simulator for silicon based on carrier distribution function evolutions. Solid State Electron. **33**, 27 (1990)
83. S. Jin, M.V. Fischetti, Ting-wei Tang, Theoretical study of carrier transport in silicon nanowire transistors based on the multisubband Boltzmann transport equation. IEEE Trans. Electron Devices **55**, 2886 (2008)
84. B.R. Nag, *Electron Transport in Compound Semi-Conductors* – Springer Series in Solid-State Sciences, vol. 11 (Springer, Berlin, 1980)

# Appendix A
# Review of Time-Independent and Time-Dependent Perturbation Theory

## A.1 Stationary (Time-Independent) Perturbation Theory

In most applications we treat in this text we must solve the following type of problems: a physical system described by a total Hamiltonian $\widehat{H}_{\mathrm{tot}}$ which can be decomposed into a term $\widehat{H}_0$—which we can somehow treat "exactly"—and a term $\alpha\ \widehat{H}(t)$ (in general dependent on time $t$) which usually renders the problem intractable; that is, we do not know how to diagonalize the full Hamiltonian $\widehat{H}_0 + \alpha\ \widehat{H}(t)$. Almost invariably, with a few exceptions usually worth a Nobel prize, the problem is handled by limiting ourselves to the case in which the troublesome term $\alpha\ \widehat{H}(t)$ is "small" (a concept to be somehow quantified) compared to $\widehat{H}_0$. This is reflected by the statement that the dimensionless quantity $\alpha$, which we have factored out of the term $\alpha\ \widehat{H}(t)$, is small compared to 1. In this case, we treat $\alpha\ \widehat{H}(t)$ as a small "perturbation" and solve the problem by expressing the solutions (eigenvalues and eigenvectors) as series of powers of $\alpha$. Typically, we shall be satisfied with the first term, or, occasionally, the first two terms.

Let's consider first the case in which $\widehat{H}(t)$ does not depend on time. As we just said, we are assuming that the problem we must solve

$$(\widehat{H}_0 + \alpha\widehat{H})\ |\phi_n\rangle = W_n\ |\phi_n\rangle \tag{A.1}$$

is too complicated. However, we have assumed that $\alpha$ is small and that we are able to solve the "unperturbed" problem $\widehat{H}_0|\psi_n\rangle = E_n|\psi_n\rangle$. Then, one can show that, to second order in $\alpha$, the perturbed eigenvalues $W_n$ are given by

$$W_m = E_m + \alpha\, H_{mm} + \alpha^2 \sum_{n \neq m} \frac{|H_{nm}|^2}{E_m - E_n} + o(\alpha^3)\ , \tag{A.2}$$

where $H_{mn} = \langle\psi_n|\widehat{H}|\psi_m\rangle = \int \psi_n^*(\mathbf{r})\widehat{H}\psi_m(\mathbf{r})d\mathbf{r}$, and, to first order, the perturbed eigenfunctions are

$$|\phi_m\rangle = |\psi_m\rangle + \alpha \sum_{n \neq m} \frac{H_{nm}}{E_m - E_n}\ |\psi_n\rangle \tag{A.3}$$

The proof of these statements proceeds as follows: Let us expand $|\phi_n\rangle = |\psi_n\rangle + \alpha|\phi_n^{(1)}\rangle + \alpha^2|\phi_n^{(2)}\rangle + \cdots$ and $W_n = E_n + \alpha E_n^{(1)} + \alpha^2 E_n^{(2)} + \cdots$. Inserting these expressions into the initial Schrödinger



M. Fischetti, W.G. Vandenberghe, *Advanced Physics of Electron Transport in Semiconductors and Nanostructures*, Graduate Texts in Physics, DOI 10.1007/978-3-319-01101-1

equation we have

$$
\begin{aligned}
&(\widehat{H}_0 + \alpha\,\widehat{H})\,(|\psi_n\rangle + \alpha|\phi_n^{(1)}\rangle + \alpha^2|\phi_n^{(2)}\rangle + \cdots) \\
&\quad = (E_n + \alpha E_n^{(1)} + \alpha^2 E_n^{(2)} + \cdots)\,(|\psi_n\rangle + \alpha|\phi_n^{(1)}\rangle + \alpha^2|\psi_n^{(2)}\rangle + \cdots)
\end{aligned}
\tag{A.4}
$$

Equating now the terms of the same order in $\alpha$ appearing at both sides of the equation, we have

$$
\widehat{H}_0|\psi_n\rangle = E_n|\psi_n\rangle \text{ to order } \alpha^0\text{; this is the unperturbed problem} \tag{A.5}
$$

$$
\widehat{H}|\psi_n\rangle + \widehat{H}_0|\phi_n^{(1)}\rangle \;=\; E_n^{(1)}|\psi_n\rangle + E_n|\phi_n^{(1)}\rangle \text{ to order } \alpha\,. \tag{A.6}
$$

Writing now $|\phi_n^{(1)}\rangle = \sum_i a_{ni}|\psi_i\rangle$, this equation becomes

$$
\widehat{H}|\psi_n\rangle + \widehat{H}_0\sum_i a_{ni}|\psi_i\rangle \;=\; E_n^{(1)}|\psi_n\rangle + E_n\sum_i a_{ni}|\psi_i\rangle\,. \tag{A.7}
$$

Multiplying this by $\langle\psi_l|$ and integrating

$$
H_{ln} + E_l a_{nl} \;=\; E_n^{(1)}\delta_{ln} + E_n a_{nl}. \tag{A.8}
$$

For $l = n$, we obtain immediately $E_n^{(1)} = H_{nn}$. For $l \neq n$, instead, we have

$$
a_{nl} \;=\; \frac{H_{ln}}{E_n - E_l} \tag{A.9}
$$

Extending to second order, extensions that we shall need in later sections, if we equate terms of order $\alpha^2$ at the left- and right-hand sides of Eq. (A.4), we have

$$
\widehat{H}_0|\phi_n^{(2)}\rangle \;+\; \widehat{H}|\phi_n^{(1)}\rangle \;=\; E_n|\phi_n^{(2)}\rangle \;+\; E_n^{(1)}|\phi_n^{(1)}\rangle \;+\; E_n^{(2)}|\psi_n\rangle\,. \tag{A.10}
$$

Expanding now $|\phi_n^{(1)}\rangle$ and $|\phi_n^{(2)}\rangle$ over the (complete) basis of the states $|\psi_m\rangle$, we have

$$
\phi_n^{(1)}\rangle \;=\; \sum_m a_{nm}^{(1)}\,|\psi_m\rangle\,, \tag{A.11}
$$

and

$$
\phi_n^{(2)}\rangle \;=\; \sum_m a_{nm}^{(2)}\,|\psi_m\rangle\,. \tag{A.12}
$$

Inserting now these expansions into Eq. (A.10), we have

$$
\begin{aligned}
\widehat{H}_0|\sum_m a_{nm}^{(2)}\,|\psi_m\rangle \;+\; \widehat{H}\sum_m a_{nm}^{(1)}\,|\psi_m\rangle &= E_n\sum_m a_{nm}^{(2)}\,|\psi_m\rangle + E_n^{(1)}\sum_m a_{nm}^{(1)}\,|\psi_m\rangle \\
&\quad + E_n^{(2)}|\psi_n\rangle\,.
\end{aligned}
\tag{A.13}
$$

Taking now the inner product of this equation with $\langle\psi_n|$:

$$
c_{nn}^{(2)}E_n \;+\; \sum_m a_{nm}^{(1)}H_{nm} \;=\; c_{nn}^{(2)}E_n \;+\; a_{nn}^{(1)}E_n^{(1)} \;+\; E_n^{(2)}\,. \tag{A.14}
$$

The first term on the left-hand side (lhs) is equal to the first term on the right-hand side (rhs). Using now the fact that $a_{nm}^{(1)} = H_{mn}/(E_n - E_m)$ and $E_n^{(1)} = H_{nn}$, we have

$$E_n^{(2)} = \sum_{m \neq n} \frac{H_{mn} H_{nm}}{E_n - E_m} = \sum_{m \neq n} \frac{|H_{mn}|^2}{E_n - E_m} \,. \tag{A.15}$$

In typical situations, the perturbation $\widehat{H}$ breaks a symmetry of the unperturbed Hamiltonian $\widehat{H}_0$. For example, we may consider an isolated atom (so, a spherically symmetric Hamiltonian $\widehat{H}_0$) in the presence of a weak electric or magnetic field (giving rise, respectively, to the Stark and Zeeman effects). These fields may be viewed as perturbations which break the spherical symmetry. In these cases, looking at the expressions above for the second-order correction to the eigenvalue $W_m$ and the first-order corrections to the eigenstate $|\phi_m\rangle$, we see that we may run into problems. Indeed the spherical symmetry of $\widehat{H}_0$ implies that it will commute with the angular momentum operators $\widehat{\mathbf{L}}$ and $\widehat{L}_z$. Therefore, each eigenstate $|n>$ of $\widehat{H}_0$ will also be labeled by the angular momentum quantum numbers $l$ and $m$ (so, as a mater of notation specific to this problem, the $|\phi_n\rangle \rightarrow |\phi_{n,l,m}\rangle$). However, the energy $E_n$ will not depend on $l$ or $m$, thanks to the spherically symmetry. Since the perturbed eigenvalues are given by a simple extension of the equation given above:

$$W_n = E_n + \alpha \, H_{nlm,nlm} + \alpha^2 \sum_{n',l',m' \neq n,l,m} \frac{|H_{n'l'm',nlm}|^2}{E_n - E_{n'}} + o(\alpha^3) \,, \tag{A.16}$$

problems will occur in the sum whenever we consider terms with $l \neq l'$, $m \neq m'$, but $n = n'$, since the denominator will vanish. In these cases, we have to modify slightly our approach following what is known under the name of "degenerate perturbation theory." Indeed we could bypass the difficulty we have just emphasized if we could modify the matrix elements $H_{n'l'm',nlm} \rightarrow H'_{n'l'm',nlm}$, so that

$$H'_{nl'm',nlm} = \langle \phi'_{nl'm'} | H | \phi'_{nlm} \rangle = \lambda_{nlm} \, \delta_{ll'} \, \delta_{mm'} \,. \tag{A.17}$$

Clearly, this amounts to a diagonalization of the perturbation Hamiltonian $\widehat{H}$ in each eigenspace corresponding to the same eigenvalue $E_n$ of the unperturbed Hamiltonian. Thus, the states $|\phi'_{nlm}\rangle$ constitute the new basis of eigenvectors in each subspace with eigenvectors $\lambda_{nlm}$. Therefore the new expression for the second-order corrections to the energy,

$$W_{nlm} = E_n + \alpha \, \lambda_{nlm} + \alpha^2 \sum_{n' \neq n; l'm'} \frac{|H'_{n'l'm',nlm}|^2}{E_n - E_{n'}} + o(\alpha^3) \,, \tag{A.18}$$

is now free from singularities. Obviously, this diagonalization also removes the singularities appearing in the first-order corrections to the eigenstates.

## A.2 Time-Dependent Perturbation Theory

Let us consider now the more general case in which the perturbation term is time-dependent:

$$[\widehat{H}_0 + \widehat{H}(t)] \, |\Phi\rangle = i\hbar \frac{\partial |\Phi\rangle}{\partial t} \,. \tag{A.19}$$

As in the previous case, we know that $\widehat{H}_0|\psi_n\rangle = E_n|\psi_n\rangle$ and, from the time evolution of the eigenstates of the unperturbed Hamiltonian, $|\Psi_n\rangle = |\Psi_n(t)\rangle = \mathrm{e}^{-iE_nt/\hbar}|\psi_n\rangle$. Writing $|\Phi(t)\rangle = \sum a_n(t)|\Psi_n\rangle$, we have $\partial|\Phi(t)\rangle/\partial t = \sum_n[\dot{a}_n - (\mathrm{i}/\hbar)E_n a_n]|\Psi_n\rangle$. Inserting this into the original equation, we obtain

$$[\widehat{H}_0 + \widehat{H}(t)] \sum_n a_n(t)|\Psi_n\rangle = \sum_n [\mathrm{i}\hbar\dot{a}_n(t) + E_n a_n(t)]|\Psi_n\rangle \,, \tag{A.20}$$

or:

$$\sum a_n(t)(\widehat{H}_0 - E_n)|\Psi_n\rangle + \sum_n [a_n\widehat{H}(t) - \mathrm{i}\hbar\dot{a}_n]|\Psi_n\rangle = 0 \,. \tag{A.21}$$

The first term vanishes. Multiplying by $\langle\Psi_l|$, integrating, and recalling that $\langle\Psi_l(t)|\Psi_n(t)\rangle = \delta_{ln}$, we have

$$\sum_n [a_n(t)H_{ln}(t)\mathrm{e}^{\mathrm{i}(E_l-E_n)t/\hbar} - \mathrm{i}\hbar\dot{a}_n\delta_{ln}] = 0 \,. \tag{A.22}$$

Defining the quantity $\hbar\omega_{ln} = E_l - E_n$, this equation can be rewritten as

$$\mathrm{i}\hbar\dot{a}_l = \sum_n a_n(t)\, H_{ln}\, \mathrm{e}^{\mathrm{i}\omega_{ln}t} \,. \tag{A.23}$$

We now assume that at $t = 0$ the system was in the state $|\psi_j\rangle$, so that $a_j(0) = 1$, all other $a(0)$'s vanish. We also assume that the term $\widehat{H}$ is "weak," so that the coefficients $a_n$ in the rhs can be taken as constant. Then:

$$a_j(t) \approx 1 - \frac{\mathrm{i}}{\hbar}\int_0^t H_{jj}(t)\,\mathrm{d}t \to 1 - \frac{\mathrm{i}}{\hbar} H_{jj}\, t \quad \text{for constant } H \text{ turned on at } t=0 \tag{A.24}$$

$$a_k(t) \approx -\frac{\mathrm{i}}{\hbar}\int_0^t H_{kj}(t)\,\mathrm{e}^{\mathrm{i}\omega_{kj}t}\,\mathrm{d}t \to \frac{H_{kj}}{\hbar\omega_{kj}}(1 - \mathrm{e}^{\mathrm{i}\omega_{kj}t}) \quad \text{for } k \neq j \,. \tag{A.25}$$

Now, the probability of finding the system in the state $k \neq j$ at time $t$ will be $P_k(t) = |a_k(t)|^2$. The "transition rate" (that is, probability per unit time) will be $P_k(t)/t$, that is

$$\frac{1}{t}\left|\frac{H_{kj}}{\hbar\omega_{kj}}(1 - \mathrm{e}^{\mathrm{i}\omega_{kj}t})\right|^2 = \left|\frac{2H_{kj}}{E_k - E_j}\right|^2 \frac{1}{t}\sin^2\left(\frac{E_k - E_j}{2\hbar}t\right)$$
$$= |H_{kj}|^2\,\frac{4\sin^2[(E_k - E_j)t/(2\hbar)]}{[E_k - E_j]^2 t} \,. \tag{A.26}$$

One can show that

$$\lim_{t\to\infty} \frac{4\sin^2[(E_k - E_j)t/(2\hbar)]}{[E_k - E_j]^2 t} = \frac{2\pi}{\hbar}\,\delta(E_k - E_j) \,, \tag{A.27}$$

where $\delta(x)$ is the "Dirac delta function(al)" that has the property:

$$\int_{-\infty}^{\infty} f(x)\,\delta(x)\mathrm{d}x = f(0) \,. \tag{A.28}$$

Saying that the Dirac delta is a functional, and not a function, means that this expression has a rigorous mathematical meaning only under integration over $E(k)$. It is indeed by multiplying both sides of Eq. (A.27) by an arbitrary, well-behaved function $f[E(k)]$ that one proves its validity. In more correct mathematical terms, the functional $\delta$ is a linear mapping between a space of complex-valued functions and the field of complex numbers $\mathbb{C}$ such that for any function $f$ in the functional space, $\delta[f] = f(0)$. Therefore:

$$\text{transition rate}\, j \rightarrow k \; = \; \frac{2\pi}{\hbar} \, |H_{kj}|^2 \, \delta(E_k - E_j) \,, \tag{A.29}$$

which is the famous "Fermi's Golden Rule." For a harmonic perturbation, $\widehat{H}(t) = \widehat{H}(0)\mathrm{e}^{-\mathrm{i}\omega t}$, we obtain a similar expression:

$$S(j \rightarrow k) \; = \; \frac{2\pi}{\hbar} \, |H_{kj}|^2 \, \delta(E_k - E_j - \omega) \,, \tag{A.30}$$

## A.3 The First Born Approximation

The problem we wish to consider now is slightly different. We consider, as before, a problem whose solutions $|\psi_n\rangle$ are known, $\widehat{H}_0|\psi_n\rangle = E_n|\psi_n\rangle$, considering specifically free particles in three dimensions, so that $|\psi_n\rangle$ is a plane wave,

$$\langle \mathbf{r}|\psi_{\mathbf{k}}\rangle \; = \; \frac{1}{\Omega^{1/2}} \, \mathrm{e}^{\mathrm{i}\mathbf{k}\cdot\mathbf{r}} \,, \tag{A.31}$$

having replaced the "quantum number" $n$ with the wavevector $\mathbf{k}$ and having normalized the wavefunction to the volume $\Omega$. As in Eq. (A.1), we consider adding a weak, time-independent perturbing scattering perturbation $\widehat{H}(\mathbf{r})$ that is "localized"; that is, nonzero only in a small region around the origin and vanishes at infinity. Assuming that $|\psi_n\rangle$ represents a wave with energy $E = \hbar^2k^2/(2m)$ incident from infinity, we wish to calculate the scattering cross section.

The scattered wave, $|\phi\rangle$, is a solution of Eq. (A.1) that we can write as

$$\left[\frac{\hbar^2}{2m}\nabla^2 \, + \, H(\mathbf{r})\right] \phi(\mathbf{r}) \; = \; E \, \phi(\mathbf{r}) \,, \tag{A.32}$$

or

$$\left[\nabla^2 \, + \, k^2\right] \phi(\mathbf{r}) \; = \; \frac{2m}{\hbar^2} \, \phi(\mathbf{r}) \,. \tag{A.33}$$

This is the Helmholtz equation that can be solved using Green's functions:

$$\phi(\mathbf{r}) \; = \; \psi(\mathbf{r}) \; - \; \frac{2m}{\hbar^2} \int \mathrm{d}\mathbf{r}' \, \frac{\mathrm{e}^{\mathrm{i}\mathbf{k}\cdot(\mathbf{r}-\mathbf{r}')}}{4\pi|\mathbf{r}-\mathbf{r}'|} \, H(\mathbf{r}') \,. \tag{A.34}$$

Considering the scattered wave $\phi(\mathbf{r})$ very far from the region where the scattering potential is nonzero, $r >> r'$, to first order in $r'/r$ we can approximate $|\mathbf{r}-\mathbf{r}'|$ with $\mathbf{r} - \hat{\mathbf{r}}\cdot\mathbf{r}'\cdot\hat{\mathbf{r}}$, where $\hat{\mathbf{r}}$ is the unit vector along the direction of $\mathbf{r}$. Looking at the scattered wavevector along the direction of $\mathbf{r}$, that is, setting $\mathbf{k}' = k\hat{\mathbf{r}}$, we can rewrite Eq. (A.34) as

$$\phi(\mathbf{r}) = \frac{1}{\Omega^{1/2}} \left[ e^{i\mathbf{k}\cdot\mathbf{r}} + \frac{e^{i\mathbf{k}'\cdot\mathbf{r}}}{r} f(\mathbf{k},\mathbf{k}') \right] , \tag{A.35}$$

where:

$$f(\mathbf{k},\mathbf{k}') = \frac{m\Omega^{1/2}}{2\pi\hbar^2} \int d\mathbf{r}' \, e^{-i\mathbf{k}'\cdot\mathbf{r}'} H(\mathbf{r}') \, \phi(\mathbf{r}') . \tag{A.36}$$

In Eq. (A.35), the first term inside the square bracket represents the un-scattered wave, while the second term represents a wave scattered in the direction of $\mathbf{r}$ with amplitude given by the function $f(\mathbf{k},\mathbf{k}')$. Note that scattering is elastic, so $k = k'$.

Equations (A.35) and (A.36) give us a solution only formally, since the unknown function $\phi(\mathbf{r})$ appears inside the integral at the right-hand side of Eq. (A.36). The Born approximation consists in a series of approximations: at the lowest order, we assume that the scattered wave is very weak, so that as $r \to \infty$ it is negligible. Then, we can replace the unknown wave $\phi(\mathbf{r})$ with the incident wave $\psi(\mathbf{r}) = \langle \mathbf{r}|\psi\rangle$ inside the integral Eq. (A.36). The Born series of approximations can continue: calculating the scattered wave by replacing the unknown wave $\phi(\mathbf{r})$ with the incident wave $\psi(\mathbf{r})$ in Eq. (A.36) and inserting the result into Eq. (A.35), we obtain a better approximation to the scattered wave. We can now repeat the process, replacing this new expression for the scattered wave into Eq. (A.36) and repeat, every iteration being known as the $n$-th Born approximation.

The differential cross section can be obtained as follows. The incident flux is $\mathbf{j} = (\hbar/m)\mathrm{Im}[\psi^*\nabla\psi] = \hbar\mathbf{k}/(mV)$. The outgoing flux is $\mathbf{j}' = (\hbar/m)\mathrm{Im}[\phi^*\nabla\phi] = [\hbar\mathbf{k}'/(mV)]|f(\mathbf{k},\mathbf{k}')|^2/r'$. Therefore, the differential cross section per infinitesimal solid angle $d\omega$ is $d\sigma/d\omega = r'^2 j'/j = |f(\mathbf{k},\mathbf{k}')|^2$. We can write the total scattering cross section as

$$\sigma = \int_{k=k'} d\omega \, |f(\mathbf{k},\mathbf{k}')|^2 . \tag{A.37}$$

Assuming now that we have many scattering centers, say $N_{sc}$, sufficiently "diluted," so that each of them can be treated as independent of all other scattering centers, we can define a frequency, $1/\tau(\mathbf{k})$, at which incident particles with wavevector $\mathbf{k}$ will experience collisions as

$$\begin{aligned}\frac{1}{\tau(\mathbf{k})} &= jN_{sc}\sigma = N_{sc}\frac{\hbar k}{mV}\int_{k=k'} d\omega \, |f(\mathbf{k},\mathbf{k}')|^2 \\ &= \frac{N_{sc}}{\Omega}\frac{2\pi}{\hbar}\int d\mathbf{k}' \left| \int d\mathbf{r}' \, e^{-i\mathbf{k}'\cdot\mathbf{r}'} H(\mathbf{r}') \, \phi(\mathbf{r}') \right|^2 \delta[E(\mathbf{k}) - E\mathbf{k}'] ,\end{aligned} \tag{A.38}$$

where the last step is a bit "tricky": it involves use of the Dirac delta-function to express the angular integral over the surface $k' = k$ as a volume integral and a change of variables, from $\mathbf{k}$ to $E$.

Employing now the first Born approximation, this equation becomes

$$\frac{1}{\tau(\mathbf{k})} = N_{sc}\frac{2\pi}{\hbar}\int d\mathbf{k}' \, |\langle \mathbf{k}'|\widehat{H}|\mathbf{k}\rangle|^2 \, \delta[E(\mathbf{k}) - E(\mathbf{k}')] . \tag{A.39}$$

Note that this equation is identical to the Fermi's Golden Rule, Eq. (A.29), integrated over all final states. Note, however, that the physical situations considered, as well as the approximations made, to obtain these equations are very different. Fermi's Golden Rule applies to weak time-dependent perturbations. The first Born approximation applies to time-independent scattering potentials that are localized and weak enough to lead to scattered waves of very small amplitude. It can also be shown that the first Born approximation becomes more accurate when the incident particle has a large energy, typically when $k >> (2m\lambda U_0)/\hbar^2$, where $U_0$ is the strength of the scattering potential (the height of a barrier, depth of a well) and $\lambda$ is a length over which the scattering potential is localized (for example, the inverse screening parameter $\beta$ for the screened Coulomb potential of an ionized impurity). Also, Eq. (A.39) is valid only when the scattering centers are sufficiently diluted. The distinction between Fermi's Golden Rule and the first Born approximation is seldom made clear in the literature.

# Appendix B
# Local Empirical Pseudopotential Parameters

## B.1 Overview

In Chap. 7 examples are given regarding the use of the supercell method to study Si inversion layers and nanowires, III–V hetero-channels, graphene sheets, and C nanotubes (CNTs). These structures have been treated by employing several forms of local empirical pseudopotentials presented in the literature. Since lack of "portability" is one of the major drawbacks of the empirical nature of the technique, care must be taken in employing the model pseudopotential best suited to the system of interest. In this appendix we list the form of the pseudopotentials employed. Tables and portions of the text are reproduced from [1].

## B.2 Silicon, Germanium (and Hydrogen) I

Zhang et al. [2] have employed the following empirical form for the Si local pseudopotential:

$$V^{(\mathrm{Si})}(q) = \frac{b_1(q^2 - b_2)}{b_3 \mathrm{e}^{b_4 q^2} - 1}\,, \tag{B.1}$$

which results in the correct workfunction, so that it can be used to study free-standing Si layers bounded by vacuum. The surface states which may arise from the dangling bonds can be removed by terminating them with H, whose local empirical pseudopotential is given by the expression [3]:

$$V^{(\mathrm{H})}(q) = \begin{cases} b_0 + b_1 q + b_2 q^2 + b_3 q^3 & \text{for } (q \leq 2) \\ b_{-1}/q + b_{-2}/q^2 + b_{-3}/q^3 + b_{-4}/q^4 & \text{for } (q > 2) \end{cases}\,. \tag{B.2}$$

A form identical to Eq. (B.1) has been employed by Reboredo and Zunger [4] for Ge. Table B.1 lists the parameters appearing in Eqs. (B.1) and (B.2) as well as the cutoff energy $E_{\mathrm{cutoff}}$ which determines the number of plane waves (i.e., of **G**-vectors) used via the relation $\hbar^2 G^2/(2m) \leq E_{\mathrm{cutoff}}$. Note that in this and in the following tables the pseudopotential is assumed to be normalized to the atomic volume in the relaxed crystal and all parameters are expressed in Rydberg atomic units (i.e., energy in Rydberg, length in Bohr units).



M. Fischetti, W.G. Vandenberghe, *Advanced Physics of Electron Transport in Semiconductors and Nanostructures*, Graduate Texts in Physics, DOI 10.1007/978-3-319-01101-1

**Table B.1** Local pseudopotential parameters for Si and H from Zunger's group [2–4]

| | $b_0$ | $b_1$ | $b_2$ | $b_3$ | $b_4$ | $b_{-1}$ | $b_{-2}$ | $b_{-3}$ | $b_{-4}$ | $E_{\text{cutoff}}$ |
|---|---|---|---|---|---|---|---|---|---|---|
| Ge | | 0.584954 | 2.344131 | 3.24496 | 0.64970 | | | | | 5 |
| Si | | 0.53706 | 2.19104 | 2.05716 | 0.48716 | | | | | 4.5 |
| H | −0.2832 | 0.019604 | 0.12462 | −0.03790 | | 0.05796 | −0.7754 | 1.9384 | −2.044 | 8 |

The Ge and Si pseudopotentials are normalized to the atomic volume of the Ge and Si ions in Ge and Si crystals. The H pseudopotential is also normalized to the Si atomic volume

**Table B.2** Local pseudopotential parameters for Si and Ge from the Bell-Labs group [5]

| | $a_1$ | $a_2$ | $a_3$ | $a_4$ | $a_5$ | $a_6$ | $E_{\text{cutoff}}$ |
|---|---|---|---|---|---|---|---|
| Si | $106.0686/\Omega_{\text{Si}}$ | 2.2278 | 0.6060 | −1.9720 | 5.0 | 0.3 | 5 |
| Ge | $54.4512/\Omega_{\text{Ge}}$ | 2.3592 | 0.7400 | −0.3800 | 5.0 | 0.3 | 5 |

## B.3 Silicon and Germanium II

For the study of systems containing both Si and Ge atoms, Friedel et al. [5] have employed the following form for the Fourier transform of the local pseudopotential:

$$V^{(\text{Si,Ge})}(q) = \frac{a_1(q^2-a_2)}{e^{a_3(q^2-a_4)}+1}\,\frac{1}{2}\left[\tanh\left(\frac{a_5-q^2}{a_6}\right)+1\right], \tag{B.3}$$

similar to Eq. (B.1) but with an additional large-$q$ cutoff (given by the hyperbolic-tangent factor) within square brackets above. Using the parameters listed in Table B.2 one can obtain the correct band alignment of Si and Ge, but not the correct workfunction.

## B.4 Carbon and Hydrogen

The band structure of bulk C has been studied by Kurokawa et al. [6] using the following interpolation for the local pseudopotential, fully equivalent to the pseudopotential in Eq. (B.3) apart from the large-$q$ cutoff:

$$V^{(\text{C})}(q) = \frac{b_1(b_3q^2-b_2)}{e^{b_3q^2-b_4}+1}, \tag{B.4}$$

and, for H a rescaling of Eq. (B.2). These parameters have been used to study the amorphous C–H-based crystallites, but the C pseudopotential yields—quite surprisingly—reasonable results also for graphene and CNTs, as shown below. Note that, like in the cases considered above of Si and III–V compound semiconductors, several other empirical pseudopotentials have been given for C in the diamond structure, for both local [7] and nonlocal [8] models. However, Kurokawa et al. provide an expression which is both suitable for interpolation as well as yielding the correct workfunction for C in the diamond structure. In addition, in combination with the associated local empirical pseudopotential for H, also given by Kurokawa and co-workers, the electronic structure of *trans*-polyacetylene is correctly reproduced. This is significant, since these C pseudopotentials appear to reproduce both the hybridized $sp^3$ tetragonal diamond bonding coordination as well as the planar $sp^2$ (with $p_z$ out-of-plane orbitals hybridizing into $\pi^*$ orbitals) coordination found in graphene, nanoribbons, and, with the addition of curvature effects we discuss in Chap. 7, also nanotubes. Using a cutoff energy of 25 Ry the plane-wave method used here yields an indirect $\Gamma'_{25}$-$\Delta_1$ gap of about 5.0 eV and values consistent with Kurokawa's results for the direct $\Gamma'_{25}$-$\Gamma_2$, $X_4$-$X_1$, and $L'_3$-$L_3$ gaps. For numerical convenience we have employed a cutoff energy of 15 Ry which yields similar values for the direct gap, but a slightly smaller value for the indirect $\Gamma'_{25}$-$\Delta_1$ gap of 4.2 eV. We have spot-checked our results for graphene,

**Table B.3** Local pseudopotential parameters for C and H from Kurokawa et al. [6]

| | $b_0$ | $b_1$ | $b_2$ | $b_3$ | $b_4$ | $b_{-1}$ | $b_{-2}$ | $b_{-3}$ | $b_{-4}$ | $E_{\text{cutoff}}$ |
|---|---|---|---|---|---|---|---|---|---|---|
| C | | 1.781 | 1.424 | 0.354 | 0.938 | | | | | 15 |
| H | −0.397 | 0.02759 | 0.1754 | −0.0531 | | 0.0811 | −1.086 | 2.71 | −2.86 | 15 |

The pseudopotentials are normalized to atomic volume of C in diamond

**Table B.4** Local pseudopotential parameters for C from Mayer [9] normalized to the atomic volume of C in diamond

| | $A_1$ | $A_2$ | $A_3$ | $a_1$ | $a_2$ | $a_3$ | $b_1$ | $b_2$ | $b_3$ | $E_{\text{cutoff}}$ |
|---|---|---|---|---|---|---|---|---|---|---|
| C | 0.7796 | 2.1837 | −7.2698 | 0.12126 | 1.9148 | 0.60078 | −2.68574 | −0.11989 | 2.27101 | 15 |

nanoribbons, and CNTs employing a larger cutoff of 25 Ry without finding any significant difference. One such example will be shown below. Table B.3 lists the parameters appearing in the expressions above. One-electron empirical local pseudopotential for C fitted to the band structure of graphene has been proposed by Mayer [9]. The real-space form of the potential is given by the sum of three Gaussians:

$$V^{(\mathrm{C})}(r) = -\sum_{j=1}^{3} A_j \, \mathrm{e}^{-a_j r^2} , \tag{B.5}$$

so that its Fourier-transformed expression will be

$$V^{(\mathrm{C})}(q) = \sum_{j=1}^{3} b_j \, \mathrm{e}^{-q^2/(4a_j)} , \tag{B.6}$$

with $b_j = -(\pi/a_j)^{3/2} A_j$. Table B.4 lists the relevant parameters. Note that these are *one-electron* empirical forms which would provide only one valence band per diamond primitive cell, so they may be suitable for transport, but not to handle the physics of the nanostructures of interest.

## B.5 III–V Compound Semiconductors

Pseudopotentials for many group-III and group-V elements have been provided by Zunger's group: as, Al, and Ga in [10, 11], P in [11], and In in [12]. The form chosen for the $q$-dependence of the ionic pseudopotential is the sum of four Gaussians as follows:

$$V^{(\mathrm{III-V})}(q) = \sum_{j=1}^{4} a_j \mathrm{e}^{-b_j(q-c_j)^2} [1 - d_j \mathrm{e}^{-f_j q^2}] , \tag{B.7}$$

where we have added the factor $[1 - d_j \mathrm{e}^{-f_j q^2}]$ to further adjust the band alignments among $In_{0.53}Ga_{0.47}As$, InP, and $Al_{0.48}In_{0.52}As$. The parameters are chosen in order to fit the bulk band structure, of course, but also to provide the correct band alignment among the various elements, thus making it possible to study the electronic states of hetero-structures. Table B.5 lists the values of the parameters entering Eq. (B.7), as always in atomic units, and assuming that the ionic pseudopotentials are normalized to the volume of the ion in the relaxed crystal. Note that these parameters should be used with small cutoff energies (5 Ry) and with spin–orbit interaction. When the spin–orbit interaction

**Table B.5** Local pseudopotential parameters for group-III and group-V elements from Zunger's group [10–12]

| | Al | Ga | In | As (in GaAs and InAs) | As (in AlAs) | P | N |
|---|---|---|---|---|---|---|---|
| $\Omega_a$ | 111.3 | 131.4 | 169.75 | 145.2 | 145.2 | 139.2 | 75.0 |
| $a_1$ | −1.32712 | −1.24498 | −1.13393 | −1.05821 | −1.10411 | −1.02821 | −0.2262800 |
| $a_2$ | 0.158114 | 0.0464357 | 0.046658 | −0.0434312 | −0.00368081 | −0.049431 | −0.9268300 |
| $a_3$ | 0.0601648 | 0.0366517 | 0.0322835 | 0.10569 | 0.0921512 | 0.11769 | 0.2372800 |
| $a_4$ | 0.0174946 | −0.0133385 | −0.0123714 | −0.0021763 | 0.0174946 | 0.012176 | 0.208300 |
| $b_1$ | 1.59819 | 1.52748 | 2.15 | 0.959327 | 0.972439 | 0.870327 | 1.816013 |
| $b_1^*$ | | 1.51748[a] | 1.70 | 0.949327[a] | | 0.860327[b] | |
| $b_2$ | 2.10827 | 0.574047 | 0.15 | 2.94679 | 5.50601 | 3.18679 | 0.918842 |
| $b_3$ | 0.527745 | 0.959082 | 0.57 | 0.820922 | 1.18638 | 0.470922 | 0.909593 |
| $b_4$ | 11.2708 | 11.2708 | 13.75 | 6.53145 | 6.53147 | 5.81145 | 13.51921 |
| $c_1$ | 0.0 | 0.0 | 0.0 | 0.0 | 0.0 | 0.0 | 0.0 |
| $c_2$ | 1.77453 | 2.01935 | 2.25 | 0.85164 | 1.22845 | 0.88964 | 0.85177 |
| $c_3$ | 2.5955 | 2.09782 | 2.45 | 1.22436 | 1.35897 | 1.02836 | 1.55591 |
| $c_4$ | 2.93581 | 2.93581 | 2.73 | 2.46808 | 2.46793 | 2.4008 | 2.62852 |
| $d_1$ | 0.08 | 0.0 | 0.0 | 0.0 | 0.0 | 0.04 | - |
| $f_1$ | 5.00 | 0.0 | 0.0 | 0.0 | 0.0 | 5.00 | - |

The correction factor $b_1^*$ to be used in absence of spin–orbit interaction has been determined in this work. A cutoff energy $E_{cutoff}$ of 5 Ry should be employed when using these parameters. The band-alignment parameters $d_j$ and $f_i$ are all zero, except for the parameters $d_1$ and $f_1$ for Al and P shown in the last rows. The pseudopotentials are normalized to the atomic volumes $\Omega_a$ indicated (in atomic units) in the first row

[a]Only in GaAs

[b]Only in AlP and InP, not in GaP

**Table B.6** Local pseudopotential parameters for the InGaAs system from Zunger's group [13]

| | Ga | In | As |
|---|---|---|---|
| $b_1$ | 55.738 | 107.76 | 48.54 |
| $b_2$ | 2.793 | 1.915 | 2.536 |
| $b_3$ | 3.209 | 3.460 | 1.463 |
| $b_4$ | 0.238 | 0.414 | 0.4937 |

A cutoff energy $E_{cutoff}$ of 5 Ry should be employed when using these parameters. The pseudopotentials shown here have been already multiplied by the atomic volume

is ignored, small corrections should be made to the width of the deeper Gaussian (the parameter $b_1$), resulting in the parameter indicated by $b_1^*$ in the Table B.5. Alternative forms for the empirical pseudopotentials for In, Ga, and As in the InAs, GaAs, and InGaAs system have been provided by Kim et al. [13]. The form of the pseudopotentials as a function of $q$ is identical to Eq. (B.1) with the addition of strain-related terms. The coefficients are given in Table B.6. These pseudopotentials have not been used here.

## References

1. M.V. Fischetti, B. Fu, S. Narayanan, J. Kim, Semiclassical and quantum electronic transport in nanometer-scale structures: empirical pseudopotential band structure, Monte Carlo simulations and Pauli master equation, in *Nano-Electronic Devices: Semiclassical and Quantum Transport Modeling*, ed. by D. Vasileska, S.M. Goodnick (Springer, New York, 2011), pp. 183–247

2. S.B. Zhang, C.-Y. Yeh, A. Zunger, Electronic structure of semiconductor quantum films. Phys. Rev. B **48**, 11204 (1993)
3. L.-W. Wang, A. Zunger, electronic structure pseudopotential calculations of large (approx. 1000 Atoms) Si quantum dots. J. Phys. Chem. **98**, 2158 (1994)
4. F.A. Reboredo, A. Zunger, Surface-passivation-induced optical changes in Ge quantum dots. Phys. Rev. B **63**, 235314 (2001)
5. P. Friedel, M.S. Hybertsen, M. Schlüter, Local empirical pseudopotential approach to the optical properties of Si/Ge superlattices. Phys. Rev. B **39**, 7974 (1989)
6. Y. Kurokawa, S. Nomura, T. Takemori, Y. Aoyagi, Large-scale calculation of optical dielectric functions of diamond nanocrystallites. Phys. Rev. B **61**, 12616 (2000).
7. W. Saslow, T.K. Bergstresser, M.L. Cohen, Band structure and optical properties of diamond. Phys. Rev. Lett. **16**, 354 (1966).
8. L.H. Hemstreet Jr., C.Y. Fong, M.L. Cohen, Calculation of the band structure and optical constants of diamond using the nonlocal-pseudopotential method. Phys. Rev. B **2**, 2054 (1970)
9. A. Mayer, Band structure and transport properties of carbon nanotubes using a local pseudopotential and a transfer-matrix technique. Carbon **42**, 2057 (2004)
10. K. Mäder, A. Zunger, Empirical atomic pseudopotentials for AlAs/GaAs superlattices, alloys, and nanostructures. Phys. Rev. B **50**, 17393 (1994)
11. L. Bellaiche, S.-H. Wei, A. Zunger, Localization and percolation in semiconductor alloys: GaAsN vs GaAsP. Phys. Rev. B **54**, 17568 (1996)
12. L. Bellaiche, L.-W. Wang, S.-H. Wei, A. Zunger, Resonant hole localization and anomalous optical bowing in InGaN alloys. Appl. Phys. Lett. **74**, 1842 (1999)
13. J. Kim, L.-W. Wang, A. Zunger, Comparison of the electronic structure of InAs/GaAs pyramidal quantum dots with different facet orientations. Phys. Rev. B **57** R9408 (1998)

# Appendix C
# Quick Overview of Special Relativity and the "Minimal" Electron–Photon Coupling

## C.1 Principles of Special Relativity

Here we shall discuss very succinctly the basic principles of special relativity. There are two reasons for doing this.

First, when introducing optical devices, we saw that the electron–photon perturbation Hamiltonian is given by [see Chap. 16, Eq. (16.4), for example]:

$$\widehat{H}_{\text{phot}} = \frac{\text{i}e\hbar}{m_{\text{el}}c}\mathbf{A}\cdot\nabla\,. \tag{C.1}$$

(The speed of light, $c$, is within parentheses: it appears when using Gaussian units, it does not when using SI units. In this appendix, we use Gaussian units, more commonly used in the physics literature dealing with the electromagnetic field.) This interaction term can be obtained by starting from the free-electron Hamiltonian $\mathbf{p}^2/(2m)$ and replacing the electron momentum $\widehat{\mathbf{p}}$ with $\widehat{\mathbf{p}} - e\widehat{\mathbf{A}}/c$. (The factor of $c$ in the denominators appears when using Gaussian units, which are more convenient in this context and we shall use them here. Note also that here $e$ represents the charge of a particle *with sign*.) The reason behind this substitution relies on some basic principles of special relativity and it is worth understanding how this comes about.

The second reason stems from the recent popularity of graphene as a "promising" material for nanoelectronics applications. The band structure of graphene shows a unique feature: at the Fermi level the dispersions of the valence and conduction bands cross as straight lines. This $E(k)$ relation resembles the dispersion of relativistic massless particles and it has triggered interest in exploiting what is known about relativistic massless spin-1/2 particles in order to understand the electronic properties of graphene. Such particles are described by a spin-1/2 relativistic wave equation (the Dirac equation for zero-mass spin-1/2 Fermions), so that it is interesting to see how the Schrödinger equation can be generalized to satisfy the principle of relativistic invariance.

### *C.1.1 Galilean Invariance and Maxwell's Equations*

As soon as Maxwell's equation were formulated, it became clear that there was a major difference with respect to Newton's law. Let's start with Newton's second law,



M. Fischetti, W.G. Vandenberghe, *Advanced Physics of Electron Transport in Semiconductors and Nanostructures*, Graduate Texts in Physics, DOI 10.1007/978-3-319-01101-1

$$\mathbf{F} = m\mathbf{a} \quad \text{or} \quad \mathbf{F} = m\frac{\mathrm{d}^2\mathbf{x}}{\mathrm{d}t^2} . \tag{C.2}$$

and consider this same equation as could be written by somebody (called an "observer") which is moving with respect to us with uniform velocity $\mathbf{u}$. Let's assume that we and the other observer use a reference frame with parallel $x$-, $y$-, and $z$-axes, that the other observer moves along the $x$-axis, so that $\mathbf{u} = (u, 0, 0)$, and that the origin of the two reference frames coincides at a fixed instant in time which we take $t = 0$. Then, calling $(x, y, z)$ our frame and $(x', y', z')$ the other observer's frame, an object located at a point $(x, y, z)$ in our frame will be located at a point $(x', y', z')$ in the observer's frame, such that

$$\begin{aligned} x' &= x - ut \\ y' &= y \\ z' &= z \end{aligned} \quad . \tag{C.3}$$

Therefore, expressing Newton's second law in the "primed" frame,

$$\mathbf{F} = m\frac{\mathrm{d}^2\mathbf{x}'}{\mathrm{d}t^2} = m\frac{\mathrm{d}^2(\mathbf{x} - \mathbf{u}t)}{\mathrm{d}t^2} = m\frac{\mathrm{d}^2\mathbf{x}}{\mathrm{d}t^2} - \frac{\mathrm{d}\mathbf{u}}{\mathrm{d}t} = m\frac{\mathrm{d}^2\mathbf{x}}{\mathrm{d}t^2} . \tag{C.4}$$

In other words, Newton's law retains the same algebraic form in all frames which are moving with uniform velocity with respect to our frame. This principle can be generalized by saying that the laws of mechanics are valid in all "inertial frames." An observer, by performing experiments, cannot tell whether he/she is moving with respect to other inertial frames. This is the principle of Galilean relativity, since it was proposed (noted, discovered, invented?...the choice of a term is a matter of deep philosophical discussions) by Galileo. Consider now Maxwell's equations. For simplicity, let's just consider a wave equation:

$$\left(\nabla^2 - \frac{1}{c^2}\frac{\partial^2}{\partial t^2}\right)\psi = 0 . \tag{C.5}$$

Let's apply the transformation Eq. (C.3). Since in the "unprimed" frame $\psi(x, y, z, t) = \psi(x' + ut, y, z, t)$, so that $\partial\psi/\partial x' = \partial\psi/\partial x + (1/u)\partial\psi/\partial t$, we find

$$\left(\nabla^2 - \frac{1}{c^2}\frac{\partial^2}{\partial t^2} - \frac{2u}{c^2}\frac{\partial^2}{\partial x \partial t} - \frac{u^2}{c^2}\frac{\partial^2}{\partial x^2}\right)\psi = 0 . \tag{C.6}$$

What a mess! The form of the equation has been completely altered by the transformation! In hindsight, we already knew it had to be so: after all, magnetic fields caused by moving charges must disappear when we use a frame in which the charges are at rest. Therefore, the **E** and **B** fields do not transform correctly under the Galilean transformation Eq. (C.1). Moreover, the Lorentz force depends explicitly on the velocity of the particle, so that the form of the equation will differ in a different inertial frame. Historically, this is also related to the difficulty of understanding electromagnetic waves: sound waves are oscillations of the medium in which they propagate. But electromagnetic waves are oscillations of what?

In order to fix the situation we have three alternatives we can choose from:

1. Maxwell's equations are wrong. The correct equations, yet to be discovered, are invariant under Galilean transformations.

2. Galilean invariance is valid for mechanics, not for electromagnetism. This is the historical solution before Einstein: the "ether" determines the existence of the "absolute frame" in which the ether is at rest and Maxwell's equations hold.
3. Galilean invariance is wrong. There is a more general invariance—yet to be discovered—which preserves the form of Maxwell's equation. Classical mechanics is incorrect and must be reformulated so that it is invariant under this new transformation.

Having to choose between abandoning Maxwell (option 1) or Newton (option 3), physicists chose the easier option 2. Einstein, instead, decided to follow the third option, guided by two postulates:

1. *Postulate of relativity:* All physical laws must "look the same" in all frames moving with uniform velocity with respect to each other.
2. *Postulate of the constancy of the speed of light:* The speed of light is the same (numerically the same!) independent of the velocity of the observer or of its source. This stems logically from the Michelson–Morley experiment of 1887, but the result could have been explained "saving" ether and using the Lorentz–Fitzgerald contractions.

Armed with these postulates, Einstein set out to build a new set of transformations between inertial frames. Maxwell's equation is now invariant under this new set of transformations, but Newtonian mechanics has to be modified: if two frames move at a relative speed much smaller than the speed of light, the "new" transformations approach the usual Galilean transformation and Newton is approximately correct. But for relative velocities approaching the speed of light, the laws of mechanics deviate enormously from Newton's laws.
It is impossible to pay justice to special relativity in such a short time. We shall only discuss those few concepts which are required to answer our original question of why we perform the substitution $\widehat{\mathbf{p}} \rightarrow \widehat{\mathbf{p}} - (e/c)\widehat{\mathbf{A}}$.

### C.1.2 Lorentz Transformations

Consider the same two frames ("primed" $K'$ and "unprimed" $K$ frames) considered above. Assume now that at $t = 0$ (the time at which the origins of the two frames coincided... at least when looking at our clock...!) a ray of light was emitted from the origin. Since light travels at the same speed in both frames, we must have for the wavefronts of the emitted light:

$$ct^2 - x^2 - y^2 - z^2 = 0 \quad \text{and} \quad ct'^2 - x'^2 - y'^2 - z'^2 = 0\,, \tag{C.7}$$

so that, assuming that space–time is isotropic and homogeneous,

$$ct^2 - x^2 - y^2 - z^2 = \lambda(u)[ct'^2 - x'^2 - y'^2 - z'^2]\,. \tag{C.8}$$

The function $\lambda(u)$ is a possible velocity-dependent change of scale between the two frames. However, since going from $K$ to $K'$ must involve the same transformation as going from $K'$ to $K$ with a sign-flip for $u$, we must have $\lambda = 1$. Using the notation $x_0 = ct$, $x_1 = x$, $x_2 = y$, and $x_3 = z$, Eq. (C.8) is satisfied if:

$$\begin{aligned} x'_0 &= \gamma(x_0 - \beta x_1) \\ x'_1 &= \gamma(x_1 - \beta x_0) \\ x'_2 &= x_2 \\ x'_3 &= x_3 \end{aligned}\,, \tag{C.9}$$

where

$$\beta = \frac{u}{c} \quad \text{and} \quad \gamma = (1-\beta^2)^{-1/2} \,. \tag{C.10}$$

The transformations, Eq. (C.9), are called “Lorentz transformations.” Obviously the inverse transformations read

$$\begin{aligned} x_0 &= \gamma(x_0' + \beta x_1') \\ x_1 &= \gamma(x_1' + \beta x_0') \\ x_2 &= x_2' \\ x_3 &= x_3' \end{aligned} \,, \tag{C.11}$$

Note that, unlike the Galilean transformations, now time and space transform together: simultaneous events in one frame will not be simultaneous in another frame. Indeed consider two events $(ct_1, \mathbf{x}_1)$ and $(ct_2, \mathbf{x}_2)$. Thanks to Eq. (C.8), their “distance”

$$s_{12} = c^2(t_1 - t_2)^2 - |\mathbf{x}_1 - \mathbf{x}_2|^2 \tag{C.12}$$

is an “invariant” (that is, it’s the same in all inertial frames). If $s_{12} > 0$, the separation between the events is said to be “time-like”: it is always possible to find a transformation (actually with $\beta = |\mathbf{x}_1 - \mathbf{x}_2|/(c|t_1 - t_2|)$) such that in the transformed frame the two events are at the same spatial location, separated only by time. If $s_{12} < 0$, the separation between the events is said to be “space-like”: it is always possible to find a Lorentz transformation such that in the transformed frame the two events are simultaneous, separated only spatially. If, finally, $s_{12} = 0$, the separation is said to be “light-like”: one event lies on the “light-cone” of the other.

### *C.1.3 Proper Time and Time Dilatation*

Since “time” has become an observer-dependent quantity, when dealing with moving particles it is convenient to consider the time in the frame in which the particle is at rest. If $\mathbf{v}(t)$ is the velocity of the moving particle in “our” frame, consider the invariant:

$$\begin{aligned} ds^2 &= dx_\mu dx^\mu = c^2 dt^2 - dx^2 - dy^2 - dz^2 \\ &= c^2 dt^2 - v^2 dt^2 = c^2(1 - v^2/c^2) dt^2 = c^2 dt^2/\gamma^2 \,. \end{aligned} \tag{C.13}$$

Since this quantity is invariant (that is, it is numerically the same in all inertial frames), in the frame in which the particle is at rest it will be $ds^2 = c^2 d\tau^2$, where $\tau$ is the time in that frame. This time is called “proper time.” If the particle travels over a time interval $\tau_2 - \tau_1$ in its proper time, as seen by us the time interval will stretch to the interval $t_2 - t_1$ obtained by integrating Eq. (C.13) along the particle trajectory:

$$t_2 - t_1 = \int_{\tau_1}^{\tau_2} d\tau \, \gamma(\tau) = \int_{\tau_1}^{\tau_2} \frac{d\tau}{\sqrt{1 - v(\tau)^2/c^2}} \,. \tag{C.14}$$

Since $\gamma > 1$, the time interval we observe, $t_2 - t_1$, is longer than the proper time interval $\tau_2 - \tau_1$. This has been experimentally verified: the $\mu$-mesons (actually, leptons) produced by cosmic-ray hits in the upper atmosphere, often reach the ground. Since the lifetime of the $\mu$-meson is about 2.2 μs, even at the speed of light the particle could not travel more than about 660 m before decaying. Yet, they can easily be detected after having traveled distances more than two orders of magnitude longer (the thickness of the atmosphere, of the order of $10^5$ m). This is because their lifetime, as observed by us, is stretched enormously, as these particles travel at speeds approaching the speed of light.

### *C.1.4 Lorentz Contraction*

Consider a rod of length $L'$ at rest in the $K'$ frame. Let the ends of the rod be at $x' = x'_1$ and $x' = x'_2$, so that $L' = x'_2 - x'_1$. What is the length of the rod in our $K$ frame? By the Lorentz transformations, Eq. (C.9) (which we must use since when we measure the length of the rod we measure its ends at the same time in our frame), we have

$$L = x_2 - x_1 = \frac{1}{\gamma}(x'_2 - x'_1) = \frac{L'}{\gamma} < L' . \tag{C.15}$$

The rod in our frame appears shorter than in its rest frame. This is the Lorentz–Fitzgerald contraction which was postulated (without proof or arguments behind) in order to explain the Michelson–Morley experiment.

### *C.1.5 Addition of Velocities*

Let's consider the Lorentz transformation

$$\begin{aligned} \mathrm{d}x_0 &= \gamma(\mathrm{d}x'_0 + \beta \mathrm{d}x'_1) \\ \mathrm{d}x_1 &= \gamma(\mathrm{d}x'_1 + \beta \mathrm{d}x'_0) \\ \mathrm{d}x_2 &= \mathrm{d}x'_2 \\ \mathrm{d}x_3 &= \mathrm{d}x'_3 \end{aligned} . \tag{C.16}$$

In our $K$ frame, the velocity $v = c\mathrm{d}x_1/\mathrm{d}x_0$ of a particle moving with velocity $v' = c\mathrm{d}x'_1/\mathrm{d}x'_0$ in the $K'$ frame will be

$$v = c\frac{\mathrm{d}x_1}{\mathrm{d}x_0} = c\frac{\mathrm{d}x'_1 + \beta \mathrm{d}x'_0}{\mathrm{d}x'_0 + \beta \mathrm{d}x'_1} = c\frac{\mathrm{d}x'_0(\mathrm{d}x'_1/\mathrm{d}x'_0 + \beta)}{\mathrm{d}x'_0(1 + \beta \mathrm{d}x'_1/\mathrm{d}x'_0)} = \frac{v' + u}{1 + v'u/c^2} . \tag{C.17}$$

Note how the particle velocity $v'$ and the frame velocity $u$ add, as seen by us: In the limit of small velocities, $v \approx v' + u$, as usual. But as either $v'$ or $u$ approaches $c$, the denominator grows and the sum-velocity $v$ as seen by us cannot exceed $c$. This is consistent with the second postulate.
Note that the 4-vector

$$U_\mu = (\gamma c, \gamma \mathbf{u}) \tag{C.18}$$

transforms like the coordinate vector $x_\mu$.
*Note the we shall use the contravariant (e.g., $x_\mu$) and covariant (e.g., $x^\mu = g^{\mu\nu}x_\nu$) notation. Using the metric $(+,-,-,-)$ for $\mu$ = 0,3 we have $x_\mu = (x_0, \mathbf{x})$ but $x^\mu = (x_0, -\mathbf{x})$). Thus $x_\mu x^\mu = x_0^2 - \mathbf{x} \cdot \mathbf{x}$.*

### *C.1.6 4-Momentum*

In classical mechanics the momentum and energy of a particle are

$$\begin{aligned} E &= E(0) + \tfrac{1}{2}mu^2 \\ \mathbf{p} &= m\mathbf{u} \end{aligned} . \tag{C.19}$$

(We warn the reader that in this Appendix the total energy of a particle is indicated by the symbol $E$, while the electric field is indicated by the vector (bold) symbol $\mathbf{E}$. The magnitude of the electric field is never used, so no confusion should arise.) The term $E(0)$ is a constant which refers to the rest energy of the particle. It is usually ignored in non-relativistic discussions.
In order to generalize these concepts, we can start by finding arbitrary functions of the velocity, $\mathscr{E}(u)$ and $\mathscr{M}(u)$, such that

$$\begin{aligned} E &= \mathscr{E}(u) \\ \mathbf{p} &= \mathscr{M}(u)\mathbf{u} \end{aligned} , \tag{C.20}$$

with the constraints [dictated by the fact that we want to recover Eq. (C.19) in the limit $u \to 0$]:

$$\begin{aligned} \frac{\partial \mathscr{E}}{\partial u^2}(0) &= \frac{m}{2} \\ \mathscr{M}(0) &= m \end{aligned} . \tag{C.21}$$

The general expressions for the functions $\mathscr{E}$ and $\mathscr{M}$ can be obtained by analyzing the elastic collision of two identical particles and require momentum and energy conservation in two inertial frames $K$ and $K'$. We'll skip the derivation (see Jackson, *Classical Electrodynamics*, Sect. 11.5) and simply state the result: the general form for these two functions consistent with the two postulates, and energy and momentum conservation is

$$\begin{aligned} \mathscr{E}(u) &= \gamma mc^2 \\ \mathscr{M}(u) &= \gamma m \end{aligned} . \tag{C.22}$$

Note that in the limit $u \to 0$, $E = \gamma mc^2 \to mc^2 + mu^2/2$, which is the classical result with a rest energy $mc^2$.
From this we can define an energy–momentum 4-vector, $p^\mu$:

$$p_\mu = (\gamma mc, \gamma m\mathbf{u}) = (E/c, \gamma m\mathbf{u}) = mU_\mu , \tag{C.23}$$

having used Eq. (C.18) in the last step.
The invariant length of the energy–momentum 4-vector is

$$p^\mu p_\mu = p_0^2 - |\mathbf{p}|^2 = E^2/c^2 - \gamma^2 m^2 u^2 = \gamma^2(m^2c^2 - m^2u^2) = m^2c^2 . \tag{C.24}$$

Finally, from this expression we can write the energy $E$ of a particle as

$$E = \sqrt{c^2p^2 + m^2c^4} . \tag{C.25}$$

### C.1.7 Lorentz Force

Our goal now is to re-express the Lorentz force (recall that we are using Gaussian units here):

$$\frac{\mathrm{d}\mathbf{p}}{\mathrm{d}t} = e\left(\mathbf{E} + \frac{\mathbf{u}}{c} \times \mathbf{B}\right) \tag{C.26}$$

in a manifestly covariant form and find a Hamiltonian from which Eq. (C.26) may be derived. First, let's rewrite Eq. (C.26) in terms of the proper time $\tau$ and add to it the equation expressing the power-balance $dE/\mathrm{d}\tau = (e/c)\mathbf{u} \cdot \mathbf{E}$, so that we can employ the 4-vector $p^{\mu}$:

$$\begin{aligned} \frac{\mathrm{d}\mathbf{p}}{\mathrm{d}\tau} &= \gamma\frac{\mathrm{d}\mathbf{p}}{\mathrm{d}t} = e\gamma\mathbf{E} + e\gamma\frac{\mathbf{u}}{c} \times \mathbf{B} = \frac{e}{c}(U_0 + \mathbf{U} \times \mathbf{B}) \\ \frac{dp_0}{\mathrm{d}\tau} &= \gamma\frac{dp_0}{\mathrm{d}t} = \gamma\frac{e}{c}\mathbf{u} \cdot \mathbf{E} = \frac{e}{c}\mathbf{U} \cdot \mathbf{E}. \end{aligned} \tag{C.27}$$

**Defining the electromagnetic field tensor:*

$$F^{\mu\nu} = \begin{pmatrix} 0 & -E_x & -E_y & -E_z \\ E_x & 0 & -B_z & B_y \\ E_y & B_z & 0 & -B_x \\ E_z & -B_y & B_x & 0 \end{pmatrix}, \tag{C.28}$$

*we can write Eq. (C.27) in the manifestly covariant form:*

$$\frac{dp^{\mu}}{\mathrm{d}\tau} = \frac{e}{c}F^{\mu\nu}U_{\nu}. \tag{C.29}$$

We must now find a Hamiltonian function whose dynamic equations (the Hamilton equations of motion) yield the Lorentz force equation. To do this correctly it would be necessary to develop a bit of Lagrangian theory. So, here we follow a "pragmatic approach": let's define the Hamiltonian:

$$H = \sqrt{c^2[c\mathbf{p} - (e/c)\mathbf{A}]^2 + m^2c^4} + e\Phi\,, \tag{C.30}$$

where $\Phi$ is the scalar potential. Considering that the particle velocity in terms of $\mathbf{p}$ and $\mathbf{A}$ is (as it follows from Lagrangian theory):

$$\mathbf{u} = \frac{c\mathbf{p} - e\mathbf{A}}{\sqrt{(\mathbf{p} - (\mathbf{e}/\mathbf{cA})^2 + m^2c^4}}\,, \tag{C.31}$$

with some algebra one can verify that the Hamilton equations of motion ($i = 1, 2, 3$):

$$\frac{\partial H}{\partial x_i} = \frac{dp_i}{\mathrm{d}\tau} \tag{C.32}$$

are equivalent to the Lorentz force equation [the first of the two Eq. (C.27)]. Comparing this expression with Eq. (C.25) we see that the interaction between a charged particle and the electromagnetic field has been accounted for by replacing the particle momentum $\mathbf{p}$ with $\mathbf{p} - (e/c)\mathbf{A}$. Restoring SI units and keeping in mind that here $e$ represents the charge of the particle, for an electron with negative charge $-e$, this is indeed $\mathbf{p} + e\mathbf{A}$, which is what we wanted to show.

## C.2 Relativistic Wave Equations

### *C.2.1 The Klein–Gordon Equation*

One possible way to “derive” (wrong term, but let’s ignore deep philosophical discussions!) Schrödinger’s equation is to write the energy–momentum dispersion for a free particle of mass $m$,

$$E = \frac{p^2}{2m}, \tag{C.33}$$

set $E \to \mathrm{i}\hbar\partial/\partial t$, $\mathbf{p} \to -\mathrm{i}\hbar\nabla$, and view Eq. (C.33) as operators acting on a vector $\psi$ in some Hilbert space $\mathscr{H}$. Thus:

$$E = \frac{p^2}{2m} \to -\frac{\hbar^2}{2m}\nabla^2\psi(\mathbf{r},t) = \mathrm{i}\hbar\frac{\partial}{\partial t}\psi(\mathbf{r},t), \tag{C.34}$$

which is the wave equation for non-relativistic particles. Searching for a relativistic wave equation, we should start by replacing the non-relativistic energy–momentum relation given by Eq. (C.33) with its relativistic counterpart given by Eq. (C.25). We immediately encounter a big problem: Eq. (C.25) contains a square root. As innocent as this may seem, viewing this equation in terms of operators would yield a mathematically very nasty outcome: taking the square root of a linear differential operator yields a nonlocal, unbound (so, not continuous in the operator sense) operator. As a first, alternative approach, let’s avoid this “square-root” issue by considering the square of Eq. (C.25):

$$E^2 = c^2p^2 + m^2c^4 \to -\hbar^2\frac{\partial^2}{\partial t^2}\phi(\mathbf{r},t) = -c^2\hbar^2\nabla^2\phi + m^2c^4\phi(\mathbf{r},t), \tag{C.35}$$

or

$$(c^2\hbar^2\Box^2 + m^2c^4)\,\phi(\mathbf{r},t) = 0, \tag{C.36}$$

where the d’Alembert operator is defined as $\Box^2 = (1/c^2)\partial^2/\partial t^2 - \nabla^2$ and $\phi$ represents the relativistic state vector. This is called the “Klein–Gordon equation.” In relativistic Quantum Mechanics it is customary to use units in which $c = \hbar = 1$, so the Klein–Gordon equation, Eq. (C.36), is written simply as

$$(\Box^2 + m^2)\,\phi(x) = [\partial_\mu\partial^\mu + m^2]\phi(x) = 0, \tag{C.37}$$

where $x$ is the 4-vector $(\mathbf{r},t)$, inner products have the metric $(-,-,-,+)$, and $\Box^2$ takes the form $\partial^2/\partial t^2 - \nabla^2 = \partial_\mu\partial^\mu$, the latter form having been written using the notation $\partial_\mu = \partial/\partial x_\mu$ and also the “usual” relativistic convention according to which the index $\mu$ runs over the 4 dimensions and the sum is implied over repeated indices. The Klein–Gordon equation is relativistically covariant, as desired, but its interpretation as a wave equation for a single particle (so, not as a second-quantization field equation) is problematic for two major reasons. First, plane-wave solutions of Eq. (C.37) are of the form $\phi(\mathbf{r},t) = \exp[-\mathrm{i}(\mathbf{k}\cdot\mathbf{r} - Et)]$ with both positive and negative energies, $E = \pm\sqrt{k^2+m^2}$. One may attempt to interpret the negative-energy waves as anti-particles of the positive-energy solutions, but this interpretation encounters problems as soon as interactions are introduced. For example, adding the interaction with the electromagnetic field, $\partial_\mu \to \partial_\mu - ieA_\mu$, we get an equation which admits transitions between negative- and positive-energy solutions. This problem affects all relativistic first-quantization formulations and can be bypassed only in second quantization. More serious is a second problem specific to the Klein–Gordon equation: the charge-current 4-vector

$$j_\mu(x) = i[\partial_\mu \phi^*(x)\phi(x) - \phi^*(x)\partial_\mu \phi(x)] , \tag{C.38}$$

obeys the conservation law

$$\partial^\mu j_\mu(x) = 0 , \tag{C.39}$$

as it follows directly from Eq. (C.37), but attempting to identify $\rho(x) = -ij_0(x) = -i[\dot{\phi}^*(x)\phi(x) - \phi^*(x)\dot{\phi}(x)]$ as a probability density yields negative values. For example, for an eigenstate of energy $E$, $\rho(x) = 2E\phi^*(x)\phi(x)$, which is negative for negative $E$. One may simply ignore negative-energy solutions, but, yet again, interactions can induce transitions to those and the interpretation becomes really problematic.

### C.2.2 The Dirac Equation

A second approach is quite different. We saw that squaring Eq. (C.25) results in negative-energy solutions and negative probability densities. On the other hand, retaining the square root is not an option, since it results in nonlocal, unbound operators. But looking at the expression

$$E^2 = p^2 + m^2 \tag{C.40}$$

one is reminded of the "factorization" of the form $A^2 = \alpha^2 + \beta^2$ as $AA^*$ where $A = \alpha + i\beta$. This looks like a clean way to take some sort of "square root." The quantity i (the imaginary unit) does the trick when we consider the "two-dimensional" form $AA^* = \alpha^2 + \beta^2$, but we need other mathematical objects in the four-dimensional case we are considering here. We require $E^2 = m^2 + p^2$ to be expressed in a form of the type $(\beta m + \alpha \cdot \mathbf{p})(\beta m - \alpha \cdot \mathbf{p})$, where $\alpha$ and $\beta$ are mathematical entities yet to be determined. The relation we just wrote can be satisfied only if $\beta\alpha_i + \alpha_i\beta = 0$ and $\alpha_i\alpha_j + \alpha_j\alpha_i = 0$ for $i \neq j$ (so that all the unwanted cross-terms vanish) and $\alpha_i^2 = \beta^2 = 1$. This implies that the quantities $\beta$ and $\alpha$ must be matrices. It turns out that these quantities (with algebraic properties related to the so-called quaternions) have to be $4 \times 4$ matrices and can be represented in terms of the Pauli matrices as follows

$$\gamma^i = \begin{pmatrix} 0 & \sigma_i \\ -\sigma_i & 0 \end{pmatrix} , \quad \gamma^0 = \begin{pmatrix} I & 0 \\ 0 & -I \end{pmatrix} \tag{C.41}$$

where $i = 1, 3$, $I$ is the $2 \times 2$ identity matrix, and $\sigma_i$ are the Pauli matrices:

$$\sigma_1 = \begin{pmatrix} 0 & 1 \\ 1 & 0 \end{pmatrix} \quad \sigma_2 = \begin{pmatrix} 0 & -i \\ i & 0 \end{pmatrix} \quad \sigma_3 = \begin{pmatrix} 1 & 0 \\ 0 & -1 \end{pmatrix} . \tag{C.42}$$

**In the literature one may find several different conventions regarding the definitions of the Dirac matrices, the difference originating from the different convention used for the metric: a 4-vector $x$ can be defined as $(x_0, \mathbf{x})$ with $(+,-,-,-)$ metric, as done here. Alternatively, $x$ can be defined as $(\mathbf{x}, x_4) = (\mathbf{x}, ix_0)$ with purely imaginary $x_4$ and Euclidean metric $(+,+,+,+)$.*

Note that these matrices obey the anticommutation laws:

$$[\gamma^\mu, \gamma^\nu]_+ = 2\, g^{\mu\nu} , \tag{C.43}$$

where $g^{\mu\nu} = g_{\mu\nu}$ is the diagonal metric tensor $(+,-,-,-)$. To see that these matrices accomplish what we set out to do, let's write the wave equation

$$(-\mathrm{i}\gamma^{\mu}\partial_{\mu} + m)\psi(x) = 0$$
$$\left(\text{or}\left(-\mathrm{i}\gamma^{\mu}\partial_{\mu} + \frac{mc}{\hbar}\right)\psi(x) = 0 \text{ with units restored, } x_0 = ct\right), \tag{C.44}$$

where $\psi$ is now a 4-vector $\psi_{\alpha}$, where $\alpha = 0,3$. This index can be viewed as a spin-index $\times$ a sign-index labeling the sign of the energy (positive for electrons, negative for positrons, in Dirac's original interpretation) and spin for the spin-1/2 case. If we multiply the Dirac equation, Eq. (C.44), by $(-\mathrm{i}\gamma^{\lambda}\partial_{\lambda} - m)$, we have

$$(-\mathrm{i}\gamma^{\lambda}\partial_{\lambda} - m)(-\mathrm{i}\gamma^{\mu}\partial_{\mu} + m)\psi(x) = (-\gamma^{\lambda}\gamma^{\mu}\partial_{\lambda}\partial_{\mu} - m^2)\psi(x) = 0\,. \tag{C.45}$$

Now, expressing $\gamma_{\lambda}\gamma^{\mu}\partial^{\lambda}\partial_{\mu}$ as 1/2 the sum of this quantity and itself, swapping the name of the dummy indices $\lambda$ and $\mu$ in the second term, noticing that $\partial^{\lambda}\partial_{\mu} = \partial^{\mu}\partial_{\lambda}$, and using the anticommutation properties, Eq. (C.43), we can write this as

$$\left[\frac{1}{2}(\gamma^{\lambda}\gamma^{\mu}\partial_{\lambda}\partial_{\mu} + \gamma^{\mu}\gamma^{\lambda}\partial_{\lambda}\partial_{\mu}) + m^2\right]\psi(x)$$
$$= \left(\frac{1}{2}[\gamma^{\lambda},\gamma^{\mu}]_{+}\partial_{\lambda}\partial_{\mu} + m^2\right)\psi(x) = (\partial^{\mu}\partial_{\mu} + m^2)\psi(x) = 0\,, \tag{C.46}$$

which is the Klein–Gordon equation. So, we have effectively "squared" the dispersion and Dirac's equation can be viewed as a sort of square root of the Klein–Gordon equation.

#### C.2.2.1 Solution of the Dirac Equation

Plane-wave solutions of Dirac's equation can be obtained as follows. Let's express the desired solution in the form

$$\psi(x) = w(\mathbf{k},E)\,\mathrm{e}^{\mathrm{i}(\mathbf{k}\cdot\mathbf{r} - \mathrm{i}Et)}\,, \tag{C.47}$$

where $w(\mathbf{k},E)$ is a four-component spinor which satisfies the equation

$$(\gamma^0 E - \gamma\cdot\mathbf{k} + m)w(\mathbf{k},E) = 0\,. \tag{C.48}$$

Let us now write $w(\mathbf{k},E)$ in terms of two two-component objects $w_+$ and $w_-$:

$$w(\mathbf{k},E) = \begin{pmatrix} w_+(\mathbf{k},E) \\ w_-(\mathbf{k},E) \end{pmatrix}. \tag{C.49}$$

The quantities $w_{\pm}$ must satisfy the two coupled two-component equations:

$$\begin{aligned} \sigma\cdot\mathbf{k}w_- &= (E-m)w_+ \\ \sigma\cdot\mathbf{k}w_+ &= (E+m)w_-\,. \end{aligned} \tag{C.50}$$

For $m \neq 0$ it is convenient to employ the rest frame of the particle. We have either

$$E = m \begin{cases} w_+ \neq 0 \\ w_- = 0 \end{cases}, \tag{C.51}$$

or

$$E = -m \begin{cases} w_+ = 0 \\ w_- \neq 0 \end{cases}. \tag{C.52}$$

The nonzero component $w_\pm$ can be chosen arbitrarily, so it is convenient to consider them as eigenvalues of $\sigma_z$. Thus we have the four solutions:

$$\begin{pmatrix} 1 \\ 0 \\ 0 \\ 0 \end{pmatrix} \begin{pmatrix} 0 \\ 1 \\ 0 \\ 0 \end{pmatrix} \begin{pmatrix} 0 \\ 0 \\ 1 \\ 0 \end{pmatrix} \begin{pmatrix} 0 \\ 0 \\ 0 \\ 1 \end{pmatrix}. \tag{C.53}$$

The first two solutions correspond to particles of energy $E = m$ with spin up and down, respectively, while the latter solutions correspond to particles of energy $E = -m$. The existence of negative-energy solutions constitute the problem which, as mentioned above, affects all first-quantization relativistic formulations. Dirac suggested that these solutions should be interpreted as anti-particles (i.e., positrons in the case of electrons). Note how spin, introduced as an ad hoc new degree of freedom in conventional Quantum Mechanics, now emerges naturally from the formulation of a relativistic wave equation. This constitutes a major success of Dirac's theory.

For arbitrary $\mathbf{k}$ the solutions are

$$u_{\mathbf{k},\sigma} = w_\sigma(\mathbf{k}, E_\mathbf{k}) = \left( \frac{E_\mathbf{k} + m}{2m} \right)^{1/2} \begin{pmatrix} \xi_\sigma \\ \frac{\sigma \cdot \mathbf{k}}{E_\mathbf{k} + m} \xi_\sigma \end{pmatrix}, \tag{C.54}$$

(for $\sigma$ = 1,2) and

$$v_{\mathbf{k},\sigma} = w_\sigma(-\mathbf{k}, -E_\mathbf{k}) = \left( \frac{E_\mathbf{k} + m}{2m} \right)^{1/2} \begin{pmatrix} \frac{\sigma \cdot \mathbf{k}}{E_\mathbf{k} + m} \xi_\sigma \\ \xi_\sigma \end{pmatrix}. \tag{C.55}$$

In these expressions $\xi_1 = \begin{pmatrix} 1 \\ 0 \end{pmatrix}$ and $\xi_2 = \begin{pmatrix} 0 \\ 1 \end{pmatrix}$. Considering now the case $m = 0$ (as for the early theories of neutrinos or looking at the analogy with the dispersion in graphene), we can choose the 4 independent 4-component spinors [in place of Eqs. (C.54) and (C.55)]:

$$u_{\mathbf{k},\sigma} = \frac{1}{\sqrt{2}} \begin{pmatrix} \zeta_\sigma \\ \frac{\sigma \cdot \mathbf{k}}{k} \zeta_\sigma \end{pmatrix}, \tag{C.56}$$

(for $\sigma$ = 1,2) and

$$v_{\mathbf{k},\sigma} = \frac{1}{\sqrt{2}} \begin{pmatrix} \frac{\sigma \cdot \mathbf{k}}{k} \zeta_\sigma \\ \zeta_\sigma \end{pmatrix}, \tag{C.57}$$

corresponding to the negative- and positive-energy dispersion $E = \pm k$ on the "light cone." The 2-component spinors $\zeta_\sigma$ are now chosen not as eigenstates of $\sigma_z$, but as eigenstates of the "helicity" $\sigma \cdot \mathbf{k}/k$:

$$\begin{aligned} \sigma \cdot \mathbf{n}\, \zeta_1 &= \zeta_1 \\ \sigma \cdot \mathbf{n}\, \zeta_2 &= -\zeta_2 \, , \end{aligned} \tag{C.58}$$

where $\mathbf{n} = \mathbf{k}/k$ is the unit vector along the direction of motion. This is rendered necessary by a profound difference between massless and massive particles: for massive particles one is free to select arbitrarily the rest-frame form of $w_\pm$ in Eqs. (C.51)–(C.55), since one can choose to define the polarization in the rest frame along any arbitrary direction. On the contrary, for massless particles traveling at the speed of light there is no such a thing as a "rest frame" and the polarization must be measured along the direction of motion, $\mathbf{k}/k$.

# Appendix D
# Pseudopotential Code for Silicon Band Structure

In this appendix, we give a MATLAB code yielding silicon's bandstructure using the empirical pseudopotential method. To compute the bandstructure: choose a directory, copy the code listed below into a file named "pseudo_silicon.m," and run the matlab script by typing "pseudo_silicon."

```
function E=pseudo_silicon(a, b_ep, tau, E_cut, k)
% pseudo_silicon calculates and plots the bandstructure using
% pseudopotentials
%
%   E = pseudo_silicon(a0, b, n_atoms, tau, E_cut)
%      a0 - lattice constant in atomic units
%      b_ep - 4 coefficients for pseudopotential
%       V_q = b_ep(1)*( q^2-b_ep(2) )./(b_ep(3)*exp( b_ep(4)*q^2 )-1);
%      n_atoms - number of atoms
%      tau - matrix made of columns indicating position of each atom
%      E_cut - cut-off energy in Rydbergs
%
%   E = bulk_pseud_william() calls pseudo_silicon(a0,
%   b_ep, n_atoms, tau, E_cut) with default values
%
%   @author William Vandenberghe

    %% Constants
    ryd_m = 5.2917725e-11;
    ryd_eV = 13.6;

    if nargin==5 % Calculate and plot the eigenvalues

        % Define the pseudopotential function
        V_q = @(q_in) b_ep(1)*( q_in.^2-b_ep(2) ) ...
            ./(b_ep(3)*exp( b_ep(4)*q_in.^2 ) -1);

        nk=size(k,1); % Number of k-points
        n_atoms=size(tau,2);
        %%%%

        %%%% Generate all relevant reciprocal vectors
        % Generators of the group of reciprocal vectors for FCC
        b=2*pi*a'^-1;
```



M. Fischetti, W.G. Vandenberghe, *Advanced Physics of Electron Transport in Semiconductors and Nanostructures*, Graduate Texts in Physics, DOI 10.1007/978-3-319-01101-1

```
    % Generate all linear combinations of the generators within
    % Manhattan distance 2
    nGxyz=ceil(2*sqrt(E_cut./(b.^2*ones(3,1))));

    cx=(-nGxyz(1):nGxyz(1))';
    cy=(-nGxyz(2):nGxyz(2))';
    cz=(-nGxyz(3):nGxyz(3))';
    G=[ kron(kron(cx,cy.^0),cz.^0) ...
        kron(kron(cx.^0,cy),cz.^0) ...
        kron(kron(cx.^0,cy.^0),cz) ]*b;

    % Retain all reciprocal vectors within the cut-off
    G = G(G.^2*ones(3,1)<E_cut,:);
    % Number of G-vectors
    nG = size(G,1);
    %%%%


    %%%% Generate the potential part of the Hamiltonian
     % Matrix containing differences between G and G' (size 3*nG^2)
    GmG2=(kron(ones(1,nG),G)-kron(ones(nG,1),reshape(G',1,3*nG)));
    VGG=zeros(nG); % Initialize potential part of the Hamiltionian
    for i_atom=1:n_atoms
        VGG=VGG+b_ep(5)/abs(det(a)) ...
            *exp(-1i*(GmG2)*kron(eye(nG),tau(:,i_atom))) ...
            .*V_q(sqrt(abs(GmG2).^2*kron(eye(nG),ones(3,1))));
    end

    nE=3*n_atoms;
    E=zeros(nk,nE);  % Initialize the eigenvalues
    for i_k=1:nk
        % Generate the kinetic part of the Hamiltonian for k_i
        T=spdiags(sum(abs(G+ones(nG,1)*k(i_k,:)).^2,2),0,nG,nG);
        % Calculate the eigenvalues
        E(i_k,:)=sort(real(eigs(T+VGG,nE,-18/ryd_eV)));
    end
    %%%%


    %%%% Plot the bandstructure in electronvolt
    plot((E-max(E(:,2*n_atoms)))*ryd_eV); axis([1 nk -15 6]);
    xlabel('k'); ylabel('E (eV)');
    fprintf('The band gap is %1.2feV\n', ...
        min(E(:,2*n_atoms+1)*ryd_eV)-max(E(:,2*n_atoms)*ryd_eV))
    %%%%

elseif nargin==0 % Call the default values

    a0=5.43e-10/ryd_m;

    b_ep=[ 0.53706 2.19104 2.05716 0.48716 a0^3/8];
      % Values from Zhang et al., Phys. Rev. B 48, 11204 (1993)

    a=a0*[[.5,.5,0];[.5,0,.5];[0,.5,.5]]; % Unit cell
    tau=a0/8*[[-1 -1 -1];[1 1 1]]'; % Atom positions

    E_cut = 10.5;

    kmax=2*pi/a0;
    dk  =.05*kmax;
```

```
        %%%% Generate array of k-values for which the bandstructure
        %  has to be solved
        k   =[(kmax*sqrt(3)/2:-dk:0)'*[1 1 1]/sqrt(3) ; ...
              (0:dk:kmax)'*[1 0 0] ;  \ldots
              (0:dk:kmax/2/sqrt(2))'.^0*[kmax 0 0] + ...
                (0:dk:kmax/2/sqrt(2))'*[0 1 1]/sqrt(2) ;  ...
              (kmax*3*sqrt(2)/4:-dk:0)'*[0 1 1]/sqrt(2)];

        E = pseudo_silicon(a, b_ep, tau, E_cut, k);

    end
return
```

# Index




M. Fischetti, W.G. Vandenberghe, *Advanced Physics of Electron Transport in Semiconductors and Nanostructures*, Graduate Texts in Physics, DOI 10.1007/978-3-319-01101-1

**D**



**E**

**F**



**G**

**L**



**M**



**N**



**O**

**T**



**U**

Printed in the United States
By Bookmasters